RENEWALS 458-4574
DATE DUE

WITHDRAWN
UTSA Libraries

American Geophysical Union | ANTARCTIC RESEARCH SERIES

Antarctic Research Series Volumes

1 Biology of the Antarctic Seas I *Milton O. Lee (Ed.)*
2 Antarctic Snow and Ice Studies *M. Mellor (Ed.)*
3 Polychaeta Errantia of Antarctica *O. Hartman (Ed.)*
4 Geomagnetism and Aeronomy *A. H. Waynick (Ed.)*
5 Biology of the Antarctic Seas II *G. A. Llano (Ed.)*
6 Geology and Paleontology of the Antarctic *J. B. Hadley (Ed.)*
7 Polychaeta Myzostomidae and Sedentaria of Antarctica *O. Hartman (Ed.)*
8 Antarctic Soils and Soil Forming Processes *J. C. F. Tedrow (Ed.)*
9 Studies in Antarctic Meteorology *M. J. Rubin (Ed.)*
10 Entomology of Antarctica *J. L. Gressit (Ed.)*
11 Biology of the Antarctic Seas III *G. A. Llano, W. L. Schmitt (Eds.)*
12 Antarctic Bird Studies *O. L. Austin, Jr (Ed.)*
13 Antarctic Ascidiacea *P. Kott (Ed.)*
14 Antarctic Cirripedia *W. A. Newman, A. Ross (Eds.)*
15 Antarctic Oceanology I *L. Reid (Ed.)*
16 Antarctic Snow and Ice Studies II *A. P. Crary (Ed.)*
17 Biology of the Antarctic Seas IV *G. A. Llano, I. E. Wallen (Eds.)*
18 Antarctic Pinnipedia *W. H. Burt (Ed.)*
19 Antarctic Oceanology II: The Australian-New Zealand Sector *D. E. Hayes (Ed.)*
20 Antarctic Terrestrial Biology *G. A. Llano (Ed.)*
21 Recent Antarctic and Subantarctic Brachiopods *M. W. Foster (Ed.)*
22 Human Adaptability to Antarctic Conditions *E. K. Eric Gunderson (Ed.)*
23 Biology of the Antarctic Seas V *D. L. Pawson (Ed.)*
24 Birds of the Antarctic and Sub-Antarctic *G. E. Watson*
25 Meterological Studies at Plateau Station, Antarctica *J. Businger (Ed.)*
26 Biology of the Antarctic Seas VI *D. L. Pawson (Ed.)*
27 Biology of the Antarctic Seas VII *D. L. Pawson (Ed.)*
28 Biology of the Antarctic Seas VIII *D. L. Pawson, L. S. Kornicker (Eds.)*
29 Upper Atmosphere Research in Antarctica *L. J. Lanzerotti, C. G. Park (Eds.)*
30 Terrestrial Biology II *B. Parker (Ed.)*
31 Biology of the Antarctic Seas IX *L. S. Kornicker (Ed.)*
32 Biology of the Antarctic Seas X *L. S. Kornicker (Ed.)*
33 Dry Valley Drilling Project *L. D. McGinnis (Ed.)*
34 Biology of the Antarctic Seas XI *L. S. Kornicker (Ed.)*
35 Biology of the Antarctic Seas XII *D. Pawson (Ed.)*
36 Geology of the Central Transantarctic Mountains *M. D. Turner, J. F. Splettstoesser (Eds.)*
37 Terrestrial Biology III *B. Parker (Ed.)*
38 Biology of the Antarctic Seas XIII [crinoids, hydrozoa, copepods, amphipoda] *L. S. Kornicker (Ed.)*
39 Biology of the Antarctic Seas XIV *L. S. Kornicker (Ed.)*
40 Biology of the Antarctic Seas XV *L. S. Kornicker (Ed.)*
41 Biology of the Antarctic Seas XVI *L. S. Kornicker (Ed.)*
42 The Ross Ice Shelf: Glaciology and Geophysics *C. R. Bentley, D. E. Hayes (Eds.)*
43 Oceanology of the Antarctic Continental Shelf *S. Jacobs (Ed.)*
44 Biology of the Antarctic Seas XVII [benthic satiation, brittle star feeding, pelagic shrimps, marine birds] *L. S. Kornicker, (Ed.)*
45 Biology of the Antarctic Seas XVIII, Crustacea Tanaidacea of the Antarctic and the Subantarctic 1. On Material Collected at Tierra del Fuego, Isla de los Estados, and the West Coast of the Antarctic Peninsula *L. S. Kornicker (Ed.)*
46 Geological Investigations in Northern Victoria Land *E. Stump (Ed.)*
47 Biology of the Antarctic Seas XIX [copepods, teleosts] *L. S. Kornicker (Ed.)*
48 Volcanoes of the Antarctic Plate and Southern Oceans *W. E. LeMasurier, J. W. Thomson (Eds.)*
49 Biology of the Antarctic Seas XX, Antarctic Siphonophores From Plankton Samples of the United States Antarctic Research Program *L. S. Kornicker (Ed.)*

50 Contributions to Antarctic Research I *D. H. Elliot (Ed.)*

51 Mineral Resources Potential of Antarctica *J. F. Splettstoesser, G. A. M. Dreschhoff (Eds.)*

52 Biology of the Antarctic Seas XXI [annelids, mites, leeches] *L. S. Kornicker (Ed.)*

53 Contributions to Antarctic Research II *D. H. Elliot (Ed.)*

54 Marine Geological and Geophysical Atlas of the Circum-Antarctic to 30° S *D. E. Hayes (Ed.)*

55 Molluscan Systematics and Biostratigraphy Lower Tertiary La Meseta Formation, Seymour Island, Antarctic Peninsula *J. D. Stilwell, W. J. Zinsmeister*

56 The Antarctic Paleoenvironment: A Perspective on Global Change, Part One *J. P. Kennett, D. A. Warnke (Eds.)*

57 Contributions to Antarctic Research III *D. H. Elliot (Ed.)*

58 Biology of the Antarctic Seas XXII *S. D. Cairns (Ed.)*

59 Physical and Biogeochemical Processes in Antarctic Lakes *W. J. Green, E. I. Friedmann (Eds.)*

60 The Antarctic Paleoenvironment: A Perspective on Global Change, Part Two *J. P. Kennett, D. A. Warnke (Eds.)*

61 Antarctic Meteorology and Climatology: Studies Based on Automatic Weather Stations *D. H. Bromwich, C. R. Stearns (Eds.)*

62 Ultraviolet Radiation in Antarctica: Measurements and Biological Effects *C. S. Weiler, P. A. Penhale (Eds.)*

63 Biology of the Antarctic Seas XXIII, Antarctic and Subantarctic Pycnogonida: Ammotheidae and Austrodecidae *S. D. Cairns (Ed.)*

64 Atmospheric Halos *W. Tape*

65 Fossil Scleractinian Corals From James Ross Basin, Antarctica *H. F. Filkorn*

66 Volcanological and Environmental Studies of Mt. Erebus *P. R. Kyle (Ed.)*

67 Contributions to Antarctic Research IV *D. H. Elliot, G. L. Blaisdell (Eds.)*

68 Geology and Seismic Stratigraphy of the Antarctic Margin *A. K. Cooper, P. F. Barker, G. Brancolini (Eds.)*

69 Biology of the Antarctic Seas XXIV, Antarctic and Subantarctic Pycnogonida: Nymphonidae, Colossendeidae, Rhynchothoraxida, Pycnogonidae, Phoxichilidiidae, Endeididae, and Callipallenidae *S. D. Cairns (Ed.)*

70 Foundations for Ecological Research West of the Antarctic Peninsula *R. M. Ross, E. E. Hofmann, L. B. Quetin (Eds.)*

71 Geology and Seismic Stratigraphy of the Antarctic Margin, Part 2 *P. F. Barker, A. K. Cooper (Eds.)*

72 Ecosystem Dynamics in a Polar Desert: The McMurdo Dry Valleys, Antarctica *John C. Priscu (Ed.)*

73 Antarctic Sea Ice: Biological Processes, Interactions and Variability *Michael P. Lizotte, Kevin R. Arrigo (Eds.)*

74 Antarctic Sea Ice: Physical Processes, Interactions and Variability *Martin O. Jeffries (Ed.)*

THE ANTARCTIC RESEARCH SERIES

The Antarctic Research Series, published since 1963 by the American Geophysical Union, now comprises more than 70 volumes of authoritative original results of scientific work in the high latitudes of the southern hemisphere. Series volumes are typically thematic, concentrating on a particular topic or region, and may contain maps and lengthy papers with large volumes of data in tabular or digital format. Antarctic studies are often interdisciplinary or international, and build upon earlier observations to address issues of natural variability and global change. The standards of scientific excellence expected for the Series are maintained by editors following review criteria established for the AGU publications program. Publication of the Series is aided by a grant from the National Science Foundation, which supports much of the underlying field work. Priorities for publication are set by the Board of Associate Editors. Inquiries about published volumes, work in progress or new proposals may be sent to Antarctic Research Series, AGU, 2000 Florida Avenue NW, Washington, DC 20009 (http:/www.agu.org), or to a member of the Board.

BOARD OF ASSOCIATE EDITORS

Rodney M. Feldmann, Chairman, *Paleontology*
Robert A. Bindschadler, *Glaciology*
David H. Bromwich, *Meteorology and Upper Atmosphere Physics*
Nelia W. Dunbar, *Geology*
Stanley S. Jacobs, *Oceanography*
Jerry D. Kudenov, *Marine/Polychaete Biology*
John C. Priscu, *Terrestrial Biology*

Frontispiece. The U.S. research vessel *Nathaniel B. Palmer* in its natural habitat: Antarctic sea ice. Photograph courtesy Martin O. Jeffries.

Volume 74 | ANTARCTIC RESEARCH SERIES

Antarctic Sea Ice
Physical Processes, Interactions and Variability

Martin O. Jeffries
Editor

American Geophysical Union
Washington, D.C.
1998

	ANTARCTIC
Volume 74	RESEARCH
	SERIES

ANTARCTIC SEA ICE: PHYSICAL PROCESSES, INTERACTIONS AND VARIABILITY
Martin O. Jeffries, Editor

Published under the aegis of the
Board of Associate Editors, Antarctic Research Series

Library of Congress Cataloging-in-Publication Data
Antarctic sea ice physical processes : interactions and variability /
    Martin O. Jeffries, editor.
        p. cm. -- (Antarctic research series ; v. 74)
    Includes bibliographical references.
    ISBN 0-87590-902-7
    1. Sea ice--Antarctic Ocean.. I. Jeffries, M. O. (Martin O.)
II. Series
GB2597.A48 1998
551.34'3'09167--dc21   97-53113
                                                 CIP

ISBN 0-87590-902-7
ISSN 0066-4634

Cover. (Background) Pancake ice riding a swell in the northeastern Ross Sea. (Insets) Iceberg and new ice growing in a polynya near Peter I Island, Bellingshausen Sea; closeup of pancake ice; vertical thin section of ice between crossed polarizing filters, showing (from top to bottom) superimposed ice, frazil ice, and congelation ice. Photographs courtesy Cornelius W. Sullivan (pancake ice closeup) and Martin O. Jeffries (all others).

Copyright 1998 by the American Geophysical Union
2000 Florida Avenue, N.W.
Washington, DC 20009

Figures, tables, and short excerpts may be reprinted in scientific books and journals if the source is properly cited.

Authorization to photocopy items for internal or personal use, or the internal or personal use of specific clients, is granted by the American Geophysical Union for libraries and other users registered with the Copyright Clearance Center (CCC) Transactional Reporting Service, provided that the base fee of $01.50 per copy plus $0.50 per page is paid directly to CCC, 222 Rosewood Dr., Danvers, MA 01923. 0066-4634/98/$01.50+0.50.
This consent does not extend to other kinds of copying, such as copying for creating new collective works or for resale. The reproduction of multiple copies and the use of full articles or the use of extracts, including figures and tables, for commercial purposes requires permission from the American Geophysical Union.

Published by
American Geophysical Union
2000 Florida Avenue, N.W.
Washington, D.C. 20009
With the aid of grant OPP-9414962
from the National Science
Foundation

Printed in the United States of America.

CONTENTS

Preface
Martin O. Jeffries — xi

Snow Cover on Sea Ice

The Winter Snow Cover of the West Antarctic Pack Ice: Its Spatial and Temporal Variability
Matthew Sturm, Kim Morris, and Robert A. Massom — 1

Snow Depth Distribution Over Sea Ice in the Southern Ocean From Satellite Passive Microwave Data
Thorsten Markus and Donald J. Cavalieri — 19

Ice Formation, Thickness and Drift in the Pack Ice

East Antarctic Sea Ice: A Review of Its Structure, Properties and Drift
Anthony P. Worby, Robert A. Massom, Ian Allison, Victoria Lytle, and Petra Heil — 41

Late Winter First-Year Ice Floe Thickness Variability, Seawater Flooding and Snow Ice Formation in the Amundsen and Ross Seas
Martin O. Jeffries, Shusun Li, Ricardo A. Jaña, H. Roy Krouse, and Barbara Hurst-Cushing — 69

Deriving Modes and Rates of Ice Growth in the Weddell Sea From Microstructural, Salinity and Stable-Isotope Data
Hajo Eicken — 89

Temporal and Regional Variation of Sea Ice Draft and Coverage in the Weddell Sea Obtained From Upward Looking Sonars
Volker H. Strass and Eberhard Fahrbach — 123

Sea Ice Drift and Deformation Processes in the Western Weddell Sea
Cathleen A. Geiger, Stephen F. Ackley, and William D. Hibler III — 141

Satellite Microwave Observations of Pack Ice Characteristics and Processes

Oscillatory Behavior In Antarctic Sea Ice Concentrations
Per Gloersen and Alena Mernicky — 161

Length of the Sea Ice Season in the Southern Ocean, 1988-1994
Claire L. Parkinson — 173

Active Microwave Remote Sensing Observations of Weddell Sea Ice
Mark R. Drinkwater — 187

Sea Ice Characteristics and Seasonal Variability of ERS-1 SAR Backscatter in the Bellingshausen Sea
Kim Morris, Martin O. Jeffries, and Shusun Li — 213

Interactions Between Ice, Ocean and Atmosphere

Antarctic Ocean-Ice Interaction: Implications From Ocean Bulk Property Distributions in the Weddell Gyre
Douglas G. Martinson and Richard A. Iannuzzi ... 243

Ice Formation in Coastal Polynyas in the Weddell Sea and Their Impact on Oceanic Salinity
Thorsten Markus, Christoph Kottmeier, and Eberhard Fahrbach ... 273

Interannual Variability in Summer Sea Ice Minimum, Coastal Polynyas and Bottom Water Formation in the Weddell Sea
Josefino C. Comiso and Arnold L. Gordon ... 293

Marginal Ice Zone Characteristics and Processes

Mesoscale Ice Features in the Summer Marginal Ice Zone off East Queen Maud Land Observed in NOAA AVHRR Imagery
Yasushi Fukamachi, Kay I. Ohshima, and Takayuki Ishikawa ... 317

Wave Damping in Compact Pancake Ice Fields due to Interactions Between Pancakes
Hayley H. Shen and Vernon A. Squire ... 325

Landfast and Marine Ice Characteristics

Some Features of the Growth, Structure and Metamorphism of East Antarctic Landfast Sea Ice
Vladimir I. Fedotov, Nikolay V. Cherepanov, and Konstantin P. Tyshko ... 343

Physical and Structural Properties of Landfast Sea Ice in McMurdo Sound, Antarctica
Anthony J. Gow, Stephen F. Ackley, John W. Govoni, and Wilford F. Weeks ... 355

Linking Landfast Sea Ice Variability to Marine Ice Accretion at Hells Gate Ice Shelf, Ross Sea
Jean Louis Tison, Reginald D. Lorrain, Ariane Bouzette, Michela Dini, Aldino Bondesan, and Michel Stiévenard ... 375

PREFACE

In a 1971 Scientific Committee on Antarctic Research report that reviewed polar contrasts in sea ice, Lyn Lewis and Willy Weeks made the following observation: "People who study sea ice in the Arctic Basin are commonly asked if they have ever studied ice in Antarctica, and they answer 'why bother, it's the same old stuff.'" Noting this was "fortunately true to a considerable extent," they added "It is clear that future work will depend critically on the logistics facilities available to allow surface observations beyond the fast ice edge at all seasons of the year. Of almost equal importance will be the development of instruments and recording equipment suited for use in the polar environment" (Lewis, E. L., and W. F. Weeks, Sea Ice: Some Polar Contrasts, in, *Antarctic Ice and Water Masses*, edited by G. Deacon, Scientific Committee on Antarctic Research, Cambridge, 23–34, 1971).

Lewis and Weeks made no specific mention of Earth-orbiting satellites, on which the first passive microwave sensor became operational in December 1972. Less than a year later the giant Weddell Polynya was observed for the first time. Perhaps more than any other development, this unexpected feature illustrated the potential to greatly expand our knowledge of sea ice through the application of spaceborne remote sensing. Simultaneously, it acted as a catalyst for a significant increase in the level of research.

The contributions to this volume demonstrate that advances in technology and improvements in logistics since the early 1970's have indeed increased our understanding of Antarctic sea ice processes, interactions, and variability. Satellite imaging allows us to study variability in the extent and area of the pack ice, along with its motion and deformation. Field experiments that enable in situ observations and measurements are enhanced by the availability of real-time remote data and accurate positioning systems aboard icebreaking research vessels. These ships can now operate during all seasons in the pack ice, providing essential samples for laboratory analysis, and validation of satellite-derived products. Both satellite and in situ data contribute to the forcing and validation fields that are vital for realistic numerical modeling and computer simulations of ice growth, drift, and decay at a variety of spatial and temporal scales.

With the benefit of hindsight the Lewis and Weeks paper reveals how little was known at the time about Antarctic pack ice. From field experiments and remote sensing to modeling of the pack ice and fast ice, the papers in this volume should dispel any lingering notion that Antarctic sea ice is "the same old stuff." At the same time, although observations have been made "beyond the fast ice edge at all seasons of the year" for some time now, these papers also demonstrate that much remains to be learned. Perhaps sea ice geophysicists, atmospheric scientists, and oceanographers of both polar persuasions will find herein the inspiration to pursue the interdisciplinary studies that are vital to improving our understanding of the complex Antarctic sea ice physical environment. Believed to be sensitive to climate change, that environment also influences sea ice biology, the subject of a companion volume, *Antarctic Sea Ice: Biological Processes, Interactions and Variability*, edited by M. Lizotte and K. Arrigo.

Many people have helped to make this book possible. I would like to thank the authors who responded so positively to my original solicitation, sent by e-mail from the R.V. *Nathaniel B. Palmer*, deep within the pack of the western Ross Sea. It was there that I received a message from the ARS Board of Associate Editors inquiring as to my interest in being the editor of this book. In the subsequent journey to this final product, I was ably assisted by many reviewers and by ARS editor Stan Jacobs. Syun–Ichi Akasofu (Director, Geophysical Institute, University of Alaska Fairbanks) provided unqualified encouragement and support for my involvement in this project. NSF grants DPP 9117211, OPP 9316767, and OPP 9614844, which have funded my Antarctic sea ice research in recent years, also provided partial support for my editorial activities.

Martin O. Jeffries
Geophysical Institute
University of Alaska Fairbanks

Editor

THE WINTER SNOW COVER OF THE WEST ANTARCTIC PACK ICE: ITS SPATIAL AND TEMPORAL VARIABILITY

Matthew Sturm[1], Kim Morris[2], and Robert Massom[3]

The snow cover on the sea ice of the Bellingshausen, Amundsen and Ross seas was examined during one autumn and two winter cruises in 1994-95. The snow was extremely heterogeneous, being composed of depth hoar, soft slabs, icy layers, slush, and new snow, often all present at a single location. These dissimilar snow types resulted from cycling between cold, calm periods and warm, windy periods with rain-on-snow and melt events. Local snow heterogeneity also resulted from sea water flooding. At virtually every location, the bottom 9 cm of snow pack was saline and lay on snow-ice as thick as the snow. Snow depth variations over hundreds of meters were as large as variations between floes up to 100 km apart, but across the cruise region, snow was deeper, less icy, and contained less depth hoar at higher latitudes. The addition of snow decreased the higher-frequency roughness of the ice, leading to a more gently undulating surface. Measurements of the snow/ice interface temperature (n=8051), snow thermal conductivity, snow depth, and surface temperature have been used to calculate the conductive heat flux through the snow (3 to 9 W m^{-2}). In autumn, the flux could be divided into a high and low class based on ice thickness, but later in the winter only one class was present. A diagenetic model of snow pack development is proposed in which snow-ice formation produces locally heterogeneous conditions in the snow pack, but at a regional scale tends to produce homogeneous conditions for the combined ice and snow system. The homogeneity is manifested in regional heat flux measurements from the ice surface.

1. INTRODUCTION

The transfer of energy and mass between an ice-covered ocean and the atmosphere is altered in a number of critical ways when snow covers the ice. Snow has a higher albedo than ice, absorbing 10 to 30% of the incoming solar radiation compared to about 40 to 80% for bare saline ice [*Grenfell and Maykut*, 1977; *Barry*, 1996]. This has regional and global climate implications [*Kellogg*, 1975; *Ledley*, 1991]. Snow is also a much better insulator than ice, reducing conductive heat losses from the ocean and therefore the rate at which the ice thickens by congelation growth [*Maykut and Untersteiner*, 1971; *Maykut*, 1986; *Eicken et al.*, 1995]. Snow can reduce the aerodynamic roughness of the ice surface [*Andreas*, 1987] by drifting in and around ice ridges and rubble fields, smoothing out higher-frequency surface roughness. This affects the turbulent transfer of energy. If a sufficient amount of snow accumulates on the ice, it will be depressed below the sea surface, creating the possibility of sea water flooding. When the resulting slush freezes, snow-ice will be produced [*Lytle and Ackley*, 1992; *Eicken et al.*, 1994; *Jeffries and Adolphs*, 1997; *Jeffries et al.*, 1997 and this volume] and latent heat released at the ice surface.

[1]U.S.A. Cold Regions Research Laboratory, Fort Wainwright, Alaska
[2]Geophysical Institute, University of Alaska, Fairbanks, Alaska
[3]Antarctic Cooperative Research Centre, University of Tasmania, Hobart, Tasmania, Australia

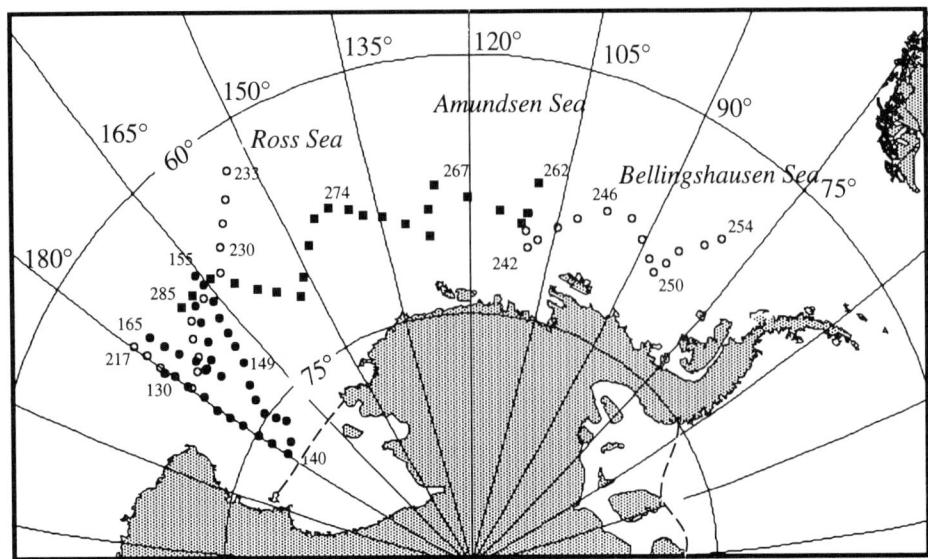

Figure 1: Track lines for Cruise 94-5 (solid squares), 95-3 (solid circles), and 95-5 (open circles), showing station locations and the day of the year the station was occupied.

Snow on sea ice also alters the electro-magnetic signature, thereby making the retrieval of sea ice properties by satellite remote sensing difficult. Since snow consists of multiple layers, each composed of grains with different size and shape, its impact on the electromagnetic radiation as measured by both passive [*Comiso*, 1983; *Lohanick*, 1990] and active systems [*Drinkwater et al.*, 1995] can be highly variable. If the snow is wet with either fresh or saline water, which it frequently is in the Antarctic, interpretation can be even more difficult [*Morris et al.*, this volume].

The pervasive impact of snow on the sea ice energy and mass balance makes it a critical part of sea ice models, but its adverse impact on remote sensing makes obtaining the regional snow distribution information needed to drive models a difficult task. Without reliable remote sensing, direct measurements must be used, but in many regions such measurements are sparse or non-existent. This is particularly true for the West Antarctic pack ice, where there have been relatively few studies (particularly in winter) compared to the vast size of the region [see *Haas and Viehoff*, 1994; *Jeffries et al.*, 1994; *Worby et al.*, 1996; *Jeffries and Adolphs*, 1997].

In this paper, we report on autumn and winter snow observations from the West Antarctic pack ice. We describe the spatial distribution of snow properties across the region and propose a diagenetic model of how the snow cover forms, including flooding, snow-ice formation, and depth hoar production. The model can be used to explain why the snow pack is so heterogeneous on a local scale. Temperature measurements from the snow are used to compute regional heat fluxes from the ice. These measurements suggest that at the regional scale, the combined snow-ice system tends to some degree of homogeneity. We speculate how local heterogenity but regional homogeneity might arise.

2. STUDY AREA

Snow and ice studies were conducted from the *R/V Nathaniel B. Palmer* in the Bellingshausen, Amundsen and Ross Seas. Results from two winter cruises and one autumn cruise are presented here (Cruise 94-5: September-October, 1994; Cruise 95-5: August-September, 1995, and Cruise 95-3: May-June, 1995). The cruise tracks (Figure 1) were designed for maximum areal coverage, and encompass an area of approximately 10^6 km^2. Due to ice conditions, the tracks were mainly in the outer 800 km of the pack, only once penetrating close to the continent. Once a day during transit in the ice, the ship stopped and a series of snow and ice measurements were made on a representative ice floe. A total of 87 stations were occupied (Figure 1).

3. DATA COLLECTION

At each station, depth, temperature, density, stratigraphy and other snow attributes were collected along 100-m traverse lines. In 1994, the lines were parallel to, but offset by 5 m from other lines on which snow and ice thickness where measured [*Jeffries et al.*, 1995a and 1995b]. On both cruises in 1995, all

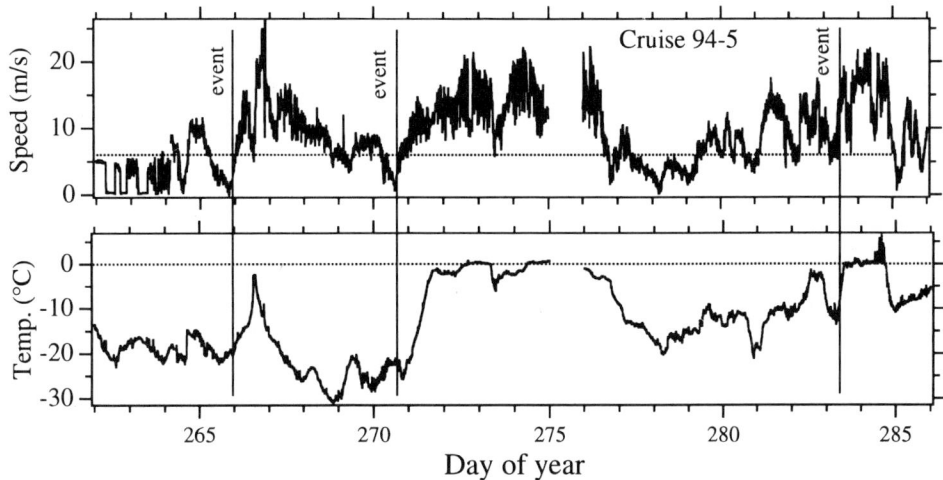

Figure 2: Near-surface wind speed and air temperature recorded aboard the R/V *Nathaniel B. Palmer* during Cruise 94-5. Vertical lines mark the onset of wind events accompanied by air temperatures near or above freezing. Note that peak wind speeds generally were not recorded until after the temperature was above freezing.

measurements were made along the same traverse lines. Lines were located so that they were representative of typical ice, though some bias towards thicker ice (>0.2-0.3 m) was present due to the inherent danger of working on thin ice.

Snow depth and temperature of the snow/ice interface were measured at 1-m intervals along each line. A thin probe marked in centimeters was pushed through the snow to the ice surface, and the depth was read from the side of the probe with an accuracy of ±1 cm. A thermistor at the tip of the probe was read using a digital reader (Omega Model 866). Temperatures (accurate to ±0.3°C) were recorded after the probe had equilibrated. Snow surface temperature was measured twice on each line. On 41 occasions, mostly during Cruise 94-5, snow depths and temperatures were also measured on ancillary lines ranging from 100 to 500 m in length. For Cruises 95-3 and 95-5, in addition to snow depth, the elevation of the snow surface was surveyed at 0.5-m intervals with an accuracy of ±1 cm using a rotating laser and automatic leveling rod.

In at least one location on each traverse line, extensive measurements of snow properties were made. A snow trench was dug and the trench wall brushed to accentuate the stratigraphy. Variations in thickness and character of strata were mapped. Vertical profiles of snow density, stratigraphy, texture, grain size, temperature, and salinity were measured in 3-cm increments. The density was measured using a 100-cm^3 cutter and a digital balance. A stereo-microscope was used to estimate the size of snow grains placed on a gridded card, as well as to estimate the degree of bonding. A temperature probe was pushed into the snow from the surface, and temperatures were recorded at 3-cm intervals. Snow samples (100 cm^3) were bagged for later determination of salinity.

The salinity samples were brought aboard the ship and allowed to melt at room temperature. The conductivity of the melt was measured using a Rosemount Analytical Conductivity Bridge (accuracy: 3% of reading) and small-volume pipette cells with cell constants of 1 or 20 depending on the salinity of the sample. From the conductivity value, the salinity was calculated [*Baker*, 1987].

Bulk snow samples from key strata were boxed and taken intact to the cold room on the ship where measurements of grain size distribution, air permeability, and thermal conductivity were made. The grain size distribution was measured by sieving [*Sturm*, 1991; *Sturm and Benson*, 1997]. The thermal conductivity was measured using a needle probe apparatus [*Sturm and Johnson*, 1992; *Sturm et al.*, 1997], and the air permeability was measured using a double-walled permeameter [*Chacho and Johnson*, 1987]. Photomicrographs of snow crystals were taken.

4. RESULTS

4.1. Stratigraphic and Textural Heterogeneity

The snow cover was notable for its heterogeneity. Three or four distinctly different types of snow were generally found inter-layered in a single stratigraphic column, and in some cases, two types (depth hoar and

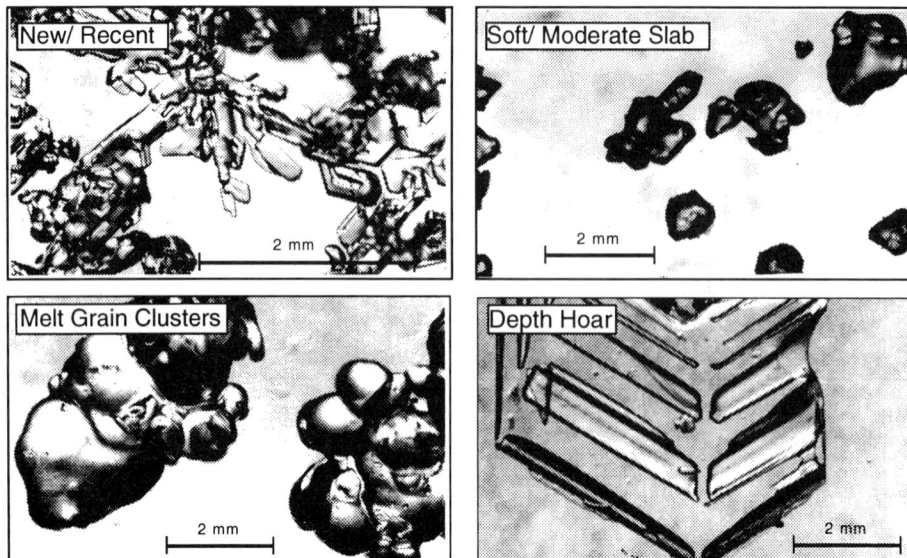

Figure 3: Photo-micrographs showing the four primary types of grains in the snow cover.

melt-grain clusters) were co-mingled within a single layer. Layers pinched and swelled over distances of less than a meter, while variations in density and grain size indicative of some (but not severe) wind transport and deposition were common. Meltwater percolation from the top of the snow pack, and wicking of brine from the bottom, were ubiquitous. This liquid water produced zones of coarse-grained snow, percolation columns, and lenses irregularly distributed both laterally and vertically in the pack. Pockets of well-developed depth hoar, the result of strong temperature gradients and an ample supply of moisture, developed between the zones of wetted snow.

A major cause of the heterogeneity was the winter weather. It cycled between periods of clear, cold, and relative calm, and periods when it was warm, windy, and wet (Figure 2). Similar weather patterns prevailed in the Bellingshausen Sea in the winter of 1993 [*Jeffries et al.*, 1994], in the Weddell Sea in 1992 [*Massom et al.*, 1997], and in East Antarctica in 1995 [*Worby et al.*, in press]. It may be common winter weather over much of the outer Antarctic ice pack.

An unusual snow feature, seen on all three cruises, suggests that this cycle of weather affected the entire cruise region. A distinctive, nubbly icy layer, previously described by *Sturm et al.* [1996] and *Massom et al.* [1997] was found both at the snow surface and within the snow pack. It consisted of small irregular lumps of icy snow, 1 to 2 cm high, spaced every few centimeters. It was observed to form when freezing rain was driven by high winds (speeds in excess of 15 m s^{-1}) into snow that was cold enough (-2°C) to cause some of the liquid water drops to freeze at the snow surface. As the rain continued, the ice nubs built into the wind, creating the characteristic nubbly texture. The texture was easily recognized in snow trenches and was traced during Cruise 94-5 for more than 1000 km.

Four primary types of snow (Figure 3), plus saline slush, were observed. Each forms under distinctly different conditions, so their close association is consistent with large cycles in weather. For example, depth hoar forms under temperature gradients in excess of a critical gradient of about 25°C m^{-1} [*Akitaya*, 1974; *Marbouty*, 1980]. Low air temperatures and high basal snow temperatures favor its growth. In contrast, melt grain clusters

TABLE 1: The percentage of types of snow in the snow cover

Cruise	94-5	95-3	95-5
Season	winter	autumn	winter
Duration (days)	262 to 285	130 to 164	216 to 254
Density (g cm^{-3})	0.36	0.35	0.39
Depth (cm)	28.5	13.6	22.4
Number of pits	164	73	45
% icy	20	26	46
% new & recent	14	7	4
% soft & moderate slabs	23	10	13
% depth hoar	31	48	29
% hard slab	6	0	0
% slush	6	9	8
k_{bulk} (W m^{-1} K^{-1})	0.124	0.112	0.138

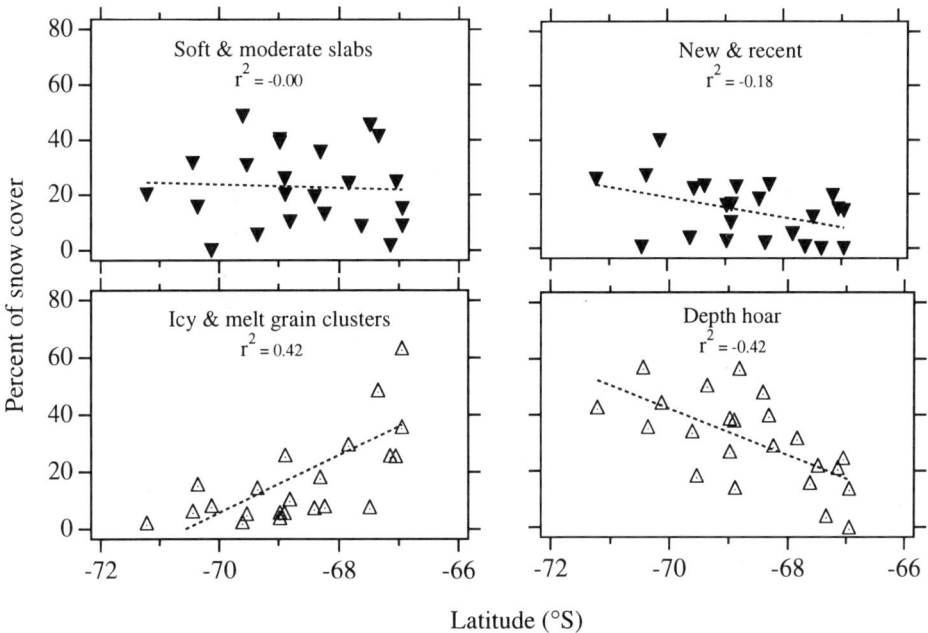

Figure 4: Variations in the percentage of the four primary types of snow with latitude, Cruise 94-5. Only depth hoar and icy snow are correlated with latitude.

form when temperatures are above freezing, or there is rain-on-snow. Undulating fine-grained layers require winds in excess of 6 m s^{-1} [*Pomeroy and Gray*, 1995].

Percentages of each snow type encountered during the cruises are listed in Table 1. The four primary types account for over 90% of the snow sampled. The amount of depth hoar was similar on the two winter cruises (94-5 and 95-5), but there was more slush and icy snow (snow with ice layers, percolation columns and coarse-grained melt-grain clusters), and slightly less soft and moderate slabs, observed on Cruise 95-5 than on Cruise 94-5. This suggests that in 1995 there was either more melting and rain-on-snow, or a greater incidence of sea water flooding, than in 1994. In the autumn (Cruise 95-3), the percentage of depth hoar was higher than later in the winter.

In 1994 (Cruise 94-5), depth hoar and icy snow varied in a systematic manner across the cruise area (Figure 4). Iciness increased with decreasing latitude and increasing proximity to the ice edge, while depth hoar varied in the opposite manner. For the other types of snow, no pattern could be discerned, and in 1995 (both Cruise 95-3 and 95-5) none of the types of snow showed a relationship with latitude or proximity to the ice edge ($r^2 < 0.1$ in all cases), though little or no icy snow was observed at the stations farther than 75° south.

One pattern that was consistent during all three cruises, however, was that layers of depth hoar or melt-grain clusters were more prevalent near the base of the snow pack, while soft and moderate slabs were more prevalent near the top (Figure 5*a*). Two conditions account for this trend. First, the temperature in the snow pack generally decreased from the base to the top (Figure 5*a*). Depth hoar development is favored by strong temperature gradients and relatively high temperatures [*Trabant and Benson*, 1972; *Akitaya*, 1974; *Sturm*, 1991], so the base of the pack was a favorable location for its growth. Second, melt-grain clusters and icy snow form after snow has been wetted and refrozen, a common cycle at the base of the pack.

As a result of the stratigraphic sequence, grain size tended to decrease from the base to the top of the snow pack (Figure 5*b*), since both depth hoar and melt grain clusters are coarse-grained types of snow, while soft and moderate wind slabs tend to be fine-grained (Table 2). The degree of sorting, on the other hand, increased with height. Wind slabs tended to be well sorted, while depth hoar, layers of depth hoar mixed with melt-grain clusters, and melt-grain clusters tended to be poorly sorted (see standard deviation in Table 2). In particular, if a layer had been subjected to any wetting, it was not only poorly sorted, but its grain size distribution also showed pronounced bimodality. For example, snow layers containing percolation columns or ice lenses contained a substantial number of grains that were many times larger than the mean grain size. These large grains have a direct impact on the passive microwave signal of the snow [*Lohanick*, 1990; *Armstrong et al.*, 1993].

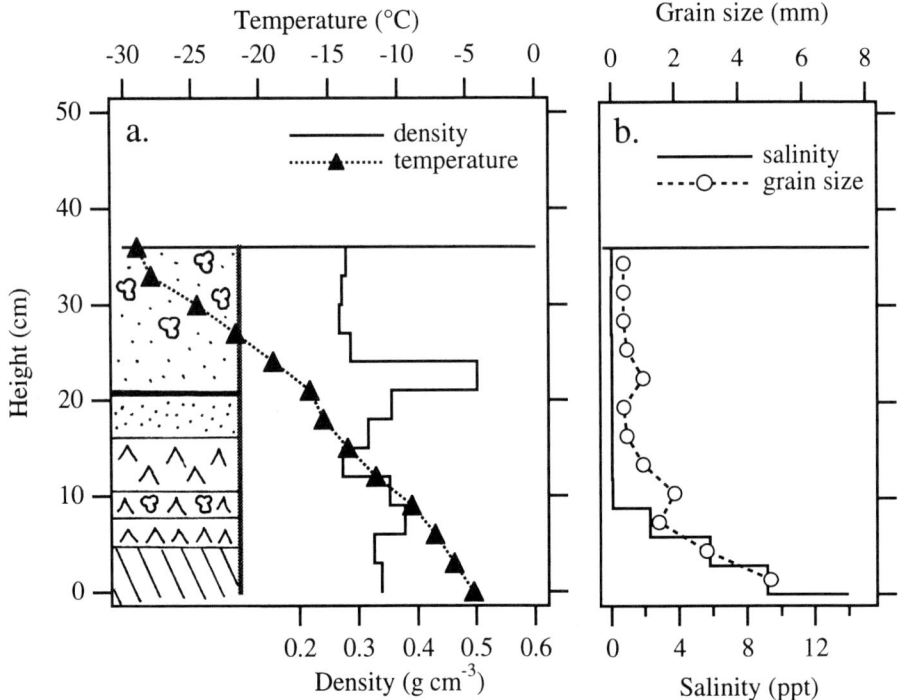

Figure 5: A typical snow pit diagram (from Station 242, Cruise 95-5). The column on the extreme left in (a) shows the snow stratigraphy (symbols are from the International Classification for Seasonal Snow on the Ground [Colbeck et al., 1990]). Other data include vertical profiles of: (a) snow density and temperature and (b) salinity and grain size.

4.2. Flooding at the Snow-Ice Interface and Snow-ice Formation

The stratigraphic record, from both the snow and sea ice, contained compelling evidence that sea water flooding at the snow/ice interface had occurred in virtually every location. Three lines of evidence show this to be the case:

1) Saline slush and standing water were commonly observed at the base of the snow. During Cruise 94-5, out of 165 snow trenches or pits, 76 had slush or standing water in the bottom; when observed, the slush averaged 4.0 cm in thickness. For Cruise 95-5, the total number of pits was 45, 23 had slush, and it averaged 3.4 cm in thickness. On Cruise 95-3, out of 73 pits, 26 contained slush; it averaged 4.6 cm in thickness. Above slush layers, salt water was often observed to have wicked up 15 cm or more in capillary columns, or spread laterally in lenses.

2) Microscopic examination of depth hoar crystals from basal layers indicated that virtually all depth hoar had undergone minor rounding, the result of minimization of surface free energy in the presence of liquid water [Colbeck, 1986a], or that melt-grain clusters had formed. Well-developed depth hoar grains were often adjacent to melt-grain clusters up to 10 mm in diameter due to capillary wicking.

3) Virtually all basal snow layers were saline (Figure 5b and Table 3), indicating that brine had been

TABLE 2: Grain size and statistics based on sieving snow

Type of Snow	No. of Samples	Size (mm)	Std. Dev. (mm)	Density (g cm⁻³)
recent snow	2	1.1	0.6	0.31
melt-grain clusters	8	1.3	0.6	0.35
soft & moderate slabs	21	1.1	0.4	0.34
hard slabs	4	0.9	0.3	0.41
depth hoar	29	2.0	0.9	0.29
depth hoar, melt clusters	14	1.5	0.8	0.30

TABLE 3: Average snow salinity, snow base to 9 cm height

Cruise	Season	Ice Surface (ppt)	0 to 3 cm (ppt)	3 to 6 cm (ppt)	6 to 9 cm (ppt)
94-5	winter	15.2	13.3	9.8	4.9
95-3	autumn	25.7	16.8	9.3	3.9
95-5	winter	21.8	16.6	9.2	3.3

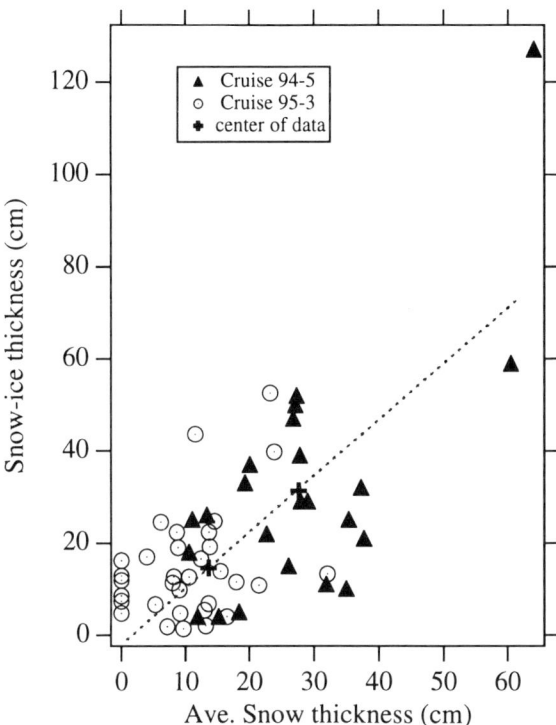

Figure 6: Thickness of the snow-ice (average of several cores per station) as a function of average snow depth for two cruises (data unavailable for cruise 95-5). The center of each data cloud is indicated by a cross, and the line connecting the crosses has a slope of 1.2 (data from Jeffries et. al, this volume and personal communication).

(28.5 cm for Cruise 94-5 vs. 22.4 cm for Cruise 95-5). During the autumn cruise (Cruise 95-3), the snow had yet to reach maximum depth, and the distribution curve was more negatively skewed. It contained more measurements on bare or nearly bare ice, and fewer measurements in snow deeper than 80 cm. On the winter cruises, depths greater than 80 cm were generally measured in snow drifts adjacent to pressure ridges (or in two cases, on multi-year floes). Interestingly, distribution curves for the winter snow of the Weddell Sea, as reported by *Eicken et al.* [1994] and *Massom et al.* [1997], are closer in form to our autumn curve than our winter curves.

Despite year-to-year regularity of the regional snow depth distribution (Figure 7), large variations in depth were observed from one station to the next (Figure 8). The pattern of these variations differed from one cruise to the next, probably the result of differences in cruise tracks introduced into the layers after their deposition. The salinity structure was remarkably similar from year to year, with the salinity decaying with height in a linear fashion.

In addition, thin sections and oxygen isotope analyses of sea ice [*Jeffries and Adolphs*, 1997, Jeffries et al., this volume] indicate that at virtually all stations there was abundant snow-ice, and in many cases the snow-ice was thicker than the snow pack itself (Figure 6). Data for Cruise 95-5 are not yet available, but the data from the other cruises show that the thickness of the snow-ice increased from autumn to winter, as the snow depth increased. However, the ratio of snow-ice to snow was nearly constant (the slope of the line connecting the centers of the two data clouds in Figure 6 is 1.2). Half or less of all snow that had been deposited on a floe was available for measurement at most locations, the rest having been converted to snow-ice.

4.3. Heterogeneity in Snow Depth Distribution

At the largest or regional scale, snow depth distribution for the two winter cruises was similar (Figure 7), though there was a difference in mean depth

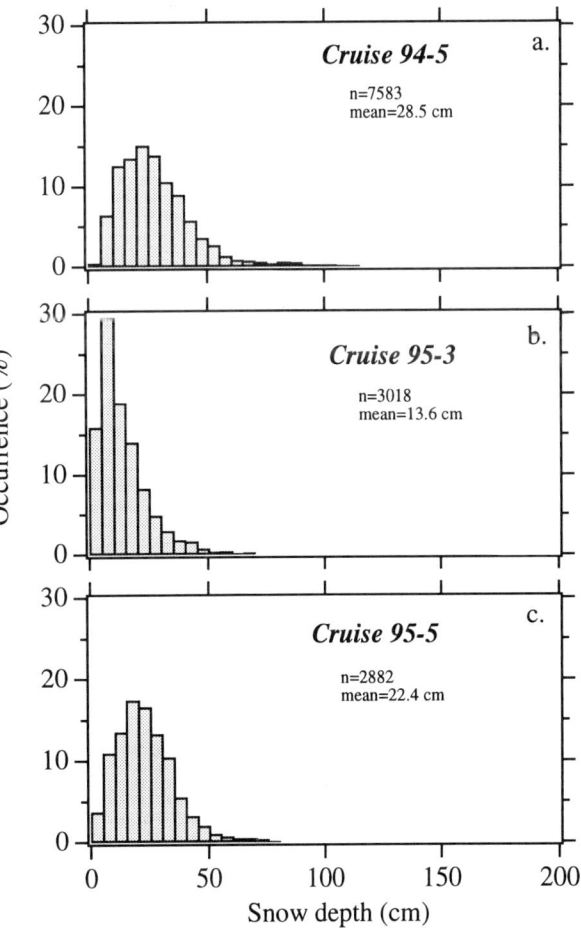

Figure 7: Depth distribution curves for all snow depth measurements, (a) Cruises 94-5, (b) 95-3, and (c) 95-5. Note the striking similarity between the two winter curves (a and c).

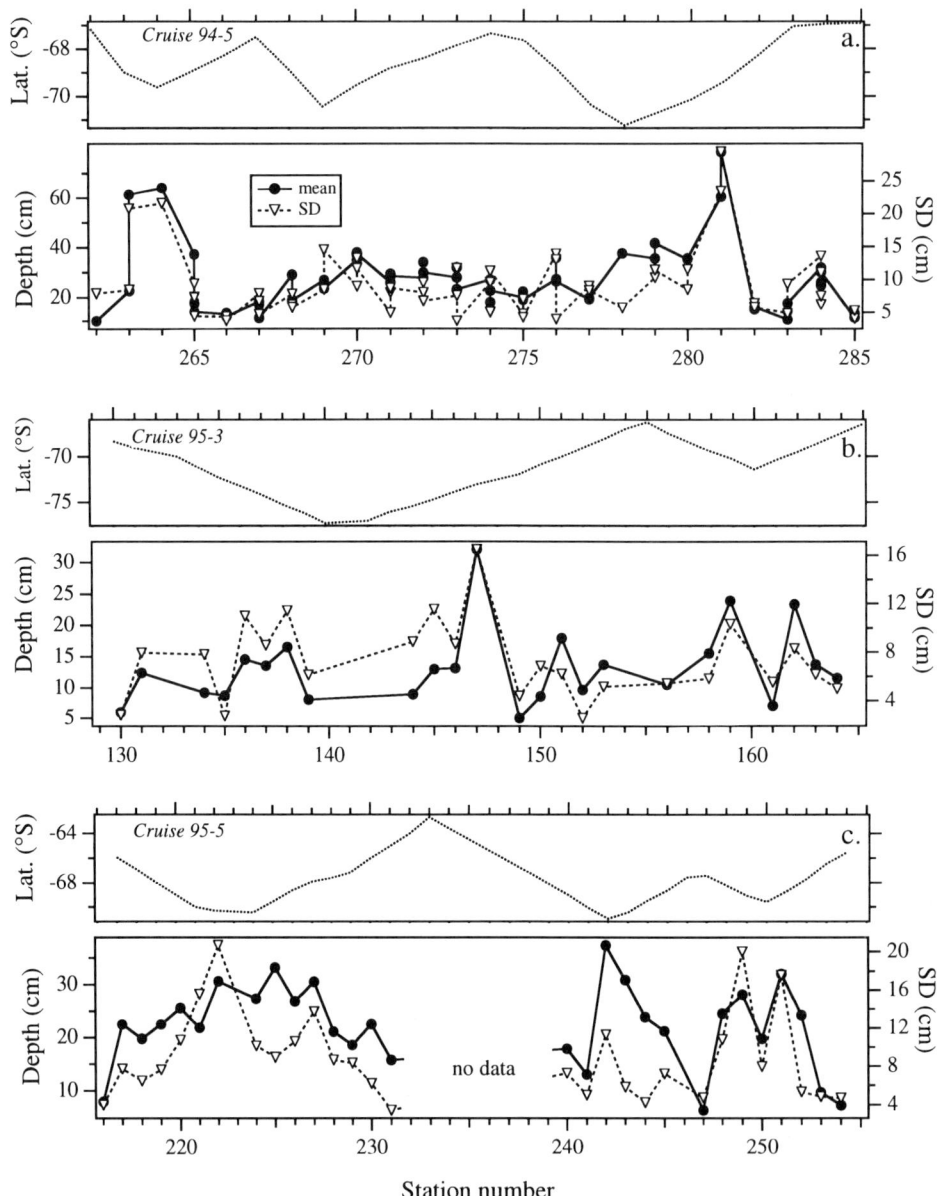

Figure 8: Variation in the mean and the standard deviation (SD) of snow depth by station number for (a) Cruises 94-5, (b) 95-3 and (c) 95-5. On Cruise 94-5, several depth traverses were done at most stations; multiple values at a single station suggest the range of local means. Overall, the results show both regional and sub-regional variability. The station latitude is also shown for each cruise. For Cruise 95-5, there is a weak correlation between depth and SD and latitude, with higher values at higher latitudes.

(Figure 1). In Figure 8, the mean snow depth and its standard deviation (SD) have been plotted as a function of station number. Stations were roughly 50 to 100 km apart. Differences in mean snow depth as large as 30 cm were observed for adjacent pairs of stations. Depth and SD were correlative, with the SD almost as large as the depth. For example, between station 280 and 281 on Cruise 94-5, a distance of 100 km, the mean depth increased by 25 cm (from 36 to 61 cm), and the SD increased from 10 to 25 cm. Both depth and SD were weakly correlated with the latitude for Cruise 95-5 (Figure 8c), with greater snow depth (and higher SD) at higher latitudes, but for the other two cruises, both values appear to be independent of latitude.

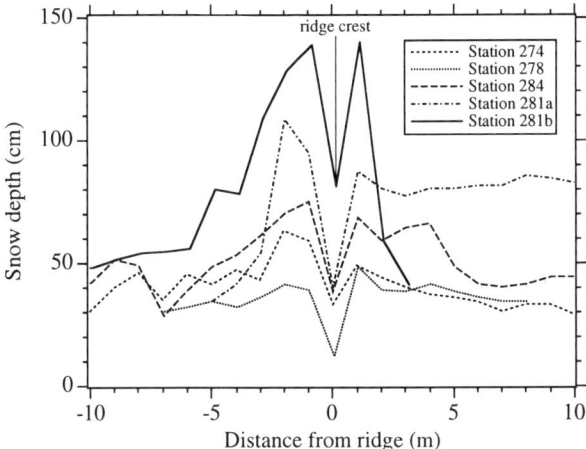

Figure 9: Snow depth across five pressure ridges. Note the thin snow at the ridge crest and the characteristic "M"-shaped pattern of snow depth across a ridge (shallow at ridge crest, deep adjacent to ridge). Drifting extends a maximum of 10 m from the ridge crest.

Snow depth on a single ice floe could vary almost as much as snow depth between different floes or stations. On Cruise 94-5, multiple depth traverse lines were measured at each station. The mean depth and SD for each traverse are plotted in Figure 8a. The range of mean values varied from 2 cm to more than 40 cm, and the average range was 7 cm. In some cases (i.e., stations 271 through 275, Cruise 94-5) the variation on a floe was greater than the variation between several stations.

A small part of the variation in the mean and SD of snow depth may have been the result of a bias introduced when traverse lines intersected ridges. Both unusually deep and shallow snow were associated with these features (Figure 9), which, unlike in the Weddell Sea [Eicken et al., 1994], tended to be symmetrical across the ridge. For a typical ridge, the area of snow with enhanced depths exceeded the area where the snow was shallow, so preferential sampling of ridges would have led to over-estimates in both mean depth and SD. However, drift snow was usually confined to narrow strips 10 to 20 m wide centered on ridges (Figure 9), and the areal extent of ridges was low (estimated at 5-10% from hourly shipboard observations), so it is unlikely the bias was large.

Variation in snow depth at scales less than 100 m was investigated using semivariograms [McBratney and Webster, 1986; Oliver and Webster, 1986; Robert and Richards, 1988; Isaaks and Srivastava, 1989]. For each 100-m cross section (Figure 10a), or depth record (Figure 10b), the semivariogram was calculated, and lines fit to it using least squares regression (Figure 10c). The intersection of these lines defined the length (R) and amplitude (S=square root of the sill, approx. equal to the SD) of structures at the snow (subscript s) and ice (subscript i) surfaces.

For Cruises 95-3 and 95-5, for which snow cross sections are available, features at the snow surface ranged from 3 to 70 m in length, with an average value of R_s equal to 23.3 and 12.7 m, respectively (Table 4). For 67% of these cases, R_s could be inferred from the snow depth alone, so for Cruise 94-5, for which only snow depths were available, a structural length of 22 m is indicated. The average amplitude (S_s) of snow surface structures was small, less than 10 cm, but the underlying ice had even more subdued structure amplitudes (S_i=5.1 cm). The addition of snow on the ice actually increased the topographic relief, but because the wavelength of the ice structures was considerably less than that of snow structures, the addition of the snow tended to "smooth" out the higher-frequency surface roughness of the ice, creating a more gently undulating surface.

The smoothing action of the snow on the ice surface can be illustrated by comparing the ratio of snow-to-ice structure lengths (R_s/R_i) to the ratio of snow-to-ice

TABLE 4: Snow and ice structure statistics from semivariograms for snow cross sections

Cruise	No. of Profiles	Ice ridges form drifts?	Snow depth stats like?	Ice S_i (cm)	R_i (m)	Snow S_s (cm)	R_s (m)	Snow to Ice S_s/S_i	R_s/R_i
95-3	24	70% Y[a] 30% N ave: 2 ridges per 100 m	ice surface: 13% snow surface: 67% neither: 16% intermediate: 4%	5.1	6.1	8.1	23.3	1.8	11.9
95-5	29	89% Y 11% N ave.: 3 ridges per 100 m	ice surface: 15% snow surface: 66% intermediate: 19%	5.1	4.2	8.8	12.7	1.9	4.1

[a] N = no, Y = yes

R = structure length s = snow surface
S = structure amplitu i = ice surface

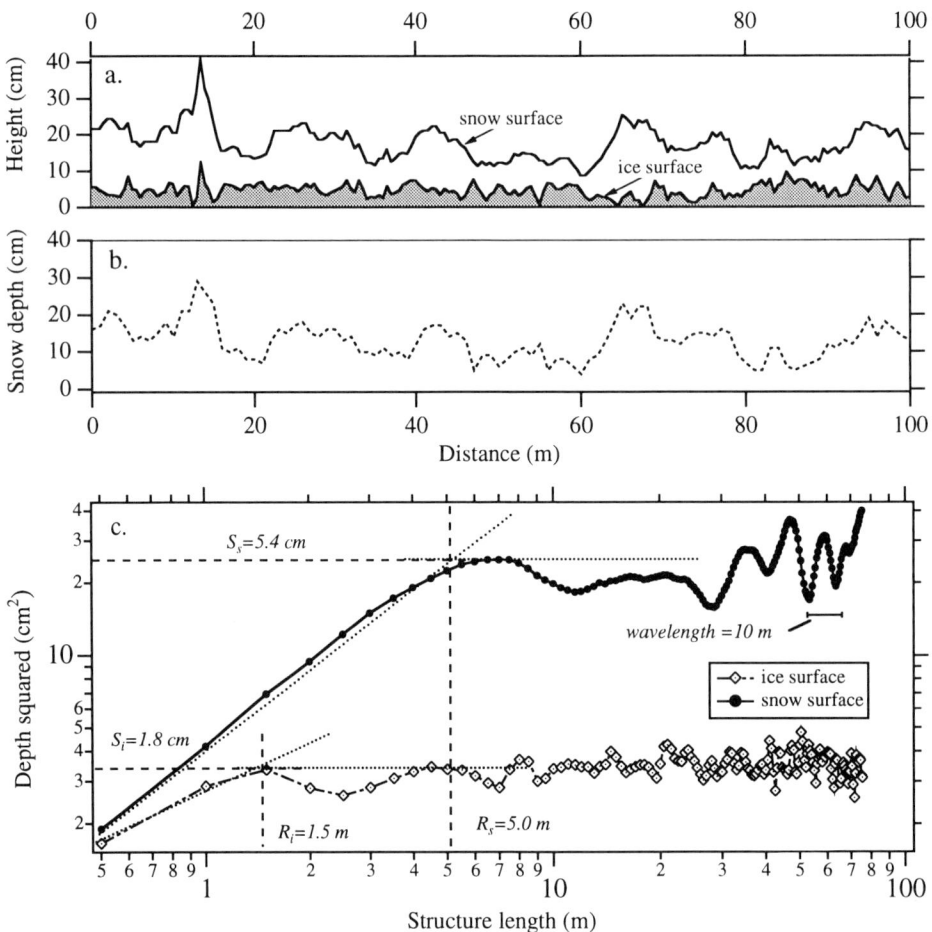

Figure 10: Typical results from a 100 m long traverse line (Station 241, Cruise 95-5) showing: (a) a cross section of the snow showing the ice and snow surface elevations, (b) the snow depth, and (c) semivariograms for the snow and ice surfaces. See text for explanation of the labels on the semivariogram.

structure amplitudes (S_s/S_i). These ratios have been tabulated in Table 4. For both cruises, the ratio for structure lengths is several times greater than the ratio for structure amplitudes.

Despite the average amplitude of ice surface structures being small, the measured cross sections indicate that ice ridges and bumps were important in creating complementary structures at the snow surface. For each cross section (Figure 10a), the number of ice ridges and bumps was counted, and it was determined by inspection whether or not there was a corresponding feature at the snow surface. The tally, listed in Table 4, shows that 70% of all ice ridges or bumps produced snow surface features on Cruise 95-3. For Cruise 95-5, 89% of all ridges and bumps initiated snow surface features. We conclude that almost any ice surface feature that has substantial relief (>20 cm) produces corresponding features at the snow surface.

4.4. Thermal Environment and Heat Flux in the Snow

4.4.1. Measurements. Here, we examine thermal conditions in the snow to see if they too were heterogeneous. Figure 5a shows a typical profile. Temperature is highest at the base and lowest at the top of the snow. The profile is nearly linear. Of 117 profiles measured, 63% were similar. These profiles were controlled by the air temperature and its recent fluctuations, the depth and thermal properties of the snow, and the ice thickness. Normally, heat moved from the ocean to the snow/ice interface by conduction through the ice with some brine convection [*Lytle and Ackley*, 1996]. When sea water flooding occurred, heat conduction through the ice was bypassed, and the temperature of the snow/ice interface was increased, as well as the rate of heat transfer.

TABLE 5: Snow/ice interface temps. and temp. gradients

Cruise	94-5	95-3	95-5
Season	winter	autumn	winter
Number of stations	25	35	27
Number of measurements	2251	3019	2781
Ave. surface temp (°C)	-10.7	-18.3	-19.0
Ave. interface temp. (°C)	-4.2	-12.1	-8.6
SD interface temp. (°C)	2.2	5.5	2.7
Ave. snow depth (cm)	28.5	13.6	22.4
Ave. temp. gradient (°C cm^{-1})	-0.23	-0.82	-0.44
SD temp. gradient (°C cm^{-1})	0.40	0.89	0.38
k_{bulk} (W m^{-1} K^{-1})	0.124	0.112	0.138
Ave. heat flux (Watts m^{-2})	2.9	9.2	6.1
SD heat flux (Watts m^{-2})	5.0	9.9	5.3

Snow/ice interface temperatures varied widely from station to station and at each station (see SD in Table 5), but average values for each cruise can be related to average air temperatures and snow depths (Table 5). Ignoring differences in cruise tracks, interface temperatures were 4.4°C higher during winter Cruise 94-5 than during winter Cruise 95-5; air temperature and snow depth were also higher. Seasonal differences are also apparent in the data. Air temperature was nearly the same on both the autumn and winter cruises of 1995 (95-3 and 95-5), but the snow was considerably thinner in the autumn (Figure 7), leading to substantially lower interface temperatures. The autumn ice may also have been thinner, with flooding more prevalent. As discussed below, the combination of these factors led to substantial differences in conductive heat flux through the snow cover.

The seasonal and inter-annual differences in interface temperature translate into important differences in the vertical temperature gradient across the snow pack [calculated by subtracting the snow/ice interface temperature (T_i) from the snow surface temperature (T_s) and dividing by the average snow depth (h_s)]. For all three cruises, average thermal gradients were equal to or greater than the critical value necessary for depth hoar growth [*Akitaya*, 1974; *Marbouty*, 1980]. However, the thinner snow and lower temperatures encountered during the autumn cruise resulted in gradients that were more than three times higher than critical (Table 5). Due to these high gradients, depth hoar comprised nearly twice as much of the snow pack in the autumn than later in the winter (Table 1).

Depth hoar is a considerably better insulator, per unit thickness, than most other types of snow [Table 6; see also *Sturm and Johnson*, 1992; *Sturm et al.*, 1997]. *Zhang et al.* [1995] have shown that the depth hoar percentage in an Arctic snow pack plays a critical role in the thermal balance of the permafrost; it may also play an important role in the thermal balance of Antarctic sea ice. A typical metamorphic change in the cruise area might convert a moderate slab into depth hoar. If this layer was 15 cm thick, the decrease in thermal conductivity (Table 6) would, for similar temperature gradients, decrease the heat flux by a factor of two.

4.4.2. Computations. Measurements of snow depth, snow type, thermal conductivity and snow/ice interface temperature are now used to derive estimates of the average vertical winter heat flux through the snow (F_a) for the cruise area. Following the "black box" model developed by *Lytle and Ackley* [1996], F_a is assumed to be equal to F_c, the conductive heat flux through the ice.

TABLE 6: Thermal conductivity of snow types measured using a needle probe

Description	ICSG Class[a]	Number of Measurements[b]	Ave. Density (g cm^{-3})	Ave. Thermal Conductivity (W m^{-1} K^{-1})
new snow	1a to 1e	10	0.133	0.070
recent snow	2a, 2b	21	0.254	0.128
small, rounded grains	3a	51	0.320	0.169
larger rounded grains	3b	9	0.345	0.163
mixed forms	3c, 4c	26	0.321	0.153
depth hoar cups in chains, weak	5b	171	0.225	0.072
depth hoar cups in chains, indurated	5a, 5b, 5c	9	0.345	0.183
melt-grain clusters	6a, 6b	21	0.416	0.247
soft to mod. slab	3a, 9d	50	0.348	0.167
hard slab	3a, 9d	16	0.379	0.237
hard drift snow	3a, 9d	78	0.440	0.355
very hard wind slab	3a, 9d	22	0.488	0.452

[a]Colbeck et al.(1990)
[b]Includes 173 measurements from Cruises 94-5, and 95-3; see also Sturm et al., in press

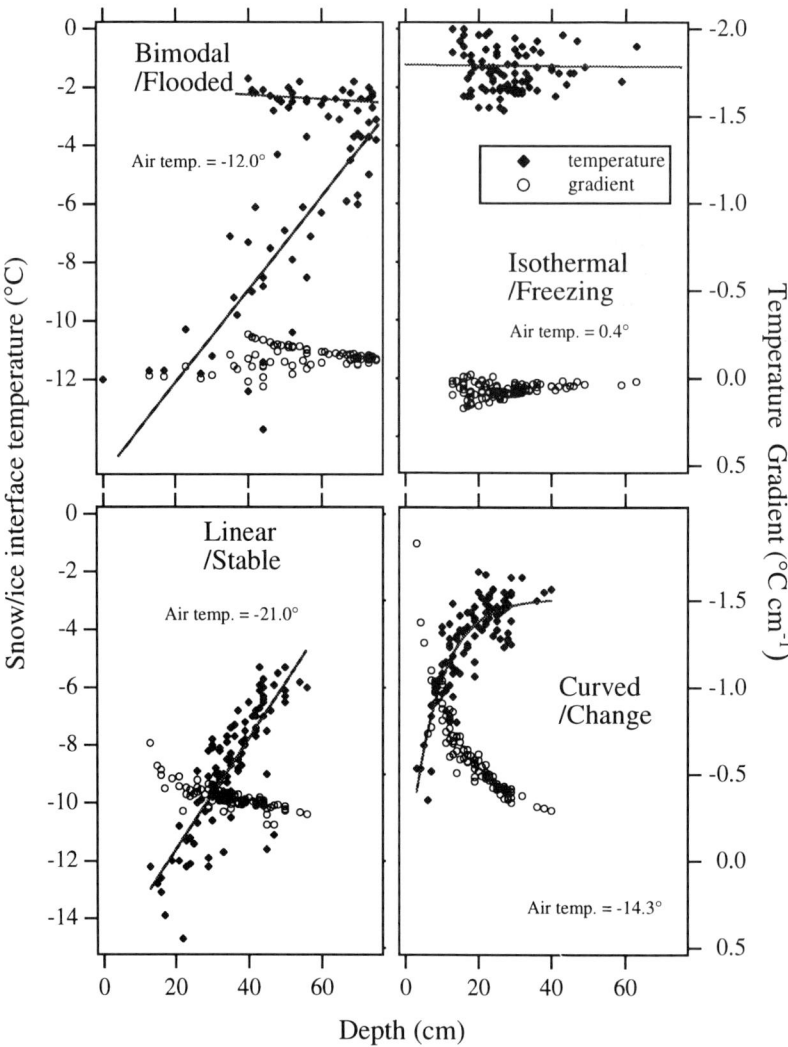

Figure 11: The four general categories into which results from snow/ice interface temperature traverses can be divided. The computed temperature gradient for each temperature measurement is also shown (open symbols).

It can be related to the rate of ice production (both snow-ice and congelation ice), if the ocean heat flux (F_w) is known.

To calculate F_a, measured snow surface and snow-ice interface temperatures, and snow depth, are used to compute the vertical temperature gradient across the snow cover ($\partial T / \partial z \approx \Delta T / h_s$) at each measurement location. Snow type percentages determined from the stratigraphic profiles are used to compute a reasonable value for the bulk effective thermal conductivity of the snow cover (k_{bulk}) for each cruise. Then the Fourier heat conduction equation:

$$F_a = -k_{bulk}\frac{\partial T}{\partial z} \approx -k_{bulk}\left\{\frac{T_s - T_i}{h_s}\right\} \quad (1)$$

is solved for F_a. By solving Equation (1) for each measurement, and averaging all measurements at a station (usually $n=101$), a spatially averaged estimate of the conductive heat flux ($\overline{F_a}$) was computed. The average of all station values gave the average for the cruise.

Four assumptions are implicit in this approach:

1) Heat fluxes through the ice and snow are vertical, probably a reasonable assumption.

2) The bulk temperature gradient is a reasonable approximation for the local gradient at the snow/ice interface. The high percentage of linear temperature profiles (63%) makes this assumption reasonable (Figure 5a).

3) The average bulk thermal conductivity (k_{bulk}) can be represented by a single value for an entire cruise. To support this assumption, we note that the relative percentages of the major types of snow varied by only a limited amount from floe to floe, and that the SD of k_{bulk} (approximately 0.03 W m^{-1} K^{-1} for all three cruises) was only about 25% of k_{bulk}. Additionally, F_a, computed using local values for k_{bulk} (derived for each station from the local snow type percentages), differed by less than 5% from values of F_a calculated using k_{bulk} for the whole cruise, with the exception of a few extreme values computed for floes on which only one or two snow pits were measured.

4) The computed temperature gradients do not reflect transient variations in temperature but instead are approximately equal to steady-state gradients and therefore proportional to the heat flux.

The final assumption warrants discussion. Results from individual snow/ice interface temperature traverses can be grouped into four categories (Figure 11). Each category arises from a particular set of thermal conditions. The *bimodal* category is observed when there is localized flooding. In flooded areas, the interface temperature may be as high as -1.8°C, depending on the salinity of the brine, while elsewhere it may be substantially lower. The *isothermal* category arises when near-freezing air temperatures prevail. At those times, the snow surface temperature approaches 0°C, and little or no drop in temperature is present across the snow pack, regardless of its depth. The third category, *near-linear* profiles, are obtained during stable (constant temperature) conditions. The last category (*curved*) results when interface temperatures are measured following an abrupt change in air temperature. In areas where the snow is thin, the interface temperature may have already equilibrated to the new conditions, but where the snow is deep, the temperature wave will not have penetrated to the base.

The last category is inconsistent with assumption 4. However, we argue that transient changes in snow surface temperature will be both positive and negative. The sheer number of measurements (n=8051) makes it likely that as many positive as negative perturbations were sampled, giving an equal number of over- and underestimates for $\overline{F_a}$.

The bulk effective thermal conductivity (k_{bulk}) of the snow pack, based on the textural percentages listed in Table 1 and the thermal conductivity values listed in Table 6, was calculated using:

$$\frac{1}{k_{bulk}} = \sum_{i=1}^{n} \frac{\varphi_i}{100 k_i} \quad (2)$$

where φ_i is the percentage of snow type i encountered during the cruise (excluding slush), and k_i is the thermal conductivity of that type of snow. For Cruise 94-5 k_{bulk} was 0.124, for Cruise 95-3 it was 0.112, and for Cruise 95-5 it was 0.138 W m^{-1} K^{-1} (Table 1). The cruise average heat flux was lower for Cruise 94-5 ($\overline{F_a}$ = 2.9 W m^{-2}) than for Cruise 95-5 ($\overline{F_a}$ = 6.1 W m^{-2}), and both of these were lower than the autumn cruise (95-3: $\overline{F_a}$ = 9.2 W m^{-2}), as could have been anticipated from annual differences in surface temperature, snow depth, bulk thermal conductivity, and average temperature gradients (Table 5).

At individual stations, interface temperatures (Figure 11) and snow depths (Figures 8, 9, and 10) varied widely, resulting in large local variations in F_a, sometimes as much as 20 W m^{-2} in 100 m. Local variations in temperature, temperature gradient, and heat flux were greater in autumn than in the winter (see SD values in Table 5). The range of these variations is suggested by the temperature gradient data in Figure 11. The local variation generally could not be predicted as a simple function of snow depth because the snow stratigraphy did not scale with snow depth.

Given the local spatial variation in F_a, it is surprising that computed values of $\overline{F_a}$ for each station, plotted as a function of the average snow surface temperature, $\overline{T_s}$, show distinct similarities (Figs 12a through 12c). For the two winter cruises, $\overline{F_a}$ increased in magnitude with $\overline{T_s}$ at a similar rate as shown by lines fit to the data ($\overline{F_a}$=0.47$\overline{T_s}$+1.54; r^2=0.71 for 94-5: $\overline{F_a}$=0.36$\overline{T_s}$-0.17; r^2=0.47 for 95-5). Some limited control by snow depth could be observed in the data, but ice thickness (data from M. Jeffries, personal communication, 1997) did not correlate with the heat flux.

In contrast, values of $\overline{F_a}$ for the autumn cruise (95-3) defined two distinct groups differentiated by ice thickness (Figure 12c). For thicker ice (> 0.65 m), the relationship between $\overline{F_a}$ and the surface temperature was similar to the relationships from the winter cruises ($\overline{F_a}$=0.451$\overline{T_s}$+5.11; r^2=0.86), but for thinner ice (≤0.65m), $\overline{F_a}$ increased with $\overline{T_s}$ at double the winter rate ($\overline{F_a}$=0.939$\overline{T_s}$+6.25; r^2=0.92). Snow depth, on the other hand, had little apparent effect on the flux.

The slope of the lines in Figure 12 is a complicated function of $\overline{T_s}$, T_i, h_s and k_{bulk}:

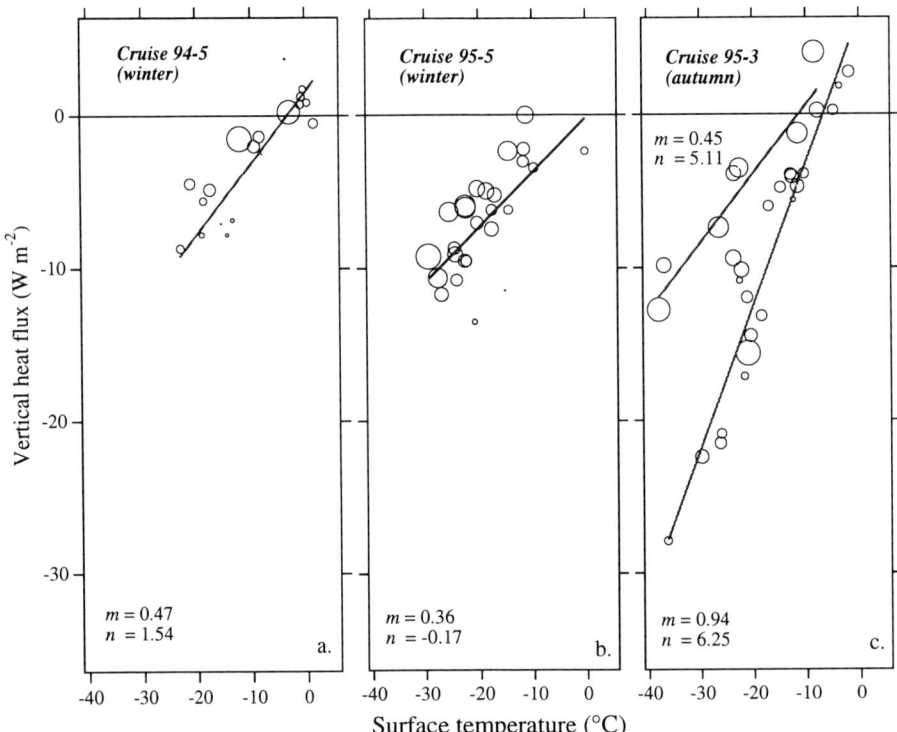

Figure 12a: The average conductive heat flux through the snow cover calculated from snow/ice interface temperatures (see text) for each ice floe station on (a) Cruise 94-5, (b) Cruise 95-5, and (c) Cruise 95-3. Each plotted value is the average of 101 measurements. For Cruise 94-5 and Cruise 95-5, the symbol size indicates the relative snow depth at each station. For Cruise 95-3, the symbol size indicates the relative ice thickness (larger circles=thicker ice). For Cruise 95-3, two heat flux classes can be differentiated based on ice thickness, but there is little sorting by snow depth. m equals the slope of the line fit to the data; n equals the intercept at $T_s=0°C$.

$$m = \frac{1}{\overline{T_s}}\left[\frac{k_{bulk}}{h_s}\left(\overline{T_s}-T_i\right) - n\right] \quad (3)$$

where m is the slope and n is the intercept of the lines. However, if $\overline{T_s}$ and T_i were constant, then the slope would be proportional to k_{bulk} divided by h_s. This ratio is the inverse of the thermal resistance of the snow pack, with the effect of the ice thickness and thermal properties manifested through $\overline{T_i}$. The similarity in the slopes suggests that there is some convergence from autumn to winter, and from winter to winter, toward a fixed value of thermal resistance for the combined sea ice and snow cover system.

5. A DIAGENETIC MODEL OF THE SNOW COVER

Some of our results, such as the latitudinal variation in snow type percentage (Figure 4), the regional snow depth distribution (Figure 7), and the regional heat flux (Figure 12 and Table 5), suggest that at the regional scale there is some degree of order or homogeneity in the snow-ice system, and that the system converges to the same state from one winter to the next. In contrast, at the local scale, results from individual ice floes, and from one floe to another (Figures 8, 9, 10 and 11, Table 4), show great heterogeneity. We now propose a diagenetic model of how the snow cover forms, and suggest one possible mechanism, based on the interaction of the snow and ice, that might work to produce local heterogeneity but regional homogeneity.

Much of the heterogeneity of the snow results from the cycling of winter weather between warm, windy storms and prolonged periods of cold with little or no wind. Snowfall comes during the storms, often at temperatures close to freezing or early in the storm development before the wind speed increases (Figure 2). One result is that the snow is "locked up" (see explanation below) before the wind is able to cause extensive drifting. This leads to an abundance of soft and moderate slabs in the snow pack (Figure 3 and Table 1) and the near-absence of hard slabs like those commonly observed in the Arctic under similar wind conditions but significantly lower temperatures [*Benson and Sturm,* 1993]. The "lock-up" also limits the formation of sastrugi and other large drift structures to a brief period just after snowfall. The two mechanisms that lock snow in place are surface melting and rain-on-snow. Several times we observed snow showers evolve into rain showers, producing nubbly ice crusts before winds were able to erode the snow. Thus, at wind speeds in excess of 15 m s^{-1}, there was little or no drifting. Three of these events are marked on Figure 2.

"Lock up" has a potentially important effect on the mass balance of the snow: it reduces sublimation and the amount of snow transported into leads, thereby maintaining snow loads at higher levels. Estimates of sublimation indicate that as much as 1/3 of the total snow pack in the Alaskan Arctic can sublime during the winter [*Pomeroy and Gray,* 1995; *Liston and Sturm,* unpublished data]. However, this rapid rate requires particles to saltate or be in suspension. The relative absence of either of these processes during high wind events over the West Antarctic pack ice suggests that sublimation losses are minimized, and therefore more of the pack is left in place loading the ice. This could contribute to the process of basal flooding and snow-ice formation and may be an important difference between Arctic and Antarctic sea ice snow covers.

The same processes that cause "lock-up" (surface melt and rain-on-snow) also cause the formation of ice lenses, percolation columns and coarse-grained zones of melt-clusters, thereby increasing the local heterogeneity of the snow. Since the bulk snow temperature is generally well below freezing, the liquid water tends to infiltrate in pipes or spread laterally at layer hydraulic discontinuities [*Colbeck,* 1990]. Percolation also forms melt-grain clusters that are manifested as bimodal grain size distributions (Table 2).

A second set of heterogeneous features (basal ice lenses, ice columns, melt-grain clusters, saline snow) form as a result of sea water flooding and wicking of brine into the snow by capillary action. Flooding produces snow-ice (Figure 6), while wicking results in the formation of saline snow (Table 3) with much of its original structure and density intact (Table 1). The flooding occurs primarily by isostatic depression of ice floes under a snow load [*Ackley et al.,* 1990; *Lytle and Ackley,* 1996; *Massom et al.,* 1997, *Jeffries and Adolphs,* 1997]. When sea water floods the base of snow, it produces slush. If the slush freezes, then layers of snow-ice are added to the ice column. Measurements made during the three cruises (Figure 6) indicate that this process is very common. Snow-ice formation consumed more than half the snow pack in some locations and was a critical part of the snow and ice mass balance.

Saline snow forms when limited amounts of brine are left in snow to freeze. Flooding, with subsequent drain-back before freezing, could produce saline snow, but particles maintained in a slush for even a short period will metamorphose into oblate spheroidal particles [*Raymond and Tusima,* 1979; *Colbeck,* 1986b]. If rafting, a common process of sea ice thickening, raised a flooded ice floe and allowed the brine to drain away, it would have to do so immediately after the flooding, or else the snow texture would indicate that the flooding had taken place. Since snow textures indicative of this process were not observed, we conclude that flooded snow must invariably have been transformed into snow-ice. *Takizawa* [1985] observed saline snow forming concurrently with snow-ice, and suggested that a process of capillary wicking was taking place. Most of the saline snow we observed was above snow ice or slush (Table 1 and Figure 6), so we also conclude that the majority formed by wicking from a slush layer.

Both snow-ice and saline snow formation contribute directly to the formation of depth hoar in two ways. First, flooding and wicking increase the temperature at the base of the snow, thereby strengthening temperature gradients (Figure 11). Second, if flooding takes place, the snow depth is reduced by an amount equal to the slush thickness. The thinner snow also increases the temperature gradient. The enhanced gradients either initiate depth hoar metamorphism or accelerate it. For example, in the Bimodal/Flooded case illustrated in Figure 11, in snow 40 to 60 cm deep, flooding increased the interface temperature 4° to 6°C, strengthening the vertical temperature gradient by -0.06 to -0.25 °C cm^{-1}. This constitutes an increase of 50% to 100%, and was sufficient in at least five locations along the 100-m line to push the temperature gradient above the critical value for depth hoar formation.

The processes listed above tend to be local and work at the grain and layer scale in the snow pack. They produce local heterogeneity. Basal flooding, with its dual impact on the snow (creating features due to wetting and features due to depth hoar growth), is particularly effective at producing small scale variations in grain and layer features. It also tends to produce heterogeneity in snow depth, as the depth is reduced only in those areas where flooding has taken place. Basal flooding is controlled by the ice thickness distribution. As ice thickness becomes more uniform, effects of flooding will also become more uniform; as ice thickness increases relative to snow, flooding effects will diminish.

Is there a way that processes which produce local heterogeneity might also produce homogeneous conditions at the regional scale? We think it is possible through the mechanism of snow-ice formation. If we ignore dynamic processes, flooding and snow-ice formation will tend to produce heterogeneous snow conditions, but their effect on ice thickness will be just the opposite. Thin floes will be preferential sites for flooding when snow is deposited. If flooded, the thin floes will thicken more rapidly than adjacent thick floes through snow-ice formation (recall that flooding increases the rate of heat loss). This selective process will repeat as many times as there are snowfalls, or until little or no difference in ice thicknesses is present in the ice distribution. The end result will be that the ice thicknesses distribution will become more uniform over time, unless dynamic processes mask the effect. As the ice becomes more uniform, the degree to which basal processes create heterogenity in the snow will also decline, and the total stratigraphic column of snow and ice will tend to become more uniform. However, the snow stratigraphy reflects the cumulative metamorphic history, and will retain heterogeneity developed earlier in the winter.

Does evidence support this hypothetical scenario? It is consistent with the following results:

1. Snow-ice was ubiquitous over the entire cruise area (Table 3), and often as thick as the snow pack itself. It increased in thickness through the winter at about the same rate as the snow (Figure 6), suggesting a balance between snow and ice thickness.

2. Depth hoar and slush comprised more of the snow pack in autumn than in winter (Table 1), consistent with higher rates of snow-ice production and concomitant depth hoar formation in the autumn than later in the winter.

3. Two distinct classes of conductive heat flux were present in the autumn but not the winter (Figure 12c). They reflected two classes of ice thickness, with thin ice exhibiting high heat fluxes consistent with rapid snow-ice formation and widespread flooding, and thicker ice exhibiting lower heat fluxes comparable to those observed later in the winter. These results suggest that the snow and ice system in autumn is more heterogeneous than later in the winter, but that even in autumn, some floes and their snow cover have already attained a "stable" configuration.

4. The slopes of lines fit to the heat flux data (Figures 12a, 12b, and 12c) are similar from one winter to the next, despite large differences in climatic conditions (Table 5). These slopes, proportional to the inverse of the thermal resistance, suggest that both year to year, as well as from autumn to winter (Figure 12c), the total snow and ice system converges toward some regional thermal resistance.

5. Variation (as reflected in SD values) in interface temperature, temperature gradient (Table 5), and heat flux all decrease from autumn to winter, suggesting increased regional homogeneity.

The convergence of the West Antarctic snow and sea ice system toward some sort of regional homogeneity, particularly for thermal resistance, has important implications for energy exchange and climate modeling. The convergence implies that the complex interaction of the many physical processes taking place in the snow and ice system are in some way self-limiting or self-regulating. As a result, year-to-year variations in energy exchange tend to be small, and potentially resistant to changes in climate. This possibility, that some sort of buffering or self-limitation is controlling the winter heat loss and perhaps the range of physical conditions that can develop in the ice pack, needs to be considered and accounted for in climate models that include sea ice. If it is not, such models will over-emphasize the climate response of the snow and ice system.

6. CONCLUSIONS

Our results highlight two aspects of the snow cover of the west Antarctic pack ice that are challenging to researchers trying to model sea ice growth and decay, or researchers who want to measure the properties of the region by remote sensing: 1) the snow is extremely heterogeneous at scales less than hundreds of kilometers, and 2) the heterogeneity is intimately linked through coupled snow and ice processes like snow-ice production and depth hoar formation.

The heterogeneity is a function of spatial scales. At the regional level, predictable stratigraphic and textural snow sequences develop each winter. For example, it can be assumed that the snow cover will consist of depth hoar, icy snow, melt-grain clusters, and soft to moderate slabs. The basal snow will be saline, often wet, and beneath it there will be snow-ice as thick as the snow. Depth hoar will increase with increasing latitude, while the iciness of the snow will increase as the edge of the pack is approached. Heat flux from the snow pack will vary from winter to winter, but the bulk thermal resistance of the ice and snow will converge from year to year.

The regional-scale variations in snow cover are predictable because the climate of the region is a reliable sequence of warm storms and cold periods. Through the action of wind and air temperature, it sets the initial ice distribution onto which the first snowfalls of autumn are deposited, and it determines how much snow will fall, how much wind redistribution will occur, and thus how much snow-ice will form. We have observed this dominance of climate on terrestrial snow covers [Sturm et al., 1995], where large inter-annual variations in weather fail to change the fundamental nature of the snow.

The same climate processes that create large-scale homogeneity produce heterogeneity at smaller scales. Snow depth can vary as much on a single floe as it can from one floe to the next or across the cruise area, and stratigraphy and thermal properties show the same sort of variation. The reason is that the snow is a cumulative product of its metamorphic history and the initial conditions when it is deposited. In autumn, the snow is deposited on a melange of floes of varying ages, thickness, histories and metamorphic states. The amount of snow, in combination with the floe thickness, produces radically different snow histories: flooding, reduction in snow depth and rapid depth hoar formation in one case, or little change in another. At any given point in space, the state of the snow pack is the result of a number of difficult-to-predict factors: 1) the age and history of the ice floe, 2) the amount of snow that has been transformed into snow-ice, 3) the proximity and size of ice roughness elements, and 4) how much snow drifting has occurred. When these factors are combined, point prediction of snow properties becomes impossible.

Is this pattern similar elsewhere in the Antarctic? We cannot be certain, but similar weather patterns have been observed in the Weddell Sea [*Massom et al.*, 1997] and in the East Antarctica [*Worby et al.*, in press]. Recent studies [*Worby et al.*, in press; *Eicken*, 1991; *Ackley et al.*, 1990; *Eicken et al.*, 1994; *Jeffries et al.*, 1997 and this volume] also indicate that flooding of the snow/ice interface is common elsewhere in the Antarctic. Combined these suggest that similar snow and ice diagenesis might be taking place, and that the ice pack may evolve toward regional homogeneity over a widespread part of the Antarctic pack ice. More striking, however, is the difference between what we observed in the West Antarctic pack ice and the Arctic pack. In the Arctic, snow ice formation is relatively rare, and depth hoar and hard wind slabs make up the bulk of the snow cover. In the snow pack, melt features and features due basal flooding are rare until spring. Presumably, similar differences exist in local vs. regional homogeneity as well. These differences are so substantial that large-scale models will need to treat snow and ice interactions differently in the Arctic and Antarctic.

Acknowledgments. Captain J. Borkowski and the crew of the R. V. *Nathaniel B. Palmer* and Antarctic Support personnel contributed to the success of this study. We also thank Martin Jeffries for his support and for providing data on snow-ice. Ted Maksym, Joanne Groves, Barbara Hurst-Cushing, Campbell Scott, Doyle Nicodemus, Greg Packard, Barney Kane, Bill Young, Erika Lawson, Gwyneth Hufford, Elizabeth Chilton, Joe Sapiano, Ricardo Jaña, André Belem, Bernhard Rabus and Ute Adolphs all deserve thanks for their willingness to collect data for us in what was often terrible weather. Jon Holmgren, Ted Maksym, Carl Benson, Gary Maykut and Steve Warren all provided valuable suggestions that greatly improved the paper. This work was supported by NSF Grant OPP-9316767. The National Science Foundation and the Antarctic Cooperative Research Centre made it possible for R. M. to participate; we thank both organizations.

REFERENCES

Ackley, S. F., M. A. Lange, and P. Wadhams, Snow cover effects on Antarctic sea ice thickness, in *Sea Ice Properties and Processes, CRREL Monograph 90-1*, edited by S. F. Ackley and W. F. Weeks, U.S. Army Corps of Engineers, Hanover, NH, 16-21, 1990.

Akitaya, E., Studies on depth hoar, *Contr. Inst. Low Temp. Science*, 26A, 1–67, 1974.

Andreas, E. L., A theory for the scalar roughness and the scalar transfer coefficients over snow and sea ice, *Boundary Layer Meteorol.*, 38, 159–184, 1987.

Armstrong, R. L., A.T.C. Chang, A. Rango, and E. Josberger, Snow depths and grain-size relationships with relevance for passive microwave studies, *Ann. Glaciol.*, 17, 171-176, 1993.

Baker, G. C., Electrical conductivity, freezing temperature, and salinity relationships for seawater and sodium chloride solutions for the salinity range of 0 to over 200 ppt., *Geophysical Institute Report UAG R-310*, 1987.

Barry, R. G., The parameterization of surface albedo for sea ice and its snow cover, *Prog. in Phys. Geog.*, 20, 1, 63-79, 1996.

Benson, C. S. and M. Sturm, Structure and wind transport of seasonal snow on the Arctic slope of Alaska, *Ann. Glaciol.*, 18, 261–267, 1993.

Chacho, E. F. and J. Johnson, Air permeability of snow, *EOS*, 68, 1271, 1987.

Colbeck, S. C., Classification of seasonal snow cover crystals, *Water Res. Res*, 22, 59S–70S, 1986a.

Colbeck, S. C., Statistics of coarsening in water-saturated snow, *Acta Metall.*, 34, 347-352, 1986b.

Colbeck, S. C., The layered character of snow covers, *Rev. of Geophys.*, 29, 81-96, 1991.

Colbeck, S. C., E. Akitaya, R. Armstrong, H. Gubler, J. Lafeuille, K. Lied, D. McClung and E. Morris, *The International Classification for Seasonal Snow on the Ground*, The International Commission on Snow and Ice of the International Association of Scientific Hydrology, and the International Glaciological Society, 23pp., 1990.

Comiso, J. C., Sea ice effective microwave emissivities from satellite passive microwave and infrared observations, *J. Geophys. Res.* 88, 7686–7704, 1983.

Drinkwater, M. R., R. Hosseinmostafa and P. Gogineni, C-band backscatter measurements of winter sea ice in the Weddell Sea, Antarctica, *Int. J. Remote Sensing*, 16(17), 3365–3389, 1995.

Eicken, H., M. A. Lange, H.-W. Hubberton and P. Wadhams, Characteristics and distribution patterns of snow and meteoric ice in the Weddell Sea and their contribution to the mass balance of sea ice, *Ann. Geophys.*, 12, 80–93, 1994.

Eicken, H., H. Fischer and P. Lemke, Effects of snow cover on Antarctic sea ice and potential modulation of its response to climate change, *Ann. Glaciol.* 21, 369–376, 1995.

Grenfell, T. C. and G. Maykut, The optical properties of ice and snow in the Arctic Basin, *J. Glaciol.*, 18(80), 445-463, 1977.

Haas, C. and T. Viehoff, Sea ice conditions in Bellingshausen/Amundsen Sea: Shipboard observations and satellite imagery during ANT XI/3, *Berichte aus dem Fachbereich Physik*, Alfred Wegner Institute für Polar und Meeresforschung, Report 51, 1994.

Isaaks, E. H., and R. M. Srivastava, *Applied Geostatistics*, Oxford University Press, New York. 561 pp, 1989.

Jeffries, M. O., R. A. Shaw, K. Morris, A. L. Veazey and H. R. Krouse. Crystal structure, stable isotopes (δ18O) and development of sea ice in the Ross, Amundsen and Bellingshausen Seas, Antarctica. *J. of Geophys. Res.*, 99(C1):985-995, 1994.

Jeffries, M. O., R. Jaña and S. Li, Sea ice and snow thickness distributions in late winter 1993 and 1994 in the Ross, Amundsen and Bellingshausen Seas. *Ant. J. U.S.*, 1995a.

Jeffries, M. O., S. Cushing and M. Porter, Sea ice development in the Ross, Amundsen and Bellingshausen Seas revealed by analysis of ice cores in late winter 1993 and 1994, *Ant. J. U.S.*, 1995b.

Jeffries, M. O. and U. Adolphs, Early winter snow and ice thickness distribution, ice structure and development of the western Ross Sea pack ice between the ice edge and the Ross Ice Shelf. *Antarct. Sci.*, 1997.

Jeffries, M. O., A. P. Worby, K. Morris and W. F. Weeks, Seasonal variations in the properties and structural composition of sea ice and snow cover in the Bellingshausen and Amundsen seas, Antarctica. *J. of Glaciol.*, 43 (143), 1997.

Jeffries, M. O., S. Li, R. Jana, H. R. Krouse, and B. Hurst Cushing, Late winter first year ice floe thickness variability, seawater flooding, and snow ice formation in the Amundsen and Ross Seas, *Antarct. Res. Series*, this volume.

Kellogg, W. W., Climatic feedback mechanisms involving the polar regions, *Climate of the Arctic*, edited by G. Weller and S. A. Bowling, Geophysical Institute, University of Alaska, 111-116, 1975.

Ledley, T. S., Snow on sea ice: competing effects in shaping climate, *J. Geophys. Res.*, 96, 17,195–17,208, 1991.

Lohanick, A. W., Some observations of established snow cover on saline ice and their relevance to microwave remote sensing, in *Sea Ice Properties and Processes*, CRREL Monograph 90-1, edited by S. F. Ackley and W. F. Weeks, U.S. Army Corps of Engineers, Hanover, NH, 61-67, 1990.

Lytle, V. I. and S. F. Ackley, Snow properties and surface elevation profiles in the Western Weddell Sea (NBP92-2). *Ant. J. U.S.*, XXVII(5): 93-94, 1992.

Lytle, V. I. and S. F. Ackley, Heat flux through sea ice in the western Weddell Sea: convective and conductive transfer processes, *J. of Geophys. Res.*, 101, (C4), 8853-8868, 1996.

Marbouty, D., An experimental study of temperature gradient metamorphism, *J. Glaciol.*, 26(94), 303–312, 1980.

Massom, R. A., M. R. Drinkwater and C. Haas, Spatial and temporal distribution of winter snowcover properties on sea ice in the Weddell Sea, *J. Geophys. Res.*, 102(C1), 1101-1117, 1997.

Maykut, G. A., The surface heat and mass balance, in *The Geophysics of Sea Ice*, N. Untersteiner, ed. NATO ASI Series B, Physics Vol. 146. Plenum Press, New York, 395-464, 1986.

Maykut, G. A. and N. Untersteiner, Some results from a time-dependent thermodynamic model of sea ice. *J. Geophys. Res.* 75, No. 6, 1550-1575, 1971.

McBratney, A. B. and R. Webster, Choosing functions for semi-variograms of soil properties and fitting them to sampling estimates, *Journal of Soil Science*, 37, 617-639, 1986.

Morris, K., M. O. Jeffries, and S. Li, Sea ice characteristics and seasonal variability of ERS-1 SAR backscatter in the Bellinghausen Sea, *Antarct. Res. Series*, this volume.

Oliver, M. A. and R. Webster, Semi-variograms for modelling the spatial pattern of landform and soil properties, *Earth Surface Processes and Landforms*, 11, 491-504, 1986.

Pomeroy, J.W. and D. M. Gray, Snowcover, *Accumulation, Relocation and Management*, NHRI Science Report No. 7, Environment Canada, National Hydrologic Research Institute, 1995, 142 p.

Raymond, C. F. and K. Tusima, Grain coarsening of water saturated snow, *J. Glaciol.*, 22, (86), 83-106, 1979.

Robert, A. and K. S. Richards, On the modelling of sand bedforms using the semivariogram, *Earth Surface Processes and Landforms*, 13, 459-473, 1988.

Sturm, M., The role of thermal convection in heat and mass transport in the subarctic snow cover, *CRREL Report 91-19*, USA Cold Regions Research and Engineering Lab., Hanover, New Hampshire, 1991.

Sturm, M. and J. B. Johnson, Thermal conductivity measurements of depth hoar, *J. Geophys. Res.*, 97, 2129–2139, 1992.

Sturm, M., J. Holmgren, and G. E. Liston, A seasonal snow cover classification system for local to global applications, *J. Climate*, 8, 5, (II), 1261-1283, 1995.

Sturm, M., K. Morris, and R. Massom, A description of the snow cover on the winter sea ice of the Amundsen and Ross Seas, *Ant. J. U.S.*, 1996.

Sturm, M., J. Holmgren, M. König, and K. Morris, The thermal conductivity of seasonal snow, *J. Glaciology*, 43, 143, 26-41, 1997.

Sturm, M. and C. S. Benson, Vapor transport, grain growth and depth hoar development in the subarctic snow, *J. Glaciology*, 43, 143, 42-59, 1997.

Takizawa, T., Salination of snow on sea ice and formation of snow ice, *Annals of Glaciology*, 6, 309-310, 1985.

Trabant, D. and C. S. Benson, Field experiments on the development of depth hoar, *Mem. Geol. Soc. of Amer.*, 135, 309–322, 1972.

Worby, A. P., M. O. Jeffries, W. F. Weeks, K. Morris, and R. Jana, The thickness distribution of sea ice and snow cover during late winter in the Bellingshausen and Amundsen Seas. *J. Geophys. Res.*, 101 (C12), 28,441-28,455, 1996.

Worby, A. P., R. A. Massom, I. Allison, V. I. Lytle, and P. Heil, East Antarctic sea ice: a review of its structure, properties and drift, this volume, *Antarc. Res. Series*, this volume.

Zhang, T., T. E. Osterkamp, and K. Stamnes, Impact of the depth hoar layer of the seasonal snow cover on the ground thermal regime, *Water Res. Res.*, 32(7), 2075-2086, 1995.

M. Sturm, USA-CRREL-Alaska, P.O. Box 35170, Ft. Wainwright, AK 99703-0170

(Received August 29, 1996; accepted April 7, 1997)

SNOW DEPTH DISTRIBUTION OVER SEA ICE IN THE SOUTHERN OCEAN FROM SATELLITE PASSIVE MICROWAVE DATA

Thorsten Markus and Donald J. Cavalieri

Laboratory for Hydrospheric Processes, Code 971, NASA Goddard Space Flight Center, Greenbelt, Maryland

A major shortcoming in polar heat budget studies is the lack of reliable large-scale information on the distribution of snow depth on sea ice. This is particularly true in the Southern Ocean where the snow depth distribution is highly variable both spatially and temporally. In this paper, we present an algorithm to calculate snow depth on sea ice using DMSP SSM/I data. In-situ snow depth measurements obtained from different expeditions to the Weddell, Bellingshausen, and Amundsen Seas are regressed on SSM/I brightness temperatures. The relationship between snow depths and microwave data is consistent for all data sets. Multi-temporal information is included to identify wet snow or melt-refreeze events which result in large snow depth retrieval errors. Using this algorithm, we calculate monthly snow depths from 1988 through 1994. Areas of deep snow cover (about 40 cm) are found in the northwestern Weddell Sea and in the Bellingshausen and Amundsen Seas. Shallower snow depths (less than 15 cm) occur in the East Antarctic region and seaward of the Ronne and Ross Ice Shelves. The latter two regions are known to have frequent coastal polynyas and therefore considerable amounts of new ice with little if any snow cover. Validation of the retrieval estimates with large-scale snow depth distributions from field experiments for various regions shows good agreement in both average snow depth and distribution. Average snow depths vary between 7 and 25 cm. The correlation coefficient between in-situ snow depths and SSM/I-derived averages is 0.81. On average, the SSM/I snow depths are lower by 3.5 cm, which results primarily from the areal integration of the SSM/I retrieval.

1. INTRODUCTION

The thermal conductivity of snow is about an order of magnitude less than that of sea ice [*Maykut and Untersteiner*, 1971]. Consequently, just a little snow on sea ice can greatly affect the heat flux between the surface and atmosphere. Although the relative importance of snow cover on single parameters, such as thermal insulation, albedo, specific heat, and ice strength are different, the overall importance of snow cover on sea ice in the climate system is well established [*Ledley*, 1991; *Eicken et al.*, 1995]. Knowledge of the distribution of snow thickness on sea ice is vital to understanding the overall heat exchange occurring in the polar regions [*Maykut and Untersteiner*, 1971]. Snow depth and accumulation rate are also important variables in the fresh water budget of the oceans. A quantitative knowledge of snow depth estimates would provide more accurate estimates of precipitation minus evaporation (P-E), which are critically needed in state-of-the-art coupled ice-ocean models [*Häkkinen*, 1995]. For all these reasons, snow depth measurements have been included in recent Antarctic field experiments [e.g. *Wadhams et al.*, 1987; *Lange and Eicken*, 1991; *Allison et al.*, 1993; *Worby and Massom*, 1995; *Jeffries*

Copyright 1998 by the American Geophysical Union

et al., 1994; Eicken et al. 1994; Worby et al., 1996; Massom et al., 1997; Sturm et al., this volume].

Despite recent efforts to improve the knowledge of the snow cover and its properties, the data are currently spotty and continuous information for the entire Southern Ocean can only be provided by satellite observations. For this reason, we developed a satellite data algorithm which provides a representative depiction of the large-scale snow depth distribution on sea ice in the Southern Ocean. A map of Antarctica and the Southern Ocean is presented in Figure 1. As no algorithm currently exists for snow depth retrieval over sea ice surfaces, this study is a first-time effort. In this study we make use of passive microwave data from the defense meteorological satellite program (DMSP) special sensor microwave/imager (SSM/I). Passive microwave sensors, unhampered by cloud cover and unaffected by darkness, are particularly well suited for monitoring changing conditions in the polar regions on a daily basis.

Current snow algorithms for use with satellite passive microwave data have been developed for land regions only [Künzi et al., 1982; Hallikainen and Jolma, 1986, Chang et al., 1987]. All of these algorithms use a linear relationship between brightness temperature difference at 19 and 37 GHz and snow depth, or rather snow water equivalent, which is calculated as a linear function of snow depth. Chang et al. [1987] and Künzi et al. [1982] found the best correlation between in-situ snow depth and brightness temperature difference at horizontal polarization, whereas Hallikainen and Jolma [1986,1992] found a higher correlation using vertical polarization data. Nevertheless, 19–37 GHz combinations were superior to combinations of other channels at either polarization.

Snow particles act as scatterers of microwave radiation. The scattering of the upwelling radiation depends on snow thickness (more scatterers) and crystal size (bigger crystals increase the scattering) and provides the physical basis for the microwave estimation of snow [Chang et al., 1987]. Rango et al. [1979] found a strong linear correlation between decreasing brightness temperature at 37 GHz (at both polarizations) with increasing snow depth. The effect of scattering decreases with increasing wavelength. Thus, brightness temperatures at 37 GHz are reduced more than brightness temperatures at 19 GHz with increasing snow depth.

All the aforementioned algorithms are limited to dry snow areas. With increasing wetness or liquid water content of the snow layer, the emissivity of snow increases more at 37 GHz than at 19 GHz [Schanda et al., 1983; Stiles and Ulaby, 1986] leading to underestimates in snow depth. Wankiewicz [1993] uses a time series of brightness temperature to detect the onset of melt and thus wet snow. Walker and Goodison [1993] use the difference between vertical and horizontal polarization at 37 GHz to distinguish areas of wet snow from snow-free land.

Passive microwave signatures of snow on sea ice have been studied in field experiments [e.g. Mätzler et al., 1984; Grenfell, 1986; Onstott et al., 1987; Comiso et al., 1989; Garrity, 1990; Lohanick, 1990; Garrity, 1992] and in experiments involving artificial sea ice grown in a tank [Grenfell and Comiso, 1986; Lohanick, 1993]. Snow cover can at times mask the signature of the underlying ice [Lohanick, 1990] so that variations in emissivity result predominantly from variations in snow cover properties. The exact relationship between snow properties and observed brightness temperatures is unclear as a result of the complex structure of snow, its layering and metamorphism [Colbeck, 1982, 1991]. For example, Grenfell and Comiso [1986] found little variation before and after snow fall at 10 GHz because of the reduced scattering of snow at lower frequencies. On the other hand, Lohanick [1993] reported a dramatic decrease

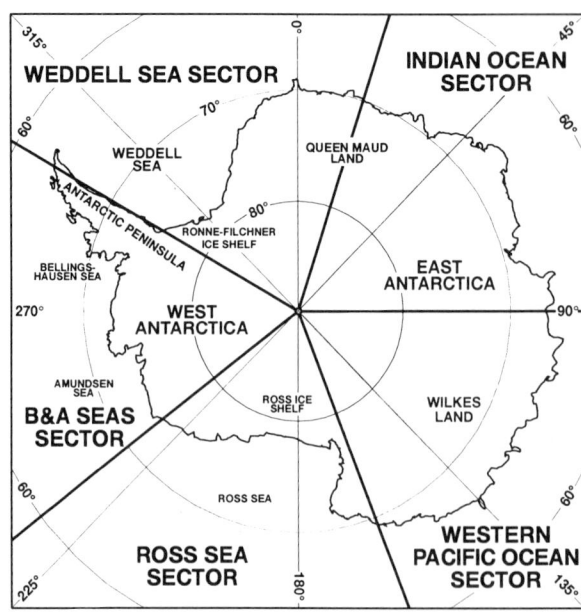

Fig.1. Map of the Antarctic region. The definition of the sectors is adapted from Gloersen et al. [1992]. "B&A Seas Sector" means Bellingshausen and Amundsen Seas Sector.

in brightness temperature at 10 GHz because of the development of a slush layer at the snow-ice interface immediately after the snow fall. The presence of depth hoar layers increases the scattering at higher frequencies as a result of larger snow grain sizes and thus reduces the brightness temperature [*Hall et al.*, 1986; *Shuman et al.*, 1993]. Layering in dry snow cover was found to strongly influence the horizontally polarized brightness temperature more than the vertical polarization [*Mätzler et al.*, 1984]. Besides these layering effects, the temporal change of snow properties such as grain size, density, and wetness affects the dielectric properties [e.g. *Hallikainen et al.*, 1986; *Onstott et al.*, 1987; *Mätzler*, 1987; *Wang et al.*, 1992; *Hallikainen and Winebrenner*, 1992; *Garrity*, 1992]. These effects are discussed in more detail in section 3.

Our approach to the retrieval of snow depth over Antarctic sea ice from passive microwave data is described in the next section. This is followed by a discussion of error sources and limitations (section 3). Monthly snow depth distributions are presented (section 4) and validated through a comparison with large-scale in-situ snow depth distributions (section 5).

2. APPROACH TO THE PROBLEM

Snow cover on sea ice is characterized by heterogeneity and variability. Snow depths can vary greatly over spatial scales as small as a meter [*Sturm et al.*, this volume]. Because of the comparatively low spatial resolution of the SSM/I sensor (69 km × 43 km footprint at 19 GHz), much of this small-scale heterogeneity and variability will be reduced in the satellite data. For this same reason attempts to model the various snow properties on microwave emission are not likely to lead to useful large-scale relationships. Instead, in this study, we use a statistical approach directly comparing in-situ snow depths with satellite radiance measurements. In-situ snow depth measurements obtained from several cruises to the Southern Ocean are compared with coincident SSM/I data. While we do not know how well these in-situ measurements represent the average snow depth over an SSM/I footprint, over 1000 snow depth measurements covering over 100 individual SSM/I pixels in different seasons and regions should provide the basis for deriving an algorithm to measure large-scale snow depth variability.

2.1. Satellite Data

The SSM/I measures microwave radiances at 19.4, 37.0, and 85.5 GHz at horizontal and vertical polarizations, and at 22.2 GHz at vertical polarization only. The sensor, operating from a near-polar orbit, conically scans the surface of the earth with a swath width of 1400 km, and thus provides near-global coverage every day. The spatial resolution of the sensor varies with frequency, ranging from 69 km × 43 km at 19.4 GHz to 15 km × 13 km at 85.5 GHz (see *Hollinger et al.* [1987] for a detailed sensor description). In this study, we use brightness temperatures mapped to a polar stereographic projection available on CD-ROM from the National Snow and Ice Data Center (NSIDC) in Boulder, Colorado [*NSIDC*, 1992]. The brightness temperature of each 25-km grid cell is a daily average of all passes during a 24 hour period.

2.2. In−Situ Data Sets

Snow depth measurements from the Bellingshausen-Amundsen Sea area were obtained from the R/V *Nathaniel Palmer* cruise in August and September 1993 [*Worby et al.*, 1996]. Snow depths were estimated when floes were tipped on their sides as the ship broke through the ice. The total number of estimates made was about 4000. Measurements within one SSM/I pixel were averaged. This led to a total of 80 averaged values of snow depth and coincident SSM/I radiances. Although these are not direct measurements, the average might well reflect the average snow depth along the route for an SSM/I pixel.

Two data sets in the Weddell Sea were available. One was snow measurements from Ice Station Weddell (ISW) [*Lytle and Ackley*, 1996]. This data set contained 11 transects near ISW made between February and May 1992. Each of the transects contains 600 individual snow depth measurements at 1 m intervals.

The other data set was wintertime data (July 1992) from an R/V *Polarstern* cruise in the Weddell Sea [*Drinkwater and Haas*, 1994; *Massom et al.*, 1997]. The data consist of between 2 and 75 individual measurements on either the same floe or on different floes. This results in snow depth averages for 20 SSM/I pixels.

Although extensive samples are taken, the areal coverage is still small compared to an area of about 625 km^2 for a single 25-km SSM/I image pixel.

Therefore, even the average of measurements for an individual location represents essentially a point measurement. The actual snow depth variability over any SSM/I pixel remains uncertain.

2.3. Method

The dominant factor affecting the observed brightness temperature of the ocean surface in polar regions is the fraction of sea ice cover within the observed area (i.e., ice concentration). This dominance results from the large contrast in microwave emission between open water and consolidated sea ice. This contrast depends on the frequency and polarization component of the measured radiation. As we are only interested in changes in brightness temperature resulting from changes in snow cover on sea ice, we exclude open water from the signal. The measured brightness temperature at frequency ν and polarization p, $T_B(\nu p)$, consists of a linear mix of brightness temperatures for open water, $T_{\text{Bow}}(\nu p)$, and sea ice, $T_{\text{Bice}}(\nu p)$:

$$T_B(\nu p) = C T_{\text{Bice}}(\nu p) + (1 - C) T_{\text{Bow}}(\nu p) \quad (1)$$

where C is the ice concentration. From this equation, $T_{\text{Bice}}(\nu p)$ is calculated and used in the snow algorithm. Changes in the resulting $T_{\text{Bice}}(\nu p)$ will reflect changes of the snow/ice system. Henceforth, the subscript "ice" indicates the use of T_{Bice} instead of T_B.

Sea ice concentration is calculated using the NASA Team algorithm [*Cavalieri et al.*, 1984; *Gloersen and Cavalieri*, 1986] with tie points for the SSM/I from *Cavalieri et al.* [1992]. This algorithm makes use of the polarization at 19 GHz

$$PR = \frac{T_B(19V) - T_B(19H)}{T_B(19V) + T_B(19H)} \quad (2)$$

and the spectral gradient ratio of the 19 GHz and 37 GHz vertically polarized components

$$GR^V = \frac{T_B(37V) - T_B(19V)}{T_B(37V) + T_B(19V)} \quad (3)$$

to derive the fraction of open water and two ice types. PR gives predominantly a measure of total ice concentration, and GR gives predominantly the fraction of the two ice types, which are first-year and multi-year ice in the Arctic. In the Antarctic, the types are referred to as type A and type B because a true multiyear ice signature cannot be unambiguously identified in the microwave data [*Cavalieri et al.*, 1992].

The ice type A signature is similar to first-year ice, whereas the ice type B signature with its negative GR presumably results from volume scattering associated with a heavy snow cover. Total ice concentration is calculated by summing these two ice type concentrations.

After correcting the observed brightness temperature for open water, the next step was to plot the ice brightness temperatures for various SSM/I channels versus in-situ snow depths. The results are presented in Figure 2. For snow depths up to 30 cm, the 19 GHz channels are independent of snow depth. For greater snow depths, the horizontal polarization data show a better linear relationship than the vertical polarization data as reflected in the higher correlation coefficient ($T_{\text{Bice}}(37H)$ yields a correlation coefficient of -0.66). Although microwave emission at 85 to 90 GHz has been found to be particularly sensitive to snow [*Grenfell*, 1986; *Cavalieri et al.*, 1986; *Comiso et al.*, 1989], the 85 GHz data are much more scattered with almost no correlation. The reasons are probably related to their greater sensitivity to weather effects and to their shallow penetration depth of only a few centimeters. Therefore, 85 GHz data may be a good indicator for determing the presence of snow, but may not be useful for deriving actual snow depths. The 85 GHz channels were therefore excluded from further consideration.

Scatterplots of snow depth versus the differences between 37 and 19 GHz at both polarizations as well as their respective gradient ratios are presented in Figure 3. The GRs have higher correlation coefficients than the differences, presumably because the ice physical temperature variability has been removed to first order. Interestingly, the correlation coefficients are almost identical to those found by *Hallikainen and Jolma* [1992] in Finland.

A key result using the 19 GHz and 37 GHz channel combination is the improvement in correlation at snow depths of less than 35 cm. This is particularly important because most snow depth measurements are below 35 cm. As noted above, the individual brightness temperatures show little correlation for lower snow depths. The channel combinations, instead, have a consistent correlation coefficient of -0.55 with a standard deviation of 5.9 cm.

Both the correlation coefficient and the standard deviation of the linear fits are quite sensitive to single outliers. By removing the three pixels in the lower left corner and the one pixel at the top right corner (Figure 3), the correlation coefficient increases to -0.77 for GR^V_{ice} and to -0.74 for GR^H_{ice}. The corre-

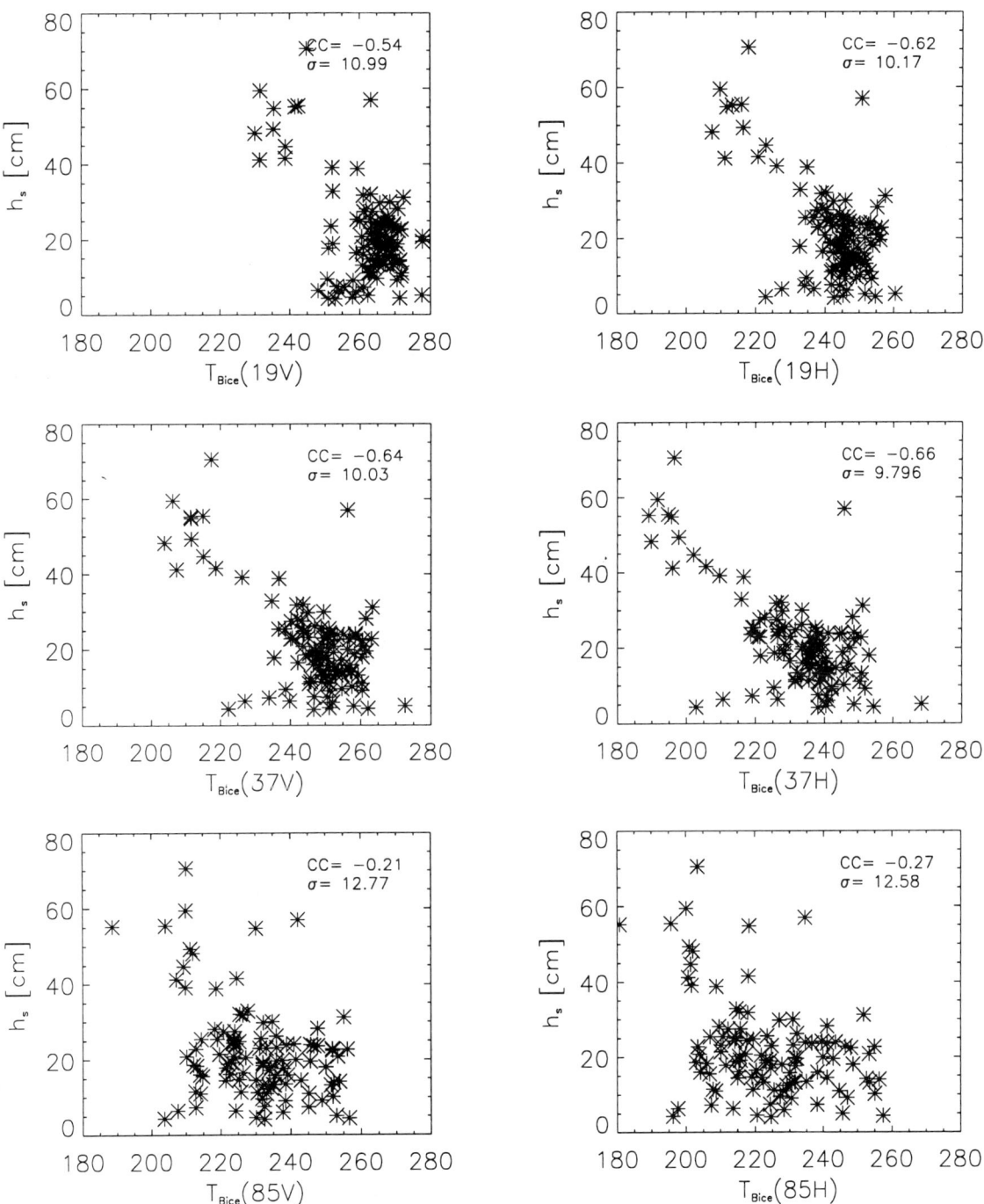

Fig.2. Scatterplots of ice-concentration-corrected brightness temperatures of the SSM/I channels versus in-situ snow depth measurements. CC is the correlation coefficient, and σ the standard deviation of a linear regression fit in cm.

lation coefficients for the T_{Bice} differences are about -0.70. Also, standard deviations decrease to 7.8 cm for GR_{ice}^V and to 8.4 cm for GR_{ice}^H. Because the eliminated points lie on both sides of the distribution, the regression coefficients vary only a little. The choice of polarization seems to be of minor importance. We selected GR_{ice}^V for use as the independent parameter, because the results are slightly better at vertical po-

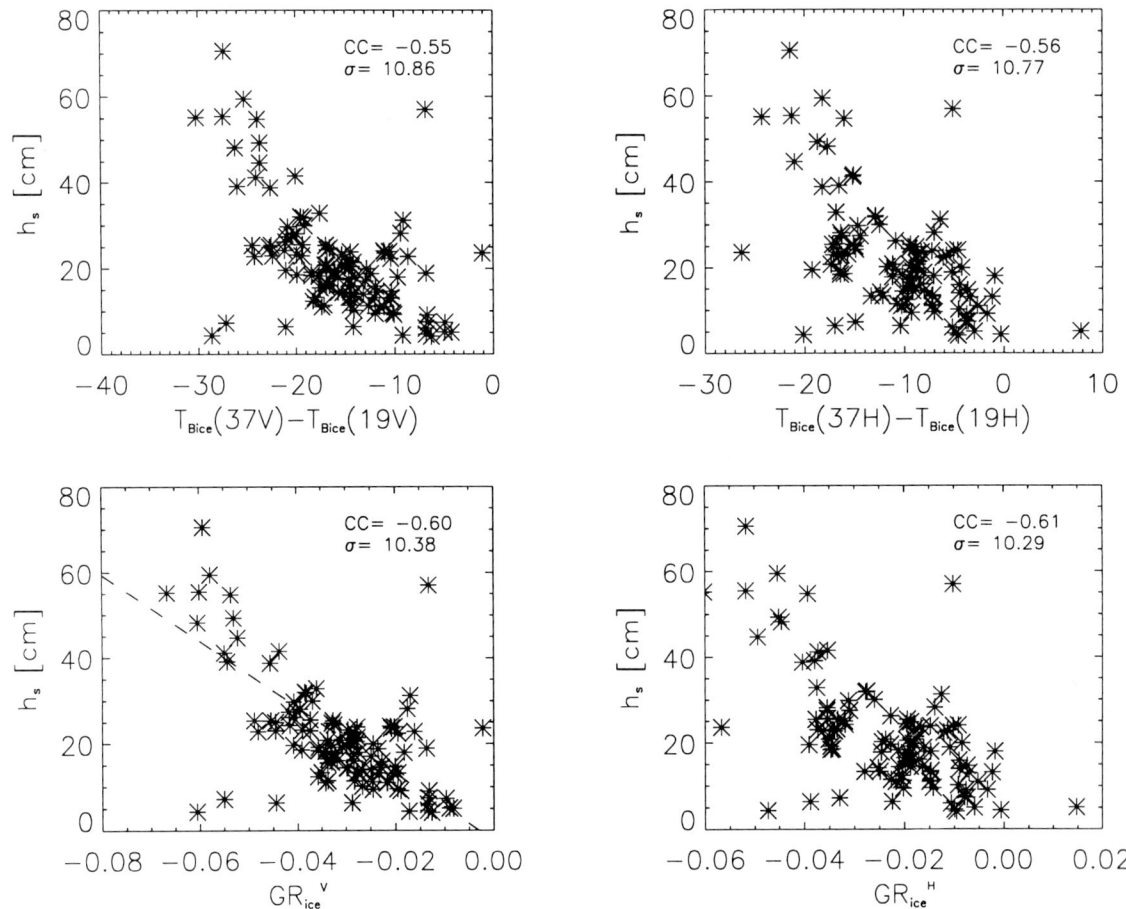

Fig.3. Scatterplots of ice brightness temperature differences of 19 and 37 GHz at both polarizations (top), and the respective spectral gradient ratios (bottom) versus in-situ snow depth measurements. CC is the correlation coefficient, and σ the standard deviation of a linear regression fit in cm. The dashed line represents the linear regression fit.

larization when four outliers are removed (less scatter especially at lower snow depths), and because the vertical polarization is expected to be less sensitive to layering in the snow [Mätzler et al., 1984].

The linear regression of in-situ snow depth and GR_{ice}^V gives the following relationship for snow depth, h_s, in centimeters:

$$h_s = -2.34 - 771 \times GR_{\text{ice}}^V. \quad (4)$$

The complete algorithm combining equations (1), (3), and (4) can be rewritten as

$$h_s = -2.34 - 771 \times \frac{T_B(37V) - T_B(19V) - k^-(1-C)}{T_B(37V) + T_B(19V) - k^+(1-C)} \quad (5)$$

where $k^- = T_{\text{Bow}}(37V) - T_{\text{Bow}}(19V)$ and $k^+ = T_{\text{Bow}}(37V) + T_{\text{Bow}}(19V)$. The open water brightness temperatures at the 19 and 37 GHz vertically polarized components are average values from open ocean areas and are used as constants (176.6 K and 200.5 K, respectively). The scatterplots (Figure 3) also indicate that the signal is saturated at a value of -0.06 for GR_{ice}^V corresponding to a snow depth of 45 cm. This saturation depth is close to the value obtained by Künzi et al. [1982] (50 cm) and higher than the value found by Sturm et al. [1993] (30 cm), both for land regions. Theoretical calculations [Ulaby et al., 1986] suggest an exponential decrease in T_B to the saturation limit. For snow thicknesses below 45 cm the standard deviation is less than 6 cm. This may be a measure of the precision of the determination of snow depth using SSM/I and/or a measure of the natural variability of snow depth at SSM/I spatial scales.

3. ERROR SOURCES AND LIMITATIONS

The regression analyses discussed in section 2 give no hint of how error sources, such as grain size, wetness, or refrozen snow alter our results. These are sources of errors which limit the general utility of the algorithm. Additional sources of error are changes in brightness temperature not directly resulting from changes in snow properties but from weather effects and ice concentration estimates. In this section, we examine the sensitivities of snow depth retrievals to ice concentration errors, weather effects, and to snow properties such as grain size and wetness. We also present a method to flag those pixels for which snow depth is indeterminable because of melt effects.

3.1. Sensitivity to Ice Concentration Errors

Input parameters to the snow algorithm are the 19 GHz and 37 GHz brightness temperatures both at vertical polarization and the ice concentration. The largest error source in ice concentration retrievals is the presence of new and young ice types. These ice types are interpreted by the algorithm as a mixture of open water and consolidated first-year ice. Ice concentration is underestimated if there is a significant amount of new and young ice within a footprint. The sensitivity of the snow depth retrievals to ice concentration variability was investigated as follows. We took a sample data set of brightness temperatures from actual observations and varied ice concentrations by ±1% at different ice concentrations ranging from 20% to 100%. The differences in snow depth per percent change in ice concentration are presented in Table 1. Except for an ice concentration of 20%, the sensitivity is always below 1 cm per 1% change in ice concentration with a minimum at 50% ice concentration. Overestimates in ice concentration result in underestimates in snow depth and vice versa. Because of the higher sensitivity at 20%, we limit the snow depth retrievals to ice concentrations between 20% and 100%. Ice concentrations less than 20% appear almost exclusively near the ice edge.

3.2. Weather Effects

Gloersen and Campbell [1988] have observed in airborne observations, and *Maslanik* [1992] has shown

TABLE 1. Changes in snow depth (Δh_s [cm]) per % variation in ice concentration (C).

C	20	30	40	50	60	70	80	90	100
Δh_s	1.3	0.7	0.5	0.3	0.4	0.6	0.7	0.8	0.8

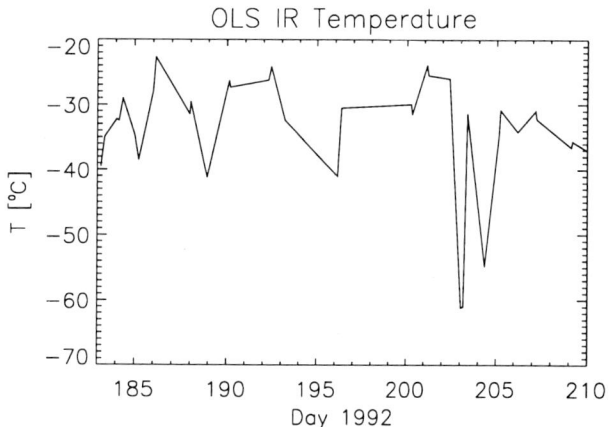

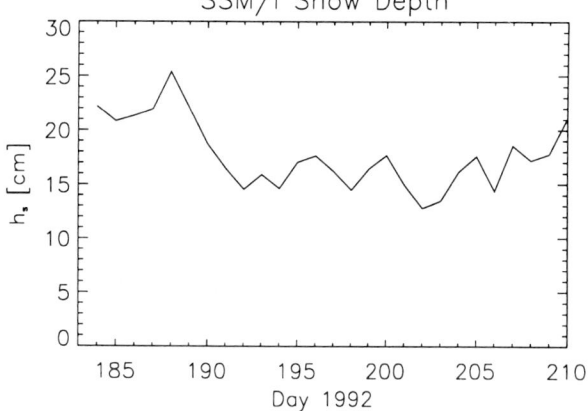

Fig. 4. Top: DMSP OLS thermal infrared temperatures for an area in the central Weddell Sea in July 1992. Bottom: Coincident SSM/I-derived daily snow depths.

in a theoretical study of weather effects on ice concentration retrievals, that at higher ice concentrations weather effects increase GR whereas PR is less affected. This increase in GR will also be reflected as an underestimate of snow depth. Thermal infrared temperatures as measured from the operational linescan system (OLS) aboard the DMSP satellites with coincident SSM/I snow depth retrievals for a pixel in the central Weddell Sea during July 1992 are presented in Figure 4. Because infrared data are very sensitive to clouds, the actual surface temperature of the earth can only be measured under cloud-free conditions. Otherwise, the measured temperature reflects the cloud top temperature. The OLS average temperature is about -30°C. Variations around this temperature might result from changes in actual surface temperature or low clouds. No cloud mask algorithm was applied. A clear low temperature peak with temperatures less than -50°C can be identified around day 203. These low temperatures presumably result from high altitude, and thus very cold, cumulus clouds. During these days, the derived snow depth drops from about 17 cm to 13 cm and reaches 17 cm again on day 205. This minimum is not significantly greater than other daily variations. There are daily variations of about 3 cm, which may be high frequency noise resulting from smaller weather effects and/or snow property variations.

3.3. Grain Size

Grain size variability affects the brightness temperatures and leads to errors in the calculation of snow depth. Larger grain sizes result in enhanced scattering and therefore decreased brightness temperatures. Because this effect is greater at 37 GHz than at 19 GHz [Chang et al., 1976], larger grain sizes will result in overestimates of snow depth.

Grain size generally increases after deposition at the surface. Temperature and humidity gradients within the snow cover result in the development of surface or depth hoar. Another mechanism which results in an increase of grain size is the refreezing of wet snow. Armstrong et al. [1993] have shown from measurements over land that the grain size effect is lower than what has been found from theoretical calculations [Chang et al., 1976]. The grain size effect is further reduced with satellite observations because extreme grain size effects will average out. Sturm et al. [this volume] report for the Bellingshausen and Amundsen Seas region average grain sizes of 1.1 mm for soft and moderate slabs and 1.3 mm to 2.0 mm for depth hoar or layers of depth hoar mixed with melt-grain clusters. Similar grain sizes are also found in the East Antarctic region with average grain sizes of about 1.6 mm [Worby and Massom, 1995]. In the Weddell Sea, grain sizes are more variable with average grain sizes increasing from 0.3 mm at the coastal region (around 10°W) to 3 mm near the tip of the Peninsula where second-year ice is present [Massom et al., 1997].

As grain sizes are not routinely obtained during the cruises in the Southern Ocean, investigation of how the grain size distribution of snow on sea ice affects our results could not be carried out. The reason is that in-situ grain size measurements require considerably more effort than the measurement of snow depths.

3.4. Thaw and Refreezing

Theoretical studies have shown that snow wetness is of much greater importance to the dielectric properties of snow than density or grain size for typical snow properties [Hallikainen et al., 1986; Ulaby et al., 1986; Hallikainen and Winebrenner, 1992]. During summer, the microwave signal can change rapidly because of the melting and refreezing of snow [e.g. Künzi et al., 1982; Schanda et al., 1983; Stiles and Ulaby, 1986; Onstott et al., 1987; Garrity, 1992; Wankiewicz, 1993]. Passive microwave algorithms for snow on land misinterpret wet snow areas as snow-free as a result of the blackbody behavior of a wet snow pack, meaning that the brightness temperature difference between 37 GHz and 19 GHz will be very small and the derived snow depth value will be close to zero. The larger increase in brightness temperature at 37 GHz than at 19 GHz resulting from snow wetness [Schanda et al., 1983] is also supported by theoretical studies [Stiles and Ulaby, 1980]. Thus, snow wetness increases GR and therefore underestimates snow depth.

wet snow may also refreeze during the night which results in a frozen crust at the top of the snow layer. This results in very large grain sizes [Colbeck, 1982]. The emissivity decreases (more with increasing frequency) because of scattering within this layer [Onstott et al. 1987; Mätzler, 1994]. A reduced emissivity at 37 GHz relative to 19 GHz would result in a decrease in GR and thus would lead to an overestimate of snow depth.

As these melting and freezing events often occur within a day, the brightness temperatures can have large diurnal variations [Stiles and Ulaby, 1980; Cavalieri et al., 1990; Wankiewicz, 1993]. This is a problem when using daily averaged brightness temperatures as input to the snow algorithm. Cyclical wettening and subsequent refreezing of the snow mass can occur at all times in the Antarctic, even in winter particularly in the marginal ice zone [Sturm et al., this volume; Massom et al., 1997].

The problem of thaw-freeze cycle effects on the retrieval of snow depth is best illustrated through an examination of the time history of snow depth over an annual cycle. Daily SSM/I ice concentrations and SSM/I snow depths for a pixel in the East Antarctic region (62.51°S, 94.25°E) are shown in Figure 5a and b respectively for 1992. The dashed line in Figure 5b represents a 7-day running mean. With the beginning of the seasonal ice growth on day 130, the snow cover begins to increase. It reaches its maximum in mid August around day 220 with a snow depth of 23 cm. After this time the snow depth decreases presumably because of both densification and less snow fall. Around day 300, the ice concentration starts its

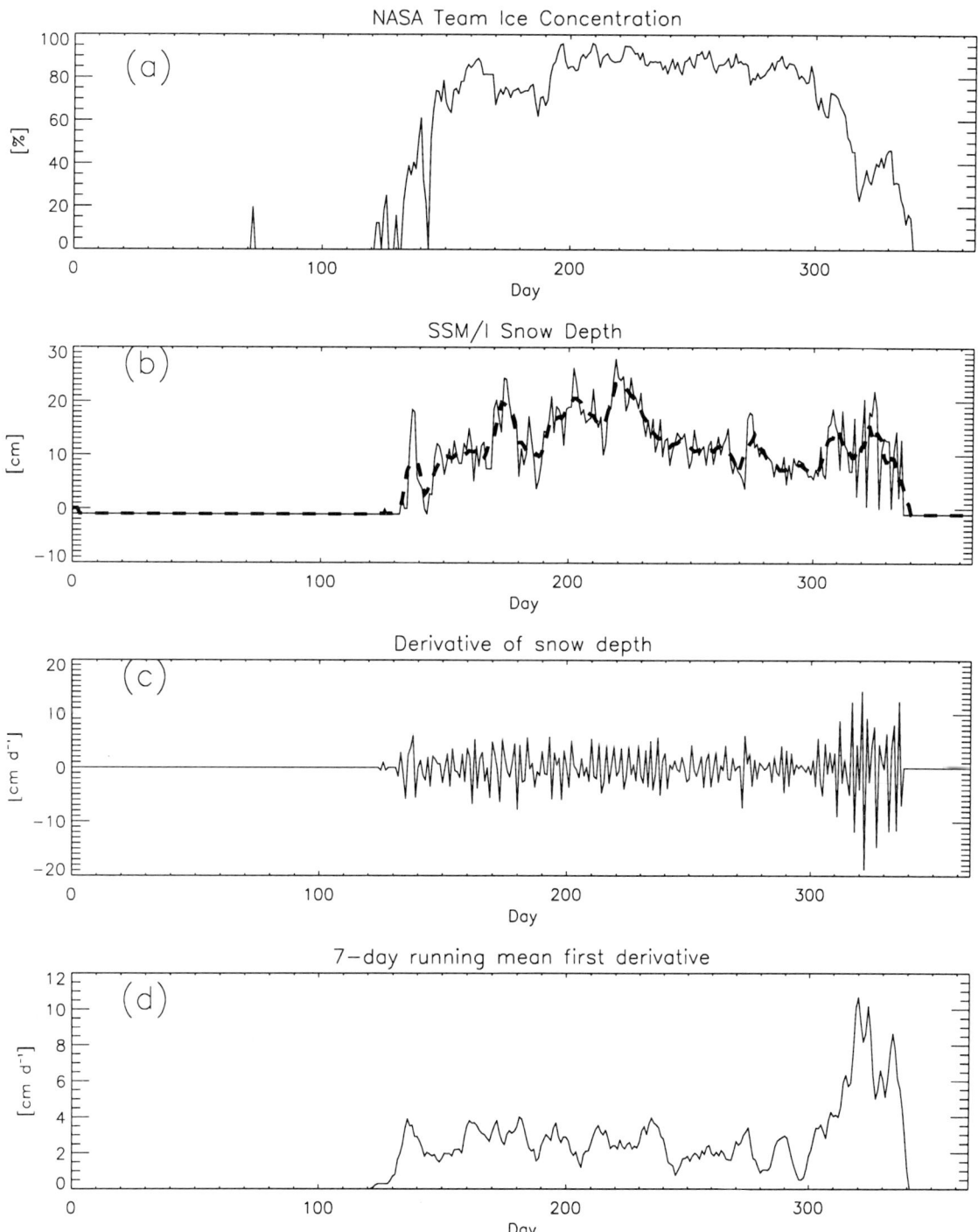

Fig.5. Time series of (a) SSM/I ice concentrations, (b) SSM/I snow depths (dashed line: 7-day running mean), (c) derivative of snow depth, and (d) 7-day running mean of the absolute value of the first derivative for an area in the East Antarctic region during 1992.

seasonal decrease indicating the beginning of melt, while the snow depth has already decreased to about 10 cm. After day 300, very high daily variations in snow depth occur. These fluctuations indicate the above mentioned thaw-refreezing period. The derivative of snow depth with respect to time (Figure 5c) shows daily variations of around 3 cm d^{-1} during the wintertime, but these increase dramatically after day 300. This daily variation gives some measure of the precision of the algorithm. Daily variation is very well described by a Gaussian distribution with an average of -0.04 cm d^{-1} and a standard deviation of 2.4 cm d^{-1}. The 7-day running mean of the absolute value of the derivative (Figure 5d) gives values of over 10 cm d^{-1} during this period. As retrieval of snow depth during melt periods will be grossly inaccurate, we flag out these days, where the 7-day running mean of the absolute value of the derivative is above 5 cm d^{-1}. The same feature can be seen for a region in the western Weddell Sea (69.44°S, 53.13°W) where perennial ice is present (Figure 6). The ice concentration is above 80% for almost the entire year (Figure 6a). Highest snow depths occur during the summer months and reach a minimum in winter (Figure 6b). In spring (November/December) and summer (January/February) daily variations are high and are of the same order of magnitude as in Figure 5. A cut-off value 5 cm d^{-1} is used to flag the melt season effects.

Subsequent to the regression analysis, we obtained snow measurements from a cruise of the *R/V Polarstern* in the Weddell Sea during September and October 1989 [*Augstein et al.*, 1991]. Highest snow depths are from the area east of the tip of the Antarctic Peninsula where the ship entered the ice pack on day 255. The snow surface was wet and air temperatures were above freezing (Eicken, unpublished data). On the following days air temperatures were around 0°C, and the snow had a surface crust layer. Melt events, detected using the derivative cut-off value of 5 cm d^{-1}, are marked as black in the SSM/I snow depths maps for every other day of September 1989 (Plate 1). This time series of snow depths demonstrates the frequent occurrence of melt events in the marginal ice zone. East of the tip of the Antarctic Peninsula, a melt event is initially observed on day 252, just before the *Polarstern* enters the sea ice in this region. A scatterplot of in-situ and SSM/I-derived snow depths is presented in Figure 7. The boxed asterisks indicate pixels which would have been flagged by the algorithm. Most of the flagged pixels either strongly underestimate (because of wet snow) or strongly overestimate (because of refrozen snow) snow depth. The correlation coefficient between in-situ and satellite snow depths of the remaining pixels is 0.7. Similar to the results in Figure 3 deeper in-situ snow depths (>45 cm) tend to be lower in the SSM/I results. Besides the possible saturation of the microwave signal, peak in-situ snow depths are smoothed over the SSM/I footprint as noted earlier.

4. MONTHLY SNOW DEPTH DISTRIBUTION

In this section we present monthly snow-depth distribution maps derived using the snow depth algorithm. Snow depth is calculated for each day of 1988 through 1994. Wet snow events are flagged following the procedure described in section 3.4. Monthly averages are calculated for only those image pixels where the ice concentration is above 20%. The 7-year mean is calculated from the monthly maps. If for more than three years the monthly averages give only flagged values, these pixels are also flagged in the climatology. 7-year mean snow depths values for a given month and pixel may represent a single monthly-averaged value (one of seven years) or an average of 2 to 7 monthly-averaged values depending on years of ice coverage. These maps are presented in Plates 2 and 3.

The monthly maps presented in Plates 2 and 3 show that the deepest snow occurs in the Weddell, Bellingshausen/Amundsen, and Ross Sea sectors. Considerably less snow occurs in the eastern sectors (0° − 180°E) in agreement with average in-situ snow depth measurements of about 10 cm in the East Antarctic region [*Worby and Massom*, 1995]. Depths range from a few cm in regions of new ice growth (e.g., in the Ross Sea polynya) to over 40 cm in the vicinity of the Antarctic Peninsula in the Weddell Sea. The spatial distribution of snow depths shown here is consistent with in-situ measurements. *Eicken et al.* [1994] and *Jeffries et al.* [1994] reported thicker snow cover in the western Weddell and Bellingshausen Seas than in other regions. *Eicken et al.* [1995] found the deepest snow occurred over perennial sea ice areas and argued that the presence of a perennial sea ice cover resulted from the thermal "snow-shielding" effect. Transects through the Weddell Sea in late winter show a decrease in snow depth from about 35 cm in the northwestern Weddell Sea near the Peninsula to 7 cm in the eastern Weddell Sea [*Lange and Eicken*, 1991; *Massom et al.*,

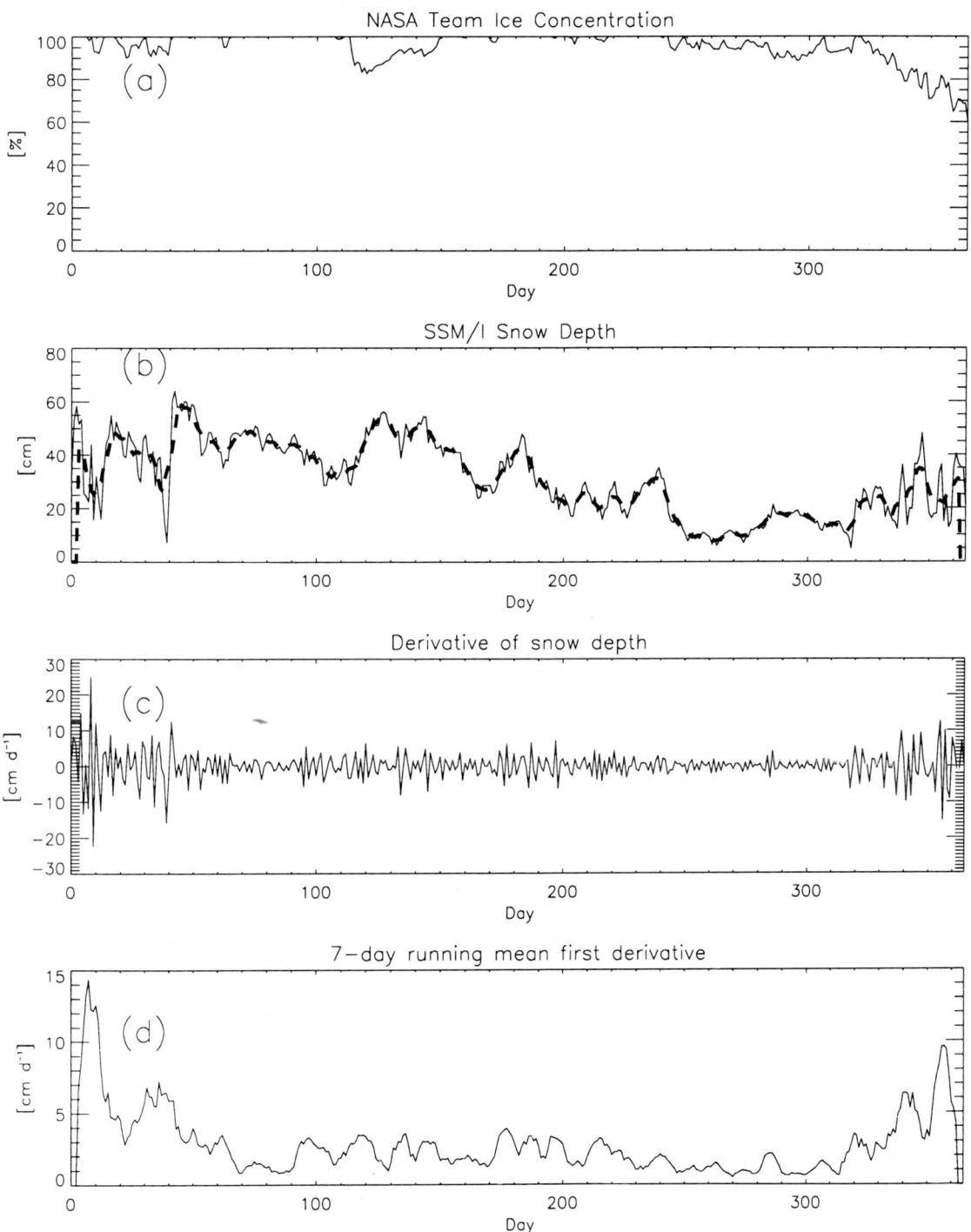

Fig.6. Time series of (a) SSM/I ice concentrations, (b) SSM/I snow depths (dashed line: 7-day running mean), (c) derivative of snow depth, and (d) 7-day running mean of the absolute value of the first derivative for an area in the western Weddell Sea region during 1992.

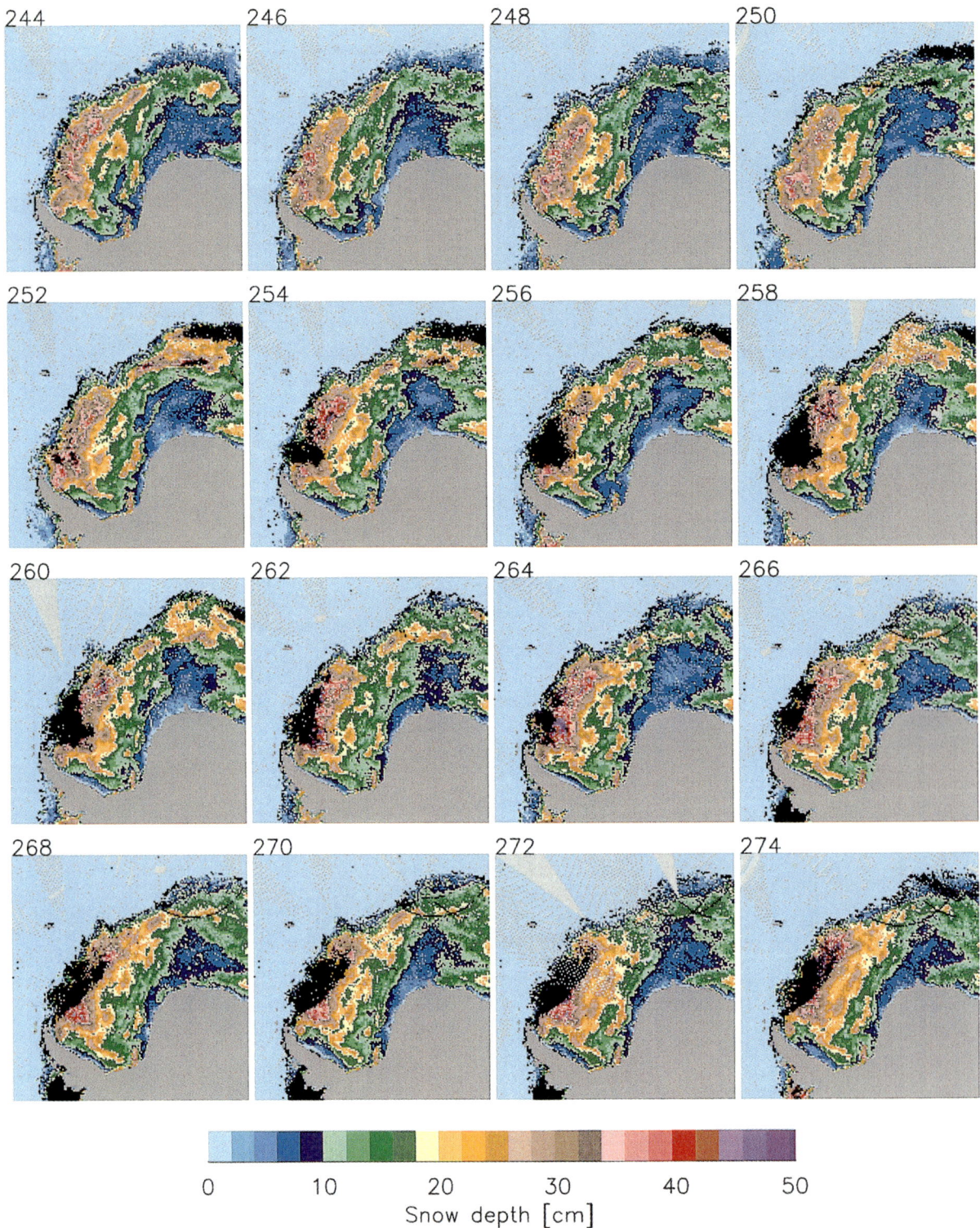

Plate 1. Identified melt events (black) during September 1989 in the Weddell Sea for every other day. The Antarctic continent is gray. Areas of missing data are indicated in light gray.

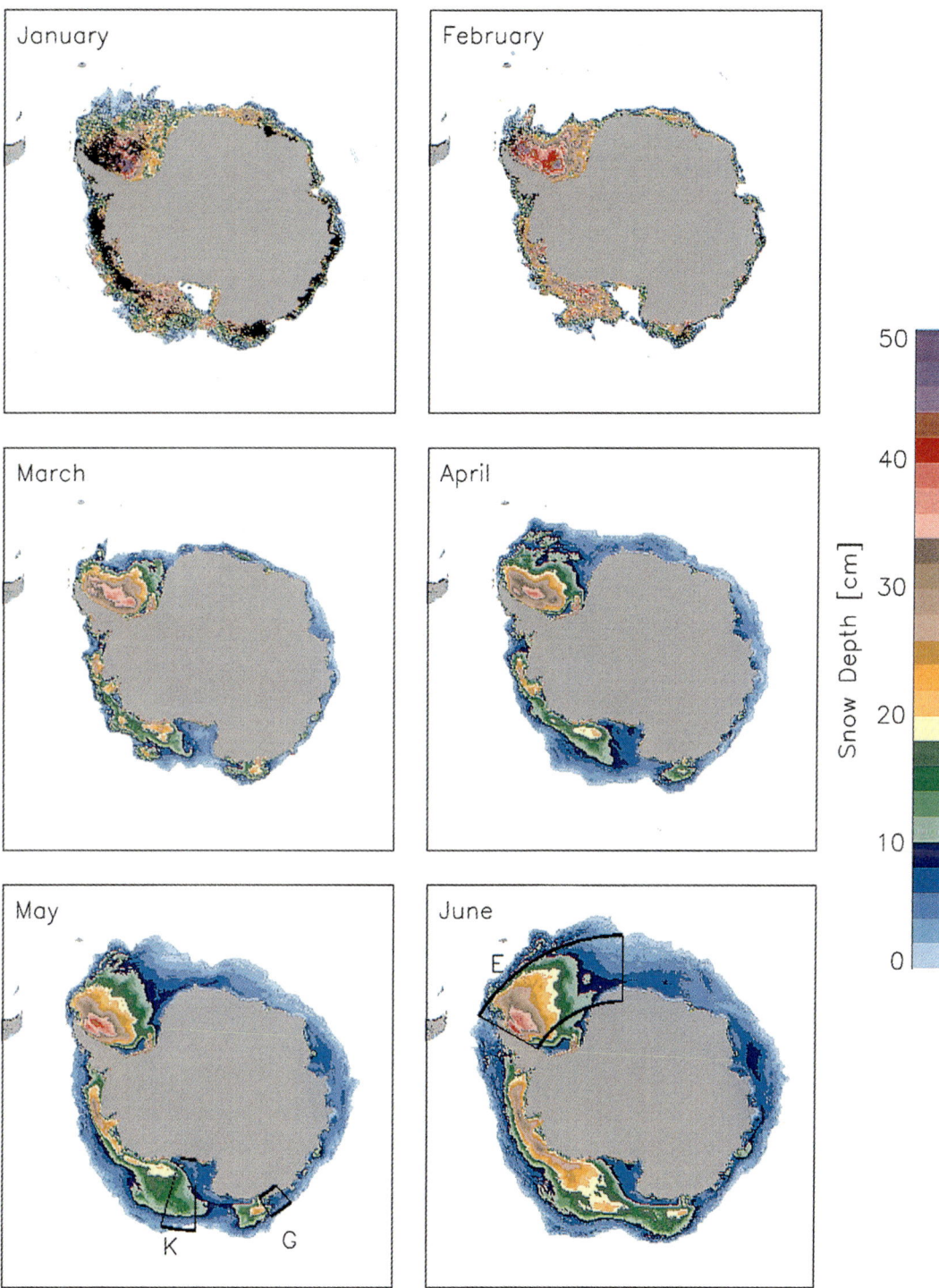

Plate 2. Mean monthly snow depths for the months January through June from the years 1988–1994. Labeled boxes indicate regions of cruises summarized in Table 2. The open ocean area is white. Black pixels indicate areas where more than 3 years of monthly averages have flagged values resulting from melting effects identified using temporal information as described in section 3.4.

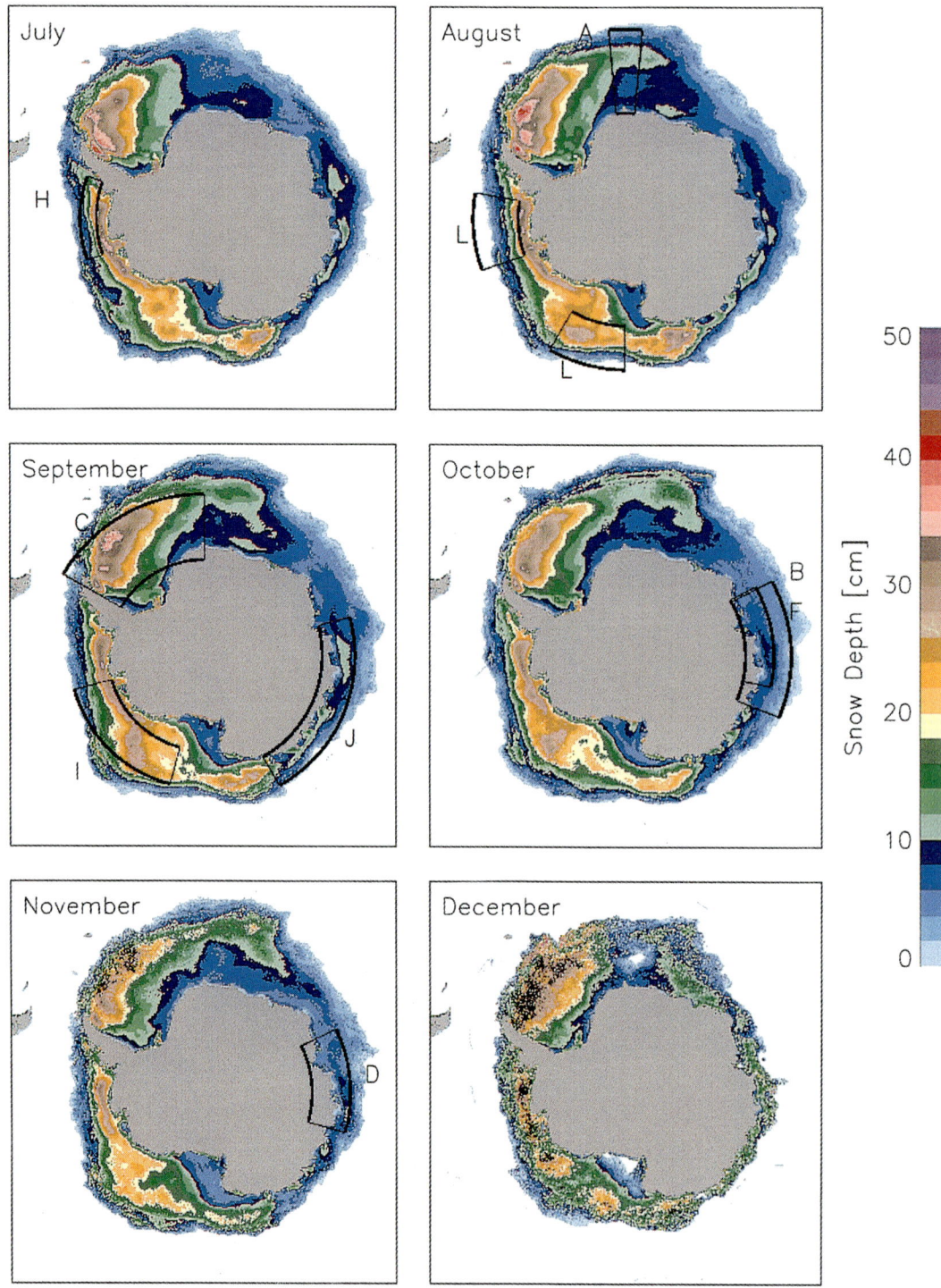

Plate 3. Mean monthly snow depths for the months July through December from the years 1988–1994. Labeled boxes indicate regions of cruises summarized in Table 2. The open ocean area is white. Black pixels indicate areas where more than 3 years of monthly averages have flagged values resulting from melting effects identified using temporal information as described in section 3.4.

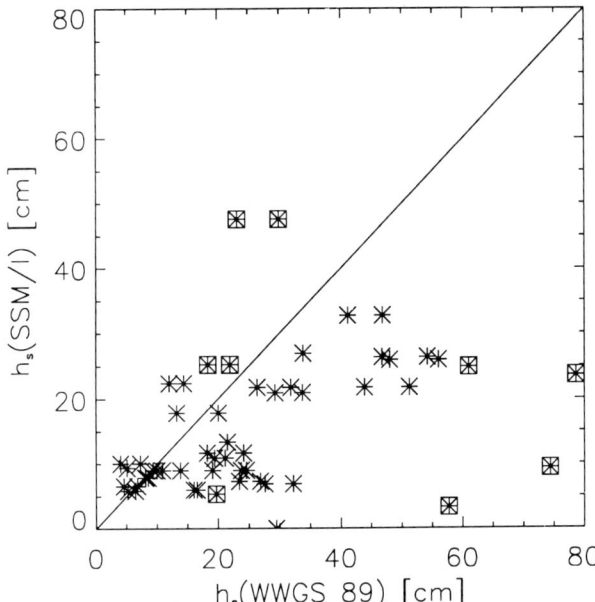

Fig.7. Scatterplot of in-situ snow depths during WWGS'89 and SSM/I-derived snow depths.

Sea embayments occur during summer. During the summer months, there is greater variability in the number of values averaged for each pixel because of rapidly changing ice concentrations giving the maps a speckle appearance.

5. VALIDATION OF RESULTS; COMPARISON WITH SNOW DEPTH DISTRIBUTIONS

Quantitative validation of the results is difficult because there are no measurements of average snow depth over an area as large as an SSM/I footprint. Nevertheless, several cruises through the Southern Ocean sea ice measured snow depth along their routes providing snow depth distributions [Wadhams et al., 1987; Allison et al., 1993; Drinkwater and Haas, 1994; Eicken et al., 1994; Jeffries et al., 1995; Worby and Masson, 1995; Worby et al., 1996; Massom et al., 1997; Sturm et al., this volume]. The different cruises which report snow depth distributions are summarized in Table 2. These data cover a variety of regions and seasons. Except for the 1986 data set (A), all measurements are made during the SSM/I period and comparison can be made with monthly averages of the respective years. Data sets from cruises E and H have already been used in the algorithm development, but here our aim is to validate the large-scale snow depth distribution. The snow depth distributions from the satellite data are calculated from pixels within the latitude-longitude range of each cruise. These regions are indicated in

1997]. The SSM/I results are in good agreement with these published values. The greatest snow depths are found in the western Weddell Sea and in the eastern Ross Sea during February (Plate 2). In-situ summer values from Ice Station Weddell confirm these snow depths [Lytle and Ackley, 1996]. Bromwich [1988] reported that the highest precipitation rates based on coastal station data in the Weddell Sea and Ross

TABLE 2. Time and location of different cruises in the Antarctic which report snow depth distributions. Some data from cruises E and H have already been used in the development of the algorithm.

	Time	Location	Reference
A	July – September 1986	Weddell Sea (5°W–5°E)	Wadhams et al., 1987
B	October – December 1988	East Antarctic (62°E–112°E)	Allison et al., 1993
C	September – October 1989	Weddell Sea (60°W–0°E)	Eicken et al., 1994
D	November 1991	East Antarctic (64°E–106°E)	Worby and Massom, 1995
E	June – July 1992	Weddell Sea (60°W–0°E)	Drinkwater and Haas, 1994; Massom et al., 1997
F	October–November 1992	East Antarctic (62°E–102°E)	Worby and Massom, 1995
G	March – May 1993	East Antarctic (139°E–149°E)	Worby and Massom, 1995
H	August – September 1993	Bellingshausen Sea (70°W–110°W)	Worby et al., 1996
I	September – October 1994	Amundsen Sea (105°W–165°W)	Sturm et al., this volume
J	September – October 1994	East Antarctic (75°E–150°E)	Jeffries et al., 1995; Worby and Massom, 1997
K	May – June 1995	Ross Sea (165°W–180°W)	Sturm et al., this volume
L	August – September 1995	Bellingshausen Sea (80°W–110°W) Ross Sea (150°W–180°W)	Sturm et al., this volume

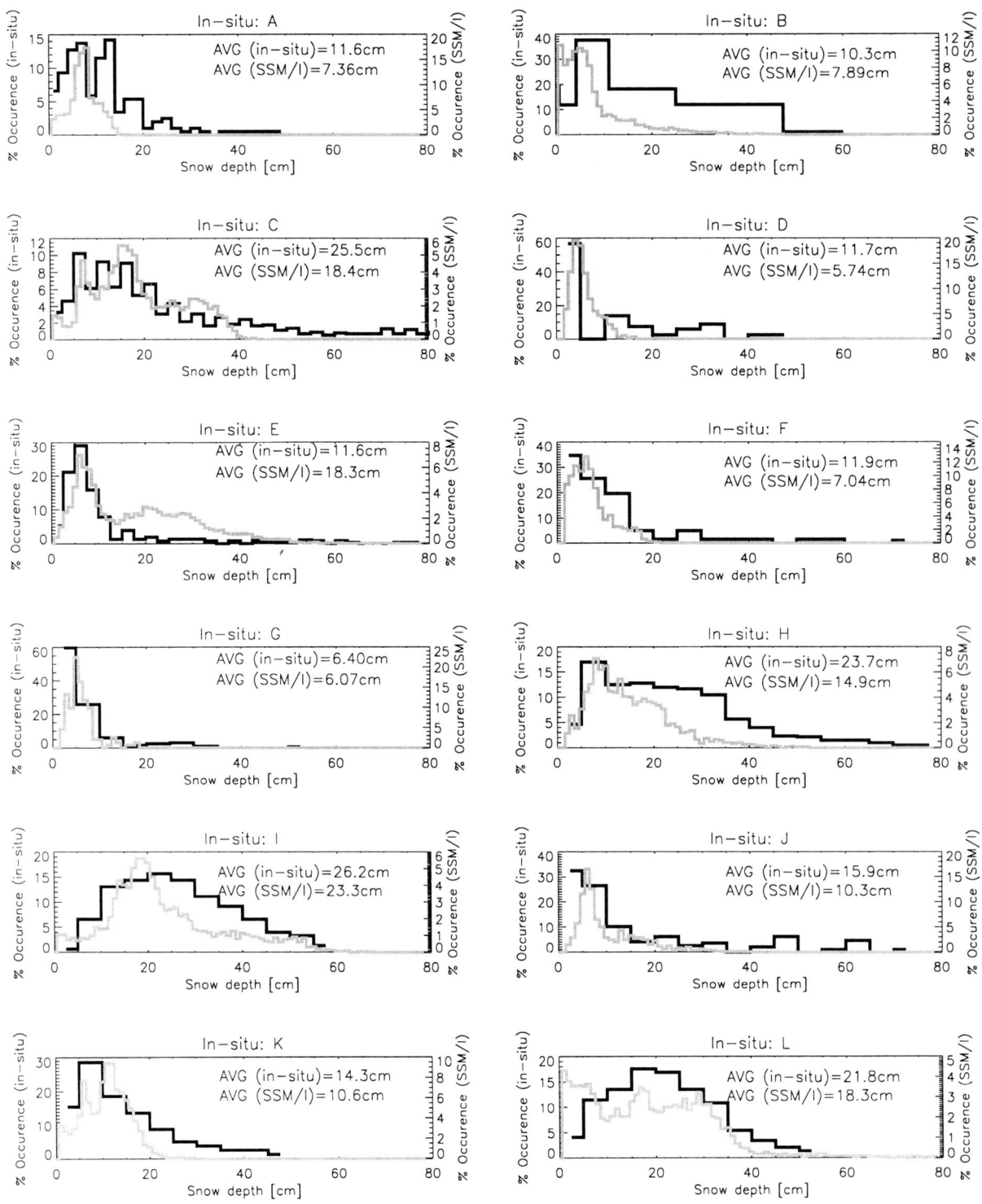

Fig.8. Snow depth distributions from field experiments (black) and from the SSM/I snow depth algorithm (gray). The letters indicate the cruises summarized in Table 2.

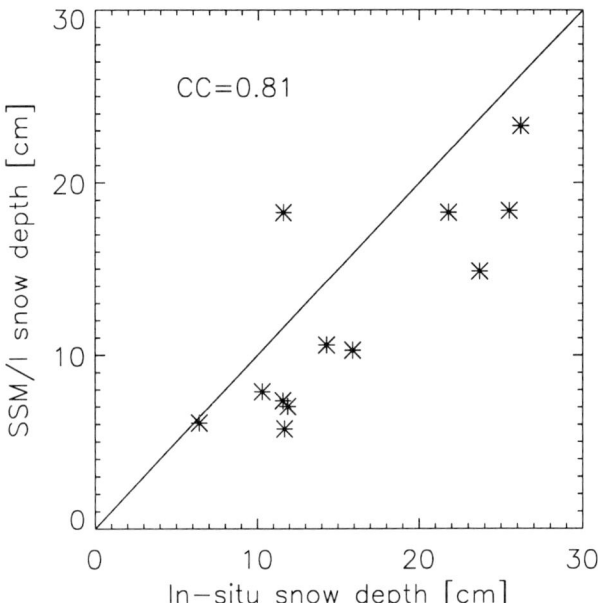

Fig.9. Average snow depths from the cruises summarized in Table 2 versus SSM/I-derived monthly averages.

Plates 2 and 3 by black line segments. Although the in-situ data are an average of point measurements made at different times and the satellite data are large-scale measurements averaged over one or two months and are coincident with the ships' exact routes, the average snow depths as well as the shape of the distributions are in reasonable agreement (Figure 8). The SSM/I average snow depths are well correlated (correlation coefficient of 0.81) with the in-situ average values (Figure 9). On average, the SSM/I snow depths are 3.5 cm below the in-situ values. There are primarily two reasons why this bias is expected. First, as mentioned earlier, high in-situ snow depths are smoothed by the large SSM/I footprint (monthly averages further increase this smoothing). Second, during most cruises, the ship passes the marginal ice zone twice and makes the majority of measurements in the interior ice, whereas the SSM/I segments include the whole marginal ice zone.

Clearly, there is a need for in-situ measurements that are representative of snow depths at SSM/I spatial scales to provide an estimate of the accuracy of the SSM/I retrievals on a pixel-by-pixel basis. In summary, these results are encouraging and suggest that the satellite-derived monthly snow depths provide a fair representation of the in-situ snow depth distribution.

6. INTERANNUAL VARIABILITY AND REGIONAL DIFFERENCES

For the purpose of looking at regional differences and interannual variability in more detail, mean monthly snow depths for each year and each sector, indicated in Figure 1, are calculated (Figure 10). Highest mean snow depths occur during the summer months in all sectors because most of the seasonal ice pack has melted, leaving only thicker ice with greater snow depth close to the continent. The highest wintertime snow depths of about 15 cm are found in the Weddell, Ross, Bellingshausen, and Amundsen Seas. These sectors also have the highest interannual variability. The interannual variability in the Weddell Sea decreases as the season progresses from summer to winter, whereas the variability in the Bellingshausen-Amundsen Seas sector stays high year round with a range of 10 cm to 20 cm. The lowest winter snow depths (6 cm) and variability are found for the Indian Ocean sector. Interestingly, the interannual variation of the whole Southern Ocean is very low which suggests that greater snow depths in one region in a certain year are compensated by lower snow depths in another region. For example, during September 1988, an unusually high average snow depth of over 20 cm is found in the Bellingshausen-Amundsen Seas sector. For the same month, the Weddell Sea and Ross Sea sectors have lower than average snow depths. The same can be seen in September 1991, when the Ross Sea sector has greater than average snow depths, whereas the Bellingshausen-Amundsen Seas have the lowest snow depth of the seven years.

7. CONCLUSIONS

An algorithm to calculate snow depths from passive microwave radiance data has been developed through the regression of in-situ snow depth measurements on microwave spectral gradient ratios defined using the 19 GHz and 37 GHz vertical polarization channels of the DMSP SSM/I. The choice of polarization seems to be of minor importance, but because of the expected higher sensitivity to layering at horizontal polarization [Mätzler et al., 1984] we use the vertical polarization. The resulting algorithm is a simple linear relationship between the spectral gradient ratio, corrected for ice concentration variations, and snow depth. Multi-temporal information is additionally used to identify wet snow, or rather melt-refreeze events, which makes snow depth signatures

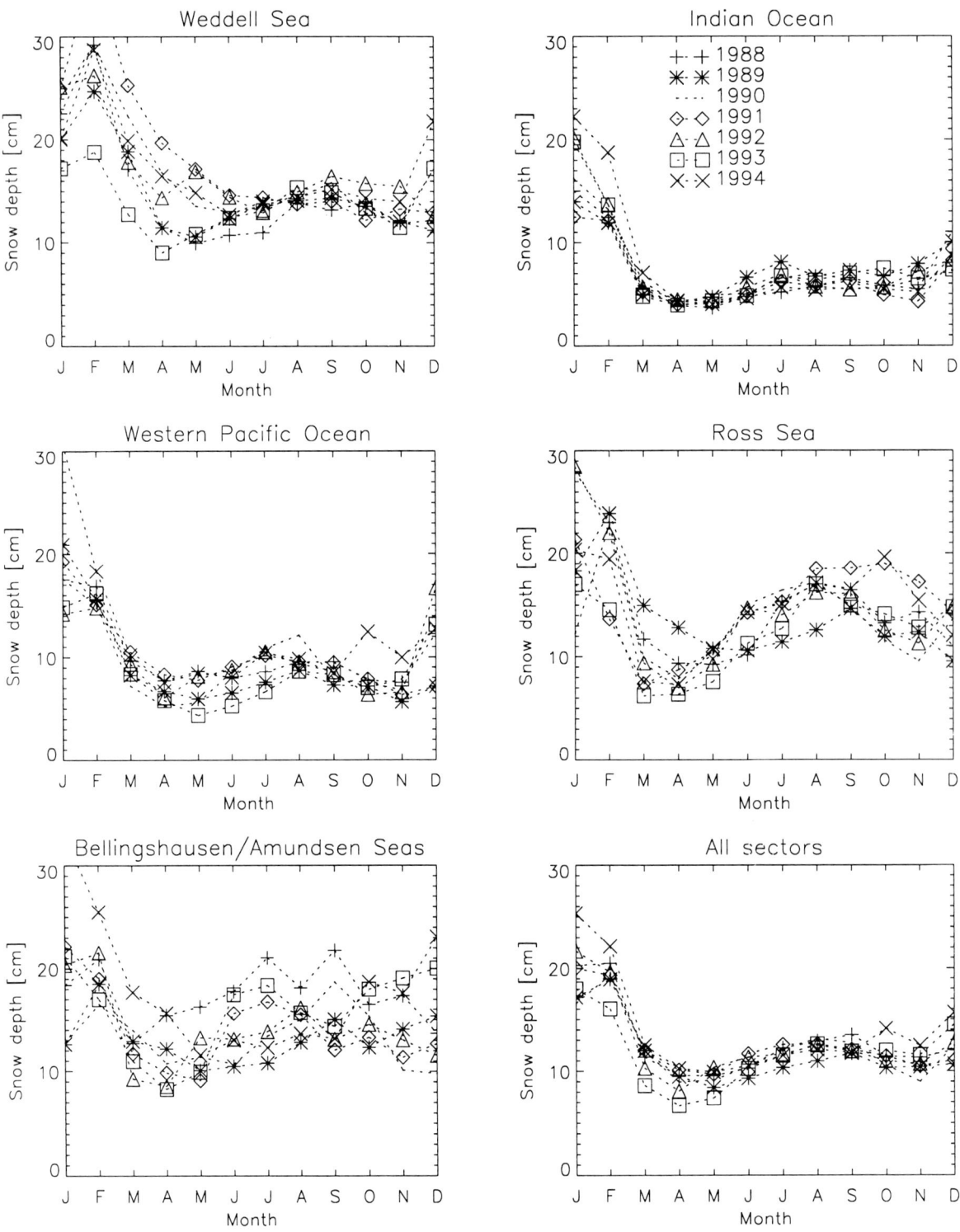

Fig.10. Mean monthly snow depths for each sector and each year.

in the microwave data indeterminable. The use of single-day values is currently hampered by spurious weather effects and melt events, which occur even in winter. Thus, monthly snow depths on sea ice are presented for the years 1988 through 1994. Large-scale snow depth distributions from shipborne measurements and monthly SSM/I snow depths agree reasonably well for different regions. The average snow depths for these regions are highly correlated (correlation coefficient of 0.81). The average difference is 3.5 cm. A quantitative measure of the accuracy of the algorithm on a pixel-by-pixel basis must await further validation studies. Validation data will be difficult to come by, because the algorithm using satellite measured radiances provides an integrated measure of snow depth over areas of approximately 3000 km^2 (assuming a 55 km footprint at 19 GHz). The only currently available data sets are point or short transect measurements which are inadequate, given the high spatial and temporal variability of the snow cover in the Antarctic. Only a dedicated snow depth survey with the aim of obtaining representative measurements in the SSM/I resolution could provide sufficient information. In addition to snow depth, information on grain size, flooding, layering, and slush at the snow/ice interface would be very helpful. This is not an easy task because information other than snow depth cannot be obtained from a running ship. Nonetheless, these data would enable the investigation of errors on a pixel by pixel basis and would aid the study of using additional channels (for example the use 85 GHz data) to improve the accuracy. The use of lower frequency channels such as 6.6 GHz and 10 GHz on the upcoming AMSR sensor should improve the retrieval of very deep snow covers. Also the combination of passive microwave data and scatterometer data for the mapping of wet snow [Mätzler et al., 1982] should be investigated.

Acknowledgments. The authors appreciate receiving digital snow depth data from Martin Jeffries, Hajo Eicken, Christian Haas, and Steve Ackley. We also acknowledge receiving the DMSP SSM/I radiance data through the National Snow and Ice Data Center in Boulder, CO. Furthermore, we thank Robyn Kelly, who spent the summer of 1996 at the NASA Summer Institute for Atmospheric and Hydrospheric Sciences, for her support. The reviewers have, with their constructive criticism, significantly improved the paper. TM is funded by the National Research Council through a Resident Research Program tenured at the NASA Goddard Space Flight Center and DJC is supported by the NASA Polar Program office

(RTOP 578-32-20) and by the NASA EOS Project (229-04-15).

REFERENCES

Allison, I., R.E. Brandt, and S.G. Warren, East Antarctic sea ice: Albedo, thickness distribution, and snow cover, *J. Geophys. Res.*, *98*, 12,417-12,429, 1993.

Armstrong, R.L., A. Chang, A. Rango, and E. Josberger, Snow depths and grain-size relationships with relevance for passive microwave studies, *Ann. Glaciol*, *17*, 171-176, 1993.

Augstein, E., N. Bagriantsev, and H.W. Schenke (eds.), The expedition ANTARKTIS VIII/1-2, 1989 with the Winter Weddell Gyre Study of the research vessels "Polarstern" and "Akademik Fedorov", *Rep. Polar Res.*, *84*, 134 pp., 1991.

Bromwich, D.H., Snowfall in high southern latitudes, *Rev. Geophys.*, *26*, 149-168, 1988.

Cavalieri, D.J., A microwave technique for mapping thin sea ice, *J. Geophys. Res.*, *99*, 12,561-12,572, 1994.

Cavalieri, D.J., P. Gloersen, and W.J. Campbell, Determination of sea ice parameters with the NIMBUS 7 scanning multichannel microwave radiometer, *J. Geophys. Res.*, *89*, 5355-5369, 1984.

Cavalieri, D.J., P. Gloersen, and T.T. Wilheit, Aircraft and satellite passive microwave observations of the Bering Sea ice cover during MIZEX West, *IEEE Trans. Geosc. Rem. Sens.*, *GE-25*, 368-377, 1986.

Cavalieri, D.J., B.A. Burns, and R.G. Onstott, Investigation of the effects of summer melt on the calculation of sea ice concentration using active and passive microwave data, *J. Geophys. Res.*, *95*, 5359-5369, 1990.

Cavalieri, D.J., J. Crawford, M.R. Drinkwater, D.T. Eppler, L.D. Farmer, R.R. Jentz, and C.C. Wackerman, Aircraft active and passive microwave validation of sea ice concentration from the defense meteorological satellite program special sensor microwave imager, *J. Geophys. Res.*, *96*, 21,989-22,008, 1991.

Cavalieri, D.J. and 16 others, NASA sea ice validation program for the DMSP SSM/I: Final report, *NASA Tech. Memo. 104559*, 126pp., 1992.

Chang, A.T.C., P. Gloersen, T. Schmugge, T. Wilheit, and H.J. Zwally, Microwave emission from snow and glacier ice, *J. Glaciol.* *16*, 23-39, 1976.

Chang, A.T.C., J.L. Foster, and D.K. Hall, Nimbus-7 SMMR derived global snow cover parameters, *Ann. Glaciol.*, *9*, 39-44, 1987.

Colbeck, S.C., An overview of seasonal snow metamorphism, *Rev. Geophys. Space Phys.*, *20*, 45-61, 1982.

Colbeck, S.C., The layered character of snow covers, *Rev. Geophys.*, *29*, 81-96, 1991.

Comiso, J.C., T.C. Grenfell, D.L. Bell, M.A. Lange, and S.F. Ackley, Passive microwave in situ observations of winter Weddell Sea ice, *J. Geophys. Res.*, *94*, 10,891-10,905, 1989.

Davis, R.E., J. Dozier, and A.T.C. Chang, Snow property measurements correlative to microwave emission at 35 GHz, *IEEE Trans. Geosc. Rem. Sens.*, 6, GE-25, 751-757, 1987.

Drinkwater, M.R., and C. Haas, Snow, sea-ice and radar observations during ANT X/4: Summary Data Report, *AWI Berichte aus dem Fachbereich Physik*, 53, 51pp., 1994.

Eicken, H., M.A. Lange, H.W. Hubberten, and P. Wadhams, Characteristics and distribution patterns of snow and meteoric ice in the Weddell Sea and their contribution to the mass balance of sea ice, *Ann. Geophysicae*, 12, 80-93, 1994.

Eicken, H., H. Fischer, and P. Lemke, Effects of the snow cover on Antarctic sea ice and potential modulation of its response to climate change, *Ann. Glaciol.*, 21, 369-376, 1995.

Garrity, C., Electrical, physical, and microwave properties of snow-covered floating ice, in *Sea Ice Properties and Processes*, edited by S.F. Ackley and W.F. Weeks, *CRREL Monogr.*, 90-1, pp. 57-61, U.S. Army Corps of Eng., Hanover, N.H., 1990.

Garrity, C., Characterization of snow on floating ice and case studies of brightness temperature changes during the onset of melt, in *Microwave Remote Sensing of Ice, Geophys. Monogr. Ser.*, vol. 68, edited by F. Carsey, pp. 313-328, AGU, Washington, D.C., 1992.

Gloersen, P., and D.J. Cavalieri, Reduction of weather effects in the calculation of sea ice concentration from microwave radiances, *J. Geophys. Res.*, 91, 3913-3919, 1986.

Gloersen, P., W.J. Campbell, D.J. Cavalieri, J.C. Comiso, C.L. Parkinson, and H.J. Zwally, Arctic and Antarctic sea ice, 1978-1987: Satellite passive microwave observations and analysis, *NASA SP-511*, Washington, D.C., 1992.

Grenfell, T.C., Surface-based passive microwave observations of sea ice in the Bering and Greenland Seas, *IEEE Trans. Geosc. Rem. Sens.*, GE-24, 378-382, 1986.

Grenfell, T.C., and A.W. Lohanick, Temporal variations of the microwave signatures of sea ice during the late spring and early summer near Mould Bay NWT, *J. Geophys. Res.*, 90, 5063-5074, 1985.

Grenfell, T.C., and J.C. Comiso, Multifrequency passive microwave observations of first-year sea ice grown in a tank, *IEEE Trans. Geosc. Rem. Sens.*, GE-24, 826-831, 1986.

Grenfell, T.C., M.R. Wensnahan, and D.P. Winebrenner, Passive microwave signatures of simulated pancake ice and young pressure ridges, *Rem. Sens. Rev.*, 9, 51-64, 1994.

Häkkinen, S., Seasonal simulation of the Southern Ocean coupled ice-ocean system, *J. Geophys. Res.* 100, 22,733-22,748, 1995.

Hall, D.K., A.T.C. Chang, and J.L. Foster, Detection of the depth-hoar layer in the snow-pack of the Arctic coastal plain of Alaska, U.S.A., using satellite data, *J. Glaciol.*, 32, 87-94, 1986.

Hallikainen, M.T., and P.A. Jolma, Retrieval of the water equivalent of snow cover in Finland by satellite microwave radiometry, *IEEE Trans. Geosc. Rem. Sens.*, GE-24, 855-862, 1986.

Hallikainen, M.T., F.T. Ulaby, and M. Abdelrazik, Dielectric properties of snow in the 3 to 37 GHz range, *IEEE Trans. Ant. Prop.*, AP-34, 1329-1340, 1986.

Hallikainen, M.T., and P.A. Jolma, Comparison of algorithms for retrieval of snow water equivalent from Nimbus-7 SMMR data in Finland, *IEEE Trans, Geosc. Rem. Sens.*, 30, 124-131, 1992.

Hallikainen M.T., and D.P. Winebrenner, The physical basis for sea ice remote sensing, in *Microwave Remote Sensing of Ice, Geophys. Monogr. Ser.*, vol. 68, edited by F. Carsey, pp. 29-46, AGU, Washington, D.C., 1992.

Hollinger, J., R. Lo, G. Poe, R. Savage, and J. Pierce, Special Sensor Microwave/Imager User's Guide, *Naval Research Laboratory*, pp.120, Washington, D.C., 1987.

Jeffries, M.O., K. Morris, A.P. Worby, and W.F. Weeks, Late winter sea-ice properties and growth processes in the Bellingshausen and Amundsen Seas, *Ant. J. of U.S.*, 29, 9-10, 1994.

Jeffries, M.O., R. Jaña, S. Li, and S. McCullars, Sea-ice- and snow-thickness distributions in late winter 1993 and 1994 in the Ross, Amundsen, and Bellingshausen Seas, *Ant. J. of U.S.*, 30, 18-21, 1995.

Josberger, E.G., P. Gloersen, A. Chang, and A. Rango, The effects of snowpack grain size on satellite passive microwave observations from the Upper Colorado River Basin, *J. Geophys. Res.*, 101, 6679-6688, 1996.

Künzi, K.F., S. Patil, and H. Rott, Snow-cover parameters retrieved from Nimbus-7 scanning multichannel microwave radiometer (SMMR) data, *IEEE Trans. Geosc. Rem. Sens.*, GE-20, 452-467, 1982.

Lange, M.A., and H. Eicken, The sea ice thickness distribution in the northwestern Weddell Sea, *J. Geophys. Res.*, 96, 4821-4837, 1991.

Ledley, T.S., Snow on sea ice: Competing effects in shaping climate, *J. Geophys. Res.*, 96, 17,195-17,208, 1991.

Lohanick, A.L., Some observations of established snow cover on saline ice and their relevance to microwave remote sensing, in *Sea Ice Properties and Processes*, edited by S.F. Ackley and W.F. Weeks, *CRREL Monogr.*, 90-1, pp. 61-67, U.S. Army Corps of Eng., Hanover, N.H., 1990.

Lohanick, A.L., Microwave brightness temperatures of laboratory-grown undeformed first-year ice with an evolving snow cover, *J. Geophys. Res.*, 98, 4667-4674, 1993.

Lytle, V.I. and S.F. Ackley, Heat flux through sea ice in the western Weddell Sea: Convective and conductive transfer processes, *J. Geophys. Res.*, 101, 8853-8868, 1996.

Maslanik, J.A., Effects of weather on the retrieval of sea ice concentration and ice type from passive microwave data, *Int. J. Rem. Sens., 13*, 34-57, 1992.

Massom, R.A., M.R. Drinkwater, and C. Haas, Winter snow cover on sea ice in the Weddell Sea, *J. Geophys. Res., 102*, 1101-1117, 1997.

Mätzler, C., Applications of the interaction of microwaves with the natural snow cover, *Rem. Sens. Rev., 2*, 259-387, 1987.

Mätzler, C., Passive microwave signatures of landscapes in winter, *Met. Atm. Physics, 54*, 241-260, 1994.

Mätzler, C., E. Schanda, and W. Good, Towards the definition of optimum sensor specifications for microwave remote sensing of snow, *IEEE Trans. Geoscience Rem. Sens., GE-20*, 57-66, 1982.

Mätzler, C., R.O. Ramseier, and E. Svendsen, Polarization effects in sea-ice signatures, *IEEE J. Oceanic Eng., OE-9*, 333-338, 1984.

NSIDC, DMSP SSM/I Brightness Temperature and Sea Ice Concentration Grids for the Polar Regions on CD-ROM, User's Guide, *National Snow and Ice Data Center*, Boulder, CO, 1992.

Onstott, R.G., T.C. Grenfell, C. Mätzler, C.A. Luther, and E.A. Svendsen, Evolution of microwave sea ice signatures during early summer and midsummer in the marginal ice zone, *J. Geophys. Res., 92*, 6825-6835, 1987.

Rango, A., A.T.C. Chang, and J.L. Foster, The utilization of space-borne microwave radiometers for monitoring snow pack properties, *Nordic Hydrol., 10*, 25-40, 1979.

Shuman, C.A., R.B. Alley, and S. Anandakrishnan, Characterization of hoar-development episode using SSM/I brightness temperatures in the vicinity of the GISP2 site, Greenland, *Ann. Glaciol, 17*, 183-188, 1993.

Stiles, W.H., and F.T. Ulaby, The active and passive microwave response to snow parameters, 1. wetness, *J. Geophys. Res., 85*, 1037-1044, 1980.

Sturm, M., T.C. Grenfell, and D.K. Perovich, Passive microwave measurements of tundra and taiga snow covers in Alaska, U.S.A., *Ann. Glaciol., 17*, 125-130, 1993.

Sturm, M., K. Morris, and R. Massom, The winter snow cover of the West Antarctic pack ice: Its spatial and temporal variability, *Antarctic Research Series*, this volume.

Ulaby, F.T., R.K. Moore, and A.K. Fung, Microwave Remote Sensing, Active and Passive, Vol. III, 2162 pp., Artech House, 1986.

Wadhams, P., M. Lange, and S.F. Ackley, The ice thickness distribution across the Atlantic sector of the Antarctic Ocean in midwinter, *J. Geophys. Res., 92*, 14,535-14,552, 1987.

Walker, A.E., and B.E. Goodison, Discrimination of a wet snow cover using passive microwave satellite data, *Ann. Glaciol., 17*, 307-311, 1993.

Wang, J.R., A.T.C. Chang, and A.K. Sharma, On the estimation of snow depth from microwave radiometric measurements, *IEEE Trans. Geosc. Rem. Sens., 30*, 785-791, 1992.

Wankiewicz, A., Multi-temporal microwave satellite observations of snowpacks, *Ann. Glaciol., 17*, 155-160, 1993.

Wensnahan, M., T.C. Grenfell, D.P. Winebrenner, and G.A. Maykut, Observations and theoretical studies of microwave emission from thin saline ice, *J. Geophys. Res., 98*, 8531-8545, 1993.

Worby, A.P., M.O. Jeffries, W.F. Weeks, K. Morris, and R. Jaña, The thickness distribution of sea ice and snow cover during late winter in the Bellingshausen and Amundsen Seas, Antarctica, *J. Geophys. Res., 101*, 28,441-28,455, 1996.

Worby A.P. and R.A. Massom, The structure and properties of sea ice and snow cover in East Antarctic pack ice, *Antarctic CRC, Research Report, 7*, 191pp., Hobart, Tasmania, 1995.

T. Markus, Code 971, NASA Goddard Space Flight Center, Greenbelt, MD 20771.

(Received March 12, 1997;
accepted August 6, 1997.)

EAST ANTARCTIC SEA ICE: A REVIEW OF ITS STRUCTURE, PROPERTIES AND DRIFT

A. P. Worby[1,2], R. A. Massom[1], I. Allison[1,2], V. I. Lytle[1], and P. Heil[1,3]

Data on the physical characteristics of East Antarctic pack ice (20°–160°E) are presented for the period 1986–1995. The ice in much of this region is confined to a narrow band that extends only 300 km from the continent at maximum extent in some locations, and retreats close to the coast in most places during summer. The ice is highly mobile and data from 32 drifting buoys show the mean drift speed to be 0.22 m s^{-1}, but highly variable on a daily basis. The net drift is divergent, but frequent periods of convergence cause floe deformation. Ice core textural analyses show this process to be a major contributor to the thickening of floes. The pack ice comprises, on average, 39% columnar ice, 47% frazil ice and 13% snow-ice, with other ice types comprising 1%. Ice salinity is shown to decrease with ice thickness, and mean core salinities are combined with monthly ice thickness distribution curves to estimate the total salt flux to the ocean over the growth season. This is estimated to be 8.0 x 10^{13} kg annually for the defined region. The mean undeformed, area-weighted ice thickness (including the open water fraction) varies seasonally from 0.31 m in December to 0.52 m in August, and the mean area-weighted snow thickness (including the open water and snow free fractions) varies from 0.02 m in March to 0.12 m in September. The mean snow density is 360 kg m^{-3} and the mean grain size is 1.6 mm. The constantly changing growth and deformation environment, coupled with high ice drift speeds, result in highly variable physical properties. Due to the observed variability it is not possible with current data to identify regional or interannual trends in ice and snow properties in the East Antarctic pack ice.

1. INTRODUCTION

Field investigations of Antarctic sea ice have, over the past decade, focussed primarily on the Weddell Sea. These include studies of the thickness and properties of the ice and snow by drilled measurements and ship-based observations [e.g., *Ackley*, 1979; *Casarini and Massom*, 1987; *Drinkwater and Haas*, 1994; *Eicken et al.*, 1994; *Gow et al.*, 1987, 1992; *Lange et al.*, 1990; *Lange and Eicken*, 1991; *Massom et al.*, 1997; *Meese et al.*, 1991; *Wadhams et al.*, 1987], acoustic and laser profiling studies of surface topography [e.g., *Lytle and Ackley*, 1991; *Dierking*, 1995] and upward-looking sonar studies to determine the ice thickness distribution [*Strass*, 1995]. Many of these studies have found that frazil ice growth is the dominant mechanism for ice formation, and that the mean unridged first-year ice thickness is rarely greater than 0.6 m. Ridging, rather than thermodynamic growth, is the predominant mechanism for increasing floe thickness beyond about 0.6 m. The Weddell Sea has also been the focus of several large interdisciplinary programs such as the Winter Weddell Sea Project [*Schnack-Schiel*, 1987], Winter Weddell Gyre Study [e.g., *Andreas et. al.*, 1993], Ice Station Weddell [e.g., *Gordon*, 1993] and the Antarctic Zone Flux Experiment [e.g., *McPhee et al.*, 1996]. These studies have further described the sea ice physical prop-

---

[1] Antarctic Cooperative Research Center, University of Tasmania, Hobart, Tasmania, Australia
[2] Australian Antarctic Division, Hobart, Tasmania, Australia
[3] Institute of Antarctic and Southern Ocean Studies, University of Tasmania, Hobart, Australia

erties, and the latter two in particular have concentrated on longer-term studies of atmosphere/ice/ocean interactions. In contrast, much of the remainder of the Antarctic pack has received relatively little attention. Studies in the Ross Sea include investigations of ice structure and properties [e.g., *Jeffries and Weeks,* 1992; *Jeffries and Adolphs*, 1997] and surface roughness [*Weeks et al.*, 1989]. These have shown that the ice in this region is structurally diverse, with frazil ice formation an important process of ice growth, and ridging an important mechanism for increasing ice thickness. In the Bellingshausen and Amundsen Seas, studies by *Haas and Viehoff* [1994], *Jeffries et al.* [1994; 1997], *Sturm et al.* [this volume] and *Worby et al.* [1996b] have been conducted since 1993.

The sea ice in different sectors of the Antarctic pack may exhibit significantly different characteristics. The Weddell and Ross Sea regions for example both contain large embayments of ice with cyclonic ocean currents influencing the drift and distribution of the ice. The Weddell Sea pack extends further north than anywhere else around the continent, and reaches a maximum of 2200 km from the coast at maximum extent. The Weddell Sea also accounts for 80% of the multi-year ice around Antarctica [*Gloersen et al.*, 1992]; in contrast the Ross Sea region is virtually ice free during summer. The Bellingshausen and Amundsen Seas have the second largest areas of summer ice, and *Jeffries et al.* [1997] have shown a higher percentage of superimposed ice than in other parts of the Antarctic pack. This sector is also atypical in that the minimum ice extent usually occurs in March, in contrast to the other sectors where minimum extent occurs in February [*Gloersen et al.*, 1992]. The East Antarctic region is different again in that it consists of a narrow and highly mobile band of sea ice, which is typically only several hundreds of kilometers wide at maximum extent. Multi-year ice is only found in small amounts, as the outer limit of the pack retreats close to the coast in most places during February. In this paper we present data on the drift and physical characteristics of the East Antarctic pack in order to develop a climatology of the region, and compare our findings with the work of other authors in different sectors around the Antarctic continent.

Published data on sea ice thickness in the East Antarctic region are sparse, and limited to a few seasons only. *Jacka et al.* [1987] presented some of the first systematic observations of sea ice characteristics in East Antarctica from a voyage in October–December, 1985, covering the region 50–70°E. These observations showed the pack ice during late spring to comprise a broad mixture of ice types, thicknesses and floe sizes, with a total ice concentration generally higher than 80%. *Allison* [1989b] made similar observations over a larger area of the East Antarctic pack (60°–120°E) during the late spring of 1986, and *Allison and Worby* [1994] presented ship-based observations of sea ice thickness and its spatial and temporal variability for the region 60°–150°E, between October and May for the years 1986–1993. *Worby and Massom* [1995] compiled ice and snow statistics from the same region for 1991–1994. These studies highlighted the high degree of variability in ice and snow thickness that is common in the spring, and the high percentage of thin ice present in all areas of the pack. The first winter study of East Antarctic pack ice characteristics and ice/ocean interaction was conducted near 140°E in August 1995. This focussed on the growth, deformation and drift of sea ice and the development of the ocean mixed layer due to ice growth [*Worby et al.*, 1996a].

Allison [1989a] reported on the drift of five buoys which were deployed in the Prydz Bay region in the autumns of 1985 and 1987. These buoys initially moved westward along the coast in the East Wind Drift (EWD). Some of them then drifted northward and eventually moved into the eastward flow of the Antarctic Circumpolar Current (ACC), while others continued westward around the continent. *Allison* [1989a] reported average daily drift speeds between 0.05 and 0.13 m s^{-1}, similar to those in the Weddell Sea (0.06–0.18 m s^{-1}) reported by *Massom* [1992]. An intermittent program of buoy deployments has continued in the East Antarctic pack since 1985.

2. DATA AND METHODS

In this paper, we use data from field investigations and satellites to develop a climatology of the sea ice conditions in the East Antarctic pack between 20°–160°E. These data have been collected over the period 1986–1995 and include:

• Ship-based observations of sea ice concentration, thickness, floe size, topography and snow cover from 18 voyages between 1986–1995. These observations cover the range 60°–160°E, for all months of the year except January–February and May–July.

• Ice cores sampled during six different voyages from 1991–1995 and analysed for ice structure, salinity and oxygen isotope composition.

• Snow cover samples from snow pits on 3 voyages from 1993–1995, analysed for grain size, density, salinity and oxygen isotope composition.

• Daily mean positions from drifting buoys deployed within the pack to estimate the drift rate of the ice. These data extend between 20°–170°E.

• Passive microwave data from the Special Sensor Microwave/Imager (SSM/I), to estimate variations in ice extent over the period 1990–1995.

3. SEA ICE EXTENT

Figure 1 shows the mean monthly ice edge locations at 5° longitudinal increments from 1991-1995 for January to August, and from 1991-1994 for September to December. These have been extracted from daily maps of sea ice concentration derived from SSM/I data using the NASA TEAM algorithm. The ice edge has been taken as the isoline of 15% ice concentration. The maps show the seasonal cycle of ice advance and retreat, as well as the large interannual variability in different seasons and in different regions of the pack. The data follow on from the studies of *Zwally et al.* [1983] and *Gloersen et al.* [1992] in examining data from the Indian Ocean sector (20°–90°E) and western Pacific Ocean sector (90°–160°E).

The growth and decay cycle of the ice in all sectors around Antarctica is characterised by 7 months of growth and 5 months of decay. Figure 1*b* shows the minimum East Antarctic pack ice extent in February, with a significant degree of interannual variability in the regions 20-35°E, 60-70°E, 80-100°E and 140-160°E. At this time, the outer margin of the East Antarctic sea ice cover retreats close to the coast in many locations, and is typically of low concentration (<50%).

The annual expansion of the pack begins in March and continues until maximum ice extent is reached in September or more often October. The ice edge advance is northward from the coast, which is consistent with other regions except the Weddell Sea where much of the advance is in a east-northeasterly direction [*Massom*, 1992]. The most rapid advance usually occurs in July and a high degree of interannual variability is observed in ice edge location at this time. This is particularly true in the regions near Davis (78°E) where the ice edge extended 450 km further offshore in July 1993 than in July 1995, and near Syowa (40°E). During advance, the most extensive zonal variability occurs during June to the west of 60°E (Figure 1*f*).

Ice edge retreat begins in November, accelerates in December and continues throughout February along the entire East Antarctic coastline (Figure 1). By November, areas of open water within the pack are no longer sites of enhanced ice production; rather they become a focus for the uptake of solar radiation and contribute to the rapid decay of the ice. By this process, the pack ice decays "from within" as well as by the retreat of the ice edge from north to south. Particularly rapid retreat occurs in regions where the ice edge extended furthest north at maximum extent, e.g., to the west of 80°E.

An important feature determining sea ice extent off East Antarctica is the northward extent of the coastline, which extends further equatorward than in any other area around the continent (with the exception of the Antarctic Peninsula). Based on climatological temperature data, *Gloersen et al.* [1992] suggested that the Pacific sector is the warmest around the Antarctic continent. As a result, the mean maximum distance between the continent and the ice edge is only about 300 km at 130°E, which is as broad as the marginal ice zone in other sectors. Further west, in the Indian Ocean sector (20°–90°E), the ice extends much further north, to a maximum of 1500 km from the coast between 20–25°E. This area is thought to contain the eastern, southward-flowing limb of the Weddell Gyre [*Deacon*, 1979] which advects ice into this region from the west. The large and abrupt southward indentation in the ice edge of 300–400 km near 25°E identifies the limit of eastward ice advection by the Weddell Gyre. This feature usually persists until December, when the supply of ice from the Weddell Sea is halted due to the retreat of the pack. At maximum extent, the distance from the coast to the mean ice edge is least between 120°–135°E, where the pack ice may extend only 300 km offshore. However, the position of the ice edge in any area of the pack may vary by tens of kilometers on a daily basis due to local wind forcing.

The monthly mean areal coverage of sea ice is shown in Figure 2, and numerically in Table 1. Total ice concentration values are not shown, as these are biased by anomalously low values in the marginal ice zone where there is a high proportion of brash ice. Over the two sectors (20°–90°E and 90°–160°E), the minimum ice extent over the 5 years shown in Table 1 is 0.8×10^6 km^2 in February 1993, with a maximum of 6.7×10^6 km^2 in October the same year. The latter compares with a lowest maximum in October 1991 of 5.8×10^6 km^2. The East Antarctic sector contains approximately 39% of the coastline of Antarctica, but accounts for only 30–35% of the pack ice area at maximum ice extent and 20–25% at minimum extent. In any given month the combined area of the pack in the Indian and Pacific sectors is roughly equivalent to the amount of ice in the Ross Sea sector alone, which covers only 50 degrees of longitude.

4. POLYNYAS

4.1 *Latent heat polynyas*

The East Antarctic coastline is characterised by a number of recurring polynyas, and several studies using satellite data have characterised and monitored their locations and behaviour. These coastal polynyas are maintained in part by the removal of nearshore ice as quickly as it forms by katabatic winds draining off the ice sheet [*Parish and Bromwich*, 1987]. *Adolphs and Wendler* [1995] used AVHRR imagery to investigate

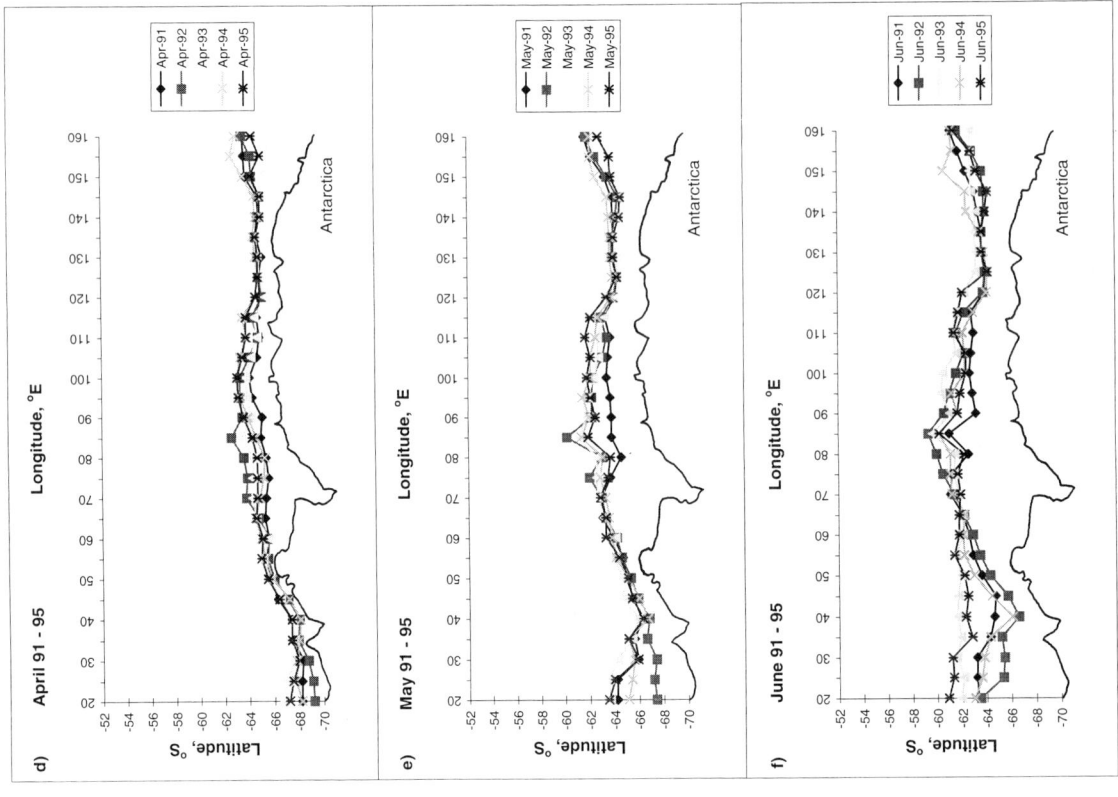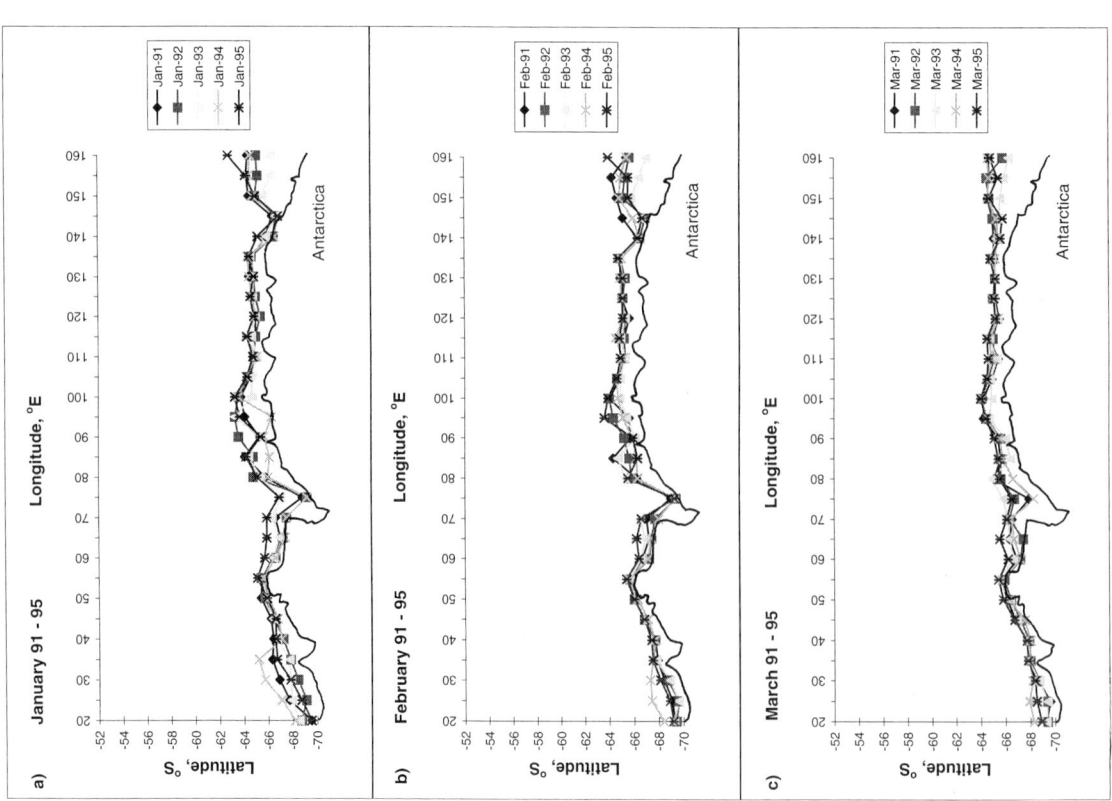

Fig. 1. Mean monthly ice edge locations in East Antarctica (*a-h*) January-August for the years 1991–1995, and (*i-l*) September-December for the years 1991–1994. The locations are at 5-degree longitudinal intervals, derived from mean monthly maps of SSM/I ice concentrations. The concentration threshold for determining ice extent is 15%.

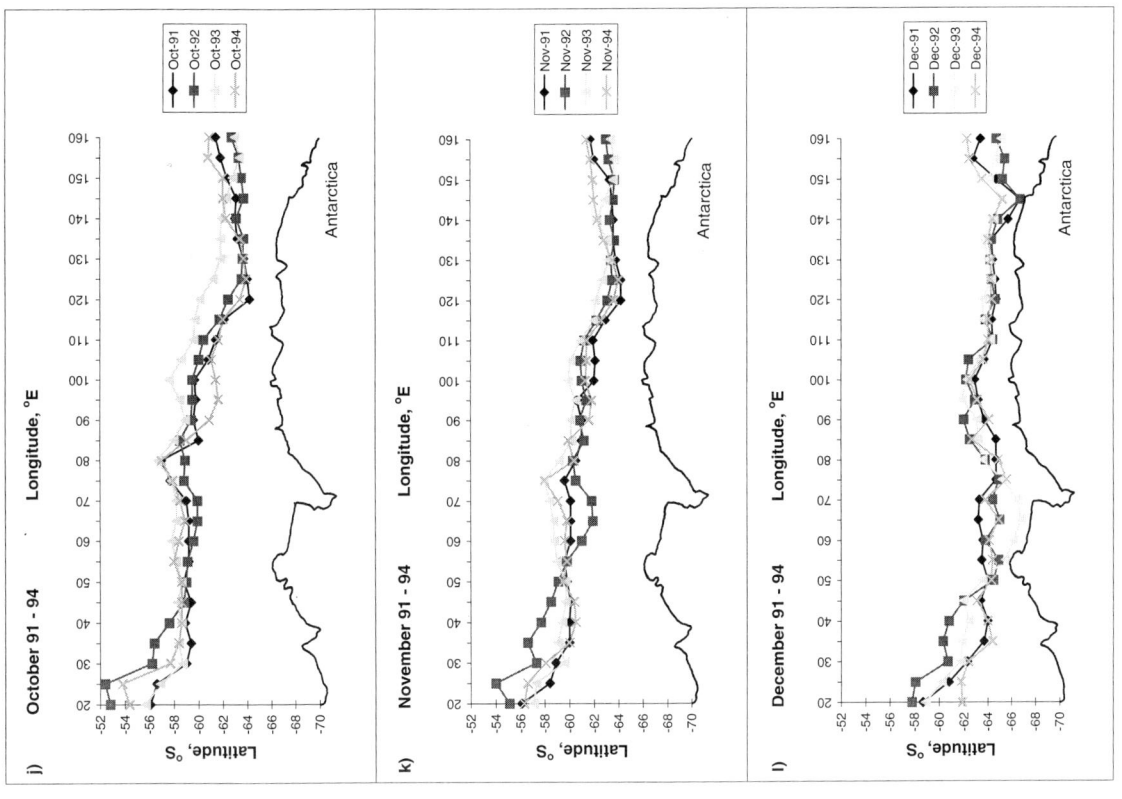
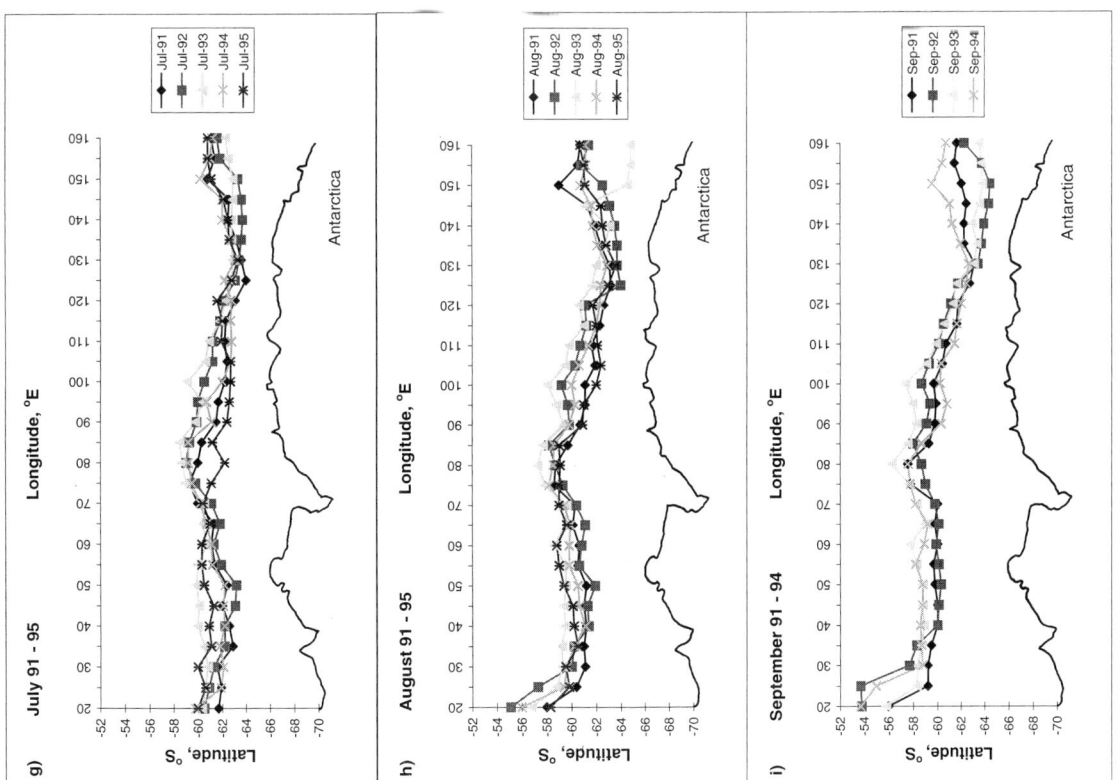

Figure 1 (continued)

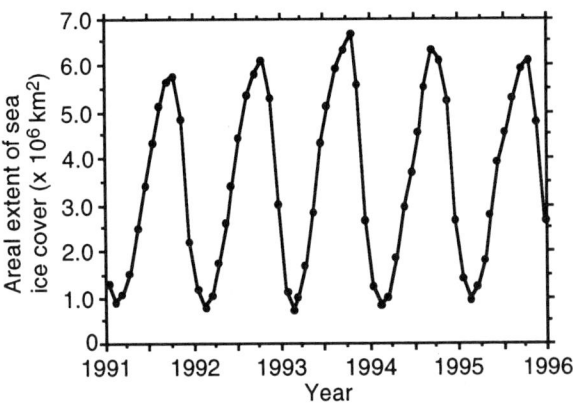

Fig. 2. Sea ice areal extent (i.e., the area of ocean covered by sea ice plus open water within the pack ice) in East Antarctica for the period January 1991 to August 1995 inclusive, recorded from mean monthly SSM/I maps of sea ice concentration. The concentration threshold for determining ice extent is 15%.

polynya behaviour off Terre Adélie and King George V Land, and suggested that both katabatic and synoptic winds play a significant role in ice removal. The contribution of ice growth in coastal polynyas was estimated by *Cavalieri and Martin* [1985] to be about 10 m per year, compared with 1 m for adjacent fast ice. The outflow of cold air from the continent can lead to air-water temperature differences of more than 20°C, with subsequent ocean-atmosphere heat fluxes in excess of 500 W m^{-2} [*Fahrbach et al.*, 1994]. As a result, polynyas may dominate the regional heat flux in winter [*Adolphs and Wendler*, 1995], and the salt flux to the ocean resulting from enhanced ice growth is significant in the production of Antarctic bottom water [*Gordon and Tchernia*, 1972; *Foster*, 1995].

Potter [1995] used SSM/I data to identify and monitor polynyas along the entire East Antarctic coast. Polynyas were identified in mean monthly images as areas of significantly reduced (<75%) ice concentration during the winter months, and for the period June–October 1987-1994, 31 coastal polynyas were identified between 20°-160°E. The polynya locations are shown in Figure 3, along with the time-averaged stream lines of the Antarctic continent surface wind field as modelled by *Parish and Bromwich* [1987]. *Potter* [1995] observed a high degree of variability in polynya persistence and recurrence over the study period, with 19 (61%) recurring every year, while others are not observed at all in some years. For example, 8 occur in four or less of the eight years. One of the more persistent polynyas studied is the Vincennes Bay polynya near 110°E, which is present in 37 of the 40 study months, reaching a maximum size of 14,000 km^2. In general the persistent polynyas tend to be larger than their more ephemeral counterparts.

Polynyas are also common to the west of coastal protrusions, such as glacier and iceberg tongues, that block the westward motion of sea ice around the continent in the East Wind Drift. Hence the removal of ice from the polynyas by wind is more efficient due to the lack of ice being advected from the east. These protrusions may also cause a heavy build up of ice on their eastward sides, which may be extensively deformed. We consider such areas to be the origin of 5-10 m thick, high freeboard floes that are occasionally observed in the pack. These are discussed further in section 6.4.

Other areas along the coast have been regularly observed to have a high percentage of thin ice and open water, suggesting that these might also be areas of high ice production. *Allison and Worby* [1994] showed a decrease in mean springtime ice thickness between the central pack ice zone and the coast in the region 60°-90°E, due to a persistent flaw lead along the fast ice edge. Similarly, voyages to the area near 140°E in May 1993 and August 1995 both reported vast areas of nilas adjacent to the coast or fast ice, with much thicker pack ice further north. These thin ice areas are also the result of katabatic or synoptic winds advecting ice from coastal regions to the central pack. An AVHRR image from August 1995 (Figure 4) clearly shows the linear flaw lead feature that is common in this region. The Soviet Antarctic Expedition *Atlas of Antarctica* [1966] also shows that during September, the southern part of the pack between 90°-95°E is typically grey to grey-white ice, with much heavier pack, including second-year floes, 100-200 km north of the coast.

TABLE 1. Mean Monthly Areal Coverage of Sea Ice (x 10^6 km^2) from 20°–160°E for the Period 1991–1995

Year	Jan	Feb	Mar	Apr	May	Jun	Jul	Aug	Sep	Oct	Nov	Dec
1991	1.333	0.931	1.074	1.521	2.551	3.440	4.353	5.138	5.707	5.778	4.894	2.210
1992	1.185	0.825	1.038	1.758	2.619	3.443	4.474	5.390	5.862	6.160	5.309	3.025
1993	1.146	0.772	1.005	1.698	2.863	4.343	5.156	5.945	6.400	6.728	5.636	2.676
1994	1.265	0.884	1.035	1.866	2.958	3.750	4.600	5.558	6.400	6.119	5.283	2.705
1995	1.426	0.971	1.235	1.837	2.801	3.949	4.590	5.330	5.953	6.147	4.842	2.717
Mean	1.271	0.877	1.077	1.736	2.758	3.785	4.635	5.472	6.064	6.186	5.193	2.667

4.2 Sensible heat polynyas

Only one recurrent sensible heat polynya occurs regularly in the East Antarctic region. This open-ocean polynya, which is usually centred near 65°S, 52°E in the Cosmonaut Sea (Figure 3), was identified in satellite imagery by *Comiso and Gordon* [1987]. Although it can be as large as 114,000 km^2 and has occurred in satellite imagery every year since 1972, it seldom persists for longer than one month at a time. *Comiso and Gordon* [1996] show that the centroid of the polynya varies only slightly with each formation (both during a given year and interannually), suggesting that its location is controlled by ocean bottom topography, and initial formation is probably induced by wind. They suggest that the formation of this sensible heat polynya may not involve convective overturning, but rather locally-enhanced upwelling of warm deep water induced through local topography and circulation interactions. Similarly, *Wakatsuchi et al.* [1994] associated notable decreases in ice thickness and concentration with upwelling and eddy formation where north-south trending ocean ridges intersect the ACC between 50°–130°E.

5. SEA ICE DRIFT

The sea ice is highly mobile. South of the Antarctic Divergence (AD) the meridional net drift is westward along the coast, while further north the net ice drift is eastward in the flow of the ACC. The motion of the ice is generally divergent as there is no constraining boundary at the northern limit of the sea ice zone. The East Wind Drift around the continent is not continuous, and there are diversions in the current at various locations along the coast that carry ice northward [*Tchernia and Jeannin*, 1984]. Off East Antarctica this occurs near 25°E, 35°E, 70°E, 95°E, and to some extent near 130°E.

Data from 32 drifting buoys deployed in the East Antarctic pack between 1985 and 1996 have been analysed to determine the drift characteristics in this region. All the buoys were deployed on sea ice floes or in leads between floes in the autumn or early winter, between 64°–68°S and 65°–145°E. The position of each buoy was obtained regularly via the Argos system with an accuracy of better than 250 m [*CLS/Argos*, 1988]. An average of 17 position fixes per day were received from each buoy; however in some cases there are data gaps due to transmission problems. The lifetime of individual buoys ranged from 5 days to 15 months.

The buoy drift tracks are plotted in Figure 5. The drift tracks only show periods when the buoys were within the pack ice; periods of open water drift, identified using the measured water temperature, have been excluded from both the figures and calculations of drift speed. The buoys were all deployed south of the AD,

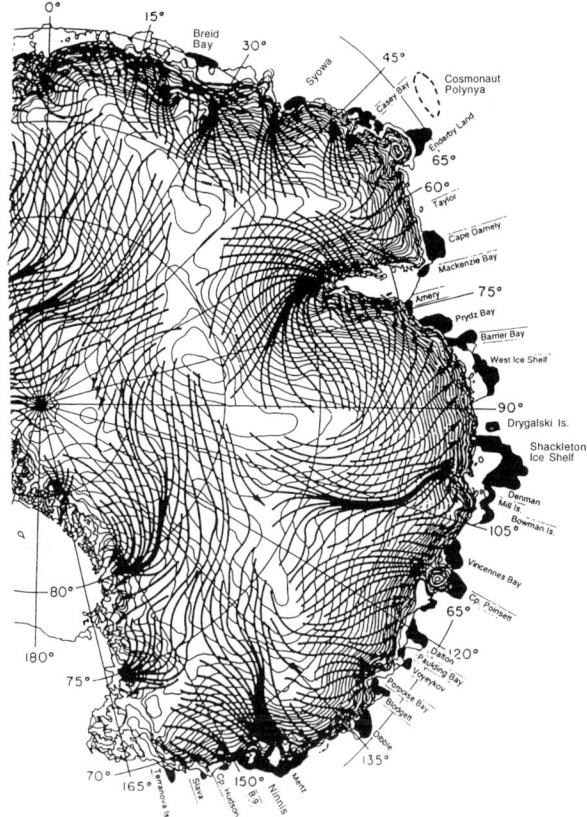

Fig. 3. Locations of recurrent coastal polynyas in East Antarctica. Each of the polynyas is identified by name, and approximate maximum areal extents observed over the period June–October from 1987–1994 are indicated by dark shading. The modelled time-averaged stream lines of the surface wind field over the Antarctic continent are after *Parish and Bromwich* [1987]. After *Potter* [1995].

and initially drifted to the west. Ten of the buoys eventually turned northward into the eastward-flowing ACC. The buoys crossed the AD at various locations; one at 20°E, two near 70°E, two near 95°E and the remainder near 130°E. One additional buoy (AAD-20) may have crossed between 105° and 120°E, although this is difficult to verify due to a large data gap. In some cases, the buoys transmitted data after they had drifted out of the ice, and in five cases the buoys survived the summer and were refrozen into the pack ice the following autumn.

Daily drift speeds have been calculated for each buoy. These are highly variable, ranging from less than 0.05 m s^{-1} (4.3 km d^{-1}) up to 0.9 m s^{-1} (78 km d^{-1}), with an average of 0.22 m s^{-1} (19 km d^{-1}). The large and rapid variations occur in response to the passage of synoptic systems [*Worby et al.*, 1996a]. The drift data are further subdivided into locations north and south of the AD, based on their average drift direction. The average

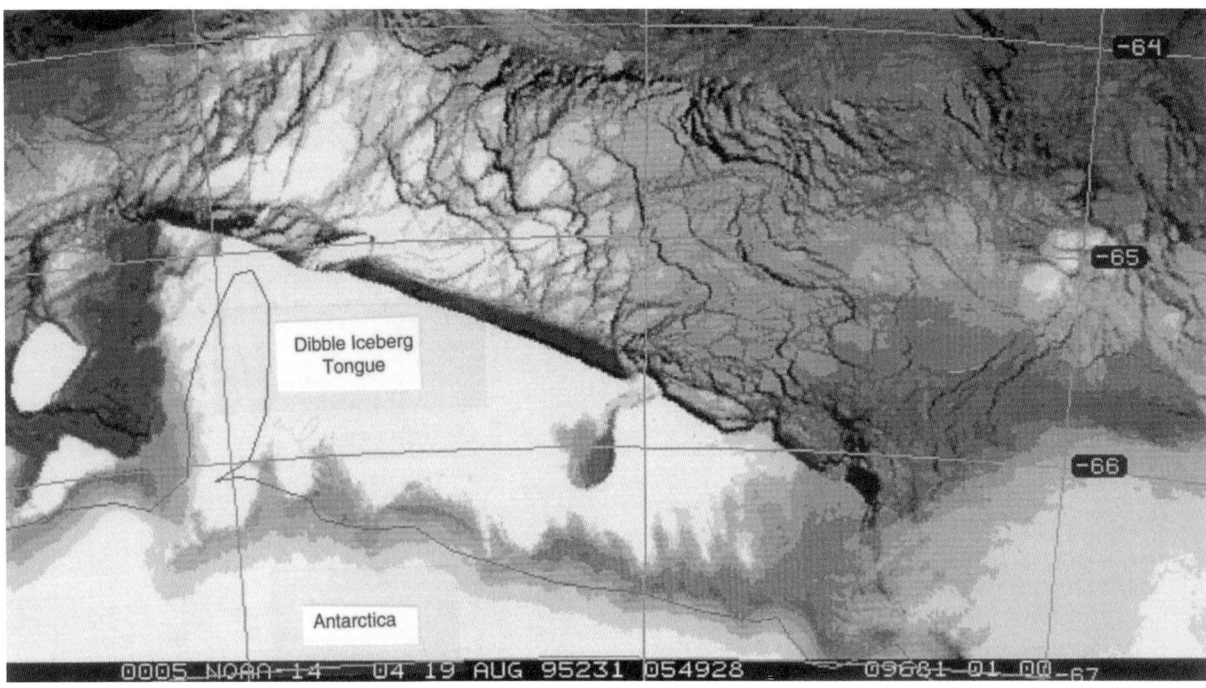

Fig. 4. An Advanced Very High Resolution Radiometer (AVHRR) image (channel 4) of the flaw polynya adjacent to fast ice along the Antarctic coast in the vicinity of 65°S, 130–135°E. The image was collected by the receiving facility of the Antarctic Meteorology Centre of Casey Base (courtesy of the Australian Bureau of Meteorology, Hobart).

speed north of the divergence was 0.23 m s^{-1}, and south of the divergence, 0.21 m s^{-1}.

South of the divergence, although the predominant drift direction was west, there were considerable daily variations in both drift direction and speed. Two buoys (AAD-18 and AAD-17) are examined in detail to demonstrate the variable drift behaviour. Figure 6 shows the drift track of buoy AAD-18 from day 116 to day 315 in 1995. Figure 7 is a plot of the daily mean speed and direction during the same time period. The average speed was 0.19 m s^{-1} which varied from 0.01 m s^{-1} to as much as 0.5 m s^{-1}. South of the AD, the average speed was 0.17 m s^{-1}, slightly less than the eastward drift speed of 0.24 m s^{-1} north of the AD. Locations where the buoy changed its drift characteristics are marked in Figures 6 and 7 by the letters A to F. The buoy track initially shows several loops (A to B), with frequent changes in direction, travelling either northeast (40°) or west-northwest (280°). Frequency analysis of these data show peaks in the energy at around 12 hours which is about the inertial period at these latitudes. From about day 172 to day 238 (B to C) the direction was fairly constant as it travelled primarily westward; however the speed was highly variable. From day 238 to day 250 (C to D) the buoy slowed considerably, never reaching speeds greater than 0.2 m s^{-1}. Although its direction was still primarily westward, it showed more variability, and on several occasions drifted eastward. After day 250 the drift speed increased as the buoys travelled northeast, crossing the Antarctic Divergence (D to E). From day 265 to day 282 (E to F) the drift speeds, on average, decreased slightly, and the drift direction again became highly variable. After day 282 (F), the buoy speed showed a marked increase and the drift direction became predominantly eastward.

In contrast, buoy AAD-17 (Figure 8), which was deployed around 66°S, 62°E, showed no loops in its drift, and followed the coastline generally westward. Its average speed of 0.34 m s^{-1} (Figure 9) is higher than most of the other buoys, although it still shows large variations from 0.05 to greater than 0.8 m s^{-1}. Periods of high speed drift are denoted by A and C in Figures 8 and 9, and periods of slow drift are denoted by B and D. The drift direction is remarkably constant, varying only 80° from 230°T (southwest) to 310°T (northwest) from day 90 to day 130. The average speed slows to less than 0.2 m s^{-1} as the buoy drifts out of Lutzow Holm Bay, (D in Figures 8 and 9) and travels predominantly north, before drifting west again.

Allison [1989a] presented data from five drifting buoys deployed near Prydz Bay and showed that the

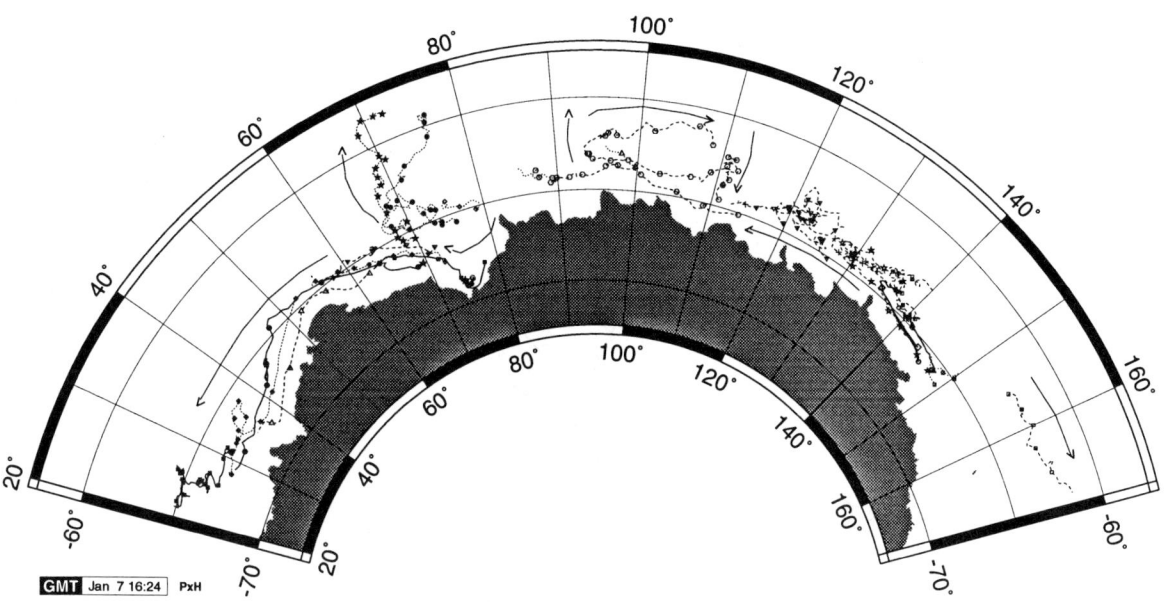

Fig. 5. Tracks of drifting sea ice buoys for 1985 (short dashed line with solid symbols), 1987 (solid line with solid symbols), 1992/93 (solid line with open symbols), 1995/96 (dotted lines) and 1996 (dashed lines). The symbols are plotted at the time of deployment and every ten days thereafter. Only periods of drift within the ice are shown. Arrows show the net drift.

variations in the drift speed are primarily influenced by the winds. The average drift of the buoys ranged from 0.12 to 0.23 m s^{-1} within the ice, with the average speed increasing significantly within the marginal ice zone. More recent observations in the pack ice near 140°E [*Worby et al.*, 1996a] also showed ice drift to be strongly correlated with local winds, with changes in wind speed and direction accelerating the ice from near zero drift to more than 0.8 m s^{-1} within hours. *Allison* [1989a] concluded that the northward divergence of the ice is an important feature in determining the location of the ice edge, and that once north of the AD the motion of the pack ice is generally divergent, with little stress between the floes. Additionally, buoy tracks have shown that the location of the AD is influenced by oceanic eddies [*Wakatsuchi et al.*, 1994], the positions of which are partly determined by the location of submarine ridges extending northward from the continental shelf.

In general, the average drift speeds are higher than those reported by *Allison* [1989a], indicating that drift speeds within the EWD and the ACC are greater than in the Prydz Bay Gyre. Two of the buoys deployed in Prydz Bay in the present study (AAD-10 and AAD-11) have mean drift speeds of 0.11 and 0.06 m s^{-1} respectively, similar to those reported in the same region by *Allison* [1989a]. A detailed analysis of the drift is beyond the scope of this paper; however, these data illustrate the highly mobile nature of the sea ice.

6. SEA ICE PROPERTIES

Ice cores were collected during six voyages to the East Antarctic pack ice between 1991 and 1995. These voyages covered a broad range of longitude from 60°–145°E, in the months of May and August–December. The cores were collected from a vast range of ice and floe types, from nilas, pancake ice and level first-year floes to heavily ridged first-year floes, multi-year floes and fast ice. The thickness of these floes varied from less than 0.1 m to as much as 10 meters. In general, a core was collected throughout the entire thickness of the ice; however, on 7 floes, where the ice was in excess of 2 m thick, we were not able to sample the entire floe, either because of logistical constraints, or because the bottom part of the floe was unconsolidated. Such cores are excluded from the analysis (e.g., Figure 10).

Data from 165 cores are presented; 153 from first-year floes, 4 from high freeboard (multi-year) floes, and 8 from the coastal fast ice. The cores are 0.1 m in diameter and were collected using either a stainless steel or fibreglass corer. For each core, a 2–3 mm-thick longitudinal section was cut, and examined between cross-polarised filters to determine its crystal structure. This initial examination of the core differentiates between columnar and granular ice [*Weeks and Ackley*, 1982], and subsequent oxygen isotope analysis of melted core sections further differentiates between granular ice of frazil and snow-ice origin. The vertical core sections

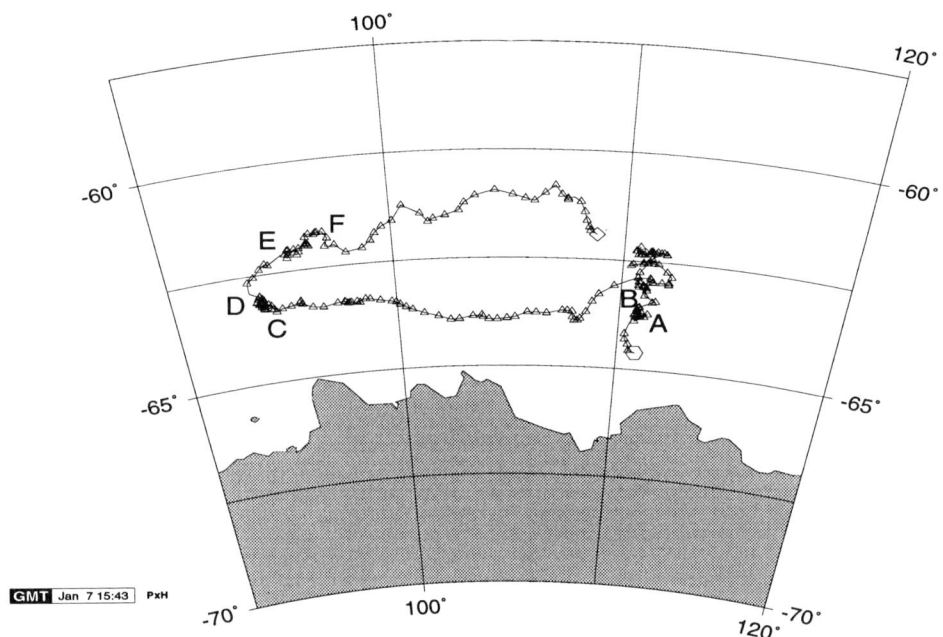

Fig. 6. Part of the drift track of buoy AAD-18, initially showing several loops before drifting predominantly westward south of the Antarctic Divergence and then eastward within the ACC. The hexagonal symbol indicates buoy deployment, and the diamond the end of the buoy drift. See text for explanation of letters.

were cut along crystal structure boundaries observed in the thin section. The majority of these were 0.01–0.05 m long and were also analysed for salinity by determining either the density or conductivity of the melted sample [*Worby and Massom*, 1995]. The accuracy of the salinity measurements is ±0.1 psu.

6.1 *Pack ice crystal structure*

Pack ice is the drifting sea ice which is usually less than one year old. A summary of the cores taken on each voyage is shown in Table 2. These have an average length of 0.66 m. The crystal structure is an indication of the conditions under which the ice formed. Initially an ice cover will have a layer of granular crystals, but subsequent growth under calm conditions will result in a columnar ice structure, or under more turbulent conditions, granular frazil ice will form. Photographs of thin sections of ice cores showing these (and other) structural differences can be found in *Weeks and Ackley* [1982]. Granular ice can also result from deformation and crushing of the columnar crystals, or from snow-ice formation caused by flooding and refreezing the base of the snow layer [e.g., *Lange et al.*, 1990].

For all the pack ice cores, 39% of the ice is columnar, ranging from 24% to 51% for individual voyages (Table 2). In the Weddell Sea *Lange and Eicken* [1991] report a slightly higher mean value of 43.1%, and a range of values for individual voyages of 30.5% to 58.1%. There is no systematic change with season, and it is not possible to determine whether these variations are representative of the ice conditions during different voyages, or are the result of the sampling strategies (one voyage had only 3 cores). The proportion of columnar ice in individual cores generally decreased with length, varying from 0% to as much as 92% in some of the shorter cores. The core stratigraphy often shows multiple layers, as reported by *Allison and Worby* [1994] and *Worby and Massom* [1995] in this region, and by numerous other authors [e.g., *Lange*, 1988; *Jeffries and Weeks*, 1992; *Jeffries et al.*, 1997] in other areas of the Antarctic pack. For cores longer than about 0.6 m the columnar ice is typically found in several different sections caused by rafting. In the thinner ice, although rafting is still apparent in the cores, the columnar ice is often found in sections as long as 0.5 m. In some cores there are sections of mixed columnar and granular ice, probably the result of deformation. Such layers have also been observed in other areas of the Antarctic pack ice [*Jeffries and Weeks*, 1992; *Lange and Eicken*, 1991]. Cores longer than 1.6 m have less than 10% columnar ice, also primarily due to deformation which can destroy the columnar crystal structure as the ice is being ridged. This ice is classified as "fragmented" if it

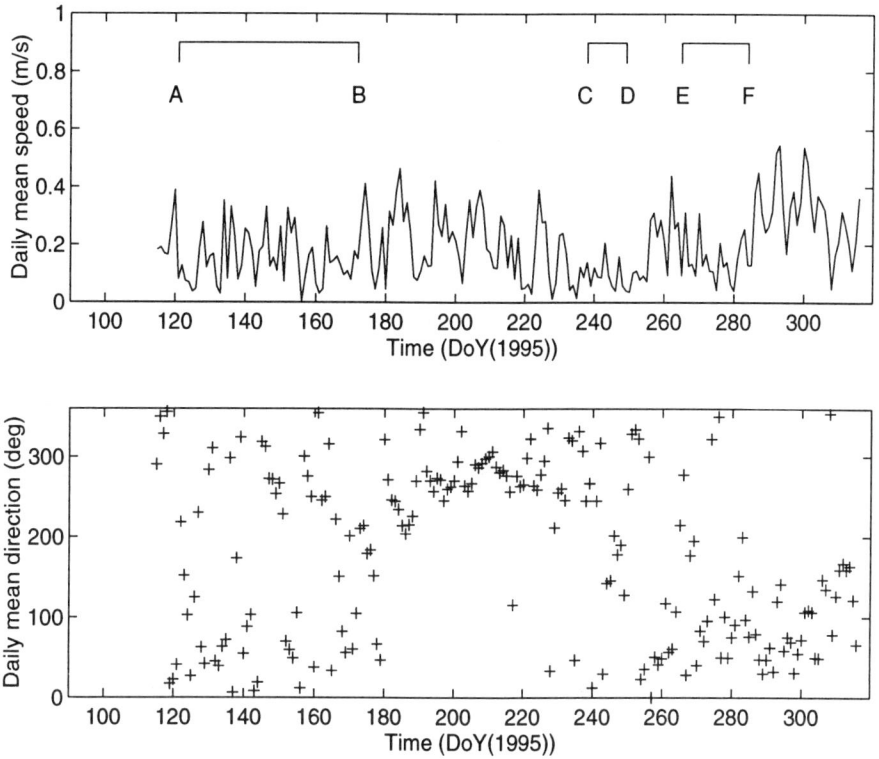

Fig. 7. Daily mean speed (*a*) and direction (*b*) of buoy AAD-18. See text for explanation of letters.

is clearly deformed congelation ice; it contributes to the 1% of "other" ice types within the pack.

Snow-ice is formed as a result of snow loading, wave induced flooding, or deformation processes forcing the ice/snow interface below sea level and flooding the base of the snow cover. The incorporation of snow into the sea ice reduces the $\delta^{18}O$ value of the sample, making it possible to differentiate between frazil ice (sea water only origin) and snow-ice (mixture of sea water and snow). Various authors have used different $\delta^{18}O$ cutoff values to identify snow-ice [e.g., *Lange et al.*, 1990; *Jeffries et al.*, 1994], with 0.0 psu being the most commonly used. For this study, snow-ice is classified as ice with a granular structure and a $\delta^{18}O$ value less than 0.0 psu. We do not attempt to define the percentage of snow present within each sample.

$\delta^{18}O$ data are available for 70 pack ice cores collected on 4 of the 6 voyages. Of these cores, 40 (57%) have no snow-ice, and 4 have sections of more than 0.40 m of snow-ice. In total, 13% of the volume of ice sampled is snow-ice. This fraction, and the fractions of other ice types, have been calculated from the 153 individual cores, not by averaging the results of each voyage. Consequently, different ice regimes, which may result from regional and seasonal variations in pack ice characteristics, cannot be identified. The fraction of snow-ice is slightly less than the 16% reported by *Lange and Eicken* [1991] in the Weddell Sea which was also derived from data collected aboard numerous voyages. The much higher value of 24–27% reported in the Bellingshausen and Amundsen Seas by *Jeffries et al.* [1997] is not directly comparable with these results as it is representative of conditions for the duration of only one voyage and does not average over potentially different ice regimes. Individual voyages to the East Antarctic (Table 2) show similarly large percentages of snow-ice.

Two of the cores contained superimposed ice, which is formed by fresh water (from snow melt) freezing onto the surface of the ice. Interestingly, this ice was observed only in first-year floes in late May, indicating that snow melt had occurred since the start of the growth season. This is consistent with recent observations by *Worby et al.* [1996a] that surface melt can occur within the pack ice in any season. Superimposed ice comprises only 0.2% of the total ice core volume sampled in East Antarctica. Significantly greater proportions of this ice type (almost 5% of the pack in summer 1992) have been reported by *Jeffries et al.* [1997] in the Bellingshausen and Amundsen Seas.

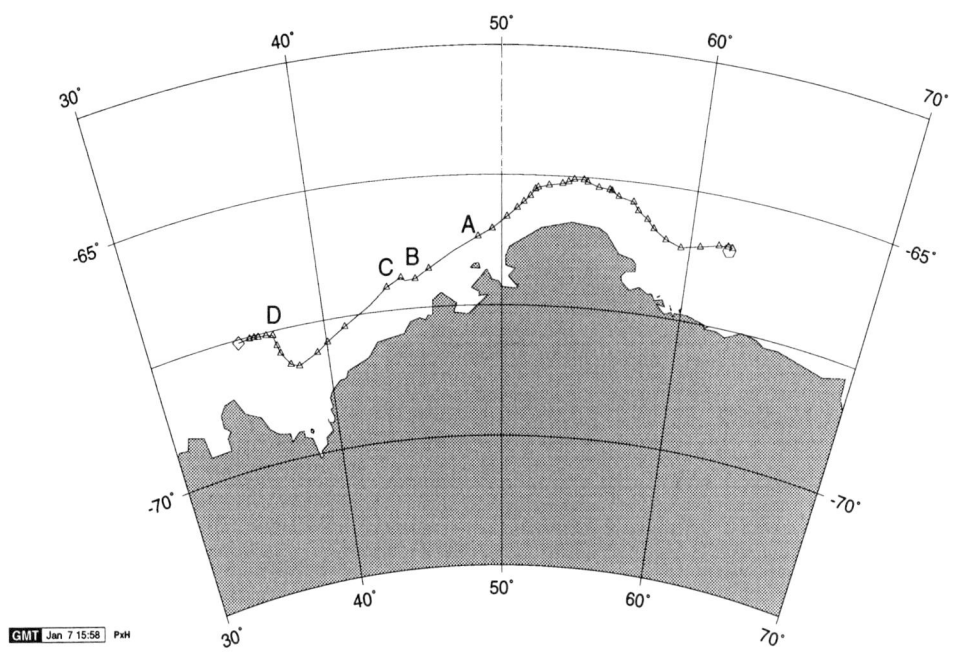

Fig. 8. Drift track of buoy AAD-17, which travelled westward along the coast. The hexagonal symbol indicates buoy deployment, and the diamond the end of the buoy drift. See text for explanation of letters.

Kawamura et al. [1993] have also reported superimposed ice in the region near Syowa (40°E). No platelet ice was observed in either the fast or pack ice cores reported in this paper. This ice type is most commonly found near the coast, and may form as a result of snow- and ice-melt water runoff from the continent, or adiabatic cooling of low density water masses beneath ice shelves [*Jeffries et al.*, 1993].

6.2 Pack ice salinity

The average salinity by volume of all the pack ice cores is 6.6 psu, with the average salinity of individual cores ranging from 3.1 to 28.2 psu. In general, the average bulk salinity decreases with increasing ice thickness (Figure 10). The salinity can be divided into two distinct categories; ice less than about 0.05 m thick has salinities ranging from 9 to 28 psu, and ice greater than 0.05 m thick shows an approximate linear decrease in salinity with thickness from 8 to 4 psu. This decrease in salinity with ice thickness indicates that the ice continues to desalinate after it is deformed, providing a salt flux to the ocean over a period of time, not just during the initial growth phase. This salinity variation with thickness is similar to that found in other regions of the Antarctic [*Gow et al.*, 1987], and with Arctic sea ice, despite structural differences in the ice [*Kovacs*, 1996]. The salinity values are used in conjunction with the ice thickness distribution to determine a net salt flux to the ocean (see Section 7.0).

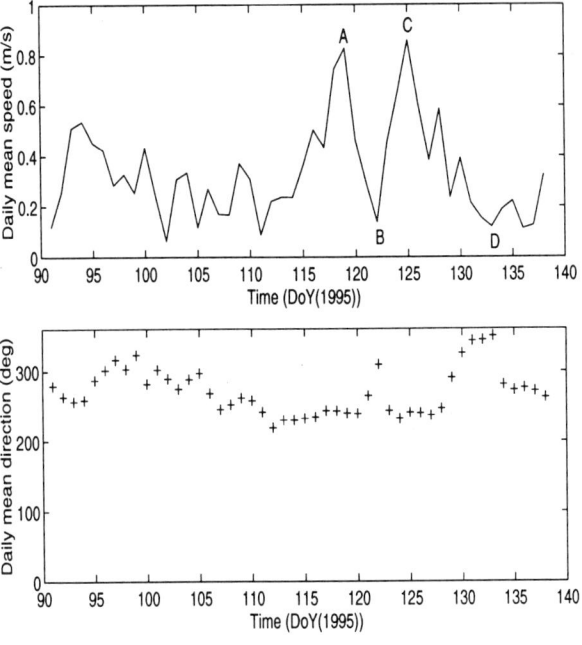

Fig. 9. Daily mean speed (*a*) and direction (*b*) of buoy AAD-17. See text for explanation of letters.

TABLE 2. Summary of the Mean Core Structure and Salinity for Each of Six Voyages Between 1991 and 1995

Voyage	Number of Cores	Average Length (m)	Average Salinity (psu)	Snow-ice (%)	Columnar Ice (%)	Total Granular Ice (%)
V1 91/92	3	0.69	5.9	45.6	30.7	69.3
V2 91/92	8	0.90	6.5	18.3	50.0	50.0
V1 92/93	20	0.65	7.6	26.0	39.8	60.2
V9 92/93	39	0.26	15.1	11.9	34.5	65.5
V1 94/95	34	1.14	6.6	...	24.0	76.0
V1 95/96	49	0.61	8.1	...	51.1	48.9
Mean	...	0.66	6.6	13.0	39.0	61.0

The snow-ice percentage is also included in the granular category, as the snow-ice data are not available for the two most recent voyages, and the two ice types can not be separated. The mean values are calculated by volume, and do not represent the mean of the data presented in the table for individual voyages. One percent of the pack is classified as "other ice types" (see section 6.1).

6.3 Fast Ice

Land fast ice is immobile and extends in a continuous sheet northward from the continent. The cores presented here were sampled near the Australian bases Davis (78°E) and Mawson (63°E), at varying distances offshore. The maximum fast ice extent near Davis is approximately 15 km, but may exceed 80 km offshore near Mawson where a series of islands assist in trapping the ice. Persistent katabatic winds keep the nearshore ice surface relatively snow free near Mawson but at Davis a substantial snow cover may accumulate on the fast ice.

The fast ice typically maintains a slow thermodynamic growth from March to late September [Heil et al., 1996], with surface ablation and bottom melt occurring before the ice breaks out between December and February. Eight fast ice cores were collected; one core was 0.65 m long, the remaining 7 ranged from 1.05 to 1.68 m in length. The shorter core was taken within 20 m of the northern edge of the fast ice zone near Davis, and may have been growing for a shorter time period or been subjected to bottom melting. Six of the cores had granular ice overlaying columnar ice that was comprised of large, long crystals typical of slow thermodynamic growth. Two of the cores were deformed, with several layers of granular ice interspersed with columnar ice. In total, 87% of the ice was columnar and only 2% snow-ice. The average salinities varied little from core to core (4.7 to 6.2 psu), with an average of 5.3 psu.

6.4 High Freeboard Floes

High freeboard floes are a characteristic of the pack in the region 140°–150°E, and range in thickness from 3 to 10 m, with a 1–2 m thick snow cover. The origin of these floes is uncertain, although they are often encountered in "swarms" suggesting that they are all of the same origin, and result from the break up of larger floes. We speculate that coastal protrusions, such as grounded icebergs or glacier tongues, impede the westward flow of ice along the coast, causing sea ice to become trapped and highly deformed.

In regions where these floes are common, they may account for a sizeable percentage of the volume of ice within the pack. Ship-based observations from August

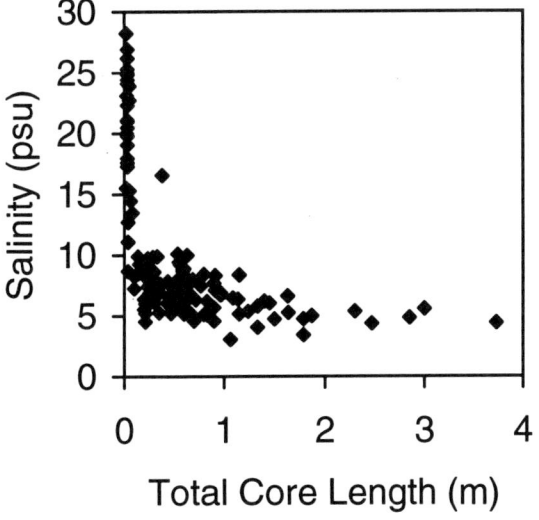

Fig. 10. Core length versus salinity (expressed in practical salinity units, psu) for 157 cores from first- and multi-year pack ice floes. Data from three cores greater than 4.0 m long are not shown.

TABLE 3. Dates and Locations of Each of 18 Voyages to the East Antarctic Pack Ice Between 1986 and 1995

Year	Month	Ship	Region
1986	Nov	Icebird	60° – 120°E
1987	Oct	Icebird	60° – 120°E
1988	Nov	Icebird	60° – 120°E
1989	Mar	Icebird	60° – 100°E
1989	Nov	Icebird	60° – 120°E
1990	Oct	Aurora Australis	60° – 120°E
1990	Dec	Aurora Australis	60°E
1991	Oct	Aurora Australis	140°E
1991	Nov	Aurora Australis	60° – 100°E
1992	Oct	Aurora Australis	60° – 100°E
1992	Dec	Icebird	110°E
1993	Apr	Aurora Australis	140°E
1993	Oct	Aurora Australis	60° – 100°E
1994	Sep	Aurora Australis	75° – 140°E
1994	Nov	Aurora Australis	75° – 120°E
1995	Mar	N. B. Palmer	140° – 170°E
1995	Apr	Aurora Australis	60° – 150°E
1995	Aug	Aurora Australis	140°E

1995 (V1 95/96) indicate that over an area of 100 km^2 with a variable concentration of thick floes between 1 and 10%, the effective ice thickness was increased by 0.24 m. The floes varied in size from tens of meters up to about 400 m in diameter. Four of these were sampled; on one of these floes, we were able to collect a core throughout the entire thickness (8.5 m), and on the other three, cores were collected from the top 1.3 to 2.3 meters of ice. Multiple ice layers were found within the snow cover, indicative of past melt events, and indicating that the ice is multi-year. Twenty five percent of the ice was columnar and 75% granular. The cores comprised many layers, and columnar ice was present at many different depths. Often the columnar ice crystals were tilted from the vertical, or partially fragmented, indicating a high degree of deformation. The floes also have a deeper draft and higher freeboard than the surrounding thinner pack ice; hence they can be expected to drift at a different velocity to other floes depending on surface winds and currents. This may affect the thickness distribution of the remaining pack ice, by both deformation of thinner floes and the creation of additional open water.

A similar ice type was reported by *Wadhams et al.* [1987] for ice in the southern Weddell Sea. They surmise that these floes originate from embayments near the edge of ice shelves which appeared to be full of highly ridged, multi-year ice. *Jeffries et al.* [1994] also reported similarly sized and shaped floes with high, flat-topped surfaces and deep snow covers in the eastern Ross and Amundsen Seas.

6.5 *Sea ice thickness distribution*

Ship-based sea ice observations have been collected on 18 voyages to East Antarctica between October 1986

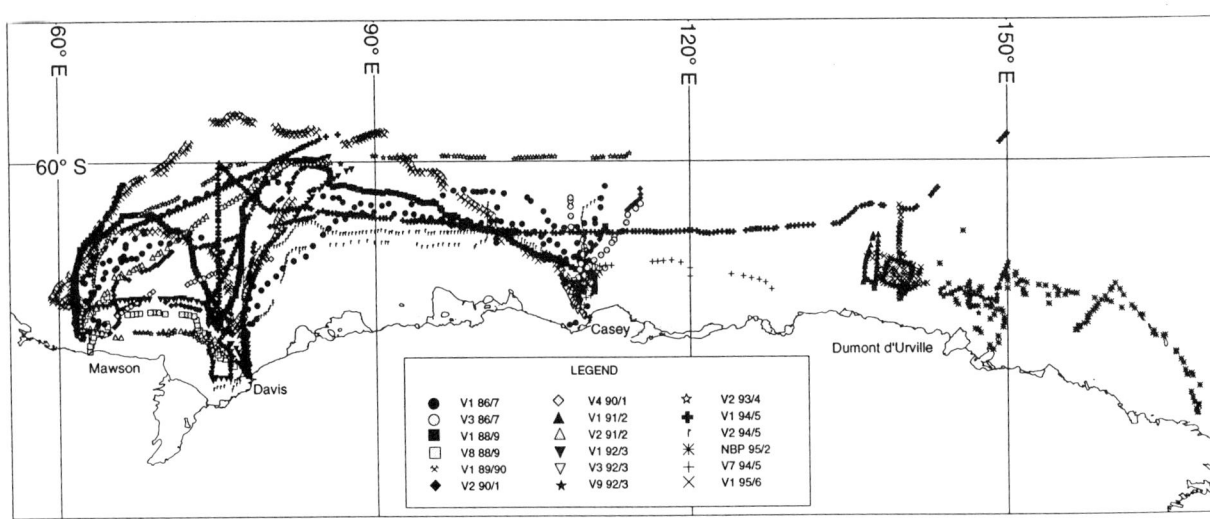

Fig. 11. Ship tracks for 18 voyages to East Antarctica over the period 1986–1995. Each symbol represents a location where ship-based sea ice observations were made.

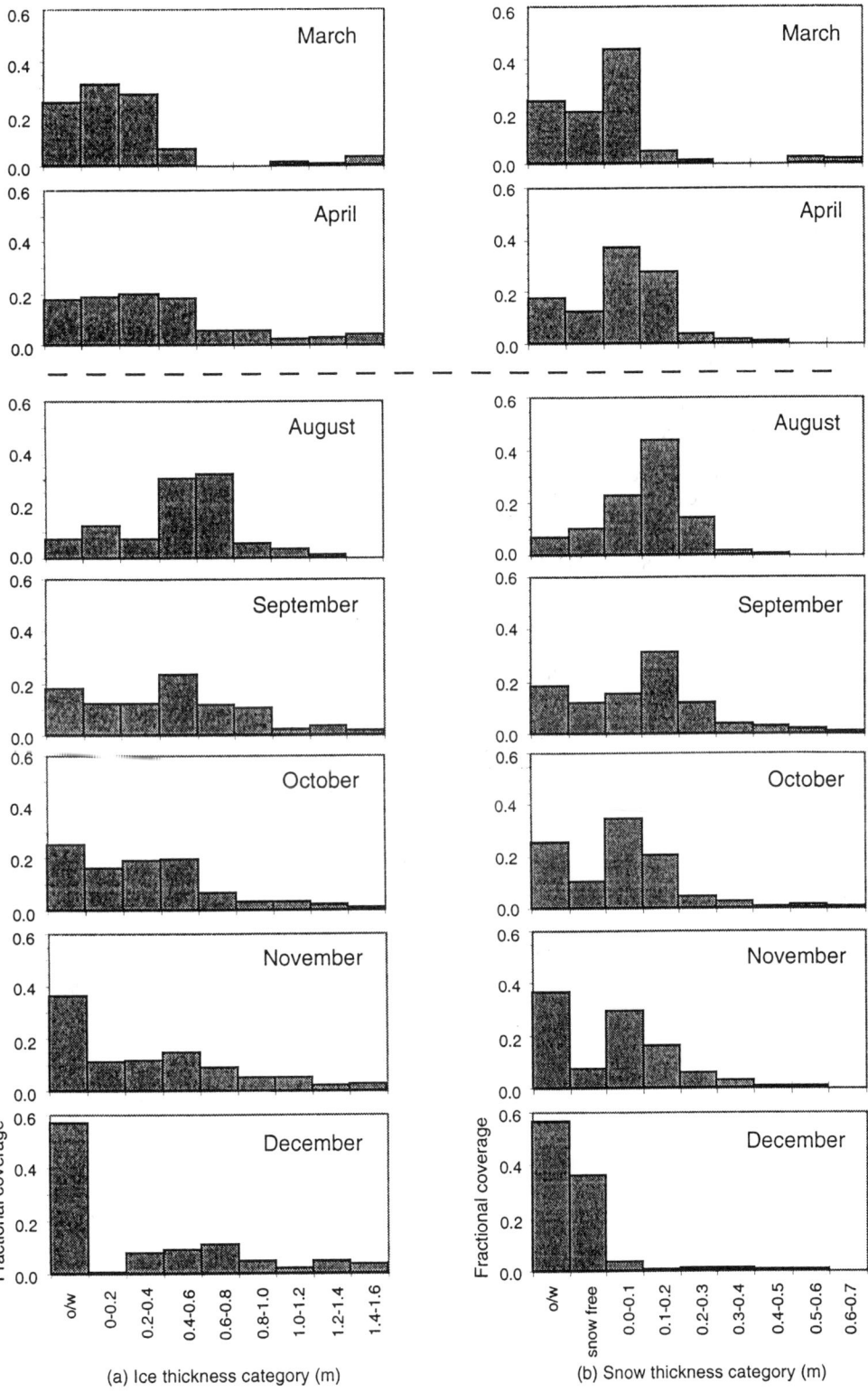

Fig. 12. Monthly thickness distributions from ship-based observations for (a) sea ice and (b) snow.

TABLE 4. Summary of the Mean Ice Concentration, and Undeformed Ice and Snow Thickness Values from Ship-Based Observations

Month	Number of Voyages	Number of Observations	Mean Ice Concentration (%)	Mean Ice Thickness (m)	Mean Snow Thickness (m)
March	3	92	76	0.36	0.02
April	3	129	83	0.48	0.11
August	1	165	93	0.52	0.11
September	1	246	82	0.47	0.12
October	10	595	75	0.35	0.07
November	8	1129	64	0.36	0.07
December	4	63	43	0.31	0.07

The mean values are calculated over the entire pack ice, including the open water fraction. Note that the mean ice and snow thickness values exclude the anomalously thick floes observed in the region 140°–150°E (described in Section 6.4).

and August 1995 using the technique described by *Allison and Worby* [1994]. The dates and locations are summarised in Table 3 and the voyage tracks shown graphically in Figure 11. Most voyages are within the longitudinal band 60°–150°E. The observations are made hourly from the ship's bridge and include the total concentration of ice and the distribution of open water. For each observation the pack is divided into three thickness categories, and the concentration, mean thickness, floe size, topography and snow cover is estimated for each category. In this way, the distribution of ice thickness and open water within the pack is recorded, as well as the mean thickness at each observation point. Only the thickness of the level areas of floes is recorded, since the thickness of ridged ice cannot be reliably determined from the ship. The level ice may have been rafted but remains sufficiently undeformed that its thickness can reliably be estimated. Additional observations of surface ridging and sail height provide a record of the extent of deformation, and these data are used to estimate the mean floe thickness as discussed in Section 6.6. The thickness distribution data presented in this section represent only the mean undeformed (or level) ice thickness.

The data presented by *Allison and Worby* [1994] have been combined with similar data from 6 additional voyages (1993–1995) to describe the seasonal cycle of the sea ice thickness distribution around East Antarctica. The complete data set (1986–1995) comprises 2419 observations, with the highest concentration of observations en route to, and in the location of, the three Australian Antarctic stations (Figure 11). The majority of the data have been collected during spring, and most years have observations in October and November. Additionally there are observations in March, April, May, September and December in a number of years (see Table 3). The data have been categorised by month, and binned into 0.2 m thickness categories. The mean monthly ice and snow thickness distribution curves are shown in Figures 12a and 12b and a summary of the data is shown in Table 4. There are sufficient data in seven months (March–April and August–December) to draw statistically meaningful conclusions about the thickness distribution of the sea ice and snow cover in this region of the Antarctic ice pack. Currently, there is still a large gap in the data set during the early winter months, with very little or no data in May, June and July.

By far the greatest seasonal changes in the ice thickness distribution are in the open water and thin ice categories. The amount of open water decreases from almost 60% in December to little more than 10% in August, and the thinnest ice thickness category (0–0.2 m) shows a 30% seasonal change between December and March. In contrast, the amount of ice greater than 1.0 m shows very little seasonal variability. Although these observations are of the undeformed component of the pack, there are occasions when thick consolidated floes can quite accurately be estimated and these are invariably included in the observations. While these floes are undoubtedly deformed ice, they comprise a small fraction of the total pack that exhibits very little sea-

TABLE 5. Mean Ice Core Salinities Binned Into 0.2 m Thickness Categories

Ice Thickness (m)	Mean Ice Core Salinity (psu)
0.0–0.2	16.9
0.2–0.4	7.2
0.4–0.6	7.1
0.6–0.8	6.5
0.8–1.0	6.1
>1.0	5.2

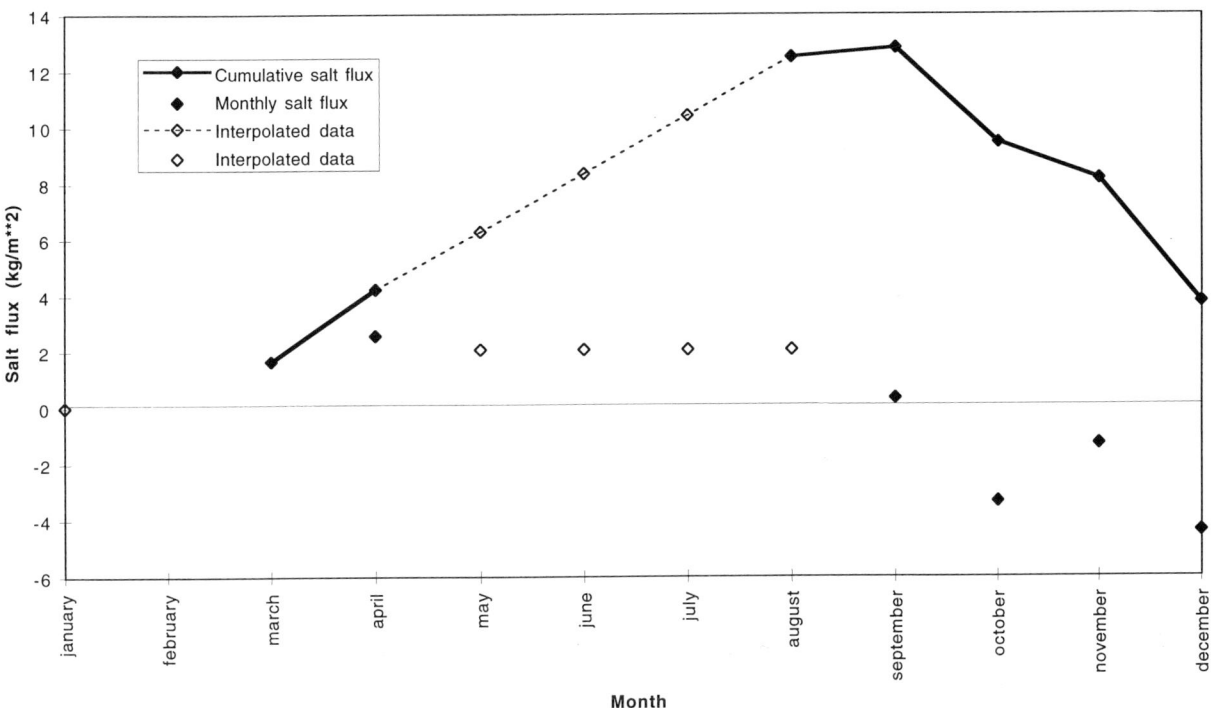

Fig. 13. Salt fluxes based on the undeformed ice thickness distributions for each month shown in Figure 12a, and the mean ice salinities shown in Table 5. The monthly fluxes and cumulative flux over the year are shown.

sonal variability, and their inclusion in the observations does not affect our conclusions.

The discussion of Figures 12a and 12b focuses on the months of March, August, October and December. In March, there is approximately 25% open water and an additional 60% of ice less than 0.4 m, indicative of rapid new ice growth over large areas of the Southern Ocean. Most of this ice has a thin snow cover with less than 10% greater than 0.1 m. A close look at the March data shows considerable variability in the composition of the pack ice in difference locations. In the area around Prydz Bay (60–90°E) there is predominantly new ice in March, while the area near 140°–150°E has a higher fraction of multi-year ice, with a substantially thicker snow cover. This thicker, multi-year ice has been excluded from the mean values presented in Table 4 which represent only first year ice.

In August, the pack is quite consolidated, and the open water fraction averages only 12%. There is only a small percentage of thin ice due to cold air temperatures at this time of year quickly refreezing leads to ice thicknesses greater than 0.4 m, and also due to the effects of deformation. This is supported by observations in the winter pack showing that ice may quickly grow to more than 0.4 m [*Worby et al.*, 1996a]. Hence, only a small fraction of the pack is comprised of open water and thin ice, the opposite of the March distribution, but the snow cover is predominantly less than 0.2 m.

By October, two changes in the ice growth regime contribute to the flattening of the thickness distribution curve. First, leads do not refreeze as quickly as observed in August, increasing the relative amount of ice less than 0.7 m thick. Second, the ice does not grow to the same thickness, primarily because of increased radiation and warmer air temperatures. As a result, there is more open water and thin ice within the pack, typically with a thinner snow cover. As the pack diverges, ice is slower to form, leading to an increase in the open water fraction, and a subsequent warming of the surface ocean water. This is a positive feedback which further limits ice production, and may result in some ice melt. The distribution curve for December reflects this, showing the greatest open water fraction, no ice thinner than 0.2 m, and a considerable decrease in ice types thinner than 0.6 m thick.

The ice thickness distribution curves for the intervening months are consistent with the discussion above. April shows a flatter distribution than March as a result of thickness increases in existing ice by both dynamic and thermodynamic growth, and new ice growth only in restricted open water areas. The September curve flattens between the thinnest category and

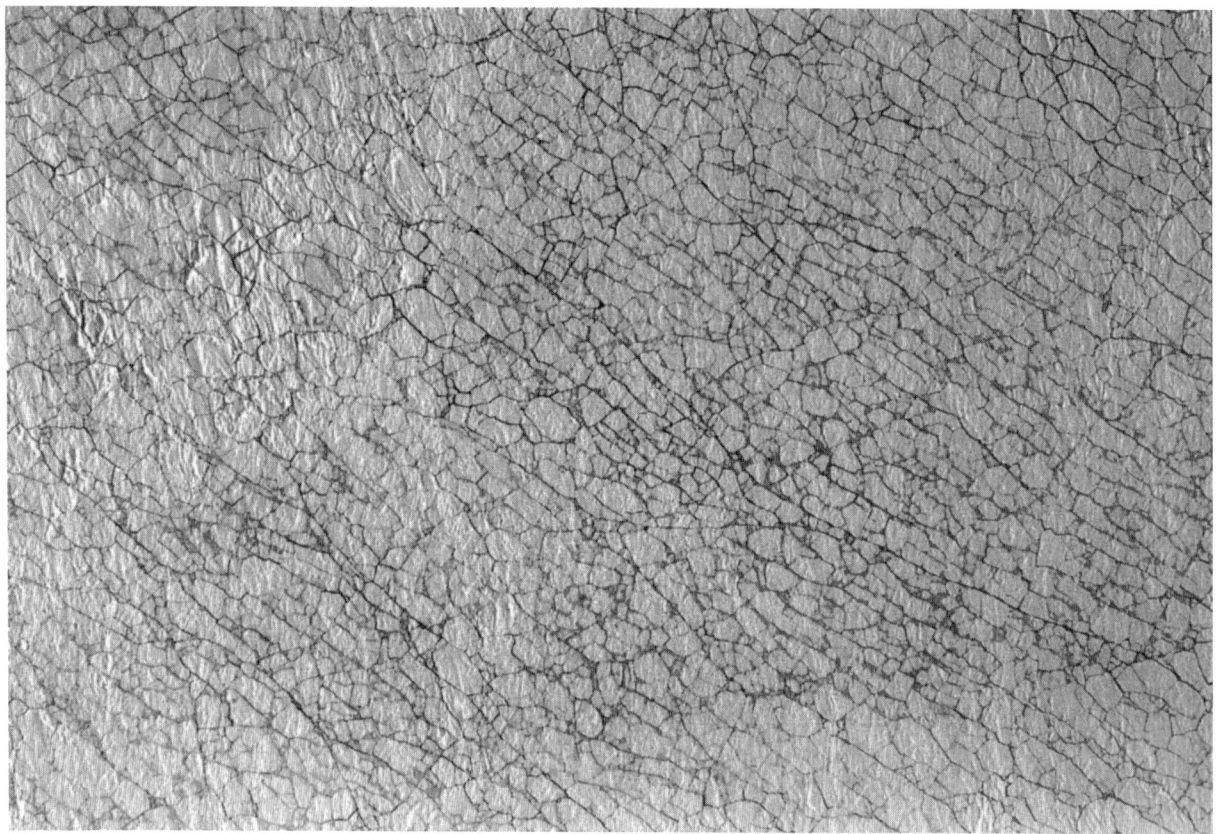

Fig. 14. An aerial photograph showing the effect of swell penetration on floe size within the pack. The image covers a region of approximately 1170 x 785 m, and the average floe size is approximately 20–40 m. Also visible is the extent of flooding at floe edges caused by wave action (discussed in Section 8.0). The photograph was taken over the pack ice near 140°E from an altitude of 8000 ft (2440 m).

the 0.4–0.6 m category, which is consistent with the trend between August and October. November, in turn, shows an increase in open water fraction and further flattening of the distribution curve in response to the divergence of the pack and limited new ice growth.

6.6 Sea ice ridging

The differential drift of ice floes within the pack may result in mechanical stacking of the floes, thereby increasing their thickness. This is extremely common in the Antarctic where the ice is highly mobile. The structural analysis of ice cores from East Antarctica, as well as other locations within the pack, indicate that dynamic processes may play an important role in the evolution of sea ice floes from the earliest stages of their development. *Worby et al.* [1996b] suggested that rafting is the dominant dynamic mechanism by which floes reach 0.4–0.6 m thick, whereas beyond that, converging floes are more likely to form ridges. There will obviously be exceptions to this, e.g., crushing new lead ice into ridges.

Observations of ridged sea ice thickness are not easily obtained. Drilled profiles across floes are only localised, but provide valuable information on the small scale spatial variability in ice thickness. *Allison and Worby* [1994] showed a number of profiles from this region, and used these to help develop a simple model for calculating the effective mean thickness of ridged ice from ship-based observations of sail height and the extent of surface ridging. A comparison of these results with laser and acoustic studies was presented by *Worby et al.* [1996b], who concluded that the corrected thickness may overestimate the real mean thickness of the pack. Hence, the model formulation has subsequently been revised to account for possible errors caused by the inclusion of snow drifts in the observations of surface ridging, and the ratio of snow and ice above sea level to ice below, observed in drilled transects. The empiri-

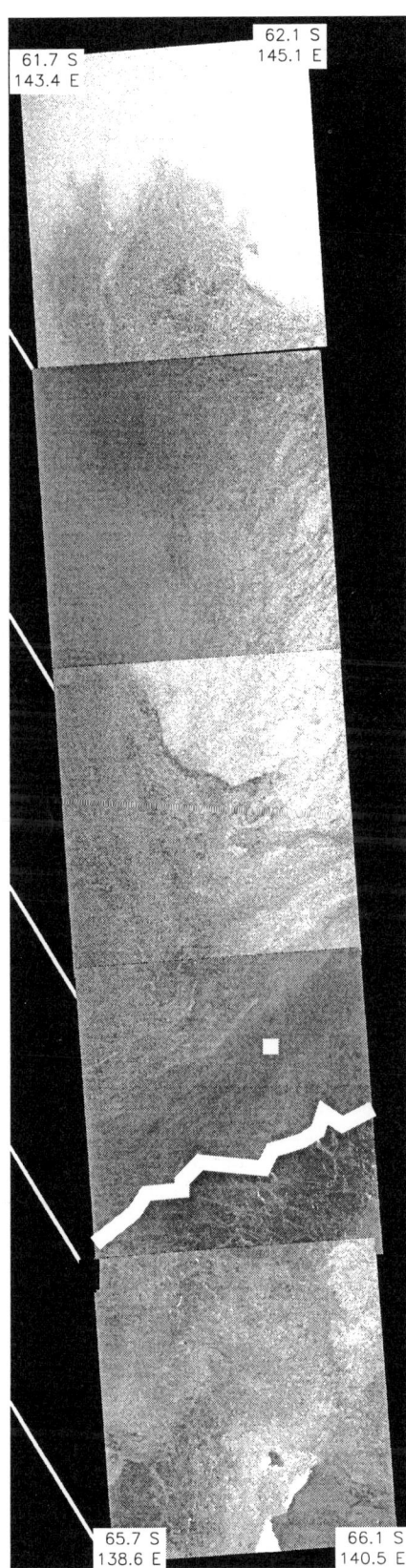

cally revised expression to calculate the average thickness of ridged floes (z_r) has the form:

$$z_r = 6(0.5RS) + z_u \qquad (1)$$

where R is the areal extent of surface ridging; S is the average sail height of ridges, and z_u is the thickness of the level ice in the floe.

Eight voyages to the East Antarctic pack between 1992 and 1995 have observations of surface topography that can be incorporated into the revised *Allison and Worby* [1994] model. These show that by incorporating the ridged ice, the mean thickness increases, on average, by 1.8 times the observed mean undeformed ice thickness. Individual voyages show increases of between 1.3 and 2.4 times.

7. SALT FLUX

The ship-based observations of sea ice thickness are combined with the measured sea ice core salinity values to estimate the salt flux to the ocean. The cores have been binned into the same 0.2 m thickness categories as the ice thickness data, and the average salinity for each category has been calculated (Table 5). The fractional coverage of each ice thickness category is derived from the ship-based observations (Figure 12a), and the total area of the pack for each month has been determined from seven complete years of SSM/I data from 1988–1994. A combination of these data is then used to calculate the total volume of ice within each 0.2 m ice thickness category, and the total salt flux to the ocean as a result of its formation and continued desalination. An initial ocean salinity (sea water from which the ice formed) of 33.0 psu is assumed [*Smith et al.*, 1984].

Figure 13 shows the cumulative salt flux over the entire sea ice growth and decay cycle, and the individual monthly values. The total salt input (from the undeformed component of the pack only) over the growth season (February–September) for the longitude band 60–150°E is 5.14 x 10^{13} kg, which represents a mean salt flux of 12.53 kg m^{-2} from a total ice volume of 2.06 x 10^{12} m^3. While the ice extent usually reaches a maximum in October, the total ice volume attains a maximum in September.

The salt flux estimate is low because it does not include the contribution of ridged ice. To correct the salt flux for the mass of ice in ridges we use the estimate of

---

Fig. 15. Sequential SAR images collected from the European Research Satellite ERS-1 on August 2, 1995, orbit 21174, © ESA, 1995. The broad white line indicates the approximate northern limit of floes greater than 100 m in diameter as determined from ship-based observations. The location of *RSV Aurora Australis* at the time of the overpass is indicated by a white square.

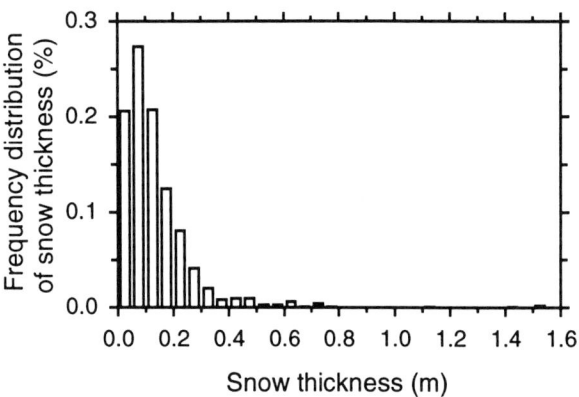

Fig. 16. The frequency distribution of all snow thicknesses measured in situ during 6 voyages to East Antarctica, October 1991–August 1995.

mean total thickness calculated in Section 6.6. Based on topography data from 8 voyages, the average extent of surface deformation is estimated to be 11%, and the effective thickness of the ice in this ridged area over the entire pack is 0.36 m (or 3.27 m over the 11% of the area that is ridged). This increases the total ice volume within the pack to 3.54×10^{12} m^3, and provides an additional salt flux to the ocean (assuming that all ridges are greater than 1 m thick and provide maximum salt flux to the ocean) of 4.27×10^{13} kg. The revised estimate of the total salt flux to the ocean as a result of sea ice formation is therefore 9.41×10^{13} kg over the entire growth season, for the region 60°–150°E.

8. FLOE SIZE AND SWELL PENETRATION

Floe size varies markedly within the Antarctic pack, from small pancakes to vast first-year floes several kilometers across. Floes continuously change in thickness and shape due to the thermal and mechanical forces acting upon them. One of the primary mechanisms affecting floe size is the penetration of swell into the pack ice. *Robin* [1963] showed in the Weddell Sea that the horizontal dimensions of ice floes are an important factor controlling wave penetration at shorter wavelengths; at longer wavelengths, the thickness of the floes is also important. *Squire et al.* [1986] confirmed that swell sufficient to cause ice breakup has been observed up to 600 km south of the ice edge in the Weddell Sea.

The open ocean to the north of the East Antarctic pack has, on average, the highest waves in the southern ocean [*Josberger and Mognard*, 1996], and these can penetrate hundreds of kilometers into the pack ice. Swell penetration tends to break large floes into smaller rectilinear floes as shown in Figure 14. Subsequent floe-floe collisions result in more rounded edges and a high percentage of brash ice between the floes. The breakup of the ice is significant insofar as it affects the ocean-atmosphere heat flux by creating open water within the pack. The addition of large quantities of saturated brash to the pack affects the microwave properties of the surface and *Massom* [unpublished data, 1995] suggests that this results in underestimates of the total ice concentration from passive microwave data. Within the swell-affected region, surface flooding may also be caused by water squirting up between floes as they are compressed in the troughs of swells. This may result in a significant area of the floes' surface becoming saturated, affecting snow-ice formation rates and ocean-atmosphere heat exchange. This effect is clearly observable as grey-white areas near the cracks in Figure 14.

A series of ERS-1 synthetic aperture radar (SAR) images collected along a single orbit near 140°E on August 2, 1995 is shown in Figure 15. The images extend from the fast ice zone in the south to the ice edge in the north, a total of about 450 km. In the southern portion of the image, large individual floes and leads can be identified, and in the northern portion of the image the floes are much smaller and can no longer be identified. Ship observations collected in the region at the same time confirm these features; in particular the northward limit of floes greater than 100 m in diameter (shown in the image).

A composite data set of floe sizes from 6 voyages between 1992 and 1995 shows no discernible change in mean floe size with distance from the ice edge. These data show that at any distance from the ice edge it is possible to find floes of any size, including vast (>2000 m) floes near the ice edge and much smaller floes in the central and southern pack. This indicates that swell is not the only mechanism for determining floe size, and that other forces such as shear or compres-

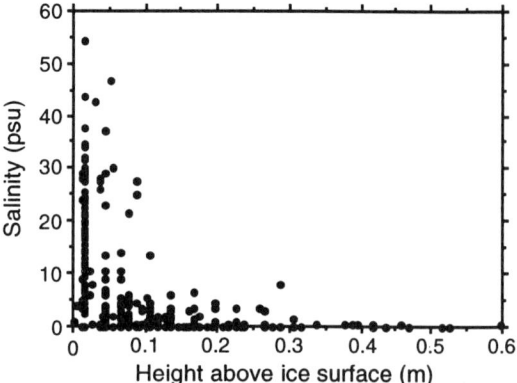

Fig. 17. A scatter diagram of snow salinity as a function of height in the snow column above the sea ice surface (n = 293), as measured during 3 cruises (in March–May 1993, September–October 1994 and August 1995).

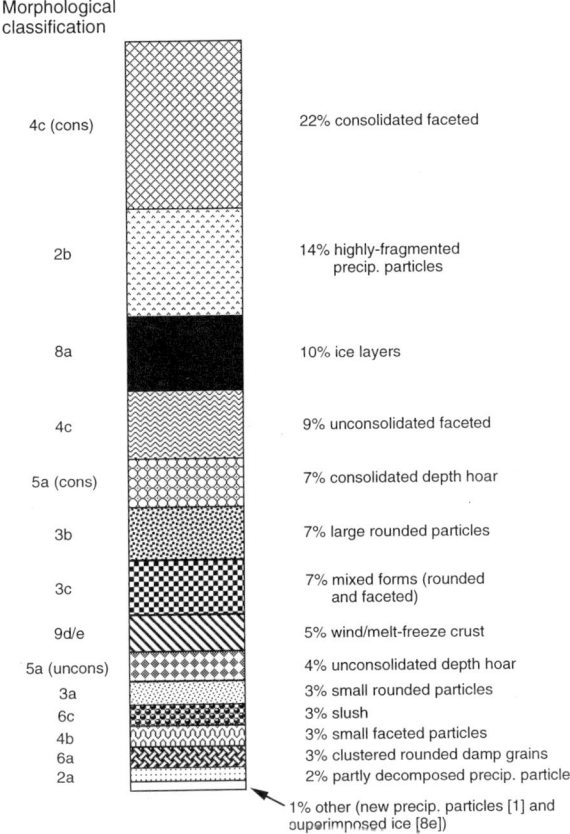

Fig. 18. A composite snow stratigraphy diagram combining all measurements from 3 cruises (in March–May 1993, September–October 1994 and August 1995). The morphological classification is based upon that of *Colbeck et al.* [1990].

sion may be equally important, possibly in different regions of the pack, or in different seasons.

9. SNOW PROPERTIES

Snow cover thickness and properties have been measured on 52 floes on 3 voyages between 1993 and 1995. The majority of snow pit data are from the same floes where ice cores were collected (see Section 6.1). The snow characterisation on each floe usually included vertical profiles of snow density, grain size and shape, salinity, $\delta^{18}O$ and temperature. Samples were collected using standard 3 cm-high cutters with a volume of 100 cm^3. Snow classifications have been made according to the morphological and process-oriented classification of *Colbeck et al.* [1990], modified to include superimposed ice (snow melt refrozen onto the sea ice surface). Grain size and shape characterisations were performed in the field or in a freezer laboratory soon after the sample was collected. Temperatures were measured by probes with thermistors in their tips, connected to a digital thermometer.

The snow cover on sea ice is important by virtue of its unique optical and insulative properties that affect both air-sea ice-ocean interaction and the microwave properties of the surface as measured by satellites [*Massom et al.*, 1997]. The rate of snow accumulation is determined not only by the precipitation rate but also by snow redistribution by wind. Meteorological data from the voyage in August 1995 shows that mean half-hourly wind speeds of >5 m s^{-1} were recorded for 83% of the time. This is the threshold proposed by *Ackley et al.* [1990] above which aeolian transport of unconsolidated snow occurs. These conditions result in an uneven distribution of snow, even across individual floes, depending on surface roughness. In the vicinity of ridges, measurements show an increase in snow thickness by a factor of 2–5 compared with adjacent regions. These results are consistent with the findings of *Eicken et al.* [1995] in the Weddell Sea. Over more level floes, the variation in snow thickness is 1.5–3, due to the presence of low dunes and scoured regions.

9.1 *Snow thickness*

A total of 1205 snow thickness measurements were made at snow pits, ice core sites and along drilled thickness transects on 6 voyages. The relative frequency distribution of these data is shown in Figure 16, and is consistent with that of the ship-based observations of snow thickness on individual floes (Figure 12b). The mean in situ thickness was 0.13±0.14 m, over the range of 0.0–2.01 m, with a mode of 0.05–0.10 m. It is not possible to accurately relate the age of floes to snow thickness due to a lack of data on the precipitation rate and aeolian redistribution. Snow-ice formation also effectively reduces the snow thickness on floes which further complicates this relationship.

9.2 *Snow density, salinity and grain size*

Snow density depends on a complex interaction of age, metamorphism and exposure to moisture, and varies with changing meteorological conditions. Snow density measurements were made on two voyages (May 1993 and August 1995) and the combined data set of 225 measurements from 2 voyages has a mean value of 360±110 kg m^{-3} over the range of 120–760 kg m^{-3}. The modal density is 300–350 kg m^{-3} with an approximately normal distribution. Only a very small percentage of the measured densities were greater than 600 kg m^{-3} and these are underestimated due to problems sampling an accurate volume of the denser snow. However, assuming an average density of 850 kg m^{-3} for these layers, which constitute 10% of the total observed

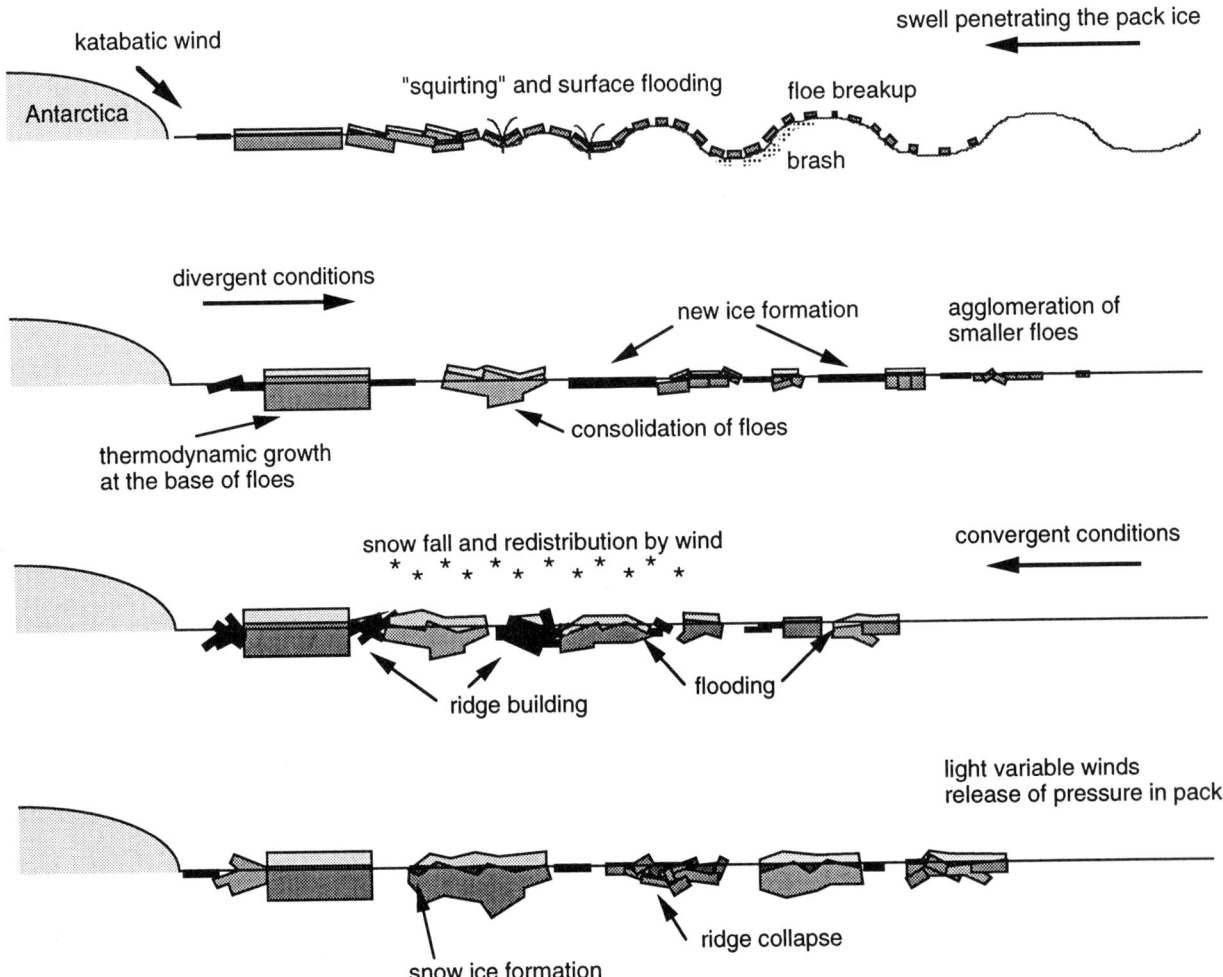

Fig. 19. A conceptual model of the thermodynamic and dynamic forcing parameters that determine the thickness distribution and physical characteristics of sea ice and snow cover as observed in the East Antarctic pack. See text for details.

snow mass, the actual mean snow density would approximate 400 kg m^{-3}.

The snow cover on Antarctic pack ice floes is quite saline, with the salt content affecting the surface melting point and the dielectric properties of the snow. The associated wetness due to sea water flooding affects the mass balance of the ice pack through snow-ice formation, biological production [*Ackley and Sullivan*, 1994] and the microwave signature of the surface [*Lohanick*, 1993; *Drinkwater*, 1995]. The snow cover may become saline by a number of processes including snow loading (see Section 6.1), deformation and wave action (see Section 8.0) or by the upward rejection of brine by newly-forming sea ice [*Perovich and Richter-Menge*, 1994]. The mean salinity for the 293 samples is 8.5± 11.5 psu, with a range between 0.05 and 66.4 psu.

Sixty percent of samples fall in the range 0.0–5.0 psu, although more than half of these have values less than 0.5 psu. Between 5 and 10% of samples fall in each of the categories between 5–10 psu and 25–30 psu. Only a small percentage of salinities are greater than 34 psu, and these are caused by the concentration of brine during the ice formation or the incorporation of frost flowers into the snow [*Drinkwater and Crocker*, 1988]. Snow salinity as a function of height of the snow cover above the ice surface is presented in Figure 17, and shows that salinities greater than 10 psu occur mostly in the 0.1 m thick layer closest to the ice surface. Similar findings have been reported by *Massom et al.* [1997] and *Sturm et al.* [this volume].

The mean grain size in the snow pack is 1.6± 1.2 mm with a range of 0.1 mm for fragmented spicules of new snow to 10.0 mm for well developed depth hoar cups in older snow. The mode is 0.20–0.25 mm.

9.3 Snow texture and stratigraphy

The snow cover on ice floes is typically composed of a complex and evolving assemblage of different snow types as a result of the variability in meteorological conditions during deposition and subsequent redistribution and metamorphism. Each of these snow types may have a significantly different thermal conductivity depending on its texture and density, and the bulk thermal conductivity of the snow pack depends upon the assemblage of snow types present [*Sturm et al.*, this volume].

A composite snow stratigraphy diagram, comprising all the snowpit measurements from 3 voyages, is presented in Figure 18. Windy conditions typically occur during and immediately following precipitation events, and this is reflected in the fragmented nature of newly-deposited snow (accounting for 14% of the total samples), consisting of spicules with a mean size of 0.2–0.3 mm. Small (0.2–0.5 mm) and large (0.5–1.0 mm) rounded grains, which together account for 10% of the snow volume, reflect equilibrium-type metamorphism [*Colbeck*, 1991], caused by relatively high air temperatures. Mixed rounded and faceted forms (7% of the snow volume) reflect intermediate stages of temperature-gradient metamorphism, and occur during the transition from warm to cold air temperatures as the snow temperature gradient increases. Faceted and depth hoar crystals (42% of the snow volume) are a feature of the cold snow cover. They form during prolonged periods of large temperature gradient (approximately 0.25°C cm^{-1} for low density snow [*Colbeck*, 1982]), when the snow surface temperature is significantly colder than the ice/snow interface temperature. Surface crusts only constitute a small proportion of the snow cover (5% of the snow volume), although they play a very significant role in reducing aeolian snow loss and redistribution.

Regular cyclical melt-refreeze activity combined with the burial of surface crusts results in a high proportion of icy layers within the snow volume which were common on all voyages to the East Antarctic pack. Buried icy layers, which account for 10% of the snow sampled, were observed up to 0.05 m thick but were typically less than 0.01 m thick. The complex stratigraphy of older, multi-year snow covers indicate that such cyclical changes occur many times during the lifetime of the snow cover. The snow cover on one 10 m thick multi-year floe (see Section 6.4) contained 10% icy layers in at least 10 distinct horizons.

10. CONCLUSIONS

The characteristics of the sea ice pack in East Antarctica are largely determined by the continual passage of synoptic weather systems through the region. *Worby et al.* [1996a] showed that ice growth in the Antarctic pack occurs sporadically in response to these synoptic systems and that alternating periods of divergence and convergence result in significant changes in the thickness distribution of the ice. In particular, new ice which forms in leads during divergent periods when the air temperature is cold, may be deformed into thicker ice during periods of convergence. *Andreas and Makshtas* [1985] show that a similar process is important in the Weddell Sea in spring, when southerly winds remove roughly 100 W m^{-2} more heat from the surface than northerly winds. This highlights the importance of changing wind direction, related to synoptic systems, on ice growth rate, snow metamorphism, precipitation and ocean-atmosphere heat exchange within the pack. The alternating warm and cold periods constantly modify the snow pack on the ice surface, resulting in a variation of grain size and densities within the snow.

The textural characteristics of the East Antarctic pack ice are similar to those found in the Weddell Sea [e.g., *Lange and Eicken*, 1991], suggesting that despite higher drift rates and a generally thinner ice cover in the East Antarctic pack, the formation mechanisms of the ice may be the same. At least 39% of the ice volume is columnar ice. This is a conservative estimate as, using our analysis it is not possible to distinguish columnar ice once its columnar structure has been destroyed by deformation and crushing. The remaining fraction of the ice volume is predominantly frazil ice formed in turbulent conditions (47%), snow-ice (13%), or other ice types (1%). A decrease in the average salinity with increasing ice thickness results in a steady salt flux to the ocean, which is estimated to be 9.41 x 10^{13} kg annually for the region 60°–150°E.

The drifting buoy data confirm the results of previous studies that the ice is highly mobile and constantly changing speed and direction. Although the ice drift is, on average, divergent, there are frequent periods of convergence when the ice is deformed, as indicated by the structural analysis of ice cores. Individual layer thicknesses in cores indicate that 40% of the layers are less than 0.1 m thick, and that the ice rarely grows thicker than about 0.5 m before being rafted or ridged. This is in good agreement with recent findings in the Bellingshausen and Amundsen Seas [*Worby et al.*, 1996; *Jeffries et al.*, 1997].

The thickness distribution of sea ice in the East Antarctic pack is determined by a complex interaction of thermodynamic and dynamic processes, which may be summarised by a conceptual model presented in Figure 19. Ocean swell may penetrate several hundred kilometers into the pack ice, and break floes into smaller segments. Depending on surface winds and currents these may have a net northward drift, resulting in a higher percentage of open water within the pack, or be compacted and subsequently pushed together by the passage of the swell. In either case, snow flooding is possible, caused by the surging or squirting of sea water as floes are knocked together by wave action, particularly in swell troughs. This may result in enhanced snow-ice formation or may completely wash away the snow cover on floes. Floes being pushed together may also result in brash ice being formed. These are processes that are important primarily near the ice edge.

At the southern limit of the pack, near the Antarctic continent, katabatic and synoptic winds maintain coastal polynyas and flaw leads which are areas of high ice production, ocean-atmosphere heat exchange, and salt input to the ocean. During periods of divergent conditions, the ice expands northwards, creating open water within the pack. At these times, the air mass over the pack is typically of continental origin and leads and areas of brash ice quickly freeze, increasing the mass of ice within the pack and consolidating small floes into larger ones. Thermodynamic growth at the base of floes is also important during these periods. As synoptic weather systems pass through the pack, the winds eventually shift, compacting the ice and pushing the ice edge south. During these periods, newly refrozen leads and older floes may raft or ridge to form thicker ice. The onshore winds are often of maritime origin, and the relatively warm, moist atmospheric conditions can result in widespread snowfall, and limit thermodynamic growth. On occasions, rain and surface melt may occur, even in the middle of winter. Wind redistributes snow across the surface of floes, creating drifts in the lee of ridge sails and blowing snow into leads. The increase in snow loading may result in flooding of the base of the snow on some floes, creating a slush layer at the ice/snow interface. Subsequent cold temperatures will transform this slush into snow-ice, and further modify the grain size and density of the remaining snow pack. When pressure within the pack is relieved by a change in wind speed or direction, newly formed pressure ridges may collapse creating a rubble of broken ice blocks and saturated snow. This may later reconsolidate with the return of colder conditions.

Any number of these processes may occur simultaneously (or sequentially), and the net effect of this continually changing growth and deformation environment is reflected in the thickness, topography, crystal structure, salinity, $\delta^{18}O$ values, and snow cover of a floe. Sea ice will invariably drift many hundreds of kilometers from the location where it first started to form and at every stage throughout its drift will be affected by the constantly changing ocean swell, air temperature, winds and precipitation. This high variability means that regional and long-term changes in many ice and snow properties are difficult to resolve. The ice and snow thickness distributions (which incorporate changes in the ice extent and concentration) are currently the only sea ice parameters in which we can confidently identify seasonal variations, albeit only in some months of the year. Numerical modelling efforts are making steady progress towards improving our understanding of many of the processes within the pack; in particular considerable progress has been made recently in modelling the net precipitation and accumulation using data from operational numerical analyses [*Budd et al.*, 1995; *Cullather et al.*, in press]. Additional field investigations, in conjunction with modelling efforts, will further improve this climatology of the East Antarctic sea ice zone, and the Antarctic pack as a whole.

Acknowledgments. The data presented in this paper have been collected over many field seasons, primarily aboard MV *Icebird* (1986–1990) and RSV *Aurora Australis* (1991–1995). The officers, crew, voyage management and support staff of both vessels are thanked for their support of this research. Thanks also to Ian Knott who has provided dedicated and thoroughly professional technical assistance; Anton Rada and Michael Wall for their contributions to data processing; and Vito Dirita for computer programming support. Other volunteer observers and field assistants are too numerous to name; their contribution has been invaluable and the authors are grateful to each one. Steve Pendlebury, Roger Lurz, Richard Jardine and Neil Adams from the Australian Bureau of Meteorology are thanked for their efforts in collecting AVHRR imagery at the Antarctic Meteorology Centre, Casey. Thanks are also extended to the National Snow and Ice Data Center, University of Colorado, for supplying the SSM/I data. The GMT public domain software [*Wessel and Smith*, 1995] was used to prepare Figures 5, 6 and 8.

REFERENCES

Ackley, S. F., Mass balance aspects of the Weddell Sea pack ice, *J. Glaciol., 24*(90), 391–406, 1979.

Ackley, S. F., and C. W. Sullivan, Physical controls on the development and characteristics of Antarctic sea ice biological communities – a review and synthesis, *Deep-Sea Res. I, 41*(10), 1583–1604, 1994.

Ackley, S. F., M. A. Lange, and P. Wadhams, Snow cover effects on Antarctic sea ice thickness, in *Sea Ice Properties and Processes, CRREL Monograph 90-1*, edited by S. F. Ackley and W. F. Weeks, pp. 16–21, 1990.

Adolphs, U. and G. Wendler, A pilot study on the interactions between katabatic winds and polynyas at the Adelie coast, eastern Antarctica, *Ant. Science, 7*(3), 307–314, 1995.

Allison, I., Pack-Ice Drift off East Antarctica and some implications, *Ann. Glaciol., 12*, 1–8, 1989a.

Allison, I., The East Antarctic sea ice zone: Ice characteristics and drift, *GeoJournal, 18*(1), 103–115, 1989b.

Allison, I., and A. P. Worby, Seasonal changes of sea-ice characteristics off East Antarctica, *Ann. Glaciol., 20*, 195–201, 1994.

Allison, I., R. E. Brandt, and S. G. Warren, East Antarctic sea ice: albedo, thickness distribution and snow cover, *J. Geophys. Res., 98*(C7), 12,417–12,429, 1993.

Andreas, E. L., and A. P. Makshtas, Energy exchange over Antarctic sea ice in the spring, *J. Geophys. Res.*, *90*, 7119–7212, 1985.

Andreas, E. L., M. A. Lange, S. F. Ackley, and P. Wadhams, Roughness of Weddell Sea ice and estimates of the air-ice drag coefficient, *J. Geophys. Res.*, *98*(C7), 12,439–12,452, 1993.

Budd, W. F., P. A. Reid, and L. J. Minty, Antarctic moisture flux and net accumulation from global analyses, *Ann. Glaciol.*, *21*, 149–157, 1995.

Casarini, M. P., and R. A. Massom, *Winter Weddell Sea Project; Sea Ice Observations, Leg 2: June–Sept. 1986*, Scott Polar Research Institute Special Report, Cambridge, 162 pp., 1987.

Cavalieri, D. J., and S. Martin, A passive microwave study of polynyas along the Antarctic Wilkes Land coast, in *Oceanology of the Antarctic Continental Shelf*, edited by S. S. Jacobs, *Antarctic Research Series 43*, pp. 227–252, American Geophysical Union, Washington, DC, 1985.

CLS/Argos, User manual, Toulouse, France, 242 pp., 1988.

Colbeck, S. C., An overview of seasonal snow metamorphism, *Rev. Geophys. Space Phys.*, *20*(1), 45–61, 1982.

Colbeck, S. C., The layered character of snow covers, *Rev. Geophys.*, *29*(1), 81–96, 1991.

Colbeck, S., E. Akitaya, R. Armstrong, H. Gubler, J. Lafeuille, K. Lied, D. McClung and E. Morris, *The International Classification for Seasonal Snow on the Ground*, The International Commission on Snow and Ice of the International Association of Scientific Hydrology, and the International Glaciological Society, 23 pp., 1990.

Comiso, J. C., and A. L. Gordon, Recurring polynyas over the Cosmonaut Sea and the Maud Rise, *J. Geophys. Res.*, *92*(C3), 2819–2834, 1987.

Comiso, J. C., and A. L. Gordon, The Cosmonaut Polynya in the Southern Ocean: structure and variability, *J. Geophys. Res.*, *101*(C8), 18,297–18,313, 1996.

Cullather, R. I., D. H. Bromwich, and M. L. Van Woert, Interannual variations in Antarctic precipitation related to El Nino-Southern Oscillation, *J. Geophys. Res.*, in press.

Deacon, G. E., The Weddell Gyre, *Deep-Sea Res. A*, *26*(9), 981–995, 1979.

Dierking, W., Laser profiling of the ice surface topography during the Winter Weddell Gyre Study 1992, *J. Geophys. Res.*, *100*(C3), 4807–4820, 1995.

Drinkwater, M. R., Applications of SAR measurements in ocean-ice-atmosphere interaction studies, in *Oceanographic Applications of Remote Sensing*, edited by M. Ikeda and F. W. Dobson, pp. 381–396, CRC Press, Boca Raton, Florida, 1995.

Drinkwater, M. R., and G. B. Crocker, Modelling changes in the dielectric and scattering properties of young snow-covered ice at GHz frequencies, *J. Glaciol.*, *34*, 274–282, 1988.

Drinkwater, M. R., and C. Haas, Snow, sea-ice, and radar observations during ANT X/4: summary data report, *Berichte aus dem Fachbereich Physik*, *53*, Alfred-Wegener-Institut für Polar- und Meeresforschung, Bremerhaven, Germany, 51 pp., 1994.

Eicken, H., H. Fischer, and P. Lemke, Effects of the snow cover on Antarctic sea ice and potential modulation of its response to climate change, *Ann. Glaciol.*, *21*, 369–376, 1995.

Eicken, H., M. A. Lange, H.-W. Hubberton, and P. Wadhams, Characteristics and distribution patterns of snow and meteoric ice in the Weddell Sea and their contribution to the mass balance of sea ice, *Ann. Geophysicae*, *12*, 80–93, 1994.

Fahrbach, E., E. Augstein, and D. Olbers, Impact of shelf and sea ice on water mass modifications and large-scale oceanic circulation in the Weddell Sea, *Antarctic Science: Global Concerns*, edited by G. Hempel, pp. 167–187, Springer Verlag, Berlin, 1994.

Foster, T. D., Abyssal water mass formation off the eastern Wilkes Land coast of Antarctica, *Deep-Sea Res. I*, *42*(4), 501–522, 1995.

Gloersen, P., W. J. Campbell, D. J. Cavalieri, J. C. Comiso, C. L. Parkinson, and H. J. Zwally, *Arctic and Antarctic Sea Ice, 1978–1987: Satellite Passive-Microwave Observations and Analysis*, NASA SP-511, 290 pp., NASA, Washington, D.C., 1992.

Gordon, A. L., and the Ice Station Weddell Group of Principal Investigators and Chief Scientists, Weddell Sea exploration from ice station, *EOS Trans. AGU*, *74*(11), 121, 124–6, 1993.

Gordon, A. L., and P. Tchernia, Waters of the continental margin off Adélie Coast, Antarctica, in *Antarctic Oceanography II: The Australian-New Zealand sector*, edited by D. E. Hayes, Washington, D.C., American Geophysical Union, pp. 59–69, 1972.

Gow, A. J., S. F. Ackley, K. R. Buck, and K. M Golden, Physical and structural characteristics of Weddell Sea pack ice, *CRREL Report*, *87-14*, 75 pp., 1987.

Gow, A. J., S. F. Ackley, V. I. Lytle, and D. Bell, Ice-core studies in the western Weddell Sea (Nathaniel B. Palmer 92-2), *Ant. J. U.S.*, *XXVII*(5), 89–90, 1992.

Haas, C., and T. Viehoff, Sea ice conditions in the Bellingshausen/Amundsen Sea: shipboard observations and satellite imagery during ANT XI/3, *Berichte aus dem Fachbereich Physik*, *51*, Alfred-Wegener-Institut für Polar-und Meeresforschung, Bremerhaven, Germany, 1994.

Heil, P., I. Allison, and V. I. Lytle, Seasonal and interannual variations of the oceanic heat flux under a landfast Antarctic sea ice cover, *J. Geophys. Res.*, *101*(C11), 25,741–25,752, 1996.

Jacka, T. H., I. Allison, R. Thwaites, and J. C Wilson, Characteristics of the seasonal sea ice off East Antarctica and comparison with satellite observations, *Ann. Glaciol.*, *9*, 85–91, 1987.

Jeffries, M. O., and U. Adolphs, Early winter snow and ice thickness distribution, ice structure and development of western Ross Sea pack ice between the ice edge and the Ross ice shelf, *Ant. Science*, 1997, in press.

Jeffries, M. O., and W. F. Weeks, Structural characteristics and development of sea ice in the western Ross Sea, *Ant. Science*, *5*(1), 63–75, 1992.

Jeffries, M. O., R. A. Shaw, K. Morris, A. L. Veazy, and H. R. Krouse, Crystal structure, stable isotopes ($\delta^{18}O$), and development of sea ice in the Ross, Amundsen, and Bellingshausen Seas, Antarctica, *J. Geophys. Res.*, *99*(C1), 985–995, 1994.

Jeffries, M. O., W. F. Weeks, R. Shaw, and K. Morris, Structural characteristics of congelation and platelet ice and their role in the development of Antarctic landfast sea ice, *J. Glaciol.*, *39*(132), 223–238, 1993.

Jeffries, M. O., A. P. Worby, K. Morris, and W. F. Weeks, Seasonal variations in the properties and structural composition of sea ice and snow cover in the Bellingshausen and Amundsen Seas, Antarctica, *J. Glaciol.*, 1997, in press.

Josberger, E. G., and N. M. Mognard, Southern Ocean monthly wave fields for austral winters 1985-1988 by Geosat radar altimeter, *J. Geophys. Res.*, *101*(C3), 6689–6696, 1996.

Kawamura, T., and K. I. Ohshima, Sea ice growth in Ongul Strait, Antarctica, *Ann. Glaciol.*, *18*, 97–101, 1993.

Kovacs, A., Sea Ice, Part 1. Bulk salinity versus ice floe thickness, *CRREL Report 96-7*, 16 pp., 1996.

Lange, M. A., Basic properties of Antarctic sea ice as revealed by textural analysis of ice cores, *Ann. Glaciol.*, *10*, 95–101, 1988.

Lange, M. A., and H. Eicken, Textural characteristics of sea ice and the major mechanisms of ice growth in the Weddell Sea, *Ann. Glaciol.*, *15*, 210–215, 1991.

Lange, M. A., P. Schlosser, S. F. Ackley, P. Wadhams, and G. S. Dieckman, ^{18}O concentrations in sea ice in the Weddell Sea, *J. Glaciol.*, *12*, 92–96, 1990.

Lohanick, A.W., Microwave brightness temperatures of laboratory-grown undeformed first-year ice with an evolving snow cover, *J. Geophys. Res.*, *98*(C3), 4667–4674, 1993.

Lytle, V. I., and S. F. Ackley, Snow properties and surface elevation profiles in the Western Weddell Sea, (NBP 92-2), *Ant. J. U.S.*, *XXVII*(5), 93–94, 1992.

Lytle, V. I., and S. F. Ackley, Heat flux through sea ice in the western Weddell Sea: convective and conductive transfer processes, *J. Geophys. Res.*, *101*(C4), 8853–8868, 1996.

McPhee, M., S. F. Ackley, P. Guest, J. Morrison, D. Martinson, R. Meunch, L. Padman, and T. Stanton, The Antarctic Zone Flux Experiment, *Bulletin of the American Meteorological Society*, *77*(6), 1221–1232, 1996.

Massom, R. A., Observing the advection of sea ice in the Weddell Sea using buoy and satellite passive microwave data, *J. Geophys. Res.*, *97*(C10), 15,559–15,572, 1992.

Massom, R. A., M. R. Drinkwater, and C. Haas, Winter snowcover on sea ice in the Weddell Sea, *J. Geophys. Res.*, *102*(C1), 1107–1117, 1997.

Meese, D. A, J. W. Govoni, V. Churun, B. Ivanov, V. Komarovskii, V. Shilnikov, and A. Zachek, Sea ice observations from the Winter Weddell Gyre Study '89, *CRREL Report 91-2*, 161 pp., 1991.

Parish, T. R., and D. H. Bromwich, The surface wind field over the Antarctic Ice Sheet, *Nature*, *327*(6125), 51–54, 1987.

Parkinson, C. L., and P. Gloersen, Global sea ice coverage, in *Atlas of Satellite Observations Related to Global Change*, edited by R. J. Gurney, J. L. Foster, and C. L. Parkinson, pp. 371–383, Cambridge University Press, Cambridge, England, 1993.

Perovich, D. K., and J. A. Richter-Menge, Surface characteristics of sea ice, *J. Geophys. Res.*, *99*(C8), 16,341–16,350, 1994.

Potter, M. J., *An Evaluation of Polynyas in East Antarctica*, Unpubl. Honours thesis, 168 pp., Inst. Ant. and Southern Ocean Studies, University of Tasmania, Australia, 1995.

Robin, G. de Q., Wave propagation through fields of pack ice, *Philosophical Transactions of the Royal Society of London A*, 225, 313–339, 1963.

Schnack-Schiel, S., The winter expedition of RV Polarstern to the Antarctic (ANT V/1-3), *Berichte zur Polarforschung*, *39*, Alfred-Wegener-Institut für Polar- und Meeresforschung, Bremerhaven, Germany, 1987.

Smith, N.R., D. Zhaoqian, K.R. Kerry, and S. Wright, Water masses and circulation in the region of Prydz Bay, Antarctica, *Deep-Sea Res. I*, *31*(9), 1121–1147, 1984.

Soviet Antarctic Expedition, *Atlas of Antarctica, 1*, Main Administration of Geodesy and Cartography of the Ministry of Geology USSR, Moscow, 1966.

Squire, V. A., P. Wadhams, and S. C. Moore, Surface gravity wave processes in the winter Weddell Sea, *EOS Trans. AGU*, *67*(44), 1005, 1986.

Strass, V. H., Sea ice draft and coverage in the Weddell Sea recorded with moored upward looking sonar, in *Report of the Third Session of the WCRP ACSYS Scientific Steering Group*, Göteborg, Sweden, 11-14 November, 1994, Appendix J, 1995.

Sturm, M., K. Morris, and R. Massom, A description of the snow cover on the winter sea ice of the Amundsen and Ross Seas, *Ant. J. U.S*, *XXX*(1–4), 21–24, 1996.

Sturm, M., K. Morris, and R. Massom, Diagenesis of the winter snow cover of the West Antarctic pack ice, 1994-1995, this volume.

Tchernia, P., and P. F. Jeannin, Circulation in Antarctic waters as revealed by iceberg tracks 1972-1983, *Polar Record*, *22*(138), 263–269, 1984.

Wadhams, P., M. A. Lange, and S. F. Ackley, The ice thickness distribution across the Atlantic sector of the Antarctic Ocean in mid-winter, *J. Geophys. Res.*, *92*(C13), 14,535–14,552, 1987.

Wakatsuchi, M., K. I. Ohshima, M. Hishida, and M. Naganobu, Observations of a street of cyclonic eddies in the Indian Ocean sector of the Antarctic Divergence, *J. Geophys. Res.*, *99*(C10), 20,417–20,426, 1994.

Weeks, W. F., and S. F Ackley, The growth, structure and properties of sea ice, *CRREL Monograph*, *82-1*, 130 pp., 1982.

Weeks, W. F., S. F. Ackley, and J. Govoni, Sea ice ridging in the Ross Sea, Antarctica, as compared with sites in the Arctic, *J. Geophys. Res.*, *94*(C4), 4984–4988, 1989.

Wessel, P., and H. F. Smith, New version of the generic mapping tool released, *Eos Trans. AGU*, *76*, 379, 1995.

Worby, A. P., and R. A. Massom, The structure and properties of sea ice and snow cover in East Antarctic pack ice, *Antarctic CRC Research Report 7*, 191 pp., Hobart, Tasmania, 1995.

Worby, A. P., N. L. Bindoff, V. I. Lytle, I. Allison, and R. A. Massom, Winter ocean/sea ice interactions in the East Antarctic pack ice, *EOS Trans. AGU*, *77*(43), 453, 456–7, 1996a.

Worby, A. P., M. O. Jeffries, W. F. Weeks, K. Morris, and R. Jaña, The thickness distribution of sea ice and snow cover during late winter in the Bellingshausen and

Amundsen Seas, Antarctica, *J. Geophys. Res.*, *101* (C12), 28,441–28,455, 1996b.

Zwally, H. J., J. C. Comiso, C. L. Parkinson, W. J. Campbell, F. D. Carsey, and P. Gloersen, Antarctic Sea Ice, 1973–1976: Satellite Passive-Microwave Observations, *NASA SP-459*, 206 pp., NASA, Washington, D.C., 1983.

___

I. Allison, P. Heil, V. I. Lytle, R. A. Massom, and A. P. Worby, Antarctic Cooperative Research Center, University of Tasmania, P. O. Box 252-80, Hobart, Tasmania, 7001, Australia.

(Received August 30, 1996; accepted February 11, 1997)

LATE WINTER FIRST–YEAR ICE FLOE THICKNESS VARIABILITY, SEAWATER FLOODING AND SNOW ICE FORMATION IN THE AMUNDSEN AND ROSS SEAS

M. O. Jeffries[1], S. Li[1], R. A. Jaña[2], H. R. Krouse[3], and B. Hurst-Cushing[1]

In September and October 1994 in the western Amundsen and eastern Ross seas, snow depth, ice thickness, draft and freeboard data were collected by drilling along transects on ice floes, and ice core samples were obtained for analysis of ice structure/stratigraphy, stable isotopic composition and brine volume. Dynamic thickening by deformation had affected most of the ice cover. Three types of floe, X, Y and Z, were identified according to their coefficient of ice thickness variation and described as rafted, moderately ridged and strongly ridged, respectively. Types X, Y and Z had distinctive ice thickness probability density functions, and mean values of 0.68 m, 0.82 m and 1.17 m, respectively. Flooding was widespread throughout the study area, but the areal extent ranged from 66% on the Type X floes to 38% on the Type Z floes. Flooding occurred primarily at sites where the snow depth was 30% of the ice thickness, and this snow load effect was probably enhanced by the fact that much of the ice cover had temperatures > 5‰ throughout the entire ice thickness, thereby promoting brine exchange between the base and the snow/ice interface. The total amount of snow ice, and the thickness of individual snow ice layers showed that, by late winter, snow ice formation had made a greater contribution (32-39% of the total mass of each floe type) than either congelation ice or frazil ice formation to the thermodynamic thickening of all the floes.

INTRODUCTION

A knowledge of the spatial and temporal variability of the thickness distribution of the Antarctic sea ice cover is essential for understanding the role of the pack ice in modifying atmosphere–ocean interactions and exchanges of heat, mass and momentum, their influence on ocean and climate variability from the local to global scale, and the biological productivity of the ice and ocean. Fundamental to understanding the ice thickness distribution is a knowledge of the dynamic (ridging and rafting) and thermodynamic (freezing and melting) processes that contribute to the development of the sea ice cover under the influence of atmospheric and oceanic forcing. Observations and measurements of the ice thickness distribution and formation processes are necessary for deriving forcing and validation fields for numerical models of air–ice–ocean interactions and climate, to determine the factors controlling the biology and ecology of sea ice biota, and for understanding remote sensing signatures of the ice cover.

The first comprehensive Antarctic sea ice thickness data set, obtained by drilling along transects on floes in the eastern Weddell Sea, showed that unridged first–year ice had a preferred thickness of 0.5–0.6 m [*Wadhams et al.*, 1987]. Using the same technique in the western Weddell Sea, even ridged first–year ice was found to be thinner than similar Arctic ice [*Lange and Eicken*, 1991a]. Although caution must be exercised due to differences in

[1]Geophysical Institute, University of Alaska Fairbanks, Fairbanks, Alaska
[2]Instituto Antártico Chileno, Santiago, Chile
[3]Department of Physics and Astronomy, University of Calgary, Calgary, Alberta, Canada

sampling strategy, i.e, estimates of ice thickness made while the ship was underway rather than direct measurements by drilling on floes, the first–year ice thickness characteristics off East Antarctica are similar to those of the eastern Weddell Sea [*Allison et al.*, 1993; *Allison and Worby*, 1994; *Worby et al.*, this volume].

Drilling investigations of the ice thickness distribution in the Pacific sector of the Antarctic pack ice indicate that the late winter first–year ice in the Bellingshausen and eastern Amundsen seas is thicker than that of the eastern Weddell Sea and the East Antarctic pack ice [*Worby et al.*, 1994, 1996]. A large area of the western Ross Sea first–year ice pack attains a greater thickness by mid–winter than the late winter first–year ice cover elsewhere [*Jeffries and Adolphs*, 1997].

The snow depth distribution on sea ice is also of some importance because, for example, of its role in the surface energy balance and ice thermal regime [*Maykut*, 1986] and in microwave emissivity and backscatter variability and the retrieval of geophysical parameters from satellite data [see *Carsey*, 1992, for numerous papers on the subject]. The snow cover also plays an important role in the mass balance of the sea ice, and it is now apparent that one of the key roles of the snow cover on Antarctic sea ice, and one which sets it apart from Arctic snow/sea ice interactions, is to contribute to seawater flooding at the snow/ice interface and the subsequent thermodynamic thickening of floes by snow ice formation.

The widespread occurrence of seawater flooding at the snow/ice interface was first documented in the Weddell Sea, where subsequent snow ice formation made a moderate, but nevertheless important, contribution to the sea ice mass balance in the absence of basal congelation ice growth [*Ackley et al.*, 1990; *Lange et al.*, 1990; *Eicken et al.*, 1994, 1995a]. A large amount of snow ice occurs also in the western and eastern Pacific sectors of the Antarctic sea ice zone, and it makes a greater contribution to the thermodynamic thickening of the ice cover than either frazil ice or congelation ice [*Jeffries and Adolphs*, 1997; *Jeffries et al.*, 1997].

The studies described above have greatly increased the knowledge of the thickness distribution and formation of sea ice in the Weddell Sea, the East Antarctic pack ice, and in the western and eastern Pacific sectors of the Antarctic sea ice zone. However, a gap remains in the knowledge of these sea ice characteristics in the central Pacific sector, i.e., the western Amundsen Sea and the eastern Ross Sea. In September and October 1994, the R.V. *Nathaniel B. Palmer* operated in the central Pacific sector of the Antarctic sea ice zone (Figure 1) in support of oceanographic and sea ice research. The sea ice program included observations and measurements of snow depth, ice thickness, draft, and freeboard, and ice core analysis of ice structure/stratigraphy, stable isotopes, and brine volume. The objective of the investigation was to document the snow and ice thickness variability, the extent of seawater flooding at the snow/ice interface and amount of snow ice, and to identify the relative importance of dynamic and thermodynamic processes in the formation of ice floes.

This paper presents the results of the late winter 1994 investigation of first–year ice floe thickness and formation in the western Amundsen and eastern Ross seas. After the methods are described, the paper is organized as follows: (1) a general description of the study area and sea ice conditions is given; (2) the general characteristics of the entire snow depth, ice thickness, freeboard and draft data sets and some inter–relationships are presented; (3) three types of ice floe are classified according to their ice thickness variability, and their snow depth, ice thickness, draft and freeboard characteristics are summarized; and (4) the oxygen isotopic composition, structure and stratigraphy, and brine volume variability of the three floe types are described. The paper concludes with a discussion of the results and the implications with regard to the dynamic and thermodynamic processes that contribute to ice floe formation, seawater flooding and snow ice formation in the central Pacific sector of the Antarctic sea ice zone.

METHODS

During the period 18 September to 13 October 1994, 23 ice floe stations (Figure 1) were occupied for snow and ice thickness measurements and retrieval of ice cores. On each floe, snow depth (z_s), ice thickness (z_i), freeboard (z_f) and draft (z_d) were measured at mechanically drilled holes (5–cm diameter) spaced at 2–m intervals along 100 m long transects. On four floes, measurements were made along a single transect. On the other floes, measurements were made along two 100–m transects oriented perpendicular to each other, primarily in L–shapes but also as a + or a T (Table 1). Every effort was made to sample a floe that was representative of the sea ice cover on a particular day and, where appropriate, the sampling transects on floes were deliberately placed across ridges. Thin ice (< 0.3 m thick) is poorly represented, due to the hazards of working on such ice, which occurred primarily in leads, and open water is not represented at all. Measurements were made at a total of 2227 holes. These data, Set A, are the focus of this study.

On each floe, 3–5 pairs of 0.1–m diameter ice cores were obtained at different sites along the transects, prima-

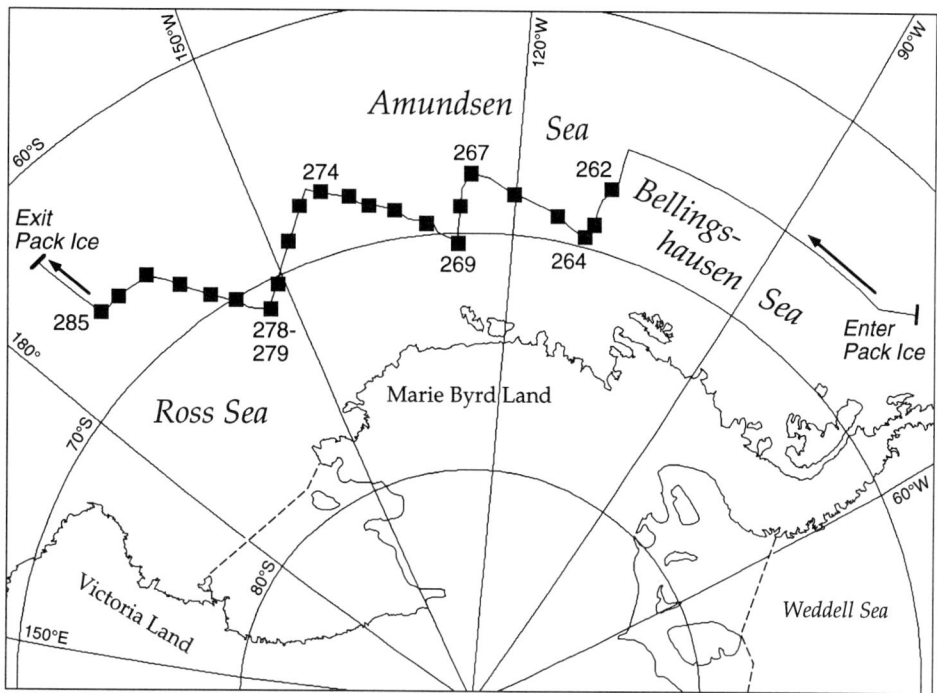

Fig. 1. Map of the study area showing the track of the R.V. *Nathaniel B. Palmer* in the pack ice of the Pacific sector of the Southern Ocean. The solid squares indicate the locations of ice floes that were investigated for this study along the saw-tooth track beginning in the central Amundsen Sea and ending in the north-eastern Ross Sea. The three figure numbers are Julian Days (e.g., 262:19 September; 285:12 October 1994). They are used as identification numbers for the floes that were investigated.

rily at their ends and the intersection when two transects were investigated. The first core of each pair was returned to the ship and kept frozen until its structure was determined later in the day in the science freezer at a temperature of −15°C. After splitting the cores longitudinally using a band saw, vertical thick (1–2 mm) sections were cut and then illuminated between crossed–polarizing filters to reveal the crystal texture. The core stratigraphy and textural variability were recorded and each core was then sampled for later oxygen isotope ($^{18}O/^{16}O$ ratio) analysis. Depending on core length and stratigraphic and textural variability, between 4 and 26 contiguous samples were taken per core for isotope analysis. The oxygen isotopic composition of a total of 735 samples was determined using standard procedures on a mass spectrometer and the values are expressed as $\delta^{18}O$ in parts per thousand (‰), as described in *Jeffries et al.* [1994]. Seventy–three ice cores, a total length of 61.02 m, were analyzed for textural, stratigraphic and isotopic variability.

The second of each core pair was obtained within 0.3 m of the first core and was used for temperature and salinity measurements, and brine volume calculations. Temperatures were measured immediately to a precision of 0.1°C using a digital thermometer and needle probe inserted into holes drilled in the core at 0.1–m intervals. The cores were then cut into 0.1 m long sections, which were sealed in plastic buckets and returned to the ship to melt. The salinity of the melted samples was measured using a Beckmann salinity/conductivity meter (Model RB5-349A Solubridge) calibrated with Standard Seawater. Seventy–three ice cores, total length 59.47 m, were analyzed. Assuming an ice density of 920 kg m^{-3}, the brine volume of each sample was calculated from the salinity and temperature data using the method of *Cox and Weeks* [1983].

A general description of the sea ice conditions was compiled from two data sets, B and C, that were acquired while the ship was underway. The first of these data sets, Set B, was obtained by making 25 snow and ice thickness estimates per hour of individual floes that were tipped on their side by the ship. Set B does not adequately represent ridges, because most are broken up by the ship, but its

TABLE 1. Summary of mean (± 1 s.d.) values of snow depth (z_s), ice thickness (z_i), freeboard (z_f) and draft (z_d) and other characteristics for ice floes investigated for this study.

Station No.[a]	Transect Config.[b]	n[c]	Mean z_s m	Mean z_i m	Mean z_d m	Mean z_f m	z_i Coeff. Variation[d]	Floe Type[e]
262	I	51	0.10±0.07	0.49±0.18	0.48±0.18	0.01±0.02	0.36	Y
263	L	109	0.23±0.09	0.91±0.47	0.88±0.44	0.03±0.07	0.51	Z
264	I	55	0.56±0.19	1.76±0.72	1.76±0.73	0.00±0.00	0.41	Z
265	L	104	0.24±0.10	0.88±0.25	0.86±0.25	0.02±0.02	0.28	Y
266	T	107	0.15±0.05	0.53±0.04	0.52±0.04	0.01±0.00	0.07	X
267	L	104	0.21±0.07	0.76±0.49	0.74±0.48	0.01±0.03	0.64	Z
268	+	103	0.32±0.08	0.67±0.13	0.73±0.12	-0.06±0.06	0.19	X
269	L	112	0.27±0.06	0.78±0.20	0.78±0.19	-0.01±0.03	0.25	Y
270	L	102	0.39±0.08	1.09±0.22	1.10±0.21	-0.01±0.06	0.20	Y
271	I	53	0.29±0.04	0.72±0.10	0.72±0.10	0.00±0.01	0.13	X
272	L	101	0.29±0.07	0.96±0.15	0.95±0.15	0.00±0.02	0.15	X
273	L	114	0.30±0.15	1.41±0.81	1.37±0.77	0.03±0.01	0.57	Z
274	L	107	0.32±0.09	1.34±0.44	1.33±0.40	0.01±0.05	0.32	Y
275	I	52	0.18±0.03	0.78±0.08	0.78±0.08	0.00±0.01	0.10	X
276	T	102	0.24±0.03	0.76±0.08	0.79±0.05	-0.03±0.01	0.10	X
277	T	107	0.17±0.05	0.61±0.24	0.59±0.25	0.02±0.06	0.39	Y
278	+	126	0.35±0.08	0.80±0.15	0.83±0.15	-0.03±0.04	0.18	X
280	T	104	0.37±0.11	0.81±0.21	0.84±0.17	-0.03±0.07	0.25	Y
281	T	102	0.63±0.23	1.28±0.58	1.34±0.56	-0.06±0.07	0.45	Z
282	L	102	0.13±0.05	0.41±0.06	0.44±0.05	-0.02±0.02	0.14	X
283	L	102	0.13±0.05	0.43±0.08	0.44±0.07	-0.02±0.03	0.18	X
284	L	102	0.34±0.10	0.87±0.10	0.92±0.11	-0.05±0.04	0.11	X
285	L	106	0.11±0.04	0.45±0.13	0.45±0.12	-0.01±0.04	0.28	Y

[a]The Station No. is also the day of year, e.g. 262 is 19 September and 285 is 12 October.
[b]Transect Config. is the configuration of the ice thickness transects. The symbol I denotes a single transect and the other symbols are self explanatory.
[c]Number of measurements of each variable at each floe.
[d]z_i Coeff. Variation is the coefficient of ice thickness variation, i.e., the standard deviation of the mean of z_i divided by the mean of z_i (σ_i/z_i).
[e]Floe Type is based on the coefficient of ice thickness variation.

advantage over Set A is that it gives more continuous geographic coverage of the thickness of unridged ice and of lead ice < 0.3 m thick. The second of the underway data sets, Set C, includes estimates made each hour of the thickness, concentration, snow cover and topographic variability of the three thickest ice categories in the ship's vicinity [*Allison et al.*, 1993; *Allison and Worby*, 1994]. We report on the Set C ice concentration and topography data.

STUDY AREA AND GENERAL DESCRIPTION OF SEA ICE CONDITIONS

The R. V. *Nathaniel B. Palmer* crossed the ice edge in the north–eastern Bellingshausen Sea at 64.5°S, 74°W on 14 September and, after a transit through the marginal ice zone along latitude 65.5°S, longitude 109°W in the central Amundsen Sea was reached on 18 September (Figure 1). Set B and Set C data were obtained during this transit, but are not reported here. Between 18 September and 12 October the ship operated entirely in first–year ice between 109°W and 171°W, following a saw–tooth track that went as far south as 71.2°S in the eastern Ross Sea (Figure 1). Set A, Set B and Set C data were collected during this period. Southward progress was limited by time and fuel constraints, and the occurrence of thicker ice and deeper snow cover at the northern margin of the perennial pack ice. The ice edge at 63.5°S, 172.5°W in the northern Ross Sea was crossed on 13 October. A general description follows of the ice conditions in the main body of the first–year pack ice where floes were sampled, i.e., the region

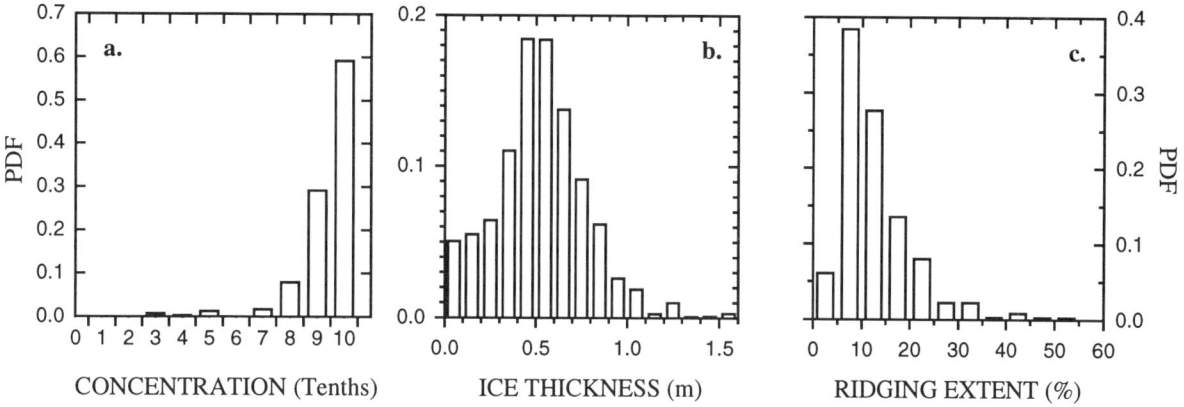

Fig. 2. Probability density functions of pack ice characteristics compiled from hourly observations made while the ship was underway in the region south of 65.5°S and between longitudes 109°W and 171°W: (a) ice concentration from Set C, (b) ice thickness from Set B, and (c) extent of ridging from Set C.

south of 65.5°S, and between longitudes 109°W and 171°W.

The ice concentration was high throughout the study area; 88% of the hourly Set C ice observations reported an ice concentration ≥ 9/10 (Figure 2a). Most of this high concentration of ice was quite thick; 71% of the Set B estimates were in the range 0.3–0.8 m (Figure 2b). The Set C data revealed that 74% of the pack ice showed evidence of ridging activity. The ridges were typically the randomly oriented, point type roughness elements described by *Lytle and Ackley* [1991: Figure 3]. Five per cent of the Set C observations reported phenomena such as finger rafting, pancake rafting and brash; these too are the result of deformation. Only 21% of the ice cover was described as level, undeformed ice, which occurred primarily as nilas, grey ice and grey–white ice in leads. The majority of floe sizes reported in Set C were in the medium (100–500 m) to vast (> 2 km) range.

Within the 74% of the pack ice that showed evidence of ridging, the areal extent of ridges was mainly in the range 5–15% (Figure 2c). Weighted according to the concentration of each ice thickness category in which they occurred, ridges accounted for an average of 15% of the area of ice that had been subject to ridging, and 12% of the total area of ice in the study area. The ridges that were sampled for Set A made up 16% of the total number of holes drilled. In view of the similarity between the Set A and Set C ridging statistics, and the fact that open water and thin ice < 0.3 m thick were in the minority, we believe that Set A provides an adequate representation of the thickness distribution of most of the ice cover in the study area.

The absence of an open water fraction in Set A is of no consequence with respect to the use of the z_s, z_i, z_f and z_d data and their relationship to seawater flooding and snow ice formation. Also, where appropriate, we can compare Set A with similar data sets obtained elsewhere by drilling, since they too lack the thin ice and open water fractions. We recognize the difficulties in making comparisons with data sets from elsewhere in the Antarctic sea ice zone, but it remains important to place the observations and measurements in the larger context in order to compile climatologies of sea ice thickness variability. We do not know if 1994 was a normal or an anomalous year, but, since we are presenting the first data set of its kind for the eastern Amunden and western Ross seas region, we can establish a benchmark for future reference.

RESULTS OF ICE THICKNESS MEASUREMENTS AND ICE CORE ANALYSIS

General Characteristics of the Snow Depth, Ice Thickness, Draft and Freeboard Data

Probability density functions. The z_s, z_i, z_f and z_d data for the individual floes are summarized in Table 1. Probability density functions (PDFs) for the entire z_s, z_i, z_f and z_d data sets are illustrated in Figure 3. The majority (68%) of the snow depth values were in the range 0.1–0.35m and there was a tail of higher values. The ice thickness distribution was bi-modal with peaks in the 0.4–0.5m and 0.7–0.9 m categories, and it too had a tail of higher values. Both the snow depth and ice thickness had

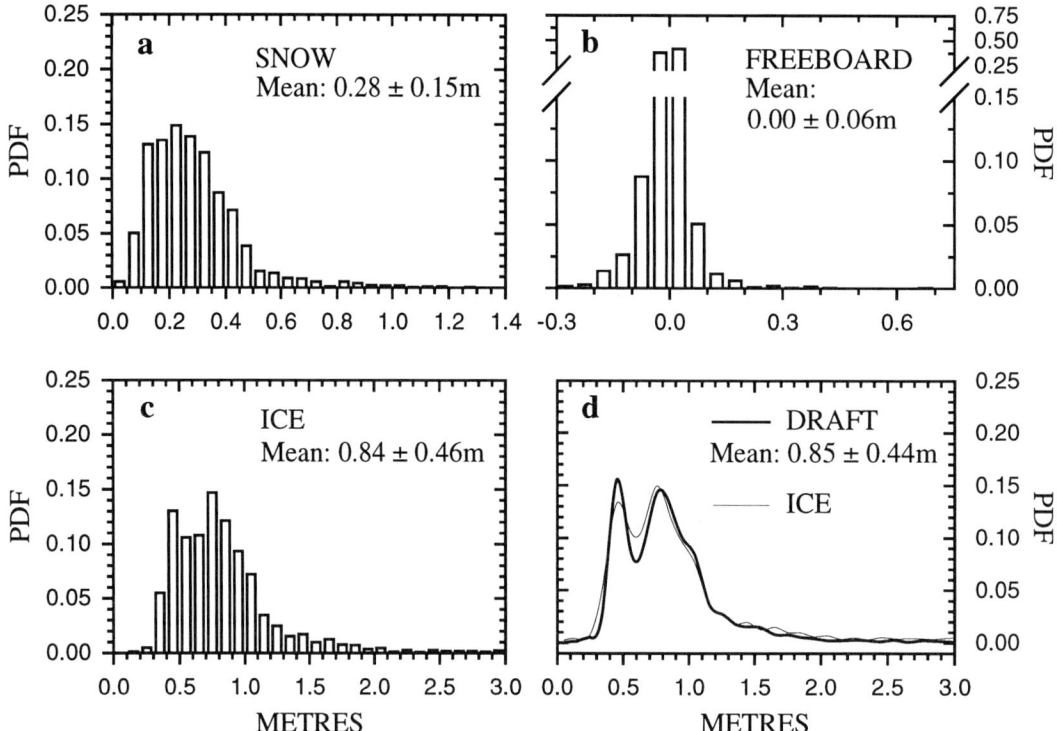

Fig. 3. Probability density functions of ice floe snow depth, ice thickness, draft and freeboard for all the Set A data. In these and all subsequent probability density functions, the draft and ice thickness distributions are shown together as curves for ease of comparison, the snow depth and freeboard data are binned at 5 cm intervals, and the ice thickness and draft data are binned at 10 cm intervals. The scale of the ice thickness distribution has been shortened and does not include a few very thick ice values.

high standard deviations with respect to their means, reflecting their high variability. The ice draft distribution closely resembled the ice thickness distribution, but the peak of higher draft values fell to the right of the peak of higher ice thickness values, i.e., draft exceeded ice thickness. The significant quantity of ice with its surface below sea level was reflected in the freeboard distribution, where 51% of the values were < 0 m, i.e., those sites were flooded with seawater. A further 42% of the freeboard values occurred in the 0-0.5–m category alone.

Isostatic balance and relationships between z_s, z_i, z_f and z_d. To be in isostatic balance, floes must satisfy the following,

$$\rho_{sw} z_d = \rho_i (z_d + z_f) + \rho_s z_s \quad (1)$$

where ρ is the density of snow (s), ice (i) and seawater (sw), and z_s, z_f and z_d are snow depth, freeboard and draft respectively [*Eicken et al.*, 1994]. For this study we used ρ_i = 920 kg m^{-3} and ρ_{sw} = 1030 kg m^{-3} [*Eicken et al.*, 1994]. Snow density measurements (n = 255) were made on 20 different floes, yielding a mean ρ_s value of 0.36±0.04 kg m^{-3} [*Sturm et al.*, this volume]. The relationship between the left and right side of (1) for all the floes is shown in Figure 4. The differences between the left and right sides of (1) vary between only –3.3% and +4.4% (a positive value indicates that the left side is greater than the right side of the equation). The mean difference for all floes is +0.1%. The data indicate that each of the floes was very close to being in isostatic balance.

There was considerable scatter in the relationship between individual snow depth and ice thickness values, particularly at higher values (Figure 5a). There was much less scatter and a high correlation between the mean snow depth and the mean ice thickness of each floe (Figure 5c) due to the fact that each floe, as a whole, was in isostatic balance, unlike the individual points across each floe [*Lange and Eicken*, 1991a; *Worby et al.*, 1996]. On aver-

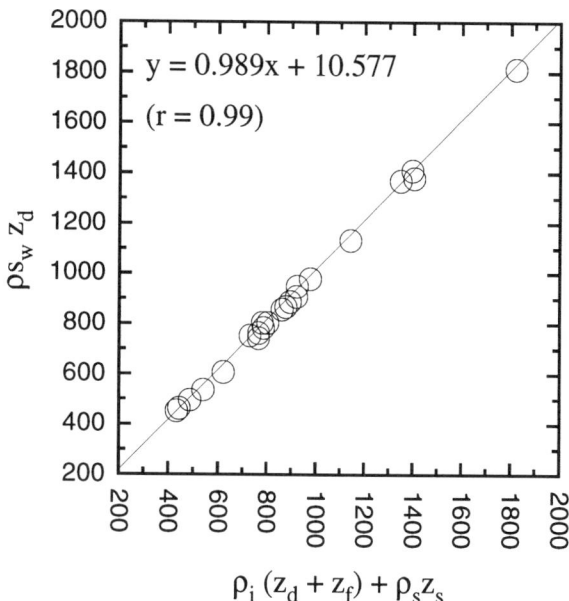

Fig. 4. All the ice floes investigated in this study were close to being in isostatic balance as represented by $\rho_{sw} z_d = \rho_i (z_d + z_f) + \rho_s z_s$. The right and left sides of the equation are the x–axis and y–axis of the graph. The straight line is represented by the regression equation and the correlation coefficient for this and all subsequent regression equations is significant at the ≥ 95% confidence level.

age, the snow depth was about one–third of the ice thickness at each floe (Figure 5c).

There was minimal scatter in the relationship between individual ice thickness and ice draft values (Figure 5b) due to the fact that, at the majority of sampling sites, the ice surface was close to sea level and the difference between ice thickness and draft was very small. Because each floe was in isostatic balance, there was even less scatter and a high correlation between the mean ice draft and the mean ice thickness of each floe (Figure 5d) compared to the individual points across each floe.

Figure 6 shows that the mean freeboard of the floes was strongly dependent on the ratio of mean snow depth to mean ice thickness, z_s/z_i. The regression equation indicates that, on average, the mean freeboard would have become negative when the mean snow depth was ≥ 30% of the mean ice thickness. According to *Eicken et al.* [1994], solving (1) for the ratio of snow depth to ice thickness, z_s/z_i, gives

$$z_s/z_i > (\rho_{sw} - \rho_i)/\rho_s \quad (2)$$

as the critical value for flooding, which occurs when $z_i < z_d$. Using the ρ_{sw}, ρ_i and ρ_s values given above, and the z_s and z_i values in Table 1, the critical value on the right side of (2) is 0.3 (close to that derived from Figure 6), while the twelve floes with a zero or positive mean freeboard (Table 1) had z_s/z_i ratios ranging from 0.204 to 0.402 (mean 0.27). For each of those twelve floes, the critical value was greater than the z_s/z_i value; hence their mean z_f values ≥ 0. Our critical value is lower than the value of 0.38 determined by Eicken et al. [1994], mainly because they used a lower snow density value than we used.

Ice Floe Classification and z_s, z_i, z_f and z_d Characteristics

The PDFs illustrated in Figure 3 provide a general summary of the z_s, z_i, z_f and z_d characteristics of most of the ice in the study area. This composite data set contains information on the ridged and unridged components of the pack ice, and separating these components from each other could provide further insight into the ice thickness variability and the processes that contribute to the development of the ice cover. To decompose the entire data Set A, the coefficient of ice thickness variation, σ_i/z_i, where σ_i is the standard deviation of the mean ice thickness z_i, has been calculated for each of the 23 ice floe stations (Table 1). The z_s, z_i, z_f and z_d data for each station were placed in three groups, X, Y and Z, according to whether the station σ_i/z_i value occurred in the ranges 0–0.19, 0.2–0.39 and ≥ 0.4, respectively. The higher the σ_i/z_i value, the greater the variability of ice thickness. Data from 10, 8 and 5 floe stations fell into the Type X, Y and Z floe categories, respectively (Table 1).

Representative profiles of snow depth and ice thickness for Type X, Y and Z floes are shown in Figures 7, 8 and 9. PDFs and descriptive statistics summarizing z_s, z_i, z_f and z_d for Types X, Y and Z are shown in Figure 10. These are described in more detail below.

Snow depth and ice thickness profiles. Station 266 (Figure 7a) had the lowest σ_i/z_i value (Table 1) of all the profiles and was thus a Type X floe. Although this floe had a very low σ_i/z_i value, it was not level; it had continuously rough bottom and snow/ice interface surfaces, and the bottom surface was rougher than the snow/ice interface (Figure 7a). This was characteristic of all Type X floes, e.g., station 268 (Figure 7b), which was rougher than station 266 and also had extensive seawater flooding at the snow/ice interface. Type Y and Type Z floes also had continuously rough bottom and snow/ice interface surfaces, but they also included locally and significantly thicker ice in ridges (Figures 8 and 9). A characteristic feature of the ridges was that the sails had a very small cross section and almost their entire mass lay below the

76 ANTARCTIC SEA ICE: PHYSICAL PROCESSES, INTERACTIONS AND VARIABILITY

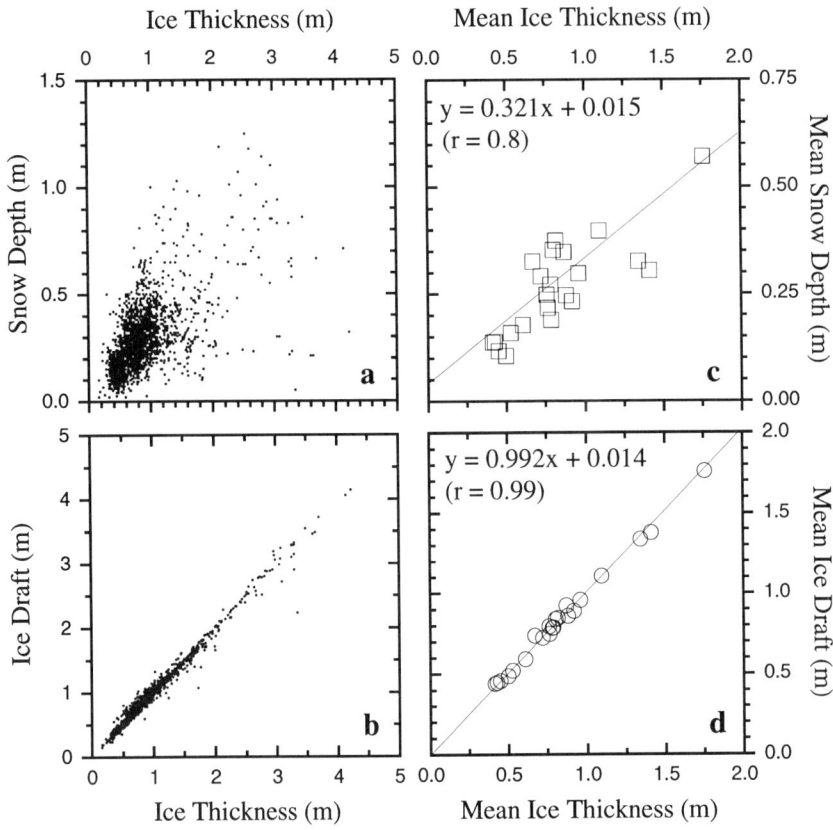

Fig. 5. Scatter plots of (a) individual snow depth and ice thickness data points, and (b) individual ice draft and ice thickness data points, and regression equations/curves and correlation coefficients of the relationships between (c) mean snow depth and mean ice thickness for individual floes, and (d) mean ice draft and mean ice thickness for individual floes.

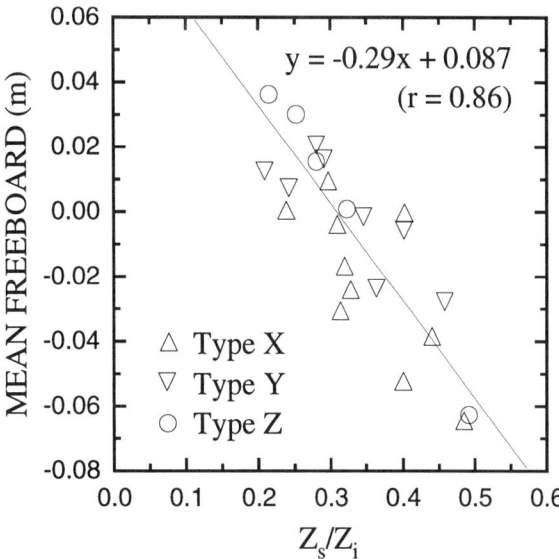

Fig. 6. Scatter plot of the relationship between the mean freeboard and the ratio z_s/z_i (mean snow depth/mean ice thickness) of individual ice floes. The straight line is represented by the regression equation in the plot. Types X, Y and Z are explained in the text.

water line. The unridged parts of some of the Type Y and Z floes sometimes had σ_i/z_i values characteristic of Type X floes, e.g., the ice to the right of the ridge in Figure 9 has a σ_i/z_i value of 0.06. Likewise, the unridged parts of some of the Type Z floes often had σ_i/z_i values characteristic of Type Y floes, e.g. the ice to the left of the ridge has a σ_i/z_i value of 0.33. In the latter case, it is noteworthy that the snow/ice interface and bottom surfaces are continuously rough, but there is no ridge evident.

Snow depth distribution characteristics. On Type X floes, most of the snow was 0.1–0.35 m deep (Figure 10*a*), while on Type Y floes the snow depth distribution was flatter with a larger number of values > 0.35 m (Figure 10*e*) compared to Type X. Type Z floes had many values in the range 0.15–0.3 m (Figure 10*i*), but the distribution had a much longer tail of high values than both Type X and Y.

Ice thickness distribution characteristics. Most of the ice in Type X floes was 0.3–0.9 m thick, with a mode of 0.7–0.9 m (Figure 10*b*). The thickness distribution of Type Y floes (Figure 10*f*) was flatter, bi-modal and had a longer tail of high values than Type X. The main charac-

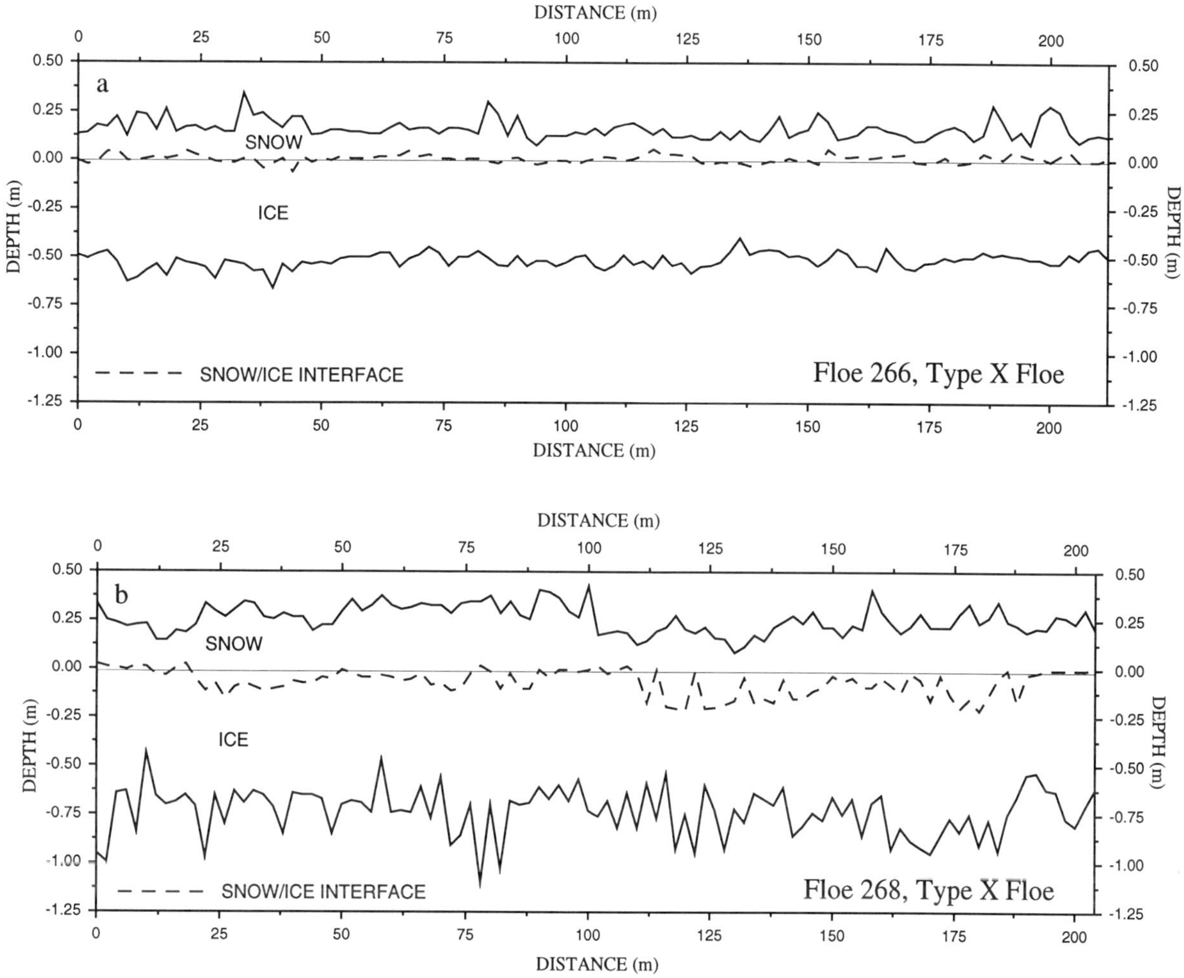

JEFFRIES ET AL.: FIRST YEAR ICE FLOE THICKNESS AND FORMATION 77

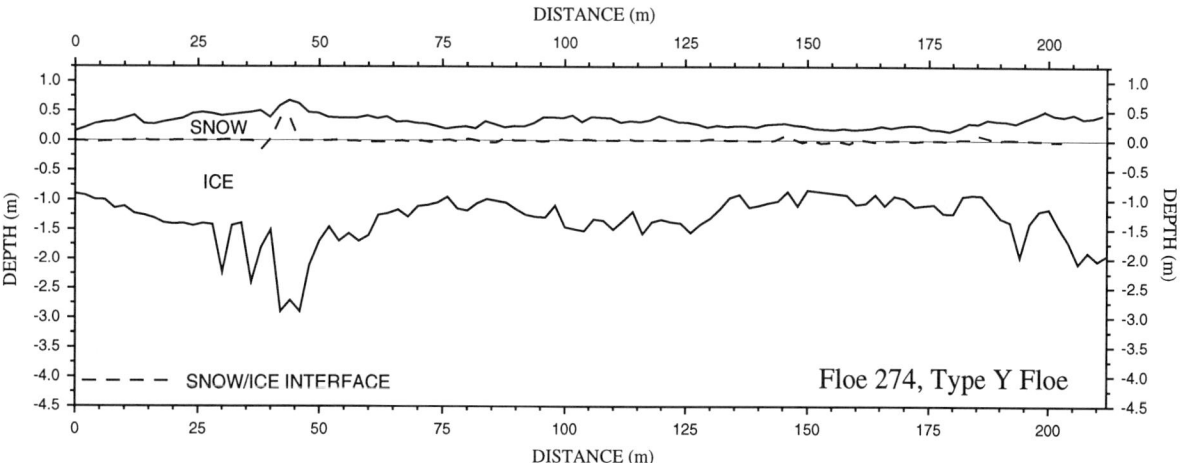

Fig. 7. Snow depth and ice thickness profiles for two Type X floes. The thin, solid line at 0cm represents sea level.

Fig. 8. Snow depth and ice thickness profiles for a Type Y floe. The vertical scale is exaggerated 3.3x relative to the vertical scale in Figure 7. The thin, solid line at 0 cm represents sea level.

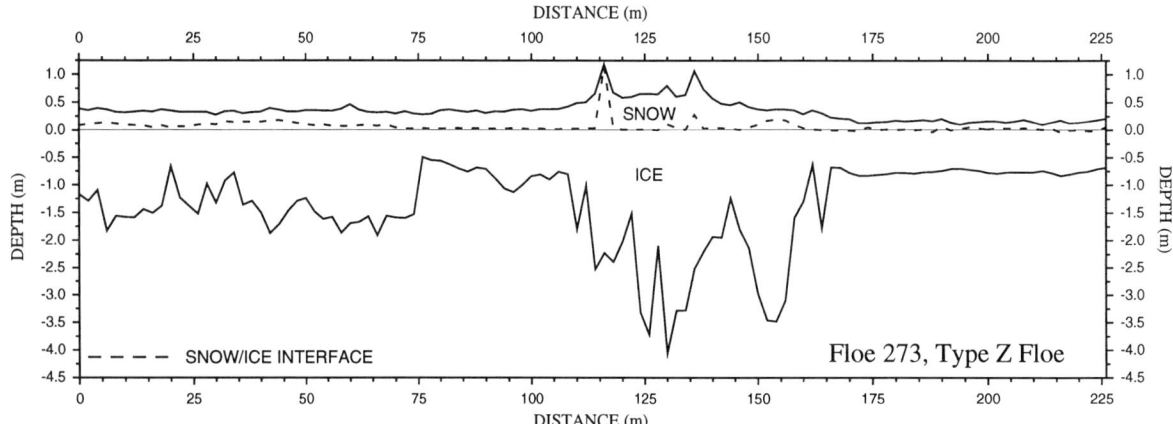

Fig. 9. Snow depth and ice thickness profiles for a Type Z floe. The vertical scale is the same as that in Figure 8, i.e., exaggerated 3.3x relative to the vertical scale in Figure 7. The thin, solid line at 0 cm represents sea level.

teristic of the Type Z thickness distribution (Figure 10*j*) compared to the other distributions was the rightward shift to higher values exemplified by its very flat appearance, the very small number of values < 0.5 m and the long tail of high values. The Type Z distribution had a number of minor peaks but no obvious modes.

Draft and freeboard distribution characteristics. The draft distribution of Type X floes fell almost entirely to the right of the thickness distribution (Figure 10*c*), i.e., a large proportion of the surface of Type X floes lay below sea level. This was reflected in the freeboard distribution, where 66% of the values were < 0 (Figure 10*d*).

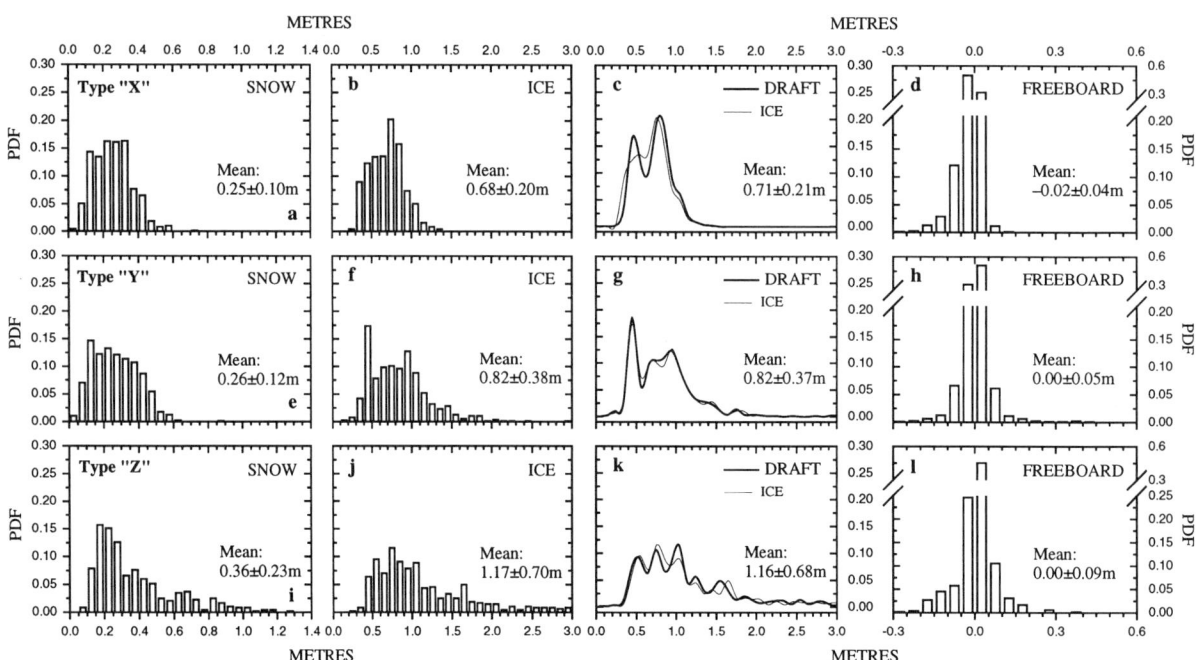

Fig. 10. Probability density functions of snow depth, ice thickness, draft and freeboard for Type X, Y and Z floes.

The draft and ice thickness distributions of Type Y floes were closely matched, with neither one leading the other over a broad range of values. The Type Y freeboard distribution (Figure 10h), with 60% of values ≥ 0, indicates that large areas of the surface of these profiles were at or above sea level. The Type Z draft distribution trailed the thickness distribution in most cases (Figure 10k) and 62% of the freeboard values were ≥ 0 (Figure 10l).

The freeboard PDFs for each type of floe (Figures 10d, 10h, 10l) indicated that, as ice thickness variability and ice thickness itself increased, the number of negative freeboard values decreased and there was less extensive flooding. This is illustrated in Figure 11, which shows the mean freeboard of individual floes as a function of their σ_i/z_i values; as the thickness increased and became more variable, the mean freeboard became more positive.

Ice Core Structure, Stratigraphy, Stable Isotope and Brine Volume Characteristics

General Description. Granular ice is easily distinguished from the columnar texture of congelation ice. However, distinguishing granular ice of snow ice origin from granular ice of frazil origin on textural grounds alone is difficult, and stable isotope analysis is required.

A mean oxygen isotope profile for all the cores, regardless of floe type, shows that the uppermost ice layers had moderately negative $\delta^{18}O$ values and the greatest variability, while the lower ice layers had slightly positive $\delta^{18}O$ values and lower variability (Figure 12). The $\delta^{18}O$ values of the lower ice layers were similar to those of the seawater (–0.8±0.1‰, n = 23) sampled from the upper 20 m of the water column adjacent to the floes that were studied. Differences between the $\delta^{18}O$ values of the lower ice layers and the seawater can be attributed to a combination of the variability of the seawater composition and the degree of isotopic fractionation that occurred during freezing.

The mean oxygen isotope profile is similar to those reported previously in the Weddell, western Bellingshausen and eastern Amundsen seas [*Lange et al.*, 1990; *Eicken et al.*, 1994; *Jeffries et al.*, 1997]. The negative $\delta^{18}O$ values in the uppermost ice layers reflect the entrainment of snow with negative $\delta^{18}O$ values into floes by seawater flooding and snow ice formation. The isotopic criterion that is generally used for differentiating between snow ice and frazil ice is that any granular ice layer with a $\delta^{18}O$ value ≤ 0‰ is snow ice, while any granular layer with a $\delta^{18}O$ value > 0‰ is frazil ice [*Lange et al.*, 1990; *Eicken et al.*, 1994;

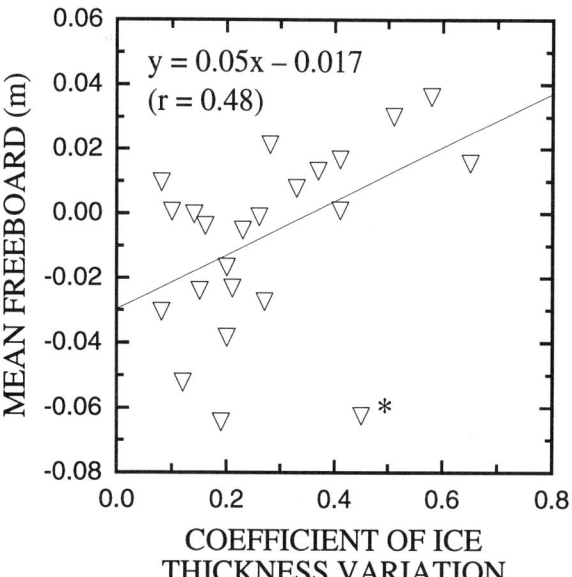

Fig. 11. Scatter plot of the relationship between the mean freeboard and the coefficient of ice thickness variation (σ_i/z_i) of individual ice floes. The straight line is represented by the regression equation in the plot. Excluding the outlying asterisked value improves the correlation coefficient to 0.65.

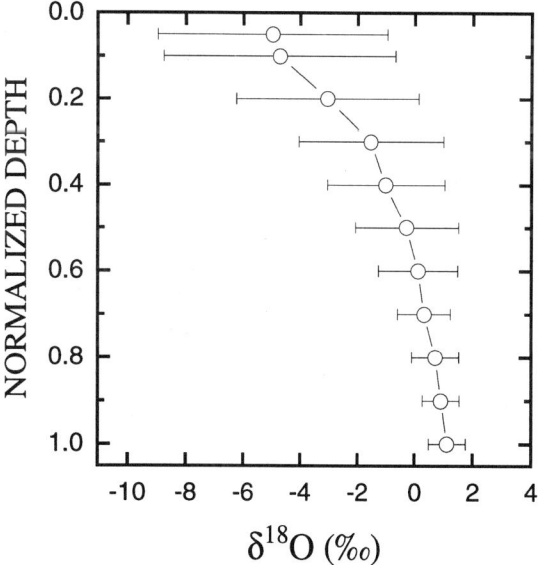

Fig. 12. Composite (mean) $\delta^{18}O$ profile for all ice cores. Because the ice cores were sampled for stable isotopes at irregular intervals according to the structural variability, the following procedure was adopted for compiling the profiles: the depth for each $\delta^{18}O$ value was normalized by dividing by the ice core length; the $\delta^{18}O$ values were binned between the surface and the base of the ice at 11 normalized depth intervals of 0.05, 0.1, 0.2 and so on at intervals of 0.1; and a mean and standard deviation $\delta^{18}O$ value was calculated for each bin. Each data point represents the mean $\delta^{18}O$ value for each bin and the horizontal bar the standard deviation value.

Worby and Massom, 1995; *Jeffries et al.*, 1997]. We use the same criterion in this study.

Snow ice was found in 69 cores. All of those cores had snow ice layers at the surface, and 10 also had snow ice layers that had been buried below the surface. This phenomenon has been observed elsewhere and is a consequence of deformation [*Lange and Hubberten*, 1992; *Jeffries et al.*, 1997]. Cavities, i.e., water– and slush–filled spaces between ice blocks and slabs, are also created during deformation. They were found in 6 cores, and although they are not ice *per se*, they are part of the ice thickness sampled and evidence of a particular process; therefore, they are included in the analysis.

Frazil ice was found in 55 cores. The majority of frazil layers were composed of many sub–layers that were distinguished from one another by sharp boundaries and differences in grain size (see *Jeffries et al.*, 1997, for additional details). No record was kept of the thickness of the individual frazil ice sub–layers, which appeared to have been stacked on top of each other.

Congelation ice was found in 66 cores. A few congelation ice layers were composed of stacked sub–layers that were separated from each other by sharp discontinuities and changes in column width (see *Jeffries et al.*, 1997, for additional details). However, most were single layers with continuous columns that indicated uninterrupted, thermodynamic thickening. In 54 cores the lowermost layer was congelation ice, but none had a skeletal layer indicating active ice growth. In fact, basal melting was indicated by the occurrence of scalloping at the bottom of some cores.

Numerous cores contained multiple congelation and frazil ice layers that appeared to have been stacked on top of each other. Similar multiple layering of different ice types in cores has been illustrated by, for example, *Jeffries et al.* [1994: Figures 7 and 8; 1997: Figure 2]. The structural complexity of cores within a single floe was often quite variable; one core might be comprised of only snow ice overlying frazil or congelation ice, while another core obtained only tens of metres away might be comprised of a snow ice layer overlying multiple layers of frazil and congelation ice. The majority of the congelation ice layers were < 0.2 m thick (Figure 13). In contrast, most of the snow ice layers were as much as 0.4 m thick and consequently had a greater mean thickness than the congelation ice layers (Figure 13).

The amount of each ice type, including cavities, as a function of the total length of all the cores that were examined is illustrated in Figure 14. Snow ice was the predominant ice type, and there was slightly more frazil than congelation ice. On the whole, the contribution from cavities was small. While few cores contained cavities, there

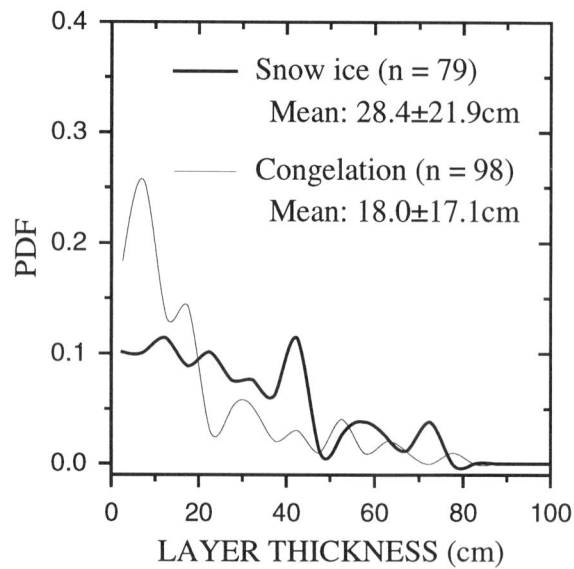

Fig. 13. Probability density functions and descriptive statistics of the thickness of layers of snow ice and congelation ice in all the ice cores. These data have been derived using a combination of isotopic and crystal structure criteria to identify the layers of the different ice types. The number of layers (n) of each ice type is also shown.

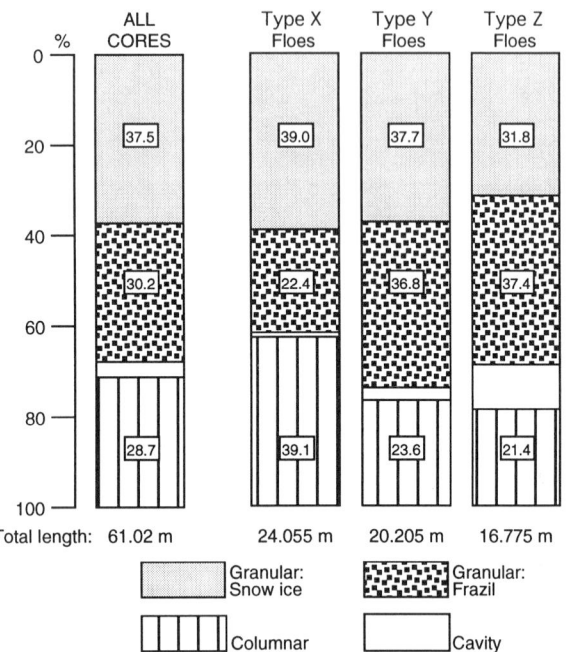

Fig. 14. Proportional representation of the amount of snow ice, frazil ice, congelation ice and cavities as a function of the total length of all the ice cores that were examined in the study area, and of all the ice cores from each ice floe type. The values in boxes are the actual amounts (%) of each ice type.

were also few instances when snow ice, frazil ice and congelation ice did not occur in the same core. Consequently, we are satisfied that the data presented in Figure 14 are a fair representation of the structural composition of the ice, and are not biased by data from only a small number of cores or floes.

The brine volume variability of all the ice cores is summarized in Figure 15a. Only 2.5% of the values were < 5%. This brine volume value is significant because, when brine volumes exceed 5%, brine pockets coalesce, gravity drainage occurs and, as brine drains out of the ice at the base, it may be replaced by seawater, which will reach the snow/ice interface if the entire ice thickness has brine volumes > 5% [*Cox and Weeks*, 1975, 1988; *Eicken et al.*, 1995]. Fifty-nine of the 73 cores analyzed had brine volumes > 5% along their entire length.

Crocker [1988] observed that flooding occurred when the snow/ice interface temperature was ≥ −8°C. Of the 59 cores with brine volumes > 5% along their entire length, 53 had snow/ice interface temperatures ≥ −8°C. At those sites with lower snow/ice interface temperatures (−10 to −8.2°C), brine volumes remained > 5% along the entire length of core because the uppermost ice layers had sufficiently high salinity values (12.5 to 18.2‰).

Variability as a function of floe types. The stratigraphic characteristics described in the previous section apply equally to the sets of cores from each type of floe, i.e., the Type X, Y and Z floes were neither more nor less stratigraphically complex than each other. However, they did differ in terms of their bulk structural composition. The results of the structural/isotopic analysis and the determination of the amount of each ice type according to the total length of core examined from each floe type are illustrated in Figure 14. Snow ice was a significant component of each floe type; the least amount of snow ice occurred in Type Z. Type X floes had significantly more congelation ice and less frazil ice than Type Y and Z. The least congelation ice and the most frazil ice occurred in Type Y and Z. The contribution of cavities increased as the ice thickness and variability increased from Type X to Type Z.

PDFs of snow ice and congelation ice layer thickness variability were derived for the entire ice core data set (Figure 13), but it is too small to be divided further for the derivation of meaningful PDFs for each floe type. However, there are a sufficient number of values to calculate mean snow ice and congelation ice layer thicknesses for each floe type (Table 2). The results show that snow ice layers were thicker than congelation ice layers in all floe types. Snow ice layers in Type Z were thicker than those in Type X and Y. Congelation ice layers in Type X were thicker than those in Type Y and Z.

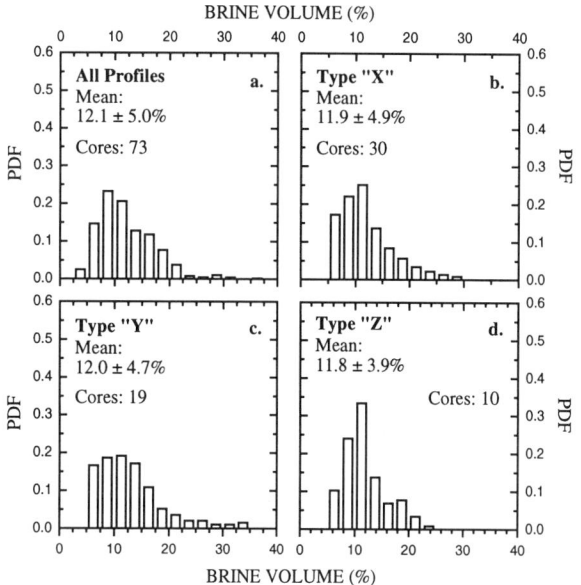

Fig. 15. Probability density functions of (a) all brine volume values in all the ice cores in the study area, and all brine volume values for those cores with brine volumes >5% along their entire length from (b) Type X floes, (c) Type Y floes, and (d) Type Z floes.

For each type of floe, most of the cores had brine volumes > 5% along their entire length. The brine volume variability in these cores is summarized by floe type in Figures 15b, c and d. The Type Z brine volume distribution was more peaked than those of Type X and Y, due to slightly different temperature and salinity characteristics, but otherwise the PDFs are fairly similar and the mean brine volume didn't vary significantly between floe types.

DISCUSSION

Ice Thickness Variability and the Role of Frazil Ice and Congelation Ice in Floe Formation

Using the coefficient of ice thickness variation as an objective means of classification, we have identified three types of first-year floe. Each of these floe types has characteristic z_s, z_i, z_f and z_d PDFs (Figure 10) and significant differences in their bulk structural composition (Figure 14). *Lange and Eicken* [1991a] identified two classes of first-year floe in the north-western Weddell Sea and described them as strongly deformed ice (Class I) and less deformed ice (Class II). Because of different analytical techniques and classification methods, a direct comparison of the Weddell Sea floe classes and our floe types is not straightforward. However, subjectively, on the basis

TABLE 2. Descriptive statistics (mean±1s.d.) for the thickness of snow ice and congelation ice layers in Type X, Y and Z floes.

	Snow Ice	Congelation Ice
Type X Floes	25.4±18.4cm	19.5±19.5cm
Type Y Floes	25.4±18.9cm	14.5±16.3cm
Type Z Floes	33.4±33.4cm	14.3±13.2cm

of the ice thickness PDFs, the Weddell Class I floes are equivalent to our Type Z floes and the Weddell Class II floes appear to be a combination of the Type X and Type Y floes. The first–year ice investigated in this study is 0.1–0.2 m thinner than the first–year ice investigated in the north–western Weddell Sea.

Frazil ice is the predominant ice type in the Weddell Sea and the East Antarctic pack ice, where amounts typically exceed 50% [*Lange and Eicken*, 1991b: Table 1; *Worby and Massom*, 1995; *Worby et al.*, this volume]. These values are somewhat higher than those observed in this study, although the excess of frazil ice over congelation ice in the Type Y and Z floes is similar to observations elsewhere. On the other hand, the excess of congelation ice over frazil ice in the Type X profiles is unusual and has only been reported in the western Ross Sea [*Jeffries and Weeks*, 1992; *Jeffries and Adolphs*, 1997] and once in the Weddell Sea as a consequence of sampling rafted nilas in coastal polynyas [*Lange and Eicken*, 1991b].

The large amount of frazil ice observed in Antarctic floes is generally attributed to the turbulent, wind– and wave–affected environment that promotes frazil ice growth and subsequent pancake ice formation [*Wadhams et al.*, 1987; *Lange et al.*, 1989]. As the pancake cycle progresses, significant thickening of the ice cover occurs by rafting [*Lange et al.*, 1989] and even ridging [*Lytle and Ackley*, 1991] of the pancakes. Notwithstanding the relatively small amount of frazil ice observed in this study, we believe that much of frazil ice with multiple stacked layers in the ice cores from all three floe types originated in the pancake cycle.

The ice thickness PDF of the Type X floes strongly resembles the first–year ice PDF in the eastern Weddell Sea [*Wadhams et al.*, 1987], but the Type X floes are 0.1–0.2 m thicker than the eastern Weddell Sea ice. The eastern Weddell Sea ice was described as undeformed inasmuch as it did not include ridges, but it was noted that the floes were characterized by much small scale roughness on the upper and lower surfaces due to pancake rafting at the time of consolidation [*Wadhams et al.*, 1987]. The small–scale roughness at the snow/ice interface and greater bottom roughness that characterize the Type X floes, and the unridged portions of the Type Y and Z floes, are typical of Antarctic ice floes in general [*Lange and Eicken*, 1991b; *Andreas et al.*, 1993] and reflect the role that rafting plays in thickening the ice cover [*Worby et al.*, 1996].

The bi–modal ice thickness distribution of the Type Y floes (Figure 10f) is common in first–year ice, and while the thicker ice peak and the tail of high values are considered to represent the ridged ice, the thinner ice peak is often described as undeformed ice [*Lange and Eicken*, 1991a; *Worby et al.*, 1996]. The term *undeformed* is often applied to the unridged areas of Antarctic ice floes, perhaps because the surface traces of deformation by rafting, and sometimes even of ridging, are effectively concealed by the snow cover [*Eicken et al.*, 1991]. We consider the use of the term *undeformed* to be misleading and prefer the term *unridged*, since it allows for the existence of undeformed ice, but also recognizes the fact that, in this study for example, much of the ice located between ridges had been deformed by rafting. As noted by *Eicken et al.* [1991] in the Weddell Sea, "It may be difficult, if not impossible, to obtain core samples in the pack–ice that have not undergone some deformation at some stage of development."

While the Type X floes owe their thickness variability primarily to deformation by rafting, the PDFs and profiles of Type Y and Type Z floes clearly indicate that deformation by ridging contributed to the thickening of parts of these floes. In terms of the thickness and the cross–sectional area of the ridges, Type Y floes can be described as moderately ridged and Type Z floes as strongly ridged. The increase in the contribution of cavities to the total ice mass of Type X, Y and Z floes (Figure 14) reflects the progressively greater degree of deformation that has contributed to the development of these floes. While some of the ridges probably were created during more turbulent episodes in the pancake cycle, it is likely that others were created by ridging of ice in leads. Regardless of the origin of the ridges, additional frazil ice might have been generated during the ridge–building events themselves [*Tucker et al.*, 1987] and thus contributed to the relatively large amount of frazil ice in the Y and Z floes.

If the frazil ice in the cores indicates that much of the ice cover formed initially as a consequence of the turbulent conditions associated with the pancake cycle, then the thickness of the congelation ice that occurred as the lowermost layer of the majority of cores indicates that there was subsequently moderate thermodynamic thickening by basal columnar ice growth. However, not all the congelation ice grew undisturbed at the base of the floes, as indicated by the occurrence of numerous cores containing rafted, multiple layers of congelation ice and frazil ice.

Since congelation ice occurred at the base of so many cores, the evidence for bottom melting observed during the course of the study indicates that it would have been removing primarily congelation ice. Thus, the observed amounts of congelation ice are probably conservative and not a true representation of the total contribution of congelation ice to floe development. Bottom melting might explain why the X, Y and Z floes contained generally less congelation ice than has been reported in other Antarctic sea ice zones: typically ≥ 40% in the Weddell Sea [*Lange and Eicken*, 1991*b*: Table 1] and the East Antarctic pack ice [*Worby and Massom*, 1995; Worby et al., this volume], and over 60% in the inner Ross Sea pack ice [*Jeffries and Adolphs*, 1997].

If we make the reasonable assumption that bottom melting rates were the same at the base of the Type X, Y and Z floes, then another explanation must be sought for the smaller amount of congelation ice in Types Y and Z. Bearing in mind that the congelation ice thickness values presented in Table 2 represent uninterrupted columnar ice growth, the thinner congelation ice layers in Type Y and Z compared to Type X floes (Table 2) might be an indication that there simply was not as much congelation ice growth at the base of the Y and Z floes. These thinner congelation ice layers could have been a consequence of the thicker snow cover and greater insulation of the Y and Z floes, which would have reduced the heat conduction between the base and the surface and thus the basal ice growth rates. Basal columnar ice growth rates also would have been lower if, after initial formation in the pancake cycle, the unridged ice in the Y and Z floes had been thicker than that in the X floes.

If, as *Lytle and Ackley* [1991] suggest, ridging occurs during the pancake cycle, then it is possible that the Y and Z floes owe their initial formation to more turbulent conditions than those that contributed to the initial formation of the rafted Type X floes. This would explain why the unridged, but nevertheless rafted, portions of Type Y and Z floes were often thicker than the Type X floes. Such variability in the conditions at the time of initial formation would explain the flattening and rightward shift of the Type X, Y and Z floe thickness distributions (Figure 10), and the tails of Type Y and Z floe thickness distributions would include not only ridged ice but also thicker rafted ice.

Snow, Seawater Flooding and the Formation of Snow Ice

The form of the snow depth PDFs and the relationship between snow depth and ice thickness can be attributed, in part, to the variable topography of the floes. On Type Y and Z floes, the ridges act as fences, where snow accumulates on their flanks while being eroded from the tops of the sails [*Lange and Eicken*, 1991*a*; *Eicken et al.*, 1994; *Worby et al.*, 1996; Sturm et al., *this volume*]. This leads to extremes of snow depth, and the more ridged the ice the greater the potential for significant snow drift accumulation and a longer tail of deep snow values in the PDFs (Figures 10*e* and 10*i*). Also, the more ridged the ice, the greater the ice thickness (Figures 10*f* and 10*j*) and thus more scatter in the snow depth–ice thickness relationship for individual points in the thicker ice categories (Figure 5*a*). On the other hand, on Type X floes, there are no ridges to act as snow fences; consequently, the snow is more evenly distributed on the floes and the less variable snow thickness is reflected in the short tail of the snow depth distribution (Figure 10*a*) and lower scatter in the snow depth–ice thickness relationship for individual points in the thinner ice categories (Figure 5*a*).

The greater total amount and greater mean thickness of snow ice layers compared to those of congelation ice layers indicates that, by the time the Type X, Y and Z floes had been sampled, snow ice had made a greater contribution than congelation ice to the thermodynamic thickening of the floes. Although we have no data on the thickness of the frazil ice sub–layers (the original, thermodynamically–thickened building blocks of the frazil ice layers observed in the cores), it is probable that they too were thinner than the snow ice layers. Comparisons of the thickness of snow ice layers and frazil ice sub–layers elsewhere in the Pacific sector of the Antarctic sea ice zone clearly show that snow ice can make a greater contribution than frazil ice to the thermodynamic thickening of the ice cover [*Jeffries and Adolphs*, 1997; *Jeffries et al.*, 1997].

The presence of snow ice layers at the top of almost all the cores, and the far greater number of surface than buried snow ice layers, indicates that seawater flooding and snow ice formation occurred primarily after the Type X floes, and the unridged portions of the Type Y and Z floes, had been deformed by rafting. The widespread occurrrence of flooding and snow ice formation probably explains, in part, why the roughness of the snow/ice interface is less than that of the bottom surface, since the seawater preferentially floods the lowest elevations, e.g., profile 268 (Figure 7*b*). Once snow ice has formed in these flooded areas, a previously rough snow/ice interface would be smoother than the original surface [*Worby et al.*, 1996; *Sturm et al.*, this volume]. Snow ice formation also helps to maintain isostatic balance [*Worby et al.*, 1996]. It also reduces the depth of the snow cover; consequently, the snow depths measured do not represent the total amount of snow deposited on the floes.

The amount of snow ice contributing to the mass of each type of floe indicates that there must have been considerable seawater flooding of the snow/ice interface prior to the time of this investigation. The amount of snow ice, both as a function of the entire set of cores or of the different profile types (Figure 14), is greater than most observations in first–year ice elsewhere in the Antarctic: 3–4% in the Weddell Sea [*Lange et al.*, 1990; *Eicken et al.*, 1994]; 13.5% in the East Antarctic pack ice [*Worby et al.*, this volume]; 15–24% elsewhere in the Pacific sector of the Antarctic sea ice zone [*Jeffries and Adolphs*, 1997; *Jeffries et al.*, 1997].

At the time of this investigation, the floes remained flooded to a considerable extent and the amount of flooding exceeded almost all previous reports elsewhere in the winter Antarctic pack ice: 15–38% in the Weddell Sea [*Wadhams et al.*, 1987; *Ackley et al.*, 1990; *Lange and Eicken*, 1991a]; 40–53% in the East Antarctic pack ice [*Worby et al.*, this volume]; 18–30% elsewhere in the Pacific sector of the Antarctic sea ice zone [*Jeffries and Adolphs*, 1997; *Jeffries et al.*, 1997]. Each of the floe types had a large number of freeboard values in the 0-5 cm range that indicates a high flooding potential at many other sites (Figures 3 and 10).

Currently, it is understood that three factors contribute to seawater flooding at the snow/ice interface. The first factor is the snow load, where the weight of snow is sufficient to depress the ice surface below sea level causing seawater to infiltrate the snow/ice interface [*Ackley et al.*, 1990; *Eicken et al.*, 1994]. The second factor is the ice load, where it is the weight of ice in a ridge or a raft that locally depresses the ice surface below sea level and/or it fractures the ice creating pathways for seawater to reach the snow/ice interface [*Ackley*, 1986; *Ackley and Sullivan*, 1994]. The third factor is vertical brine exchange between the base of the ice and the snow/ice interface [*Fritsen et al.*, 1994; *Lytle and Ackley*, 1996].

The occurrence of a large amount of snow ice and extensive seawater flooding on these late winter floes corroborates numerical model simulations of these phenomena, which indicate that the probability of flooding and snow–ice formation increases as the length of the ice growth season increases [*Eicken et al.*, 1995]. This probability increases primarily due to the increasing snow load on the ice, which depresses the snow/ice interface below sea level. It is clear from the data obtained during this investigation that flooding was strongly dependent on the snow load and that, on average, flooding occurred when the snow depth was $\geq$ 30% of the ice thickness in late winter (Figure 6 and Equation 2).

The flooding potential represented by the individual freeboard values $\geq$ 0 (Figures 3 and 10), or by the mean floe freeboard values $\geq$ 0 (Table 1), can be realized by additional snow accumulation or redistribution of the existing snow cover [*Ackley et al.*, 1990], i.e., the snow load increases to the point that the critical value, 0.3 in this investigation, is exceeded. The flooding potential can be achieved also by melting ice off the bottom, while the snow depth remains constant or increases, and continued bottom melting might maintain the flooding at the snow/ice interface. Bottom melting, which was evident during this study, would have altered the z_s/z_i ratios and caused an effective increase in the snow load and thus the extensive seawater flooding that was observed. This was probably particularly true of the final two legs of the cruise between days 274 and 285 (Figure 1) when the majority of floes had a negative mean freeboard (Table 1). Concurrent physical oceanographic measurements revealed that temperatures in the shallow mixed layer were above the *in situ* freezing point, possibly due to warm water upwelling in the Ross Sea Gyre (H. Hellmer and S. Jacobs, personal communications, 1994 and 1995, cited in *Jeffries et al.*, 1995). A 0.1–0.2-m decrease in the mean ice thickness in the vicinity of 69.5°–71°S, 150°–156°W was attributed to this oceanographic phenomenon [*Jeffries et al*, 1995].

The occurrence of bottom melting suggests that the ice floe thickness values reported here are conservative and underestimate the true ice thickness. However, the underestimate might not be too great. The process of melting ice off the bottom leading to further flooding and snow ice formation at the snow/ice interface amounts to a 'conveyor belt' (S.F. Ackley and V.I. Lytle, personal communication, 1995, cited in *Jeffries et al.* [1997]), where the basal losses are balanced by the surface gains. Consequently, the thickness of an ice floe might be close to being in an equilibrium state and the measured thickness values good estimates of the true thickness. The greater amount of snow ice compared to most previous reports can be explained by the conveyor belt, which would have caused an effective increase in the contribution of snow ice to the thermodynamic thickening of the floes and the total ice mass, as ice, mainly congelation ice, was lost at the base of the floes.

The ridging of the Type Y and Z floes has been described as moderate to strong, respectively, and in a number of cases flooding was observed in the vicinity of ridges.

The degree of ridging of Type Y and Z floes is defined by a proxy measure, the coefficient of ice thickness variation, where the higher the coefficient the more variable and greater is the overall thickness of a floe due to ridging. Figure 11, which shows that the higher the coefficient of variation the higher the freeboard of a floe, suggests that, although there might have been flooding due to a locally depressed snow/ice interface and/or due to seawater flow through fractures within and around ridges at the time of the study, the overall effect of the ridges appears to have been to reduce the extent of flooding on a floe as a whole. The snow load, represented by the z_s/z_i ratio, on the ridged and rafted Type Y and Z floes is not significantly different from that on the rafted–only Type X floes (Figure 6), yet the latter have more extensive flooding. This suggests that the ridges might reduce flooding on the Y and Z floes, where the snow load is insufficient to overcome the considerable buoyancy of the ridges represented by their large mass below sea level.

Nevertheless, there is a significant amount of snow ice at the surface of the Y and Z floes (Figure 14), indicating that the floes were flooded more extensively at one time. Most of the flooding and snow ice formation probably occurred prior to ridge formation, although we don't discount the possibility that some occurred after ridging. Any post–ridging snow ice formation that did occur might have compounded the buoyancy effect of the ridges and contributed to raising the snow/ice interface above sea level across much of the individual ridged floes.

Most of the ice sampled in this investigation met the temperature and brine volume criteria for flooding by vertical brine exchange. The similarity between the brine volume distributions for each floe type (Figure 15) suggests that each had a similar, high potential for flooding by this process. Vertical brine exchange would be independent of the extent of ridging and buoyancy effects and might explain the flooding that was observed on the Y and Z profiles.

SUMMARY AND CONCLUSION

The variability of snow depth and ice thickness, of draft, freeboard and seawater flooding, and of the structure and stratigraphy of first–year ice floes in late winter in the western Amundsen and eastern Ross seas has been described for the first time. The main findings of the study are summarized below.

1. Dynamic processes, i.e., rafting and ridging, played an important role in the thickening of much of the ice cover. In view of the high concentration of ice > 0.3 m thick throughout the study area, as represented by the Set B and C data, the PDF for the entire Set A data (Figure 3b) was probably a good representation of the consequences of deformation for the first–year ice thickness distribution in the central Pacific sector of the Antarctic sea ice zone.

2. Using the coefficient of ice thickness variation, the large–scale ice thickness variability represented by all the Set A data has been decomposed into subsets, each with characteristic z_s, z_i, z_f and z_d distributions representing three different floe types. The simple form of the ice thickness distribution of Type X floes suggests that rafting efficiently transformed large amounts of thin ice into thicker ice categories. Two thicker floe types, Y and Z, are moderately and strongly ridged, respectively, and their ice thickness distributions are more complex than that of the Type X floes. This is because Type Y and Z floes contain ridged ice and a significant quantity of heavily rafted ice. Together, the ice thickness distributions of the three types of floe profile reflect the complex nature of the dynamic thickening and consolidation of individual floes and the pack ice as a whole.

3. The sets of ice cores from individual floes and for each of the floe profile types ranged from simple to complex in terms of the number, disposition and thickness of the layers of snow ice, frazil ice, congelation ice and cavities, and the total amount of each ice type, they contained. This structural/stratigraphic complexity was partly due to the fact that, by late winter, the medium to vast floes that were investigated were composed of many, once smaller floes, each of which had had slightly different growth histories prior to their consolidation into larger floes. However, despite the variability of individual ice cores, there were significant differences in the bulk structural composition and the proportions of each ice type observed in the X, Y and Z floes. This suggests that the ice floe classification based on the ice thickness variability, which resulted in distinctive z_s, z_i, z_f and z_d probability density functions and profiles of snow depth and ice thickness, was not simply identifying different portions of a continuum of ice thickness variability, but that the thickening of each floe type differed even after consolidation and formation of the larger floes. These differences included variations in the extent of seawater flooding of the snow/ice interface and subsequent snow ice formation, which provided some insight into the factors controlling these processes.

4. The initial formation of all floes was dominated by frazil ice growth and rafting, even ridging, of pancakes. Subsequent thermodynamic thickening of all floes included moderate congelation ice growth, but by late winter the thermodynamic thickening of all floe profiles was dominated by snow ice formation. It is evident from the amount of snow ice in the cores that seawater flooding of

the snow/ice interface had been widespread prior to our arrival, and the draft/freeboard data showed that it continued to be widespread during the study, particularly on the Type X floes. As the ice cover developed during the course of the winter, the cause and extent of flooding and subsequent snow ice formation on all profiles was a complex function of the interplay between deformation, ice and snow loading, and the ice thermal regime and its effect on brine volumes. Ridges might have provided sufficient buoyancy to counter the snow load and lift much of the snow/ice interface of the thicker Type Y and Z floes above sea level. Thus, the snow load, coupled with vertical brine exchange promoted by high ice temperatures due to the deep snow cover, might have been the major cause of flooding by late winter.

Acknowledgements. This research was supported by NSF grants OPP–9117721 and OPP–9316767. Jaña's participation in the cruise was made possible by NSF, Office of Polar Programs, and Instituto Antartico Chileno. The data were acquired with the assistance of the following individuals; Science Party personnel: Ute Adolphs, Dave Crane, Stephanie Cushing (supported by a Research Experience for Undergraduates [REU] supplement), Rob Massom, Shawn McCullars (NSF Young Scholar Program), Kim Morris, Janet Nonelly, Marjorie Porter (NSF High School Teacher Enhancement Program), Bernard Rabus, Jane Stevens and Matthew Sturm; Antarctic Support Associates personnel: Barney Kane, Rod McCabe, Doyle Nicodemus, Russ Nilson, Buzz Scott and Bill Young. We are indebted to Captain J. Borkowski, the officers and crew of the R. V. *Nathaniel B. Palmer* for their support and assistance, which contributed to a productive and enjoyable voyage. B. H.–C. was supported in part by an NSF REU supplement. H.R.K. acknowledges the support of the Natural Sciences and Engineering Research Council of Canada for the Stable Isotope Laboratory, University of Calgary. Kim Morris and Ted Maksym read early versions of this paper and we appreciate their comments and suggestions on its content and organization. Stan Jacobs had editorial responsibility for this paper. His chosen reviewers, Terry Tucker, Greg Crocker, and one individual who wished to remain anonymous, helped us to make further improvements.

REFERENCES

Ackley, S. F., Sea ice pressure ridge microbial communities, *Antarct. J. of the U.S.*, 21, 172–174, 1986

Ackley, S. F., and C.W. Sullivan, Physical controls on the development and characteristics of Antarctic sea ice biological communities – a review and synthesis, *Deep Sea Res.*, 10(11), 1583–1604, 1994.

Ackley, S. F., M. A. Lange, and P. Wadhams, Snow cover effects on Antarctic sea ice thickness, in *Sea Ice Properties and Processes*, edited by S. F. Ackley and W. F. Weeks, CRREL Monogr. 90–1, 16–21, 1990.

Allison, I., R. E. Brandt, and S. G. Warren, East Antarctic sea ice: Albedo, thickness distribution and snow cover, *J. Geophys. Res.*, 98(C7), 12417–12429, 1993.

Allison, I., and A. P. Worby, Seasonal changes in sea ice characteristics off East Antarctica, *Ann. Glaciol.*, 20, 195–201, 1994.

Andreas, E. A., M. A. Lange, S. F. Ackley, and P. Wadhams, Roughness of Weddell Sea ice and estimates of air–ice drag coefficients, *J. Geophys. Res.*, 98(C7), 12,439–12,452, 1993.

Carsey, F. D. (editor), *Microwave Remote Sensing of Sea Ice*, Geophysical Monograph 68, AGU, Washington, D.C., 1992.

Cox, G. F. N., and W. F. Weeks, Brine drainage and initial salt entrapment in sodium chloride ice, *CRREL Res. Rep.* 354, 1975.

Cox, G. F. N., and W. F. Weeks, Equations for determining gas and brine volumes in sea ice, *J. Glaciol.*, 29(102), 306–316, 1983.

Cox, G. F. N., and W. F. Weeks, Numerical simulations of the profile properties of undeformed first–year sea ice during the growth season, *J. Geophys. Res.*, 93(C10), 12,449–12,460, 1988.

Crocker, G, Physical processes in Antarctic landfast sea ice, Ph.D. thesis, University of Cambridge, 1988.

Eicken, H., M. A. Lange, and G. S. Dieckmann, Spatial variability of sea–ice properties in the northwestern Weddell Sea, *J. Geophys. Res.*, 96(C6), 10,603–10,615, 1991.

Eicken, H., M. A. Lange., H.–W. Hubberten, and P. Wadhams, Characteristics and distribution patterns of snow and meteoric ice in the Weddell Sea and their contribution to the mass balance of sea ice, *Ann. Geophys.*, 12, 80–93, 1994.

Eicken, H., H. Fischer, and P. Lemke, Effects of the snow cover on Antarctic sea ice and potential modulation of its response to climate change, *Ann. Glaciol.*, 21, 369–376, 1995.

Fritsen, C. H., V. I. Lytle, S. F. Ackley, and C. W. Sullivan, Autumn bloom of Antarctic pack–ice algae, *Science*, 266(5186), 782–784, 1994.

Jeffries, M. O., and W. F. Weeks, Structural characteristics and development of sea ice in the western Ross Sea, *Antarct. Sci.*, 5(1), 63–75, 1992.

Jeffries, M. O., R. A. Shaw, K. Morris, A. L. Veazey, and H. R. Krouse, Crystal structure, stable isotopes ($\delta^{18}O$) and development of sea ice in the Ross, Amundsen and Bellingshausen Seas, Antarctica, *J. Geophys. Res.*, 99(C1), 985–995, 1994.

Jeffries, M. O., R. J. Jaña, S. Li, and S. McCullars, Sea ice and snow thickness distributions in late winter 1993 and 1994 in the Ross, Amundsen and Bellingshausen seas, *Antarct. J. of the U.S.*, 30(1-4), 18–21, 1995.

Jeffries, M. O., and U. Adolphs, Early winter snow and ice thickness distribution, ice structure and development of the western Ross Sea pack ice between the ice edge and the Ross Ice Shelf, *Antarct. Sci.*, 9(2), 188-200, 1997.

Jeffries, M. O., A. P. Worby, K. Morris, and W. F. Weeks, Seasonal variations in the properties and structural composition of sea ice and snow cover in the Bellingshausen and Amundsen seas, Antarctica, *J. Glaciol.*, 43(143), 138-151, 1997.

Lange, M. A., P. Schlosser, S. F. Ackley, P. Wadhams, and G. S. Dieckmann, ^{18}O concentrations in sea ice of the Weddell Sea, Antarctica, *J. Glaciol.*, 36(124), 315-323, 1990.

Lange, M. A., and H. Eicken, The sea ice thickness distribution in the northwestern Weddell Sea, *J. Geophys. Res.*, 96(C3), 4821–4837, 1991a.

Lange, M. A., and H. Eicken, Textural characteristics of sea ice and the major mechanisms of ice growth in the Weddell Sea, *Ann. Glaciol.*, 15, 210–215, 1991b.

Lange, M. A., and H.-W. Hubberten, Isotopic composition of sea ice as a tool for understanding sea ice processes in the polar regions, in, *Physics and Chemistry of Ice*, edited by N. Maeno and T. Hondoh, Sapporo, Japan, Hokkaido University Press, 399–405, 1992.

Lange, M. A., S. F. Ackley, P. Wadhams, G. S. Dieckmann, and H. Eicken, Development of sea ice in the Weddell Sea, *Ann. Glaciol.*, *12*, 92–96, 1989.

Lytle, V. I., and S. F. Ackley, Sea ice ridging in the eastern Weddell Sea, *J. Geophys. Res.*, *96*(C10), 18,411–18,416, 1991.

Lytle, V. I., and S. F. Ackley, Heat flux through sea ice in the western Weddell Sea: Convective and conductive transfer processes, *J. Geophys. Res.*, *101*(C4), 8853–8868, 1996.

Maykut, G. A., The Surface Heat and Mass Balance, in *The Geophysics of Sea Ice*, edited by N. Untersteiner, Plenum Press, New York, 395–463, 1986.

Sturm, M., K. Morris, and R. A. Massom, The winter snow cover of the West Antarctic pack ice: Its spatial and temporal variability, *Antarctic Research Series*, this volume.

Tucker, W. B., III, A. J. Gow, and W. F. Weeks, Physical properties of summer sea ice in the Fram Strait, *J. Geophys. Res.*, *92*(C7), 6787–6803, 1987.

Wadhams, P., M. A. Lange, and S. F. Ackley, The ice thickness distribution across the Atlantic sector of the Antarctic ocean in midwinter, *J. Geophys. Res.*, *92*(C13), 14,535–14,552, 1987.

Worby, A. P., and R. Massom, The structure and properties of sea ice and snow cover in East Antarctic pack ice, *Antarctic CRC Res. Rep.*, 7, 1995.

Worby, A. P., W. F. Weeks, M. O. Jeffries, K. Morris, and R. Jaña, Late winter sea ice and snow thickness distributions in the Bellingshausen and Amundsen Seas, *Antarct. J. of the U.S.*, *29*(1), 13–15, 1994.

Worby, A. P., M. O. Jeffries, W. F. Weeks, K. Morris, and R. Jaña, The thickness distribution of sea ice and snow cover during late winter in the Bellingshausen and Amundsen Seas, *J. Geophys. Res.*, *101*(C12), 28,441-28,455, 1996.

Worby, A. P., R. A. Massom, I. Allison, V. I. Lytle, and P. Heil, East Antarctic sea ice: A review of its structure, properties and drift, *Antarctic Research Series*, this volume.

B. Hurst–Cushing, M. O. Jeffries and S. Li, Geophysical Institute, University of Alaska Fairbanks, P. O. Box 99775–7320, 903 Koyukuk Drive, Fairbanks, AK 99775–7320.

R. Jaña, Instituto Antartico Chileno, Luis Thayer Ojeda 814, Providencia, Santiago, Chile.

H. R. Krouse, Department of Physics and Astronomy, University of Calgary, 2500 University Drive, N.W., Calgary, AB, Canada T2N 1N4.

(Received August 27, 1996; Accepted March 11, 1997.)

DERIVING MODES AND RATES OF ICE GROWTH IN THE WEDDELL SEA FROM MICROSTRUCTURAL, SALINITY AND STABLE-ISOTOPE DATA

Hajo Eicken

Alfred Wegener Institute for Polar and Marine Research, Bremerhaven, Germany

The dependence of texture, salinity and $H_2^{18}O$ concentration of sea ice on the mode and rate of ice growth was studied based on ice, brine and seawater samples from the Weddell Sea. Mean salinities and $\delta^{18}O$ values were 6.0±1.1 and 1.04±1.00 ‰ for first-year and 4.8±1.4 and 0.40±0.71 ‰ for second-year ice, respectively. Deviations between ice of different textural composition and between averaged core profiles are mainly a result of snow- and superimposed ice formation. Growth history and properties of exemplary cores of young ice, frazil and congelation first- and second-year ice are studied through analysis of grain-size profiles, $\delta^{18}O$ and salinity. The mean $\delta^{18}O$ of brine, –2.8 ‰ (min. –10 ‰), is explained through Rayleigh fractionation in a half-closed system and entrainment of meteoric water. The near-constant isotopic composition of surface seawater in the study area allows for determination of effective fractionation coefficients from the bottom ice layers and the parent water mass, with a maximum value close to 2.70 ‰. A stagnant boundary-layer/seawater-entrainment fractionation model is derived to explain growth-rate dependent fractionation of $H_2^{18}O$ for sea ice. The model is validated through comparison of $\delta^{18}O$ profiles with simulation results from an ice-growth model. Through an exponential approximation, growth rates have been derived from the ice-core data, yielding a time-integrated growth rate of 0.22 mm h^{-1}. The data indicate significantly lower oceanic heat fluxes in the western as compared to the eastern and central Weddell Sea. Drawing on grain-size and salinity data and ice-growth simulations, temporal and spatial variability of ice accretion and the major modes of sea-ice growth are discussed.

1. INTRODUCTION

The contrasts between Arctic and Antarctic sea ice are manifold, extending from the growth environment through the properties and structure of the ice to the history of its study. The long tradition and wealth of sea-ice studies carried out in the Arctic have shaped our understanding of sea-ice growth and decay, beginning with the classical works of *Weyprecht* [1868] and *Malmgren* [1927], continuing with investigations from drifting and near-shore stations [*Zubov*, 1945, *Cherepanov*, 1957, *Weeks and Lee*, 1958 and 1962, *Schwarzacher*, 1959, *Untersteiner*, 1968] and culminating in the extensive field programs of the past two decades. Comparatively few studies of Antarctic sea ice had been carried out up until the 1970's. The subsequent increase in the number of expeditions and satellite imagery [*Gloersen et al.*, 1992] soon demonstrated that Antarctic sea ice differs considerably from its Arctic counterpart.

Most (≥80 % by area) of the Antarctic ice cover does not survive austral summer, with expanses of perennial sea ice present only in the Western Weddell, the Bellingshausen, Amundsen and Ross Seas. As established through ice-thickness measurements, the ice cover is much thinner than in the Arctic, with level ice not exceeding 0.5 to 0.7 m thickness in most areas [*Jacka et al.*, 1987, *Wadhams et al.*, 1987]. Thickening is mostly a result of dynamic growth processes, in particular the formation of frazil and pancake ice in the swell-dominated regime of the advancing ice edge, resulting in high proportions of granular ice [*Gow et al.*, 1982 and 1987, *Lange et al.*, 1989, *Jacka et al.*, 1987, *Lange and Eicken*, 1991a, *Jeffries et al.*, 1994]. In the regime of the so-called "pancake cycle" [*Lange et al.*, 1989], deformation in a convergent regime is characterized by rafting of individual pancakes or floe segments [*Wadhams et al.*, 1987, *Lange et al.*, 1989]. The oceanic heat flux, higher by up to an order of magnitude in the Southern Ocean as compared to the Arctic seas, is instrumental in controlling the maximum ice thickness grown [*Gordon and Huber*, 1990, *McPhee and Martinson*, 1994, *Martinson*, 1994, *Lytle and Ackley*, 1996]. In addition, the ice-growth regime in the Southern Ocean is strongly affected by snow accumulation. Apart from its insulating properties resulting in a decrease of ice growth rates [*Eicken et al.*, 1995a], snow contributes positively to the mass balance of sea ice through flooding and the formation of snow ice [*Lange et al.*, 1990, *Ackley et al.*, 1990, *Eicken et al.*, 1994a and 1995a, *Jeffries et al.*, 1994, 1997 and this volume, *Worby and Massom*, 1995] or formation of superimposed ice [*Panov and Fedotov*, 1977, *Kawamura et al.*, 1993, *Jeffries et al.*, 1997].

A thorough understanding of the role of sea ice as an important component of the Southern Ocean, no matter whether from an oceanographic, climatological or biological perspective, requires quantitative estimates of the relative importance of different ice-growth processes. While remote sensing is an important tool in this context [*Gloersen et al.*, 1992, *Comiso et al.*, 1992, *Drinkwater*, 1995], at present only rudimentary information about ice growth and characteristics can be derived. Given the spatial and temporal restrictions of most field studies, the analysis of sea-ice cores provides important data in this context. Each core represents an archived time series of the interaction between ocean, ice and atmosphere from the initiation of ice formation until the time of sampling. During icebreaker expeditions traversing wider areas of the Antarctic pack ice it is thus possible to obtain a data set with a comparatively high spatial and temporal resolution.

The growth regime of a sea-ice sample is commonly inferred from its textural stratigraphy or stable-isotope composition [*Weeks and Ackley*, 1986, *Gow and Tucker*, 1990, *Souchez et al.*, 1988, *Tison and Haren*, 1989, *Lange et al.*, 1990]. The former is easily determined from thick or thin sections, though at present mostly on a qualitative (identification of textural classes) or semi-quantitative basis. Nevertheless, the simple distinction between granular ice of frazil origin and columnar ice formed through freezing of seawater at the base of the ice [*Gow et al.*, 1987, *Lange and Eicken*, 1991a] provides valuable information about the relative importance of dynamic versus thermodynamic processes. Due to the easily identifiable signature of precipitation (depleted in the heavier deuterium and oxygen-18 isotopes), stable isotopes have been employed to quantify the contribution of snow ice to total ice thickness, since the latter may not be identified based on structural evidence alone [*Lange et al.*, 1990, *Eicken et al.*, 1994a, *Jeffries et al.*, 1994, *Worby and Massom*, 1995]. Furthermore, Souchez and co-workers have demonstrated that the deuterium fractionation upon freezing of freshwater is dependent on the ice growth rate [*Souchez et al.*, 1987] and have extended this approach to sea ice for the analysis of a fast ice core from the eastern Antarctic [*Souchez et al.*, 1988]. Whereas the isotopic signal appears to be preserved in the bulk of the ice, salinity is not as unequivocally interpreted in terms of an ice floe's growth history. Apart from the initial segregation and entrapment of brine at the advancing ice-water interface [*Wettlaufer*, in press], desalination and small-scale variability complicate the picture considerably [*Untersteiner*, 1968, *Cox and Weeks*, 1988, *Eicken*, 1992].

This study aims to characterize the sea-ice growth regime (i.e. modes and rates of growth) in the Weddell Sea based on the integral analysis of ice cores and corresponding upper-ocean salinity and stable-isotope data. The core samples were obtained along a traverse across the Weddell Gyre and a meridional transect through the eastern Weddell Sea during late winter and early spring of 1989 (Figure 1). Ice floes of different ages (young, first-year and second-year), growth history, snow cover and regional origin were analysed. The following issues will be addressed in more detail.

(1) The co-evolution of ice texture, salinity and stable-isotope composition as well as their utility as a record of growth conditions will be addressed through intercomparison of parameters for the same sample. This aspect has received comparatively little attention in the past, since most studies focussed on a single particular process or parameter, e.g., deduction of snow-ice formation from stable-isotope data alone [*Lange et al.*, 1990, *Eicken et al.*, 1994a, *Jeffries et al.*, 1994 and 1997] or derivation of the growth mode from textural analysis [*Gow et al.*, 1982 and 1987, *Lange and Eicken*, 1991a]. In addition to the standard textural classification, high-resolution vertical grain-size profiles have been derived through image analysis of thin sections.

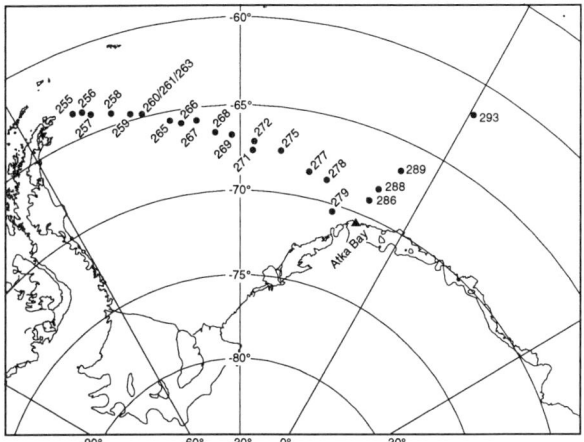

Fig. 1. Study area and sea-ice sampling locations during the Winter Weddell Gyre Study (WWGS 89) with RV *POLARSTERN* in September and October of 1989. The site numbers correspond to the day of the year during which sampling took place.

(2) The second-year ice cores obtained from the western branch of the Weddell Gyre allow for a closer scrutiny of first- and second-year ice contrasts with respect to structure and different properties. While the distinction between ice of different age is at the heart of sea-ice nomenclature [*World Meteorological Organization*, 1985] and an important component of Arctic field and remote-sensing programs, it has not been addressed in great detail in the Antarctic.

(3) The stable-isotope composition seems to hold considerably more promise for the deduction of ice growth modes and growth rates, than either ice texture or salinity. *Souchez et al.* [1987, 1988] demonstrated a growth rate-fractionation dependence for HDO in freshwater and sea ice. Here, this analysis is extended and refined in order to explain and model fractionation of oxygen-18 and deuterium as a function of growth rate for sea ice in general. Isotope and salinity data from Antarctic pack ice and the underlying water column at the end of the growth season are of considerable value in this context. Wind mixing and thermo-haline convection have homogenized the composition of the surface mixed layer. In contrast with conditions near the Antarctic coast or throughout the Arctic, with steep gradients in the isotopic and salinity structure of under-ice waters due to freshwater input from melting ice shelves or rivers, the offshore Weddell Sea mixed layer exhibits only minimal variability [*Schlosser et al.*, 1989, *Fahrbach et al.*, 1995, *Weppernig et al.*, 1996].

(4) The integrated analysis of stable-isotope and salinity data of these samples allows for an extension of the fundamental studies carried out on segregation processes at the advancing ice-water interface and the distribution of salt between ice and water either in the laboratory [*Weeks and Lofgren*, 1967, *Cox and Weeks*, 1975, *Wettlaufer*, in press] or in Arctic landfast ice [*Nakawo and Sinha*, 1981] to the Antarctic ice pack. It will be shown, that isotopic fractionation may be considered a touchstone for the applicability of the classical Burton-Prim-Slichter (BPS) theory [*Burton et al.*, 1953, *Weeks and Lofgren*, 1967] to the segregation of sea-salt ions at the advancing ice-water interface.

(5) Finally, based on relationships derived between fractionation coefficient and growth rate, a one-dimensional thermodynamic growth and salt-flux model will be extended to simulate the isotopic composition of growing sea ice. The combined analysis of field data and model simulations allows for an assessment of the growth modes and rates as well as the relevant heat fluxes controlling the growth of ice in the Weddell Sea.

Thus, this paper is one of a growing number of contributions, some of which have been referred to above or can be found in this volume, demonstrating the complexity of Antarctic sea-ice growth and decay. Yet, apart from emphasizing hemispheric differences, the shortcomings of this study may also be regarded as a plea for a closer, microscopic look at the boundary zone where seawater turns into sea ice.

2. MATERIAL AND METHODS

The data presented here were collected during the Winter Weddell Gyre Study 1989 (WWGS 89) aboard the icebreaker *POLARSTERN* [*Augstein et al.*, 1991]. The ship followed a transect across the Weddell Gyre and along the 0° meridian during September and October 1989 (Figure 1). The sea-ice program comprised ice observations, measurements of ice thickness and snow depth, core analysis, remote sensing, and biological studies. Other aspects of parts of the data set have already been discussed in previous publications by *Lange and Hubberten* [1992] and *Eicken et al.* [1994a]. In this study, data from 31 ice cores (22 first-year ice, 9 second-year ice) obtained from level ice at 28 different sites (Table 1) as well as measurements on liquid brine separated from the ice will be analysed. The cores were drilled with a fiberglass-barrel corer of 0.1 m diameter. Ice thickness, snow depth and freeboard were measured at and in the vicinity of the sites. Ice temperatures were recorded by drilling holes into the core contained in an insulated tube and inserting a Testotherm temperature sonde (accuracy 0.1 K). The main cores were immediately transferred to a cold laboratory (< −25 °C) aboard the ship for textural analysis on thick and thin sections. After being cut according to the textural stratigraphy, segments were split lengthwise. By aligning segments between breaks within the core the azimuthal orientation of vertical sections was kept the same throughout the core. Vertical and horizontal thin sections were recorded with a Hamamatsu C2400 high-resolution video camera under standardized conditions between crossed polarizers for the derivation of textural parameters through automated image analysis (width of images 50 mm for horizontal and 100 mm for vertical sections, see *Eicken* [1993] for details). After digitization and low-pass filtering, pore signals were removed through non-linear morphological filters. Grain boundaries were delineated by a Sobel edge-detection filter. Here, downcore profiles of the horizontal intercept length, i.e., the apparent cross-section of grains in a vertical section, are reported (vertical resolution of 0.2 mm).

Salinity was measured with a WTW LF2000 conductivity sensor on melted sections (roughly half of core cross-section, accuracy ±0.5 % of the measured value). Approximately one quarter to one third of each section was shipped in sealed 0.2 mm polyethylene bags to the Alfred Wegener Institute (AWI) at –30 °C for stable-isotope analysis. Brine samples were obtained either from core holes drilled to specific depths or by centrifuging samples under in-situ temperatures as described by *Weissenberger et al.* [1992]. Samples were transferred in wax-sealed glass bottles at 4 °C for measurements of salinity and $H_2^{18}O$ at AWI. Salinity measurements with a CTD sonde and stable-isotope data of water from the uppermost 10 to 50 m of the water column are also reported (for details with respect to measurements see original report by *Fahrbach et al.* [1995]).

At the AWI, $^{18}O/^{16}O$ ratios were determined on ice and water samples, employing an automatic water-carbon dioxide equilibration system, connected on-line to a Finnigan MAT Delta S mass spectrometer. Measurements were carried out against a laboratory standard, calibrated against IAEA Vienna Standard Mean Ocean Water (VSMOW) and Standard Light Antarctic Precipitation (SLAP). For routine measurements the accuracy is <0.05 ‰ for ice and <0.03 ‰ for water samples in the δ-notation. The δ-notation indicates the ratio of $^{18}O/^{16}O$ in the sample s relative to VSMOW as

$$\delta^{18}O = \left[\frac{\left(^{18}O/_{16}O\right)_s}{\left(^{18}O/_{16}O\right)_{VSMOW}} - 1 \right] \cdot 1000‰.$$

3. TEXTURAL STRATIGRAPHY, SALINITY AND δ¹⁸O OF SEA-ICE SAMPLES

3.1. General Overview of Ice Properties

Mean values of textural parameters, salinity and $\delta^{18}O$ integrated over the entire length of each of the cores are shown in Table 1. The mean core lengths of 0.73 and 1.55 m for first- and second-year ice correspond closely to thickness measurements along 100-m profiles [*Wadhams et al.*, 1987, *Lange and Eicken*, 1991b, *Eicken et al.*, 1994a]. A detailed outline of the distinction between first- and second-year ice is given by *Eicken et al.* [1991, section 3.1] and *Eicken et al.* [1994a]. The main criteria are level ice thickness and snow depth as determined along longer thickness profiles (to exclude deformed ice of greater local thickness), the internal structure of the snow cover [*Eicken et al.*, 1994a], the presence of clear ice layers at the ice surface (see section 3.2 below), and the occurrence of internal chlorophyll maxima within the ice cover along with other biological indicators (for details see *Gleitz et al.* [in press]). As is shown for six cores taken in second-year ice (sampling location 260, Figure 1) by *Eicken et al.* [1991], the bottom of the ice layer grown in the first year in most cases corresponds to a pronounced local maximum in the chlorophyll concentration and a corresponding discontinuity in the salinity profile just below. This local maximum is the result of algal growth at the bottom of the ice during the summer months. The same is apparent for core 25501 (Figure 4), where an increase in salinity and a slight decrease in $\delta^{18}O$ occur just below a local maximum in chlorophyll concentration at 1.23 m depth (not shown in Figure). Except where specifically indicated (i.e. Tables 5 and 6), the term "second-year ice" implies that the sampled ice floe had survived one summer; an ice core taken from such a floe consists of an upper ice layer dating from before the summer and a lower layer formed after the summer.

The fraction of columnar ice within cores is slightly higher (10 to 20 %) compared to previous observations in the Weddell Sea [*Lange and Eicken*, 1991a]. This is partly a result of intensive sampling of an area composed predominantly of congelation ice in the eastern Weddell Sea (cores 28601 to 28661, with an average fraction of 74 % columnar in all cores; for nomenclature of ice types refer to *Weeks and Ackley* [1986], *Gow et al.* [1987] and *Eicken and Lange* [1989]). In addition, this may reflect interannual variability, as is also suggested by the differences between first- and second-year cores. With a mean salinity of 6.0 ‰, the first-year ice is somewhat more saline than second-year cores from the western Weddell Sea (significant at the 5-% level) and slightly less saline than Arctic sea ice of comparable thickness sampled during winter [*Cox and Weeks*, 1974, *Weeks and Ackley*, 1986]. The salinity-depth plot of all individual measurements shown in Figure 2a displays high values at the top surface, with near-constant salinities around 4 to 5 ‰ in deeper layers, and a slight increase in the bottom layers, i.e., the C- or S-shaped salinity profile typical of Antarctic sea ice [*Eicken*, 1992]. Most of the values deviating significantly from the clustered data points are composed of granular ice.

These deviations are also apparent in the isotopic composition of the ice (Figures 2 and 3, Table 2). The values found at the top of cores up to a depth of 0.9 m,

TABLE 1. WWGS 89 Sea-Ice Core Parameters (Length L, Fraction of Columnar Ice f_c, Mean Thickness of Stratigraphic Units z_{su}, Mean Salinity S, Mean $\delta^{18}O$ δ_m, Salinity of Bottom Ice Layer S_{bt}, $\delta^{18}O$ of Bottom Ice Layer δ_{bt}), $\delta^{18}O$ of Surface Seawater at Corresponding Sampling Location δ_w, and Derived Fractionation Coefficient ε_{eff}

Core	L, m	f_c, %	z_{su}, m	S, ‰	δ_m, ‰	S_{bt}, ‰	δ_{bt}, ‰	δ_w, ‰	ε_{eff}, ‰
First-year Sea Ice									
25901	0.82	71	0.09	5.9	0.94	3.8	2.14	−0.28	2.42
26501	0.96	0	0.96	5.1	−2.72	3.3	−0.76	−0.29	(−0.47)
26601	0.90	70	0.30	6.8	0.00	5.5	0.86	−0.32	1.18
26801	0.42	32	0.14	5.4	−0.34	4.8	1.84	−0.32	2.16
26901	0.64	10	0.13	4.5	1.91	3.6	2.39	−0.27	2.66
27101	0.56	30	0.11	4.3	1.49	2.6	2.30	−0.32	2.62
27111	0.66	14	0.22	6.1	1.13	1.8	2.03	−0.32	2.35
27121	0.88	44	0.18	5.4	0.56	2.3	2.16	−0.32	2.48
27201	2.27	18	0.16	4.3	0.89	5.3	1.87	−0.32	2.19
27501	0.52	54	0.09	6.2	1.78	4.9	2.27	−0.26	2.53
27701	0.60	83	0.15	6.6	1.56	18.9	−0.96	−0.24	(−0.72)
27801	0.93	75	0.19	6.0	1.16	4.8	2.08	−0.30	2.38
27901	0.54	56	0.11	8.0	1.39	6.5	2.13	−0.22	2.35
28601	0.43	72	0.22	4.8	1.57	4.6	2.01	−0.28	2.29
28602	0.44	86	0.22	6.7	1.20	7.1	1.32	−0.28	1.60
28611	0.50	66	0.12	7.0	1.37	6.0	1.84	−0.28	2.12
28622	0.56	59	0.07	8.3	1.27	4.3	1.67	−0.28	1.95
28631	0.68	65	0.10	7.1	1.44	2.5	2.05	−0.28	2.33
28661	0.43	98	0.22	7.4	1.57	4.2	1.99	−0.28	2.27
28801	1.03	71	0.52	4.9	1.58	3.1	1.26	−0.28	1.54
28901	0.68	82	0.34	5.7	1.42	3.2	1.83	−0.28	2.11
29301	0.67	90	0.22	5.3	1.72	5.4	1.98	−0.28	2.26
Mean	0.73	57	0.22	6.0	1.04	4.9	1.65	−0.29	2.19
σ	0.39	28	0.19	1.1	1.00	3.4	0.89	0.03	0.37
Second-year Sea Ice									
25501	2.07	8	0.26	2.4	−0.39	2.4	1.80	−0.30	2.10
25701	2.40	32	0.16	4.2	1.80	3.6	2.28	−0.27	2.55
25801	1.50	9	0.15	5.0	−0.55	4.4	1.32	−0.13	1.45
26001	1.39	72	0.15	6.2	0.07	2.7	2.16	−0.28	2.44
26003	1.19	73	0.20	5.1	0.54	2.4	2.07	−0.28	2.35
26005	1.24	58	0.21	7.0	0.62	2.4	2.24	−0.28	2.52
26101	1.24	69	0.16	5.5	0.19	2.5	2.07	−0.28	2.35
26301	1.63	39	0.08	3.2	0.30	3.6	1.77	−0.29	2.06
26311	1.29	74	0.26	5.0	1.01	6.6	1.49	−0.29	1.78
Mean	1.55	48	0.18	4.8	0.40	3.4	1.91	−0.27	2.18
σ	0.42	27	0.06	1.4	0.71	1.4	0.34	0.05	0.37
All Cores									
Mean	0.97	54	0.21	5.7	0.86	4.5	1.73	−0.28	2.19
σ	0.54	28	0.17	1.3	0.96	3.0	0.77	0.04	0.37

which are significantly lower than the composition of the surface seawater with a $\delta^{18}O$ of −0.28 ‰ (Table 1), indicate a substantial snow component in the ice. The same is true for the few samples at the bottom, which all represent snow ice rafted to greater depth in core 27201. The isotopic composition indicates to what extent a sample is of marine or meteoric origin. Due to fractionation upon evaporation and precipitation, snow or rain precipitating from an air mass is progressively depleted in the heavy oxygen isotope, ^{18}O, with respect

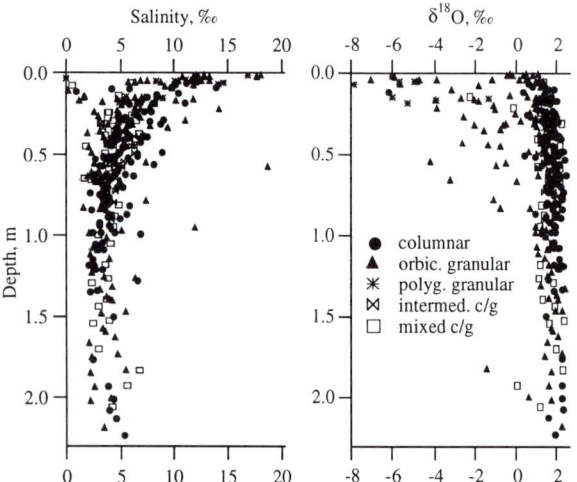

Fig. 2. Salinity and $\delta^{18}O$ plotted vs. mid-depth of individual core segments, with different symbols denoting the five main textural classes; four data points with $\delta^{18}O < -8$ ‰ have been omitted to increase the clarity of presentation, two of these correspond to the upper layers of core 25501 shown in Figure 4.

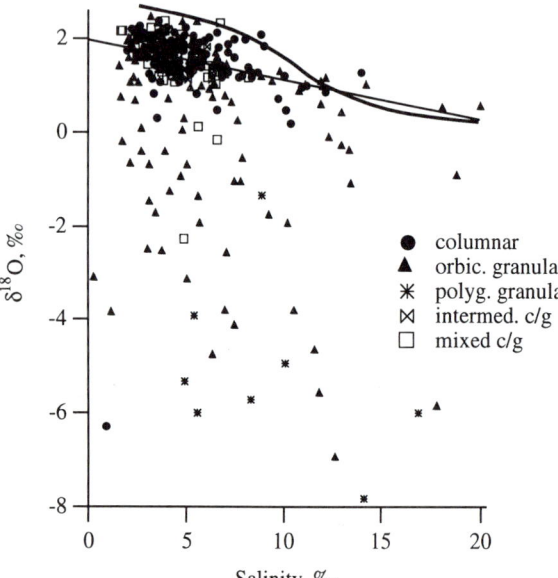

Fig. 3. Plot of $\delta^{18}O$ vs. salinity for all core segments, with different symbols denoting the five main textural classes (four data points with $\delta^{18}O < -8$ ‰ have been omitted to increase the clarity of presentation, two of these correspond to the upper layers of core 25501 shown in Figure 4). The linear least-squares fit for all data points with $\delta^{18}O > -0.28$ ‰ is shown by the thin, solid line ($\delta^{18}O = 1.94 - 0.085\ S$, $r^2 = 0.19$). Note that most of the segments exhibiting larger deviations from the mean are composed of granular ice. The thick upper line corresponds to the initial salinity S_0 and isotopic composition of sea ice computed from the fractionation model (derived from equations (19), (20) and (22)).

to the oceanic source area at lower latitudes. Owing to reservoir effects, the ocean's composition does not deviate significantly from a value of 0 ‰ (or roughly –0.3 ‰ for Antarctic surface waters), whereas snow deposited in the Weddell Sea region varies in $\delta^{18}O$ between –3 and –28 ‰ (mean –16 ‰) depending on latitude and season [*Eicken et al.*, 1994a]. Consequently, snow ice formed through flooding of the ice surface and the snow cover can be identified through a corresponding isotopic signal [*Lange et al.*, 1990, *Eicken et al.*, 1994a, *Jeffries et al.*, 1994, *Worby and Massom*, 1995]. With the exception of one anomalous data point, all samples with $\delta^{18}O$ values <0 ‰ are composed of polygonal or orbicular granular ice, including a granular component in samples of deformed mixed columnar/granular ice (Figure 3 and Table 2). The largest snow component (roughly half of the total ice volume, Table 2) is found in polygonal granular ice

TABLE 2. Sample Depth, Thickness, Salinity and $\delta^{18}O$ of Individual Stratigraphic Units of WWGS 89 Sea Ice Samples

Textural class, Parameter	Mean	σ	Min.	Max.	n
Columnar					
Depth (m)*	0.64	0.44			141
Thickness (m)	0.25	0.22			52
Salinity (‰)	5.0	2.2	0.9	13.9	141
$\delta^{18}O$ (‰)	1.61	0.78	–6.28	2.39	141
Orbicular granular					
Depth (m)*	0.60	0.58			115
Thickness (m)	0.17	0.23			59
Salinity (‰)	6.2	4.0	0.4	20.1	115
$\delta^{18}O$ (‰)	0.21	1.98	–6.99	2.43	115
Orbicular granular samples with $\delta^{18}O > -0.28$ ‰					
Salinity (‰)	5.7	3.7	1.7	20.1	81
$\delta^{18}O$ (‰)	1.28	0.60	–0.26	2.43	81
Polygonal granular					
Depth (m)*	0.09	0.06			11
Thickness (m)	0.11	0.09			10
Salinity (‰)	9.1	5.0	0.03	16.9	11
$\delta^{18}O$ (‰)	–7.00	4.14	–17.32	–1.34	11
Mixed columnar/granular					
Depth (m)*	0.75	0.57			42
Thickness (m)	0.07	0.06			37
Salinity (‰)	4.4	1.6	0.6	8.2	42
$\delta^{18}O$ (‰)	1.07	2.39	–13.24	2.37	42
Intermediate columnar/granular					
Depth (m)*	0.54	0.15			21
Thickness (m)	0.21	0.11			9
Salinity (‰)	4.6	0.9	3.2	6.3	21
$\delta^{18}O$ (‰)	1.52	0.23	1.12	1.86	21

* Mean depth in core of centerpoint of textural segment

(i.e. isometric grains with planar boundaries contrasting with the rounded shape of orbicular granular, see also more detailed discussion in section 3.2). Since the determination of the meteoric ice fraction for the same data set was the subject of a study by *Eicken et al.* [1994a], the topic will not be addressed in detail here.

In the salinity-$\delta^{18}O$ plot, most data points fall within the salinity range of 3 to 8 ‰ and corresponding $\delta^{18}O$ values of 1 to 2.5 ‰ (Figure 3). The differences in salinity and isotopic composition for ice belonging to different textural classes are apparent from the compilation in Table 2. Of particular interest are the comparatively small differences between columnar and orbicular granular ice. The set of granular ice includes snow ice with very low $\delta^{18}O$ and comparatively high salinities as a result of salt retainment after flooding [Figures 2 and 3, see also *Eicken*, 1992, *Eicken et al.*, 1994a]. Upon removal of data points below the mean seawater composition with $\delta^{18}O < -0.28$ ‰, which still leaves a fraction of snow ice undetected, the contrast between columnar and granular ice diminishes even further (Table 2).

3.2. Properties of Four Exemplary Sea-Ice Cores

In this section, four exemplary sea-ice cores typifying second-year ice (Figure 4), first-year ice predominantly of granular (Figure 5) and of columnar texture (Figure 6) as well as young ice (Figure 7) will be discussed in more detail.

Second-year ice (core 25501, Figure 4). The core was obtained from a second-year floe in the northwestern Weddell Sea (Figure 1), with a mean ice thickness of 1.93 ± 0.57 m and a mean snow depth of 0.58 ± 0.20 m. The core consists almost exclusively of granular and mixed columnar/granular ice, with a thin layer of columnar ice just below the upper surface. The extremely low $\delta^{18}O$ values and salinities (< −10 and <1 ‰, respectively), indicate that a significant fraction of the uppermost ice originates from meteoric water. Unlike ordinary snow ice, however, the topmost layer of clear, mostly bubble-free polygonal granular ice (Figure 4) most likely formed through congelation of snow meltwater at the base of the snow pack. This is also indicated by the isotopic composition ($\delta^{18}O = -17.3$ ‰), which corresponds closely to the average composition of pure Weddell Sea snow of −16.2 ‰ [*Eicken et al.*, 1994a], and the low salinity of 0.03 ‰. Similarly, the congelation ice below may represent a meltwater layer that refroze at the surface of the ice cover. Downward percolation of meltwater within the snow, formation of internal melt layers and refreezing at the base of the snow pack (referred to as superimposed ice) has been reported for different fast-ice sites in the eastern Antarctic [*Panov and Fedotov*, 1977, *Ishikawa and Kobayashi*, 1985, *Kawamura et al.*, 1993] as well as for the Bellingshausen and Amundsen Seas [*Jeffries et al.*, 1997], i.e., in areas with high accumulation rates and extensive surface melting during the summer months. Of all 9 WWGS 89 second-year cores, superimposed ice was only encountered in this one sample, corresponding to a relative contribution of roughly 1 % to the total volume of second-year ice.

Below these surface layers, $\delta^{18}O$ and salinity increase to values typical of ordinary sea ice. Apart from direct entrainment of snow, downward percolation of snow and ice meltwater may have affected the isotopic and salinity signal down to a depth of roughly 0.5 m (see also section 5.1). The asymptotic increase in $\delta^{18}O$ with depth is a feature common to most of the cores analysed in this and other studies [e.g., *Lange et al.*, 1990, *Worby and Massom*, 1995]. Both salinity and the mean chord length are much more variable than the stable-isotopic composition in the lower half of the core. The homogeneous thickness of level ice at the site, the undisrupted isotopic profile and the absence of deformation zones in the textural stratigraphy indicate that all the ice at this site originated from accretion of frazil ice from below. The layers with large differences in grain size and shape (up to a factor of >2 in the mean chord length) may be an indication of the episodic nature of these events.

First-year ice of predominantly granular texture (core 27111, Figure 5). This core was obtained from an ice floe of mean thickness 1.49 ± 0.91 m with a mean snow depth of 0.26 ± 0.10 m, from the center region of the Weddell Gyre (Figure 1). The core is predominantly composed of granular ice with a columnar ice inclusion at roughly 0.5 m depth. The inclination of the columnar crystals (Figure 5) indicate this to be a rafted or deformed segment embedded in a matrix of granular ice. The fact that neither the salinity nor the stable-isotope profile deviate from the typical decrease or increase with depth may in part be attributed to the comparatively poor resolution of the core segments. Apart from the increase associated with the columnar inclusion, the granular ice exhibits only small variability with respect to grain size.

First-year ice of predominantly columnar texture (core 27801, Figure 6). This core was taken at the eastern margin of the Weddell Gyre near the transition to the coastal current (Figure 1), from a floe of mean thickness 0.90 ± 0.17 m with 0.27 ± 0.14 m of snow. Apart from a few centimeters of granular and mixed granular/columnar ice at the very surface, the ice is exclusively of columnar texture. Both the vertical dimensions and the horizontal grain cross sections display distinct layering. The increase in grain size from 0.56 to 0.62 m and from 0.64 to 0.8 m occurs gradually, with grains strongly interlocking in the vertical and with no corresponding changes in isotopic compo-

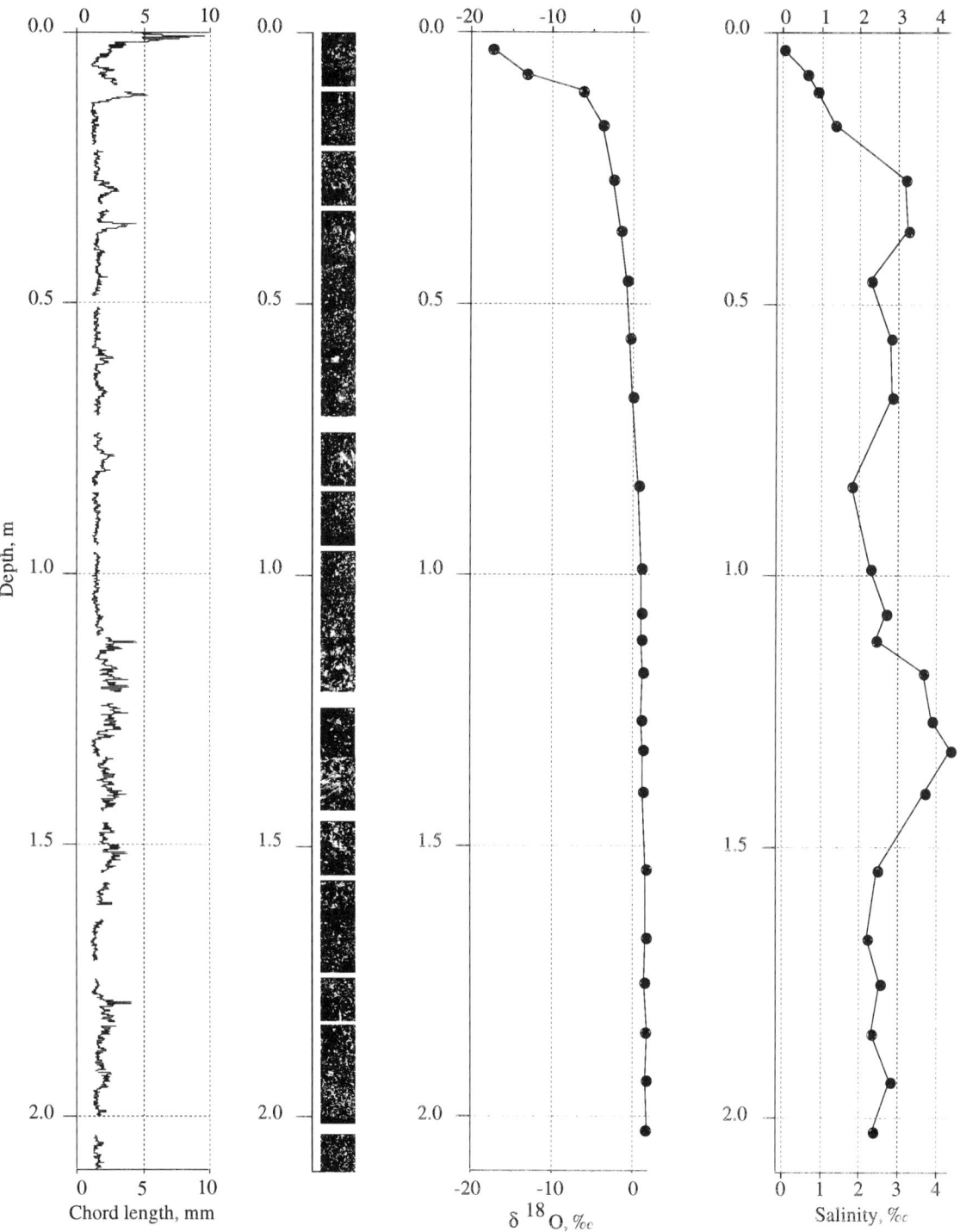

Fig. 4. Vertical profiles of mean chord (intercept) length (left) across the width of grains visible in the vertical thin sections (middle left), $\delta^{18}O$ (middle right) and salinity (right) for second-year ice core 25501. The thin sections have been recorded between crossed polarizers, providing an indication of the textural characteristics of the ice (shown to scale, gaps between sections correspond to missing data).

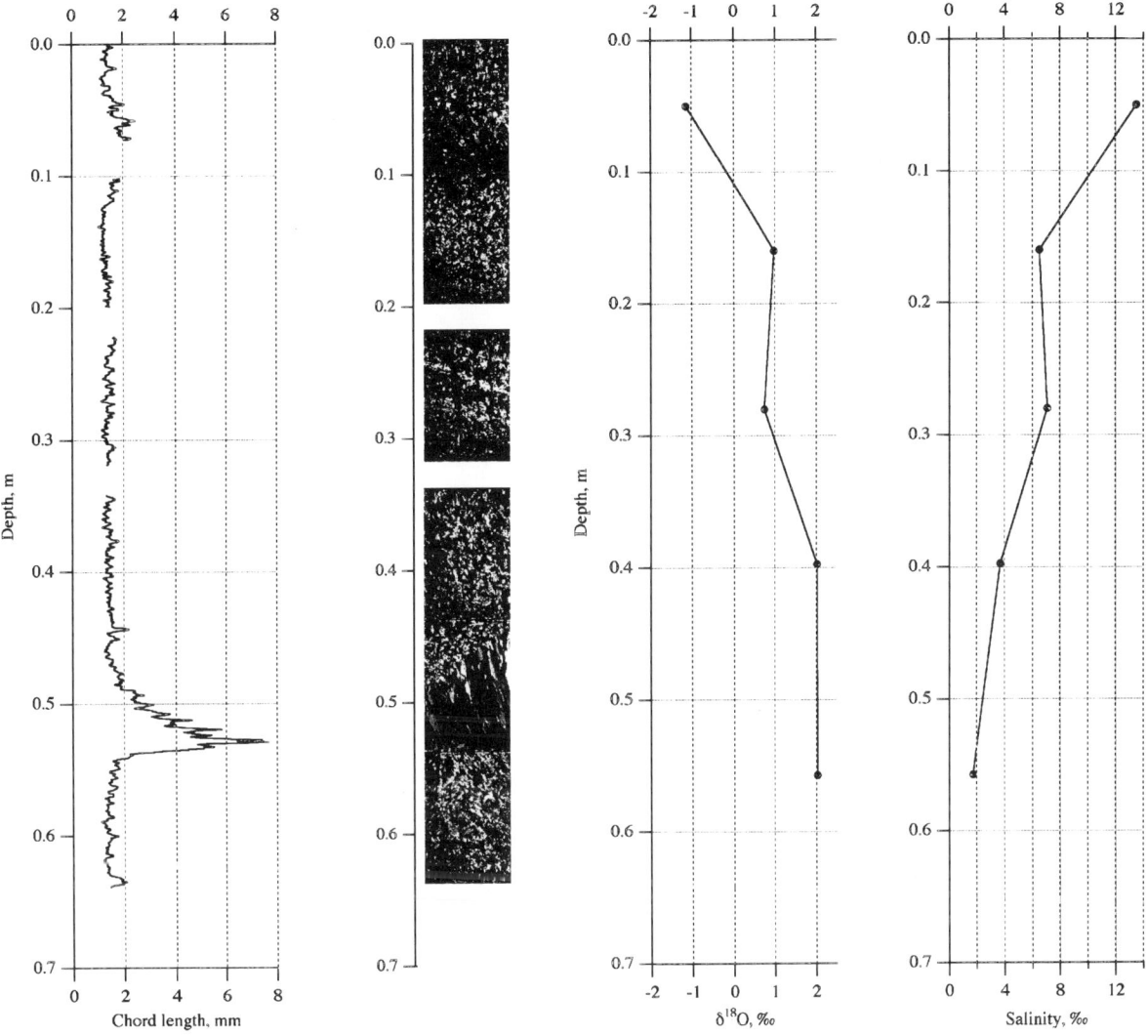

Fig. 5. Vertical profiles of mean chord (intercept) length (left) across the width of grains visible in the vertical thin sections (middle left), $\delta^{18}O$ (middle right) and salinity (right) for first-year ice core 27111. The thin sections have been recorded between crossed polarizers, providing an indication of the textural characteristics of the ice (shown to scale, gaps between sections correspond to missing data).

sition or salinity. The decrease at 0.30 m depth is a true discontinuity (coinciding with a break in the core). The vertical profiles of all three parameters, chord size, $\delta^{18}O$ and salinity, show the same pattern from 0 to 0.30 m depth and from 0.30 to 0.60 m, with an increase in chord size and $\delta^{18}O$ corresponding to a decrease in salinity. This is a clear indication of a rafting event, where two adjacent floe segments have been stacked on top one another. While the $\delta^{18}O$ values at 0.05 and 0.33 m depth correspond closely to one another (–0.68 and –0.60 ‰), the salinities show the same trend, yet, the absolute values deviate significantly (12.3 and 8.0 ‰), most likely due to subsequent desalination [*Eicken*, 1992].

Young ice (core 28661, Figure 7). Although this core obtained near the coast at Atka Bay (Figure 1) appears quite similar to the one shown in Figure 6, the ice has a shorter growth history, indicated by the smaller thickness (mean of 0.44 ± 0.03 m for the floe) and the thin snow cover (mean of 0.04 ± 0.01 m). Ice observations and the analysis of remote-sensing data indicate this to be a patch of young ice formed in a divergent regime over an area of roughly 150 km extent along the cruise track (see also detailed discussion in section 6). The cores 28601 to 28661 listed in Table 1 were all obtained from this ice field. While salinity and grain size vary considerably down-core, the increase in $\delta^{18}O$ from 1.09 ‰ at the top to 1.99 ‰ at the bottom

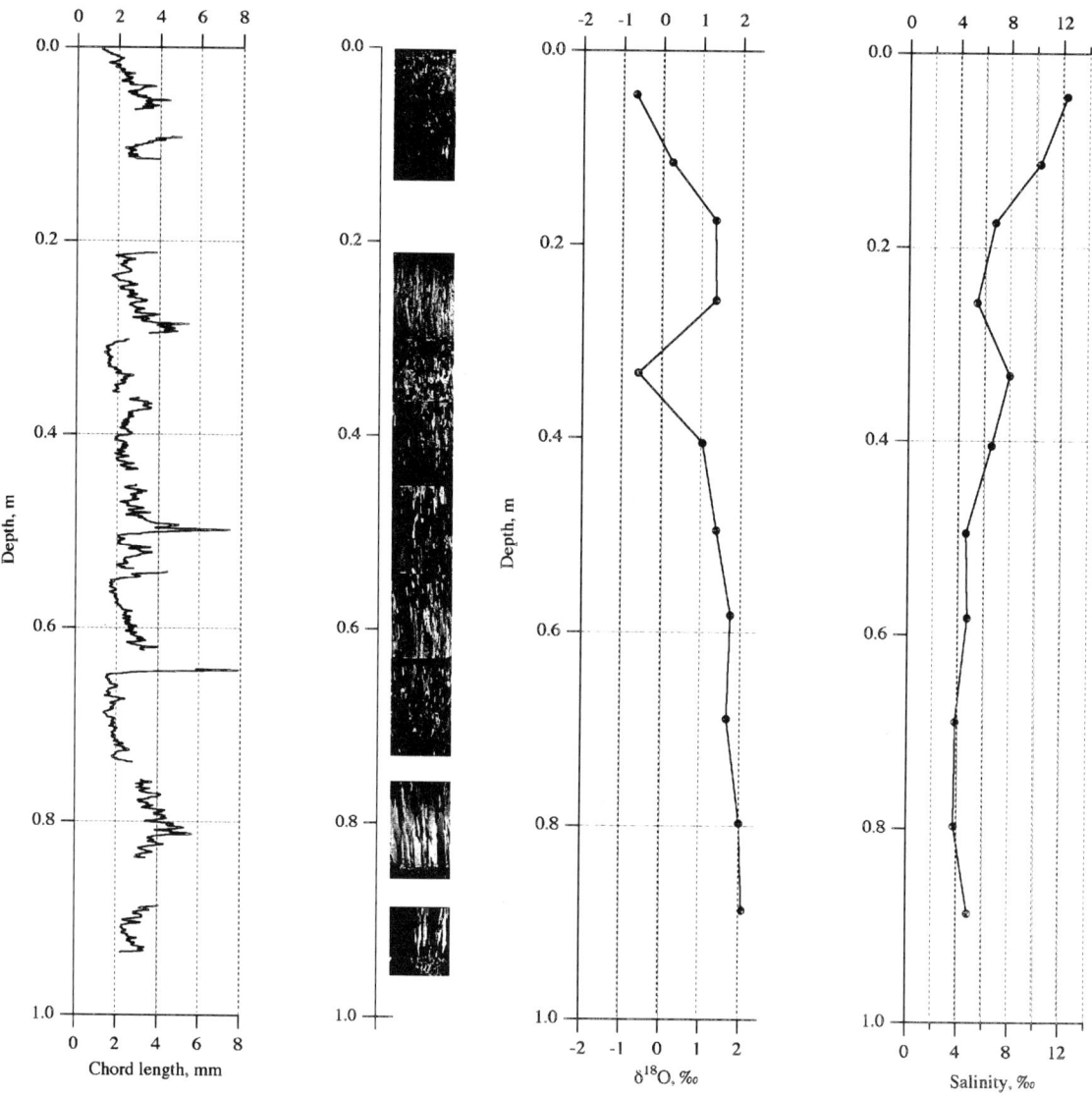

Fig. 6. Vertical profiles of mean chord (intercept) length (left) across the width of grains visible in the vertical thin sections (middle left), $\delta^{18}O$ (middle right) and salinity (right) for first-year ice core 27801. The thin sections have been recorded between crossed polarizers, providing an indication of the textural characteristics of the ice (shown to scale, gaps between sections correspond to missing data).

is comparatively small, which is common to all cores from the area.

3.3. Bulk Properties, Mean Salinity and $\delta^{18}O$ Profiles: Contrasts Between First- and Second-Year Sea Ice

While the largest differences with respect to mean core parameters for first- and second-year ice are found for ice thickness (Table 1), the contrast with respect to mean core salinity or $\delta^{18}O$ is remarkably small. The averaged salinity profiles (Figure 8) exhibit the same shape for first- and second-year ice with differences due mainly to the larger fraction of low-salinity ice in the lower sections of the latter. The mean normalized $\delta^{18}O$ profiles correspond closely below a relative depth of roughly 0.2 (Figure 9b). The major differences at the top are due to the larger fraction of snow ice present in second-year ice. The disproportionate accumulation of snow on ice that has survived a summer season enhances flooding and subsequent snow-ice formation [*Eicken et al.*, 1994a and 1995a]. These processes are also responsible for the surprisingly small differences in mean surface salinities between first- and second-year ice

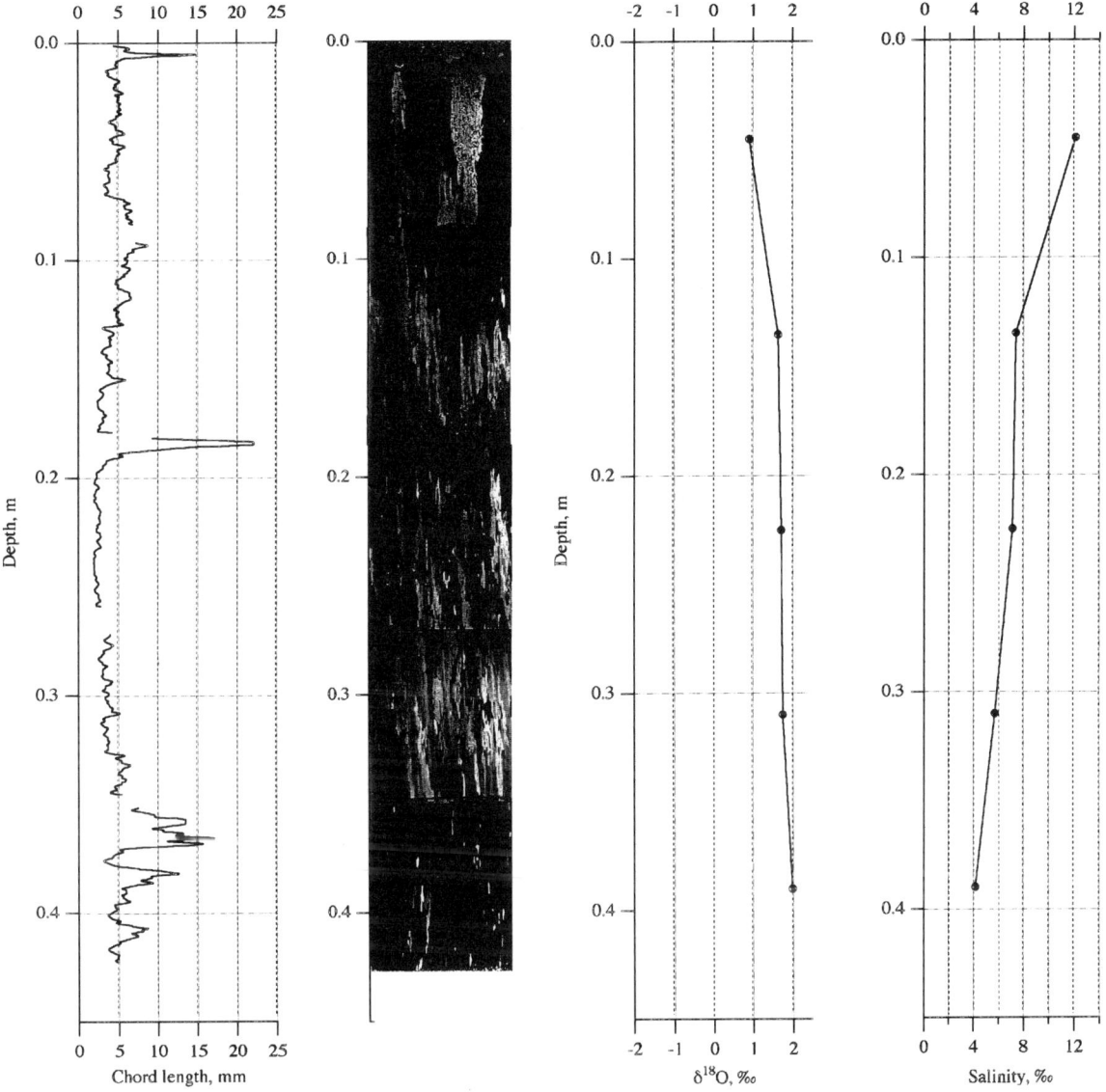

Fig. 7. Vertical profiles of mean chord (intercept) length (left) across the width of grains visible in the vertical thin sections (middle left), $\delta^{18}O$ (middle right) and salinity (right) for young-ice core 28661. The thin sections have been recorded between crossed polarizers, providing an indication of the textural characteristics of the ice (shown to scale). Note that the fine-grained layers apparent at the bottom of the sections at 0.28 and 0.34 m are artifacts due to sample preparation.

(Figure 8), as a larger fraction of salt is retained during snow-ice formation. This counterbalances the decrease in salinity due to drainage or meltwater percolation (core 25501, Figure 4) and results in a twofold increase of the standard deviation of surface salinities for second-year as compared to first-year ice (Figure 8).

While the variability of ice structure, salinity and isotopic composition is enhanced as the ice passes through the summer into the next growth season, contrasts between different textural classes for the two age groups are actually diminished (Table 3). This is true both for the upper ice layers, the only part of a second-year floe that actually dates back to before the previous summer, and the bottom part of the core, which may be as young as any first-year ice core. The correspondence of the lower 80 % of the ice cover $\delta^{18}O$ profiles (Figure 9b) also testifies to this limited contrast in the properties of different textural classes.

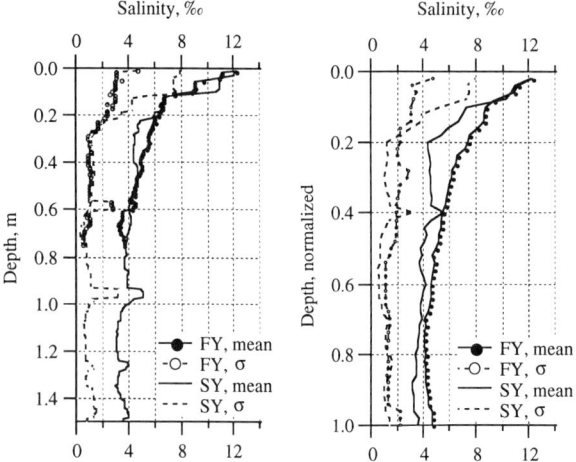

Fig. 8. Mean salinity profiles and standard deviation (σ) for all WWGS 89 cores, computed at 0.01-m intervals for the absolute depth of core segments (left) and for depths with core length normalized to 1 (right).

4. ISOTOPIC FRACTIONATION AND SALT SEGREGATION DURING SEA-ICE GROWTH: THEORY AND MEASUREMENTS

4.1. Equilibrium Isotopic Fractionation During Growth of Freshwater Ice From a Semi-Infinite Reservoir

Due to differences in the molecular vibrational energies, HDO and $H_2^{18}O$ are depleted in the liquid phase during freezing of water under equilibrium conditions. The equilibrium fractionation factor α_{eq} describes the magnitude of this effect for the isotopic ratios in the liquid and solid phase R_s and R_l:

$$\alpha_{eq} = \frac{R_s}{R_l} \quad (1).$$

Alternatively, a fractionation coefficient ε_{eq} describes the offset between the $\delta^{18}O$ values of the solid and the liquid phase:

$$\varepsilon_{eq} = (\alpha_{eq} - 1) \cdot 1000\text{‰} \quad (2).$$

In most cases, the growth of ice from the liquid phase is associated with an advance of the ice-water interface into the liquid reservoir with an interface velocity v_i (Figure 10). At positive, finite values of v_i, fractionation results in the depletion of the heavy isotopic species $H_2^{18}O$ in a layer adjacent to the phase boundary,

with a corresponding shift in the isotopic composition of the ice. The resulting effective fractionation coefficient ε_{eff} depends on the interface velocity, i.e., the ice-growth rate, and the structure of the boundary layer. The equilibrium fractionation coefficient (denoted ε_{eq}^* to distinguish it from the theoretical value) can be derived through extrapolation of laboratory measurements of ε_{eff} to zero growth rates. For fractionation of $H_2^{18}O$ during freezing of pure freshwater, ε_{eq}^* has been determined as 3.0 ± 0.1 ‰ [*O'Neil*, 1968], 2.87 ‰ [*Beck and Münnich*, 1988] and 2.91 ± 0.03 ‰ [*Lehmann and Siegenthaler*, 1991]. Through additional support from theoretical studies [*Lehmann and Siegenthaler*, 1991], a value of $\varepsilon_{eq}^* = 2.91$ ‰ is regarded as the best estimate for pure ice.

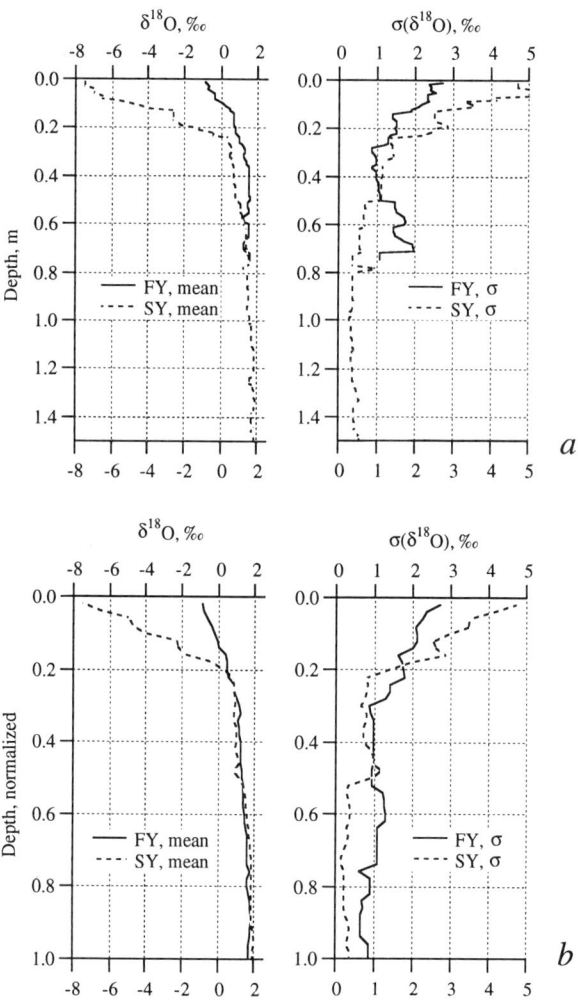

Fig. 9. Mean $\delta^{18}O$ profiles and standard deviation (σ) for all WWGS 89 cores, computed at 0.01-m intervals for (*a*) the absolute depth of core segments and (*b*) for depths with core length normalized to 1.

TABLE 3. Properties of First- (FY) and Second-year (SY) Sea Ice Stratigraphic Units (WWGS 89 Samples)*

Ice Type, Parameter	Depth (m)	Thickness (m)	Salinity (‰)	$\delta^{18}O$ (‰)	n
FY Columnar					
Mean	0.40	0.27	5.7	1.69	92
σ	0.41	0.20	2.3	0.41	(31)
SY Columnar					
Mean	0.93	0.22	3.7	1.49	51
σ	0.39	0.25	1.2	1.18	(21)
FY Granular (excluding polygonal gr.)					
Mean	0.36	0.15	7.7	-0.04	65
σ	0.47	0.20	4.5	2.08	(35)
SY Granular (excluding polygonal gr.)					
Mean	0.87	0.18	4.3	0.56	51
σ	0.60	0.25	2.3	1.79	(24)

* Numbers shown in brackets apply only to thickness measurements

4.2. Effective Isotopic Fractionation During Growth of Freshwater Ice From a Semi-Infinite Reservoir: Results From a Stagnant Boundary-Layer Diffusion Model

Purely diffusive fluxes. The depletion of $H_2^{18}O$ in the boundary layer adjacent to the ice-water interface results in a diffusive influx of the isotopic species from the reservoir (Figure 10). In the absence of other mixing processes, the flux is described according to Fick's law as

$$\frac{\partial C}{\partial t} = D \frac{\partial^2 C}{\partial z^2} + v_i \frac{\partial C}{\partial z} \quad (3)$$

for the one-dimensional case, with growth proceeding at a velocity of v_i in z-direction and C the concentration of the chemical species with a molecular diffusivity of D [*Garandet et al.*, 1994]. The boundary condition at the interface ($z = 0$) is given by

$$-D \left(\frac{\partial C}{\partial z} \right)_0 = v_i (1-k) C_0 \quad (4)$$

and in the far-field ($z \to \infty$)

$$C = C_\infty \quad (5)$$

where k represents a macroscopic segregation or partition coefficient. Generally, the usage of k implies that the chemical species under consideration is either retained within microscopic inclusions in the host phase or occurs as a true solid solution. Only in the latter case does the concept of fractionation (e.g., of water isotopes) denoted by a fractionation factor α or a corresponding coefficient ε apply. The former, general case is described by the segregation coefficient k (e.g., segregation of sea-salt ions). Under steady-state conditions ($\partial C/\partial t = 0$), the solution to the equation system is given by [*Burton et al.*, 1953, *Garandet et al.*, 1994]

$$C(z) = C_\infty \left[1 + \frac{1-k}{k} \exp\left(-\frac{zv_i}{D}\right) \right] \quad (6).$$

Convecto-diffusive fluxes and the boundary-layer concept. In natural systems, solute transport does not occur through molecular diffusion alone but is enhanced by convecto-diffusive fluxes and turbulent mixing in the melt. These processes are commonly integrated into a boundary-layer concept, such that the solute concentration in the liquid takes on the reservoir value C_∞ at a finite distance z_{bl} from the interface [*Garandet et al.*, 1994] (Figure 10). The magnitude of the boundary layer is then defined as

$$z_{bl} = \frac{C_\infty - C_0}{\left(\frac{\partial C}{\partial z} \right)_0} \quad (7),$$

which simplifies to

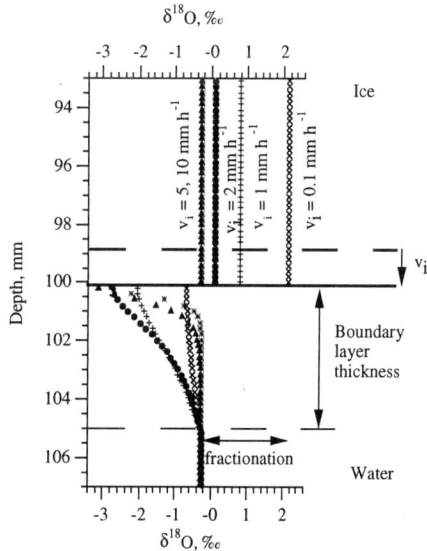

Fig. 10. Isotopic composition of growing freshwater ice and the underlying water layer as derived for different ice-growth velocities v_i from a stagnant boundary-layer fractionation model (boundary layer thickness 5 mm, $\delta^{18}O$ of reservoir water is –0.28 ‰, for details see text).

$$z_{bl} = \frac{D}{v_i} \qquad (8)$$

in the case of pure diffusion. For the computation of the boundary layer thickness, defined with $C = C_\infty$ at $z = z_{bl}$, BPS theory prescribes z_{bl} such that for the steady-state case one obtains [Burton et al., 1953]:

$$C(z) = \frac{C_\infty}{k + (1-k)\exp\left(-\frac{z_{bl} v_i}{D}\right)} \cdot \left[(1-k)\exp\left(-\frac{z v_i}{D}\right) + k\right] \qquad (9),$$

which is equivalent to equation (6) for $z_{bl} = \infty$. At finite growth velocities, the effective (macroscopic) segregation coefficient k_{eff} is given by:

$$k_{eff} = k \frac{C_0}{C_\infty} = \frac{k}{k + (1-k)\exp\left(-\frac{z_{bl} v_i}{D}\right)} \qquad (10);$$

hence, the boundary layer thickness is given by:

$$z_{bl} = \frac{D}{v_i}\left[\ln\left(\frac{1}{k}-1\right) - \ln\left(\frac{1}{k_{eff}}-1\right)\right] \qquad (11).$$

Burton et al. [1953] (referred to as BPS hereafter) introduced the concept of such a boundary layer in their analysis of the impurity concentration in germanium single crystals grown from a melt, i.e., for the case of a solid solution with a planar liquid-solid interface. In their pioneering sea-ice work, Weeks and Lofgren [1967] found that experimental data on the segregation of salt ions during the formation of sea ice are compatible with BPS theory. They derived the boundary layer thickness from a series of laboratory experiments, later extended to a field study in the Arctic [Weeks and Lofgren, 1967, Cox and Weeks, 1975, Nakawo and Sinha, 1981]. Souchez et al. [1987, 1988] based a study of the growth-dependence of deuterium fractionation for freshwater and sea ice on this approach. The aim of this work is to explain stable-isotope profiles in terms of growth-rate dependent fractionation.

The isotopic composition of freshwater ice was computed by solving equation (3) with a finite-difference approximation, similar to the approach taken by Souchez et al. [1987]. Since the diffusivities D of $H_2^{18}O$ and NaCl differ considerably from standard values valid for temperate conditions (2.66 x 10^{-9} m^2 s^{-1} at 20 °C for $H_2^{18}O$ [Gat, 1981] and 1.48 x 10^{-9} m^2 s^{-1} at 25 °C for NaCl [Weast, 1984]) all values for D have been transformed to the respective in-situ temperature based on the Stokes-Einstein relation, with brine viscosities taken from Cox and Weeks [1975]. At –2 °C, $D(H_2^{18}O)$ = 1.20 x 10^{-9} m^2 s^{-1} and D(NaCl) = 6.9 x 10^{-10} m^2 s^{-1}, employing a shape factor of 4 for $H_2^{18}O$ and 6 for NaCl [Levine, 1983]. The $\delta^{18}O$ profiles through an ice sheet and the sub-ice water layer for simulations under different growth velocities are shown in Figure 10. The dependence of the apparent fractionation coefficient ε_{eff} on growth rate and boundary layer thickness z_{bl} is summarized in Figure 11. At lower growth rates ε_{eff} increases towards the theoretical maximum, since the diffusive influx of $H_2^{18}O$ into the boundary layer is able to compensate depletion due to fractionation upon ice growth. For large values of z_{bl}, corresponding to the pure-diffusion case, ε_{eff} is close to 0 except for very small growth rates. For an ice cover of 0.5 m thickness, maximum macroscopic growth rates amount to a few mm h^{-1}. Values of 5 mm h^{-1} or more are only reached during the thickening of very thin new ice. Values well above 10 mm h^{-1}, corresponding to zero fractionation for all but very small boundary-layer thicknesses, are to be expected only for the growth of individual frazil crystals (Hobbs [1974, p. 584], see also section 8.1).

4.3. Deriving Equilibrium and Effective Isotopic Fractionation Coefficients for Sea Ice

Previous studies of equilibrium isotopic fractionation in sea ice. The question whether oxygen and hydrogen isotope fractionation coefficients are the same for freezing of fresh- and seawater often surfaces in the interpretation of sea-ice data. Friedman et al. [1964] discuss deuterium measurements on sea-ice samples and take into account the amount of trapped seawater or brine. Based on experimental work with NaCl solutions, Craig and Hom [1968] give a fractiona-

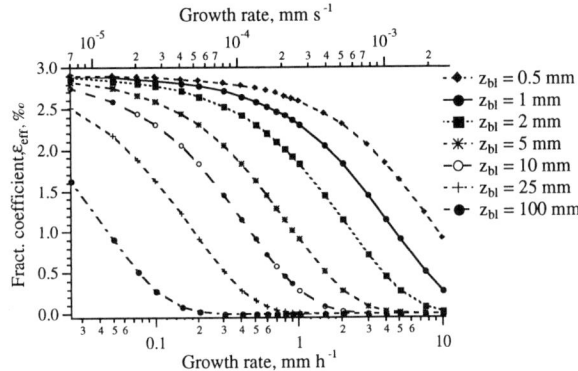

Fig. 11. Dependence of the apparent fractionation coefficient, ε_{eff}, on the growth rate and the boundary layer thickness, z_{bl}, as derived from the stagnant boundary layer model with ε_{eq} = 2.91 ‰.

tion coefficient of $\varepsilon_{eq}* = 2.70$ ‰ for ice grown from seawater.

Due to hydration effects, the activity of $H_2^{18}O$ depends on the concentration of different ionic species in the solution. In their experimental studies covering the range of concentrations encountered in seawater and brine, Taube and co-workers [*Taube*, 1954] did not find any effect of Na^+ on the fractionation of oxygen isotopes in water. *Sofer and Gat* [1972] reach the same conclusion. They do show, however, that various other ions present in seawater and brine may affect fractionation. With respect to sea ice, the hydration effects in solutions of $MgCl_2$ are of particular interest. A concentration of 0.46 M of $MgCl_2$, corresponding to sea-ice brine in equilibrium with ice at roughly –15 °C [*Richardson*, 1976], results in a $\delta^{18}O$ value lower by 0.45 ‰ as compared to pure water. Thus, the concentration of non-hydrating $H_2^{18}O$ in the liquid decreases, resulting in a corresponding effective decrease of the fractionation coefficient upon freezing. This is of little consequence for the bulk ice composition, however.

Effective fractionation coefficients derived from stable-isotope data of sea ice and parent waters. The $\delta^{18}O$ of the upper water column at the sea-ice sampling stations (10 to 50 m water depth) has an average of –0.28 ‰ (Table 1). Except for one location with a surface water $\delta^{18}O$ of –0.13 ‰, the data range between –0.22 and –0.32 ‰. Maximum values of up to –0.05 ‰ occur at water depths between 200 and 1000 m [*Fahrbach et al.*, 1995, *Weppernig et al.*, 1996]. While the $\delta^{18}O$ of surface waters may decrease to below –0.4 ‰ as a result of snow and ice melt during summer or in the vicinity of melting ice shelves, the bulk of the measurements ranges between –0.25 and –0.35 ‰ [*Schlosser et al.*, 1989, *Fahrbach et al.*, 1995, *Weppernig et al.*, 1996]. Since sampling took place at the end of the ice-growth season, the present data set is ideally suited for the derivation of effective fractionation coefficients. Before the onset of snow and ice melt, the composition of the surface waters in the interior Weddell Sea is as constant as one could hope to achieve under laboratory conditions, where vapour-flux and contamination may introduce comparable errors. In contrast, the isotopic composition of Arctic surface waters varies considerably due to continental runoff and ice-growth effects. The water underneath the pack ice varies between –3 and 0.5 ‰ in $\delta^{18}O$ in the central Arctic Basin [*Bauch et al.*, 1995] and between –1 and –16 ‰ along the Beaufort Sea coast [*Macdonald et al.*, 1995].

The fractionation coefficient ε_{eff}, derived according to

$$\varepsilon_{eff} = \delta_{bt} - \delta_w \quad (12)$$

is shown in Table 1 for all samples. Core 26501, composed exclusively of snow ice, and core 27701 with its lowermost segment consisting of rafted ice were excluded from the analysis. For first- and second-year ice ε_{eff} do not differ. Bottom-layer salinity and ε_{eff} are plotted in Figure 12. Bottom samples consisting of granular ice of frazil origin do not exhibit lower ε_{eff} values than congelation ice samples. Measurements by *Melling and Moore* [1995] on ice in the Canadian Beaufort Sea with a mean value of $\varepsilon_{eff} = 2.0$ ‰ do not deviate significantly from the bulk of the Antarctic data (Figure 12), despite different ice-growth conditions. Since finite, non-zero ice-growth rates shift ε_{eff} below the "equilibrium" value, the maximum observed fractionation coefficients are of particular interest. For the WWGS 89 samples, the maximum bottom value is $\varepsilon_{eff} = 2.66$ ‰ (the maximum of all core segments based on a seawater composition of $\delta^{18}O = -0.28$ ‰ equals 2.71 ‰), compared to a maximum ε_{eff} of 2.7 ‰ for the *Melling and Moore* dataset. From their laboratory experiments with NaCl ice, *Craig and Hom* [1968] also deduce a maximum value of $\varepsilon_{eff} = 2.7$ ‰. Summarizing, a value of $\varepsilon_{eff} = 2.70$ ‰ appears to apply to fractionation

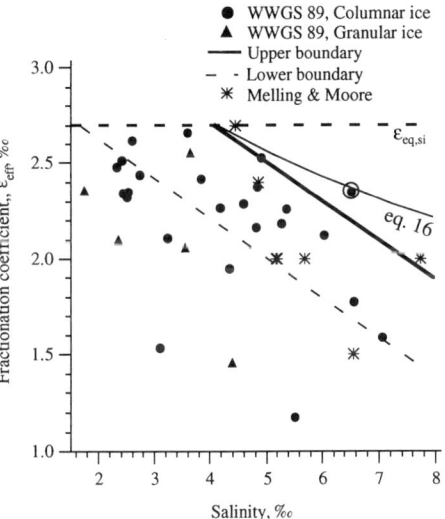

Fig. 12. Fractionation coefficient, ε_{eff}, derived from the isotopic composition of the bottom ice layer and the corresponding surface water value plotted against the salinity of the bottom ice layer. Stars denote values from *Melling and Moore* [1995] for sea ice from the Beaufort Sea. The solid line is the upper limit for the WWGS 89 data set (regression curve determined for 50-% fraction of all data points closest to upper boundary, lower boundary denoted by thin dashed line, with $r^2 = 0.25$, significant at the 5-% level; not including *Melling and Moore* data and the bottom sample consisting of intermediate columnar/ granular ice marked by a circle). The horizontal dashed line corresponds to the equilibrium fractionation coefficient of sea ice of 2.70 ‰. The thin solid line shows the relation between fractionation and segregation coefficients based on equation (16).

between seawater and sea ice for the slowest growth rates realized under natural or laboratory conditions.

4.4. Effective Isotopic Fractionation During Growth of Sea Ice From a Semi-Infinite Reservoir: Results From a Stagnant Boundary-Layer Diffusion Model

Derivation of the salt segregation coefficient and the boundary layer thickness. Determination of the growth-rate dependence of salt segregation or isotopic fractionation as outlined in section 4.2 requires an estimate of the boundary-layer thickness z_{bl}. At present, neither field nor laboratory data are available for z_{bl} with respect to HDO and $H_2^{18}O$. Pending such measurements, the approach taken by *Souchez et al.* [1988] will be followed. Thus, writing equation (11) for both compounds, NaCl and $H_2^{18}O$, incorporating an estimate of $z_{bl}(NaCl)$ based on previous laboratory and field studies, and determining k_{eq}^*, k_{eff} and ε_{eq}^*, ε_{eff}, the equations can be solved for $z_{bl}(H_2^{18}O)$ and v_i. *Souchez et al.* [1988] assumed that the salinity and δD at the bottom of the fast ice core analysed in their study corresponded to "equilibrium" zero-growth rate segregation.

Here, a different approach is taken. With the exception of the study of *Cox and Weeks* [1975], who worked with radioactive ^{22}NaCl, which they could quantitatively analyse in-situ through γ-spectroscopy, previous field and laboratory studies of the initial salt segregation during ice growth have had to cope with two problems. First, brine drainage during ice growth and sampling affects measurements of the segregation coefficient. Second, the minimum growth rates achieved in the laboratory and in the field may still be larger than required by the zero-growth rate condition for the determination of k_{eq}^*. The isotopic composition of the ice, on the other hand, is much less affected by these processes (see also section 5.1). Furthermore, both the initial salinity and the isotopic composition of an ice layer are dictated by the growth rate according to equation (10). Consequently, the ideal situation, i.e., excluding the effects of brine drainage on salinity, is given by an upper limit bounding the cluster of salinity-ε_{eff} data set shown in Figure 12. This limit has been approximated by a linear function:

$$\varepsilon_{eff} = 3.54 - 0.206 S_0 \quad (13).$$

To decrease the sensitivity of this estimate on highly desalinated samples, only those 50 % of all data points located closest to the limit have been considered for the regression model (Figure 12; note also the high fraction of granular samples among the excluded data). The choice of a linear function as delimiter is justified by the near-linear relation between ε_{eff} and k_{eff} in the salinity range from 4 to 7 ‰ (see below and the solid line plotted in Figure 3). This first approach will have to be improved through further data and direct intercomparisons between growth rate and fractionation, however.

With a zero-growth estimate for the fractionation coefficient ε_{eq}^* of 2.70 ‰ arrived at above, the corresponding zero-growth salinity S_0 can be derived from equation (13). With $S_0 = 4.1$ ‰ and a mean seawater salinity of $S_w = 34.45$ ‰ (Table 1), k_{eq}^* is given by:

$$k_{eq}^* = \frac{S_0}{S_w} = 0.12 \quad (14).$$

This estimate is equal to the value of 0.12 derived from field measurements by *Nakawo and Sinha* [1981]. In their laboratory experiments, *Cox and Weeks* [1975] determined a k_{eq}^* of 0.26. In their synthesis of existing data, *Weeks and Ackley* [1986] and *Cox and Weeks* [1988] adhere to the formulation based upon *Nakawo and Sinha* ($k_{eq}^* = 0.12$) and a slope of $z_{bl}/D = 4.2 \times 10^3$ s mm^{-1}.

This approach is only valid for a finite range of ice growth velocities. At true zero growth velocity, a planar ice-water interface is stable and sea ice is free of lamellar salt inclusions. The derivations of this work obviously only apply to the case of an unstable planar interface, with the transition suggested to occur at growth velocities of $<2 \times 10^{-7}$ mm s^{-1} (<10 mm year^{-1}) according to linear stability analysis [*Wettlaufer*, in press]. A further check on the magnitude of k_{eq}^* is provided by sea ice of several hundred years of age sampled at the bottom of the Ross Ice Shelf at J-9 [*Zotikov et al.*, 1980]. The bottommost layer of this ice grown at 10 to 20 mm year^{-1}, amounts to 4.1 ‰ which corresponds to a value of k_{eq}^* of roughly 0.12. Congelation ice at the bottom of the Ward-Hunt Ice Shelf, most likely at least several hundred years of age, shows similar values [*Jeffries*, 1991]. At present, it is not clear, whether smaller k_{eq}^* such as the value of 0.044 determined by *Souchez et al.* [1988] represent an entirely different growth regime or are affected by desalination processes (see also the study by *Tison et al.* [1993] on complex, diffusional exchange processes). In combination with the applied ratio z_{bl}/D of 724.3 s mm^{-1}, the latter estimate yields initial salinities S_0 which appear too low in comparison with other data sets (e.g., $S_0 = 1.8$ ‰ for a growth velocity of 0.8 mm h^{-1}).

Now, from the synthesis of *Weeks and Ackley* [1986] based on *Nakawo and Sinha* [1981], the boundary layer thickness can be derived as

$$z_{bl}(NaCl) = 4.2 \cdot 10^3 D(NaCl) = 2.9 mm \quad (15),$$

with $D(NaCl) = 6.9 \times 10^{-10}$ m^2 s^{-1} at -2 °C as outlined above. Based on equation (11):

$$v_i = \frac{D(NaCl)}{z_{bl}(NaCl)}\left[\ln\left(\frac{1}{0.12}-1\right)-\ln\left(\frac{1}{k_{eff}}-1\right)\right]$$

$$= \frac{D(H_2^{18}O)}{z_{bl}(H_2^{18}O)}\left[\ln\left(1-\frac{1}{1.0027}\right)-\ln\left(1-\frac{1}{\alpha_{eff}}\right)\right] \quad (16).$$

With the diffusivities listed above and $\varepsilon_{eq}^* = 2.70$ ‰ an estimate of the boundary layer thickness $z_{bl}(H_2^{18}O) = 1.3$ mm can be obtained. Thus, the growth rate (in mm s^{-1}) is given by:

$$v_i = 0.92 \cdot 10^{-3}\left[-5.92 - \ln\left(1-\frac{1}{\alpha_{eff}}\right)\right] \quad (17).$$

Lacking direct measurements of the relevant parameters as a function of growth rate, this derivation represents the best estimate, integrating the available field measurements from the Arctic as well as the laboratory results of Cox and Weeks to derive $z_{bl}(NaCl)$ and then using this singular value to determine a corresponding integral value for $z_{bl}(H_2^{18}O)$ for the Weddell Sea data. This estimate is only valid strictly for the zero-growth rate case with the established value of ε_{eq}^*. At higher growth rates, the inclusion of a seawater fraction into the ice matrix leads to increasing deviations from the model. Thus, as shown in Figure 12, equation (16) diverges towards higher values of ε_{eff} with increasing salinity. These effects as well as the sensitivity of the derivation on the estimate of the value of z_{bl} are discussed in more detail below.

Effects of growth-rate dependent inclusion of liquid phase into the ice matrix. The growth-rate dependence of isotopic fractionation formulated in equation (17) only partly takes into account the effects of liquid inclusions on the isotopic composition at zero growth rate, in that $\varepsilon_{eq}^* = 2.70$ ‰. The idealized picture shown in Figure 10 assumes that the actual process of ice growth and hence of fractionation is confined to a planar interface of zero thickness. Ice (Ih), however, does not accomodate the major sea salt ions into its crystal lattice [Hobbs, 1974, Gross et al., 1977]. During ice growth the salt is rejected such that the advancing ice-water interface becomes morphologically unstable and breaks up into a cellular or dendritic interface with intra- and intercrystalline brine layers [Figure 13, Weeks and Ackley, 1986, Wettlaufer, in press]. Thus, the relations derived for fractionation in a stagnant boundary layer only apply to the solid phase. The liquid retained within the pores has to be taken into account separately. As will be shown in section 5.1, the composition of the liquid fraction, and hence of the

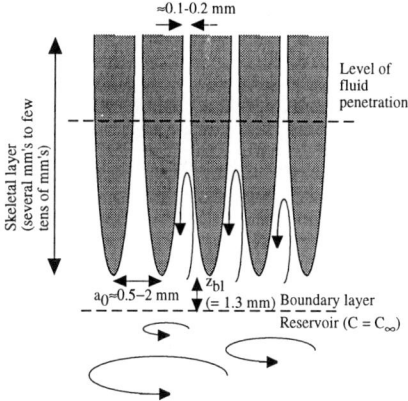

Fig. 13. Schematic drawing of the structure and exchange processes in the bottommost layers of sea ice.

bulk volume of sea ice, is essentially fixed once the porosity drops below a critical value (corresponding to a critical permeability) and the system can be considered half-closed. The derived ε_{eq}^* of 2.70 ‰ corresponds to a seawater/brine volume fraction ($\delta^{18}O = -0.28$ ‰) of roughly 0.07, the remaining 0.93 composed of pure ice with $\delta^{18}O = 2.63$ ‰. Comparisons with the value of 0.13 derived from the analysis of brine samples (section 5.1) and value of 0.12 based on the salt segregation are indicative of inherent discrepancies, which have to be resolved in future studies.

The isotopic effect of liquid inclusions could be corrected based on the bulk ice salinity, as suggested by *Friedman et al.* [1964]. Desalination processes and exchange within the skeletal layer (Figure 13) may produce significant deviations from an ideal mixing curve, however. Furthermore, such an approach may be problematic without a-priori knowledge of initial salinities S_0. To account for a growth-rate dependent inclusion of liquid, it has been assumed that this is best described by the relationship yielding the coefficient $k_{eff,s} = S_0/S_w$ for salt segregation in a newly grown ice layer (based on equation (10)):

$$\varepsilon_{eff,si} = (1-k_{eff,s})\varepsilon_{eff,i} \quad (18),$$

where $\varepsilon_{eff,si}$ and $\varepsilon_{eff,i}$ are the fractionation coefficients for bulk sea ice and for pure ice. The salt segregation coefficient $k_{eff,s}$ is determined according to the relationships indicated above. For growth velocities 1.25×10^{-5} mm s$^{-1} \leq v_i \leq 3.6 \times 10^{-4}$ mm s^{-1}, an approximation to equation (10) by *Cox and Weeks* [1988] was used:

$$k_{eff} = 0.7617 + 0.0568\ln v_i \quad (v_i \text{ in mm s}^{-1}) \quad (19).$$

Given the larger uncertainties in the determination of k_{eff} and ε_{eff} at small growth rates, this approach may not be reliable below growth velocities of 1.25 x 10^{-5} mm s^{-1} or 1.1 mm d^{-1} (corresponding to $k_{eff} = 0.12$), which is not problematic for studies such as this one, where the vast majority of analysed samples grew at faster rates (see also discussion above of segregation at very slow growth). Furthermore, except for one measurement, *Nakawo and Sinha's* data fall below a v_i of 2.0 x 10^{-4} mm s^{-1} (17 mm d^{-1}). For $v_i > 3.6$ x 10^{-4} mm s^{-1},

$$k_{eff} = \frac{0.26}{0.26 + 0.74 \exp(-724.3 v_i)} \quad (v_i \text{ in mm s}^{-1}) \quad (20)$$

based on *Cox and Weeks* [1975]. The fractionation factor for pure sea ice is computed according to equation (10), with $\alpha_{eq}^* = 1.00291$ [*Lehmann and Siegenthaler*, 1991]:

$$\varepsilon_{eff,i} = \left[\frac{\alpha_{eq}^*}{\alpha_{eq}^* + (1 - \alpha_{eq}^*) \exp\left(-\frac{z_{bl} v_i}{D}\right)} - 1 \right] \cdot 1000$$

$$= \left[\frac{1.00291}{1.00291 - 0.00291 \exp(-1083.3 v_i)} - 1 \right] \cdot 1000 \quad (21).$$

Since the pure ice phase is part of the skeletal layer, the corresponding boundary-layer thicknesses have to be taken into account. The fractionation coefficient $\varepsilon_{eff,si}$ for bulk sea ice with respect to $H_2^{18}O$ can now be derived from equations (18) and (21) in combination with either equation (19) or (20). The growth velocity can be

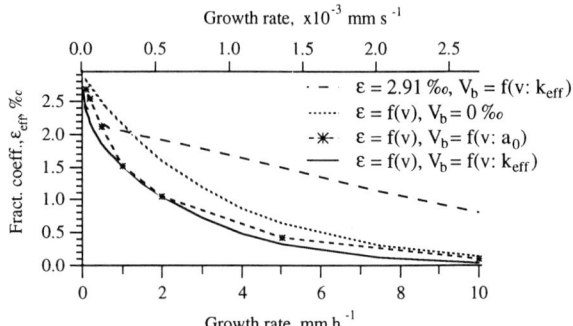

Fig. 15. Fractionation coefficient ε_{eff} for $H_2^{18}O$ plotted as a function of growth rate for different fractionation models, all based on a boundary layer thickness of $z_{bl} = 1.3$ mm. The bottom two curves represent the full model, taking into account fractionation in the pure ice phase as well as the incorporation of pure liquid fraction. Of these two curves, the solid line is based on liquid inclusions derived from a k_{eff} salinity model ($V_b = f(v:k_{eff})$) and the dashed line based on a geometric liquid inclusions model ($V_b = f(v:a_0)$). The upper two curves represent the case for pure ice ($V_b = 0$ ‰), and for ice with a constant composition ($\varepsilon_{eq} = 2.91$ ‰) and growth-rate dependent amount of liquid inclusions ($V_b = f(v:k_{eff})$).

determined through inversion of this system of equations. An exponential approximation is given by:

$$\varepsilon_{eff,si} = -7.8328 \cdot 10^{-3} + 2.2416 \exp(-1369.2 v_i)$$
$$+ 0.4918 \exp(-29447 v_i) \quad (22),$$

where v_i is in mm s^{-1} (99.2 % of variance explained by regression model). Note, however, that the validity range shown in Figure 14 is not completely covered by data points from the studies cited in *Cox and Weeks* [1988] (see equations (19) and (20)). Inversely, the growth rate can be derived from the fractionation coefficient according to

$$v_i = -1.2741 \cdot 10^{-4} + 2.1344 \cdot 10^{-3} \exp(-1.106 \varepsilon_{eff,si})$$
$$+ 1.3856 \cdot 10^{-3} \exp(-12.036 \varepsilon_{eff,si}) \quad (23)$$

for $\varepsilon_{eff,si} \leq 2.00$, and

$$v_i = -1.1139 \cdot 10^{-5} + 3.6618 \cdot 10^{-2} \exp(-2.9024 \varepsilon_{eff,si})$$
$$+ 1.5348 \cdot 10^{-3} \exp(-4.4752 \varepsilon_{eff,si}) \quad (24)$$

for $\varepsilon_{eff,si} > 2.00$. The validity range with respect to the growth velocities of the underlying data set is given above.

Figure 15 summarizes the results in a plot of the fractionation coefficient ε_{eff} as a function of growth velocity. Note the offset between the "master" curve

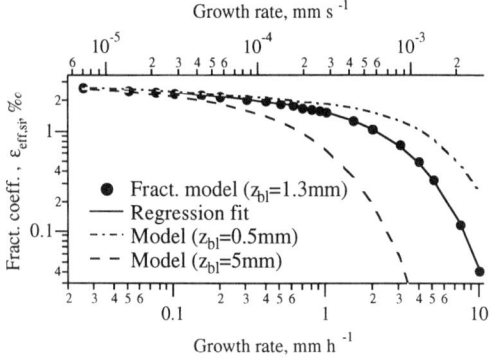

Fig. 14. Fractionation coefficient $\varepsilon_{eff,si}$ for $H_2^{18}O$ plotted as a function of growth rate based on the model assuming fractionation for the pure solid phase as well as growth-rate dependent inclusion of non-fractionated liquid (boundary layer thickness $z_{bl} = 1.3$ mm). The solid line represents a least-squares exponential fit to the data points (circles). The dashed lines are regression fits to the same model, but based on boundary-layer thicknesses z_{bl} of 0.5 and 5 mm.

(solid line) based on equations (18) to (21) and the pure ice case. The model given by equation (17) converges towards the pure ice curve with increasing growth rates (not shown), with an offset of 0.22 ‰ at a growth rate of 0.001 mm h^{-1}. Disregarding the effect of fractionation for the pure ice phase leads to even larger deviations. For comparison, the amount of non-fractionated liquid enclosed within the system has also been computed from a simple geometric model for the case of congelation ice. Based on the work of *Nakawo and Sinha* [1984], the spacing of brine layers between ice lamellae (see Figure 10) a_0 (in mm) was parameterized as a function of growth rate v_i (in mm s^{-1}) according to:

$$a_0 = 0.3324 + 2.8140 \exp(-13896 v_i) \quad (25).$$

The thickness of the brine layers t was determined as 0.116 mm, based on the assumption that for an arbitrarily selected growth rate of 2.8 mm s^{-1} (1 mm h^{-1}), the volume fraction of liquid (equal to t/a_0) corresponds to the value based on equation (19). The resulting curve for the fractionation coefficient is in reasonable agreement with the master curve.

5. POSTGENETIC CHANGES IN THE ISOTOPIC COMPOSITION OF SEA ICE

5.1. Rayleigh Fractionation of Brine in a Half-Closed System

The isotopic evolution of brine retained within the ice is controlled by sequential fractionation upon freezing of liquid within the pores. As the brine becomes increasingly depleted in the heavy isotope, its composition shifts towards lower $\delta^{18}O$ values. While this process has been invoked to qualitatively interpret brine data first reported by *Lange and Hubberten* [1992], as yet there is no quantitative analysis. *Jouzel and Souchez* [1982] and *Souchez and Jouzel* [1984] explained the isotopic composition of freshwater and ice at the base of melting-refreezing glaciers through a Rayleigh fractionation process (water and ice are in thermodynamic equilibrium with no isotopic exchange between phases). In an extension of this work, equations will be derived, describing the isotopic composition of the two major sea-ice components, solid ice and brine, after pore close-off prevents further exchange with the seawater. In contrast with freshwater systems, where the output of liquid from a constant-total-volume system due to the smaller density of ice is given by a proportionality constant [*Souchez and Jouzel*, 1984], here the derivation will consider two independent output or sink functions. The former is given by the thermal evolution of the system controlling the subsequent formation of ice, the latter is determined from the phase relationships for standard seawater based on the work of *Assur* [1960].

From the mass balance for a system containing N_s molecules in the solid and N_l in the liquid phase, with R_s and R_l denoting the ratio of the isotopes ($^{18}O/^{16}O$ in this case) in the solid and liquid phase and α_{ls} the equilibrium distribution or fractionation factor between liquid and solid phase, it follows that

$$\alpha_{ls} R_l dN_s + R_l dN_{ex} = -d(R_l N_l) = -N_l dR_l - R_l dN_l \quad (26)$$

where N_{ex} is the number of molecules removed from the system through brine expulsion. With $f_{ls} \, dN_l$ of the liquid transformed into solid ice and a corresponding amount $f_{ex} \, dN_l$ expelled to the bulk liquid, where $f_{ls} + f_{ex} = 1$, we obtain from (26):

$$(\alpha_{ls} f_{ls} + f_{ex} - 1) \frac{dN_l}{N_l} = \frac{dR_l}{R_l} \quad (27).$$

Thus, the isotopic composition of brine and pure ice resulting from a reduction in the amount of liquid within the system from its original amount of $N_{l,0}$ to N_l by freezing and expulsion can be derived by integration of equation (27):

$$\int_{N_{l,0}}^{N_l} (\alpha_{ls} f_{ls} + f_{ex} - 1) \frac{dN_l}{N_l} = \int_{R_{l,0}}^{R_l} \frac{dR_l}{R_l} \quad (28),$$

yielding for the ratio between isotopes in the solid and the liquid phase

$$R_l = R_{l,0} \left(\frac{N_l}{N_{l,0}} \right)^{f_{ls}(\alpha_{ls}-1)} \quad (29)$$

and

$$R_s = \alpha_{ls} R_l = \alpha_{ls} R_{l,0} \left(\frac{N_l}{N_{l,0}} \right)^{f_{ls}(\alpha_{ls}-1)} \quad (30).$$

Transforming isotopic notations into the δ-notation, where the isotopic composition of phase x is related to the deviation from a standard *std* as

$$\delta_x = \left(\frac{R_x}{R_{std}} - 1 \right) \cdot 1000 \quad \text{(in ‰)}, \quad (31)$$

yields:

$$\delta_l = (\delta_{l,0} + 1000) \left(\frac{N_l}{N_{l,0}} \right)^{f_{ls}(\alpha_{ls}-1)} - 1000 \quad (32)$$

and

$$\delta_s = \alpha_{ls}(\delta_{l,0} + 1000)\left(\frac{N_l}{N_{l,0}}\right)^{f_{ls}(\alpha_{ls}-1)} - 1000 \quad (33).$$

The average composition of the bulk ice volume formed through sequential freezing of a fraction $N_{l,1} - N_{l,2}$ can be determined through integration of equation (33) within the appropriate limits, yielding:

$$\delta_s = \frac{1}{\frac{N_{l,2}}{N_{l,0}} - \frac{N_{l,1}}{N_{l,0}}} \left\{ \frac{\alpha_{ls}(\delta_{l,0} + 1000)}{f_{ls}(\alpha_{ls} - 1) + 1} \left[\left(\frac{N_{l,2}}{N_{l,0}}\right)^{f_{ls}(\alpha_{ls}-1)+1} - \left(\frac{N_{l,1}}{N_{l,0}}\right)^{f_{ls}(\alpha_{ls}-1)+1} \right] - 1000\left(\frac{N_{l,2}}{N_{l,0}} - \frac{N_{l,1}}{N_{l,0}}\right) \right\} \quad (34).$$

The ratio between the frozen and the expelled fraction of liquid f_{ls} to f_{ex} can be derived from the mass balance for the liquid volume $V_{l,0}$ before and $V_{l,1}$ after freezing V_i of pure ice

$$\rho_{l,0} V_{l,0} = \rho_{l,1} V_{l,1} + \rho_i V_i \quad (35),$$

with ρ denoting the density of the components. With

$$\frac{V_{ex} \rho_{ex}}{V_i \rho_i} = \frac{f_{ex}}{f_{ls}} = \frac{1 - f_{ls}}{f_{ls}} \quad (36),$$

we obtain:

$$f_{ls} = \frac{1}{\rho_{ex}\left(\frac{\rho_{l,0} V_{l,0}}{V_i \rho_i \rho_{l,1}} - \frac{1}{\rho_{l,1}} - \frac{V_{l,0}}{V_i \rho_i} + \frac{1}{\rho_i}\right) + 1} \quad (37).$$

The respective values for f_{ls} have been derived through (numerical) integration of the above equation over the respective temperature interval (with f_{ls} averaging at 0.90 between -1.86 and -22.9 °C). The density of pure ice is [Yen et al., 1991]

$$\rho_i = 0.917 \cdot (1 - 1.17 \cdot 10^{-4} \theta) \quad (38)$$

and the density of brine ρ_l based upon the thermodynamic phase relations [Assur, 1960]:

$$\rho_l = 1 + 0.8 \cdot \left(1 - \frac{54.11}{\theta}\right)^{-1} \quad (39),$$

with θ denoting the temperature in °C.

The isotopic composition of brine retained within the ice matrix can now be computed from equations (32) and (37) to (39). Assuming that fractionation is close to but not at equilibrium, an effective fractionation coefficient $\varepsilon_{eff} = 2.5$ ‰ has been used. The resulting curves for different boundary conditions are shown in Figure 16. The initial composition of the brine ($\delta_{l,0}$ in equation (32)) corresponds to the mean $\delta^{18}O$ of surface seawater of -0.28 ‰. Fractionation of the brine only starts to take effect as the system closes off and brine is not replaced by seawater. Brine expulsion due to volume expansion is assumed to occur throughout, in accordance with laboratory observations [Cox and Weeks, 1988]. In Figure 16, fractionation curves for different close-off volumes $V_{b,0}$ (corresponding to $N_{l,0}$ in equations (32) and (33)) and bulk ice salinities S are

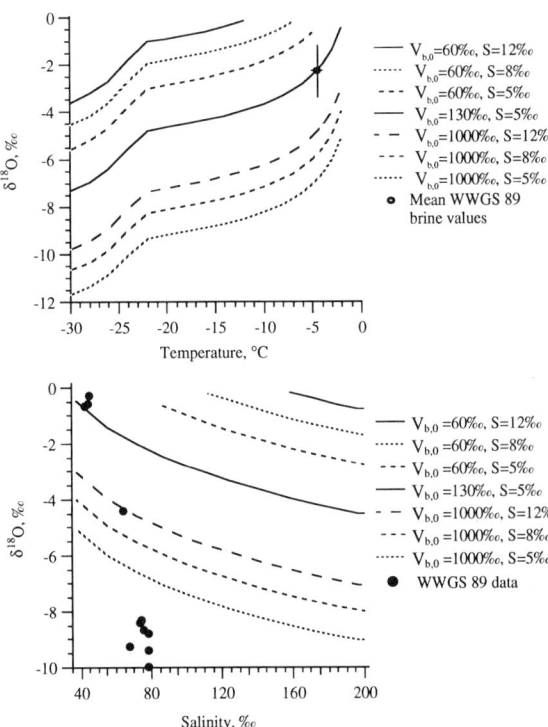

Fig. 16. Isotopic composition of brine retained within a volume of freezing sea ice as a function of temperature (top) and brine salinity (bottom), derived from a Rayleigh fractionation model. The initial composition of the liquid fraction corresponds to the average $\delta^{18}O$ of seawater (-0.28 ‰). Fractionation only starts to take effect as the system closes off, brine expulsion takes place throughout. Here, fractionation curves for different close-off porosities, $V_{b,0}$, and bulk ice salinities, S, are shown. At the top, the mean composition of all samples from first-year ice is also indicated (horizontal and vertical bar give the standard deviation of temperature and $\delta^{18}O$, respectively); at the bottom, samples for which both salinity and $\delta^{18}O$ were determined are shown.

shown. A value of 1000 ‰ is the upper limit for fractionation, since this assumes that as the first ice crystals in a volume of water start to form, all the concentrated brine is retained within the ice volume. The lower limit is represented by a value of 60 ‰, based on field and laboratory observations which indicate that the ice is effectively impermeable at smaller values [*Cox and Weeks*, 1988, *Eicken et al.*, 1995a]. More realistic values are likely to fall between these two extremes, since (1) the exchange with the underlying seawater generally drops off rapidly with distance from the ice bottom [*Lake and Lewis*, 1970, *Eide and Martin*, 1975], and (2) complete exchange of brine by seawater is limited to exceptional cases not considered here [*Lake and Lewis*, 1970, *Lytle and Ackley*, 1996]. Based on the average isotopic composition and temperature of brine collected from first-year ice (Figure 16), the effective "isotopic close-off" brine volume amounts to roughly 130 ‰. This corresponds to a temperature of –2 or –3°C for ice with a bulk salinity of 5 or 8 ‰. Based on the observed temperature gradients and ice thicknesses, close-off would thus be attained at a level 0.05 to 0.2 m above the ice bottom, a plausible estimate for the winter period.

The isotopic composition of brine samples qualitatively supports the conclusions drawn from the Rayleigh fractionation model (Table 4, Figure 16). Several methodological problems, in particular mixing of liquid from different depth levels flowing into a sampling hole and melting or refreezing in core samples from which brine is extracted, greatly increase the error compared to measurements of ice or seawater samples, however. Nevertheless, the differences between $\delta^{18}O$ vs. brine salinity curves computed from the Rayleigh model (Figure 16) and the low $\delta^{18}O$ values, originating from heavily snow laden second-year ice, shown in the Figure are significant. These cannot be explained by Rayleigh fractionation due to freezing in a single cycle but are most likely a result of downward percolation of snow meltwater. Freezing of brine containing even a small fraction of melted snow with an average $\delta^{18}O$ of –16 ‰ [*Eicken et al.*, 1994a] would explain the observed values. Snow meltwater percolation has been documented through isotopic composition and texture (e.g., core 25501, Figure 4).

Since the composition of brine cooled to lower temperatures in equilibrium with the ice matrix differs considerably from the seawater composition, one would expect the ice formed by sequential freezing of pores to differ from the bulk sea ice as well. The average composition of ice formed upon cooling brine within pores from temperature T_1 to T_2 has been computed through equation (34) (Figure 17). As more and more of the brine freezes out, the average composition of the ice tends to the value of the brine at the starting temperature, taking into account that brine expulsion described by the factor f_{ls} in equation (34) selectively reduces the amount of $H_2^{16}O$ with respect to $H_2^{18}O$ in the system, resulting in a positive deviation from the original $\delta^{18}O$ of the liquid phase. For a value of $f_{ls} = 1$, the average ice composition after complete freezing of the liquid phase would come to –0.28 ‰ in this case. Since the spread in isotopic composition shown in Figure 17 is well above the accuracy of the measurements, it should thus, at least in theory, be possible to link the thermal evolution of an ice volume to its isotopic composition resulting from pore freezing. As evident from Figure 17 (straight line changing by little more than 0.04 ‰ as an ice volume is cooled from –2 to –30 °C), the bulk isotopic composition of sea ice, i.e., a sample composed of (1) bulk ice with a composition determined by the fractionation upon freezing of seawater, (2) ice formed at a later stage through freezing of brine contained within pores and (3) the remaining unfrozen brine, does not change significantly with temperature.

5.2. Advective and Diffusive Transport Within the Ice Column

Lange and Hubberten [1992] suggested that diffusion processes may affect the isotopic composition of growing sea ice. Based on an effective diffusion coefficient of 3×10^{-11} m^2 s^{-1}, a 7 ‰ $\delta^{18}O$ layer anomaly led to a reduction in the $\delta^{18}O$ of the ice 0.05 m away from the discontinuity by roughly 1 ‰ after 5 months in their simulation. Compared with the solid-state self-diffusion coefficient of $H_2^{18}O$, amounting to 5×10^{-15} m^2 s^{-1} [*Hobbs*, 1974], the former estimate appears too high, even with a contribution by surface diffusion. While compositional gradients and diffusivities in the brine are comparatively large, diffusive or advective transport of isotopes in the liquid phase is not likely to affect ice composition during the ice-growth season. Sea ice can generally be considered impermeable at liquid volume fractions below 50 to 70 ‰ [*Cox and Weeks*, 1988, *Eicken et al.*, 1995a] which prevail in the interior of the ice throughout most of the year. Neither volume nor surface diffusion are effective in "bridging" ice layers

TABLE 4. Mean temperature and $\delta^{18}O$ of Brine from First- and Second-year Sea Ice (averages over 37 individual measurements for 14 sites)

Ice Type,	$\delta^{18}O$ (‰)		Temp. (°C)	n
	Mean	σ		
First-year	–2.3	2.3	–4.4	12
Second-year	–6.3	5.0	–3.0	2
All samples	–2.8	2.9	–4.2	14

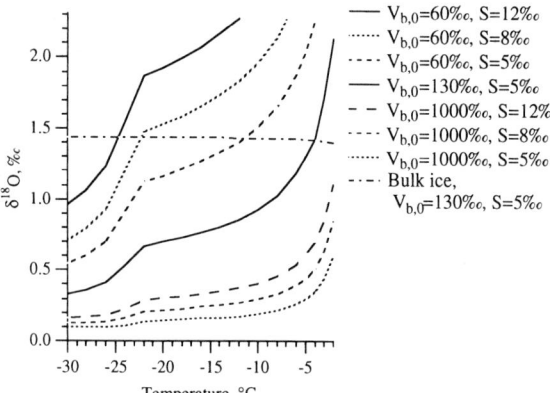

Fig. 17. Average composition of ice formed within pores through freezing of brine as a function of temperature. The near-horizontal line running from 1.40 ‰ at –2 °C to 1.44 ‰ at –30 °C corresponds to the isotopic composition of the entire volume, consisting of (1) the initial ice with $\delta^{18}O$ = 1.65 ‰ (mean value for bottom first-year ice), (2) the ice formed through freezing of brine (corresponding to the other curves shown in the Figure) and (3) unfrozen brine with an initial $\delta^{18}O$ = –0.28 ‰.

between pores under these circumstances. Even with diffusion within the brine, the volume fraction of the pores is too small to result in significant bulk isotopic fluxes. During summer, however, when penetrating shortwave radiation and seawater percolation greatly increase porosity [*Maykut*, 1986, *Lytle and Ackley*, 1996], diffusion may affect the isotopic signal, but only if strong gradients in the brine are present, e.g., due to snow meltwater. The prerequisites for such processes and their relative importance are outlined by *Tison et al.* [1993] in their study of the chemical composition of frazil layers at the base of Hell's Gate Ice Shelf (Ross Sea). With respect to the ice-growth rates derived in this study, the effect is considered to be negligible in relation to other sources of error for growing first-year ice. For second-year ice, advection and diffusion of low-$\delta^{18}O$ snow meltwater in the uppermost layers may have to be taken into account.

6. SIMULATION OF $\delta^{18}O$ PROFILES WITH A ONE-DIMENSIONAL ICE-GROWTH/SALT-FLUX MODEL: COMPARISONS BETWEEN SIMULATED AND FIELD DATA

Lacking suitable field or laboratory data, a direct comparison between measured ice-growth rates and derivations from stable-isotope fractionation was not feasible for this study. Here, a first, rough assessment was attempted on the basis of $\delta^{18}O$ data simulated with an ice-growth model. Previous studies both for Arctic [*Cox and Weeks*, 1988] and Antarctic sea ice [*Eicken*, 1992] have shown that the ice-growth/salt-flux model developed by *Cox and Weeks* is able to reproduce and predict salinity profiles of growing young sea ice. This model was extended to simulate the $\delta^{18}O$ profile of sea ice. Assuming that the growth history of the ice is, on average, well described as testified by a comparison between simulated and measured salinity data, we can assess the performance of the model and the underlying formulations through a comparison between measured and simulated $\delta^{18}O$ data.

The one-dimensional ice-growth/salt-flux model is described in detail in *Eicken et al.* [1995a]. In short, the formulation by *Maykut and Untersteiner* [1971] has been employed to derive ice-growth rates by solving the energy-balance equation for the lower and the upper ice surface with respect to the conductive heat flux. Heat transport through the ice cover has been determined with an explicit finite-difference approximation (vertical resolution 0.02 m). The evolution of the salinity profile is determined by the growth-rate dependent salt segregation coefficient k_{eff} and formulations to account for gravity drainage and expulsion of brine [*Cox and Weeks*, 1988]. The thermal properties of the ice are prognostic variables depending on temperature and salinity of an ice layer. The isotopic composition of the ice as a function of growth rate has been determined through the exponential approximation for the fractionation coefficient $\varepsilon_{eff,si}$ (equation (22)). Desalination processes have not been considered, given their negligible effect on the isotopic composition of a bulk ice volume (section 5.1, Figure 17). The meteorological and oceanographic boundary conditions for the simulations are representative of the study area in the central and eastern Weddell Sea and correspond to the detailed description in *Eicken* [1992] and *Eicken et al.* [1995a] except where indicated.

A comparison between averaged salinity and $\delta^{18}O$ profiles for 21 of the 22 WWGS 89 first-year cores (core 26501 consisting almost exclusively of snow ice has been disregarded) and corresponding model results (Figure 18) shows reasonably close agreement, in particular when taking into account spatial variability (see also standard deviations shown in Figure 8) and the fact that mean profiles are compared with simulations. To demonstrate the sensitivity of the results on the underlying fractionation model, the following simulations are also shown: (1) fractionation given by equation (17), i.e., ignoring the growth-rate dependent seawater entrainment, (2) full fractionation model with a different boundary-layer thickness z_{bl}, differing by a factor of 0.4 and 4, respectively. It is evident that ignoring the effects of seawater entrainment on isotopic fractionation leads to significant deviations, in particular at intermediate growth rates (see also Figure 15). At medium to high growth rates, z_{bl} also has a strong influence on the

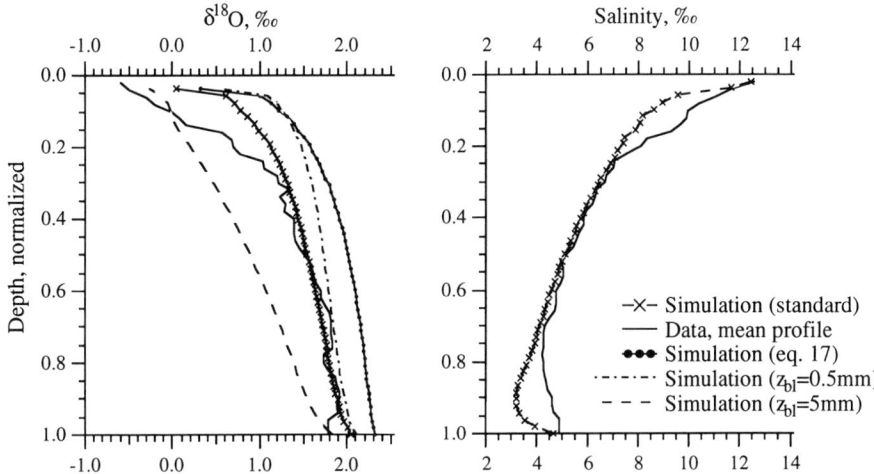

Fig. 18. Comparison between measured and simulated $\delta^{18}O$ and salinity profiles for WWGS 89 first-year ice cores. Profiles have been normalized with respect to depth. See also Figures 8 and 9 for true-depth profiles and standard deviation. Core 26501 consisting entirely of snow-ice has been disregarded. The boundary conditions for the simulation are: ice formation starting on May 15, snow accumulation rate 0.6 mm d^{-1}, oceanic heat flux 8 W m^{-2}. The solid line with dots is for a simulation with fractionation based on a simplified fractionation model (equation (17)). The dash-dotted and dashed line are for the full fractionation model (as solid curve with crosses), but based on a boundary-layer thickness z_{bl} of 0.5 and 5 mm, respectively.

simulation results (although the assumed deviations greatly surpass the expected range of uncertainty or variability).

Section 8.1 of the paper is devoted to the discussion of differences between granular and columnar ice and their effect on simulated and observed isotopic composition. The largest deviations are apparent in the upper quarter of the profiles, with the simulation over-predicting $\delta^{18}O$ values and underestimating the salinity. As discussed in section 3, this is due to the contribution of snow ice, which is not accounted for in the model. The lower stable isotope values are proportional to the amount of meteoric ice. Snow-ice salinities are generally higher than those of ordinary sea ice due to retention of seawater and brine within the snow-ice layers [*Eicken*, 1992, *Eicken et al.*, 1995a, *Lytle and Ackley*, 1996].

Additional support for the validity of the derived fractionation-coefficient/growth-rate relationship is provided by the comparison between ice-core data and model results for a regionally and temporally confined study area with a more adequate record of meteorological data. During the WWGS 89 expedition, six cores (location 286, Figure 1, Table 1) were taken from a large patch of young ice of uniform thickness (mean of 0.5 m) lacking pressure ridges or other signs of deformation. Based on the analysis of passive-microwave satellite data (SSM/I-derived ice concentrations provided by the National Snow and Ice Data Center in Boulder, Colorado) from the period prior to sampling, the origin of the ice from a larger polynya north of Atka Bay and its day of formation (day 245) could be determined. The ice-growth model was then forced with air temperature and wind speed data from Neumayer station for the period of interest (daily averages of 3-hourly observations were utilized, *König-Langlo* [1992]). Given the climatological differences between the Neumayer location near the ice-shelf margin and the pack-ice area to the North, based on an approach by *Kipfstuhl* [1991] the wind speed was reduced and air temperature (in °C) increased by 25 %, respectively. The resulting model curves reproduce most patterns observed in the data (Figure 19). The reduction in $\delta^{18}O$ at a relative depth of 0.35 and the increase at 0.65 correspond to a 6 K decrease and a 9 K increase in mean air temperature with a resulting change in ice-growth velocity. The deviations at the top correspond to the occurrence of frazil ice in the cores. In part, however, they are also due to errors in the derivation of growth rates during the initial ice growth phase.

7. DERIVING GROWTH RATES OF SEA ICE FROM $\delta^{18}O$ PROFILES

Based on the fractionation model, sea-ice growth rates have been derived from $\delta^{18}O$ measurements of the core sections (Table 5) for a surface seawater $\delta^{18}O$ of −0.28 ‰. The arithmetic mean of columnar-ice growth rates varies significantly between first- and second-year ice, even though the range of values covered by first-

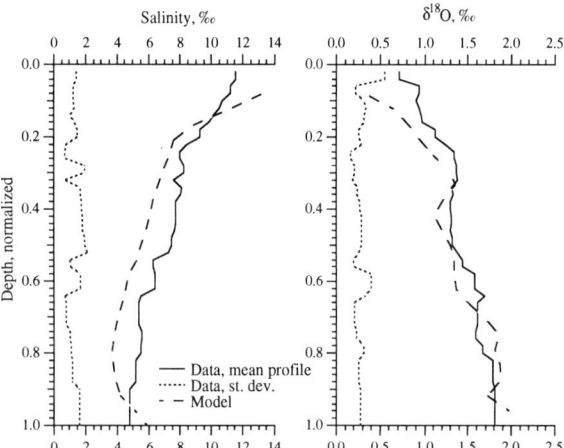

Fig. 19. Comparison between measured and simulated salinity and $\delta^{18}O$ profiles for six young-ice cores (location 286 in Figure 1) taken within a radius of <100 km north of Neumayer Station. The boundary conditions for the simulation are based on the modified meteorological record from Neumayer Station and the starting day of ice formation has been obtained from satellite data. The boundary conditions for the simulation are: ice formation starting on September 2, snow accumulation rate 1.0 mm d^{-1}, oceanic heat flux 15 W m^{-2}.

year ice is quite broad (note that in Table 5 and Figure 20 contrary to general nomenclature, second-year ice refers only to those bottom layers of an ice floe which grew during the second-year of its lifetime, whereas the upper ice layers originating from the previous year are considered to be first-year ice; see also discussion in section 3.1). This is mostly due to the decrease in growth rate with increasing ice thickness, apparent in the individual and averaged $\delta^{18}O$ profiles (Figures 6, 7 and 9). Apart from this "insulation effect", fractionation coefficients and hence growth rates vary considerably within individual cores.

The derived values of v_i are affected by three error sources. First, while all core segments with an identifiable snow-ice component ($\delta^{18}O < -0.28$ ‰) were disregarded, the growth rates may still be overestimated in cases where the contribution of snow resulted in a lowering of $\delta^{18}O$ above the threshold. Second, changes in the surface water composition affect estimates of v_i. Given the narrow range in seawater composition during most of the ice-growth season [*Fahrbach et al.*, 1995, *Weppernig et al.*, 1996] an anomaly such as a lowering of surface water $\delta^{18}O$ to -0.4 ‰ or a rise to -0.1 ‰ would result in an growth-rate overestimate by 20 % or an underestimate by 25 % for sea ice with $\delta^{18}O = 1.8$ ‰. Since the growth rate is not constant within individual core segments, linear averaging over the isotopic composition will result in an error for the growth rate derived from a non-linear model.

The time-integrated mean growth rate $v_{i,t}$ is a more meaningful parameter than the arithmetic average:

$$v_{i,t} = \frac{\sum_{seg} z_{seg}}{\sum_{seg} t_{seg}} \quad (40),$$

where t_{seg} is the time necessary to grow an ice thickness equivalent to the sample depth interval z_{seg}. The resulting values are smaller than the arithmetic average and do not deviate as much with respect to different ice types (Table 5). A mean value $v_{i,t}$ of approximately 5 mm d^{-1} for the entire data set can thus be translated directly into a mean ice thickness of 0.5 m grown over the interval of 100 d. The frequency distribution of first-year ice growth rates peaks in the interval from 2.5 to 5 mm d^{-1}, and extends further into the high growth-rate regime than second-year columnar ice (Figure 20).

Glancing at Table 5 and Figure 20, the question arises why freezing of frazil ice is included in this analysis, given that the isotopic fractionation model has been derived for the growth of congelation ice. This

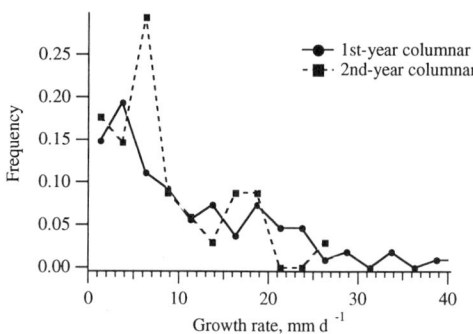

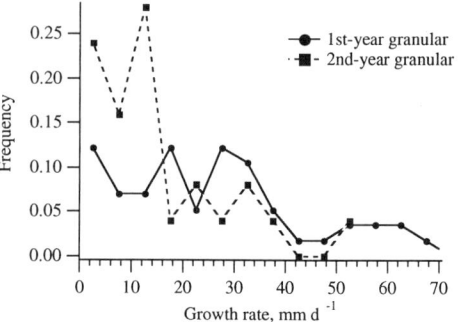

Fig. 20. Frequency distributions of growth rates for columnar (top) and granular ice (bottom) derived from $\delta^{18}O$ profiles with the model described in the text. In both Figures, a small part (<7 and 11 %, respectively) of the spectrum at growth rates larger than 40 or 70 mm d^{-1} is not shown. Second-year ice refers only to those ice layers that have been grown in the second-year of an ice floe's lifetime (i.e., the bottom layers).

TABLE 5. Arithmetic and Time-integrated Mean Growth Rates of Sea Ice from $\delta^{18}O$ Profiles of WWGS 89 Cores

Ice Type	Growth rate (mm d^{-1})			n
	Arithm. mean	σ	t-integr. mean	
First-year columnar	14.0	16.3	4.9	106
Second-year columnar*	8.3	6.1	4.0	34
First-year granular	39.4	45.2	13.0	57
Second-year granular*	14.5	13.1	2.9	25
Young columnar (286)	18.7	15.6	8.1	25
Young granular (286)	32.8	12.1	31.7	10
All samples	19.5	27.8	5.2	286

* Second-year ice only refers to the layers at the bottom of second-year floes actually originating from the second year of growth; first-year ice includes upper layers from second-year floes

will be discussed in detail in section 8.1. The overall growth rate derived for granular ice is much larger than for columnar and extends to much higher values (although the maxima with values >100 mm d^{-1} may be affected by incorporation of meteoric ice). Nevertheless, the values derived for second-year granular ice are much smaller than for first-year ice (Figure 20).

8. DISCUSSION

8.1. Contrasts Between Frazil and Congelation Ice

Overviews of the growth of Antarctic sea ice generally stress the importance of "dynamic" growth processes. Apart from the widespread occurrence of rafting and ridging events, this is manifested in the predominance of granular ice in Antarctic sea-ice cores [*Gow et al.*, 1982 and 1987, *Jacka et al.*, 1987, *Lange and Eicken*, 1991a, *Jeffries et al.*, 1994, *Worby and Massom*, 1995]. Snow-ice formation contributes to this ubiquity of granular ice [*Lange et al.*, 1990, *Eicken et al.*, 1994a, *Worby and Massom*, 1995, *Jeffries et al.*, 1997 and this volume]. A major fraction of the granular ice forms as frazil during the advance of the ice edge, eventually congealing into pancake ice in the ocean's wave field [*Wadhams et al.*, 1987, *Jacka et al.*, 1987, *Lange et al.*, 1989]. Frazil growth in the water column and in leads and polynyas furthermore contributes significantly to ice thickness in the interior pack [*Gow et al.*, 1982 and 1987, *Eicken and Lange*, 1989, *Lange and Eicken*, 1991a, *Jeffries et al.*, 1994]. The centimeter-sized ice platelets forming at greater water depth and accumulating underneath coastal sea ice can be considered an extreme form of "frazil" growth [*Dieckmann et al.*, 1986, *Eicken and Lange*, 1989, *Kipfstuhl*, 1991]. What are the differences and similarities between granular ice of frazil origin and columnar ice formed through congelation freezing?

Most important are the textural differences, with granular ice composed of more or less isometric grains, lacking the intracrystalline brine inclusions of vertically elongated columnar crystals [*Weeks and Ackley*, 1986]. Bulk salinity differences between the two ice types may not be as significant, as a comparison of mean profiles and individual segments of different textures has shown [*Eicken*, 1992]. Granular ice with comparatively high salinities and correspondingly low $\delta^{18}O$ is mostly snow ice (Figure 3). Excluding snow ice ($\delta^{18}O$ < –0.28 ‰, Table 2), the 1.3 ‰ difference between mean salinities of granular and columnar core segments is reduced to 0.7 ‰. Part of this similarity can be explained by brine drainage which obliterates original salinity contrasts, as discussed in *Eicken* [1992] and to some extent apparent in the comparison of first- and second-year ice salinities in Tables 1 and 3. The differences in isotopic composition of granular ice of frazil origin and congelation ice are more pronounced with values of 1.28 ± 0.60 and 1.61 ± 0.78 ‰, respectively. Nevertheless, the characteristic patterns in the $\delta^{18}O$ profile are also displayed by cores consisting predominantly of granular ice (Figure 5; see also cores shown by *Lange et al.* [1990] and *Worby and Massom* [1995]). Frazil embedded in congelation ice often does not exhibit any significant deviations from the adjacent layers.

To assess the validity of growth-rate determinations, frazil and granular ice have to be considered in more detail. WMO nomenclature [*World Meteorological Organization*, 1983] defines frazil ice as "fine spicules or plates of ice, suspended in water". In turbulent, supercooled freshwater, frazil forms as discs few millimeters in diameter and few micrometers to tens of micrometers thick, sometimes sintering into aggregates of small numbers of crystals [*Martin*, 1981, *Daly*, 1994]. Frazil formation in saltwater appears to be essentially similar with respect to size and morphology of crystals formed [*Martin*, 1981, *Hanley and Tsang*, 1984]. For supercoolings of 0.01 to 0.1 K, the growth velocities of freshwater frazil summarized in *Hobbs* [1974, p. 584] range between 10^{-3} and 5×10^{-2} mm s^{-1} in the direction along the basal plane. With a boundary-layer thickness of 1 mm, these velocities would correspond to a pure-ice fractionation coefficient between 1.3 and <0.1 ‰, respectively (Figure 11). Since any velocity greater than about 2×10^{-3} mm s^{-1} results in fractionation coefficients below 0.5 ‰, the frazil platelets are likely to exhibit $\delta^{18}O$ values close to the seawater composition. The large ice platelets found under coastal ice and growing at lower supercoolings at up to several hundred meters water depth exhibit fractionation coefficients around 2.2 to 2.3 ‰ [*Kipfstuhl*, 1991], indicative of slower growth rates, possibly in combination with smaller boundary layer thicknesses.

With respect to crystal morphology, consolidated granular ice generally has little in common with loose frazil ice, however. Granular-ice crystals are mostly isometric or with aspect ratios ranging between 2:1 to 5:1 (Figures 4 and 5). Individual frazil crystals are thus smaller in volume by one to two orders of magnitude. These changes in shape and volume-to-surface area come about as a result of joining of individual crystals [*Martin*, 1981] and coarsening in a liquid immersion. The latter process is mainly driven by the concurrent reduction in free surface energy as the kinetic growth form of dendritic crystals grades into an assembly of spheroidal or ellipsoidal grains [*Colbeck*, 1986, *Wettlaufer*, in press]. For the case of sea ice, these processes are likely to govern the evolution of grease and slush ice, which may remain in a semi-consolidated state for several hours to days [*Martin*, 1981, *Hanley and Tsang*, 1984, *Lange et al.*, 1989]. Frazil accumulating underneath a solid ice cover will undergo similar transformations. The original isotopic signature of individual frazil crystals is likely to be overprinted by the growth in the seawater during the coarsening process. The porosity of unconsolidated freshwater frazil layers mostly ranges between 50 and 80 % [*White*, 1992, *Daly*, 1994]. Thus, it may be assumed that the bulk isotopic signal of granular ice forming from frazil is actually generated during consolidation of the slush or grease ice layer.

Such a consolidation process would be controlled by the transfer of latent heat, released upon freezing of the liquid volume within the slush, to the atmosphere. For consolidating grease ice at the sea surface, the high turbulent fluxes will allow for rapid passage of a "freezing front" downward through the slush. Frazil accumulating at the bottom of second-year ice will consolidate at much slower rates, with the conductive flux through the overlying ice layers limiting the growth rate. The observed $\delta^{18}O$ profiles of granular ice agree at least qualitatively with such a scenario (Figure 5). Moreover, the differences in derived growth rates for first- and second-year granular ice (Table 5 and Figure 20) also strongly support this conjecture. Further, quantitative support requires an analysis of the isotopic fractionation process within the liquid immersion phase that is likely to constitute the bulk signal of granular ice.

8.2. Segregation of Sea-Salt Ions, $H_2^{18}O$ and HDO at the Ice-Water Interface

While *Souchez et al.* [1988] and this study indicate that it is possible to estimate sea-ice growth rates from isotopic fractionation data, they also point towards a major gap in the analysis. As outlined in section 4.2 (equation (10)), in BPS theory of solute segregation at the melt-solid interface, the distribution coefficient k_{eff} depends not only on the growth velocity v_i but also on the boundary-layer thickness z_{bl}. Lacking joint measurements of ice growth rate and isotopic composition, z_{bl} has only been estimated indirectly. Despite the utilization of the established interrelation between salt segregation and isotopic fractionation (section 4.4) for the broad range of conditions covered by this data set to improve upon a previous estimate by *Souchez et al.* [1988], the question remains whether z_{bl} can be considered constant for different growth rates or hydrodynamic boundary conditions. Generally, this is assumed to be the case at smaller values of v_i [*Burton et al.*, 1953, *Garandet et al.*, 1994] and has been assumed so over the entire range covered by *Weeks and Lofgren* [1967] and *Souchez et al.* [1987]. Since the convecto-diffusive fluxes are subsumed in z_{bl}, the magnitude of gravitational acceleration or large differences in seawater salinity affect the magnitude of z_{bl} and hence k_{eff}, as demonstrated experimentally by *Langhorne and Robinson* [1983]. The hydrodynamic regime is also of importance in this context. Contrasting with the original BPS problem with a microscopically smooth interface, the situation is more complex for the case of a lamellar or cellular sea-ice/water interface [*Langhorne and Robinson*, 1983 and 1986].

Hydrodynamically, the boundary zone is characterized by a laminar sub-layer (viscous effects dominating), a transition zone and a fully turbulent layer below. The thickness of the viscous sub-layer Δl is given for smooth walls by [*Schlichting*, 1982]:

$$\Delta l \approx \frac{5v}{u_{*0}} \qquad (41),$$

where v is the kinematic viscosity and u_{*0} the friction velocity. With typical friction velocities of 0.5 to 1.0 x 10^{-2} m s^{-1} [*McPhee*, 1990, *Omstedt and Svensson*, 1992] and $v = 1.9$ x 10^{-6} m^2 s^{-1}, Δl amounts to 1 to 2 mm. For a fully rough surface with all roughness elements protruding from the laminar sub-layer, the factor in the numerator of equation (41) can increase from 5 to values above 70 [*Schlichting*, 1982]. For a typical lamellar ice/water interface, the ice mass fraction increases from 0 to values above 0.7 to 0.9 within the skeletal layer of mostly several millimeters vertical extent (*Weeks and Ackley* [1986], see Figure 13). Thus, viscous sub-layer and boundary layer $z_{bl}(H_2^{18}O)$ are of the same magnitude. While these two boundary-layer concepts are not physically equivalent, the correspondence allows the preliminary conclusion that z_{bl} does not depend as much on the hydrodynamic regime as previous studies have shown for the microscopically smooth interface of pure freshwater ice [*Souchez et al.*, 1987, *Lehmann and Siegenthaler*, 1991].

The processes governing isotopic exchange and fractionation in the transition zone between melt and "closed" ice above the level of fluid penetration in Figure 13 require further testing and validation. Thus, the concept of a stagnant boundary layer needs to be reconciled with the fluid exchange observed in the skeletal layer of growing sea ice. The magnitude of z_{bl}, as derived from field measurements, may nevertheless be quantitatively correct, since it integrates the effect of fluid exchange, which has been shown to affect the lowermost 10 to 20 mm of the ice cover over characteristic time scales of minutes to hours [*Lake and Lewis*, 1970, *Eide and Martin*, 1975, *Niedrauer and Martin*, 1979, *Cota et al.*, 1987]. The replacement of interstitial liquid leads to the observed lowering of apparent fractionation coefficients of sea ice as compared to freshwater ice. It also affects fractionation of the ice growing at some depth within the skeletal layer above the tips of the ice lamellae (Figure 13). Here, the boundary-layer concept approaches its limit. This is even more true for the case of salt segregation at the interface, because, unlike the stable isotopes of water, the major sea-salt ions are only incorporated in trace amounts in the solid phase [*Gross et al.*, 1977]. In the latter case it is the interface morphology and the evolution of porosity throughout the thickness of the skeletal layer that controls the amount and salinity of brine entrained within the ice cover. These processes, however, are not described by BPS theory, which has been developed for a smooth interface, though some work suggests that the formalism still holds true (see discussion of the topic in *Weeks and Ackley* [1986]).

8.3. Deducing Modes and Rates of Ice Growth in the Weddell Sea

The relative fractions of granular and columnar ice derived from the ice cores analysed fall within the lower range of granular ice fractions in Antarctic sea ice between 40 and 70 % [*Clarke and Ackley*, 1984, *Gow et al.*, 1987, *Lange and Eicken*, 1991a, *Jeffries et al.*, 1994 and 1997, *Worby and Massom*, 1995]. As noted previously, the higher fractions of columnar ice in the WWGS 89 data set are partly due to sampling of a patch (>150 km diameter, location 286) of congelation ice formed off Atka Bay. Depending on the extent of congelation growth in coastal polynyas or open-water areas within the pack [*Eicken and Lange*, 1989], these processes may result in significant interannual as well as regional variability in ice type occurrence. Since the growth of frazil ice in open water is generally characterized by much higher ocean-atmosphere heat fluxes and hence higher ice production rates [*Smith et al.*, 1990], such variability will most likely affect the ice thickness distribution.

As shown in Figure 21a and supported by other data sets [*Gow et al.*, 1987, *Jeffries et al.*, 1994, *Worby and Massom*, 1995], the maximum ice thicknesses often correspond to high fractions of granular ice within cores. This is partly due to deformation processes, with multiple rafting events leading to ice thicknesses >2 m even in level ice such as core 27201 marked in Figure 21. As indicated by the isotopic composition (Figure 21c), snow ice is of importance as well. Oceanic heat flux, which may reach values of up to 40 W m^{-2} in certain areas [*Gordon and Huber*, 1990, *McPhee and Martinson*, 1994], and the insulating effect of ice and snow cover constrain the amount of thickening by pure congelation freezing at the base of an ice floe [*Maykut*, 1986]. In the Arctic, during a single season typically 1.5 to 2 m of ice are grown thermodynamically and the mean thickness of level ice attained after several years of growth is roughly 3 m [*Maykut*, 1986, *Eicken et al.*, 1995b]. Here, the maximum level ice thickness grown

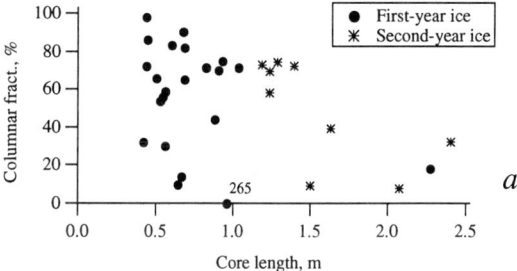

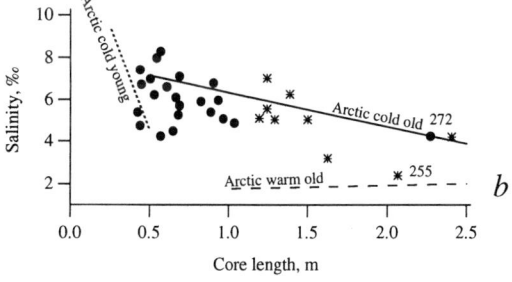

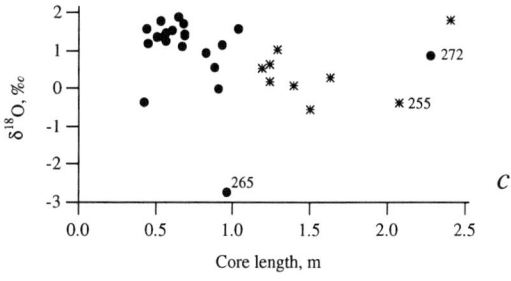

Fig. 21. Average core properties (fraction of (*a*) columnar ice, (*b*) salinity and (*c*) $\delta^{18}O$) plotted as a function of core length.

without deformation or frazil accumulation amounted to about 0.7 to 0.8 m (core 26601 or core 27801 in Figure 6, consisting almost exclusively of columnar ice with a rafted layer at the top); larger level-ice thicknesses were only achieved through rafting or contributions by snow ice (core 27201). Even when not plainly apparent through tilting of columnar grains, brine layers or stratigraphic interfaces (e.g., at 0.5 m depth in core 27111, Figure 5), rafting and deformation can be identified from the "stacked" salinity and stable-isotope profiles (Figure 6). Second-year ice sampled close to the northwestern margin of the Weddell Sea at the end of winter grew to maximum level thicknesses of 1.63 m; larger thicknesses are due to snow-ice formation and deformation (core 25501 in Figure 21).

Most snow ice, easily identified through its isotopic signature, in the Weddell Sea and other sectors of the Antarctic forms by freezing of seawater in a flooded snow cover [*Spichkin*, 1966, *Lange et al.*, 1990, *Eicken et al.*, 1994a, *Lytle and Ackley*, 1996, *Jeffries et al.*, 1997]. Less common is superimposed-ice formation through refreezing of meltwater at the base of the snow cover, resulting in clear ice with low salinity and $\delta^{18}O$ values [*Panov and Fedotov*, 1977, *Kawamura et al.*, 1993, *Jeffries et al.*, 1997]. With large accumulation rates and extensive summer melt as prerequisites to the formation of superimposed ice, it is mostly confined to coastal or warmer, high-accumulation areas [*Panov and Fedotov*, 1977, *Kawamura et al.*, 1993, *Jeffries et al.*, 1997]. This study provides a first indication of its occurrence in the western Weddell Sea (Figure 4, Table 6).

The stable-isotope profiles can provide further valuable information on ice growth processes. Thus, the second-year ice cores 25501 and 26001 to 26101 display a characteristic local minimum in the $\delta^{18}O$ curve at roughly 1.25 and 1.0 m, respectively. In all but one case, this minimum coincides with a local salinity maximum and a discontinuity in the vertical grain-size profile and is furthermore associated with an overlying chlorophyll maximum (Figure 4, see detailed discussion in section 3.1 and *Gleitz et al.*, in press). Below, salinity decreases and $\delta^{18}O$ increases to the maximum values observed within the core. This discontinuity is interpreted as the boundary between the first and second year of ice growth, corresponding to a distinct decrease in derived ice-growth rates (Table 5). The net ice accretion during the second year amounts to 0.41 ± 0.19 m (without granular ice in core 25501 0.35 ± 0.09 m). At Ice Station Weddell from February until late May, *Lytle and Ackley* [1996] observed net accretion of a few centimeters for ice with a comparable snow cover but larger initial thickness. Since the ice growth period applicable for this study is several months longer, these values are not unrealistically high.

A rough assessment of regional differences in oceanic heat flux can be based on the derived growth rates dz_i/dt (corresponding to $v_{i,t}$ in Table 5), with the accretion rate at the ice bottom given by the balance of oceanic and conductive heat fluxes F_w and F_c [*Maykut*, 1986, *Wettlaufer*, 1991]:

$$F_w = F_c - \rho_{si} L \left(\frac{dH}{dt} \right) \quad (42),$$

with sea ice density ρ_{si} and latent heat of freezing L. F_c is determined by the thickness and thermal conductivity of ice and snow, z_i, z_s, k_i and k_s, respectively:

$$F_c = \frac{k_s k_i}{k_s z_i + k_i z_s} (T_s - T_b) \quad (43),$$

with surface and bottom temperature T_s and T_b. With $k_s = 0.2$ W m^{-1} K^{-1}, $k_i = 1.7$ W m^{-1} K^{-1} [*Yen et al.*, 1991], and assuming the same upper surface boundary conditions (including T_s) in both cases, we can derive a ratio $F_{c,1}/F_{c,2}$ between the conductive heat flux for first-year ice in the central and eastern and second-year ice in the western Weddell Sea. In accordance with field measurements [*Eicken et al.*, 1994a], the snow depth has been taken to increase linearly from 0 (first-year) or 0.15 m (second-year) at day 121 to 0.15 or 0.6 m at day 273. Ice thickness increases with the root of time from 0 or 0.8 m at day 121 to 0.8 or 1.2 m at day 273. This yields

$$\frac{F_{c,1}}{F_{c,2}} = 5 \quad (44).$$

With $\rho_{si} = 915$ kg m^{-3} and $L = 290$ kJ kg^{-1} [*Yen et al.*, 1991], we obtain:

TABLE 6. Modes and Rates of Sea-Ice Growth in the Weddell Sea based on WWGS 89 Data Set

Mode/rate of ice growth	Relative contribution		
	1st-ice, east./centr. WS	2nd-year ice, west. WS	All
Meteoric ice total (%)	3	5	4
Superimposed ice (mean core %)	0	1	<0.1
Congelation ice (mean core %)	57	48	54
Annual ice accretion (m)*	0.66	0.41	0.61
T-integr. growth rate (mm d^{-1})	8.4	3.3	5.2
Oceanic heat flux	$F_{w,1}$	$0.2 F_{w,1}$-4.8	

* Ice accretion during first- and second-year of ice-floe growth history (value for first-year ice average of all non-deformed core lengths)

$$F_{w,1} = F_{c,1} - 2.65 \cdot 10^8 v_{i,1} \qquad (45)$$

and

$$F_{w,2} = 0.2 F_{c,1} - 2.65 \cdot 10^8 v_{i,2} \qquad (46).$$

With the time-integrated ice velocities v_i (averaged over frazil and congelation ice, Table 5), the oceanic heat flux under second-year ice in the western Weddell Sea can be related to that in the eastern as ($F_{w,1} \geq 24$ Wm^{-2}):

$$F_{w,2} = 0.2 F_{w,1} - 4.8 \qquad (47).$$

Given the large uncertainties in the surface boundary conditions as well as problems due to averaging both over larger regions and within sample volumes (fractionation and growth rate are exponentially related), this estimate can only be taken as a rough indication. A more accurate determination would require a full treatment of the surface energy balance and samples of smaller vertical extent. The relative measurement error (based on an accuracy of ε_{eff} of 0.05 ‰) is <10 % for $v_i \geq$ 1.5 x 10^{-4} mm s^{-1} ($\varepsilon_{\text{eff}} \leq 1.85$ ‰) and <25 % for 1.5 x 10^{-4} mm s^{-1} >$v_i \geq 1.3$ x 10^{-5} mm s^{-1} (1.85 ‰ < $\varepsilon_{\text{eff}} \leq$ 2.52 ‰). The same accuracy for the determination of F_w would require exact knowledge of F_c (e.g., from regional meteorological data) as well as a finer sample resolution. The relative measurement errors based on the mean growth rates (Table 6) are <15 % and <18 % for first- and second-year ice, respectively. Due to other sources of error (regional and vertical averaging, etc.), this estimate is probably not better than within ±50 % of the true mean F_w. Nevertheless, it clearly indicates that oceanic heat fluxes are considerably smaller in the western than in the eastern Weddell Sea. This is in line with under-ice turbulence measurements at Ice Station Weddell with F_w ranging between <1 and 4 W m^{-2} [*McPhee et al.*, 1992] or a winter value of F_w between 3 and 6 W m^{-2} from ice temperature data [*Lytle and Ackley*, 1996] compared to estimates of a minimum of 19 W m^{-2} [*Martinson*, 1994] to between 37 and 41 W m^{-2} during winter [*Gordon and Huber*, 1990] for the central and eastern Weddell Sea. For $F_{w,2} = 3$ W m^{-2}, equation (47) gives $F_{w,1} = 39$ W m^{-2}, which is in fair agreement with the cited observations. Thus, such ice-core data constitutes a useful supplement to studies confined to one particular location such as oceanographic moorings or drifting ice camps. It can also help in elucidating spatial variability in oceanic heat flux, observed even over smaller areas [*Wettlaufer*, 1991, *Omstedt and Wettlaufer*, 1992, *Lytle and Ackley*, 1996], which may also be responsible for some of the intercore variability in this study.

The down-core grain-size curves demonstrate that a quantitative textural stratigraphy contains much more information than the mere textural classification. Since grain morphology and crystallographical orientation of columnar ice crystals are affected by the under-ice current regime [*Weeks and Gow*, 1978, *Langhorne and Robinson*, 1986], the detailed analysis of grain-size variations seen on vertical scales of millimeters as in core 27801 (Figure 6) may serve as a proxy for the hydrodynamic regime and help in interpreting the stable-isotope data.

The analysis of individual core segments (Tables 1, 2 and 3) and entire core profiles confirmed previous discussions by Souchez and co-workers [*Souchez et al.*, 1988, *Tison and Haren*, 1989] that salinity provides a much less reliable record of ice growth history than the isotopic composition, given that the latter is only affected to a very small extent by desalination (Figure 17). Their interdependence is not completely obliterated, however. Both in single core profiles (Figures 4 and 6) and in a scatter plot of the entire data set, traces of the direct correlation between fractionation and salt segregation can still be discerned. This is demonstrated in Figure 3 by the limit provided by the curve relating S_0 and $\delta^{18}O$ through the growth rate-dependent relation. It is mainly entrainment of meteoric ice and desalination that shifts individual samples to positions below this curve.

The equalizing effect of "ageing" processes is particularly distinct for second-year ice. Compared to changes in ice salinity and other ice properties of Arctic sea ice during the first melt season, these changes are minor, however. This is demonstrated by the regression line shown for the salinity-ice thickness relation in warm, old ice in Figure 21b, based on a compilation by *Cox and Weeks* [1974]. The mean core salinities of Weddell Sea second-year ice are well above this line. Moreover, mean core salinities for first- and second-year ice only differ by 1.2 ‰ (Table 1). Arctic first-year ice is comparable to Antarctic first-year ice with mean salinities of 6 to 8 ‰ [*Cox and Weeks*, 1974, *Weeks and Ackley*, 1986]; the mean salinity of Arctic multi-year ice including second-year ice, on the other hand, does not exceed 3 ‰ and averages at slightly above 2 ‰ [*Cox and Weeks*, 1974, *Eicken et al.*, 1995b]. This is mostly a result of downward meltwater percolation during summer, reducing salinities in the upper 0.5 to 1 m to values <1 ‰ [*Untersteiner*, 1968, *Eicken et al.*, 1995b]. Such "meltwater flushing" is generally not observed in the Antarctic, where floe melting takes place mostly from the sides and below, with comparatively little surface ablation [*Andreas and Ackley*, 1981]. Hence, surface salinities in second-year ice in the Weddell Sea can be quite high, in particular in conjunction with flooding events [*Eicken et al.*, 1991, *Lytle and Ackley*, 1996]. In some areas, though not to a large extent in the Weddell Sea, formation of superimposed ice may lead to a reduction in surface salinity (Figures 4 and 21). Likewise, evidence of retexturing and significant decreases in ice density typical of Arctic second- and multi-year ice [*Weeks and Ackley*, 1986, *Gow and Tucker*, 1990, *Eicken et al.*, 1995b], were not observed in the WWGS cores, with the exception of superimposed ice (core 25501).

This has several important consequences. First, microwave remote-sensing techniques which discriminate age classes mainly from different signatures of the upper ice and snow layers [*Comiso et al.*, 1992, *Drinkwater*, 1995] are not likely to discern ice of different age based on the ice properties. In fact, since flooding is more widespread on second-year ice with its deeper snow cover [*Eicken et al.*, 1994a, *Lytle and Ackley*, 1996], classification criteria derived for Arctic sea ice cannot simply be transferred to Antarctic sea ice. Second, the contrasts in general ice properties observed for different age classes of Arctic sea ice are not as pronounced in Antarctic sea ice. Finally, the ecological importance of Antarctic sea ice as a habitat to a productive community of organisms depends critically on the habitability of the upper ice layers where irradiative fluxes are highest. Low ambient salinities with corresponding low nutrient concentrations, as found in the upper layers of Arctic sea ice, would thus eliminate a site of high biological production [*Legendre et al.*, 1992, *Fritsen et al.*, 1995].

9. SUMMARY AND CONCLUSIONS

Based upon ice-core data from the Weddell Sea, the factors controlling texture, isotopic composition and salinity of Antarctic pack ice were studied. Subsequently, the modes and rates of ice growth were derived from the core data. It was possible to explain brine $\delta^{18}O$ values down to -5 ‰ through Rayleigh fractionation within brine-filled pores. The lowermost values extending to -10 ‰ were affected by entrainment of snow meltwater during summer. As the composition of the ice sequentially frozen within brine channels and pores should exhibit a distinct zonation due to fractionation, high-resolution stable-isotope measurements in marine-ice bodies may yield valuable information about the thermal history and evolution of an ice mass. This is of particular interest for the study of features such as the marine-ice layers found in the Filchner-Ronne and the Amery Ice Shelves [*Oerter et al.*, 1992, *Hellmer and Jacobs*, 1992]. Here, a test for compositional zonation or layering may help in distinguishing between ice growth through freezing in cracks or confined zones [*Hellmer and Jacobs*, 1992, *Souchez et al.*, 1995], localized brine expulsion in "chimney" type features or other expulsion processes [*Eicken et al.*, 1994b].

In contrast with the Rayleigh fractionation processes, the bulk signal of the ice is a result of kinetic fractionation during initial freezing at the bottom of the ice cover. Utilizing the mixed layer of the Weddell Sea as a natural laboratory with a near-constant isotopic composition, a comparatively large data set of effective fractionation coefficients could be obtained. Based on the established relationships between isotopic fractionation and salt segregation at the advancing ice-water interface, the following issues were addressed:

(1) A regionally averaged estimate of the salt distribution coefficient $k_{eq}*$ was obtained for Antarctic sea ice, taking into account desalination effects, and found to correspond closely to measurements on Arctic sea ice [*Nakawo and Sinha*, 1981]. An effective boundary layer thickness was determined with respect to the transfer of NaCl and $H_2^{18}O$ from the liquid reservoir to the microscopic ice surface.

(2) The maximum stable-isotope fractionation factors observed in sea ice were explained in terms of the enclosure of pristine seawater.

(3) Based on field measurements of salinity and $\delta^{18}O$, and use of the BPS boundary-layer model, a set of equations relating the ice-growth velocity to the effective isotope fractionation coefficient for sea ice was derived and approximated by a simple exponential relation.

(4) The growth-rate dependent fractionation was incorporated into a one-dimensional ice-growth/salt-flux model for validation with field observations.

(5) Growth rates were derived for first- and second-year ice in the western and eastern Weddell Sea.

While the model simulations generally match the field data, further work is required to establish to what extent isotopic fractionation is dependent on interface morphology, ice type or the hydrodynamic regime. These open questions notwithstanding, the growth rates derived for different ice types in the Weddell Sea provide supplementary evidence of the strong contrasts in oceanic heat fluxes between the central/eastern sector and the western, perennially ice-covered sector of the Weddell Sea. Detailed analysis of exemplary ice cores demonstrated, that through a combination of isotopic fractionation data, high-resolution textural stratigraphy and supporting salinity measurements, the growth history and the relative contributions of deformation, meteoric ice, frazil and congelation thickening as well as the relevant growth rates can be derived from sea-ice cores. Based on the entire data set, a summary of the importance of these different processes is shown in Table 6. With gradual improvement of the underlying models and with an ever-increasing number of samples obtained both in the Arctic and Antarctic, sea-ice cores can provide valuable complementary data on ice-ocean-atmosphere interaction, increasing for instance the regional or temporal coverage of field experiments confined in space and time. One might thus eventually be able to think of each ice floe accessible to coring as a fail-safe multi-sensor buoy that records and archives a wide spectrum of ice-ocean-atmosphere interaction data.

Acknowledgments. This work relied on the competence of personnel aboard RV Polarstern as well as ample help from the whole of the sea-ice team in the field. I am particularly grateful for the support of M. Lange before, during and after this cruise. H.-W. Hubberten is thanked for stable-isotope measurements and E. Fahrbach for supplying oceanographic data. F. Valero Delgado was a great support in the preparation of samples and figures. Discussions with D. Wolf-Gladrow and R. Zeebe on isotopic fractionation were interesting and helpful. The comments by V. Lytle, S. Pfirman and an anonymous reviewer as well as M. Jeffries' editorial help aided in improving the manuscript. Special thanks go to J.-L. Tison, who provided a most thorough and helpful review of the manuscript. This is publication no. 1281 of the Alfred Wegener Institute for Polar and Marine Research.

10. REFERENCES

Ackley, S. F., M. A. Lange, and P. Wadhams, Snow cover effects on Antarctic sea ice thickness, in *Sea Ice Properties and Processes,* edited by S. F. Ackley and W. F. Weeks, pp. 16-21, CRREL Monogr. 90-1, 1990.

Andreas, E. L., and S. F. Ackley, On the differences in ablation seasons of the Arctic and Antarctic sea ice, *J. Atmos. Sci., 39,* 440-447, 1981.

Assur, A., Composition of sea ice and its tensile strength, *SIPRE Res. Rep., 44,* 1960.

Augstein, E., N. Bagriantsev, and H. W. Schenke, The expedition ANTARKTIS VIII/1-2, *Ber. Polarforsch., 84,* 1989.

Bauch, D., P. Schlosser, and R. G. Fairbanks, Freshwater balance and the sources of deep and bottom waters in the Arctic Ocean inferred from the distribution of $H_2^{18}O$, *Prog. Oceanog., 35,* 53-80, 1995.

Beck, N., and K. O. Münnich, Freezing of water: isotopic fractionation (Abstract), *Chem. Geol., 70,* 1-2, 1988.

Burton, J. A., R. C. Prim, and W. P. Slichter, The distribution of solute in crystals grown from the melt. Part I. Theoretical, *J. Chem. Phys., 21,* 1987-1991, 1953.

Cherepanov, N. V., Using the methods of crystal optics for determining the age of drift ice (in Russian), *Probl. Arktik, 2,* 179-184, 1957.

Clarke, D. B., and S. F. Ackley, Sea ice structure and biological activity in the Antarctic marginal ice zone, *J. Geophys. Res., 89,* 2087-2095, 1984.

Colbeck, S. C., Statistics of coarsening in water-saturated snow, *Acta metall., 34,* 347-352, 1986.

Comiso, J. C., T. C. Grenfell, M. Lange, A. W. Lohanick, R. K. Moore, and P. Wadhams, Microwave remote sensing of the Southern Ocean sea ice cover, in *Microwave Remote Sensing of Sea Ice,* edited by F. D. Carsey, pp. 243-259, Geophysical Monograph 68, American Geophysical Union, Washington, 1992.

Cota, G. F., S. J. Prinsenberg, E. B. Bennett, J. W. Loder, M. R. Lewis, J. L. Anning, N. H. Watson, and L. R. Harris, Nutrient fluxes during extended blooms of arctic ice algae, *J. Geophys. Res., 92,* 1951-1962, 1987.

Cox, G. F. N., and W. F. Weeks, Salinity variations in sea ice, *J. Glaciol., 13,* 109-120, 1974.

Cox, G. F. N., and W. F. Weeks, Brine drainage and initial salt entrapment in sodium chloride ice, *CRREL Res. Rep., 345,* 1975.

Cox, G. F. N., and W. F. Weeks, Numerical simulations of the profile properties of undeformed first-year sea ice during the growth season, *J. Geophys. Res., 93,* 12449-12460, 1988.

Craig, H., and B. Horn, Relationships of deuterium, oxygen 18, and chlorinity in the formation of sea ice, *Trans. AGU, 49,* 216-217, 1968.

Daly, S. F. (ed.), International Association for Hydraulic Research Working Group on Thermal Regimes - Report on frazil ice, *CRREL Spec. Rep., 94-23,* 1994.

Dieckmann, G. S., G. Rohardt, H. Hellmer, and J. Kipfstuhl, The occurrence of ice platelets at 250 m depth near the Filchner Ice Shelf and its significance for sea ice biology, *Deep-Sea Res., 33,* 141-148, 1986.

Drinkwater, M. R., Applications of SAR measurements in ocean-ice-atmosphere interaction studies, in *Oceanographic Applications of Remote Sensing,* edited by M. Ikeda and W. F. Dobson, pp. 381-396, CRC Press, Boca Raton, 1995.

Eicken, H., Salinity profiles of Antarctic sea ice: field data and model results, *J. Geophys. Res., 97,* 15545-15557, 1992.

Eicken, H., Automated image analysis of ice thin sections - instrumentation, methods and extraction of stereological and textural parameters, *J. Glaciol., 39,* 341-352, 1993.

Eicken, H., and M. A. Lange, Development and properties of sea ice in the coastal regime of the southeastern Weddell Sea, *J. Geophys. Res., 94,* 8193-8206, 1989.

Eicken, H., M. A. Lange, and G. S. Dieckmann, Spatial variability of sea-ice properties in the northwestern Weddell Sea, *J. Geophys. Res., 96,* 10603-10615, 1991.

Eicken, H., M. A. Lange, H. -W. Hubberten, and P. Wadhams, Characteristics and distribution patterns of snow and meteoric ice in the Weddell Sea and their contribution to the mass balance of sea ice, *Ann. Geophys., 12,* 80-93, 1994a.

Eicken, H., H. Oerter, H. Miller, W. Graf and J. Kipfstuhl, Textural characteristics and impurity content of meteoric and marine ice in the Ronne Ice Shelf, Antarctica. *J. Glaciol., 40,* 386-398, 1994b.

Eicken, H., H. Fischer, and P. Lemke, Effects of the snow cover on Antarctic sea ice and potential modulation of its response to climate change, *Ann. Glaciol., 21,* 369-376, 1995a.

Eicken, H., M. Lensu, M. Leppäranta, W. B. Tucker III, A. J. Gow, and O. Salmela, Thickness, structure and properties of level summer multiyear ice in the Eurasian sector of the Arctic Ocean, *J. Geophys. Res., 100,* 22697-22710, 1995b.

Eide, L. I., and S. Martin, The formation of brine drainage features in young sea ice, *J. Glaciol., 14,* 137-154, 1975.

Fahrbach, E., G. Rohardt, N. Scheele, M. Schröder, V. Strass, and A. Wisotzki, Formation and discharge of deep and bottom water in the northeastern Weddell Sea, *J. Mar. Res., 53,* 515-538, 1995.

Friedman, I., A. C. Redfield, B. Schoen, and J. Harris, The variation of the deuterium content of natural waters in the hydrologic cycle, *Rev. Geophys., 2,* 177-224, 1964.

Fritsen, C. H., V. I. Lytle, S. F. Ackley, and C. W. Sullivan, Autumn bloom of Antarctic pack-ice algae, *Science, 266,* 782-784, 1994.

Garandet, J. P., J. J. Favier, and D. Camel, Segregation phenomena in crystal growth from the melt, in *Handbook of Crystal Growth, vol. 2,* edited by D. T. J. Hurle, pp. 659-707, Elsevier, Amsterdam, 1994.

Gleitz, M., A. Bartsch, G. Dieckmann, and H. Eicken, Composition and succession of sea ice diatom assemblages in the Weddell Sea, Antarctica. *AGU Antarctic Research Series,* in press, 1997.

Gloersen, P., Campbell, W. J., Cavalieri, D. J., Comiso, J. C., Parkinson, C. L., Zwally, H. J., *Arctic and Antarctic sea ice, 1978-1987: Satellite passive-microwave observations and analysis,* NASA SP-511, National Aeronautics and Space Administration, Washington, 1992.

Gordon, A. L., and B. A. Huber, Southern Ocean winter mixed layer, *J. Geophys. Res., 95,* 11655-11672, 1990.

Gow, A. J., S. F. Ackley, W. F. Weeks, and J. W. Govoni, Physical and structural characteristics of Antarctic sea ice, *Ann. Glaciol., 3,* 113-117, 1982.

Gow, A. J., S. F. Ackley, K. R. Buck, and K. M. Golden, Physical and structural characteristics of Weddell Sea pack ice, *CRREL Rep., 87-14,* 1987.

Gow, A. J., and W. B. Tucker III, Sea ice in the polar regions, *in Polar Oceanography, Part A, Physical Science,* edited by W. O. Smith Jr., pp. 47-122, Academic Press, San Diego, 1990.

Gross, G. W., P. M. Wong, and K. Humes, Concentration dependent solute redistribution at the ice-water phase boundary. III. Spontaneous convection. Chloride solutions, *J. Chem. Phys., 67,* 5264-5274, 1977.

Hanley, T. O'D., and G. Tsang, Formation and properties of frazil in saline water, *Cold Reg. Sci. Technol., 8,* 209-221, 1984.

Hellmer, H. H., and S. S. Jacobs, Ocean interaction with the base of Amery Ice Shelf, Antarctica, *J. Geophys. Res., 97,* 20305-20317, 1992.

Hobbs, P. V., *Ice Physics,* Clarendon Press, Oxford, 1974.

Ishikawa, N., and S. Kobayashi, On the internal melting phenomenon (puddle formation) in fast sea ice, East Antarctica, *Ann. Glaciol., 6,* 138-141, 1985.

Jacka, T. H., I. Allison, R. Thwaites, and J. C. Wilson, Characteristics of the seasonal sea ice of East Antarctica and comparisons with satellite observations, *Ann. Glaciol., 9,* 85-91, 1987.

Jeffries, M. O., Massive, ancient sea-ice strata and preserved physical-structural characteristics in the Ward Hunt Ice Shelf, *Ann. Glaciol., 15,* 125-131, 1991.

Jeffries, M. O., R. A. Shaw, K. Morris, A. L. Veazey, and H. R. Krouse, Crystal structure, stable isotopes ($\delta^{18}O$), and development of sea ice in the Ross, Amundsen, and Bellingshausen seas, Antarctica, *J. Geophys. Res., 99,* 985-995, 1994.

Jeffries, M. O., S. Li, R. Jaña, H. R. Krouse, and B. Hurst-Cushing, Late winter first-year ice floe thickness variability, seawater flooding and snow ice formation in the Amundsen and Ross Seas, *AGU Antarct. Res. Ser.,* this volume.

Jeffries, M. O., A. P. Worby, K. Morris, and W. F. Weeks, Seasonal variations in the properties and structural composition of sea ice and snow cover in the Bellingshausen and Amundsen Seas, Antarctica, *J. Glaciol., 43,* in press, 1997.

Jouzel, J., and R. A. Souchez, Melting-refreezing at the glacier sole and the isotopic composition of the ice, *J. Glaciol., 28,* 35-42, 1982.

Kawamura, T., K. I. Ohshima, S. Ushio, and T. Takizawa, Sea-ice growth in Ongul Strait, Antarctica, *Ann. Glaciol., 18,* 97-101, 1993.

Kipfstuhl, J., Zur Enstehung von Unterwassereis und das Wachstum und die Energiebilanz des Meereises in der Atka Bucht, Antarktis, *Ber. Polarforsch., 85,* 1991.

König-Langlo, G., The meteorological data of the Georg-von-Neumayer-Station (Antarctica) for 1988, 1989, 1990 and 1991, *Ber. Polarforsch., 116,* 1992.

Lake, R. A., and E. L. Lewis, Salt rejection by sea ice during growth, *J. Geophys. Res., 75,* 583-597, 1970.

Lange, M. A., and H. Eicken, Textural characteristics of sea ice and the major mechanisms of ice growth in the Weddell Sea, *Ann. Glaciol., 15,* 210-215, 1991a.

Lange, M. A., and H. Eicken, The sea ice thickness distribution in the northwestern Weddell Sea, *J. Geophys. Res., 96,* 4821-4837, 1991b.

Lange, M. A., and H.-W. Hubberten, Isotopic composition of sea ice as a tool for understanding sea ice processes in the polar regions, *in Physics and Chemistry of Ice,* edited by N. Maeno and T. Hondoh, pp. 399-405, Hokkaido University Press, Sapporo, 1992.

Lange, M. A., S. F. Ackley, P. Wadhams, G. S. Dieckmann, and H. Eicken, Development of sea ice in the Weddell Sea, Antarctica, *Ann. Glaciol., 12,* 92-96, 1989.

Lange, M. A., P. Schlosser, S. F. Ackley, P. Wadhams, and G. S. Dieckmann, ^{18}O concentrations in sea ice of the Weddell Sea, Antarctica, *J. Glaciol., 36,* 315-323, 1990.

Langhorne, P. J., and W. H. Robinson, Effect of acceleration on sea ice growth, *Nature, 305,* 695-698, 1983.

Langhorne, P. J., and W. H. Robinson, Alignment of crystals in sea ice due to fluid motion, *Cold Reg. Sci. Technol., 12,* 197-214, 1986.

Legendre, L., S. F. Ackley, G. S. Dieckmann, B. Gulliksen, R. Horner, T. Hoshiai, I. A. Melnikov, W. S. Reeburgh, M. Spindler, and C. W. Sullivan, Ecology of sea ice biota, 2. Global significance, *Polar Biol, 12,* 429-444, 1992.

Lehmann, M., and U. Siegenthaler, Equilibrium oxygen- and hydrogen-isotope fractionation between ice and water, *J. Glaciol., 37,* 23-26, 1991.

Levine, I. A., *Physical Chemistry,* 2nd edition, McGraw-Hill, New York, 1983.

Lytle, V. I., and S. F. Ackley, Heat flux through sea ice in the western Weddell Sea: Convective and conductive transfer processes, *J. Geophys. Res., 101,* 8853-8868, 1996.

Macdonald, R. W., D. W. Paton, and E. C. Carmack, The freshwater budget and under-ice spreading of Mackenzie River water in the Canadian Beaufort Sea based on salinity and $^{18}O/^{16}O$ measurements in water, *J. Geophys. Res., 100,* 895-919, 1995.

Malmgren, F., On the properties of sea-ice, *Norweg. North Pol. Exped. "Maud" 1918-1925,* Vol. 1, No. 5, 1-67, 1927.

Martin, S., Frazil ice in rivers and oceans, *Ann. Rev. Fluid. Mech., 13,* 379-397, 1981.

Martinson, D. G., Ocean heat and seasonal sea ice thickness in the Southern Ocean, *in Ice in the climate system,* edited by W. R. Peltier, pp. 597-609, Springer-Verlag (NATO ASI Ser., vol. I 12), Berlin, 1994.

Maykut, G. A., The surface heat and mass balance, *in The Geophysics of Sea Ice,* edited by N. Untersteiner, pp. 395-463, Martinus Nijhoff Publ., Dordrecht (NATO ASI B146), 1986.

Maykut, G. A., and N. Untersteiner, Some results from a time-dependent thermodynamic model of sea ice, *J. Geophys. Res., 76,* 1550-1575, 1971.

McPhee, M. G., Small-scale processes, *in Polar Oceano-*

graphy, Part A, Physical Science, edited by W. O. Smith Jr., pp. 287-334, Academic Press, San Diego, 1990.

McPhee, M. G., and D. G. Martinson, Turbulent mixing under drifting pack ice, Science, 263, 218-221, 1994.

McPhee, M. G., D. G. Martinson, and J. H. Morison, Upper-ocean measurements of turbulent flux in the western Weddell Sea, Antarct. J. U. S., 27, 103-105, 1992.

Melling, H., and R. M. Moore, Modification of halocline source waters during freezing on the Beaufort Sea shelf: evidence from oxygen isotopes and dissolved nutrients, Cont. Shelf. Res., 15, 89-113, 1995.

Nakawo, M., and N. K. Sinha, Growth rate and salinity profile of first-year sea ice in the high Arctic, J. Glaciol., 27, 315-330, 1981.

Nakawo, M., and N. K. Sinha, A note on brine layer spacing of first-year sea ice, Atmosphere-Ocean, 22, 193-206, 1984.

Niedrauer, T. M., and S. Martin, An experimental study of brine drainage and convection in young sea ice, J. Geophys. Res., 84, 1176-1186, 1979.

Oerter, H., J. Kipfstuhl, J. Determann, H. Miller, D. Wagenbach, A. Minikin, and W. Graf, Evidence for basal marine ice in the Filchner-Ronne Ice Shelf, Nature, 358, 399-401, 1992.

Omstedt, A., and U. Svensson, On the melt rate of drifting ice heated from below, Cold Reg. Sci. Technol., 21, 91-100, 1992.

Omstedt, A., and J. S. Wettlaufer, Ice growth and oceanic heat flux: models and measurements, J. Geophys. Res., 97, 9383-9390, 1992.

O'Neil, J. R., Hydrogen and oxygen isotope fractionation between ice and water, J . Phys. Chem., 72, 3683-3684, 1968.

Panov, V. V., and V. I. Fedotov, Pripay vostochnoy Antarktidy, Gidrometeoizdat, Leningrad, 1977.

Richardson, C., Phase relationships in sea ice as a function of temperature, J. Glaciol., 17, 507-519, 1976.

Schlichting, H., Grenzschicht-Theorie, Braun, Karlsruhe, 1982.

Schlosser, P., R. Bayer, A. Foldvik, T. Gammelsrod, G. Rohardt, and K. O. Münnich, ^{18}O and helium as tracers of ice shelf water and water/ice interaction in the Weddell Sea, J. Geophys. Res., 95, 3253-3263, 1989.

Schwarzacher, W., Pack-ice studies in the Arctic Ocean, J. Geophys. Res., 64, 2357-2367, 1959.

Smith, S. D., R. D. Muench, and C. H. Pease, Polynyas and leads: an overview of physical processes and environment, J. Geophys. Res., 95, 9461-9479, 1990.

Sofer, Z., and J. R. Gat, Activities and concentrations of oxygen-18 in concentrated aqueous salt solutions: analytical and geophysical implications, Earth Planet. Sci. Lett., 15, 232-238, 1972.

Souchez, R. A., and J. Jouzel, On the isotopic composition in δD and δ^{18}O of water and ice during freezing, J. Glaciol., 30, 369-372, 1984.

Souchez, R., J.-L. Tison, and J. Jouzel, Freezing rate determination by the isotopic composition of the ice, Geophys. Res. Let., 14, 599-602, 1987.

Souchez, R., J.-L. Tison, and J. Jouzel, Deuterium concentration and growth rate of Antarctic first-year sea ice, Geophys. Res. Lett., 15, 1385-1388, 1988.

Souchez, R., J.-L Tison, R. Lorrain, C. Fléhoc, M. Stiévenard, J. Jouzel, and V. Maggi, Investigating processes of marine ice formation in a floating ice tongue by a high-resolution isotopic study, J. Geophys. Res., 100, 7019-7025, 1995.

Spichkin, V. A., Seawater accumulation on the Antarctic fast ice, Sov. Antarct. Exped. Inform. Bull., 59 (in translation, 6(3), 235-236), 1966.

Taube, H., Use of oxygen isotopes in the study of hydration of ions, J. Phys. Chem., 58, 523-528, 1954.

Tison, J.-L., and J. Haren, Isotopic, chemical and crystallographic characteristics of first-year sea ice from Breid Bay (Princess Ragnhild Coast - Antarctica), Antarct. Sci., 1, 261-268, 1989.

Tison, J. -L., D. Ronveaux, and R. D. Lorrain, Low salinity frazil ice generation at the base of a small Antarctic ice shelf, Antarctic Sci., 5, 309-322, 1993.

Untersteiner, N., Natural desalination and equilibrium salinity profile of perennial sea ice, J. Geophys. Res., 73, 1251-1257, 1968.

Wadhams, P., M. A. Lange, and S. F. Ackley, The ice thickness distribution across the Atlantic sector of the Antarctic ocean in midwinter, J. Geophys. Res., 92, 14535-14552, 1987.

Weast, R. C. (ed.), CRC Handbook of Chemistry and Physics (64th edition), CRC Press, Boca Raton, Fla., 1984.

Weeks, W. F., and S. F. Ackley, The growth, structure and properties of sea ice, in The Geophysics of Sea Ice, edited by N. Untersteiner, pp. 9-164, Martinus Nijhoff Publ., Dordrecht (NATO ASI B146), 1986.

Weeks, W. F., and A. J. Gow, Preferred crystal orientations along the margins of the Arctic Ocean, J. Geophys. Res., 84, 5105-5121, 1978.

Weeks, W. F., and O. S. Lee, Observations on the physical properties of sea ice at Hopedale, Labrador, Arctic, 11, 134-155, 1958.

Weeks, W. F., and O. S. Lee, The salinity distribution in young sea ice, Arctic, 15, 92-108, 1962.

Weeks, W. F., and G. Lofgren, The effective solute distribution coefficient during the freezing of NaCl solutions, in Physics of Snow and Ice: International Conference on Low Temperature Science. Proceedings, vol. 1, part 1, edited by H. Oura, pp. 579-597, Inst. Low Temp. Sci., Hokkaido, 1967.

Weissenberger, J., G. Dieckmann, R. Gradinger, and M. Spindler, Sea ice: a cast technique to examine and analyze brine pockets and channel structure, Limnol. Oceanogr., 37, 179-183, 1992.

Weppernig, R., P. Schlosser, S. Khatiwala, and R. G. Fairbanks, Isotope data from Ice Station Weddell: Implications for deep water formation in the Weddell Sea, J. Geophys. Res., 101, 25723-25739.

Wettlaufer, J. S., Heat flux at the ice-ocean interface, J. Geophys. Res., 96, 7215-7236, 1991.

Wettlaufer, J., Introduction to crystallization phenomena in sea ice, in Physics of Ice-Covered Seas, edited by M. Leppäranta, IAPSO Advanced Study Institute, in press.

Weyprecht, K., Die Metamorphosen des Polareises, Perles, Wien, 1879.

White, K. D., Determining the intrinsic permeability of frazil ice. Part 1. Laboratory investigations, CRREL Rep., 91-23, 1991.

Worby, A. P., and R. A. Massom, The structure and properties of sea ice and snow cover in East Antarctic pack ice, Res. Rep. No. 7, Antarctic CRC, Hobart, 1995.

World Meteorological Organization, WMO sea-ice nomenclature, terminology, codes and illustrated glossary, *WMO/DMM/BMO 259-TP-145*, Secretariat of the WMO, Geneva, 1985.

Yen, Y. C., K. C. Cheng, and S. Fukusako, Review of intrinsic thermophysical properties of snow, ice, sea ice, and frost, *in Proceedings 3rd International Symposium on Cold Regions Heat Transfer, Fairbanks, AK, June 11-14, 1991*, edited by J. P. Zarling and S. L. Faussett, pp. 187-218, University of Alaska, Fairbanks, 1991.

Zotikov, I. A., V. S. Zagorodnov, and J. V. Raikovsky, Core drilling through the Ross Ice Shelf (Antarctica) confirmed basal freezing, *Science, 207,* 1463-1465, 1980.

Zubov, N. N., *Arctic ice,* Izd. Glavsevmorputi, Moscow, 1945 (translated by U.S. Navy Oceanogr. Office Springfield, 1963).

Hajo Eicken, Alfred-Wegener-Institut für Polar- und Meeresforschung, Postfach 12 01 61, D-27515 Bremerhaven, Germany.

(Received August 16, 1996; accepted May 10, 1997.)

TEMPORAL AND REGIONAL VARIATION OF SEA ICE DRAFT AND COVERAGE IN THE WEDDELL SEA OBTAINED FROM UPWARD LOOKING SONARS

Volker H. Strass and Eberhard Fahrbach

Alfred-Wegener-Institut für Polar- und Meeresforschung, Bremerhaven, Germany

Up to two years–long time series of ice draft and coverage were obtained between 1990 and 1994 with moored Upward Looking Sonars (ULSs) at six locations in the Weddell Sea. The six mooring locations, between the tip of the Antarctic Peninsula and Kapp Norvegia, cover the flow into and out of the southern Weddell Sea. The measurements are analyzed for the temporal and regional variations, and used to estimate the relative contribution of thermodynamic growth and deformation to the ice draft distribution in the Weddell Sea. Whereas the ice draft in the gyre centre undergoes a distinct annual cycle, temporal changes occur rather irregularly in the boundary regions, especially in the east. The mean ice draft varies between 0.8 m in the central Weddell Gyre, 2.2 m in the eastern inflow and 2.8 m in the western outflow. In opposition to this, the mode of the ice draft is highest (0.7 m) in the gyre centre and lowest (0.2 m) in the boundary regions. In the western outflow a second maximum at 1.1 m results from advection of deformed ice. During circulation with the Weddell Gyre mainly ice with a draft of less than 1 m is transformed into thicker ice; the major transformation occurs from drafts of 0.2 m to drafts of 1.2 m. It is concluded that deformation (rafting and ridging) contributes a minimum of 30% to the total amount of ice formed in the inner Weddell Sea and a minimum of 50% in the boundary regions.

INTRODUCTION

Air-sea interaction in the polar and subpolar seas is largely controlled by the ice cover. The ice cover governs the exchange of heat, light, water, gases, and momentum between the atmosphere and the ocean. It thus strongly influences the state and the dynamics of the mixed layer of the ocean [e.g. *Martinson*, 1990], as well as the boundary layer of the atmosphere [e.g. *Hartmann et al.*, 1994]. The sea ice also contributes to the haline buoyancy fluxes. Freezing creates a process of salt and freshwater separation, thereby generating significant density contrasts. Whereas the ice drift accomplishes a large-scale transport of light freshwater, the saline brine rejected during freezing contributes substantially to the formation of the dense ocean deep and bottom waters, driving the global-scale thermohaline circulation [*Mosby*, 1934; *Broecker*, 1991]. As the circulating deep water masses generally are formed within the polar and subpolar seas, the related transports of heat, salt, and any other constituent of natural and anthropogenic origin, are influenced by the sea ice extent which controls their exchange with the atmos-phere.

Sea ice also provides a habitat for polar and subpolar organisms, ranging from unicellular algae to birds and mammals [e.g. *Spindler and Dieckmann*, 1994]. By virtue of its influence on the penetration of light and on the buoyancy fluxes, sea ice also controls pelagic primary production [e.g. *Smith*, 1987; *Sakshaug and Skjoldal*, 1989; *Strass and Nöthig*, 1996]. Both ice-related and pelagic primary production determine the associated input of carbon to the food chain, and thus affect the so-called biological pump of carbon from the atmosphere to the deep ocean and sediment [e.g. *Eppley and Peterson*, 1979; *Knox and McElroy*, 1984].

While the ice cover is of significance for the physical, chemical, and biological processes in the polar and subpolar seas, it is likewise considered one of the geophysical variables that is most sensitive to climate variability [e.g. *Wadhams*, 1994a] and an essential part of a feedback mechanism which enhances the effect of changing concentrations of atmospheric greenhouse gases on climate [*Manabe and Stouffer*, 1980]. The sensitivity to variations in the seasonal forcing is indicated by the natural seasonal change of the ice covered area surrounding Antarctica, which exceeds the area of the continent itself. In the Arctic the seasonal variation is somewhat smaller.

The two primary variables of state of the sea ice cover are its areal percent coverage and its thickness. Whereas the areal distribution of the ice cover can be assessed reasonably well by satellite remote sensing, the capabilities of remote sensing to yield useful data on ice thickness remain rather limited [*Carsey et al.*, 1992]. The ice thickness is still best measured either by drilling or by detecting the ice draft from below by use of Upward Looking Sonars in moorings or on submarines [e.g. *Wadhams*, 1994b]. Drilling can provide only poor spatial and temporal coverage, and possibly introduces a bias arising from the fact that thin ice as well as heavily ridged ice is poorly represented. Moored Upward Looking Sonars (ULS) have the potential to obtain ice draft data, which can be converted to the ice thickness, with high temporal resolution (minutes) over periods of years. These instruments, however, have been introduced only quite recently. Consequently, ULS data are rather sparse and mainly restricted to the Arctic [*Vinje and Berge*, 1989; *Kvambekk and Vinje*, 1992; *Allison and Moritz*, 1995; *Melling et al.*, 1995; *Vinje et al.*, 1997].

The severeness of the gap in draft/thickness data has been recognized [*Allison and Moritz*, 1995; *OOSDP*, 1995]. To quote *Allison and Moritz* [1995]: "The outstanding monitoring problem is the determination of the ice thickness. Knowledge of the thickness distribution is of the highest priority to understanding the role of sea ice in the coupled atmosphere-ice-ocean system, and to detecting any variation in the ice due to climate change. However, there are very few systematic data on ice thickness, and for the Antarctic there is little information even on the broad spatial and seasonal climatology of ice thickness that can be used as a basis for model validation."

With the aim of filling part of this data gap we deployed Upward Looking Sonars at six locations in the Weddell Sea, Antarctica. The data collected represent the first ever multiyear time series of sea ice draft obtained from the Weddell Sea. The time series quantify the annual cycles, and how they vary regionally. The data are used to estimate the contribution of thermodynamic growth and deformation to the formation of sea ice and its large scale distribution.

INSTRUMENTS, DATA, AND METHODS

The measurements at the six locations were made during the period 1990 to 1994 (Table 1 and Figure 1) with moored Upward Looking Sonars (ULS) type ES-300V, manufactured by the Christian-Michelsen-Institute in Bergen, Norway [*Johannessen*, 1990]. Originally, six instruments were deployed at the end of 1990 with R/V Polarstern. Five were recovered successfully in late 1992 and early 1993. Four had worked properly during more than 2 years of deployment (Table 1). At all 6 mooring locations new instruments were deployed, three of which were recovered in March 1995. Three instruments were lost. Two of the three recovered instruments had collected data during the first part of the deployment period, as indicated in Table 1.

The ULSs were mounted at the top of the moorings, at a depth of about 100 to 150 m below the sea surface (Figure 2). This depth represents roughly the maximum limit for proper operation, and was chosen to keep the instruments clear of passing icebergs. Nevertheless, most of the lost instruments (4 of 12 originally deployed) were probably torn off their moorings, or displaced together with their moorings, by icebergs. Also, some of the recovered instruments were noticeably scratched, and the data subsequently gave clear evidence of iceberg encounters.

The basic ULS principle of operation is transmitting pulses of sound towards the surface and measuring the time until return of the echo. From the sound pulse travel time the distance between the instrument and the reflecting target can be derived, provided information on the sound velocity distribution above the ULS exists. By subtracting from the echo-derived range the depth of the instrument, which is inferred from the readings of a high-precision pressure sensor in the ULS, an estimate of the depth of the sound reflecting target (the ice draft when the echo originates from a water/ice interface) is obtained. The pressure readings, however, need to be previously corrected for interference signals from air pressure variations.

The acoustic frequency of our ULSs is 300 kHz, which compares to a wavelength of roughly 0.5 cm. The pulse width is 66 ms. The sound pulse travel time is measured with a resolution 13 ms, yielding a resolution of the acoustic distance of approximately 1 cm. The opening angle of the acoustic beam is 2° (-3 dB drop) which, in combination with an instrument depth varying between 100 and 150 m, results in a surface window diameter of approximately 10 m. Pressure is measured with a resolution of 5 hPa. Also measured is the tilt of the instruments (with 1° resolution), the temperature at the depth of the instrument (resolution 0.1 K), and a further variable indicative of the target strength; the latter proved very valuable in discrimina-

Table 1: Alfred-Wegener-Institut ULS mooring locations and deployment periods in the Weddell Sea

ULS Ser.-No.	ULS Part.-No.	AWI Mooring	Position	Deployment Period	Measurement Period
06	09/90	207-2	63° 45.1' S, 50° 54.3' W	23.Nov.90 - 10.Jan.93	20.Nov.90 - 26.Nov.92
07	10/90	210-2	69° 39.6' S, 15° 42.9' W	11.Dec.90 - 22.Dec.92	10.Dec.90 - 16.Dec.92
08	11/90	208-2			instrument failure
09	12/90	212-2	70° 54.7' S, 11° 57.8' W	14.Dec.90 - 19.Dec.92	12.Dec.90 - 18.Dec.92
10	13/90	217	64° 25.1' S, 45° 51.0' W	24.Nov.90 - 08.Jan.93	20.Nov.90 - 26.Nov.92
11	14/90	209-2		not recovered	
23	28/91	212-3		not recovered	
24	29/91	208-3	65° 37.6' S, 36° 29.4' W	04.Jan.93 - 13.Mar.95	04.Jan.93 - 25.Jul.94
25	02/92	209-3	66° 37.4' S, 27° 07.2' W	31.Dec.92 - 12.Mar.95	31.Dec.92 - 10.Nov.93
26	03/92	217-2			instrument failure
27	04/92	207-3		not recovered	
28	05/92	210-3		not recovered	

ting between echoes returned from ice or open water surfaces. The sampling interval was set to either 4 or 8 minutes; each sample was composed of a burst of 4 pulses, one pulse per second.

The processing of the ULS measurements to yield data of ice draft and coverage encompassed the following steps.

1. Information on the vertical sound velocity profile and its variation in time was obtained with the aid of a mixed layer model [*Lemke et al.*, 1990] tuned to the salinity and temperature profiles measured during the deployment and recovery cruises, and to the temperature record of the instrument. Using this sound velocity model the sound pulse travel times were converted to acoustic distances.

2. The air pressure variations were eliminated from the pressure sensor records using the analyses of the European Centre of Medium Range Weather Forecast (ECMWF) for the mooring sites.

3. Using the corrected pressure readings and the acoustic distances the raw draft values were subsequently calculated.

4. Draft data were rejected as obviously erroneous when either the measured temperature or pressure was out of range, the echo was uncertain, the tilt angle exceeded 5°, the two most similar draft measurements obtained from one burst of pulses deviated by more than 2 m, or the draft was either larger than the ULS depth or less than -10 m. Nevertheless, the amount of rejected data is rather small, less than 0.03% of the total data set.

5. Drafts associated with either ice or open water were discriminated using the measure of target strength and the short term draft variations encountered during one burst of pulses. Relating the number of ice cycles to the total yields an estimate of the percent ice coverage.

6. The drafts of identified open water were used to adjust the surface level.

7. The bias (footprint error), which is likely to result from the ULS principle of operation as the combined effect of beam spreading (causing an extended surface window) and echo detection, was determined and subsequently removed. For determination of the bias, horizontal drill hole profiles collected in vicinity of our mooring sites were used.

The uncertainty in an ice draft measurement which remains after the final stage of the data processing is composed of an error in the surface level adjustment (estimated as 3.4 cm), an instrumental error (approximately 1 cm), and an uncertainty in the determination of the bias (roughly 2 cm). Assuming Gaussian error

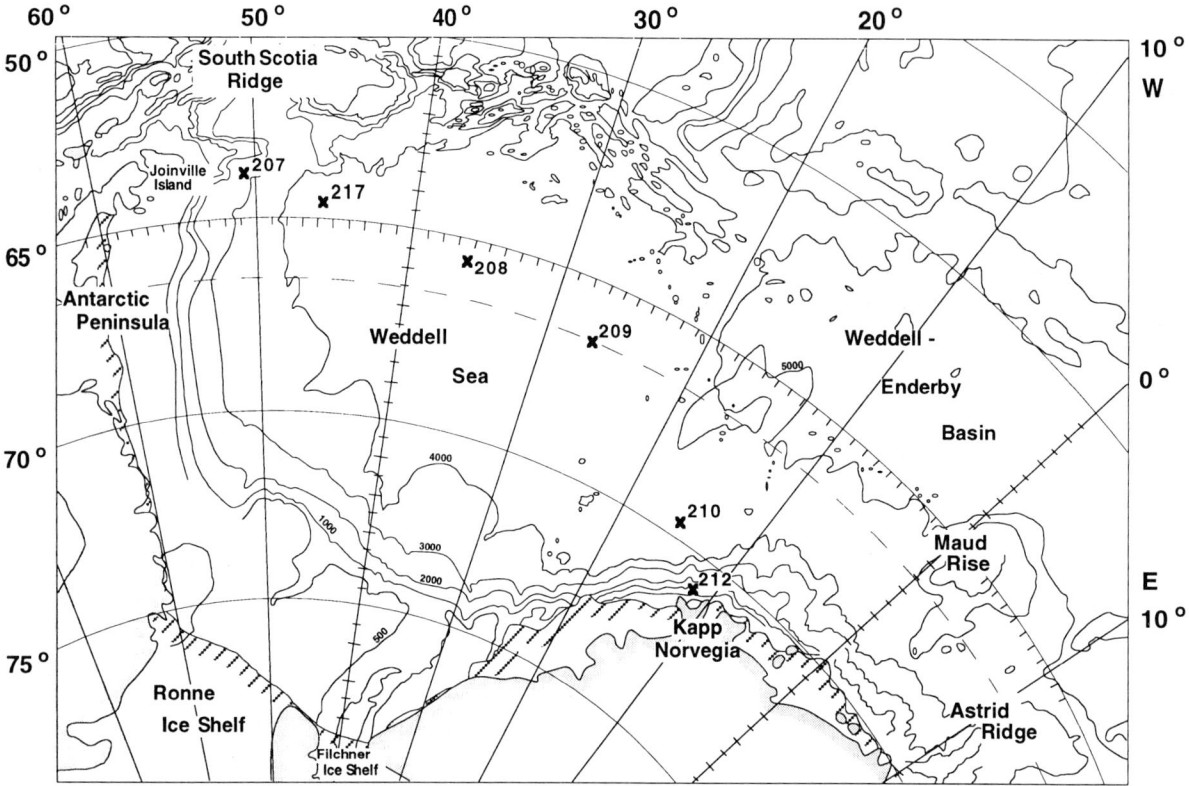

Fig. 1. Map of ULS moorings maintained by the Alfred-Wegener-Institut (AWI) in the Weddell Sea. The periods covered by measurements at the six mooring positions are given in Table 1.

propagation, these combine to a total measurement uncertainty of about 4 cm. The error in the ice coverage is estimated as 1.5%. Indications of other sources of measurement error, like shifts of the open water mode due to air bubbles beneath heavily wave-agitated surfaces or a loss of echoes from certain classes of sea ice (e.g. thin level ice), reported from ULS measurements in the Arctic [*Melling et al.*, 1995], were not found. Although sea ice and icebergs cannot be clearly distinguished in the ULS measurements, it is unlikely that icebergs and their fragments have a significant effect on the results (see the Discussion at the end of the paper). More details of the data processing and error estimation are provided in *Strass* [1997].

Given the nature of the data set, namely Eulerian time series at fixed positions, statistical analysis results in time referenced quantities:

Ice coverage indicates the fraction of time that ice is present during a given period (a day or a month, for instance) at a ULS mooring position. It would be equal to a spatial referenced statistic (for instance, ice concentration as obtained from satellite imagery) under certain conditions, e.g. an ice field that moves over the ULS at constant speed, or, if moving at variable speed, an ice field that is homogeneous over the length scale by which it drifts during the interval used in the time referenced statistic.

The mean ice draft is calculated as the sum of all draft measurements of ice, divided by the number of data cycles identified as ice, in a given time interval. To make a distinction, we use the expression "effective draft" for the mean draft calculated from all data cycles, including both ice and open water. For calculation of the means we normally include all valid measurements, without any restrictions to a certain range of drafts.

The standard error of the mean is calculated as the ratio of the standard deviation to the square root of the number of independent observations in the particular time interval; the number of independent observations is estimated by relating the product of sampling rate and surface window to the ice drift speed. The actual drift speed at the mooring sites is estimated from the ECMWF winds, using the empirical relationship determined by *Kottmeier and Sellmann* [1996] for the Weddell Sea; usually, the drift speed is high enough to accomplish a complete exchange of the ice between samplings in the surface window illuminated by the sonar. The statistical error is of typically 1% of the

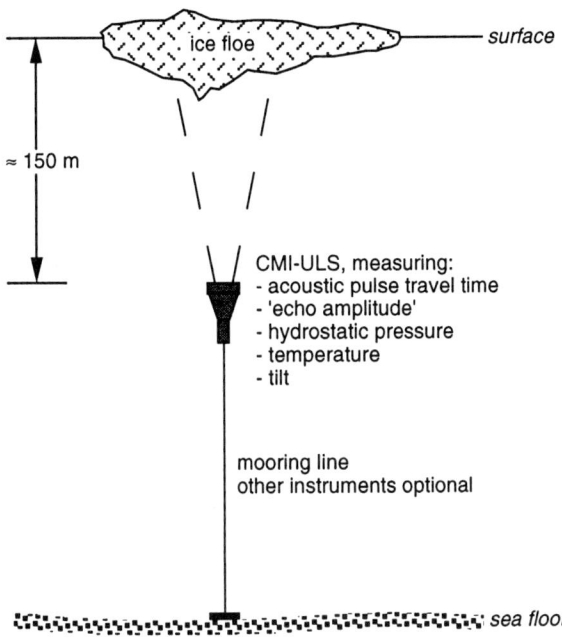

Fig. 2. Schematic representation of a ULS deployment.

monthly means (as shown later in Table 2), and is in addition to the measurement uncertainty of 4 cm indicated above.

In order to describe how the relative frequency of occurrence of ice draft measurements is distributed over the range of possible values, we use the discrete form of the probability density function,

$$p(d) = P(z<d<z+Dz)/Dz.$$

$P(z<d<z+Dz)$ is the probability that a draft measurement, d, lies between the values z and z+Dz. In the practical application it is estimated by dividing the number of draft measurements between z and z+Dz by the total number of data cycles recorded during a given period of time. Relating $P(z<d<z+Dz)$ to the bin width Dz (which we choose to be 0.1 m) results in the probability density p(d). The integral of p(d) over the whole range of possible values would equal 1. The maximum of p(d) indicates the draft mode, the most frequent (or most probable) draft.

RESULTS

The ULS mooring positions were chosen to cover the inflow as well as the outflow of ice [e.g. *Kottmeier and Sellmann*, 1996; *Emery et al.*, 1997] and water masses [e.g. *Fahrbach et al.*, 1994] to and from the southern Weddell Sea, as determined by the wind field and the Weddell Gyre. The Weddell Gyre is characterized by a basin-scale cyclonic circulation following the isobaths; the mean flow is southwestward at the eastern margin (moorings AWI-210 and -212) and northward at the western margin (moorings AWI-207 and -217) of the Weddell Sea. The mean flow in the centre of the gyre, where moorings AWI-208 and -209 were placed, is rather weak and of no clearly defined direction. Contrasts between the interior of the Weddell Gyre and the boundary currents are clearly reflected in the distribution of the measured ice drafts.

The Major Temporal Developments - A Qualitative Description

An overview of the ice draft measurements is presented in Figure 3. Most obvious is the annual cycle in the central Weddell Sea, e.g. at the location of mooring AWI-208 (Figure 3c). The time series from this mooring extends over one and a half years, including two austral summers which are marked by the frequent occurrence of open water. During the austral winter, the draft measurements are dominated by ice, and the draft values are clustered in a rather narrow band varying between 0 and 1 m. The obvious temporal trend of this band suggests that the ice draft increases almost monotonically during the first six months of the freezing period (April to September), then levels off, and at the end decreases (during December and January) much more rapidly than it grew previously. Beginning around mid-winter (July) the scatter in the ice drafts tends to increase, with higher drafts being observed more frequently during the subsequent months. Open water occurs from about October onwards, and becomes more abundant during January. During this season the vertical displacements of the open water surface, most likely caused by short gravity waves (see caption of Figure 3), also increase. The temporal increase of the surface displacements can be attributed to a wider opening of the ice, allowing for a larger wind fetch and the associated development of a wind sea. Open water observed during austral winter on the other hand is related to leads, the area of which is too limited to allow for substantial growth of the wind waves.

The other time series from the interior of the gyre (mooring AWI-209), of shorter duration, shows virtually the same behavior (Figure 3d) as the first one (AWI-208). Also at the position of AWI-210, located in the eastern part of the deep Weddell basin, a similarity can be identified (Figure 3e). These three time series have in common the fact that the little ice still remaining during the summer has a draft typically less than 1 m.

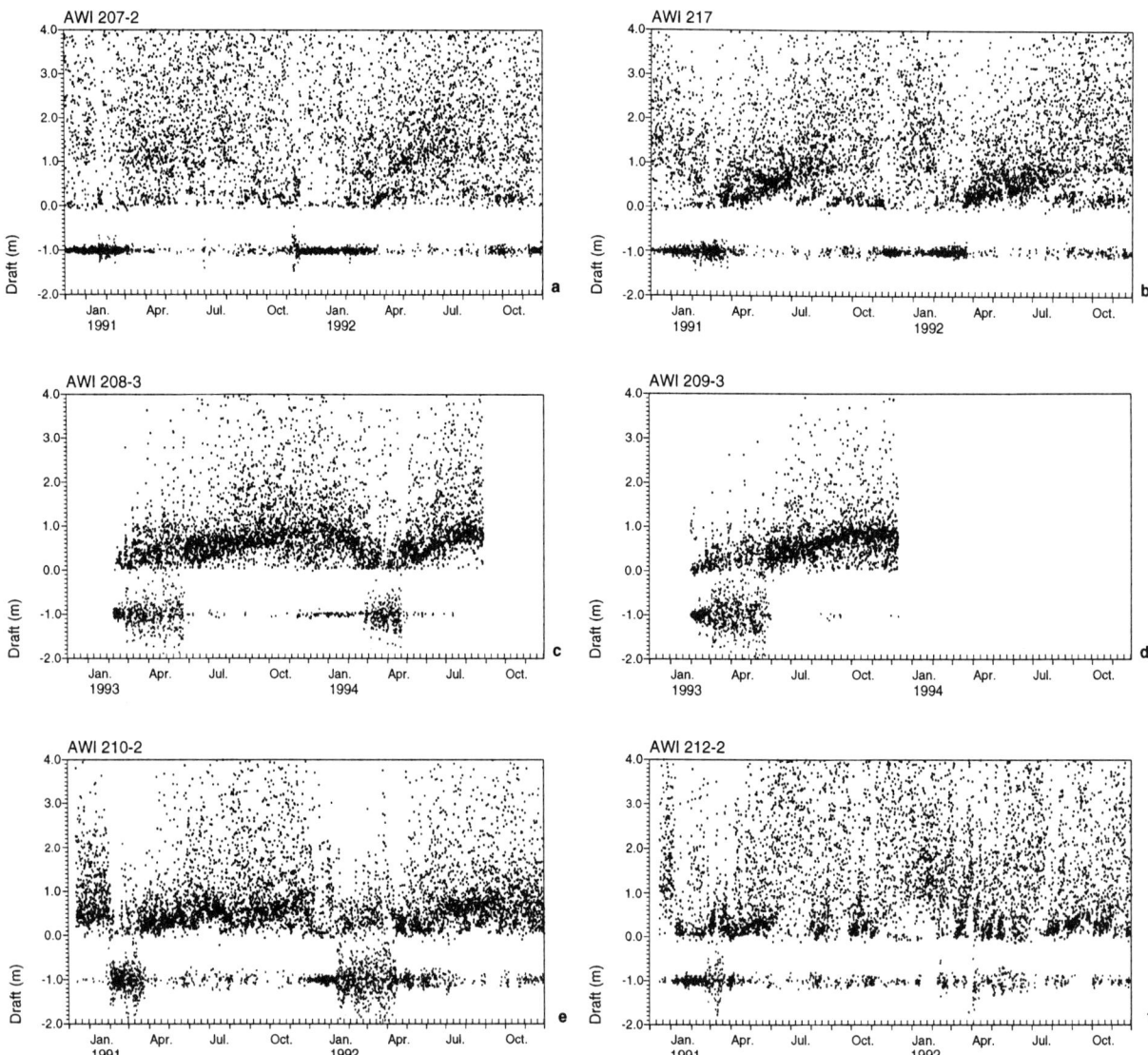

Fig. 3. Time series of drafts of ice and open water as determined from the ULS measurements. The drafts of identified open water are offset by -1 m for clarity of visualisation. The original open water surface displacements are scattered symmetrically around the zero level. This scatter results from short surface gravity waves, the pressure field of which, unlike that of long waves, does not extend deep enough to be measured by the ULS pressure sensor and hence to be eliminated during the data processing. The scatter due to surface waves also affects the ice draft data, but is averaged out if means are taken. These time series are subsamples, extracted at two-hourly intervals from the complete data set, which was collected with sampling intervals of 4 or 8 minutes. The individual panels in the figure, a - f, are organized so as to reflect the mooring positions across the Weddell Gyre from north-west to south-east.

At the western side of the deep Weddell basin similarities to the interior are less pronounced. During both freezing seasons covered by the two-year time series from AWI-217, the ice draft tends to increase monotonically only during the first four months, April to July (Figure 3b). Later, from around August onwards, the draft becomes very variable with an increase in the number of larger draft values. Open water and very thin ice of typically less than 0.3 m draft, probably new ice formed in the leads, also are observed more frequently during the second half of winter.

Closer to the coast, within the boundary currents on

either side of the gyre, periods of monotonic growth are of even shorter duration and the ice drafts are highly variable. At AWI-207 in the western boundary region (Figure 3a), only the second winter (early 1992) begins with monotonic growth in ice draft, and this period of monotonic growth lasts for only about two months (April and May). Newly grown thin ice (of less than 0.5 m draft) contributes little to the total ice cover. A substantial amount of thick ice (above 1 m) which survived the preceding austral summers is still present at the beginning of the new freezing seasons. Besides the initial phase of monotonic growth in early winter, new ice formation is observed during several additional periods, which follow partial ice break up. Such sporadic openings of the ice cover followed by phases of new ice formation dominate the temporal development on the opposite, eastern boundary region, at mooring AWI-212 (Figure 3f). Because of the sporadic behavior, no clear annual cycle can be identified at this location. The periods of opening and refreezing apparently occur on time scales of days to weeks. When the reclosure of the ice cover is accompanied by a rapid increase of the ice draft (as in early June 1991), the reclosure is more likely due to ice convergence and not to freezing.

Ice Coverage, Ice Draft Mode and Mean

In the central Weddell Sea (moorings AWI-208 and 209) the annual cycle is evident from three statistical quantities: the mode of the ice draft, the mean ice draft, and the ice coverage. The daily ice draft mode, representing the most frequently observed ice draft, varies around 0.3 m at the end of summer (Figures 4c and d). With superimposed short-term variations, it increases

Table 2: Monthly Mean Ice Drafts (m)
The monthly means given for AWI-207, 217, 210, and 212 are averages taken from two years of observation, those of 208 are from either one or two years, and those of 209 are from one year only. The lower values in each row (±) indicate the standard errors of the means.

	AWI 207-2	AWI 217	AWI 208-3	AWI 209-3	AWI 210-2	AWI 212-2
Jan	2.64 ± 0.03	1.80 ± 0.02	0.64 ± 0.01	0.23 ± 0.01	0.86 ± 0.01	2.22 ± 0.03
Feb	2.27 ± 0.02	1.33 ± 0.01	0.68 ± 0.01	0.43 ± 0.01	0.76 ± 0.03	1.79 ± 0.02
Mar	1.93 ± 0.02	0.71 ± 0.01	0.64 ± 0.01	0.52 ± 0.01	0.70 ± 0.01	1.36 ± 0.03
Apr	2.17 ± 0.02	0.85 ± 0.01	0.77 ± 0.02	0.48 ± 0.01	0.57 ± 0.02	1.12 ± 0.03
May	2.56 ± 0.02	1.09 ± 0.02	0.95 ± 0.03	0.57 ± 0.01	0.89 ± 0.01	1.48 ± 0.02
Jun	2.34 ± 0.02	1.12 ± 0.01	1.22 ± 0.02	0.77 ± 0.01	1.16 ± 0.01	2.63 ± 0.02
Jul	3.14 ± 0.03	1.79 ± 0.02	1.41 ± 0.02	0.87 ± 0.01	1.01 ± 0.01	3.56 ± 0.04
Aug	3.21 ± 0.03	2.06 ± 0.02	1.37 ± 0.02	1.01 ± 0.01	1.34 ± 0.01	1.32 ± 0.03
Sep	3.28 ± 0.04	2.67 ± 0.02	1.48 ± 0.03	1.02 ± 0.01	1.17 ± 0.01	3.13 ± 0.03
Oct	3.32 ± 0.03	2.29 ± 0.02	1.47 ± 0.02	0.93 ± 0.01	1.44 ± 0.01	1.30 ± 0.02
Nov	3.22 ± 0.03	2.06 ± 0.02	1.33 ± 0.02	0.99 ± 0.01	1.09 ± 0.01	2.91 ± 0.03
Dec	3.23 ± 0.03	2.68 ± 0.03	1.29 ± 0.03		0.94 ± 0.01	2.92 ± 0.03

Table 3: Monthly Ice Draft Modes (m)
The mode indicates that draft, analyzed in classes of 0.1 m width, which is observed most frequently during each month. The modes given for AWI-207, 217, 210, and 212 are averages taken from two years of observation, those of 208 are from either one or two years, and those of 209 are from one year only.

	AWI 207-2	AWI 217	AWI 208-3	AWI 209-3	AWI 210-2	AWI 212-2
Jan	0.7	0.9	0.4	0.1	0.4	0.8
Feb	0.1	0.2	0.2	0.3	0.3	0.7
Mar	0.2	0.3	0.5	0.4	0.4	0.2
Apr	0.7	0.3	0.5	0.4	0.3	0.1
May	0.8	0.5	0.6	0.5	0.3	0.2
Jun	0.8	0.6	0.6	0.6	0.6	0.3
Jul	1.4	0.8	0.7	0.7	0.6	0.1
Aug	0.3	0.6	0.7	0.8	0.7	0.3
Sep	0.1	0.6	0.8	0.8	0.7	0.3
Oct	0.2	0.2	0.8	0.8	0.6	0.3
Nov	0.6	0.4	0.8	0.8	0.6	0.1
Dec	0.8	1.2	0.8		0.3	1.2

130 ANTARCTIC SEA ICE: PHYSICAL PROCESSES, INTERACTIONS AND VARIABILITY

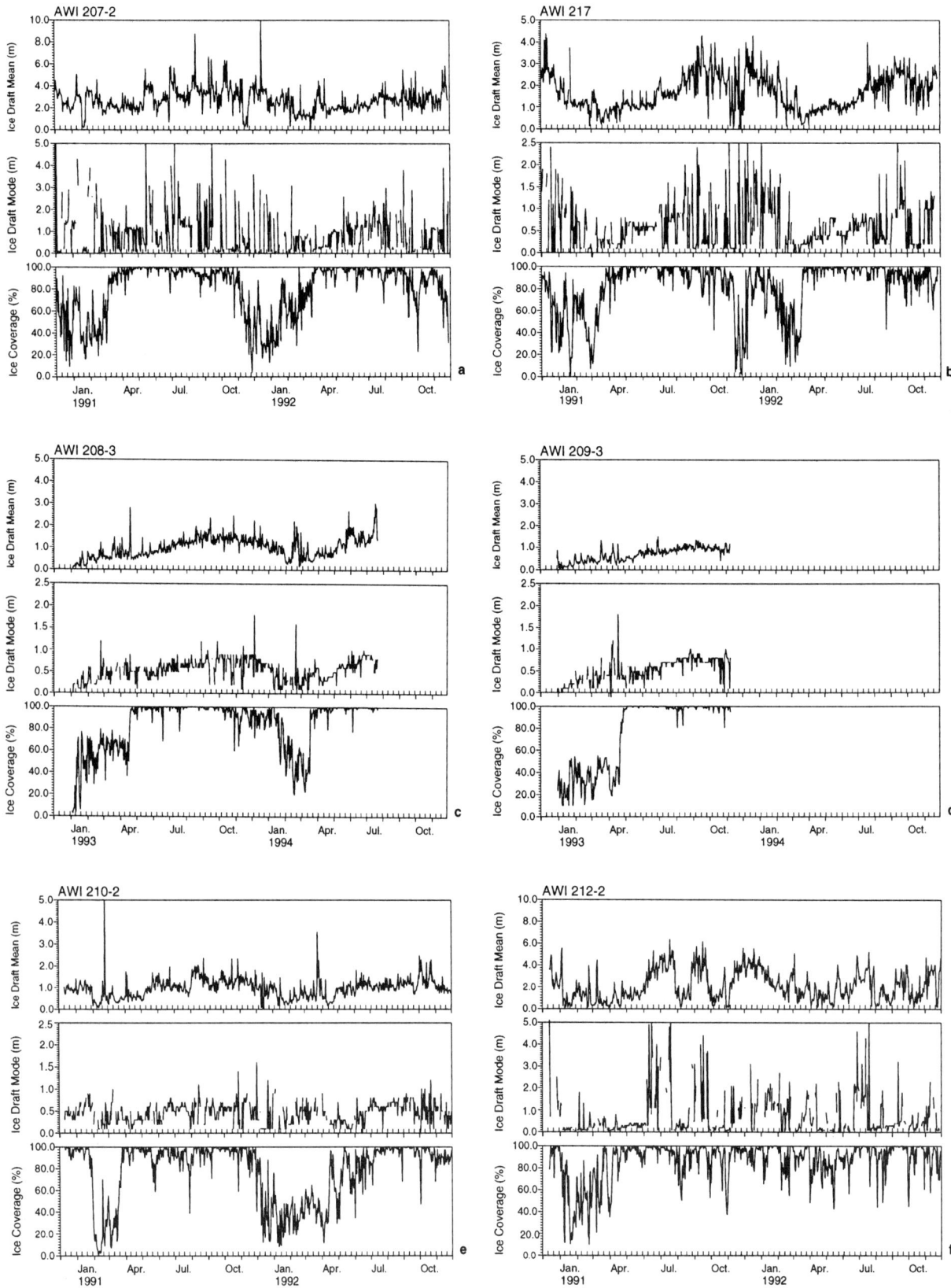

gradually during winter to 0.8 m, and this value persists for the period August/September through November (Table 3). The monthly mean ice draft (Table 2) at AWI-208 varies annually between a minimum of 0.64 m in March and a maximum of 1.48 m in September. The freezing season in the central Weddell Sea starts with a rapid closure of the ice cover. The coverage increases during only one week (either in March or April, depending on the year) from around 50% or even less to more than 95% (Figures 4c and d). This rapid closure seems to be caused by the formation of new ice, as suggested by a simultaneous decrease of the ice draft mean and mode. Rapid autumnal closure of the ice cover is also observed in the boundary regions, but not that regularly; it characterizes the start of the first of the two freezing seasons contained in the AWI-207 time series (Figure 4a), the second freezing season at AWI-217 (Figure 4b), and the first freezing season at AWI-210 (Figure 4e).

In the boundary regions, an annual cycle is not consistently evident from all three statistical quantities, ice draft mean and mode and coverage. At the westernmost mooring, AWI-207, the annual cycle is clearest in the ice coverage (Figure 4a); as given by the monthly means (Table 4) the coverage is highest in May (97.3%), and lowest in December (45.7%). Only a weak annual cycle appears in the mean ice draft, with a minimum of 1.93 m in March, and a maximum of 3.32 m in October (Table 2). Further towards the central Weddell Sea at AWI-217, both the mean ice draft and the coverage reveal a clear annual cycle, although the broad winter maximum of draft is interrupted by intermediate minima. The broad summer minimum of ice coverage also is interrupted by intermediate maxima (Figure 4b). The minimum monthly mean ice draft of 0.71 m is found in March, and the maximum of 2.68 m in December, but the September value of 2.67 m is very close. At AWI-217 the ice draft mode also is subject to an annual cycle, in that distinctly higher daily modes (Figure 4b) occur in late winter and spring. However, very small modes (i.e., much thin ice) are also found during these times, with the result that the monthly draft mode (Table 3) does not follow an annual cycle. At AWI-210, east of the gyre centre, the ice

Table 4: Monthly Mean Ice Coverage (%)
The monthly means given for AWI-207, 217, 210, and 212 are averages taken from two years of observation, those of 208 are from either one or two years, and those of 209 are from one year only.

	AWI 207-2	AWI 217	AWI 208-3	AWI 209-3	AWI 210-2	AWI 212-2
Jan	51.7	65.0	61.0	27.5	63.9	73.4
Feb	53.2	52.0	53.3	35.2	27.7	69.1
Mar	85.1	61.2	62.2	39.9	47.5	81.4
Apr	96.0	94.1	85.2	46.8	74.6	82.7
May	97.3	98.2	98.8	99.4	82.4	88.8
Jun	96.9	96.7	98.4	99.9	89.0	93.6
Jul	95.7	96.1	99.0	99.4	91.3	92.3
Aug	95.0	94.3	99.8	97.8	95.8	86.5
Sep	87.3	91.8	99.2	99.6	97.4	96.0
Oct	88.8	91.0	96.9	98.6	93.9	88.3
Nov	68.9	78.1	90.4	98.9	91.5	90.2
Dec	45.7	71.7	89.8		71.6	94.3

coverage again reveals the clearest annual cycle (Figure 4e), with a minimum monthly mean value of 27.7% in February, and a maximum of 97.4% in September (Table 4). The mode and mean of the draft are also subject to an annual cycle which, however, is largely masked by shorter-term variations and is of only small amplitude. The minimum monthly mean draft of 0.57 m is found in April, and the maximum of 1.44 m in October (Table 2). The respective values of the monthly modes (Table 3) are 0.3 m (found in December

Fig. 4. Time series of the ice draft (top), the mode of the ice draft (middle), and the ice coverage (bottom), determined on the basis of daily intervals. The daily mean ice draft (top) is calculated exclusively from measurements obtained from ice, i.e., data cycles identified as open water are not considered. The mode (middle) is defined as the most frequent ice draft occurring in the preset time interval of 1 day, and is determined in classes of 0.1 m width; difficulties in uniquely determining the mode with 0.1 m resolution from daily intervals result from the wide scatter of the drafts, and cause the time series of the mode to appear as spiky with many gaps. The ice coverage (bottom) is determined as the fraction of ice data in percent related to the total amount of data cycles made up by ice and open water.

as well as in February, April and May) and 0.7 m (August and September).

Whereas the summer minima of the ice coverage, the draft mode and the mean draft typically occur simultaneously in February, the time of occurrence of the winter maxima differs between the three variables: the maximum coverage is found in June/July, the maximum draft mode in September, and the maximum mean draft in September/October. In other words, the ice draft mean and mode continue to increase for two to four further months after the ice cover started to open in advance of spring.

In the above analysis of the annual cycle we simply investigated summer-winter differences, and ignored the possibility that the phase may vary significantly between years. For instance, the first summer minimum of the monthly mean (not shown) ice coverage in the record from AWI-207 occurred in February, while the minimum of the next summer occurred in December of the same year. Or, at AWI-210, the ice coverage during the first summer was below 50% for only two months, but below that threshold for four months during the next summer. At the westernmost mooring, AWI-212, no clear annual cycle is revealed by any of the variables. An annual minimum of the ice coverage does occur in the first summer covered by measurements, but is lacking in the second (Figure 4f).

The lack of a minimum ice coverage during summer 1991/1992 is confirmed by satellite passive microwave (SSM/I) data (T. Martin and H. Schottenmüller, personal communication, 1995). These also show that the ice cover at the position of AWI-212 opened during summer 1990/1991, but that an almost closed ice field, aligned along the shelf ice edge and extending seaward to about halfway between moorings AWI-212 and 210, persisted throughout the summer of 1991/1992.

So far we have considered mean drafts determined exclusively from data cycles identified as ice. To investigate for the variation of the ice volume in a particular region we have to take into account varying degrees of ice coverage, i.e., we have to consider the effective draft, calculated from both ice and open water. The time series of the effective draft (Figure 5) for the central Weddell Sea, from moorings AWI-208, 209, and 210, show rather similar features and the most regular annual cycles. The minimum of the monthly mean effective draft in the first austral summer in the record from AWI-208 is 0.1 m, and the maximum in the first winter is 1.5 m. In the second summer the minimum is 0.4 m, and the maximum in the second winter which, however, was not completely covered by measurements, equals 1.6 m. The summer minimum contained in the time series from AWI-209 is 0.1 m, and the winter maximum equals 1.0 m. The record from AWI-210 indicates minima of 0.2 m in both summers, and the maxima of the two winters are 1.5 m and 1.4 m, respectively.

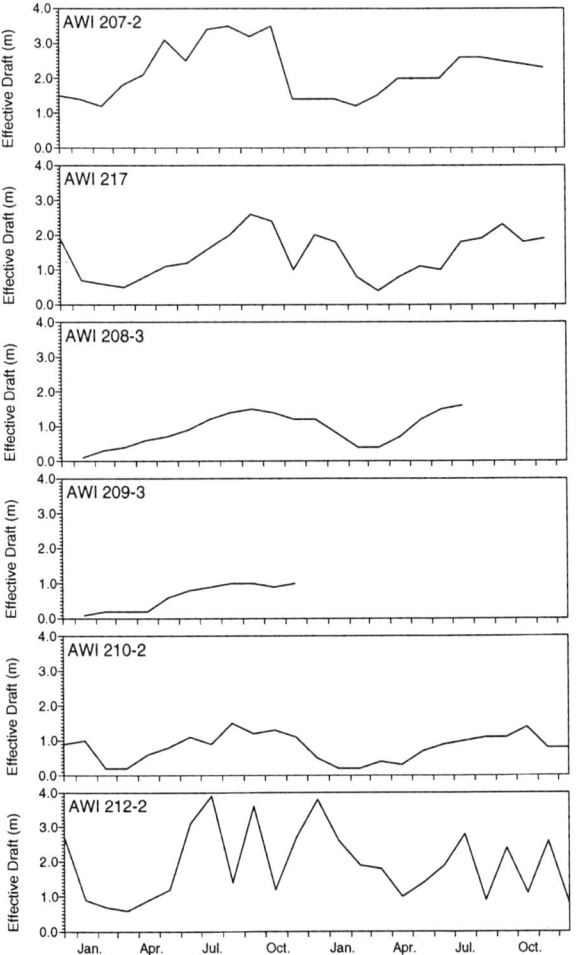

Fig. 5. Time series of the monthly mean effective draft, including ice as well as open water. From top to bottom, the six time series follow the mooring positions from west to east across the Weddell Sea. The two-year long records from mooring positions AWI-207, 217, 210, and 212 start in December 1990, those from AWI-208 and 209 in December 1992.

The differences between the respective winter maxima and summer minima are 1.4 m, 1.2 m, 0.9 m, 1.3 m, and 1.2 m. The scatter of these values is rather small, taking into account that they are based on observations made with a time delay of up to three years, and at locations up to 1000 km apart. The average winter-summer difference is 1.2 m.

Estimating a meaningful annual draft amplitude from the other three time series is more difficult, because of the larger irregularities in the annual cycle. Therefore we compare three-monthly (seasonal) averages of consecutive summers and winters. The first summer (January, February, March) in the record from AWI-217 indicates an average effective draft of 0.6 m, and the first

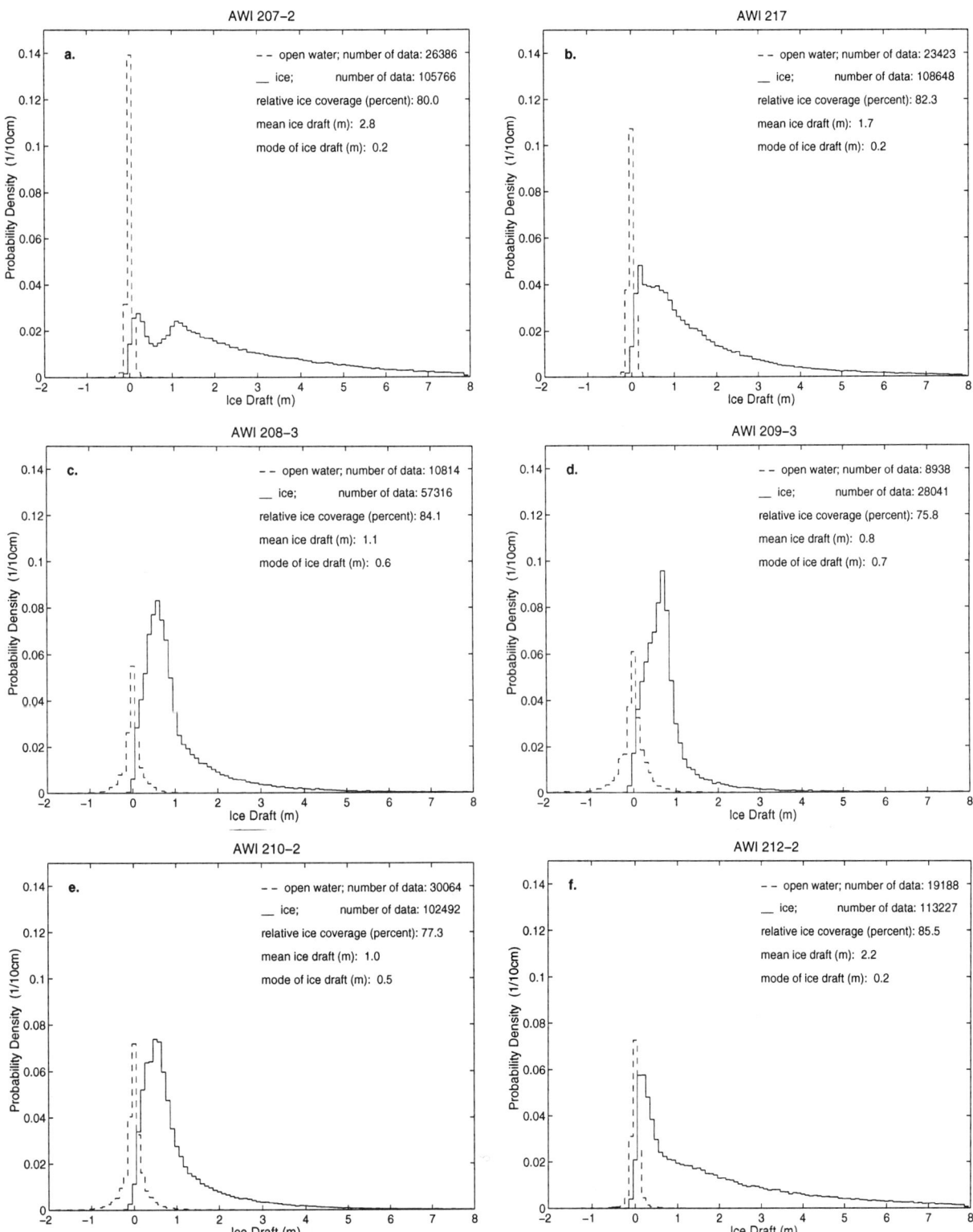

Fig. 6. Probability of occurrence of a draft measurement (ice or open water) in 0.1 m bins in the range -2 to 8 m. The sequence of graphs, a - f, is arranged so as to cross the Weddell Gyre from north-west to south-east.

winter (August, September, October) an effective draft of 2.3 m. The second summer (February, March, April) and winter (August, September, October) give values of 0.7 m and 2.0 m, respectively. From AWI-207 we obtain 1.4 m for the first summer (December, January, February) and 3.4 m for the first winter (August, September, October), 1.3 m for the second summer (December, January, February) and 2.6 m for the second winter (July, August, September). The winter-summer differences here are 1.7 m, 1.3 m, 2.0 m, and 1.3 m, with an average of 1.6 m. For the eastern boundary region (AWI-212) the annual draft amplitude cannot be assessed since the interannual variations appear to be of the same magnitude as the differences between the winter and summer seasons.

Probability Density Distributions

The probability density distributions determined from the six mooring records reveal some more regional differences of the ice cover. All the ice draft distributions shown in Figure 6 are non-Gaussian. However, those from the central Weddell Sea (moorings AWI-208, 209, and also 210; Figures 6c-e) are the least skewed and the most narrow; here the majority of the drafts is found in the interval 0.1 to 1 m. At the other sites the measured drafts are spread over a wider range. At the westernmost mooring location (AWI-207, Figure 6a) the probability density function of the ice draft is bi-modal, with one peak at 0.2 m and a second one of comparable magnitude at 1.1 m. Some weak evidence of a bi-modal distribution also can be found at AWI-217 (Figure 6b) immediately east of AWI-207, and also located in the western outflow though not in its core. At AWI-217, in addition to the primary mode at 0.2 m, there is a shoulder between 0.6 and 0.9 m. No indications of a bi-modal distribution exist at the other

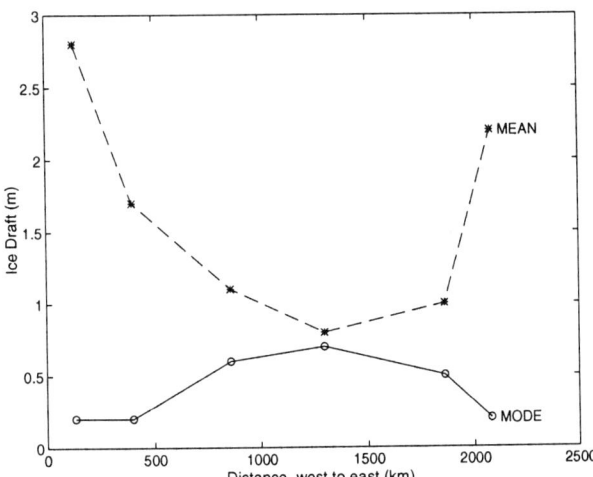

Fig. 7. Horizontal profile of the ice draft mode (continuous) and mean (dashed), crossing the Weddell Sea along the line of moorings from north-west to south-east. Mode and mean are based on the complete time series. Distance km 0 is located at Joinville Island at the northern tip of the Antarctic Peninsula.

side of the Weddell Gyre, in the inflow at the eastern boundary (AWI-212; Figure 6f). Here the probability density distribution shows a rather narrow maximum between 0 and 0.5 m, centered at 0.2 m. But larger drafts are abundant at AWI-212. Table 5 shows that more than 46% of the measured ice drafts in the eastern inflow are between 1.6 and 10 m. That is comparably close to the almost 59% large drafts found in the western outflow at AWI-207. In contrast, less than 20% of the drafts are in the class 1.6 to 10 m in the central Weddell Sea, at AWI-208, AWI-209, and AWI-210.

The systematic variation of the statistical properties across the Weddell Sea is displayed in Figure 7. The ice draft mode of the complete time series is highest (0.7 m at AWI-209) in the central Weddell Sea, and lowest in the boundary regions (0.2 m on both sides). The mean ice draft changes across the gyre inversely to the mode. The overall mean draft is lowest in the centre (0.8 m), and highest in the boundary regions: 2.8 m in the west and 2.2 m in the east.

The increase of the mean ice draft from 2.2 m to 2.8 m between inflow in the east and outflow in the west is mainly due to conversion of ice with drafts of less than 1 m to thicker ice. Figure 8 suggests that almost 75% of the thin ice with a draft of 0.2 m, which makes up the mode at AWI-212 in the inflow, is transformed to thicker ice. On average, most of the thin ice is converted to ice of 1.2 m draft during circulation through the southern Weddell Sea. For the larger drafts,

Table 5: Classification by draft ranges; numbers are in % of total drafts < 10 m observed at each mooring location.

	0.0 - 0.5 m	0.6 - 1.5 m	1.6 - 10 m
AWI 207-2	15.88	24.93	58.98
AWI 217	26.35	35.63	37.82
AWI 208-3	32.48	48.42	19.05
AWI 209-3	38.47	52.63	08.49
AWI 210-2	38.69	42.52	18.42
AWI 212-2	29.86	23.41	46.25

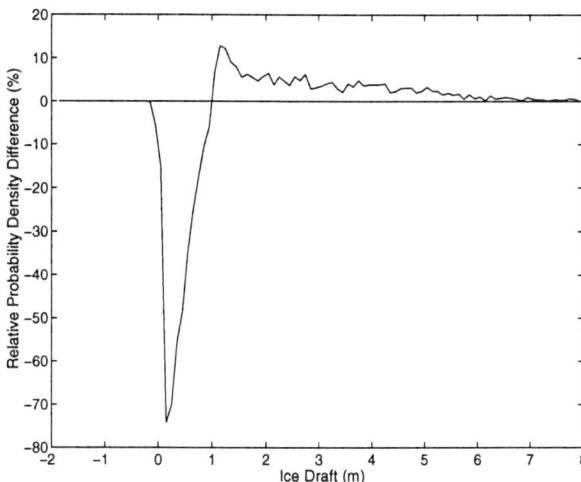

Fig. 8. Distribution of the difference of the probability density of drafts between moorings AWI-212 and AWI-207, elucidating the transformations which occur during circulation with the Weddell Gyre from the eastern inflow to the western outflow. The relative difference in %, related to the common mean of the two probability densities, is shown.

the gain in probability density gradually decreases. Virtually no increase takes place for drafts larger than 6 m, i.e., the same amount of very large drafts as occurs in the outflow at mooring AWI-207 already exists in the inflow region.

Also interesting in the probability density distributions of Figure 6 are the regional differences which exist in the open water surface displacements resulting from short surface gravity waves. All the open water displacement distributions are approximately Gaussian. But they differ significantly in width among the various locations. In the boundary regions (moorings AWI-207 and 212, as well as 217) displacements of the free surface are generally small, confined to the interval ±0.2 m. The distributions in the central Weddell Sea (at mooring locations AWI-208, 209, and 210) are much broader, with displacements to beyond ±1 m. The relation between the sea state and the wind fetch indicates that the ice cover at the central moorings opens in summer on a wide scale. In contrast, opening of the ice cover in the boundary regions occurs in the form of polynyas, leaving some ice between the mooring locations and the open ocean.

We also have made an attempt to investigate the data for the absolute maximum of the sea ice draft and the typical draft of icebergs. However, neither of these could be identified. The probability density distributions continue to tail off beyond the 8 m depicted in Figure 6, without giving evidence of a clear drop, a further maximum or even a shoulder which could be attributed to the existence of a cut-off in sea ice drafts or to a typical draft of icebergs. However, the distributions appear as more ragged at the far end approaching 100 m, due to the overall rareness of drafts in that range. From the data as they stand, sea ice and icebergs are indistinguishable by their draft distributions.

DISCUSSION AND CONCLUSIONS

Using moored Upward Looking Sonars we have collected a data set which contributes substantially to filling a gap in Weddell Sea ice draft observations. The new data set enabled us to quantify the annual cycle and to obtain well-resolved probability density distributions. That allows conclusions on the governing processes of ice formation to be drawn.

In the inner Weddell Gyre (moorings AWI-208, 209, and 210), where the monthly ice draft mode was observed to increase monotonically during the freezing season, the mode reaches maximum values at the end of winter that vary regionally from 0.7 to 0.8 m. These values compare with the 0.71 m mean ice draft of the first year level ice as obtained by drilling [*Lange and Eicken*, 1991]. This indicates that the modal draft values relate to level ice.

From the monotonic increase of the modal draft during winter in the inner Weddell Gyre we conclude that rafting, which would appear as step-like events, contributes to the formation of the level ice usually at a very early stage (corresponding to the observations of *Worby et al.* [1996] in the Bellingshausen and Amundsen Seas) when the ice thickness is still in the range of the resolution of our measurements (0.1 m bins for determination of the mode). The continuation of the monotonic increase until the end of the freezing season can then be explained by thermodynamic growth. The 0.8 m end-of-winter maximum draft mode thus indicates the limit of thermodynamic ice growth in the Weddell Sea which, as suggested by models [e.g. *Harder and Lemke*, 1994], amounts to 1 m per year.

The total ice formation during one freezing season in the central Weddell Sea is notably higher, as indicated by the 1.2 m winter-summer difference of the mean drafts. The 0.4 m surplus, compared to the 0.8 m modal draft of the end-of-winter ice, suggests that at least one third of the annually formed ice draft results from the processes of deformation, i.e. rafting and ridging. In the western Weddell Sea the difference between the mean drafts of winter and summer indicates an annual ice formation of 1.6 m. Comparing this value to the end-of-winter ice mode of the central Weddell Sea suggests that deformation contributes a minimum of 50% to the total amount of ice formed.

At the western and eastern margin of the Weddell Gyre the mode of the ice draft, with typical values of

0.2 m, is much less than in the central Weddell Sea. In the west (mooring AWI-207), advection of ice from the south [e.g. *Kottmeier and Sellmann*, 1996] limits the formation of new ice. The little new ice which is formed in the divergent ice field at the northern edge of the Weddell Sea is rapidly discharged into the Antarctic Circumpolar Current. Therefore, the mean draft of the newly formed ice remains as low as 0.2 m. In the east (mooring AWI-212), the low value results from the short duration of single freezing periods, which are associated with the opening and closing of the coastal polynya on time scales of days to weeks. However, many phases of new ice formation may take place during one particular freezing season, ultimately leading to the thick ice in the western outflow, with an overall mean draft of 2.8 m. This thick ice has been piled up by rafting and ridging from thin ice layers that originally resulted from thermodynamic growth. The effect of the processes of deformation is, while conserving the volume of ice, to reopen the surface and thus to pave the way for freezing of new ice.

Only in the core of the western outflow (at AWI-207) are young level ice and deformed ice clearly distinguished by their modal drafts, 0.2 m and 1.1 m, respectively (Figure 6a). The probability density distributions obtained at the other locations show a continuous transition between the draft mode, normally made up by the level ice, and the drafts of the deformed ice, with the thinnest deformed ice being the most frequent. At AWI-212 in the eastern inflow, for instance, a change of slope in the probability density distribution (Figure 6f) is noted between drafts of 0.5 m and 0.6 m, which obviously marks the transition between the level ice and the deformed ice. Hence, the most frequent draft of the deformed ice here is as small as 0.6 m, but is 3 times the level ice mode of 0.2 m.

The data show that large ice drafts are also readily produced at the entrance of the gyre flow into the Weddell Sea. The mechanism which makes this region one of massive ice production is probably associated with the formation of new ice in the coastal polynya, new ice which is then rafted and ridged at the seaward ice edge of the polynya when the winds are offshore, or rafted and ridged at the shelf ice edge when the winds veer onshore and the polynya recloses, as reported, for example, by *Eicken and Lange* [1989].

The difference between the mean ice draft of 2.2 m measured in the eastern inflow and the 2.8 m observed in the western outflow is only 0.6 m. As revealed by comparison of the respective probability density distributions (Figure 8), that change in mean draft is due mainly to transformation of the thinner ice, i.e. rafting and ridging of the level ice formed in the northeast. For the really large drafts, 6 m or more, virtually no further increase was observed between the inflow and the outflow. This does not necessarily mean that no further large drafts are produced, although thick ice floes certainly are more resistant to further rafting and ridging. What it reveals is an equilibrium of formation and destruction of the really large drafts in the southern Weddell Sea.

An interesting feature of the vernal disappearance of the ice in the western Weddell Sea (moorings AWI-207 and 217) is the delay between the opening of the ice cover and the decrease of the ice draft. While the average percent ice coverage begins to decrease as early as between June and September (Table 4), the monthly mean ice draft remains at high values close to the annual maximum until December (Table 2). This phase shift indicates that the initial opening of the ice cover is related more to divergence in the ice field than to melting of the ice. Melting contributes substantially to the decrease of the ice volume later in summer when the ice coverage already is reduced significantly.

The importance of advection for the local ice cover is underlined by the simultaneous increase of the ice draft and coverage in early summer 1991 (mid December through January) in the western Weddell Sea, which occurred distinctly after the spring opening in October and November (Figures 3b and 4b). This increase is also paralleled by large modal values that vary around 1.2 m. Most likely, this summer increase of ice draft and coverage is due to advection from the southern Weddell Sea. The advection occurred after the ice cover of the northern Weddell Sea had opened and more rapid northerly drift was possible.

Whereas our modal draft values are within the range hitherto reported from drillings [e.g. *Ackley*, 1979; *Lange and Eicken*, 1991; *Eicken et al.*, 1994], our mean values are notably higher. For instance, *Lange and Eicken* [1991] obtained a mean draft of 1.3 m in the northwestern Weddell Sea, where our closest ULS moorings (AWI-207 and 217) gave mean drafts of 2.8 m and 1.7 m, respectively. As an average value for a wider area of the Weddell Sea *Eicken et al.* [1994] estimated a mean ice thickness of 0.9 m, while our overall mean draft from all six ULS mooring sites is 1.6 m. Our higher mean drafts result from the fact that our probability density distributions tail off less steeply than those derived from the data sets obtained by drilling.

It is unlikely that the longer tails in our ULS-based probability density distributions originate from spurious echoes, e.g., from air bubbles or frazil ice beneath heavily wave-agitated sea surfaces, as reported from ULS measurements in the Arctic [*Melling et al.*, 1995]. Figure 3 shows that during longer periods of open water with increased surface wave amplitudes (notably the summer/early autumn seasons contained in the records from moorings AWI-208, 209, and 210), i.e. when the probability for the occurrence of bubble clouds is highest, large drafts are least frequently measured, and the average draft is far less than 1 m (Table 2).

Icebergs, of course, may have contributed to our mean draft values. The simplest approach to exclude icebergs from affecting our results is to restrict the calculation of means to ice of less than a maximum draft. If we take, for convenience, 15 m as the threshold, the resulting means are, however, not much lower than those obtained from the full range given in Figure 6. Restriction to 15 m maximum draft results in a mean of 2.70 m (instead of 2.75 m for the full range) at AWI-207, 1.69 m (instead of 1.72 m) at AWI-217, 1.08 m (instead of 1.09 m) at AWI-208, 0.78 m (0.78 m) at AWI-209, 1.03 m (instead of 1.04 m) at AWI-210, and 2.17 m (instead of 2.24 m) at AWI-212. The difference between the two approaches to estimate the mean is as low as 0.03 m, on average. That the difference is that low results from the fact that only small percentages of the measured drafts were beyond 15 m: 0.20% at AWI-207, 0.05% at AWI-217, 0.02% at AWI-208, less than 0.01% at AWI-209, 0.02% at AWI-210, and 0.13% at AWI-212. From this comparison icebergs thus appear to make only a minor contribution to the comparatively large mean drafts obtained from our ULS measurements.

However, icebergs, and in particular their fragments, may also have drafts less than 15 m. An approach to estimate the possible contamination by iceberg fragments is to consider the iceberg volume transports. The study by *Determann* [1991] suggests an annual iceberg calving rate of 180×10^9 m^3 a^{-1} at the Filchner-Ronne Ice Shelf, which is a major source of icebergs in the Weddell Sea. Another estimate of the iceberg calving rate is given by *Jacobs et al.* [1992]. If we set their estimate for the whole Antarctic in proportion to the Weddell Sea sector, we obtain a volume input of icebergs of 329×10^9 m^3 a^{-1}. In a worst case scenario, with maximum impact on our results, all icebergs entered annually would disintegrate entirely inside the Weddell Sea. If distributed evenly over the area south of our mooring transect (1.571×10^{12} m^2), icebergs would, on average, add between 0.12 m (Determann calving rate) and 0.21 m (Jacobs et al. rate) to the ice thickness, or 0.09 to 0.17 m to the ice draft. Even these comparatively small values certainly overestimate the iceberg effect because losses due to melting and, more important, the export of intact icebergs to the north are neglected.

The degree of overestimation due to neglecting the iceberg export can, however, be estimated only very roughly from our measurements. The time series from AWI 207-2 contains clear evidence of an iceberg encounter, during which the ULS was depressed to 190 m depth for 10 hours during June 31, 1991. The iceberg thickness and freeboard suggested by the draft of 190 m are 229 m and 39 m, respectively. The freeboard agrees with the 40 m height of the most frequently observed icebergs reported from earlier studies [*Weeks and Mellor*, 1978]. The ice drift during that time, as implied by the ECMWF winds, was 0.15 m s^{-1} to the north, and the current measured by a current meter at 300 m was 0.10 m s^{-1} northward. If we take the average to indicate the drift of the iceberg, it's meridional dimension would have been 4.5 km. An idea of the probability of the passage of icebergs of comparable size can be obtained from our mooring recovery/loss statistics (Table 1), as icebergs are the most likely reason for mooring losses. During the first deployment period we had the encounter mentioned above at the one mooring, and lost one other mooring, giving a ratio of 2/6. During the second deployment period we had a loss ratio of 3/6. Multiplying the total length of our mooring section which runs approximately 2000 km through the Weddell Sea by 2/6 indicates the horizontal dimension along the section of the sum of icebergs which pass across during a two years period, the deployment period. As we are interested in the iceberg export we have, however, to divide by 2 to account for the gyre circulation with inflow in the east and outflow in the west. If we assume that all icebergs which caused mooring losses have (at least) the dimension of the one recorded during the encounter, and divide again by 2 to scale the deployment period to the time basis of one year, we obtain an iceberg export rate of 177×10^9 m^3 a^{-1}. This estimate indicates that the export of berg ice volume through large icebergs is able to compensate almost completely the iceberg volume influx given by *Determann* [1991], and to compensate more than half of the influx given by *Jacobs et al.* [1992]. The high compensation rate results from the considerable volume of large icebergs, though their probability of occurrence at a particular time at a given position is indeed very low, 6×10^{-4}, as calculated from the encounter in the time series from AWI 207-2. Nevertheless, our estimate of the iceberg export rate can be considered as a conservative estimate for various reasons, first because we have taken the encounter/loss rate from the first deployment period with the lowest losses, and second because a mooring can be lost only once. Moreover, our mooring array does not cover the continental shelf and the upper part of the continental slope. The observation of an increase of the iceberg frequency towards the coast [*Morgan and Budd*, 1978; *Weeks and Mellor*, 1978] as well as observed drift tracks [*Swithinbank et al.*, 1977; *Birkenmajer et al.*, 1980] which reveal that icebergs circulate along the rim of the Weddell Sea close to the shelf ice edge, suggest that a substantial iceberg export occurs shoreward of our mooring array. The view of substantial export of intact icebergs to the northern periphery of the Weddell Sea into open water is consistent with the frequent observation of icebergs in that region [*Weeks and Mellor*, 1978], and the finding that icebergs stay largely intact as long as they are surrounded by pack ice [*Swithinbank et al.*, 1977; *Weeks and Mellor*, 1978;

Birkenmajer et al., 1980], and tend to disintegrate and melt rapidly only after being exposed to the rough seas and higher temperatures in the open water to the north [*Morgan and Budd*, 1978; *Birkenmajer et al.*, 1980; *Hamley and Budd*, 1986]. If the export of berg ice through large icebergs, of a size which is naturally excluded from our data analysis, is taken into account, the contamination by berg ice will be significantly lower than the 0.17 m draft calculated from the worst case scenario. The error introduced through icebergs may be a few cm, but, even if as large as 0.17 m, too low to explain the discrepancy of our mean drafts from those determined earlier by drilling.

What remains is the possibility that the larger sea ice drafts. which indeed occur in the Weddell Sea. are underrepresented in hitherto published distributions obtained from drillings, mainly because ice fields incorporating heavily deformed ice with the very deep keels are of limited accessibility for ship- (and ship plus helicopter) based observations. Although the above comparison of our mean drafts with those obtained earlier from drillings is not strictly appropriate due to discrepancies in location and time of the different data sets, and because of inherent methodical discrepancies (e.g. with respect to the treatment of voids in pressure ridges), the comparison might nevertheless indicate significant differences of the estimates of mean drafts in the climatological sense.

The new data set has permitted description of complete annual cycles of sea ice draft at 6 locations in significantly different regimes. This is the basis for validating modelled sea ice thicknesses. However, we are still unable to draw conclusions on the longer-term (e.g. interannual, decadel, and climatic) variability. These goals can only be achieved by continued measuring efforts.

Acknowledgements. The Upward Looking Sonars were deployed as part of the Weddell Gyre Study and are component of the German contribution to WOCE. The success of the mooring program is attributable to the skills of Gerd Rohardt and Ekkehard Schütt as well as to the valuable assistance of the captain and crew of R/V Polarstern. The ice drilling profiles used for bias correction were kindly provided by Hajo Eicken. Ralph Timmermann assisted in the data processing. The paper has benefitted greatly from the comments of Ernst Augstein and the constructive criticism of three reviewers. Contribution number 1263 of the Alfred-Wegener-Institut für Polar- und Meeresforschung.

REFERENCES

Ackley, S.F., Mass-balance aspects of Weddell Sea pack ice, *J. Glaciol.*, 24(90), 391-405, 1979.

Allison, I., and R.E. Moritz, Sea Ice in the Global Climate System: Requirements for an Ocean Observing System, *Report of the Ocean Observing System Development Panel*, 28 pp., Texas A&M University, College Station, TX, U.S.A., 1995.

Birkenmajer K., L. Kumoch, and K. Zubek, The last stages of Trolltunga drift in the Weddell Sea, Antarctica, *Polish Polar Res.*, 1(2-3), 235-237, 1980.

Broecker, W.S., The great ocean conveyor, *Oceanography*, 4, 79-89, 1991.

Carsey, F.D., R.G. Barry, D.A. Rothrock, and W.F. Weeks, Status and future directions for sea ice remote sensing, in *Microwave Remote Sensing of Sea Ice*, edited by F.D. Carsey, Geophysical Monograph 68, Chapter 26, pp. 443-446, AGU, Washington, 1992.

Determann, J., Numerical modelling of ice shelf dynamics, *Antarctic Science,* 3(2), 187-195, 1991.

Eicken, H., and M.A. Lange, Development and properties of sea ice in the coastal regime of the southeastern Weddell Sea, *J. Geophys. Res.*, 94(C6), 8193-8206, 1989.

Eicken, H., M.A. Lange, H.-W. Hubberten, and P. Wadhams, Characteristics and distribution patterns of snow and meteoric ice in the Weddell Sea and their contribution to the mass balance of sea ice, *Ann. Geophysicae*, 12, 80-93, 1994.

Emery, W.J., C.W. Fowler, and J.A. Maslanik, Satellite-derived maps of Arctic and Antarctic sea ice motion, *Geophys. Res. Lett.,* 24(8), 897-900, 1997.

Eppley, R.W., and B.J. Peterson, Particulate organic matter flux and planktonic new production in the deep ocean, *Nature*, 282, 677-680, 1979.

Fahrbach, E., G. Rohardt, M. Schröder, and V. Strass, Transport and structure of the Weddell Gyre, *Ann. Geophysicae*, 12, 840-850, 1994.

Hamley, T.C., and W.F. Budd, Antarctic iceberg distribution and dissolution, *J. Glaciol.*, 32(111), 242-251, 1986.

Harder, M., and P. Lemke, Modelling the extent of sea ice ridging in the Weddell Sea, in *The Polar Oceans and Their Role in Shaping the Global Environment: The Nansen Centennial Volume.*, edited by O.M. Johannesen, R.D. Muench, and J.E. Overland, Geophysical Monograph 85, pp. 187-198, AGU, Washington, 1994.

Hartmann, J., C. Kottmeier, C. Wamser, and E. Augstein, Aircraft measured atmospheric momentum, heat and radiation fluxes over Arctic sea ice, in *The Polar Oceans and Their Role in Shaping the Global Environment: The Nansen Centennial Volume*, edited by O.M. Johannesen, R.D. Muench, and J.E. Overland, Geophysical Monograph 85, pp. 443-454, AGU, Washington, 1994.

Jacobs S.S., H.H. Hellmer, C.S.M. Doake, A. Jenkins, and R.M. Frohlich, Melting of ice shelves and the mass balance of Antarctica, *J. Glaciol.*, 38(130), 375-387, 1992.

Johannessen, A.A., CMI ES-300 User's Guide, 21 pp., Christian-Michelsen Institut, N-5036 Fantoft, Bergen, Norway, 1990.

Knox, F., and M.B. McElroy, Changes in atmospheric CO_2: Influence of the marine biota at high latitude, *J. Geophys. Res.*, 89(D3), 4629-4637, 1984.

Kottmeier, Ch., and L. Sellmann, Atmospheric and oceanic forcing of Weddell Sea ice motion, *J. Geophys. Res.*, *101*(C9), 20809-20824, 1996.

Kvambekk, Å.S., and T. Vinje, Ice draft recordings from upward looking sonars (ULSs) in the Fram Strait and the Barents Sea in 1987/88 and 1990/91. *Norsk Polarinstitutt Rapportserie*, *79*, 43 pp., Oslo, Norway, 1992.

Lange, M.A., and H. Eicken, The sea ice thickness distribution in the northwestern Weddell Sea, *J. Geophys. Res.*, *96*(C3), 4821-4837, 1991.

Lemke, P., W.B. Owens, and W.D. Hibler, A coupled sea-ice mixed-layer pycnocline model for the Weddell Sea, *J. Geophys. Res.*, *95*(C6), 9513-9526, 1990.

Manabe, S., and R.J. Stouffer, Sensitivity of a global climate model to an increase of CO_2 concentration in the atmosphere, *J. Geophys. Res.*, *85*(C10), 5529-5554, 1980.

Martinson, D.G., Evolution of the Southern Ocean winter mixed layer and sea ice: open ocean deep water formation and ventilation, *J. Geophys. Res.*, *95*(C7), 11641-11654, 1990.

Melling, H., P.H. Johnston, and D.A. Riedel, Measurements of the underside topography of sea ice by moored subsea sonar, *J. Atmos. Oceanic Technol.*, *12*, 589-602, 1995.

Morgan, V.I., and W.F. Budd, The distribution, movement and melt rates of Antarctic icebergs, in *Iceberg Utilization, Proceedings of the First International Conference, Iowa State University, Ames, Iowa*, edited by A.A. Husseiny, pp. 220-228, Pergamon Press, New York, 1978.

Mosby, H., The waters of the Atlantic Antarctic Ocean, *Scientific results from the Norwegian Antarctic Expedition 1927-1928*, *1(11)*, 1-131 pp., Norske Videnskaps-Akademi, Oslo, 1934.

OOSDP (The Ocean Observing System Development Panel), Scientific Design for the Common Module of the Global Ocean Observing System and the Global Climate Observing System: An Ocean Observing System for Climate, 265 pp., Department of Oceanography, Texas A&M University, College Station, TX, U.S.A., 1995.

Sakshaug, E., and H.R. Skjoldal, Life at the ice edge, *Ambio*, *18*(1), 60-67, 1989.

Smith, W.O., Phytoplankton dynamics in marginal ice zones, *Oceanogr. Mar. Biol. Ann. Rev.*, *25*, 11-38, 1987.

Spindler, M., and G. Dieckmann, Ecological significance of sea ice biota, in *Antarctic Science*, edited by G. Hempel, pp. 60-68, Springer-Verlag, Berlin and Heidelberg, 1994.

Strass, V.H., Measuring sea ice draft and coverage with moored Upward Looking Sonars, *Deep-Sea Res.*, in press, 1997.

Strass, V.H., and E.-M. Nöthig, Seasonal shifts in ice edge phytoplankton blooms in the Barents Sea related to water column stability, *Polar Biol.*, *16*, 409-422, 1996.

Swithinbank, C., P. McClain, and P. Little, Drift tracks of Antarctic icebergs, *Polar Rec.*, *18*, 495-501, 1977.

Vinje, T., and T. Berge, Upward looking sonar recordings at 75°N - 12°W from 22 June 1987 to 20 June 1988, *Norsk Polarinstitutt Rapportserie*, *51*, 66 pp., Oslo, Norway, 1989.

Vinje, T., N. Nordlund, S. Østerhus, R. Korsnes, Å. Kvambekk, and E. Nøst, Monitoring ice thickness in Fram Strait, *J. Geophys. Res.*, in press, 1997.

Wadhams, P., The Antarctic sea ice cover, in *Antarctic Science*, edited by G. Hempel, pp. 45-59, Springer-Verlag, Berlin and Heidelberg, 1994a.

Wadhams, P., Sea ice thickness changes and their relation to climate, in *The Polar Oceans and Their Role in Shaping the Global Environment: The Nansen Centennial Volume*, edited by O.M. Johannesen, R.D. Muench, and J.E. Overland, Geophysical Monograph 85, pp. 337-362, AGU, Washington, 1994b.

Weeks, W.F., and M. Mellor, Some elements of iceberg technology in *Iceberg Utilization, Proceedings of the First International Conference, Iowa State University, Ames, Iowa*, edited by A.A. Husseiny, pp. 45-98, Pergamon Press, New York, 1978.

Worby, A.P., M.O. Jeffries, W.F. Weeks, K. Morris, and R. Jaña, The thickness distribution of sea ice and snow cover during late winter in the Bellingshausen and Amundsen Seas, Antarctica, *J. Geophys. Res.*, *101*(C12), 28441-28455, 1996.

V. H. Strass and E. Fahrbach,
Alfred-Wegener-Institut für Polar- und Meeresforschung,
Postfach 120161, D-27515 Bremerhaven, Germany.

(Received July 29, 1996;
accepted September 16, 1997)

SEA ICE DRIFT AND DEFORMATION PROCESSES IN THE WESTERN WEDDELL SEA

Cathleen A. Geiger

Thayer School of Engineering, Dartmouth College, Hanover, New Hampshire

Stephen F. Ackley

Snow and Ice Branch, U.S. Army Cold Regions Research and Engineering Laboratory, Hanover, New Hampshire

William D. Hibler III

Thayer School of Engineering, Dartmouth College, Hanover, New Hampshire

Data from Ice Station Weddell (ISW) during 1992 are used to examine sea ice drift and deformation activity to identify relevant external forces responsible for driving specific sea ice processes. Power spectra results from wind, sea ice, and ocean current measurements together with deformation analysis of sea ice reveal the following. First, the drift of sea ice in the western Weddell region is a low frequency (<1 cycle day $^{-1}$) dynamic process driven primarily by low frequency forcing in the form of moderate steady ocean currents and intermittent strong winds from high energy storm activity. Second, higher frequencies, specifically diurnal and semi-diurnal tidal/inertial oscillation frequencies, are the main contributors to sea ice deformation in this region. Shear deformation has large high and low frequency components, with elongation (normal deformation) oriented parallel to the shelf break being the main form of deformation at low frequencies. The observed higher frequency processes are driven by 12 and 24 hour ocean oscillations (2 and 1 cycle day $^{-1}$) with a 12 hour peak contributing the most to the total shear activity and the 24 hour peak contributing more to the solid body rotation (vorticity) of ice on scales as large or larger than the ISW array (150 km). East to west rising ocean bottom topography of the continental slope is also believed to play a major role in the directional preference of both observed ice drift and deformation in this region. Ice drift and deformation seem particularly sensitive to the forcing caused by topographic change as enhanced by ocean currents.

OVERVIEW: DRIFT AND DEFORMATION

Movement of sea ice is often referred to as ice "drift" because of the historic notion that ice is a passive recipient of the surrounding air and water motion. Today, however, we recognize that sea ice moves with a velocity resulting from both internal and external forces governed by the momentum balance in Eq. (1) [*Hibler*, 1979],

$$m\frac{D\vec{v}}{Dt} = -m f \hat{k} \times \vec{v} + \vec{\tau}_a + \vec{\tau}_w - m g \nabla H + \vec{F}_{ice}. \quad (1)$$

The external air ($\vec{\tau}_a$) and ocean ($\vec{\tau}_w$ and $-m g \nabla H$) forces, the effect of the earth's rotation ($-m f \hat{k} \times \vec{v}$)

and the internal ice force ($\vec{F}_{ice}$) make this a complicated system, particularly the internal ice force which resists alteration of the shape of the ice in a non-linear manner.

When the shape of the ice changes, deformation ensues and the shape is said to "yield" to the stresses and forces imposed on it. Stress, strain and strain-rate are terms used to quantify the processes of resistance, yielding and deformation. To clarify these terms, it is assumed at geophysical scales that sea ice can be treated in a two-dimensional plane with stress being the internal force per unit length and strain being relative displacement which is a dimensionless quantity. Strain-rate is the relative displacement per unit time which has the same units as frequency (s^{-1}) and is the spatial derivative (gradient) of the velocity field. Stress is linked to strain and strain-rate through a constitutive relation (or constitutive law). Since two-dimensional stress is force per unit length, the gradient of the resultant stress from the constitutive relation gives the internal force of the ice ($\vec{F}_{ice}$) as a force per unit area.

While considerable progress has been made in understanding the rheology of sea ice [see e.g., Hibler et al., in press], the precise non-linear (and likely anisotropic) character of sea ice rheology is still uncertain. This is primarily because of its complex multi-scalar, multi-fractured nature. Strain-rate is physically observable and as will be shown here, statistically calculable on a large scale. While understanding the geophysical strain-rate of the sea ice field does not provide us with the answer to the ice rheology problem, it does give very important information about the deformational properties of the ice which brings us one step closer to a clearly definable constitutive relation. Selecting the Ice Station Weddell (ISW) large scale array from 1992 as the data set, the focus here is to provide the reader with a series of statistical tools for analyzing large scale sea ice deformation as well as describing the limitations of the data and how to improve on the method. Finally we make use of this information to identify geophysical scale events occurring in the deformation field of the western Weddell Sea ice pack.

DATA PROCESSING

During the 1992 Ice Station Weddell (ISW) experiment [Gordon et al., 1993], a large scale drifting Argos buoy array spanning 150 km was set up in the western Weddell Sea. Five Argos buoys at remote sites located 25 to 100 km from the central camp recorded geographical position about every three hours, on average, via satellite, while a Global Positioning System (GPS) unit at the camp recorded geographical positions about 20 times per day. An overview of the buoy names and their deployment dates is given in Table 1. Figure 1 shows the general track of the buoys, their relative configuration within the array and the local bathymetry.

Data Filtering

Processing of the buoy data included the conversion of the temporally asynchronous geographical positions into a linearly interpolated hourly time series database containing the instrument number, date, time, and geographical position from each site. Using the time frame when all buoys were active but located on the camp floe, just prior to deployment, an estimate of instrument error was found to be about one half kilometer radially, with the largest relative errors being between the GPS and Argos units [Geiger, 1996]. Error in distance due to instrument accuracy propagates to error in velocity which, because of its presence at every measured time step, shows up as a high frequency signal. In order to minimize the measurement error, position data were passed through the 4 pole Butterworth low pass filter by Roberts and Roberts [1978] with the squared transfer function of

$$\mid H_B(j\omega) \mid^2 = \left[1 + \frac{\tan(\frac{\omega T}{2})}{\tan(\frac{\omega_c T}{2})}\right]^{2n} \quad (2)$$

where j is the complex value $\sqrt{-1}$, ω_c is the cutoff frequency ($T_c = 2\pi\omega_c^{-1}$ is the cutoff period), T is the discrete sampling time interval and n is the number of poles in the complex plane located at the polar coordinate angles $\pm\theta_1 = \pi/8$ and $\pm\theta_2 = 3\pi/8$. A Z-transform allows us to produce the discrete filtered signal y_k for any given time k through a two-step recursive formula to remove a phase shift in the transfer function signal [Roberts and Roberts, 1978]. Since the data set used must be synchronous and at regular sampling intervals, the one hour linearly interpolated data define the initial data sampling interval. The sampling interval T adequate for this data set is three hours, as this is approximately the average sampling time of the Argos buoys (the lesser sampled data set). Cutoff periods of T_c =7, 9, 11, 15, 18, 21, 25, 30 and 35 hours are selected for the analysis. The critical frequency in interpreting the

Table 1: Overview of Large Scale Array Program

Instrument Site (No.)	Activated (Jul. Day)	Deployed (Jul. Day)	Stopped (Jul. Day)	Total (hours)
Camp (1400)	43.44	43.44	158.66	2766.17
Alex (1430)	41.58	48.71 to 48.92	151.55	2640.45
Brent (1431)	41.72	61.75 to 62.04	153.71	2688.57
Dimitri (1432)	41.57	61.46 to 61.75	153.52	2687.92
Ed (1433)	41.57	50.67 to 51.00	182.95	3395.10
		63.46 to 63.83		
Chris (1435)	62.05	62.05	182.95	2902.73

filtered information is the Nyquist frequency which equates to a Nyquist period of $2T$, which is six hours in this case. Time periods below T (or frequencies above ω) are not computed in the spectral algorithm used. Examination of the error resulting from the selected cutoff time is considered further in the error analysis section.

Velocity Data

Velocity is computed from filtered geographical positions using a local Cartesian projection parallel to and centered at 69°S (y origin) and 54°W (x origin) and the linear calculation $u_i^{(k)} = \Delta x_i^{(k)}/\Delta t$, $v_i^{(k)} = \Delta y_i^{(k)}/\Delta t$ for velocity components at each of the $i = 1$ to 6 sites and k time intervals. The orientation of the x,y grid is chosen such that the y axis is parallel to the local shelf break topography with positive y values oriented northward in the general direction of the buoy drift.

Meteorological stations were located at three of the large scale array sites: site Chris (west of camp), site Dimitri (east of camp) and the main camp (Figure 1). From the data recorded, hourly averaged wind magnitude and direction were used. Because of the high correlation in wind data between recorded sites (Ed Andreas, personal communication 1995), the longest most continuous record, from site Chris, was chosen to represent surface meteorological readings for the array. Ocean current meter arrays were deployed at sites Alex, Chris and Dimitri each at 50 m and 100 m depths, at site Ed at 50 m and at the main camp at 25 m, 50 m and 200 m depths [*Muench and Gordon*, 1995]. Due to the strong barotropic nature of the current in this area, the presence of only one site with current measurements less than 50 m deep and the presence of the most accurate position system on the camp (a GPS unit), the main camp data plus results from *Muench and Gordon* [1995] and *Levine et al.* [1997] were chosen to represent the overall local current structure.

The ocean currents reported here are expressed as absolute velocity current as received in a pre-processed form (Muench, personal communication 1996). The complex demodulation filter by *McPhee* [1988] had already been applied to obtain absolute current from the difference between GPS, on the camp, velocities and the ocean current meter velocities from the ice-suspended current meters at 25 m and 200 m. An extensive tidal analysis done by *Levine et al.* [1997] confirms that this method accurately reflects the tidal activity in the area. We note that these data were derived from ice-mounted

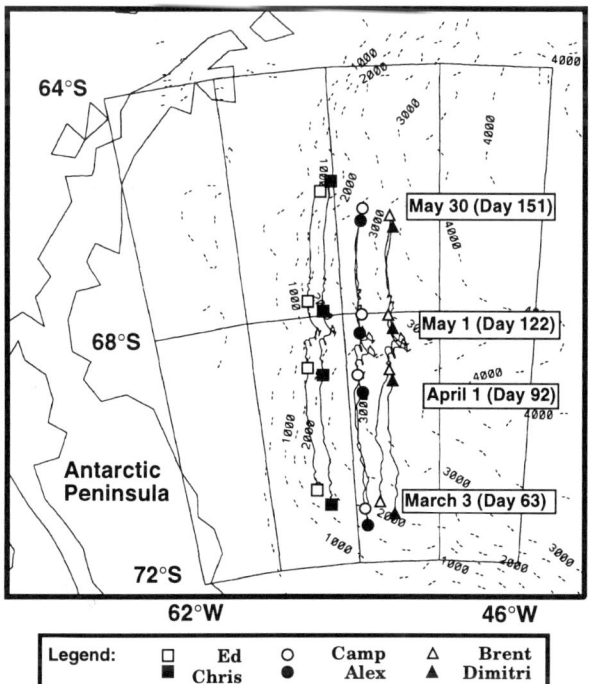

Fig. 1. Western Weddell Sea region with Ice Station Weddell (ISW) buoy tracks and local bathymetry.

current meters with ice motion statistically removed and hence not totally independent of the ice motion. The wind velocity and the 25 m and 200 m ocean current data were processed by converting them into the same format as the buoy information. To be consistent, the wind and ocean current data were additionally subjected to the same 3 hour cutoff low pass filter as the buoy data for comparison of spectral records. Results from *Muench and Gordon* [1995] show the ocean current data to be accurate to well within 1 cm s^{-1}, which will be used as a bound on the error for ocean current information presented here.

Strain-Rate Analysis

Multiple linear regression is used to statistically solve for the unknown centroid velocity components u and v and deformation tensor components $\partial u/\partial x$, $\partial v/\partial x$, $\partial u/\partial y$ and $\partial v/\partial y$ of the buoy array. The regression model is based on a two dimensional Taylor expansion of the velocity about the centroid of the array at the position (x, y). For ease of numerical computation, the model is expressed in matrix form (Note: Indicial notation used here, repeated indices sum.)

$$Z_{in} = X_{ij}\, \beta_{jn} + Err_{in} \qquad (3)$$

where

$$Z_{i1} \Rightarrow \{u_1, u_2, \cdots, u_N\} \qquad (4)$$
$$Z_{i2} \Rightarrow \{v_1, v_2, \cdots, v_N\}$$

$$X_{ij} \Rightarrow \begin{Bmatrix} 1 & x_1 - x & y_1 - y \\ 1 & x_2 - x & y_2 - y \\ \vdots & \vdots & \vdots \\ 1 & x_N - x & y_N - y \end{Bmatrix} \qquad (5)$$

$$\beta_{j1} \Rightarrow \left\{u, \frac{\partial u}{\partial x}, \frac{\partial u}{\partial y}\right\} \qquad (6)$$
$$\beta_{j2} \Rightarrow \left\{v, \frac{\partial v}{\partial x}, \frac{\partial v}{\partial y}\right\}$$

$$Err_{i1} \Rightarrow \{error_{u_1}, \cdots, error_{u_N}\} \qquad (7)$$
$$Err_{i2} \Rightarrow \{error_{v_1}, \cdots, error_{v_N}\}$$

for $i = 1, \cdots, N$ (Number of buoy sites)
$j = 1, 2, 3$ (Number of β's per component)
$n = 1, 2$ (Velocity components) $\Rightarrow u, v$.

The unknown β_{jn} matrix is solved using the approximation [*Hines and Montgomery*, 1990, chapter 15]

$$\hat{\beta}_{jn} = (X_{ji}\, X_{ij})^{-1}\, X_{ji}\, Z_{in} \qquad (8)$$

where X_{ji} is the transpose of the X_{ij} matrix and $\hat{\beta}_{jn}$ is the least squares best fit solution for β_{jn}. Examination of the Err_{in} term is presented in the error analysis section below. This procedure largely follows *Hibler et al.* [1974] except that the data are in x, y coordinates based on absolute satellite positioning rather than polar based coordinates used by Hibler to eliminate relative position rotation rates, which contaminated strain-rate estimates earlier. For additional information on multiple linear regression see, for example, *Geiger* [1996] and *Hines and Montgomery* [1990].

The above regression procedure allows us to estimate the following strain-rates and differential kinematic parameters (DKPs).

Non-invariant strain-rates:

$$\dot{e}_{xx} = \frac{\partial u}{\partial x} \qquad (9)$$

$$\dot{e}_{yy} = \frac{\partial v}{\partial y} \qquad (10)$$

$$\dot{e}_{xy} = \left(\frac{\partial u}{\partial y} + \frac{\partial v}{\partial x}\right)/2 \qquad (11)$$

Invariant strain-rates:

$$\dot{e}_I = \frac{\partial u}{\partial x} + \frac{\partial v}{\partial y}$$
$$= \dot{e}_1 + \dot{e}_2 \qquad (12)$$

$$\dot{e}_{II} = \sqrt{\left(\frac{\frac{\partial u}{\partial x} - \frac{\partial v}{\partial y}}{2}\right)^2 + \left(\frac{\frac{\partial u}{\partial y} + \frac{\partial v}{\partial x}}{2}\right)^2}$$
$$= \frac{\dot{e}_1 - \dot{e}_2}{2} \qquad (13)$$

Differential kinematic parameters (DKPs):

$$\dot{e}_I = \mathrm{DV} = \frac{\partial u}{\partial x} + \frac{\partial v}{\partial y} \qquad (14)$$

$$\mathrm{ND} = \frac{\partial u}{\partial x} - \frac{\partial v}{\partial y} \qquad (15)$$

$$\mathrm{SD} = \frac{\partial v}{\partial x} + \frac{\partial u}{\partial y} \qquad (16)$$

$$\mathrm{VT} = \frac{\partial v}{\partial x} - \frac{\partial u}{\partial y} \qquad (17)$$

$$\dot{e}_{II} = \mathrm{MS} = \sqrt{\left(\frac{\mathrm{ND}}{2}\right)^2 + \left(\frac{\mathrm{SD}}{2}\right)^2}. \qquad (18)$$

$\dot{e}_1$ and $\dot{e}_2$ are known as the principal strain-rates and are defined as the maximum and minimum extensional strain-rates respectively [*Fung*, 1977]. The DKPs can be illustrated as follows [*Massom*, 1992]: Divergence (DV) is the rate of change of area and also the first invariant of the system. The two quantities of normal (ND) and shear (SD) deformation denote change in shape but are not invariant. Normal deformation is the change in shape due to stretching in one direction and shrinking in the other without area change (i.e. elongation) while shear is due to stretching without area change of cross component terms. Finally, vorticity (VT) is an invariant for a non-rotating coordinate system and does not contribute to the deformation of the system; hence, the DKPs contain the three invariants of divergence (DV), maximum shear (MS) and vorticity (VT), with the remaining two non-invariant terms (SD and ND) being components of the second invariant. When examining the general deformation characteristics of the ice field, the invariant terms are the most useful because they show results independent of the chosen coordinate system. The non-invariant descriptions are dependent on the chosen coordinate system and often useful when trying to discern deformation responses due to a particular orientation in the field. In the current study, topography seems to have an important impact on the ice deformation field. Because of this, a local coordinate system has been chosen which more or less aligns with the topography in order to examine these effects further.

Spectral Analysis

A robust power spectrum algorithm has been constructed by *Press et al.* [1992] and used here with some modification. In addition to the inclusion of data windows, multiple periodograms and restriction between the Nyquist frequency ($-f_c$ to f_c), this algorithm uses an overlapping scheme to increase the number of periodograms κ by a factor of 2 (i.e. without overlap $\kappa = N/(4*M)$, and with overlap $\kappa = ((N/M)-1)/2$ where N is the total number of data selected and M is the number of desired frequencies). For the data presented here, $M = 64$ frequency bins is chosen, producing 5 periodograms from a sampling rate of three hours over a 2100 hour time series. This produces about 700 points of data which are padded with zeros at the end of the last periodogram to complete the 128 time samples needed for a 64 frequency periodogram. Modifications to the *Press et al.* [1992] algorithm include computation of spectra for multiple input variables and normalizing per band width (divide by Δf) to obtain the power spectral density. Output of the normalized power spectral density is plotted as a function of frequency (in cycles day^{-1}).

Time series of the array's drift and deformation give a considerable amount of information about the processes involved, but we gain even more insight by examining the power spectra of these terms. We note, however, that while analysis of the power spectrum of velocity is relatively straightforward, the analysis of the strain-rate is not, as we are dealing with a tensor.

One important property we wish to know is the power of the total strain-rate which is defined in 2D Cartesian space by

$$(\dot{e}_{ij})^2 \equiv \dot{e}_{ij}\dot{e}_{ij} = \dot{e}_{xx}^2 + \dot{e}_{yy}^2 + 2\dot{e}_{xy}^2. \qquad (19)$$

This expression is invariant as are the divergence and maximum shear encountered earlier. Through Parseval's Theorem [*Spiegel*, 1968], we can compute the power density of the divergence by Fourier transformation of the divergence quantity, i.e. $\dot{e}_1 + \dot{e}_2$, squaring the transformed result and dividing by the frequency bin width Δf. A similar procedure can be done with the maximum shear, taking into account the fact that it is a quantity whose magnitude is sign-dependent in a given coordinate system. Consequently, a plot of the magnitude of the maximum shear only represents a portion of the information contained in this invariant. To quantify the total strain-rate in terms of these other two invariant terms, we note the following relationship in the frequency domain:

$$\begin{aligned}2(\dot{e}_{ij}^2) &= (\dot{e}_{xx}+\dot{e}_{yy})^2 + (\dot{e}_{xx}-\dot{e}_{yy})^2 + (2\dot{e}_{xy})^2 \\ &= \mathrm{DV}^2 + \mathrm{ND}^2 + \mathrm{SD}^2 \\ &= (\text{divergence})^2 + (2\text{ max. shear})^2 \\ &= (\dot{e}_1+\dot{e}_2)^2 + (\dot{e}_1-\dot{e}_2)^2. \end{aligned} \qquad (20)$$

Hence, the total strain-rate power is equal to half the sum of the square of the Fourier components of the three deformational DKPs (vorticity excluded) or half the squared sum of both the sum and difference of the maximum and minimum strain-rates. Using this information, we define the following useful invariant power spectra terms to describe the deformational power of the system:

- Total Strain-Rate Power

$$(\dot{e}_{ij})^2 = \frac{DV^2 + ND^2 + SD^2}{2}$$
$$= \frac{(\dot{e}_1 + \dot{e}_2)^2 + (\dot{e}_1 - \dot{e}_2)^2}{2} \quad (21)$$

- Total Strain-Rate Divergence Power

$$\frac{DV^2}{2} = \frac{(\dot{e}_1 + \dot{e}_2)^2}{2} \quad (22)$$

- Total Strain-Rate Shear Power

$$\frac{ND^2 + SD^2}{2} = \frac{(\dot{e}_1 - \dot{e}_2)^2}{2}$$
$$= \frac{(\dot{e}_{ij})^2 - (\dot{e}_1 + \dot{e}_2)^2}{2} \quad (23)$$

RESULTS

General Ice Drift Features

An overview of the buoy drift pattern is shown in Figure 1. Within the time frame of March 3 (day 63) to May 29 (day 150) for leap year 1992, the following features are noted. The general drift pattern in the western Weddell is primarily northward and very slightly eastward. Relative to the bathymetry, the array configuration and ice drift proceed northward along the continental shelf slope with the westernmost sites (Ed and Chris) tracking along the 1500 to 2000 m isobaths and the remaining sites tracking closer to the 2500 to 3500 m isobaths to the east. The depth of the water column below the array decreases by about half from east to west. At the beginning of the drift period a large convergence event moves site Alex from east to south of the camp (some of this activity occurs before day 63 while the array is being deployed). Between April 1 (day 92) and May 1 (day 123), the entire array undergoes at least two complete cyclonic (clockwise) loops. The largest of these, between day 116 and 121, coincides with the passage of a 5 day atmospheric low pressure system also seen in the surface wind data and noted during the field experiment. The wind shifts from northeast to southeast to northwest during that time, producing a corresponding sea ice response, as seen in Figure 1. A final feature to note is that the relative position of the buoys contains a significant amount of drift perturbation activity resulting from local deformation processes in the ice. Investigation of this activity is considered in detail below.

Time Series of Drift and Deformation

A series of multiple linear regressions were made using different low pass filter (LPF) cutoff times from 7 to 35 hours. Time series plots of the centroid velocity components from these runs are shown in Figure 2. The general trend (i.e. the low frequency signal) is similar for each low pass filter cutoff time between 7 and 35 hours. In contrast, three different ranges of low pass filter cutoff times show a decreasing amount of velocity perturbation activity from less than 12 hours (7, 9, and 11 hour LPF), between 12 and 24 hours (15, 18, and 21 hour LPF), and greater than 24 hours (25, 30, and 35 hour LPF). We also note from Figure 2 that there is negligible difference in the LPF cutoff times within these three ranges (e.g. 15, 18 and 21 hour LPF are indistinguishable in the plots), but considerable differences exist between each range. Based on these results, the deformation tensor components from three representative cutoff times (9, 15 and 30 hours) were selected and subjected to strain-rate analysis. Examples from the 9 and 30 hour results are presented here in Figure 3. In comparing both directionally-dependent deformation

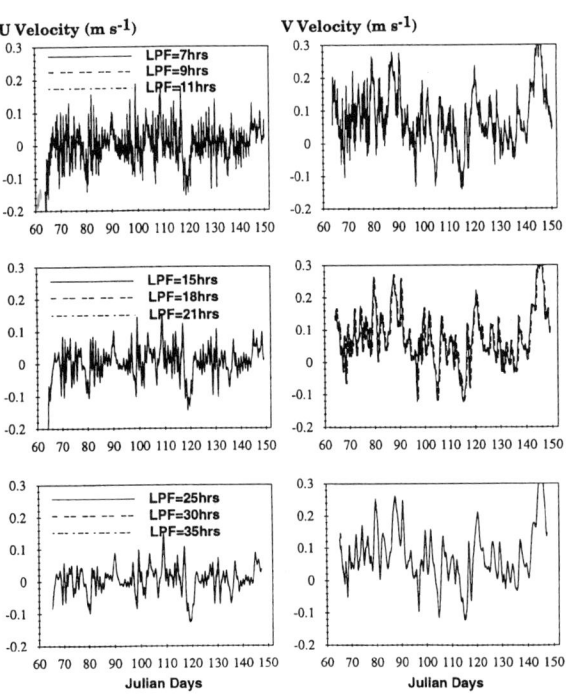

Fig. 2. Buoy array's centroid velocity components from multiple linear regression with low pass filter cutoff times of 7, 9 and 11 hours (upper panels), 15, 18 and 21 hours (middle panels) and 25, 30 and 35 hours (lower panels).

tensor components (Figure 3) and invariant components of velocity magnitude, divergence and maximum shear (Figure 4), we note a significant change in scale from $\pm 4 \times 10^{-6}$ s^{-1} for the 9 hour LPF versus $\pm 1.5 \times 10^{-6}$ s^{-1} for the 30 hour LPF deformation terms. In contrast, the velocity magnitude (upper panels in Figure 4) retains the same range of magnitude but is significantly smoother at the 30 hour LPF than for the 9 hour LPF. An intermediate decrease in deformation amplitude occurs from the 9 to 15 hour and 15 to 30 hour LPF cases.

To obtain a general idea of the trends in these time series, the average value, rms (root-mean-square) and average magnitude for each of the velocity, deformation tensor and DKP components for 9, 15, and 30 hour LPF cutoff times are presented in Table 2. From Table 2 and Figure 2 we see that the centroid has a u velocity, in the x direction, which oscillates close to zero, but a large average magnitude (4.15 cm s^{-1}) at the 9 hour LPF. At the 30 hour LPF, the value decreases by a third to 2.85 cm s^{-1} as does the rms, indicating strong eastward (positive) to westward (negative) fluctuations, which increase as higher frequency information is included. The y direction, on the other hand, centers close to 7.7 cm s^{-1} at the 9 and 15 hour LPFs and only decreases to 7.5 cm s^{-1} at the 30 hour LPF. This reduction concurs with the 6-7 cm s^{-1} average northward drift computed by *Muench and Gordon*[1995] using a 40 hour LPF. The basic pattern that emerges from this is a fairly steady strong northward drift parallel to the bottom topography with strong east-west oscillations in the current, in the direction normal to the topography, which are about a third stronger in the x direction when the higher sub-daily information is included.

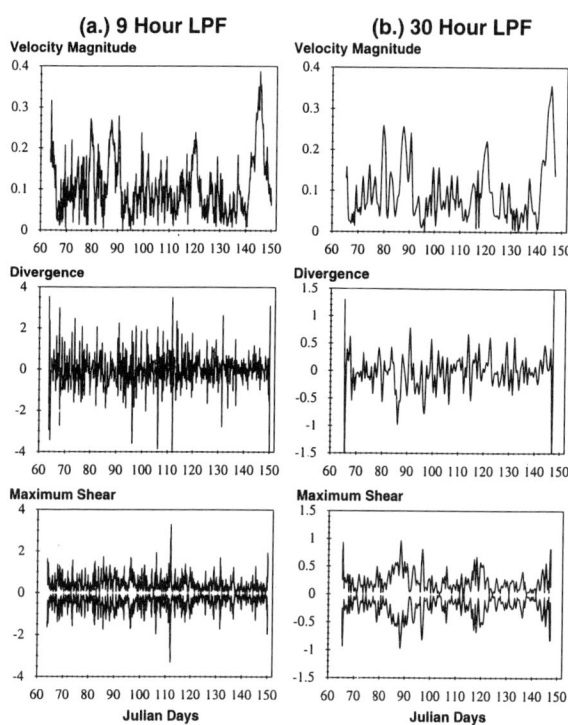

Fig. 4. Buoy array invariants of centroid velocity magnitude, divergence and maximum shear computed from multiple linear regression analysis using 9 (left) and 30 (right) hour low pass filters (LPF). Velocities are in units of m s^{-1}. Divergence and maximum shear are in units of $\times 10^{-6}$ s^{-1}. Since maximum shear is positive definite, both positive and negative values of quantity shown.

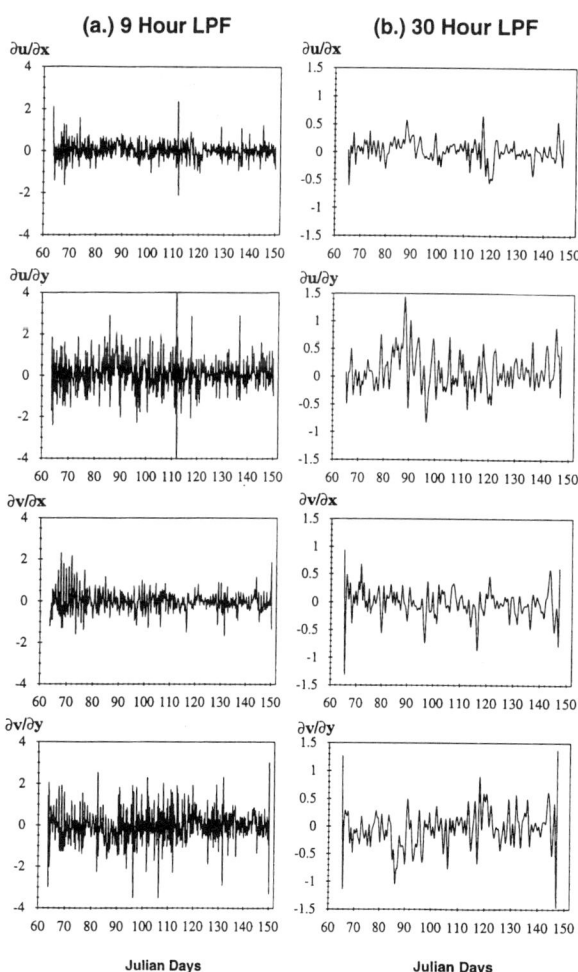

Fig. 3. Spatial differentials of velocity resulting from multiple linear regression analysis using 9 (left) and 30 (right) hour low pass filters (LPF). Components are in units of $\times 10^{-6}$ s^{-1}.

Table 2: Overview of Regression Results

	9 Hour LPF			15 Hour LPF			30 Hour LPF		
	Avg.	rms	Mag.	Avg.	rms	Mag.	Avg.	rms	Mag.
u (cm s^{-1})	0.32	5.49	4.15	0.38	4.84	3.72	0.44	3.85	2.85
v (cm s^{-1})	7.73	8.71	9.15	7.70	8.52	9.02	7.50	8.41	8.72
Vel. Mag. (cm s^{-1})	10.79	7.03	10.79	10.39	6.89	10.39	9.68	6.95	9.68
$\partial u/\partial x$ ($\times 10^6$ s^{-1})	0.02	0.37	0.27	0.03	0.26	0.20	0.03	0.18	0.14
$\partial u/\partial y$ ($\times 10^6$ s^{-1})	0.10	0.73	0.52	0.10	0.47	0.36	0.10	0.33	0.25
$\partial v/\partial x$ ($\times 10^6$ s^{-1})	-0.02	0.44	0.32	-0.02	0.36	0.26	-0.03	0.23	0.17
$\partial v/\partial y$ ($\times 10^6$ s^{-1})	-0.04	0.72	0.52	-0.04	0.45	0.35	-0.05	0.31	0.24
DV ($\times 10^6$ s^{-1})	-0.02	0.78	0.55	-0.01	0.49	0.36	-0.02	0.31	0.22
Max. Shear ($\times 10^6$ s^{-1})	---	---	0.95	---	---	0.66	---	---	0.47
ND ($\times 10^6$ s^{-1})	0.06	0.83	0.62	0.06	0.56	0.43	0.08	0.41	0.30
SD ($\times 10^6$ s^{-1})	0.08	0.82	0.58	0.08	0.56	0.41	0.07	0.41	0.30
VT ($\times 10^6$ s^{-1})	-0.12	0.89	0.66	-0.12	0.62	0.49	-0.12	0.39	0.32

Abbreviations are as follows: rms is the root-mean-square statistic, avg. is the average value and mag. is the magnitude of the value (vel. mag. being the velocity magnitude), DV is the divergence, ND is the normal deformation (also known as elongation), SD is the local shear deformation and VT is the vorticity.

Except for the maximum shear, which is a positive definite quantity, averages for the deformational components are centered close to zero at all LPF times (Table 2), with average magnitudes and rms values being a factor of 5 to 10 (and more) larger than this mean. The $\partial u/\partial y$ average seems to be particularly large at about 0.1×10^{-6} s^{-1} with the next largest at 0.04×10^{-6} s^{-1} in the $\partial v/\partial y$ term, which is also in the y direction. From Table 2 we also note that there is, on average, a convergence (i.e. negative divergence) of about 0.01 to 0.02×10^{-6} s^{-1} which does not change much for the different LPF times. However, both the magnitude and rms spread of the divergence and maximum shear decrease by a factor of 2 from the 9 hour to the 30 hour LPF showing a much greater effect of the sub-daily frequencies on the deformation versus the velocities. While the normal and shear deformation components are not invariant, they are components of the maximum shear and are greater than the average divergence by a factor of 3 and 4 respectively. This estimate suggests that shear is contributing about three to four times as much to the deformation process as divergence. With regard to the rms spread (or variability) in these averages, we see from Table 2 that variations in the y direction are about twice those in the x direction even for velocity. This information, combined with Figures 2, 3 and 4, strongly suggests that the deformation activity changes much more than the velocity within the 9 to 30 hour period range. Additionally, the magnitude of the divergence is about half that of the maximum shear, which appears to be dominated by a strong y directional preference of motion. The y-orientation of the shelf slope also seems to contribute to these directional preferences.

As a final observation, the difference between the 9, 15, and 30 hour LPF deformations need to be considered in light of the total amount of deformation accomplished. The energy dissipated during ice deformation is a positive definite quantity which is a function of strain-rate and so the integral, in time, of energy dissipated is guaranteed to be underestimated if it is completed from time-averaged stress and strain-rate, i.e. the average product is not equal to the product of the averages. The main difference is that the net divergence, including the higher frequency sub-daily oscillations, represents much more deformational activity and energy expenditure (twice) than that derived from the daily or greater time scale measurement. Such differences have a considerable impact on processes affecting the open water fraction.

Power Spectra

Results of a power spectra for the average and individual buoy velocities are shown in Figures 5 and 6. The plots show power density (normalized power per frequency bin width) resolved to 64 frequency bins ranging from the DC signal at 0 cycles day^{-1} down to the Nyquist frequency of 4 cycles day^{-1} (6 hour period). The frequency resolution is $\Delta f = 0.06$ cycles day^{-1}. The shaded region is the 90% confidence interval from χ^2 analysis, which is discussed further in the error analysis section.

Beginning with Figure 5, we note two general trends. First, the largest power signal is at the lowest frequencies (<0.8 cycles day^{-1}) with a linear log (i.e. exponential) increase in power density (so called "red noise") towards these lower frequencies. Second, there are prominent power spikes near 1, 2, and 3 cycles day^{-1} (24, 12, and 8 hour periods, respectively). The 12 and 24 hour power spikes are believed to be due to ocean tides with the 12 hour spike also being associated with inertial oscillations from the ice [*Rowe et al.*, 1989; *Foldvik et al.*, 1990; *Levine et al.*, 1997]. The power spikes near 3 cycles day^{-1} are believed to be associated with high frequency non-linear ice interaction. However, because these high frequency, 3 cycles day^{-1} signals are quite close to the cutoff Nyquist frequency of 6 hours (4 cycles day^{-1}), we are less confident of their significance, so the impact of the 3 cycles day^{-1} peak is not being considered in detail in this investigation.

In addition to these general trends, there is a considerable amount of regional variability. At sites Alex (southeast), Brent (northeast) and Dimitri (east), the 12 hour power spike is greater than the 24 hour spike, while at the camp (center), Chris (west) and Ed (northwest), both 12 and 24 hour power spikes are quite pronounced and about equal in magnitude. Relative to the bottom topography in the area, the eastern sites are in the deep basin and have a noisy signal, while the western sites located along the continental slope (Figure 1) have stronger, clearer peaks. The increased power density in both the diurnal (24 hour) and semi-diurnal (12 hour) frequencies in the western section corresponds with the rise in bottom topography discussed in the preceding section. More specifically, there is an apparent increase in sub-daily ice activity along the continental slope region due to topographically enhanced ocean circulation patterns, which, although primarily tidal in origin, may be enhanced by inertial oscillations in the sea ice at the semi-diurnal peak, which is the more pronounced of the two.

The power spectra of the individual velocity components (u (east) and v (north)) differ in that the low frequency signal of the u component (Figure 6) is much lower than the velocity magnitude ($u^2 + v^2$) and is, at most, the same order (between 10^2 and 10^3 (m s^{-1})^2) as the power spikes at 12 and 24 hours. In contrast, the low frequency signal seen in the velocity magnitude (Figure 5) is also seen in the v component of the power spectra (Figure 6). The strong 12 hour power spike seen in the velocity magnitude at the eastern sites is also quite visible in the v component, but indistinguishable from the rest of the signal in the eastern sites for the u component. On the other hand, strong 12 and 24 hour spikes (2 and 1 cycles day^{-1}, respectively) are seen in the three westernmost sites (Ed, Chris, Camp) in the u component but are very difficult to distinguish in the v component. The exception is at site Ed, the site furthest up the shelf slope, where strong 12 and 24 hour signatures in both the u and v components are observed.

Plots of the buoy array's centroid velocity and the defined invariant strain-rate power density are shown in Figure 7. The power signal in the centroid velocity contains many of the features seen in the individual buoy spectra, including the large low frequency signature in the y direction, the two peaks at 12 and 24 hours (2 and 1 cycles day^{-1}) in the x direction, and

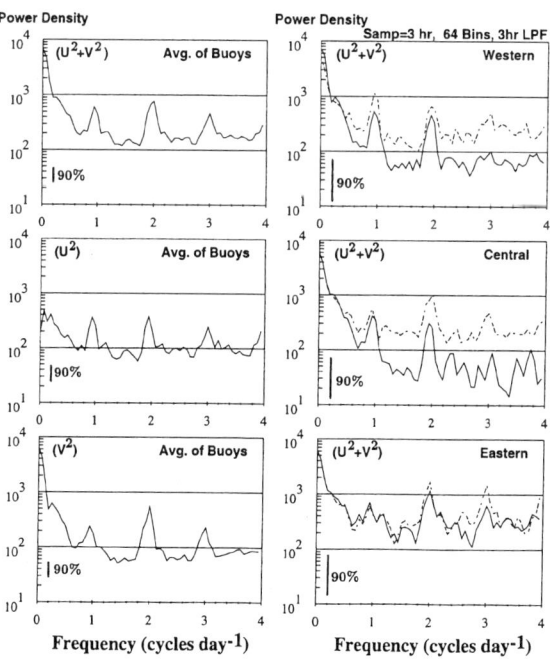

Fig. 5. Power density spectra of average buoy speed (upper left), U, V and regional distribution of buoy speed (right). The spectra are resolved to 64 frequency bins with amplitudes in units of m^2 s^{-1} and an applied low pass filter cutoff time of three hours (3 LPF). Western sites are Ed (dashed) and Chris (solid), central sites are Alex (dashed) and Camp (solid) and eastern sites are Brent (dashed) and Dimitri (solid). The 90% confidence interval of χ^2 results is shown by the vertical bar in each spectral plot.

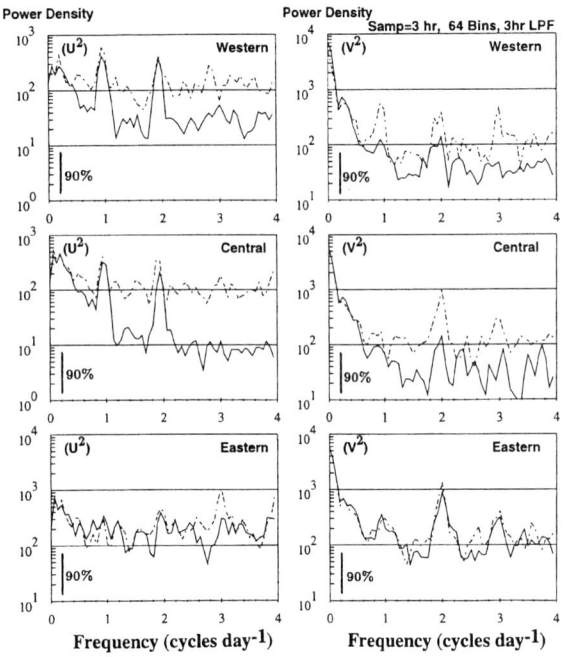

Fig. 6. Power density spectra of regionally distributed U (normal to shelf break on left) and V (parallel to shelf break on right) buoy velocity components. The spectra are resolved to 64 frequency bins with amplitudes in units of m^2 s^{-1} and an applied low pass filter cutoff time of three hours (3 LPF). Western sites are Ed (dashed) and Chris (solid), central sites are Alex (dashed) and Camp (solid) and eastern sites are Brent (dashed) and Dimitri (solid). The 90% confidence interval of χ^2 results is shown by the vertical bar in each spectral plot. Note the reduction by a factor of 10 in the upper two U spectral amplitudes versus velocity magnitude and V amplitudes.

the lack of a 24 hour signature (1 cycle day^{-1}) in the y direction. The signals absent from these velocity plots are the prominent high frequency peaks near 8 hours, which have been purposely filtered out (see error analysis section for clarification). With regard to the deformation, we see that the total strain-rate power density is, as suspected from the time series plots, dominated by high frequency peaks at 12 and 24 hours. The new information gained from the total strain-rate plot is that the 12 hour peak is stronger than the 24 hour peak and there is a fairly strong low frequency signal which was not evident before. Looking at the two invariant sub-components of the total strain-rate, divergence and invariant shear (bottom of Figure 7), we see that the shear is more than twice the divergence and, therefore, the greater contributor to the total strain-rate power. We also see that the low frequency signal is present in the shear but missing from the divergence, while the 12 and 24 hour signatures are present in both.

ERROR ANALYSIS

The multiple linear regression analysis discussed in the last section included strain-rates on the order of 10^{-6} s^{-1}, which were computed from velocities on the order of 10 cm s^{-1}. On-site instrument calibration [*Geiger*, 1996] indicates an average distance error of 530 ± 88m radially and a corresponding error of 375 ± 62m in the x and y directions (manufacturers' specification of Argos receivers is typically 500 m radially). Given the very small strain-rate values, it is important to have an understanding of the errors involved in the regression procedure in order to justify the results obtained in the last section. Three types of error analysis are conducted in this section in order to determine the quality of the data: truncation error, assessment of the signal to noise ratio, and χ^2 analysis of the power spectra [*Geiger*, 1996; *Hines and Montgomery*, 1990; *Hibler et al.*, 1974].

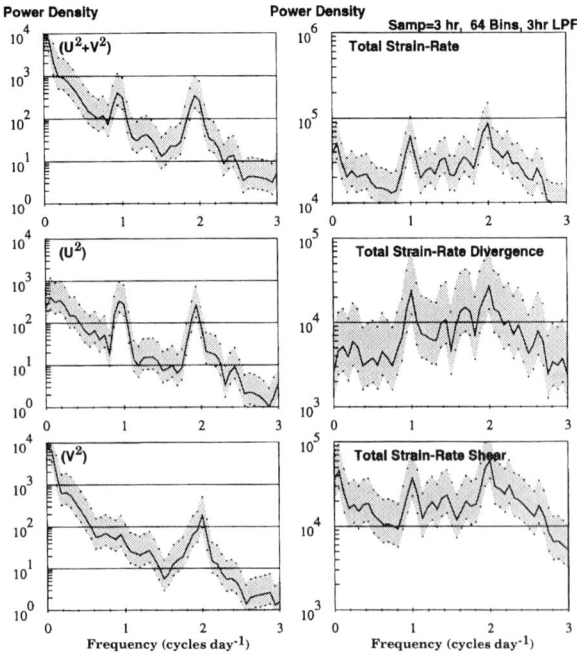

Fig. 7. Power density spectra of the buoy array's centroid velocity and invariant strain-rate components resolved to 64 frequency bins. The amplitudes of velocity spectra are in units of m^2 s^{-1} and the strain-rates are in units of $(\times 10^{-6})^2$ s^{-1}. The 90% confidence interval of χ^2 results is shown in the shaded region.

Buoy velocities are calculated using $\Delta x_i/\Delta t$, where Δx_i is the distance traversed by a buoy in a given direction (x or y) over a given amount of time Δt. An added distance to this is the instrument error which, when divided by Δt, also results in a velocity. Because the instrument error occurs at every time step, it shows up as a high frequency signal. Specifically, Δx_{err} produced by the instrument error is the propagated error [Beers, 1957]

$$\Delta x_{\text{err}} = \sqrt{(x[t]_{\text{err}})^2 + (x[t-1]_{\text{err}})^2}$$
$$= \sqrt{2}\bar{x}_{\text{err}} \quad (24)$$

where $\bar{x}_{\text{err}}$ is the average instrument error. As we increase the cutoff time in the low pass filter, the high frequency signals decrease. Defining Δt for this case to be the low pass filter time we can estimate the propagated instrument error velocity (u_{err_i}, v_{err_i}) as follows.

$$u_{\text{err}_i} = v_{\text{err}_i} = \frac{\Delta x_{\text{err}}}{\Delta t_{\text{LPF}}}$$
$$= \frac{\sqrt{2}\bar{x}_{\text{err}}}{\Delta t_{\text{LPF}}}$$
$$= \frac{\bar{r}_{\text{err}}}{\Delta t_{\text{LPF}}}. \quad (25)$$

The truncation error Err_{in} in the regression analysis is computed by

$$Err_{in} = Z_{in} - \hat{Z}_{in} \quad (26)$$

where Z_{in} is the actual velocity component value at each site and $\hat{Z}_{in}$ is the best fit value for $i = 1$ to N buoy sites and $n = 1$ to 2 velocity components. This truncation error is a measure of the difference between the estimated linear values from the truncated Taylor expansion model selected ($\hat{Z}_{in}$) and the true values (Z_{in}). Their differences account for influences not associated with the linear model. The two dominant influences suspected are 1) non-linear strains and 2) instrument error. Since the sum of both of these must be less than or equal to the total truncation error estimate, the instrument error should be less than the truncation error for the model to be reliable. This truncation error is computed at each site such that a system average is compared to the average propagated instrument error velocity. The averaging scheme used in this case is as follows. The truncation error at each site at each point in time is computed for each velocity component. Then, the temporal average of these values is computed at each site as is its average rms spread. Finally, the average and rms from each site are averaged for the whole system for each of the u and v velocity components.

Plots of the u and v truncation error averages are plotted together with the estimated instrument error velocity (Eq. 25) in Figure 8. We see that both the regression truncation error average and instrument error decrease with increased low pass filter times, as do their variability, i.e. rms. More importantly, we see that the estimated instrument error velocity for no low pass filter (1 hour) is much greater than the truncation error and well beyond the limits of the average range of that error. Hence, without low pass filtering, the multiple linear regression yields results which are not statistically reliable. With a 3 hour low pass filter, the instrument error is at least within the upper bound of the average truncation er-

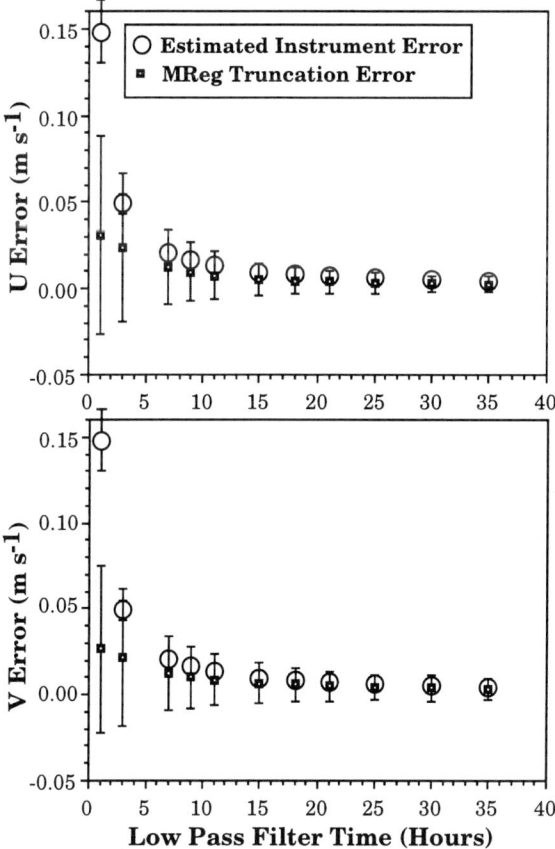

Fig. 8. Truncation error of multiple linear regression (MReg) analysis and estimated instrument error. Error bars are the average rms computed from all sites at given low pass filter cutoff times indicated in hours.

Table 3: Average Signal to Noise Ratio

	9 Hour LPF	15 Hour LPF	30 Hour LPF
u (cm s^{-1})	12.67	18.14	28.81
v (cm s^{-1})	27.23	40.24	73.35
$\partial u/\partial x$ ($\times 10^6$ s^{-1})	2.17	2.66	3.76
$\partial u/\partial y$ ($\times 10^6$ s^{-1})	1.54	1.76	2.61
$\partial v/\partial x$ ($\times 10^6$ s^{-1})	2.49	3.17	4.07
$\partial v/\partial y$ ($\times 10^6$ s^{-1})	1.55	1.47	1.85

ror, but substantially higher than the average truncation error. Proceeding with larger LPF cutoff times, we see that the instrument error slowly moves closer to the average truncation error. These results indicate that the model is statistically reliable at higher cutoff times, but should be interpreted with caution because the average of the instrument error is still greater than the average error for the model proposed. Because of this result, cutoff times of less than 7 hours are not used and cutoff times of 9, 15 and 30 hour regression results are subjected to additional testing to further quantify the bounds of the errors involved.

For the 9, 15, and 30 hour regression results, average magnitudes for all 6 regression values (u, v, $\partial u/\partial x$, $\partial u/\partial y$, $\partial v/\partial x$, $\partial v/\partial y$) are known at each point in time over the three month period of day 63 to 150. Additionally the standard deviation ($StdD$) for each of these values can be computed from the regression results using

$$StdD_{jn} = \sqrt{(Var_n)\,(C_{jj})/(N-1)} \qquad (27)$$

where

$$C_{jj} = \text{Main diagonal of } (X_{ji}\,X_{ij})^{-1}$$
$$Var_n = \frac{1}{DoF}\sum_{i=1}^{N}(Z_{in}-\hat{Z}_{in})^2$$
$$DoF = 2N - (\text{number of } \beta \text{ terms} + 1)$$

where DoF is the degrees of freedom.

Regarding the regression magnitude at each time step as the signal and the standard deviation as an estimate of the noise, a signal to noise ratio at each time interval is computed. An average of this ratio is given in Table 3. From this table we see that the velocity terms are estimated quite well from the regression model, i.e. a high number means high signal to low noise ratio, but the deformation tensor components have much more noise. Additionally, there is more noise in the y direction components than in the x direction. Considering the fact that we are dealing with very small quantities, the error estimates show that there is more signal than noise and so the deformation results are reliable. Hence, the velocity results have very little noise and are quantitatively very significant, but the deformation values must be regarded with some caution since they have more noise in their signal, particularly for the shorter low pass filter times.

The final error analysis of this section is an estimate of the power spectra amplitude error. The probability of a squared sample data value (such as the power of an FFT value) within a specified two-sided confidence interval range is given by the Chi-Squared statistic (χ^2) [Montgomery and Hines, 1990]

$$P\{\frac{\nu\hat{\phi}^2}{\chi^2_{\alpha/2,\nu}} \leq \Gamma^2 \leq \frac{\nu\hat{\phi}^2}{\chi^2_{1-\alpha/2,\nu}}\} = 1-\alpha \qquad (28)$$
$$\alpha => (1-\alpha)100 = \text{percent confidence}$$

where the values expressed in {} give the confidence interval, Γ^2 is the true squared value, $\hat{\phi}^2$ is the estimated squared value, ν is the degrees of freedom and α is chosen according to the probability range determined by $100\%(1-\alpha)$. Welch [1967] has determined the degrees of freedom (ν) for the overlapping power spectra scheme to be

$$\nu \approx 2\frac{9n}{11} \qquad (29)$$

where n is the number of periodograms.

The three month data set is broken down into segments containing 128 three hour interval time samples. Each of these segments produces a 64 bin periodogram over the desired frequency range. Employing an overlapping scheme, the first 128 samples, i.e. $t = 0$ to $t = 128$, produces the first periodogram, $t = 64$ to $t = 192$ produces the second periodogram, and so on. This sampling scheme produces 5 periodograms containing 64 frequency bins from the three month data set. The resulting degrees of freedom are 8.2 for all but the average buoy case, which is 49.1. For a 90% confidence interval, this produces a lower and upper bound of 0.52 and 2.83 (for a 6 buoy average the values are 0.74 and 1.44) times the estimated value to obtain the confidence interval. As seen from the spectral plots in Figure 9, the peaks of the velocity spectra are statistically well above the noise of the data, since the lower bound of these peaks is clearly above the upper bound of the background signal. Hence, in concurrence with the previous results, we find that the velocity peaks found from these data are statistically significant to within a 90% confidence interval. The low frequency, diurnal and semi-diurnal peaks in the strain-rates, however, are more suspect. While the lower bound of many of the identified strain-rate peaks rises above the mean value of the power density amplitude data, and in many cases above the background level of corresponding lower frequency signals (to their left on the graphs), in a number of cases the strain-rate peaks of interest fail to rise above all of the 90% confidence background signal, i.e., some of the upper bounds of background information are still higher than the lower bound of the peaks of interests. This information agrees with the signal-to-noise ratio results such that a statistically quantitative result regarding the strain-rate results can not be established at 90% confidence, but the strain-rate peaks do rise enough above the rest of the signal such that they are significant somewhere below 90%. Hence, as with the signal-to-noise ratio result, the reader is cautioned about the strain-rate results. It is for this reason that interpretation of the strain-rate data has been restricted in this study to the understanding of the presence of the peaks of interest and their location within the frequency domain (low frequency, diurnal and semi-diurnal peaks) rather than delve into the specific quantitative value they have at these frequencies.

The capability of improving strain-rate measurements is already available as more reliable GPS units and, in particular, differential GPS units are used. Additionally, longer time series with as many buoys as possible are encouraged for future experiments to both increase the number of periodograms one can create and increase the degrees of freedom, both of which work to reduce the range in the χ^2 confidence interval.

DISCUSSION

Wind and Ocean Forcing

We can identify potential causes of the observed ice motion by examining the spectra of the wind and current meter data. As seen in Figure 9, the same strong exponential increase toward the low frequencies is seen in both wind and ocean data for frequencies longer than 1 cycle day^{-1}. Taking into account the fact that the ocean is 1000 times more dense (3 orders of magnitude) than air, the low frequency contribution from both ocean and air in terms of kinetic energy is comparable (upper right panel of Figure 9). In the ocean current data, the presence of two prominent power spikes at 1 and 2 cycles day^{-1} are seen, as they are in the ice drift and deformation data. The kinetic energy at these higher frequencies, 24 and 12 hour periods, is about half an order of magnitude greater in the ocean current data than in the wind forcing at the same frequencies. From this

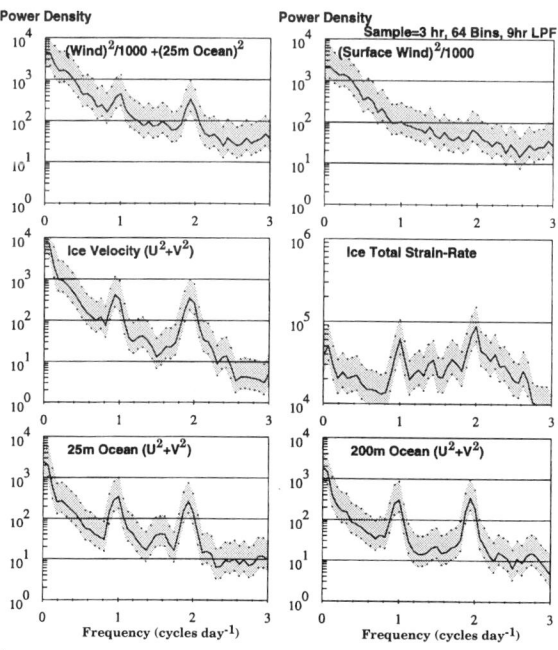

Fig. 9. Power density spectra for surface wind, ice, 25 m and 200 m water velocities and total ice strain-rate resolved to 64 frequency bins. The power density of wind, ice and ocean velocity magnitudes are in units of m^2 s^{-1} and strain-rate is in units of $(\times 10^{-6})^2$ s^{-1}. The wind power (velocity squared) is divided by 1000 to compare it with the kinetic energy in the ocean by compensating for the differences in density of the two media. The 90% confidence interval of χ^2 results is shown in the shaded region.

result we conclude that both the low frequency wind and ocean current are comparable external forces on the ice, while at diurnal and semi-diurnal frequencies the ocean dominates.

The difference between u and v velocity components seen in the buoy data is not pronounced in the current data (and therefore not shown). There is only a slightly lower signal in the v component at 12 and 24 hours than in the u component (with the v component oriented parallel to the shelf slope). One difference that does exist in the ocean current is the reduced low frequency signal in the u component compared to the strong low frequency signal in the v component, reflecting the western boundary current along the shelf-slope break [*Muench and Gordon*, 1995]. This same situation was observed in the buoy data, indicating that the dominantly northward ocean current is contributing substantially to the northward drift of the array.

With respect to regional variability of ocean currents, an overview of the ocean circulation around ISW is given in *Muench and Gordon* [1995] where three relevant circulation conditions are described. First, all four current meter sites indicate a primarily barotropic flow locally, which is reflected in the similarity in the signal between the 25 m and 200 m ocean spectra in Figure 9. Second, a western boundary intensification is observed in the 50 m measurements, which average from 5 cm s^{-1} at the westernmost site to at least 1 cm s^{-1} at the easternmost site. Third, fluctuations of 10-20 cm s^{-1} superimposed on the mean flow consist of semi-diurnal, diurnal, inertial and low frequency mesoscale signals. This information coincides very closely with the power spectra information described above. The mean northward ice drift described by *Muench and Gordon* [1995] corresponds to the strong low frequency power signature in the v velocity component. The orientation of the continental slope is apparently responsible for the western boundary intensification and also the main cause of increased diurnal and semi-diurnal activity at the western end of the array. Finally, the diurnal and semi-diurnal power spikes from the spectral plots correspond to the 10-20 cm s^{-1} superimposed fluctuations observed in the current meter data, which further supports the assumption that ocean circulation patterns contribute significantly to these high frequency features.

Comparing the wind and ocean current results with the buoy spectra, we find that the mean ice drift is being driven by both low frequency wind and ocean circulation patterns. At first glance, the ocean appears to have the most direct influence on the northward drift of the ice. However, caution must be taken in this interpretation since the internal resistance of the ice must also be acting in such a way that the ice resists southward compaction due to the presence of the continent south and west of the array. When the wind blows northward it provides a force complementary to the ocean current. Comparing results here with the general drift results shown in Figure 1, we also find that a considerable amount of low frequency response must be coming from the variable passage of storms, such as the one identified around day 120, and that this storm activity contributes substantially to deviations in drift trajectory. There are also a number of wind-related processes at frequencies higher than those considered here, which are not being included, such as form drag and other high frequency wind effects reported in *Andreas and Claffey* [1995]. Additionally, it is important to realize that the large scale ocean circulation responsible for the mean ocean current is itself driven by the large scale wind circulation from above and steered by the bottom topography from below.

Cross-spectra between the u and v components in the ice and ocean confirm (Figure 10) that the 12 and 24 hour power spikes seen in the ice drift are due to the diurnal and semi-diurnal ocean oscillations. In particular, the strong coherence (80%) between the ice and ocean u component supports the hypothesis that the diurnal and semi-diurnal tidal signals are being amplified normal to the shelf break due to the rise (by a factor of two) in topography and that this is being directly transmitted to the ice. In the cross-spectra, the frequency band of 0.2 to 0.6 cycles day^{-1} (2.5 to 5 day periods) shows a strong (> 80%) coherence between wind and ice magnitudes as well. Additionally, strong coherence exists between the low frequency northward drift of the ice and the u component of the ocean, as well as a notable coherence ($\approx$ 65%) at the semi-diurnal frequency between the northward drift of the ice and the northward motion of the 25 m ocean current.

In terms of a strong coherence between the wind and ocean velocities with the deformation of the field, the only clear signal that can be obtained with the data set here is at low frequencies (93% in ocean example shown in upper right panel of Figure 10). However, the presence of strong peaks in the deformation at diurnal and semi-diurnal frequencies matching similar-frequency oceanic peaks (Figure 9)

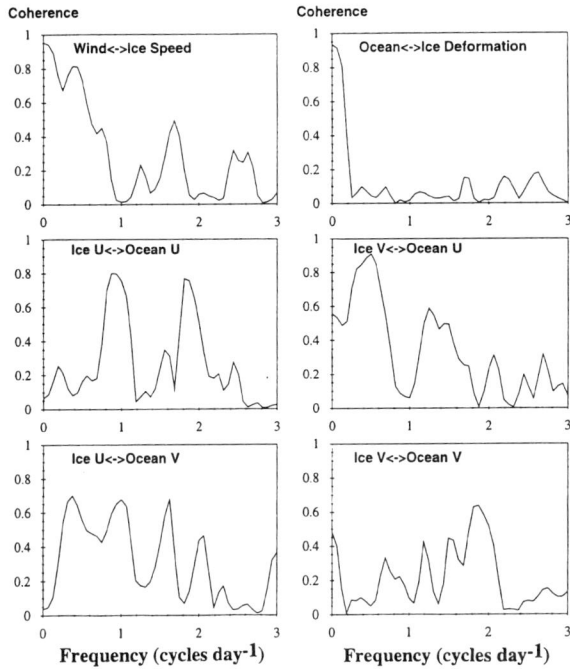

Fig. 10. Examples of results of normalized cross spectral coherence between ice speed and deformation versus wind and ocean speeds (upper panels), respectively, and between individual velocity components in the ice versus the ocean. Spectra is resolved to 64 frequency bins.

indicates that a connection should exist between the tidal motion of the ocean and the deformation of the ice. Because of the highly non-linear behavior of the ice, it is not currently possible to identify exactly how the ocean tidal frequencies correlate with the deformation. Inertial oscillations within the ice are probably the main cause, as such motion is very close to the semi-diurnal frequency in the Weddell Sea region, where the critical latitude (74° 28'S) of the two frequencies are resonant. Since the ice response contains a considerable amount of high frequency activity and directional variability that neither the wind nor the ocean exhibit, through process of elimination this additional activity must be attributable to internal ice interaction. Further examination, through modeling studies, to identify the non-linear responses of the ice at these sub-daily frequencies is one proposed method to consider in further understanding these results.

Directional Preference

Looking to the non-invariant components, we examine the matter of directional preference in the velocity and deformation components. Power spectra for each of the deformation tensor components and half the power density of each DKP term are shown in Figure 11. As was the case with the velocity, the individual deformation tensor components show significant differences in power distribution. The main contributors to the total power are the y differential terms, which have a number of power spikes, the largest being the 12 hour peak (2 cycles day^{-1}) in the $\partial v/\partial y$ term. In the x direction, the signal is significantly less, but there are two very clear 12 and 24 hour signatures in both x differential components, with the $\partial v/\partial x$ term having a very strong 24 hour (1 cycle day^{-1}) peak. Looking at this same information in terms of DKPs (part (a) in Figure 11), we see that the large 24 hour peak from the $\partial v/\partial x$ term shows up primarily in the vorticity (VT). The divergence (DV) is identical to that shown in Figure 7, but in this context we see that its 12 and 24 hour peaks are coming from both $\partial u/\partial x$ and $\partial v/\partial y$ terms. However, the low frequency signal from both of these work against each other to cancel out in the divergence.

In contrast, the normal deformation (ND, which is the difference between these same two strain-rate components) contains both 12 and 24 hour peaks and the low frequency peak. The large spike at 12 hours from the $\partial v/\partial y$ term is clearly the main contributor to this signal. The normal deformation is also a greater contributor to the total strain-rate than the shear deformation (SD), particularly at the 12 hour period. The shear deformation is primarily contributing an assortment of power peaks over the whole range of the spectra shown.

The presence of general low power activity across the spectral distribution is a classic "symptom" associated with non-linear activity (*Schuster*, 1988). The error analysis described earlier showed that non-linear behavior of the ice may be a source of error in addition to the measurement error. These results support such an error to a greater extent in the deformation than in the velocity. The amount of background signal in the y direction versus x direction further supports such a claim since a large measurement error would manifest itself in both directions. As seen in part (b) of Figure 11, however, the cleanest signals come from the x direction where there is a low background signal and very pronounced peaks (particularly in the $\partial v/\partial x$ component). The high energy spectra in the y direction, however, contains many "noisy" peaks with an overall greater back-

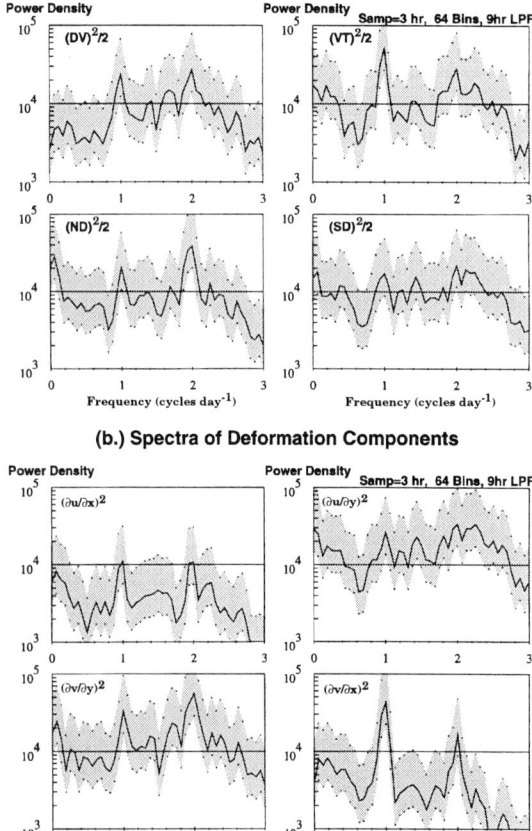

Fig. 11. Power density spectra of differential kinematic parameters (DKP) and deformation tensor components resolved to 64 frequency bins. The power density is expressed in units of $(\times 10^{-6})^2$ s^{-1}. The 90% confidence interval of χ^2 results shown as the shaded region. The upper two DKP components are invariant while the lower two are not. The deformational components corresponding to the DKP components are positioned directly below in the corresponding column.

ground signal. Such a signal suggests considerable non-linear ice activity in a preferred direction rather than a large inherent measurement error.

In terms of forcing, it appears that the 12 hour tidal/inertial oscillations are generating most of the shear activity in the form of elongation parallel to the shelf-slope, while the 24 hour oscillations are working on the ice as a solid body to turn the whole system. These findings are corroborated by *Foldvik et al.* [1990] who, using bottom moored current meters, found that diurnal ($\approx$ 24 hours) tides in this region move barotropically (as a solid body at the surface),

while the semi-diurnal ($\approx$ 12 hour) tides are influenced by the depth. Their findings also show that all four major tidal components (O_1, K_1, M_2 and S_2) move as a tidal wave around the edge of the Weddell basin in a clockwise direction. This information further supports the findings that most of the sub-daily signal we see here is tidal and that there is a difference in response between the diurnal and semi-diurnal signals. In close proximity to the research area, the latitude 74° 28'S is known as a critical latitude where the inertial period matches the M_2 (semi-diurnal) tidal component [*Foldvik et al.*, 1990]. Thus, there is also a strong likelihood that damping of inertial oscillations by ice interaction is working either in combination with or in response to the semi-diurnal tidal signal. Such an influence is one mechanism which may be responsible for the differences in diurnal and semi-diurnal ice responses. Additionally, we see that shear is affected more by high frequency inputs (12 hour tides), while translation and rotation are affected more by lower frequency forcing (one day and longer). The bathymetry, particularly the shelf-slope and surrounding land masses, is clearly acting as a counter force in all the motion described above, such that a directional preference in both the velocity and deformation of the system is primarily regulated by the local topography.

Temporal Effects

We can sum up the power from a specified band width ($n \Delta f$) over a single periodogram to obtain the integrated power density and divide that by the band width to obtain information about frequency band changes with time. We will now examine results from such integrated power spectra to understand the time evolution of the main frequency bands we have been examining; namely, low frequencies (D.C. to <0.9 cycles day^{-1}) and sub-daily frequencies (0.9 to 2.7 cycles day^{-1}). The integrated spectra computed from 32 frequency bins over 15 day time periods for the wind, ocean currents, centroid ice velocity and invariant strain-rate components are presented in Figure 12.

As seen in the previous power spectra results, the low frequencies contribute the most to the overall average power density for the velocities. The overall change in time of low frequency ice velocity is an initial rise in power from mean day 70 to 84 followed by relatively moderate power from mean day 92 to 132 and finally a large increase followed by a decrease in power from mean day 140 to 148. The

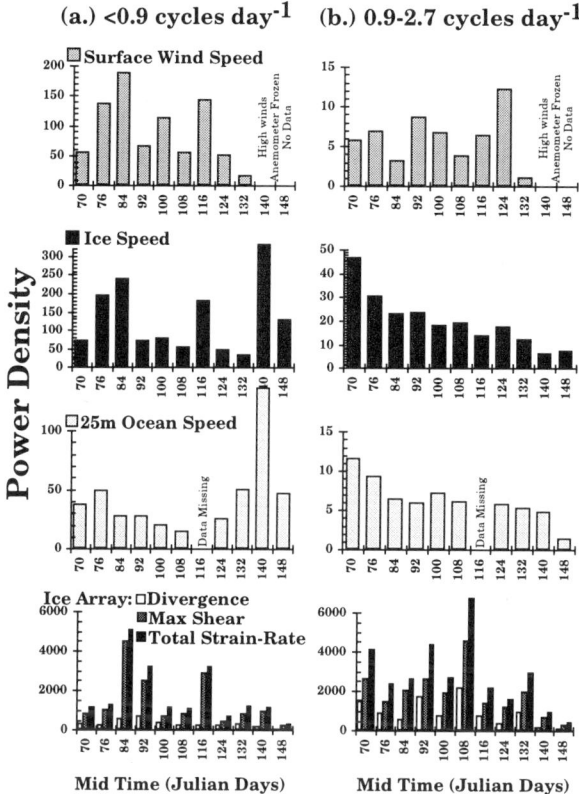

Fig. 12. Fifteen day integrated power spectra for air, ice and ocean speeds and invariant strain-rate variables of divergence, maximum shear and total strain-rate (see text for description). The integrated power density for speed is in units of $m^2\ s^{-1}$ and the strain-rate is in units of $(\times 10^{-6})^2\ s^{-1}$. Time is labeled by the mid-point (mid time) of each 15-day temporal integration in Julian days.

smaller rise and fall in power from mean day 108 to 124 corresponds to the passage of a large storm system around day 120 as discussed earlier. Overall decreasing power density for the sub-daily ice velocity frequencies is seen with power density values around 46 $m^2\ s^{-1}$ at the beginning near mean day 70 down to around 10 $m^2\ s^{-1}$ by mean days 140 and 148.

The wind in the upper panels of Figure 12 shows the same temporal evolution of power density as the ice speed, with the low frequency band clearly being the main source of that power. The ocean current on the other hand is more steady with time over the whole frequency range, except at mean day 140 where the current reaches a maximum of about 130 $m^2\ s^{-1}$. The average buoy velocity power density is, in nearly all cases, larger than the ocean current power density. When the wind is low, the buoy power is nearly the same as the ocean current power, but when the wind power is high, the ice drift power is correspondingly higher than the ocean. From these results we can conclude with reasonable certainty that the ocean current is providing a steady source of moderate low frequency power to the ice drift, while the wind is providing an intermittent source of very high power but at low frequencies. This supports the earlier conclusion that the ice drifts northward as a result of a topographically steered ocean current, but also that there must be a large input of intermittent power coming from the wind from variable directions.

The distribution of power density for the ocean current (Figure 12) shows the low frequency ocean power to be the main contributor to the overall power density. The low frequency band ranges between about 20 and 130 $m^2\ s^{-1}$ while the sub-daily frequency band is much less, with ocean currents only once (25 m current on day 70) exceeding 10 $m^2\ s^{-1}$. The ice drift has an even larger spread, with the low frequency band ranging from about 30 to over 300 $m^2\ s^{-1}$.

In addition to these overall features we see that the ice velocity follows the wind at the low frequency band, but in sub-daily range there is a decrease of power density with time. The reduction in power relates to a decrease in oscillation activity, including both tidal and inertial periods. Since the buoy array is both drifting in space northward and advancing in time, there are two possible scenarios. First, the ocean current in this frequency band may not be as strong at the northwest end of the western Weddell gyre as in the southwest, possibly due to topographic changes or because stronger tidal/inertial activity may be occurring at the southwest end of the basin. Second, the ice is becoming thicker and the compactness is increasing, both because the winter is progressing and the ice is flowing northward into an eastward extending land mass (Antarctic Peninsula). This creates a grid-lock at the surface, which can significantly damp the ocean current below, especially if the wind is working against the ocean current during this process. A combination of these scenarios is most likely the case.

With regard to the deformation activity at the different frequency bands, we see in the lowest panels of Figure 12 that the sub-daily frequency band makes the largest overall contribution to total power, with shear being its largest component in all cases. These results were also seen in the time series plots (Figures 3 and 4), but now, in addition, we see that during periods of high wind or shortly thereafter (e.g. near mean day 84 to 92 and 116), there is a substantial increase in low frequency strain-rate power in the shear but not in the divergence. Thus storm events are causing considerable shear deformation but not as much divergence.

The sub-daily frequency band is where the 12 and 24 hour power peaks lie and is heavily influenced by the tides. In these integrated power spectra, we do see a general decrease in the power density with time, as is the case with the ocean current. However, we do not see a section by section pattern emerging, which would clearly show that this is the case for the deformation. There are a number of possible explanations for this uncertainty. First, as seen in the error analysis, the strain-rate data do have a considerable amount of noise in their signal, either due to measurement error or the presence of non-linear motion not accounted for by the strain analysis model, such that direct correspondence to current spectra is at best only in terms of general trends. Non-linear interaction may propagate energy from both the high and low frequency bands into the sub-daily frequency band shown here. Low frequency wind and ocean activity can also induce ice processes of shear, elongation and ridge building, which then transmit non-linear deformation across the region.

SUMMARY

Data from Ice Station Weddell during 1992 has been used to examine sea ice drift and deformation activity in the western Weddell Sea and to identify external forces responsible for driving specific processes. Use of power spectra and deformation analysis on large scale drift together with spectra of wind and current measurements have shown that the velocity, or general drift, of the sea ice pack in the western Weddell region is driven primarily by low frequency forcing (i.e. greater than one day periods). In contrast, sub-daily frequencies, specifically diurnal and semi-diurnal tidal frequencies, appear to be the main driving force behind sea ice deformation in this region. The bottom topography also plays a major role by inducing a directional dependence in both ice drift and deformation. The internal ice interaction seems to be particularly sensitive to such topographic influences, such that spatial differences in the underlying ocean current due to topographical features are reflected in the non-linear interaction, possibly in conjunction with kinematic waves and inertial oscillation activity within the ice. In terms of identifying key deformation processes, comparisons using invariant quantities provide information about the total deformation process and its components of divergence and shear. The non-invariant components also provide a considerable amount of information about contributions due to deformation oriented relative to the topography. With regard to the statistical "fitness" of the deformation information, we have provided useful information about sea ice deformation, but there is ample room for improving the quantitative value of these results. Improvements to this type of data can be achieved through field experiments of longer duration than three months using an increased number of sites and instrumentation such as differential GPS with improved position measurement accuracy. Additionally, numerical studies delving into the processes of sub-daily tidal and inertial oscillation coupling and kinematic wave studies may reveal more information about the mechanisms which couple the sub-daily ocean tidal oscillations to the sub-daily peaks observed in the deformation spectra.

With regard to specific sea ice dynamic processes and the forces most responsible for driving them, we can now identify the following two key results with reasonable certainty. First, the general drift of the sea ice pack in the western Weddell region is a low frequency dynamic process which is driven primarily by low frequency forcing in the form of moderate but steady low frequency ocean currents and strong winds from intermittent high energy storm activity. Ocean eddies may act in a similar fashion to the wind, but there is insufficient evidence from this study to verify that. Second, deformation of the sea ice pack is composed of both low (<1 cycle day^{-1}) and sub-daily (1 cycle day^{-1} or more) frequency processes, with the sub-daily processes clearly dominating. The low frequency processes are strongest during and after episodes of high winds. There is also evidence that moderate low frequency ocean currents must also have an effect. For this region in particular, divergence is clearly a more sub-daily frequency process with very little low frequency contribution. Shear has large high and low frequency components, with elongation deformation being the main form of deformation at low frequencies relative to the shape of the underlying topography. The sub-daily processes are clearly driven by 12 and 24 hour ocean oscillations (2 and 1 cycles day^{-1}), with the 12 hour peak contributing most to the total shear activity, while the 24 hour peak contributes more to the solid body rotation of the ice on scales at least as large as the ISW array (150 km). *Foldvik et al.* [1990] conclude that diurnal tides produce barotropic currents, which is synonymous with the local solid body rotation seen in the ISW array, and supports the conclusions of the deformation analysis conducted here.

Additionally, the presence and orientation of the continental shelf and shelf slope is a major contributor in terms of forcing and directional orientation of the drift and deformation components. On the continental shelf, strong 12 and 24 hour tidal effects are seen in both u and v components at the westernmost site (site Ed). The three sites located just east of site Ed (sites Chris, Alex and the Camp), and in deeper water, are dominated by both 12 and 24 hour (2 and 1 cycles day^{-1}) responses in the u velocity (oriented normal to the shelf slope), while in the v direction, the 12 and 24 hour peaks are reduced to the noise level, except for a 12 hour peak at site Alex, which is the easternmost of these three sites. This spatial distribution of velocity at specific frequencies is what produces the strong field vorticity near 1 cycle day^{-1} (24 hour period) and a strong elongation (normal deformation) of the field near 2 cycles day^{-1} (12 hour period). Hence, topographic steering of tidal forces is also a major force driving the spatial variability of the velocity field and thus the differences in the non-invariant strain-rate components.

In terms of identifying changes in these processes with time we found that changes in wind intensity produce parallel changes in the ice drift at low frequencies (<0.9 cycles day^{-1}) while sub-daily (tidal) frequencies (0.9-2.7 cycles day^{-1}) change in the ice relative to the ocean and exhibit a decrease in power as the winter progresses. Deformation activity dominates at the sub-daily frequencies, but is complicated by non-linear processes and possibly unfiltered instrument error, and is therefore more difficult to relate to specific wind or ocean events in time. At low frequencies (<0.9 cycles day^{-1}), however, shear processes correlate very clearly with periods of increased wind activity, while divergence changes are much smaller.

One impact of these results is the ability to monitor sea ice drift and deformation activity. High resolution satellite imagery, e.g. from active radar, has a high spatial resolution but typically low temporal resolution (3 day pass average) so it can potentially detect ice drift adequately but underestimate deformation. Buoy arrays lack the high spatial resolution but do have high temporal resolution to record most of the ice deformation activity, especially differential GPS arrays with low measurement errors.

If the western Weddell shelf break is representative of other such topographic regions, then ice forecasting requires a combination of both satellite and *in-situ* buoys to correctly predict ice activity, at least in regions where sub-daily forcing is very strong as demonstrated here.

Acknowledgments. This work was made possible by grants from NSF OPP 90-2489, DPP 9315934, ONR Assert N00014-93-1-1221 and USRA NAS-5-32484. The Argos buoy data from remote sites of ISW have been provided courtesy of Robin Muench at Earth and Space Research (ESR) to whom we are most grateful. We are grateful to Jay Ardai, Vicky Lytle, Bruce Elder and Dave Bell for their assistance during the field and analysis program on Ice Station Weddell. Thanks also goes to Greg Flato for his particularly insightful review of this manuscript, which led to major improvements.

REFERENCES

Andreas, E.L. and K.J. Claffey, Air-ice drag coefficients in the western Weddell Sea. 2. A model based on form drag and drifting snow, *J. Geophys. Res.*, 100 (C3), 4833-4843, 1995.

Beers, Y, *Introduction to the Theory of Error* Addison-Wesley Publishing Co. Reading, Mass, pp 65, 1957.

Foldvik, A., J.H. Middleton and T.D. Foster, The tides of the southern Weddell Sea, *Deep Sea Res.*, 37 (8), 1345-1362, 1990.

Fung, Y.C, *A First Course in Continuum Mechanics* 2nd Edition. Prentice-Hall, Inc. Englewood Cliffs, New Jersey, pp 340, 1977.

Geiger, C.A, Investigation of dynamic sea ice processes in the Weddell Sea during 1992. Ph.D. Thesis. Dartmouth College, pp. 387, 1996.

Gordon, A.L. and Ice Station Weddell Group of Principal Investigators and Chief Scientists, Weddell Sea Exploration from Ice Station. *EOS Trans. AGU* 74 (11), 121,124-126, 1993.

Hibler, W.D, A dynamic thermodynamic sea ice model, *J. Phys. Oceanogr.*, 9, 815-846, 1979.

Hibler, W.D. and E.M. Schulson, On modeling sea ice fracture and flow in numerical investigations of climate, *Ann. Glaciol.*, In Press.

Hibler, W.D., W.F. Weeks, A. Kovacs and S.F. Ackley, Differential sea-ice drift. Part 1: Spatial and temporal variations in sea-ice deformation. *J. Glaciol.*, 13 (69), 437-455, 1974.

Hines, W.W and D.C. Montgomery, Chapter 15: Multiple Regression. in *Probability and Statistics in Engineering and Management Science* (3rd Edition), 487-558, 1990.

Levine, M.D., L. Padman, R.D. Muench and J.H. Morison, Internal waves and tides in the western Weddell Sea: Observations from Ice Station Weddell, *J. Geophys. Res.*, 102 (C1), 1073-1089, 1997.

Massom, R.A, Observing the advection of sea ice in the Weddell Sea using buoy and satellite passive microwave data. *J. Geophys. Res.*, 97 (C10), 15559-15572, 1992.

McPhee, M.G, Analysis and prediction of short-term ice drift. *Trans. ASME J. Offshore Mech. Arctic Eng.*, 110, 94-100, 1988.

Muench, R.D. and A.L. Gordon, Circulation and transport of water along the western Weddell Sea margin. *J. Geophys. Res.*, 100 (C9), 18,503-18,515, 1995.

Roberts, J. and T.D. Roberts, Use of the Butterworth low-pass filter for oceanographic data. *J. Geophys. Res.*, 83(C11), 5510-5514, 1978.

Rowe,M.A., C.B. Sear, S.J. Morrison, P. Wadhams, D.W.S. Limbert and D.R. Crane, Periodic motions in the Weddell Sea pack ice. *Ann. Glaciol.*, 12, 145-151, 1989.

Schuster, H.G, Chapter 1: Experiments and simple models in *Deterministic chaos: An introduction*, (2nd Edition), Verlagsgesellschaft. Weinheim. 7-19, 1988.

Spiegel, M.R, *Schaum's Mathematical Handbook*, McGraw-Hill Book Company, pp. 271, 1968.

Welch, P.D, A use of fast fourier transform for the estimation of power spectra: A method based on time averaging over short, modified periodograms, *IEEE Trans. Audio and Electroacoust.*, AU-15, 70-73, 1967.

S.F. Ackley, USACRREL, 72 Lyme Road, Hanover, NH 03755-1290; C.A. Geiger[1], Goddard Space Flight Center, Code 971, Greenbelt, MD 20771; W.D. Hibler, 8000 Cummings-Dartmouth College, Hanover, NH 03755

---

[1]Formerly at Dartmouth College

(Received July 3, 1996;
Accepted March 19, 1997)

OSCILLATORY BEHAVIOR IN ANTARCTIC SEA ICE CONCENTRATIONS

Per Gloersen

Oceans and Ice Branch, Laboratory for Hydrospheric Processes, NASA Goddard Space Flight Center, Greenbelt, Maryland

Alena Mernicky

Department of Mathematics and Statistics, University of Vermont, Burlington, Vermont

A frequency analysis methodology recently applied to Arctic sea ice [*Gloersen et al.*, 1996] is used here to explore the oscillations present in Antarctic sea ice during the 9-year lifetime of the Scanning Multichannel Microwave Radiometer on board the NASA Nimbus-7 spacecraft. The analysis includes determination of the spatial distribution of the trend in the sea ice concentrations, which was found to vary from about -40% per decade in the Weddell Sea, just east of the Antarctic Peninsula, and near the Adelie Coast to about +35% per decade in the Ross Sea. These large positive and negative trends were found earlier [*Gloersen and Campbell*, 1991] to average out to no significant overall trend in sea ice extents and concentrations. A positive trend in the Weddell sea occurs approximately in the location of the Weddell Polynya [*Zwally et al*, 1976, 1983; *Gordon*, 1978] which has not appeared since the early 1970's, and may indicate that there will be no reoccurrence in the subsequent decade. Spatial distribution of the amplitudes, phases, and confidence levels of the El Niño-Southern Oscillation (ENSO) [*Philander*, 1990] components at periods of 2.4 and 4.2 years observed earlier in sector averages of ice areas and extents in the Southern Ocean is shown to be localized to only parts of the sectors in which the earlier composite observations were made. Finally, the phase diagrams of the 2.4- and 4.2-year ENSO components in the sea ice cover of the Amundsen-Bellingshausen Seas reveal that the phases of the sea ice oscillations are about the same as those of the oscillations in scalar near-surface winds over the adjacent Southern Ocean, with the clear implication of a relationship between those winds and the sea ice.

INTRODUCTION

Variations in sea ice concentrations are known to be driven by wind stress, inertial force, and oceanic currents, but the extent to which positive feedback exists from these variations to the oceanic and atmospheric processes is less well understood. The sea ice variations include trends, seasonal cycles, and other oscillatory components, e.g., those of the El Niño-Southern Oscillation (ENSO) occurring at periods of 2.4 and 4.2 years. Examination of these variations is important both for verifying outputs of coupled ocean-atmosphere-ice models and perhaps discovery of some new aspects of these interactions. In this paper, we focus on processes in the Antarctic sea ice pack and the nearby open water in the Southern Ocean during the period of October 26, 1978 to August 20, 1987, when the Scanning Multichannel Microwave Radiometer (SMMR) on board the NASA Nimbus 7 satellite was in operation.

We found earlier that the overall decadal trend in the ice extent and area in the sea ice surrounding Antarctica is insignificant during the SMMR period, while a significant decrease of 2% per decade occurs in the Arctic [*Gloersen and Campbell*, 1991]. Further, the Arctic trend was found to be an average of widely varying local trends, ranging from -40% to +20% per decade [*Gloersen et al., 1996*]. Here, we examine the local trends in the Antarctic sea ice canopy, using a band-limited regression (BLR) analysis [*Kuo et al., 1990; Gloersen and Campbell, 1991*] for each grid point rather than ordinary least squares (OLS) as was done earlier for the Arctic. This has the advantage of producing more reliable trends and confidence levels.

Oscillatory components of the ENSO exist in the total and regional sea ice covers in both hemispheres [*Gloersen, 1995*]. The spatial distributions of these components in the Arctic indicate that the largest significant amplitudes occur in the seasonal sea ice zones [*Gloersen et al., 1996*]. Smaller, but still significant, ENSO oscillations appear in the perennial Arctic pack. Here, we examine the spatial distribution of the ENSO components in the Antarctic sea ice pack, and in the near-surface scalar wind fields in the adjacent open ocean. We also examine the spatial distribution of the seasonal cycle, synthesized as five harmonics of the annual cycle, and the spatial distribution of the phase of the annual cycle. Among other things, this permits a basis for assessing the importance of the ENSO components.

LOCAL MEANS AND DECADAL TRENDS

We apply the BLR method used earlier on time series of sea ice extent and area [*Gloersen and Campbell, 1991*] here to each grid point of the polar stereographic representation of sea ice concentrations in the Southern Ocean [*Gloersen et al., 1992*]. This is a significant improvement over the ordinary least squares technique used earlier for the Arctic [*Gloersen et al., 1996*]. The method is described in detail elsewhere [*Kuo et al., 1990; Gloersen and Campbell, 1991*], but briefly it is similar in matrix structure to weighted least squares regression [*Draper and Smith, 1996*], differing in that a truncated form of the sinc matrix replaces the variance matrix in the expression for the intercept and trend of a time series (but not in the formulation of the confidence levels). We obtain the truncated sinc matrix by first finding the components of its singular-value decomposition,

$$AA_{ij} = [\sin 2\pi W(i-j)/\pi(i-j)] = [U \cdot S \cdot V^T]_{ij} \quad (1)$$
$$i,j = 0,1,2,...,N-1$$

where W is the half-bandwidth of the filter (here 4/N), N is the number of observations in the record, U and V are unitary matrices (here equal because the sinc matrix AA is symmetric), S is a diagonal matrix of eigenvalues in decreasing order of importance and T signifies the matrix transpose. Truncation is obtained by constructing a truncated form of AA from the first eight columns of U, V, and S. This truncation, A, provides a sharp bandpass filter about a center frequency of zero with minimal leakage. With the truncation, A, the BLR solution for the data vector y is:

$$a = (D'AD)^{-1}(D'Ay) \quad (2)$$

where D is a (N x 2) design matrix containing ones in the first column to multiply the intercept and time in the second column to multiply the slope of the trend line and a is a vector containing their coefficients. Additional formulation provides also the standard deviations of the intercept and slope, from which the confidence levels are obtained by means of the Student's T-Test [*Draper and Smith, 1996*].

Applying the BLR technique to the Antarctic sea ice grids results in the means and decadal trends shown in Plate 1, and their confidence levels (not shown). The mean is obtained from the intercept and slope produced by BLR by adding the slope multiplied by half of the SMMR time interval to the intercept, and the decadal trend by multiplying the slope (in year^{-1}) by ten.

The highest of the means shown in Plate 1a occurs in the Weddell Sea east of the Antarctic Peninsula, with average concentrations ranging from 88% - 96%. This compares with the 9-year averages of sea ice concentrations shown earlier for each month which range from 88% to 100% [*Gloersen et al., 1992*]. This region is the only one resembling a perennial sea ice pack with the persistence and high concentration level typical of the perennial sea ice pack in the central Arctic. Its microwave signature, however, differs from that of multiyear ice in the Arctic in that it resembles more closely that of first-year Arctic sea ice (*unpublished data*). The average concentrations of 72% - 84% in the Amundsen-Bellingshausen Seas reflects the presence of sea ice in the 9-year average ice concentrations at the Austral minimum sea ice cover.

The trends depicted in Plate 1b range from -40% per decade off the Adelie Coast to +32% in the Amundsen Sea off the Getz Ice Shelf. Although the average of this wide variety of trends has been reported earlier as insignificant [*Gloersen and Campbell, 1991*], there are only small areas where the local trends are zero. The spatial distribution of trends is banded, with negative trends near parts of the Antarctic coastline and positive trends near other parts. Proceeding outwards from the coastal trends, the trend bands change sign one or two times before reaching open water in most places. For most of the Ross and Amundsen Seas, however, the trends are uniformly positive. In the vicinity of the Weddell Polynya that occurred in 1974-6 [*Zwally et al.*, 1983], the trends are uniformly positive, perhaps indicating that the polynya will not reoccur in the subsequent decade. As was observed in the Arctic [*Parkinson*, 1992; *Gloersen et al.*, 1996], the trend patterns in Plate 1b are similar to those of the length of sea ice season in the Antarctic observed by *Parkinson* (1994). These local trends may well be parts of slow cycles with periods longer than the 9-year period investigated here. Confidence levels for these trends are in the range 70% - 85%.

LOCAL SEASONAL CYCLES

The seasonal cycle for each grid point of the data is approximated by an OLS fit of the first five harmonics of the annual cycle to the data detrended by the BLR technique:

$$y_s = \sum_{n=0}^{5} a_n \cos 2\pi nt/\tau + b_n \sin 2\pi nt/\tau \qquad (3)$$

where $\tau = 1$ year. The amplitude of the local seasonal cycle is then defined as the rms of the coefficients a_n, b_n. The spatial distribution of these amplitudes is shown in Plate 1c. The smallest amplitudes, less than 3%, occur in the western Weddell Sea, adjacent to the Antarctic Peninsula, an area of known persistent ice. The largest amplitudes, near 50%, occur also in the Weddell Sea, in the vicinity of the Weddell Polynya that occurred in 1974-6, indicating a tendency to melt earlier there than in other parts of the Southern Ocean.

It is interesting to examine the phases of the annual cycle, the largest component of the seasonal cycle, which can be obtained from the coefficients in Equation 3:

$$\varphi = \arctan(b_1/a_1) \qquad (4)$$

Except for some small areas near the coastline in the vicinity of the Antarctic Peninsula (There is no discontinuity here; the phases wrap around the scale from $-\pi$ to $+\pi$.), the phase differences of the annual cycle given in Plate 1d are 1.5 radians (about 3 months) or less. The patterns show widespread coherency and are consistent with the sequence of monthly 9-year averages shown elsewhere [*Gloersen et al.*, 1992]. For instance, the appearance of the Ross Sea polynya early in the melt cycle is indicated in Plate 1d. A model for this entire distribution of phase is not yet developed.

SPATIAL DISTRIBUTION OF SOME SEA ICE OSCILLATIONS

A time series of the Antarctic sea ice areas, the integral of the ice concentrations over the Southern Ocean, can be represented very well by the combination of model fits given by Equations 2 and 3, with an R^2 of 0.96 [*Gloersen and Campbell*, 1991]. However, on a local basis, the fit is not nearly as good, as indicated by the standard deviation of the residuals (not shown) obtained by subtracting the local means, trends, and season cycles (Plate 1a-c) from the time sequence of ice concentration grid maps. The maximum value of the standard deviations is 24%, occurring in a small area in the Amundsen Sea. Standard deviations in the range 18%-22% are found in bands of perimeter ice facing the Pacific Ocean and in a small area off the tip of the Antarctic Peninsula in the Weddell Sea, but elsewhere the values range from 0%-18%. These standard deviations indicate that there remain significant variations in the Antarctic sea ice cover other than the trends and seasonal oscillations, as observed on a smaller scale earlier for the Arctic [*Gloersen et al.*, 1996].

We shall follow the methodology [*Thomson*, 1982; *Lindberg*, 1986; *Lindberg and Park*, 1987; *Park et al.*, 1987a & 1987b] in that earlier paper, where the details are presented and not repeated here, in searching for oscillations other than seasonal in the Antarctic residuals. The method, which we shall designate as band-limited spectral analysis (BLSA) begins by producing 8 filtered Fourier transforms, y_k, from the data, y:

$$y_k(\tau) = \sum_{t=1}^{3224} V_k(t) \cdot y(t) \cdot e^{-2\pi t/\tau} \qquad (5)$$
$$k = 1, 2, \ldots 8$$

where τ is the period of the oscillation of interest, the filters V_k are the first 8 columns of the matrix V in Equation 1, and t is the time in days taken on an interval of 6 days. Weighted sums of the y_k are then used to obtain the complex amplitude of the model function and its F-test value from which the confidence level is obtained at each grid point [*Gloersen et al.*, 1996]. In order to obtain an indication of which periods would be the most interesting to examine at each grid point, the series of grid maps was treated as a single, large matrix upon which singular-value decomposition (SVD) was applied, resulting in components in which there is a separation of spatial and temporal variability. The temporal parts of the first four principal components, typifying the entire Antarctic sea ice pack, were analyzed by BLSA, confirming that the oscillations observed in the Arctic, with ENSO periods of 2.4 and 4.2 years, and the unidentified oscillation of period 84 days, were also important in the Antarctic [*Gloersen et al.*, 1996]. Curiously, the Arctic oscillation at a period of 2.7 years, attributed to the aliasing of the S1 (or S2) tidal frequency with the revisit frequency of the satellite, appeared in the Antarctic to be shifted to 2.8 years, perhaps because the phases of the oscillation vary around the continent. Another possibility is that this oscillation is not an alias of the aforementioned tidal frequencies but is of an unknown origin. We have no conclusive explanation for this oscillation. At any rate, this new central period (of a wide peak) was chosen for further study.

The 2.8-year oscillation is the strongest of the four periods described above. Its maximum peak-to-peak variation in concentration is greater than 12% between the Ross and Amundsen Seas at about 67°S, 135°W, with a confidence level of 99% (Plate 2a and b). Elsewhere, this periodicity occurs around the continent with significant peak-to-peak values of concentration from 5% to 10% and confidence levels of at least 95%. The phases of this oscillation (Plate 2c) show a remarkable spatial coherency and vary over a range greater than 2π radians (the values wrap around on the color scale from fuchsia to light gray), perhaps contributing to the broadness of the peak of the BLSA analysis of the first four principal components described above. The weakest of the four selected oscillations is the one at a period of 84 days. The maximum peak-to-peak concentration variation of about 4% appears on the outer edge of the Ross Sea (Plate 2d) with a confidence level of greater than 98% (not shown). Lesser amplitudes appear elsewhere throughout the Antarctic ice pack, but the confidence levels of 95% or greater are restricted to the outer zones of the ice pack.

Of the ENSO components, and unlike in the Arctic, the stronger is the one with a period of 2.4 years, the amplitudes being about 50% higher in the Antarctic. The highest significant peak-to-peak variations, ranging from 11% to 15.5%, appear in the Bellingshausen and Amundsen Seas and just off the Princess Ragnhild Coast in the southern Indian Ocean, with confidence levels of 95% and higher (Plate 3a and b). Smaller patches of significant (confidence levels around 95%) variations in the 2% to 6% range appear in the Ross and Weddell Seas. The phase diagram for the 2.4-year oscillation is again spatially coherent and covers a range greater than 2π, i.e. greater than 2.4 years (Plate 3c). For comparison, we have included in Plate 3d an F-test diagram of near-surface winds (NSW) obtained also from an analysis of the SMMR data similar to one reported earlier [*Gloersen*, 1984], but using the 4.6-cm rather than the 2.5-cm polarization as the principal input to the NSW part of the algorithm and adding a correction for atmospheric opacity based on the 0.8-cm polarization. The peak-to-peak variations of the 2.4-year component of the NSW around the sea ice pack are only about 2 to 4 ms^{-1} (not shown), but the confidence levels in the Bellingshausen and Amundsen Seas are around 99%. The phases of this NSW ENSO component (not shown) are within 0.2 radians of the corresponding sea ice concentration variations in the nearby pack (Plate 3c). One might infer, then, a relationship between the atmospheric and sea ice 2.4-year oscillations in this locality. However, an equally significant 2.4-year atmospheric oscillation of a similar amplitude appearing in the Western Pacific off the Clarie Coast does not have a corresponding sea ice oscillation of significance, although some signal shows up in the amplitudes and again the phases are close (Plate 3a & b).

For the 4.2-year ENSO component, the strongest signal is in the Weddell Sea, with significant (i.e., 95% or greater confidence levels) ice concentration variations of 4% to 9% (Plate 4a and b). Lesser variations of significance occur in the Amundsen Sea near the coast and in smaller patches elsewhere around the coast of Antarctica. The amplitudes (Plate 4a) and F-test values (Plate 4b) form an interesting pattern in the Weddell Sea that approximates the Weddell gyre. Amplitudes of this component are comparable to those observed in the Arctic [*Gloersen et al.*, 1996]

The phase patterns (Plate 4c) are also spatially coherent for this component, and again cover a range greater than 2π radians, i.e., greater than 4.2 years. The corresponding F-test signal in the NSW in the South Atlantic can be only partly seen in Plate 4d, but on a Mercator projection (not shown) it is observed to extend to about 10°E and 40°S. Comparing the NSW and sea ice concentration phase patterns of this ENSO component in the South Atlantic, the NSW phases (not shown) are again within 0.2 radians of the ice concentration phases. We suggest that for the 4.2-year oscillation also, this

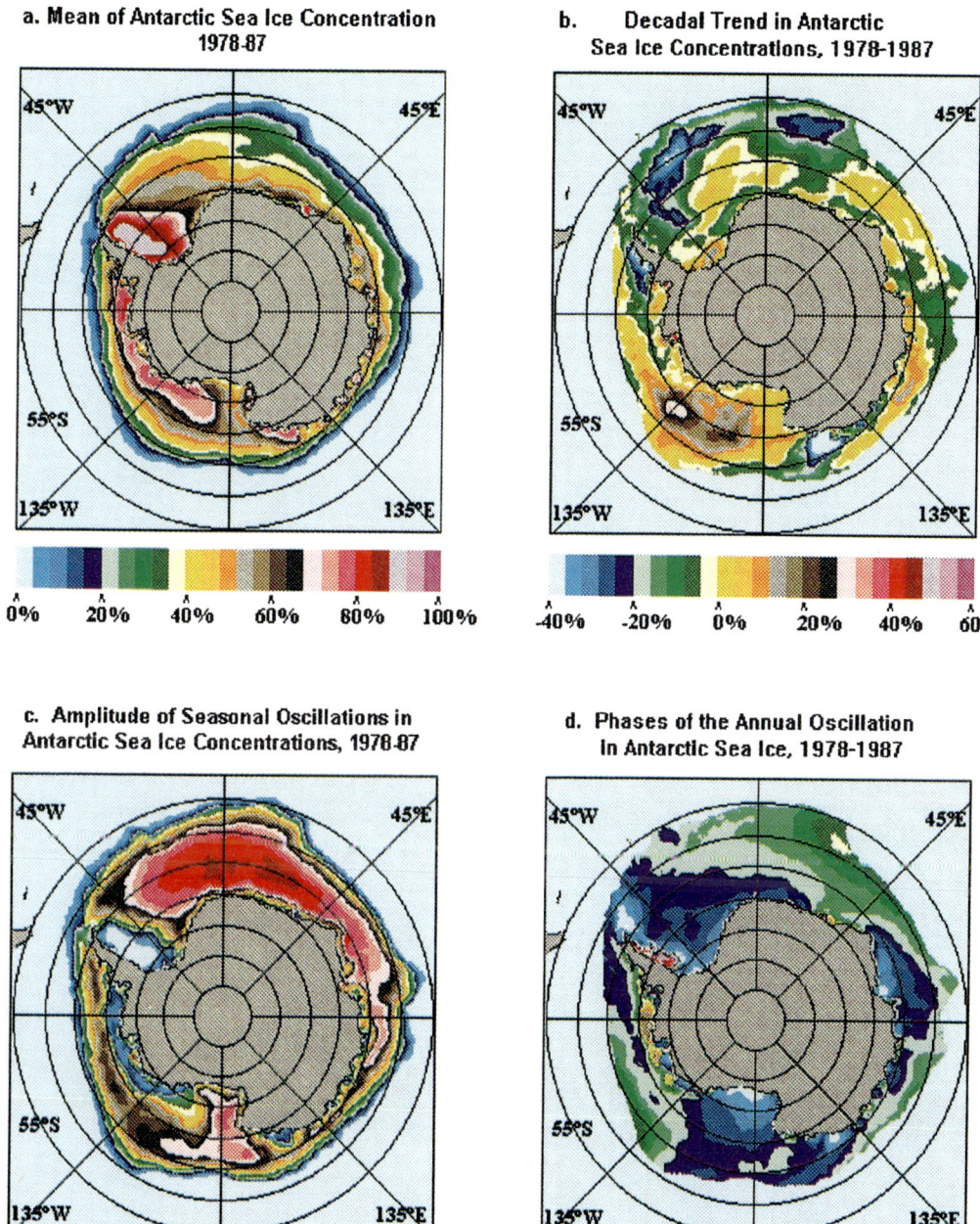

Plate 1. Results of statistical analysis of Nimbus-7 SMMR ice concentration data in the Antarctic. a. Spatial distribution of the means of the data obtained by band-limited regression (BLR). b. Decadal trends of the data obtained with BLR. c. Amplitudes of the seasonal oscillations obtained with multiple-linear regression (MLR) using an ordinary least squares fit of the first five harmonics of the annual cycle. d. Phases of the annual cycle obtained from the MLR annual cycle coefficients.

166 ANTARCTIC SEA ICE: PHYSICAL PROCESSES, INTERACTIONS AND VARIABILITY

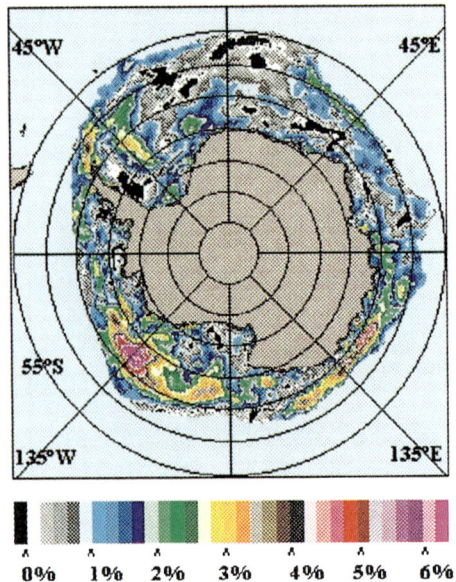

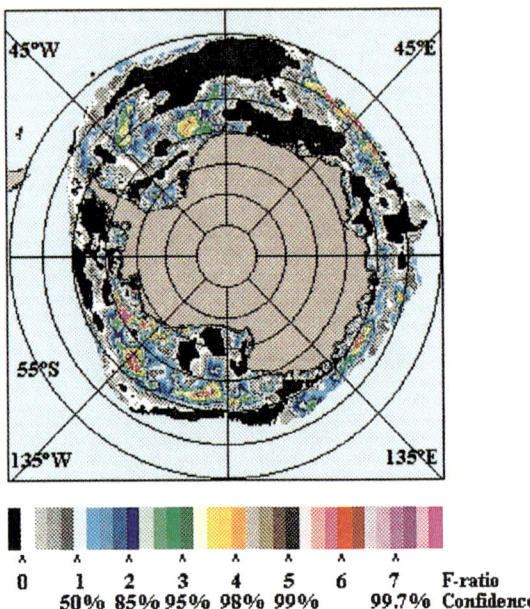

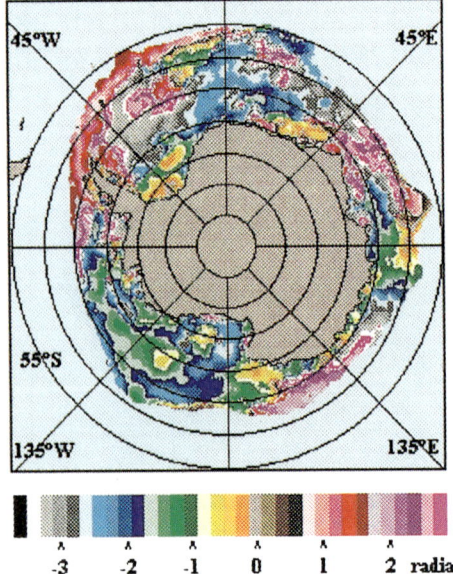

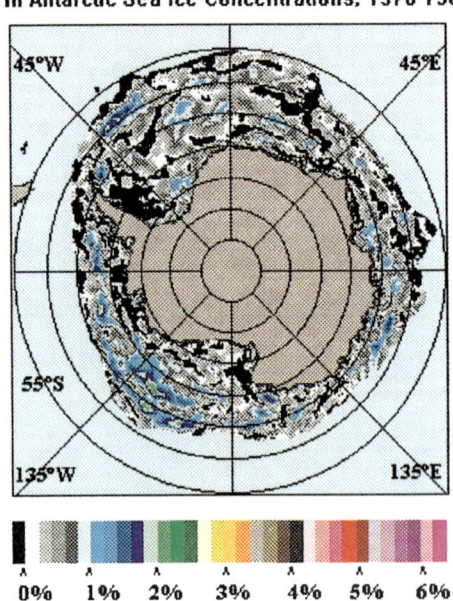

Plate 2. Results of the band-limited spectral analysis (BLSA) of SMMR ice concentrations in the Antarctic. a. Amplitudes of the 2.8-year oscillation. b. Confidence level of the amplitudes in a. c. Phases of the 2.8-year oscillation. d. Amplitudes of the 84-day oscillation.

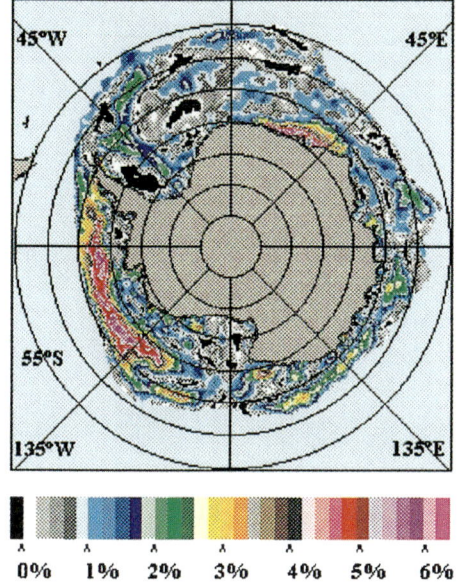
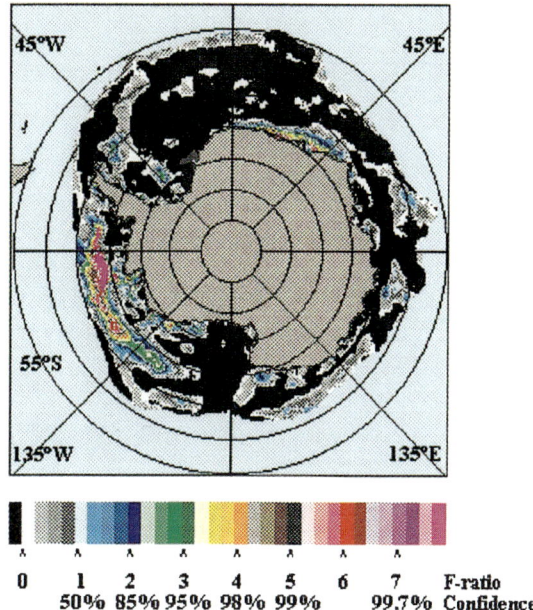

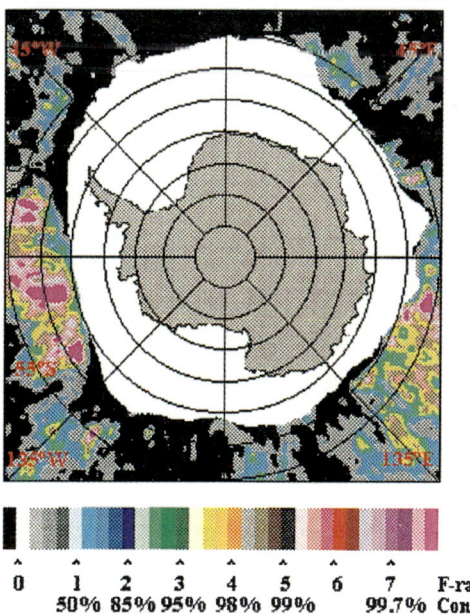

Plate 3. Results of the band-limited spectral analysis (BLSA) of SMMR ice concentrations in the Antarctic. a. Amplitudes of the 2.4-year oscillation. b. Confidence levels of the amplitudes in a. c. Phases of the 2.4-year oscillation. d. Confidence levels of the amplitudes of the 2.4-year oscillation in nearby oceanic near-surface wind speeds.

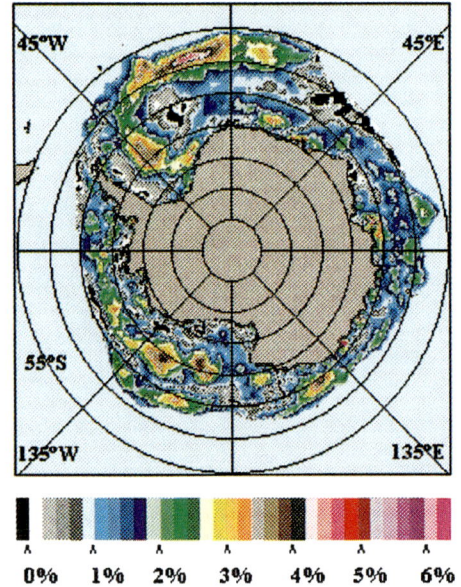
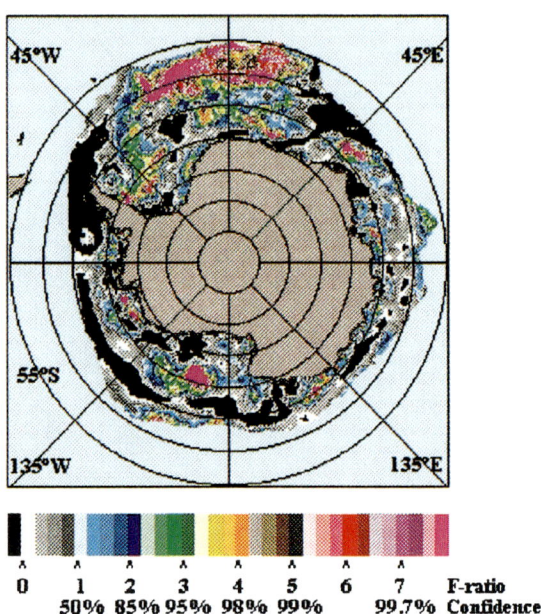
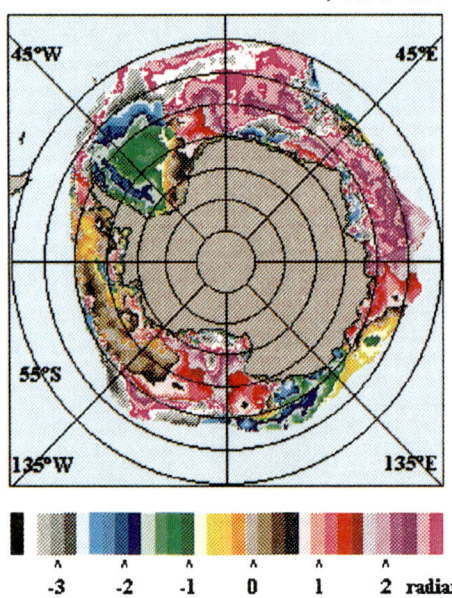
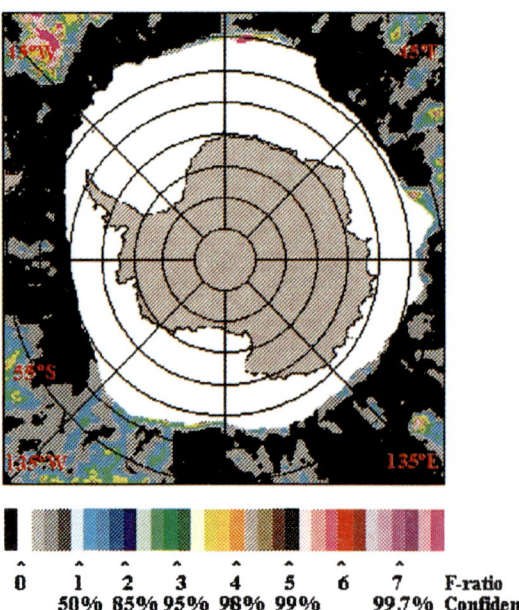

Plate 4. Results of the band-limited spectral analysis (BLSA) of SMMR ice concentrations in the Antarctic. a. Amplitudes of the 4.2-year oscillation. b. Confidence levels of the amplitudes in a. c. Phases of the 4.2-year oscillation. d. Confidence levels of the amplitudes of the 4.2-year oscillation in nearby oceanic near-surface wind speeds.

provides evidence of a relationship between the oscillations in the sea ice and nearby winds.

The ice concentration phase map also contains some evidence of the Antarctic Circumpolar Wave (ACW) with a period of 4-5 years reported recently by *White and Peterson* [1996]. Like phases are approximately 180° opposite each other in longitude in Plate 4c, indicative of two 4.2-year waves propagating around Antarctica. However, the angular extents of these two waves are unequal, one extending from about 105°E to 150°W (i.e., about 105° wide in longitude) and the diagonally opposite wave from about 150°W to about 105°W (i.e., about 255° wide in longitude).

SPECTRAL ANALYSIS OF SOME SELECTED AREAS

In the previous section, we selected four different values of τ in Equation 5 to produce the images in Plates 2 to 4. Now we shall select a number of different sea ice concentration grid elements and use Equation 5 to produce F-test spectra by running τ as a variable for each of them. These locations correspond to near-maximum signals for the 2.4-, 2.8-, and 4.2-year oscillations and are located at 67°S, 120°W; 67°S, 135°W; and 58°S, 15°W, respectively. In addition, we select a significant 4.2-year oscillation location in the Ross Sea at 72°S, 165°W. The spectra are shown in Figures 1 to 4. Curiously, these locations each show that the oscillation selected is the only significant one with a period greater

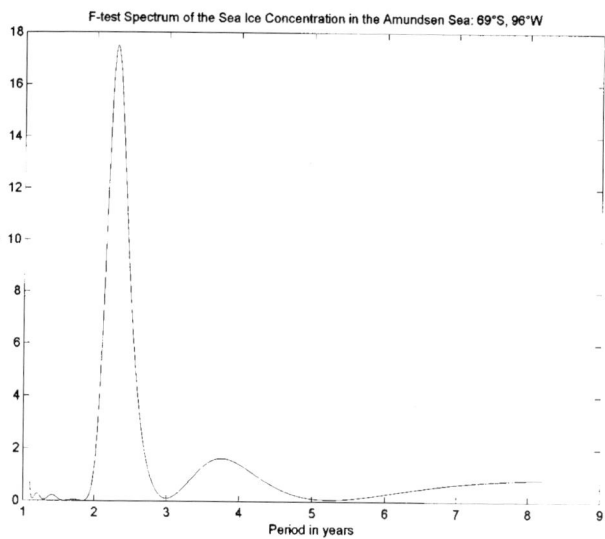

Figure 1. F-test spectrum of the sea ice concentrations in the Amundsen Sea within a grid element located at 69°S, 96°W.

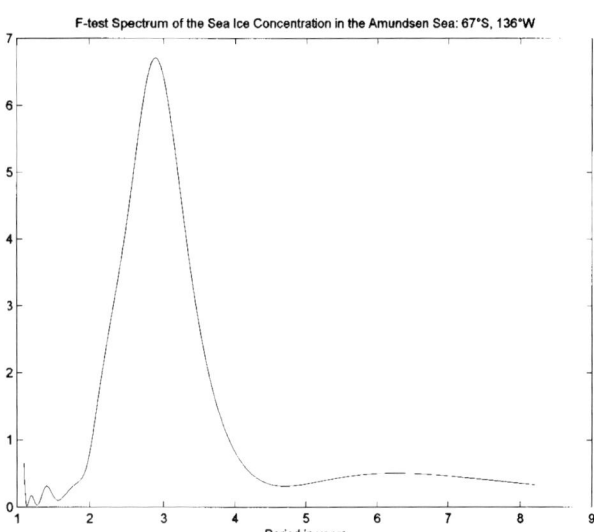

Figure 2. F-test spectrum of the sea ice concentrations in the Amundsen Sea within a grid element located at 67°S, 136°W.

than one year in its corresponding grid element. All of the oscillations appear to be detuned from the periods observed in the Arctic [*Gloersen et al.*, 1996]. The quasibiennial peak is at 2.3 years (Figure 1) rather than 2.4. The quasitriennial peak, attributed to an aliasing of the SMMR revisit time with the 12- or 24-hour tides in the Arctic, is at 2.9 years (Figure 2) rather than 2.7 as in the Arctic (and 2.8 years in the first principal component of the Antarctic sea ice concentrations). This also brings into question the earlier interpretation of this oscillation. Finally, the quasiquadrennial oscillation has a period of 3.90 years in the Ross Sea (Figure 3) and 3.95 years in the Weddell Sea (Figure 4) grid elements, in contrast with the 4.2 years observed in the Arctic. It should be kept in mind that these selected grid elements are but a small sample of the associated sectors of the Antarctic sea ice. Recalling the earlier observations of the integrated responses of the sea ice in the various Antarctic sectors which were closer to the Arctic values and the oscillations observed in the length-of-day parameter [*Gloersen*, 1995], it is clear that the periods of these oscillations must vary from one locality to another, perhaps indicative of varying basin resonance values at the various locations.

SUMMARY

We have shown that the periodic variations found in the Arctic sea ice concentrations are also present in the Antarctic, albeit with the periods in selected local areas

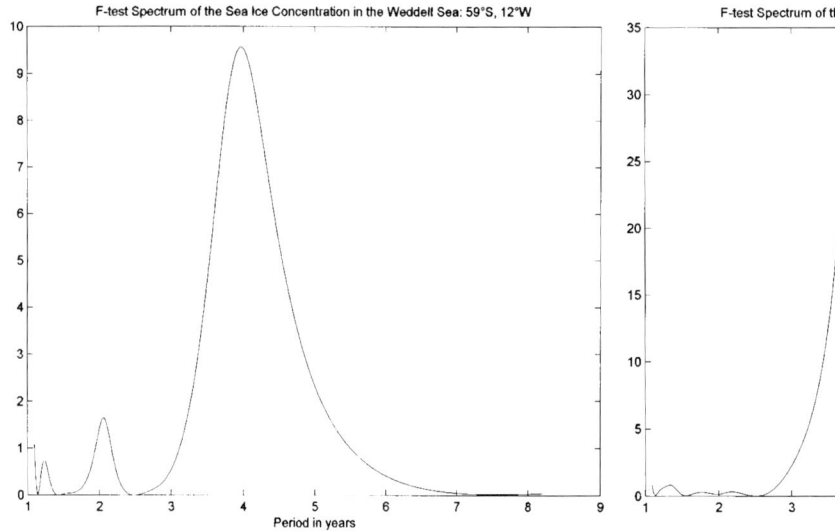

Figure 3. F-test spectrum of the sea ice concentrations in the Weddell Sea within a grid element located at 59°S, 12°W.

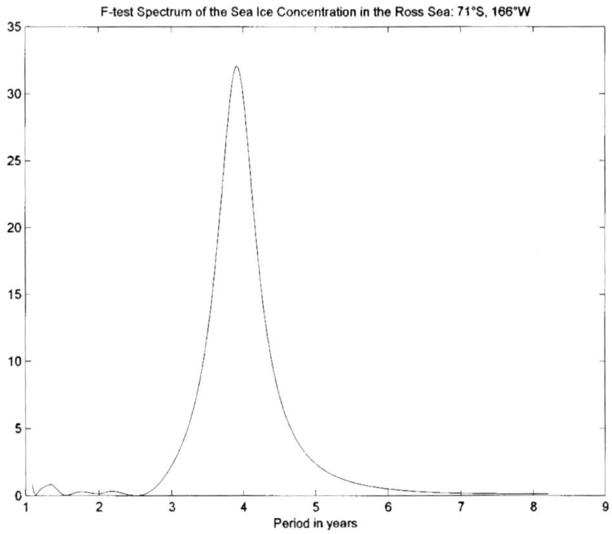

Figure 4. F-test spectrum of the sea ice concentrations in the Ross Sea within a grid element located at 71°S, 166°W.

shifted slightly from the Arctic values. The local distributions of the various Antarctic oscillations are consistent with earlier observations of the oscillations in the ice areas and extents of several Antarctic sectors [*Gloersen*, 1995]. The quasibiennial ENSO component is strongest in the Antarctic sea ice facing the Southern Pacific and in the near-surface winds in the adjacent open ocean. The quasiquadrennial components of the variation in the sea ice concentrations and nearby surface winds are strongest in the South Atlantic. Since the phase differences of the sea ice and NSW fluctuations are small, a relationship between the wind and sea ice oscillations is suggested. There is also some evidence of the recently reported ACW [*White and Peterson, 1996*] in the quasiquadrennial phase patterns of the sea ice concentrations. In the Weddell Sea, the patterns of the amplitudes and F-test values appear to follow the persistent gyre in that region, and may provide some useful information on the gyre location and fluctuations.

Acknowledgment. This work was sponsored by the Polar Programs Office of the NASA Headquarters Science Division.

REFERENCES

Draper, N., and H. Smith, *Applied Regression Analysis*, John Wiley & Sons, New York, N.Y., 1996.

Gloersen, P., Sea surface temperatures from Nimbus-7 SMMR radiances, *J. Applied Meteorology and Climate, 23,* 336-340, 1984.

Gloersen, P., and W. J. Campbell, Recent variations in Arctic and Antarctic sea ice covers, *Nature, 352,* 33-36, 1991.

Gloersen, P., W. J. Campbell, D. J. Cavalieri, J. C. Comiso, C. L. Parkinson, and H. J. Zwally, Arctic and Antarctic sea ice, 1978-1987: Satellite passive-microwave observations and analysis, *NASA SP-511,* Washington D.C. 1992).

Gloersen, P., ENSO frequency components in the global sea ice covers, *Nature, 373,* 503-506, 1995.

Gloersen, P., J. Yu, and E. Mollo-Christensen, Oscillatory behavior in Arctic sea ice concentrations, *J. Geophys. Res., 101,* 6641-6650, 1996.

Gordon, A. L., Deep Antarctic convection west of Maud Rise, *J. Phys. Oceanogr. 8,* 600-612, 1978.

Kuo, C., C. R. Lindberg, and D. J. Thomson, Coherence established between atmospheric carbon dioxide and global temperature, *Nature, 343,* 709-714, 1990.

Lindberg, C. R., Multiple taper spectral analysis of terrestrial free oscillations, Ph.D. thesis, Univ. of Calif., San Diego, 1986.

Lindberg, C. R., and J. Park, Multiple-taper spectral analysis of terrestrial free oscillations, II, *Geophys. J. R. Astron. Soc., 91,* 795-836, 1987.

Park, J., C. R. Lindberg, and F. L. Vernon III, Multiple-taper spectral analysis of high-frequency seismograms, *J. Geophys. Res., 92,* 12,675-12,684, 1987a.

Park, J., C. R. Lindberg, and D. J. Thomson, Multiple-taper spectral analysis of terrestrial free oscillations, I, *Geophys. J. R. Astron. Soc. 91,* 755-794, 1987b.

Parkinson, C. L., Spatial patterns of increases and decreases in the length of the sea ice season in the north polar region, *J. Geophys. Res., 97,* 14,377-14,388, 1992.

Parkinson, C. L., Spatial patterns in the length of the sea ice season in the Southern Ocean, 1979-1986, *J. Geophys. Res., 99,* 16,327-16,339, 1994.

Philander, S. G. H., *El Niño, La Niña and the Southern Oscillation,* Academic, San Diego, Calif., 1990.

Thomson, D. J., Spectrum estimation and harmonic analysis, *Proc. IEEE, 70,* 1055-1095, 1982.

White, W. B., and R. G. Peterson, An Antarctic circumpolar wave in surface pressure, wind, temperature, and sea-ice extent, *Nature, 380,* 699-702, 1996.

Zwally, H. J., T. T. Wilheit, P. Gloersen, and J. L. Mueller, Characteristics of Antarctic sea ice as determined by satellite-borne microwave imagers, in *Proceedings of the Symposium on Meteorological Observations from Space: Their Contribution to the First GARP Global Experiment,* Committee on Space Research of the International Council of Scientific Unions, Philadelphia, pp. 94-97, 1976.

Zwally, H. J., J. C. Comiso, C. L. Parkinson, W. J. Campbell, F. D. Carsey, and P. Gloersen, *Antarctic Sea Ice, 1973-1976: Satellite Passive-Microwave Observations,* NASA SP-459, National Aeronautics and Space Administration, Washington, D.C., 206 pp., 1983.

Per Gloersen, Code 971, NASA Goddard Space Flight Center, Greenbelt, MD 20071.
e-mail: per.gloersen@gsfc.nasa.gov

(Received October 9, 1996;
accepted February 25, 1997.)

LENGTH OF THE SEA ICE SEASON IN THE SOUTHERN OCEAN, 1988-1994

Claire L. Parkinson

Oceans and Ice Branch, NASA Goddard Space Flight Center, Greenbelt, Maryland

The length of the sea ice season in the Southern Ocean has been determined, mapped, and analyzed for the first full seven years, 1988-1994, of the DMSP Special Sensor Microwave Imager (SSM/I) data set. All years show ice season lengths decreasing northward around most of the continent except in the portion of the Weddell Sea between 0° and 45°W, where the decrease is from southwest to northeast, under the influence of the Weddell Gyre, and in the Ross Sea north of the Ross Ice Shelf, where the near-coastal Ross Sea polynya produces shorter ice seasons in the southern portion of the sea than immediately to the north. Other aspects of the spatial patterns in ice season lengths, such as the existence and distribution of perennial ice along the coasts of the Bellingshausen, Amundsen, and eastern Ross seas, vary from year to year. Maps of the differences between the yearly ice season lengths and the seven-year average ice season lengths reveal a suggestion of eastward propagation of major anomaly patterns, offering partial confirmation of the Antarctic circumpolar wave identified by *White and Peterson* [1996]. For the Southern Ocean as a whole, there is no dominant indication of either an overall shortening or an overall lengthening of the ice season, i.e., no dominant indication of widespread warming or cooling. Instead, linear trends through the seven years show the ice seasons to have shortened in the eastern Ross Sea, Amundsen Sea, far western Weddell Sea, non-coastal far eastern Weddell Sea, and the coastal regions off East Antarctica between 40° and 80°E, and to have lengthened in the western Ross Sea, Bellingshausen Sea, central Weddell Sea, and the 80°E-135°E sector off the coast of East Antarctica. Overall, the area that experienced a lengthening of the sea ice season, in terms of the linear trends, exceeded by about 20% the area that experienced a shortening of the season.

1. INTRODUCTION

The length of the sea ice season, defined locally as the number of days during the year with sea ice coverage, can be used as a measure of the season's severity. Furthermore, this measure has climatological importance beyond the sea ice cover in that, through a single number for each location for each year, it answers two critical questions for climate studies; specifically: (1) for how much of the year were the heat, mass, and momentum exchanges between the atmosphere and liquid ocean restricted by the presence of an intervening layer of ice? and (2) for how much of the year was the absorption of solar radiation by the ocean also restricted by the same layer of ice? Of course the length measure does not answer everything. Most seriously, it ignores the ice thickness, which is another critical determinant of the magnitude of the ocean/atmosphere exchanges.

By calculating the season length explicitly as the number of days with ice, not only does the resulting measure answer the above questions, but all complications regarding starting and ending dates for the ice season are avoided. For instance, in a location where ice first appears on March 20, stays for less than a day, disappears until April 1, then remains until April 3, disappears until April 7, and then stays until the spring

This paper is not subject to U.S. copyright.
Published in 1998 by the American Geophysical Union

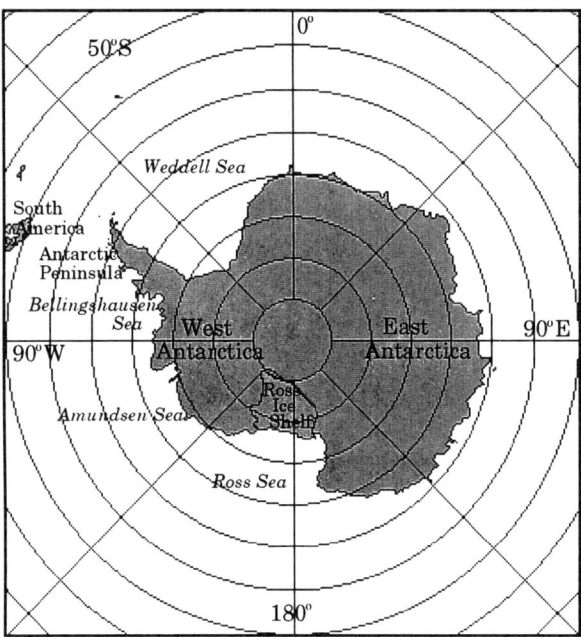

Fig. 1. Location map.

thaw, the ice season could variously be considered to have started on March 20, on April 1, or on April 7. By defining the season length as the number of days with ice, there is no need or desire to identify a season starting date, and, in this case, March 20, April 1-3, and April 7 would all be counted as ice days, and March 21-31 and April 4-6 would not be counted. Similarly, as polynyas open and close during the year, the season-length measure automatically takes them into account.

One of the advantages of the season-length measure as a climate variable is how readily it can be determined. Satellite passive-microwave data allow daily depictions of the global sea ice cover that can immediately be used to calculate and map the ice season length. As the satellite record lengthens, changes in the length of the sea ice season can be identified and used as one of many possible indicators of climate change.

Parkinson [1994] examined the length of the sea ice season in the Southern Ocean over the years 1979-1986 from the data of NASA's Nimbus 7 Scanning Multichannel Microwave Radiometer (SMMR). She found, over the 1979-1986 time frame, a general tendency toward shorter ice seasons in the northern Weddell and Bellingshausen seas and a general tendency toward longer ice seasons in the Ross Sea, around much of East Antarctica, and in a portion of the south central Weddell Sea (see Figure 1 for locations). This paper extends the earlier work to examine the length of the sea ice season over the period 1988-1994, using the more recent data from the Special Sensor Microwave Imagers (SSM/Is) on board the F8 and F11 satellites of the Defense Meteorological Satellite Program (DMSP). The next section provides a background to the data and their analysis, and section 3 presents the main results, including maps of the season lengths for each of the years 1988-1994, anomaly maps for each of the years, and maps of the trends over the 1988-1994 period. Section 4 includes a discussion of the results and comparisons with the findings of *Parkinson* [1994] for the SMMR time period.

2. DATA

The SSM/I is a seven-channel instrument receiving vertically and horizontally polarized radiation at frequencies of 19.35, 37.0, and 85.5 GHz and vertically polarized radiation at a frequency of 22.235 GHz. The 19.35 and 37 GHz data have been used to calculate ice concentrations through an algorithm based on polarization and gradient ratios, with a weather filter added that incorporates the 22.235 GHz data. The basic algorithm, often referred to as the NASA Team algorithm, is described in detail for the Nimbus 7 SMMR data in *Gloersen et al.* [1992]. Adjustments have been made for the SSM/I data as part of a group effort at Goddard Space Flight Center. These adjustments account for the replacement of the 18.0 GHz channels of the SMMR by the 19.35 GHz channels of the SSM/I and for differences in the satellite orbits. The data have also been cleaned to reduce land-to-ocean spillover and weather-related effects. The resulting SSM/I sea ice concentration data set, and documentation for it, are available from the National Snow and Ice Data Center (NSIDC) in Boulder, Colorado. The SMMR and SSM/I data sets have a six-week period of overlap in July and August 1987, allowing comparison of the ice concentrations and extents from the two instruments. Values are not identical, although they generally differ by no more than a few percentage points. Efforts continue at Goddard to minimize the differences, for instance through adjustments in tie points, and for that reason the trends calculated for this paper involve only the SSM/I data and do not extend over the two data sets. The SMMR results reported in *Parkinson* [1994] similarly involve only the data from the SMMR instrument, so that both sets of results avoid the misalignment between the instruments and should be comparable.

The first SSM/I instrument was launched on the DMSP F8 satellite on June 19, 1987 and collected data for most of the period through December 31, 1991. A second SSM/I was launched on January 12, 1989 on the DMSP F10 satellite, but data from the F10 SSM/I were of lower quality than those from the F8 and F11 satellites and did not fill in any major gaps in the F8/F11 record. Consequently, the F10 data were not used in this study. The DMSP F11 SSM/I, the third in the SSM/I series, was launched on November 28, 1991. The data stream from the F11 SSM/I began on December 3, 1991 and has been archived at NSIDC through September 1995, with data continuing to be received. Details on the processing and gridding of the SSM/I data can be found in the *DMSP SSM/I User's Guide* [*NSIDC Distributed Active Archive Center*, 1996].

Because year-round data are needed for the determination of ice-season lengths, the archived SSM/I data as of late 1996 allow the calculation of season lengths only for the seven years 1988 through 1994. To calculate the length of the sea ice season for those years, a complete set of daily ice concentration matrices was created for the January 1, 1988-December 31, 1994 period. This necessitated filling in all data gaps, which was done by spatial interpolation on the brightness temperatures for isolated points of missing data and by temporal interpolation on the ice concentrations for all larger blocks of missing data. In general, the amount of interpolation needed was small enough that the results should not be noticeably affected. However, one major span of missing data did occur during the 1988-1994 period and that covered the first 12 days of the data set. For these 12 days, values were interpolated between December 2, 1987 and January 13, 1988, resulting in the potential of an error as high as 12 days in the 1988 ice-season length in the region of the December/January retreating ice edge. The results show no obvious impact of this data gap, but the reader should be aware of its existence. For the December 1991 period of F8/F11 overlap, the F8 data were used through December 15 and the F11 data were used starting on December 16.

Once the full set of daily ice concentration matrices was obtained, the length of the sea ice season was calculated for each year at each grid point by counting the number of days with ice concentrations exceeding a preset minimum. As in *Parkinson* [1994], three preset ice concentration minima were used, 15%, 30%, and 50%, and all calculations were done with each of these three ice concentration cutoffs. Plate 1 presents the three resulting season-length maps for the first full year of the data set, 1988. Naturally, the length of the sea ice season for a lower cutoff will always be everywhere equal to or greater than the length of the season for a higher cutoff, simply because at any time and place where the ice cover is at least A%, it must also be at least B% if B is less than A. In addition to confirming that basic certainty, comparison of the three images in Plate 1 reveals that the patterns are quite similar irrespective of which cutoff is used. In view of the similarities, for the remainder of the individual years, only the results for the 30% cutoff are presented. (A 30% cutoff was selected in preference to a 15% cutoff, which is often used for ice extent studies, in order to eliminate almost all atmospheric noise, not just in the southern hemisphere images but in the corresponding northern hemisphere images. This choice is consistent with the earlier study presenting ice-season lengths from the SMMR data [*Parkinson*, 1994].) For the trends and for the 1988 anomalies, however, the results are presented for each of the three cutoffs, to confirm that the choice of the cutoff is not critical to the anomalies or trends either.

3. RESULTS

Length of the Sea Ice Season, Yearly Values

Plates 2 and 3 present maps of the length of the sea ice season, with a 30% ice concentration cutoff, for each year 1988-1994, and a map of the average length of the sea ice season over the seven years. For the seven-year average (Plate 3d), the basic overall patterns in the length of the sea ice season are as follows: (1) the season length equals or exceeds 359 days (i.e., ice remains throughout the year or very close to it) in a sizable area (496,000 km^2) of the western Weddell Sea and at scattered coastal locations around the rest of the Antarctic continent; (2) in general, ice season lengths decrease outward from the coast, with the major exception coming in the region directly north of the Ross Ice Shelf; (3) directly north of the Ross Ice Shelf, there is a region of relatively short (about 257 days) ice season lengths; (4) in much of the Weddell Sea, in contrast to most of the rest of the Southern Ocean, ice season lengths basically decrease to the northeast instead of directly north. Pattern 3 results because of the common springtime/early summer occurrence of a sizable polynya in the Ross Sea directly off the Ross Ice Shelf, and pattern 4 results because of the general cyclonic circulation in the Weddell Sea, with ice being driven southwest to northeast over much of the central portion of the sea. The reader is referred to *Gloersen et al.* [1992] and

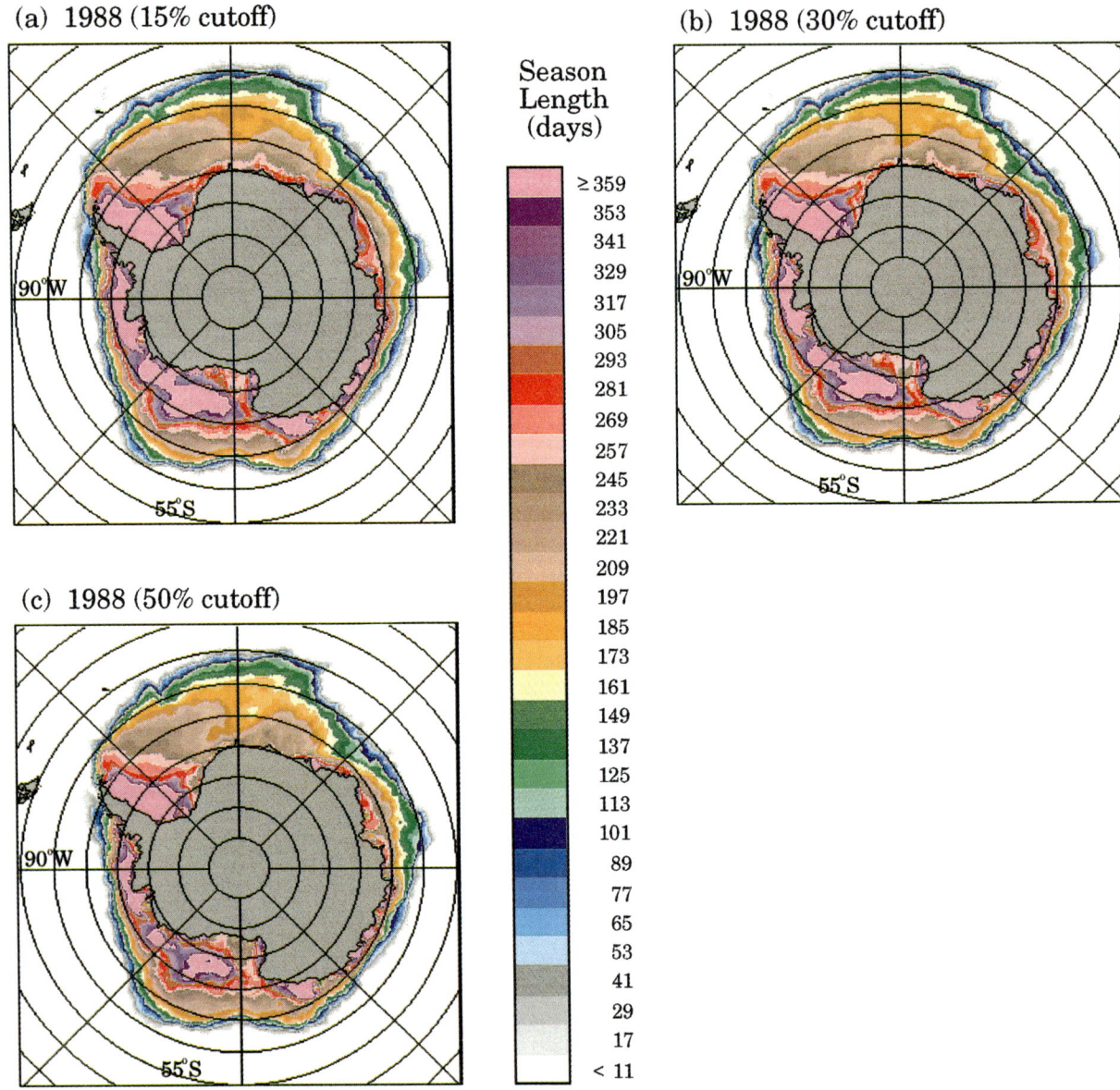

Plate 1. Maps of the length of the 1988 sea ice season in the Southern Ocean, calculated from the passive-microwave data of the DMSP Special Sensor Microwave Imager (SSM/I) using ice concentration cutoffs of (a) 15%, (b) 30%, and (c) 50%, respectively.

Jacobs and Comiso [1989] for details on the Ross Sea polynya (especially Figures 4.1.14 and 4.1.15 in *Gloersen et al.*, providing maps depicting the polynya during each of the years 1978-1986) and to *Deacon* [1979], *Kottmeier and Sellmann* [1996], and *Drinkwater* [this volume] for details on the circulation of the Weddell Sea.

Although the four basic patterns listed in the previous paragraph for the seven-year average ice season lengths (Plate 3d) also appear in each of the individual years, noticeable interannual differences exist as well (Plates 2 and 3). For instance, four of the seven years (namely, 1988-1990 and 1994) have sizable regions with ice throughout the year (for at least 359 days) in the southern Amundsen Sea and the southern portion of the far eastern Ross Sea, and two of the seven years (1988 and 1993) have ice throughout the year in the southern Bellingshausen Sea. In 1988, the year-round

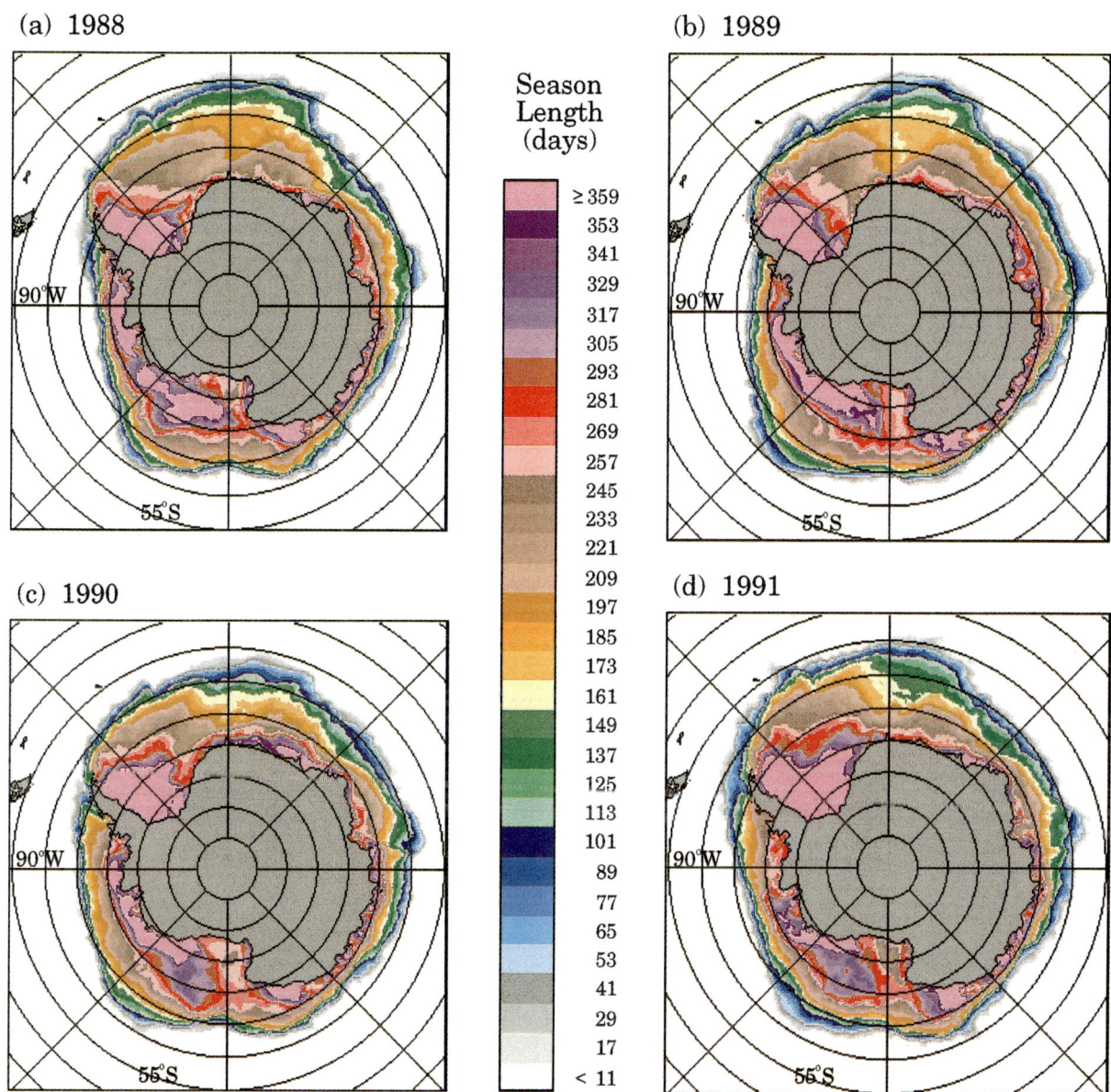

Plate 2. Maps of the length of the sea ice season in the Southern Ocean for (a) 1988, (b) 1989, (c) 1990, and (d) 1991, calculated from the passive-microwave data of the DMSP SSM/I using an ice concentration cutoff of 30%.

ice in the eastern Ross Sea is a noticeable distance from the coast, whereas in the following two years year-round ice exists adjacent to the coast. The 1988 results in the eastern Ross Sea are thereby anomalous in failing to conform with pattern 2 of the previous paragraph, instead having ice season lengths increasing outward from the coast for the first 2°-4° of latitude. Examination of the daily maps (not shown) reveals the cause to be coastal polynyas in January and February of 1988 and a coastal extension into the eastern Ross Sea of a region of open water in the western Ross Sea in the same two months, shortening the 1988 ice season lengths along the eastern Ross Sea coast.

In the Weddell Sea, 1990, 1991, and 1994 have noticeably larger regions of perennial ice than the other years, and 1992 and 1993 do not have perennial ice adjacent to the Antarctic Peninsula as the other years do (Plates 2 and 3). Defining the Weddell Sea as extending

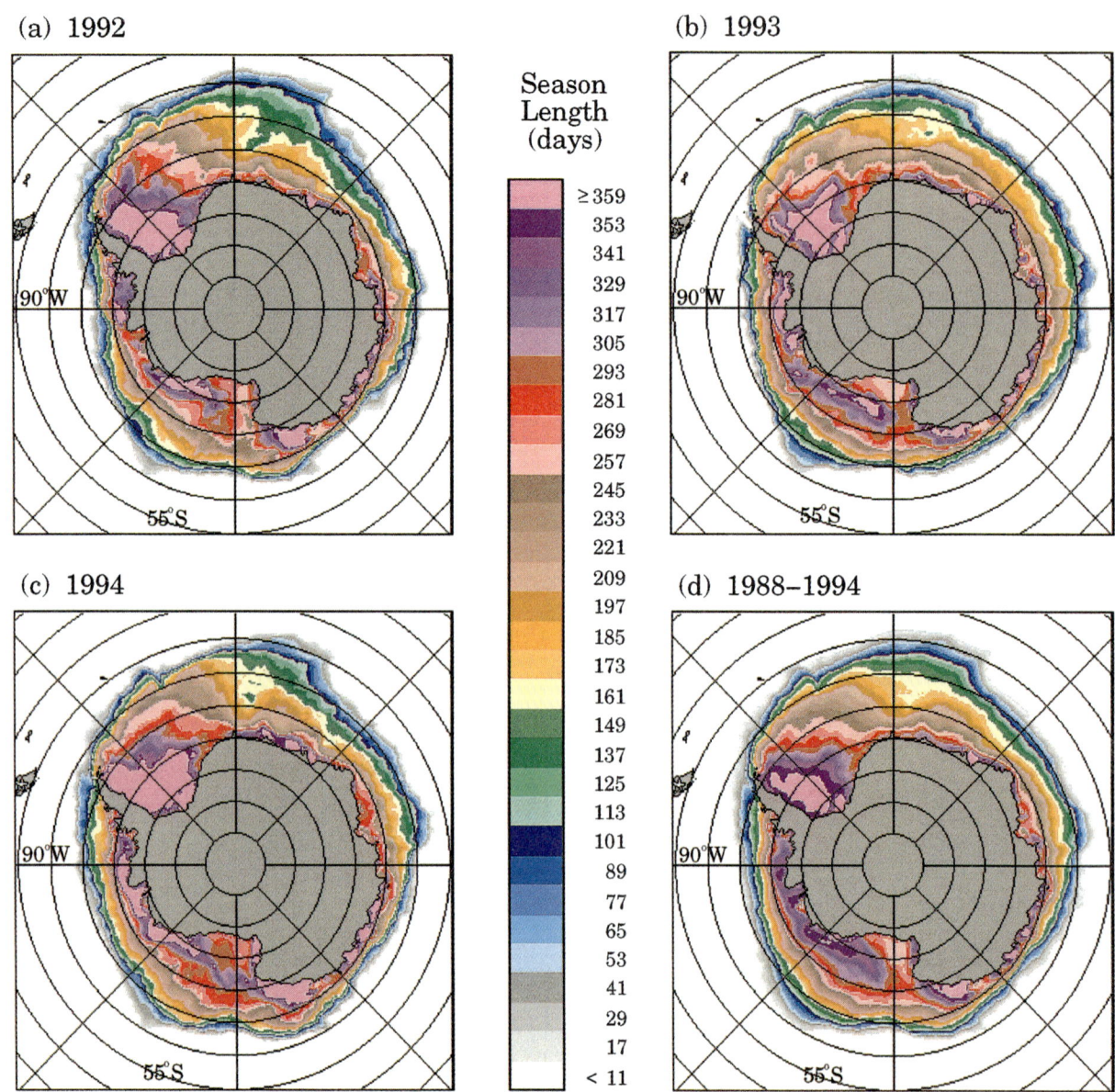

Plate 3. Maps of the length of the sea ice season in the Southern Ocean for (a) 1992, (b) 1993, and (c) 1994, calculated from the passive-microwave data of the DMSP SSM/I using an ice concentration cutoff of 30%, plus (d) the average length of the sea ice season in the Southern Ocean over the seven years 1988-1994, calculated by averaging the results depicted in Plates 2a-d and 3a-c.

from the Antarctic Peninsula eastward to 20°E, the area of perennial ice coverage in the Weddell Sea is 1.04×10^6 km^2, 1.31×10^6 km^2, and 1.12×10^6 km^2 for 1990, 1991, and 1994, respectively, with none of the other four years having perennial ice coverage as high as 0.84×10^6 km^2. Perennial ice coverages for each year for the Weddell and other seas and the Southern Ocean as a whole are plotted in Figure 2.

Interannual differences are even more pronounced off the Ross Ice Shelf, in the southwestern Ross Sea (70°S-78°S), where each year is distinctly different, reflecting the variability of the Ross Sea polynya [e.g., *Gloersen*

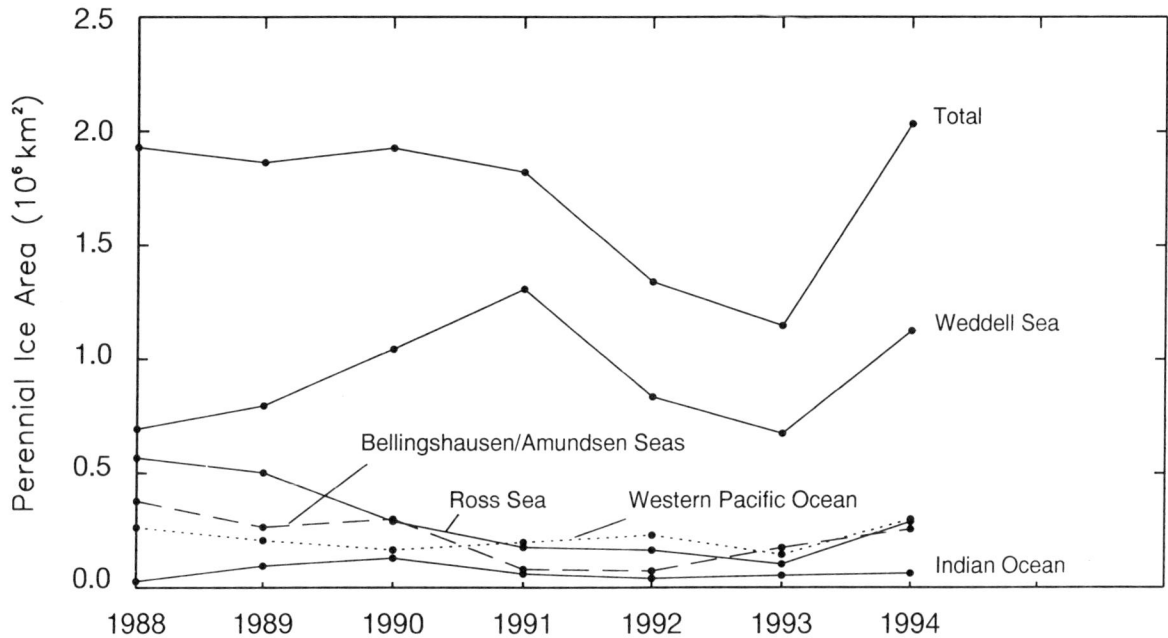

Fig. 2. Perennial ice area, defined as the area of the ocean having ice coverage of at least 30% for at least 359 days of the year, i.e., the area covered by the highest color-scale category on the maps of Plates 2 and 3. Perennial ice area is plotted for the Southern Ocean as a whole and for each of the following subdivisions: the Weddell Sea, extending eastward from the Antarctic Peninsula to 20°E; the Indian Ocean sector, extending from 20°E to 90°E; the Western Pacific Ocean sector, extending from 90°E to 160°E; the Ross Sea, extending from 160°E to 130°W; and the Bellingshausen/Amundsen Seas, extending from 130°W to 60°W, excluding the points in this range that lie to the east of the Antarctic Peninsula.

et al., 1992; *Jacobs and Comiso*, 1989]. Overall, the shortest ice seasons in this portion of the Ross Sea are in 1991, with a sizable area having season lengths under 251 days, and the longest ice seasons are in 1994, with some perennial ice at the southwestern edge of the Ross Sea (approximately 165°E, 74°S-78°S) and no ice seasons as short as 251 days anywhere within the Ross Sea south of 70°S and west of 155°W (Plates 2 and 3). Additional interannual differences, although clear on examination of the season-length maps (Plates 2 and 3), can be brought into sharper focus through mapping anomalies.

Yearly Anomaly Maps

The interannual differences in the ice season lengths, using a 30% ice concentration cutoff, are highlighted in Plates 4 and 5 with anomaly maps presented for each year. In each case, the anomaly is calculated with respect to the full 1988-1994 period, subtracting the seven-year average season lengths from the season lengths for the individual years. Spatial patterns of relatively long and relatively short ice seasons (for the individual locations, versus the seven-year average) are far clearer on the anomaly maps (Plates 4 and 5) than on the maps of season lengths (Plates 2 and 3). On the anomaly maps, relatively long seasons are colored in shades of orange and brown and relatively short seasons are colored in shades of blue and green (Plates 4 and 5).

The patterns on the anomaly maps are distinctly different from what would appear if the Southern Ocean ice cover as a whole were undergoing a marked shortening of the ice season over this period, for instance from climatic warming, or a marked lengthening, for instance from climatic cooling. In the event of a marked warming and season shortening, the earlier years would be dominated by browns and oranges and the later years by greens and blues; in the event of a marked cooling and

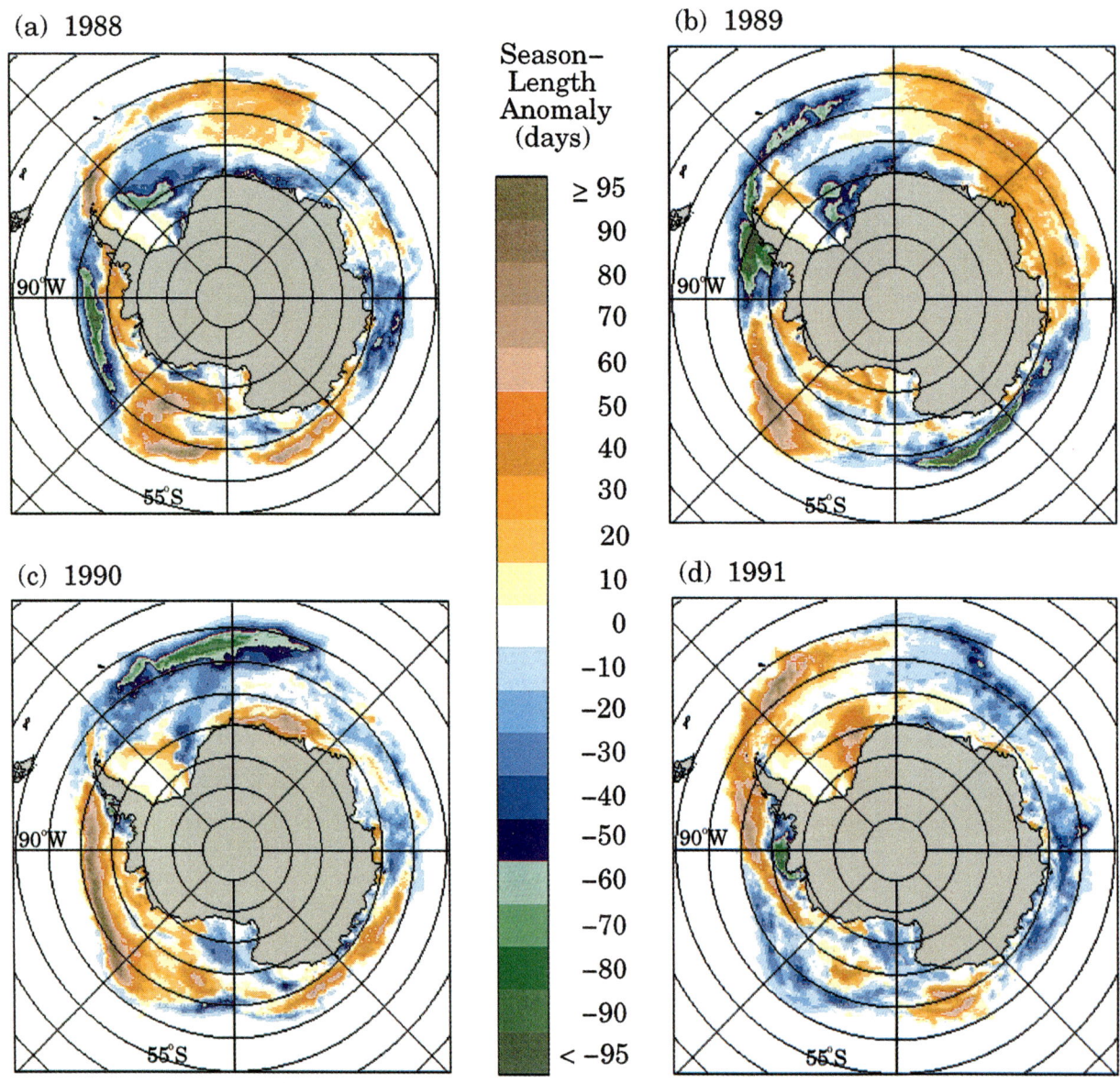

Plate 4. Anomalies in the length of the sea ice season (ice concentration cutoff of 30%) in the Southern Ocean for (a) 1988, (b) 1989, (c) 1990, and (d) 1991. These anomalies were calculated by subtracting the average season lengths of Plate 3d, for the full 1988-1994 period, from the individual annual season lengths of Plate 2.

season lengthening, the opposite would occur. Instead, each of the maps shows coherent patterns of sizable regions of relatively short ice seasons and sizable regions of relatively long ice seasons, and none is dominated by either positive or negative values.

In comparing the map sequence of Plates 4 and 5, of particular interest is the suggestion of a clockwise rotation of the regions of anomalously long and anomalously short ice seasons in the outer pack. For instance, the prominent elongated region of relatively short ice

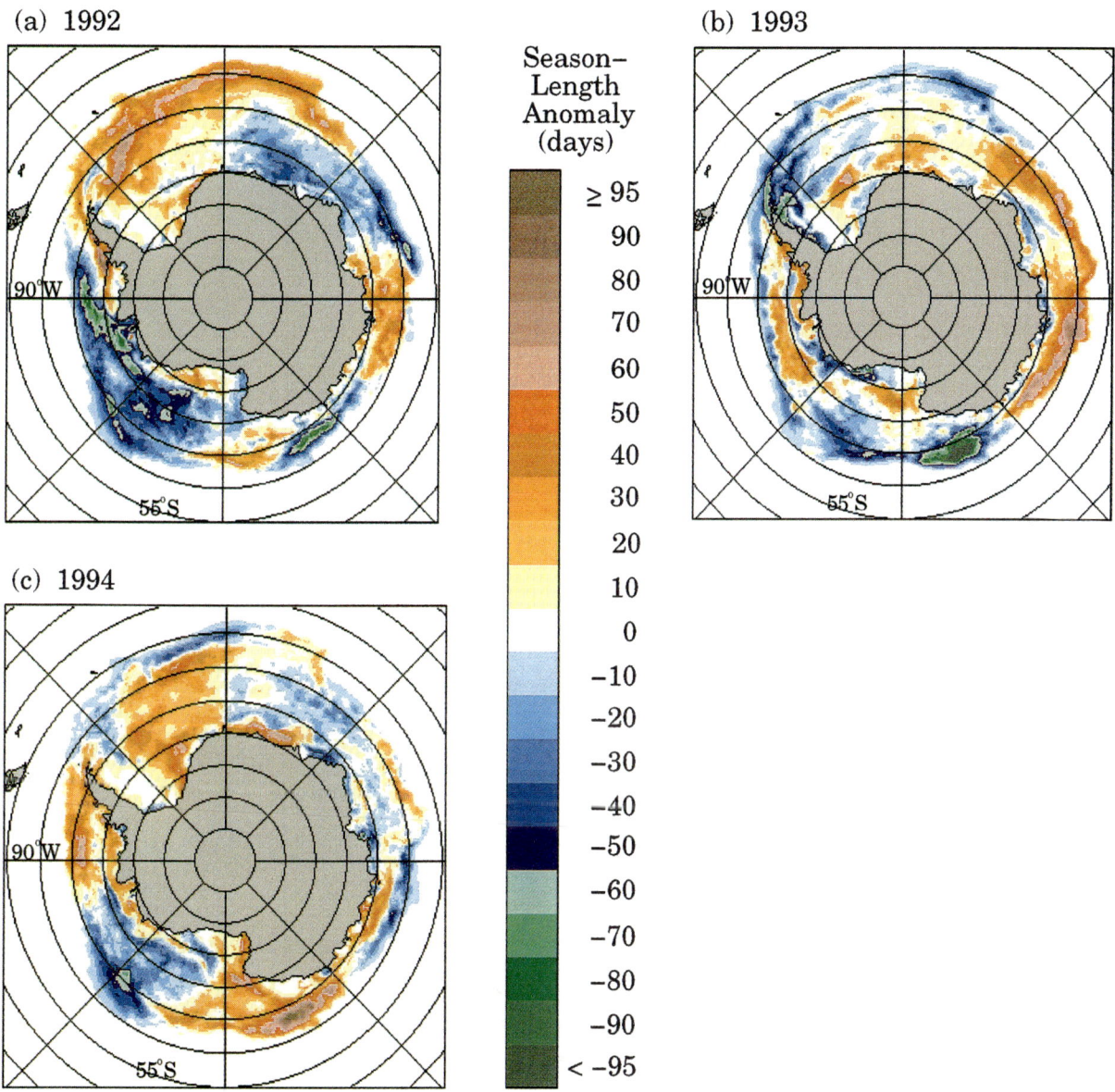

Plate 5. Anomalies in the length of the sea ice season (ice concentration cutoff of 30%) in the Southern Ocean for (a) 1992, (b) 1993, and (c) 1994. These anomalies were calculated by subtracting the average season lengths of Plate 3d, for the full 1988-1994 period, from the individual annual season lengths of Plates 3a-c.

seasons in the outer Amundsen and Bellingshausen seas in 1988 shifts to the outer Bellingshausen and western Weddell seas in 1989, then to the outer central and eastern Weddell Sea in 1990, and north of East Antarctica in 1991. Such a clockwise rotation ties in well with the Antarctic circumpolar wave identified by *White and Peterson* [1996], who found anomalies in surface pressure, wind stress, sea surface temperature, and sea-ice extent over the Southern Ocean to propagate eastward with the circumpolar flow. The results of Plates 4 and 5

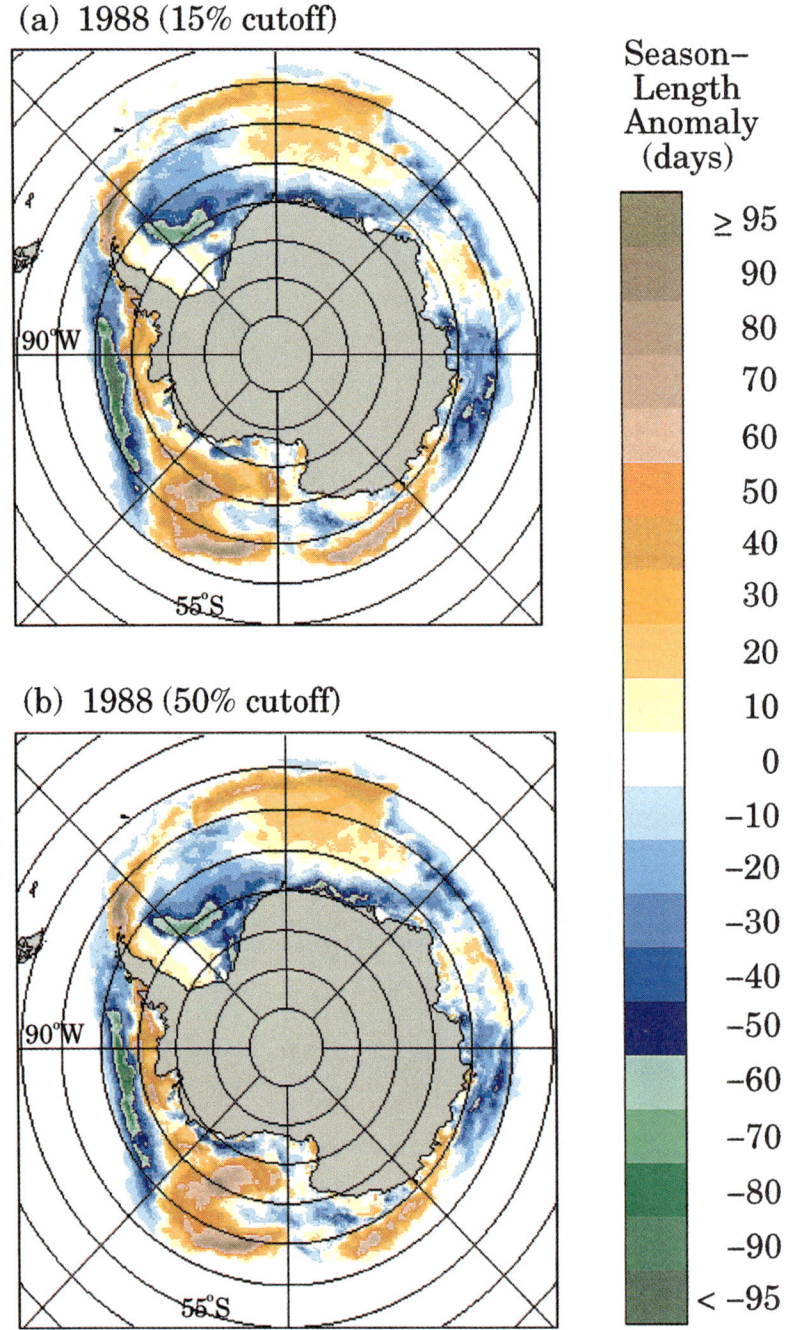

Plate 6. Anomalies in the length of the sea ice season in the Southern Ocean for 1988, using ice concentration cutoffs of (a) 15% and (b) 50%, calculated from the passive-microwave data of the DMSP SSM/I. The anomalies were calculated with respect to the average conditions over the full 1988-1994 period.

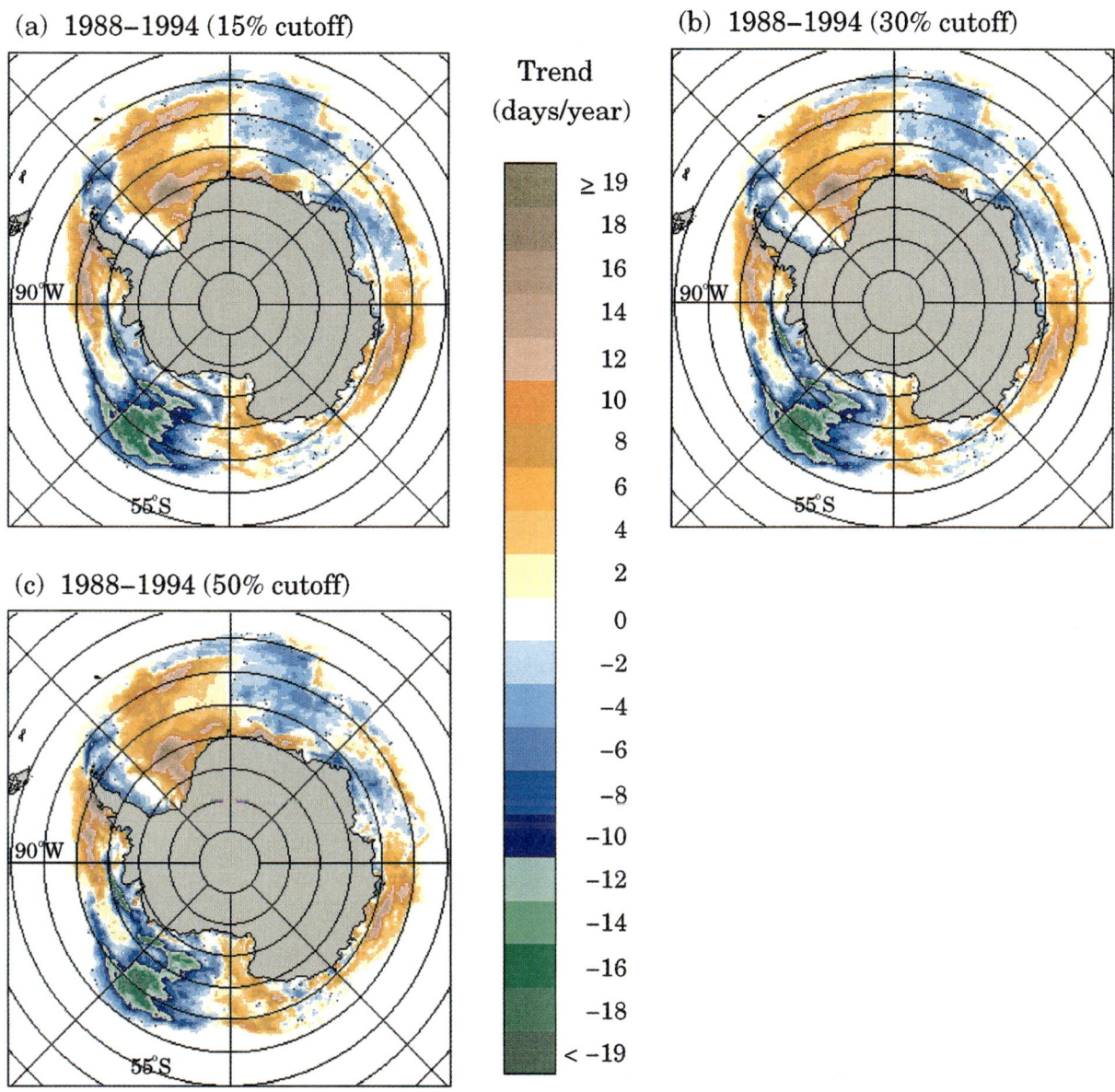

Plate 7. Trends in the length of the sea ice season in the Southern Ocean over the period 1988-1994, using ice concentration cutoffs of (a) 15%, (b) 30%, and (c) 50%, respectively. The trends are the slopes of the lines of linear least squares fit through the yearly ice season lengths of Plates 2a-d and 3a-c. Positive trends indicate a lengthening of the sea ice season, negative trends a shortening.

suggest that anomalies in the length of the sea ice season might also exhibit an eastward propagation, offering a partial confirmation (with an additional variable) of the circumpolar wave of *White and Peterson* [1996].

Plate 6 presents, for the first year of the data set, the season-length anomaly maps for ice concentration cutoffs of 15% and 50%, illustrating, by the great similarity of Plates 6a, 6b, and 4a, that the choice of ice concentration cutoff is not critical to the patterns displayed in Plates 4 and 5. Whether ice presence is defined as at least 15%, at least 30%, or at least 50% ice coverage, the season length anomaly matrices are practically iden-

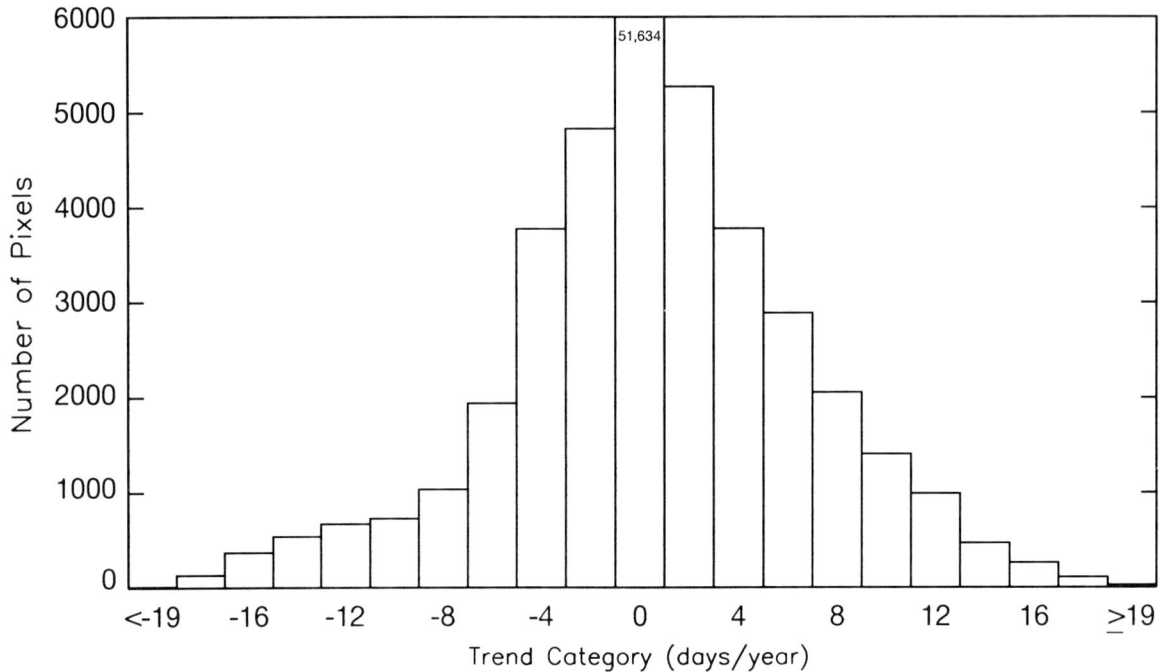

Fig. 3. Histogram of the number of pixels in each trend category of Plate 7b, for the 1988-1994 trend map using an ice concentration cutoff of 30%. Each trend category between -19 days per year and +19 days per year extends over a range of 2 days per year, centered at the value given on the x-axis. The 0 category, extending from trend values of -1 day per year to +1 day per year, contains 51,634 pixels, largely from the region outside the ice cover. This value is far off the scale of the histogram and hence is simply written near the top of the 0 category bar.

tical, not just in the patterns depicted but also in the magnitudes. Differences can be seen amongst the three 1988 anomaly maps (Plates 6a, 4a, and 6b), for instance in the outer portions of the Ross Sea ice pack at 170°E-180°E, but these differences are small in the context of the Southern Ocean ice pack as a whole. The same is also true for the anomaly maps for each of the other six years.

Trends

The data depicted in Plates 2a-d and 3a-c have been used to calculate the linear least squares fits through the time series of the seven ice season lengths at each grid point. (To adjust for different year lengths, the season lengths in leap years were multiplied by 365/366 prior to the calculation of the least squares fits.) The brevity of the record and the nonlinear fluctuations within it limit the significance that should be accorded to the individual least squares values. Nonetheless, mapping the results (Plate 7) reveals interesting patterns. Plate 7b presents a map of the slopes of the lines of least squares fit through the data of Plates 2a-d and 3a-c, i.e., using an ice concentration cutoff of 30%. Plates 7a and 7c present the corresponding maps when using ice concentration cutoffs of 15% and 50%.

The similarity of the three trend maps in Plate 7 confirms that the choice of ice concentration cutoff is not critical, at least for values of 15%, 30%, and 50% (and hence probably for the full range 15%-50%). In fact, all major patterns on each of the three maps also exist on the others, and the general appearance is of nearly identical results, although close examination reveals some differences at many locations, the strongest being in the far western Weddell Sea.

The mapped trends show that, overall, on the basis of linear least squares fits, ice season lengths decreased over the 1988-1994 period in some sizable regions of the Southern Ocean and increased in others. In particular, season lengths decreased most in the eastern Ross Sea and to lesser extents in the southern and northern Amundsen Sea, the western Weddell Sea, the non-coastal portions of the ice pack from 5°E to 40°E and the coastal regions from 40°E to 80°E and from 140°E to

160°E (Plate 7). In contrast, the season lengths increased in the western Ross Sea, throughout almost the entire Bellingshausen Sea, in the central Weddell Sea, in the coastal Weddell Sea everywhere east of 40°W, and throughout most of the 80°E-135°E sector off the coast of East Antarctica (Plate 7). The trend maps show no noticeable speckling of positive and negative values, instead containing several sizable regions with exclusively negative trends along with several sizable regions with exclusively positive trends. Furthermore, there is no strong predominance of either season-length increases or season-length decreases, although on each of the trend maps (Plate 7), the number of pixels with positive trends of at least 1 day per year exceeds the number of pixels with negative trends of less than -1 day per year. Specifically, with an ice concentration cutoff of 15%, the number of pixels with positive trends is 17,866 while the number of pixels with negative trends is 14,305. The corresponding numbers for an ice concentration cutoff of 30% are 17,249 and 14,024, and those for an ice concentration cutoff of 50% are 16,468 and 13,750. More details on the distribution of the trend values in the case of the 30% ice concentration cutoff are provided in Figure 3.

4. DISCUSSION AND COMPARISON WITH EARLIER DATA

The length of the sea ice season is relevant to such climatically important issues as the amount of solar radiation absorbed by the oceans and the magnitudes of the heat, mass, and momentum exchanges between the ocean and the atmosphere. By defining the length of the season on the basis of the calendar year, consistent comparisons can be made both from location to location and from year to year. At some locations, where the previous year's ice coverage remains into January and then melts or retreats from the location during the new year, the calendar-year definition means that the "sea ice season" for the year is composed of portions of two different geophysical ice seasons. This could be avoided for many locations by starting the year in late February, at the time of minimum ice coverage, but such a redefinition would have the major disadvantage of losing easy comparability with other geophysical variables that are presented on a calendar-year basis. For the integrated impact of the ice on the ocean and atmosphere over the course of a year, the issue of whether the impact came from a single geophysical ice season or from portions of two geophysical ice seasons is irrelevant. Instead, the critical variables are how long the ice existed at the location, i.e., the length of the sea ice season, and the ice thickness, a much more difficult variable to measure on a widescale basis.

The passive-microwave DMSP SSM/I data have allowed examination of the length of the sea ice season over the period 1988-1994, and the results presented in Plates 1-7 show that the changes in season lengths over this seven-year period exhibit interesting patterns but no dominant overall trend toward either shorter or longer ice season lengths, although the area that experienced a lengthening of the sea ice season over this period exceeded by about 20% the area that experienced a shortening of the sea ice season. Of particular interest is the apparent eastward propagation of anomalies in the season lengths in the outer reaches of the ice pack (Plates 4 and 5), suggestive of the eastward propagating circumpolar wave identified by *White and Peterson* [1996], and the spatially coherent, alternating patterns of lengthening and shortening sea ice seasons over the seven-year period as a whole (Plate 7).

Some relevant comparisons can be made with earlier results on the length of the sea ice season in the Southern Ocean, for the years 1979-1986, from the Nimbus 7 SMMR data. Those results show some suggestion of an eastward propagation of the season-length anomalies [*Parkinson*, 1994, Plates 6 and 7] but not as strong a suggestion as appears in the 1988-1994 results. The least-squares trends over the 1979-1986 period [*Parkinson*, 1994, Plate 9] show an overall lengthening of the sea ice season in the eastern Ross Sea, where the 1988-1994 period exhibits its most prominent shortening of the season (Plate 7), and an overall shortening of the sea ice season in the Bellingshausen Sea, where the 1988-1994 period exhibits a prominent lengthening. In other seas, there is a mixture of some areas where the signs of the trends in the two periods match and other areas where the signs of the trends are reversed, with the overall picture being one of an interannually varying sea ice cover with no strong indications of long-term trends toward either longer or shorter sea ice seasons.

Acknowledgments. The DMSP F8 and F11 brightness temperature data were obtained on CD-ROMs in convenient gridded format from the Earth Observing System (EOS) Distributed Active Archive Center (DAAC) at the National Snow and Ice Data Center (NSIDC) in Boulder, Colorado. The brightness temperatures were converted to sea ice concentrations as part of a group effort at Goddard Space Flight Center, the group members being, in alphabetical order, D. Cavalieri, J. Comiso, P. Gloersen, C. Parkinson, and J. Zwally. The work on the length of the sea ice season was done by the author with the assistance of Jamila Saleh of Hughes STX, who generated the images. The author also thanks Seelye Martin, two anonymous reviewers, and the volume editor Martin Jeffries for constructive comments and reviews. This work was funded by the Polar Programs office in the Science Division at NASA Headquarters.

REFERENCES

Deacon, G. E. R., The Weddell gyre, *Deep-Sea Res., 26A*, 981-995, 1979.

Drinkwater, M. R., Active microwave radar observations of Weddell Sea ice, Antarctic Research Series, this volume.

Gloersen, P., W. J. Campbell, D. J. Cavalieri, J. C. Comiso, C. L. Parkinson, and H. J. Zwally, *Arctic and Antarctic Sea Ice, 1978-1987: Satellite Passive-Microwave Observations and Analysis*, 290 pp., National Aeronautics and Space Administration, Washington, D.C., 1992.

Jacobs, S. S., and J. C. Comiso, Sea ice and oceanic processes on the Ross Sea continental shelf, *J. Geophys. Res., 94*, 18,195-18,211, 1989.

Kottmeier, C., and L. Sellmann, Atmospheric and oceanic forcing of Weddell Sea ice motion, *J. Geophys. Res., 101*, 20,809-20,824, 1996.

NSIDC Distributed Active Archive Center, *DMSP SSM/I Brightness Temperatures and Sea Ice Concentration Grids for the Polar Regions: User's Guide*, 110 pp., National Snow and Ice Data Center, University of Colorado, Boulder, Colorado, 1996.

Parkinson, C. L., Spatial patterns in the length of the sea ice season in the Southern Ocean, 1979-1986, *J. Geophys. Res., 99*, 16,327-16,339, 1994.

White, W. B., and R. G. Peterson, An Antarctic circumpolar wave in surface pressure, wind, temperature and sea-ice extent, *Nature, 380*, 699-702, 1996.

Claire L. Parkinson, Oceans and Ice Branch/Code 971, NASA Goddard Space Flight Center, Greenbelt, MD 20771.

(Received November 11, 1996; accepted April 8, 1997)

ACTIVE MICROWAVE REMOTE SENSING OBSERVATIONS OF WEDDELL SEA ICE

Mark R. Drinkwater

Jet Propulsion Laboratory, California Institute of Technology

Since July 1991, the European Space Agency's ERS-1 and ERS-2 satellites have acquired radar data of the Weddell Sea, Antarctica. The Active Microwave Instrument on board ERS has two modes; SAR and Scatterometer. Two receiving stations enable direct downlink and recording of high bit-rate, high resolution SAR image data of this region. When not in an imaging mode, when direct SAR downlink is not possible, or when a receiving station is inoperable, the latter mode allows normalized radar cross-section data to be acquired. These low bit-rate ERS scatterometer data are tape recorded, downlinked and processed off-line. Recent advances in image generation from Scatterometer backscatter measurements enable complementary medium-scale resolution images to be made during periods when SAR images cannot be acquired. Together, these combined C-band microwave image data have for the first time enabled uninterrupted night and day coverage of the Weddell Sea region at both high (25 m) and medium-scale (~20 km) resolutions.

C-band ERS-1 radar data are analyzed in conjunction with field data from two simultaneous field experiments in 1992. Satellite radar signature data are compared with shipborne radar data to extract a regional and seasonal signature database for recognition of ice types in the images. Performance of automated sea-ice tracking algorithms is tested on Antarctic data to evaluate their success. Examples demonstrate that both winter and summer ice can be effectively tracked. The kinematics of the main ice zones within the Weddell Sea are illustrated, together with the complementary time-dependencies in their radar signatures. Time-series of satellite images are used to illustrate the development of the Weddell Sea ice cover from its austral summer minimum (February) to its winter maximum (September). The combination of time-dependent microwave signatures and ice dynamics tracking enable various drift regimes to be defined which relate closely to the circulation of the sea ice in response to current and wind forcing and iceberg barriers. These are closely related to continental-shelf or central basin regimes, in which tidal forcing or barotropic circulation patterns appear to influence the sea-ice motion, respectively. These regimes provide valuable information about the regions of most prolific ice growth and influence of ice conditions upon air-sea-ice exchange processes in the Weddell Sea.

1. INTRODUCTION

During this century, a number of scientific expeditions have been carried out to study the general sea-ice characteristics and circulation patterns in the Weddell Sea [*Brenneke*, 1921; *Schnack-Schiel*, 1987; *Augstein et al.*, 1991; *Lemke*, 1994]. Direct observations have, however, been limited to the eastern and central regions due to a nearly continuous, impenetrable perennial ice cover in the southwestern and western Weddell Sea. Indeed, the early "Endurance" and "Deutschland" drift expeditions were restricted to the central part of the Weddell Sea; and the recent 1992 US-Russian driftstation experiment, though situated on a perennial ice floe in the south-western Weddell Sea, was positioned off the shelf-break over the deep ocean basin [*ISW Group*, 1993].

Scientific results to date indicate ocean circulation is dominated by the Weddell Gyre, with a southward current along the east coast and a northward flowing, western boundary current along the shelf edge at about 55°W off the Antarctic peninsula [*Orsi et al.*, 1993; *Fahrbach et al.*, 1994; *Muench and Gordon*, 1995]. Cyclonic circulation is driven by westerly winds in the north and by southerly winds in the southern Weddell Sea. Passive microwave images [*NSIDC*, 1995] together with the results of large scale sea ice models [*Fischer*, 1995] generally indicate perennial sea ice with increased thicknesses and concentration in the southwest [*Gloersen et al.*, 1992] as a result of the large scale atmospheric forcing and ice drift towards the Antarctic peninsula. But, few direct spatial or temporal observation datasets are available with which to validate such general patterns. Although buoys, for instance, provide detailed Lagrangian data [*Massom*, 1992; *Viehoff and Li*, 1995] such drifters have traditionally been placed out singly or in clusters during specific seasons and at different locations. This strategy of deployment together with the limited number of total buoy drifts in the last decade makes it impossible to characterize short-timescale variability in large-scale ice drift.

Recently, observations with active radar remote sensing techniques have enabled mapping of various sea-ice regimes within the Weddell Sea [*Drinkwater et al.*, 1993a] together with their dynamics [see papers in *Drinkwater*, 1995; *and Drinkwater and Kottmeier*, 1994]. The spatial resolution advantage which active microwave instruments have over passive microwave radiometers is the factor of 1×10^3 improvement, from 25 km to 25 m. Synthetic Aperture Radar (SAR) images available from various satellite instruments now provide information on the small-scale character of the ice cover, and on a scale which enables direct measurement of the areal fraction of open water to be quantified, together with the proportion and location of perennial ice [*Drinkwater and Lytle*, 1997]. The main limitation of SAR is the 100 km swath width and the limited temporal and spatial sampling. But a new technique of producing enhanced resolution (~ 20 km) full-polar scatterometer images now also enables large-scale dynamics of the Southern Ocean sea-ice cover to be analyzed and quantified.

This paper first introduces the C-band (5.3 GHz) sensor systems used in this study, together with some of the important and relevant sampling issues and limitations of satellite radar. Second, a broad description of the backscatter characteristics of Weddell Sea ice, together with its regional and seasonal variations is provided. Third, a high resolution description of the microwave signatures of the primary components of the ice cover is given, together with supporting information obtained from shipborne scatterometer and surface measurements. Finally, the regional and basin-wide ice dynamics are described using a combination of mesoscale SAR, buoy and Scatterometer ice kinematics products.

2. ANTARCTIC MICROWAVE RADAR IMAGING

Since 1991, two satellites of the European Space Agency (ESA), ERS-1 (July 1991–June 1996) and ERS-2 (April 1995–present), have acquired radar data of the Weddell Sea. The C-band Active Microwave Instrument (AMI) on board ERS has two alternative modes; an imaging SAR and a Scatterometer (EScat), and each allow vv-polarized backscatter cross-section ($\sigma^°_{vv}$) data to be collected (*i.e.* wavelength λ = 5.6 cm). When the AMI is not in SAR imaging mode it allows EScat backscatter values to be continuously recorded.

2.1 ERS SAR

Presently, SAR is the only instrument directly capable of acquiring all-weather (day or night), frequent repeat, high-resolution (25 m), 100 km-swath digital images of Antarctic sea-ice. High bit-rate SAR image data from these satellites require direct broadcast and recording in the region of interest [*ESA/Earthnet*, 1992]. Weddell Sea SAR data are recorded at two locations, the German Antarctic Receiving Station (GARS) situated at the Chilean General Bernardo O'Higgins station, and the Japanese Syowa station. GARS allows acquisition of SAR images of the Weddell Sea during periodic 6 week campaigns, while Syowa enables Weddell Sea data east of the Greenwich meridian to be acquired upon request. Weddell Sea image acquisition is therefore limited specifically to periods when the SAR is switched on within a receiving station mask, and which coincide with periods of operation of these stations.

2.2 ERS Scatterometer

To date, the primary focus of measurements in the ERS Scatterometer (EScat) mode of the AMI has been the estimation of ocean surface wind vectors from backscatter measurements made across a 500 km swath. Until recently, with the exception of *Drinkwater et al.* (1993a) and *Gohin and Cavanié* (1994), ERS-1 Scatterometer data had barely been exploited outside the traditional scope of measuring wind speed and direction. These low bit-rate EScat data, with an intrinsic resolution of ~ 50 km, do not require direct broadcast to a ground receiving station. Instead, when the SAR imaging mode is switched off, EScat data are continuously recorded on tape and downlinked to more accessible ESA ground stations. Scatterometer σ_{vv}° data can therefore be collected at times and in some regions where direct SAR downlink is impossible. The extensive daily coverage of the ERS-1 scatterometer is particularly valuable due to the lack of data reception in the AMI SAR mode when Antarctic receiving stations are closed. This is especially true in the Weddell Sea, because of the "campaign-style" operation of GARS and limited access to Syowa. Similarly, a large portion of the Southern Ocean sea-ice cover in the Pacific sector could not be imaged by SAR owing to the lack of a receiving station in this region prior to January 1996 when the US McMurdo receiving station opened.

Operationally, it is impossible to obtain enough SAR images in a short enough time span with which to effectively map ephemeral sea-ice conditions. Instead, a method is proposed using data acquired in the EScat mode for mapping sea-ice characteristics over the entire Southern Ocean ice cover. Recent advances in sea-ice image generation from Scatterometer data [*Long et al.*, 1993; *Drinkwater et al.* 1993a] enable complementary, medium-scale resolution images to be generated from raw EScat cell measurement data. This method allows construction of enhanced resolution (~ 20 km) scatterometer images from the backscatter data record, and, by virtue of the wider swath and larger coverage, allows mapping of the entire sea-ice cover in a few days [*Drinkwater et al.*, 1993a]. The product is a weekly average sea-ice picture, comparable to SSM/I passive microwave sea-ice data products. It is proposed that this product be used to track regional-scale spatial and temporal changes in the sea ice cover around the entire Antarctic coast. For the first time, combined C-band ERS SAR and Scatterometer microwave images therefore enable uninterrupted coverage of this geographic region at both high (25 m) and medium-scale (~ 20 km) resolution, respectively.

2.3 *Issues of Radar Time-Space Sampling Diversity*

One of the priorities for scientific studies of polar ice and snow is high spatial and temporal coverage, due to the short timescales of variability. Exploitation of radar data, particularly in studies related to the mapping and monitoring of the sea-ice cover, was one of the main driving forces behind inclusion of the SAR instrument on board ERS-1 [*Battrick*, 1985]. A large number of sea-ice studies are currently being performed using Arctic Ocean SAR data received at a number of northern hemisphere ground stations (such as Fairbanks, Alaska and Kiruna, Sweden) [see PIPOR Special Issue—*Drinkwater*, 1995]. These SAR receiving stations enable the majority of the northern hemisphere sea-ice cover to be mapped using the SAR. However, much less coverage is possible of the more extensive sea-ice cover around Antarctica, and in particular in the Weddell Sea, due to the limited operating times of GARS and Syowa. The result is a high resolution SAR image database which is at best discontinuous in space and time.

The key advantage of EScat is that it operates whenever the SAR is switched off, continuously retrieving information from the Weddell Sea without the necessity of a local receiving station. The wider EScat swath provides more frequent, and broader incidence angle ($20° \leq \theta \leq 60°$) coverage in a given location. This low-bit-rate (LBR) data source is essential to fill in areas of sparse temporal and spatial SAR coverage of the Antarctic ice cover. Until now, the intrinsic low resolution of the EScat data is the main reason why they have not been used in such applications. Nevertheless, the approach described by *Drinkwater et al.* (1993a) produces weekly backscatter maps of the entire Southern Ocean sea-ice cover at a resolution (~ 20 km) slightly higher than present conventional alternatives such as the SSM/I passive microwave radiometers. EScat images form an uninterrupted sequence of C-band backscatter (σ_{EScat}°) of Antarctic sea ice from the beginning of the ERS-1 mission. Since these backscatter images are based on multiple azimuth and incidence angle observations, the data provide greater capability for discriminating ice types and separating ice from ocean. Coupled with higher frequency passive microwave and finer resolution SAR, these data enable a better understanding of the time sequence and seasonal history of Antarctic sea-ice formation, drift, deformation and decay. Furthermore, medium-scale resolution EScat images provide the large-scale context within which the high resolution SAR images may be interpreted.

3. RADAR CHARACTERISTICS OF WEDDELL SEA ICE

Since the launch of ERS-1, C-band radar backscatter characteristics of Weddell Sea ice have been investigated during a number of surface experiments [*Drinkwater et al.*, 1993b; *Hosseinmostafa et al.*, 1995; *Drinkwater et al.*, 1995a; *Lytle et al.*, 1996]. Most of these have focused on the smaller-scale influences of sea-ice physical properties upon surface-scatterometer measurements, without placing them in the context of satellite radar measurements. In this section, a synthesis of both large- and small-scale measurements is provided. Examples are provided of satellite radar observations together with corresponding field scatterometer and surface validation data from the 1992 Winter Weddell Gyre Study (WWGS '92) [*Lemke*, 1994; *Drinkwater and Haas*, 1994].

3.1 Backscatter Signatures of Ice Floes

Field radar backscatter data were acquired in 1992 using a shipborne C-band microwave scatterometer during the WWGS '92 experiment. The C-band (4.3 GHz) frequency-modulated continuous-wave (FM-CW) radar scatterometer was operated from the port rail of R.V. *Polarstern* to obtain some of the first shipborne measurements of the C-band microwave scattering properties of Antarctic sea ice in mid-winter [*Drinkwater et al.*, 1995a]. The radar had dual polarization, enabling like- (vv) and crosspol. (hv) data to be acquired at a variety of incidence angles ($15° \leq \theta \leq 70°$). When the ship was stationary by an ice floe, the radar was scanned to obtain normalized backscatter coefficient ($\sigma°_{Ship}$) measurements as a function of incidence angle and polarization (provided the floe was large enough and uniform in extent).

The objective of the 1992 winter scatterometer measurements was to provide validation data for ERS-1 SAR observations. This was achieved by constructing a microwave backscatter signature catalogue from measurements of a variety of different ice conditions within the SAR swath. In support, detailed surface measurements were also made within the footprint of the shipborne radar, after each scan was performed. Surface information comprises snow and ice physical and chemical properties measurements together with snow and ice structural data obtained from snow pits and crystal photography [*Drinkwater and Haas*, 1994]. Overlapping data acquisitions were planned and made possible by the GARS as part of the International Space Year Project [*Lemke*, 1994].

3.1.1 *The ice margin and pancake ice.* The outer ice margin or marginal ice zone (MIZ) in Antarctica is typically characterized as a high-energy, wave-dominated environment, where the propagation of waves into the ice edge plays a dominant role in determining the style of ice growth during winter ice advance [*Tucker et al.*, 1992], and the characteristics of floes during ice-edge recession. The MIZ is typically broad in its extent during ice-edge advance. In 1992, during WWGS '92, a high-amplitude swell was experienced along the Greenwich meridian and frazil and pancake ice growth was observed for hundreds of kilometers from the ice margin into the seasonal ice zone. Figure 1a shows a photograph of pancake ice in the vicinity of Maud Rise. In contrast, after the peak in ice extent the ice edge is typically in a recessional mode, and the MIZ usually comprises brash ice and a mixture of small wave-broken floes and wave-washed piles of rubble.

In contrast to the situation for nilas and young ice forms found in calm environments, pancakes were found to be almost ubiquitous at the ice margin (along the Greenwich meridian), during ice-edge advance in 1992. The occurrence of vast expanses of pancakes and/or streamers with densely-packed sub-resolution floes results in an extremely characteristic MIZ microwave signature [*Drinkwater et al.*, 1993a; *Gohin*, 1995; *Early and Long*, in press]. Typically riding on the surface of waves, small wave-washed, porous pans often have a high salinity (Figure 1b) and sufficiently high permittivity that there is relatively strong backscatter near nadir. The rate at which $\sigma°_{Ship}$ falls with incidence angle, however, appears dependent on the size and packing density of the pancakes, the amount of open water between them, and how wet or deformed they are. Figure 1b shows the salinity characteristics of a pancake sampled on 13 June, 1992, and Figure 1c contrasts the 13 June and 14 June *in-situ* radar scatterometer signatures of pancake ice in the same locality. The 13 June backscatter coefficient is greater by 10 dB throughout the incidence angle range, with values exceeding -5 dB at 40° incidence or more, and a gradient of -0.14 dB/°. Pancake ice packing density exceeded 90 % at the location of radar measurements on 13 June and many pancakes had been rafted onto one another by the long-wavelength swell waves. Individual pancakes had an overall mean diameter of 0.7 m [*Drinkwater and Haas*, 1994] with the larger pans (up to 2.0 m in diameter) themselves comprised of several smaller pancakes. Thus, a previously frozen pancake ice cover had been broken up and reworked into larger pancakes with rough, wetted surfaces and large rims around their edges. Although the windspeed on 13 June was close to 10 m s^{-1}, wind-roughened ocean surfaces contributed little to the relatively high backscatter values, since the areas of exposed water surface between pancake edges were extremely small in extent and thus 'fetch-limited'. In comparison, on 14 June only a light swell was present and wind speeds had fallen to around 5 m s^{-1}. Photographs of the ice conditions in the location of the scatterometer beam indicate relatively small and wet, recently-formed pancakes (around 0.2 m in diameter) which were relatively smooth and having a packing density exceeding 95 %. These pancakes were separated by frazil-laden seawater, were actively growing at the time of measurement, and their surfaces had no snowcover. In this example, $\sigma°_{Ship}$ has a steeper gradient of -0.24 dB/°.

Gohin (1995, Figure 15) reports a contrasting intermediate marginal ice zone signature which appears to be a composite of signatures originating over combinations of open water and dispersed floes, or perhaps low concentration streamers or wave-herded patches of pancakes. At

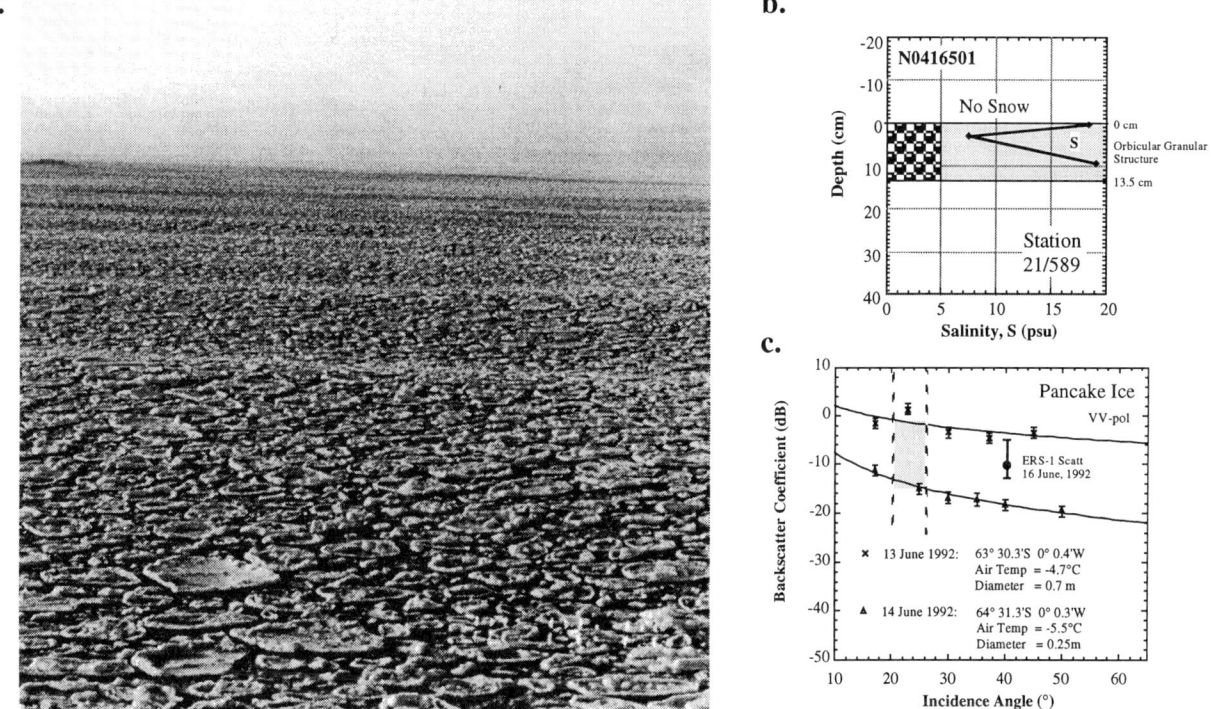

Figure 1. (a) Photograph of a pancake ice field on 13 June, 1992; (b) sea-ice sample salinity and structural characteristics at this location (site N0416501); and (c) shipborne scatterometer scan at this site location, together with contrasting scan from 14 June and mean EScat A measurement in the region around this site.

lower incidence angles (0–20°), this signature exceeds all WWGS '92 field-scatterometer measured ice backscatter values, and implies a transition between the more ocean-like near-normal-incidence signature of ice floes at low concentrations to a 40° incidence-angle signature similar to the mean EScat value plotted in Figure 1b.

Considerable azimuthal anisotropy is also noted in pancake ice margins in the EScat data (*Early and Long*, in press), and is likely due to the pancakes acting as a low-pass filter. Extensive pancake fields damp out smaller gravity and capillary waves, leaving only long wavelength swell. In some circumstances, this results in a dominant wave propagation direction (relative to the swath direction) which appears to be sensed by the scatterometer. Future attempts to fit the functional form of the wind relationship may enable the swell propagation direction to be derived.

3.1.2 *Nilas*. Large expanses of recently formed nilas (< 10 cm thick) have rarely been observed during surface shipborne experiments, largely because Weddell Sea experiments have focused on the ice margin, and swell-influenced regions of the seasonal ice pack, and not the coastal polynya regions. Consequently, the total fraction of ice formed under calm conditions is small, and the opportunity with shipborne scatterometers to sample large areas other than recently opened leads or polynyas is rare. Limited measurements of young ice were made during WWGS '92 and reported in *Drinkwater and Haas* (1994). When observed in narrow leads, nilas typically had little or no snow cover, but was periodically covered with a carpet of frost flowers. The typical signature has large σ°_{Ship} values near nadir, with a steep gradient (< -0.8 dB/°), falling to values less than -20 dB in the SAR incidence angle range. As samples of calibrated pixel values in section 3.2 testify, the overall probability of occurrence of large fractions of new ice and nilas with values below -20 dB is rare in the Weddell Sea, and these areas are generally focused in locations of long-term divergence or in polynyas along the ice shelf margins.

3.1.3 *White ice floes*. The mean thin first-year ice or white ice (in the range 30–70 cm thick) signature shown in Figure 2a is typical of ice observed in the central Weddell Sea region and is an average of 7 individual site measurements made from 1–14 July, 1992

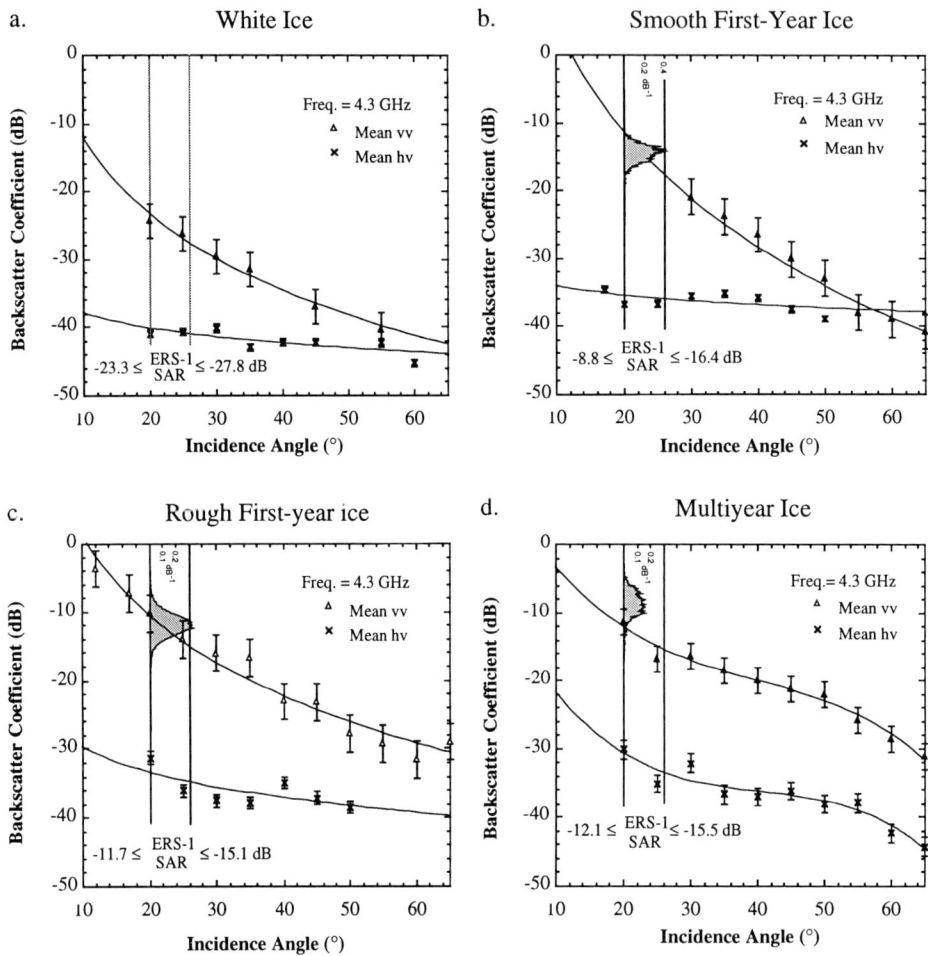

Figure 2. Mean WWGS '92 shipborne scatterometer ice signatures, from (a) white ice, (b) smooth first-year ice; (c) rough first-year ice; and (d) multiyear ice. Curves are fitted exponentially and error bars indicate the standard error of the data. Vertical lines delineate the incidence-angle range of ERS SAR (20° ≤ θ ≤ 26°) and overlapping shaded pdf's show the probability distribution from SAR pixel samples in the shipborne scatterometer measurement locations. A range of observed SAR values is shown for all sampled images for each ice type.

in the location of scene 5377_5031 shown in Figure 3. The signature of this relatively smooth, light snow-covered (1 to 6 cm snow depth) ice has $\sigma°_{Ship}$ values between -24 and -26 dB in the ERS SAR incidence angle range (20° ≤ θ ≤ 26°), an intermediate gradient of around -0.5 dB/° (between 20° and 60° incidence), and a neutral cross-polarized $\sigma°_{Ship}$ curve. Large expanses of white ice were scoured by strong winds and had snow depths of less than 5 cm [Massom et al., 1997]. Lytle et al. (1996) have shown previously that snow-free young ice (10 to 30 cm thick) may have extremely low backscatter values in this incidence angle range, but increases as the ice ages (in the absence of deformation) or roughens (with deformation).

These results nevertheless infer that smooth, level white ice with a bare ice surface or shallow snow cover alone does not play a dominant role in increasing backscatter to values shown later in the SAR pdf's (Figure 4). Alternatively, it may be that ridged portions of sea ice increase measured SAR values in this extensive region.

3.1.4 Smooth first-year ice floes. The mean signature of smooth, level (undeformed) uniform first-year ice floes (Figure 2b) is constructed from 9 individual scatterometer samples. At each site, the ice exceeded 60 cm thickness. A distinguishing feature of this signature is that $\sigma°_{Ship}$ exceeds that of the white ice signature at all angles. The surface roughness of such level ice is

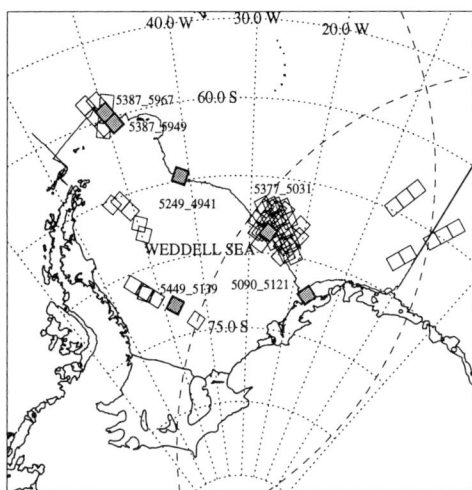

Figure 3. Map indicating ERS-1 SAR image frames used in regional backscatter analysis (see Table 1). GARS is indicated by the triangle and dashed lines delimit the intersecting receiving station coverage masks for GARS and Syowa. The WWGS '92 track of R.V. Polarstern is shown as a solid line.

generally greater than those white ice forms, while also displaying an older and deeper, layered snowcover. ERS SAR backscatter values of equivalent ice floes are expected to fall in the range $-8.8 \leq \sigma°_{Ship} \leq -16$ dB, based on the surface scatterometer measurements, but this is heavily dependent on the physical surface characteristics.

A typical medium thick (67 cm) first-year ice sample acquired on 13 July 1992 at 65.99° S 33.53° W had interleaved bands of frazil and columnar sea-ice crystal growth, and 12 cm of layered snow. The snow-ice interface temperature was -13° C, despite air temperatures lower than -30° C, and the salinity at the ice surface was 14 psu. Angular depth hoar crystals developing under the strong thermal gradient in the basal snow contained significantly more salt than the ice surface itself, and had typical salinities of 27 psu. This highly saline, rough scattering interface at the base of the snow probably accounts for extremely high backscatter values in the incidence angle range 20–26° in Figure 2b, and a relatively shallower gradient of -0.6 dB/° relative to white ice.

3.1.5 Rough first-year ice floes.
Well deformed, ridged or rubbled first-year ice floes are difficult to measure using a shipborne scatterometer, because a complete incidence angle scan is not possible from a fixed viewing position when the local topography is punctuated by piles of ice blocks. Shipborne scatterometer measurements were limited to 5 sites where surface roughness was relatively evenly distributed, and where the signature was not significantly biased by one or more features within the scan. The result is the mean signature in Figure 2c. Values of $\sigma°_{Ship}$ are lower near nadir, and range between -11.7 and -15 dB in the ERS-1 SAR incidence-angle band. The gradient of $\sigma°_{Ship}$ in the 20-60° range is -0.52 dB/°, but the small sample size and averaging of individual ridges or local roughness elements cause the greater variability in measurement points around the logarithmic fit. Cross-polarized back-scatter is greater and more variable for these rough ice floes, especially in the 20-40° range, and could be explained by second-order scattering effects such as multiple scattering from blocks.

3.1.6 Multiyear ice floes.
According to the field scatterometer data in Figure 2d, old ice floes are practically indistinguishable from rough first-year ice forms in the ERS incidence angle range. Relatively undeformed, multiyear floes were distinguished on the basis of snow depth, salinity and thickness as second-year ice. Net mean annual thermodynamic ice growth in the absence of dynamic thickening is expected to be around 1.5 m, based on the lifecycle, mean snow depth and annual net freezing rate in the Weddell Sea [*Fischer*, 1995]. However, perennial ice observed during WWGS '92 comprised massive ice floes with considerable snow accumulation. Snow depths exceeding 0.75 m were commonly measured [*Massom et al.*, 1997] while ice thickness ranged from around 2 m upwards. Shipborne scatterometer $\sigma°_{Ship}$ values from these floes fall between -12 and -15.5 dB, and the low gradient of -0.36 dB/° observed in the 30–50° incidence-angle range is caused by snow volume scattering from layers and distinctive ice lenses and pipes [*Massom et al.*, 1997]. To account for the volume scattering characteristics in Figure 2c, a modified 3rd order polynomial fit is applied instead of a single parameter exponential, to accommodate the flattening at 25° and rolloff beyond 55°. The distinctive feature is a plateau and higher backscatter value in the incidence-angle range 40–50°, and this unique characteristic may be used in 40° incidence EScat images as a method for discriminating high concentrations of thick, snow-covered perennial ice.

3.2 Mesoscale SAR Backscatter Characteristics

ERS-1 SAR data planned and acquired in conjunction with the WWGS '92 experiment are used here to illustrate comparative spaceborne radar backscatter signatures resulting from sea-ice conditions and assemblages of ice floes on scales from 100 m to 100 km. Figure 3 shows a map of the Weddell Sea region, indicating the locations of several winter SAR image frames (shaded) acquired along the WWGS '92 track of R.V. *Polarstern* (see Table 1 for

TABLE 1. List of Winter ERS-1 SAR Image Frames Used in
Regional Backscatter Data Analysis

Image (orbit/frame)	Date and Time (UTC)	Center Latitude	Center Longitude
5090_5121	6 July 92: 09:01:35.786	-72.263	343.424
5249_4941	17 July 92: 11:34:19.254	-64.706	318.550
5377_5031	26 July 92: 10:12:16.392	-68.614	333.541
5387_5949	27 July 92: 03:11:05.712	-60.406	310.928
5387_5967	27 July 92: 03:11:20.807	-59.581	310.165
5449_5139	31 July 92: 10:56:54.090	-72.934	312.629

listing). These SAR images are used for comparison with the field scatterometer data and their associated probability distribution functions (pdf's) of calibrated backscatter coefficients are shown in Figure 4. Each distribution comprises individual pixel values from "speckle-filtered" and calibrated, full-resolution images, resulting in 1×10^6 100 m pixel samples per 100×100 km frame.

3.2.1 Calibration and backscatter calibration accuracy. SAR data products received and processed by the German Processing and Archiving Facility (D-PAF) are distributed as full resolution (*i.e.* 25 m; 12.5 m pixel spacing), calibrated, 3-look images (ESA SAR.PRI products). According to a study by *Bally and Fellah* (1995), the probability that the single-pixel measured intensity lies within a ± 4.5 dB error bound is 90%, whereas the radiometric accuracy and stability errors are within specifications of a fraction of a dB [*Laur et al.*, 1996].

3.2.2 Regional C-band backscatter variability. To calculate the SAR backscattering coefficient (σ^o_{SAR}) of a distributed target which corresponds to an area of sea ice (N pixels in extent), or a group of pixels in the image, averaging is required. Intensity averaging reduces "speckle" (*i.e.* radiometric resolution errors). If "block averaging" is used to create a speckle-filtered image, the new equivalent number of looks is modified by the area of block averaging, together with the target area (*i.e.* number of pixels averaged from target), and is also a function of the incidence angle θ (since resolution is range dependent). The resulting confidence interval for an 8×8 block-averaged image with resulting 100 m pixel spacing exceeds 90 % for a ± 1 dB error bound. With further target averaging, a confidence level exceeding 90 % may be achieved for error bounds of ± 0.5 dB, with a sample box of over 250 pixels (*i.e.* 16×16 pixel box) [*Laur et al.*, 1996]. Over 99 % probability of errors less than ± 0.5 dB may be reached with samples of more than 500 pixels (*i.e.* 23×23 pixel box). Thus, for most of our purposes, having already used a box-filter to reduce the image to 100 m pixel spacing, it is only necessary to derive samples from target areas equal to or exceeding 8×8 pixels in size to satisfy the number of "equivalent" independent looks necessary to exceed a ± 0.5 dB error bound at 99 % confidence. All SAR data samples presented in this paper are from target areas exceeding this minimum threshold area.

The global SAR probability density function (pdf) shown in Figure 4, identified as the thick lined bar-histogram with 0.1 dB bin width, is a combined pdf of all image frames plotted in Figure 3. This summary pdf indicates a trimodal distribution of backscatter σ^o_{SAR}, with distinct components forming peaks at -11.5, -7.4 and -2.5 dB. Individual image pdf's indicate that the highest backscatter mode comprises pixels containing glacial ice.

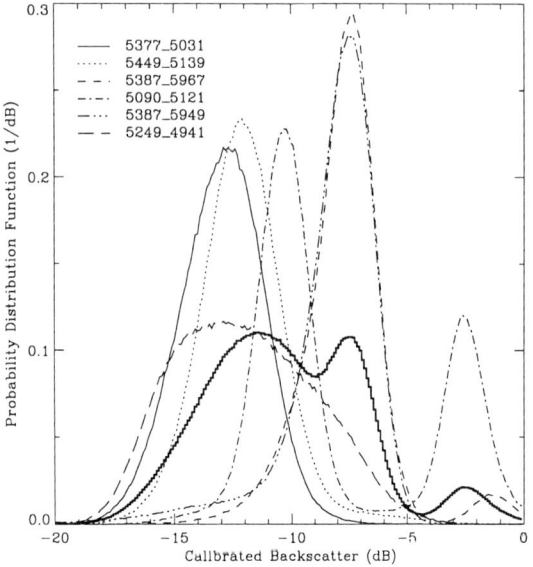

Figure 4. Backscatter pdf's for selected winter SAR scenes (highlighted in Figure 3 as shaded boxes; and listed in Table 1), and for the entire grouped SAR dataset (the thick black stepped pdf).

Image 5090_5021 was acquired on 6 July in the vicinity of R.V. *Polarstern*, as she made a transect along the periphery of the Riiser-Larsen ice shelf (see Table 1 for frame details). Consequently, this scene contains a significant number of ice-shelf pixels falling in the range -5 to +1 dB, with a mode at -2.5 dB. Similar characteristics are observed in the upper peak in the histogram of image 5387_5967. This image captures a large tabular iceberg drifting in the outer marginal ice zone, northeast of the tip of the Antarctic Peninsula. Its backscatter values slightly exceed those of the ice shelf in the previous example, resulting in a peak at -1.5 dB. In contrast, the air temperatures recorded on board R.V. *Polarstern*, in the vicinity of the iceberg were -2.0° C, and had cycled through the melting point within the previous week. It is speculated that the resulting layered snowcover on the tabular iceberg surface causes these higher backscatter values. Glacial ice, therefore, comprises a distinct portion of the upper limb of the backscatter distribution in the Weddell Sea, and may conveniently be used in winter to find tabular icebergs of significant size, either in the marginal ice zone or within the interior ice pack. In contrast, non-tabular Antarctic icebergs do not always present a uniform target area to the SAR, and rotation and drift with respect to the imaging swath often result in a reduction in contrast between the berg and its sea-ice or ocean background [see *Haykin et al.*, 1994].

The next brightest component distribution is described by the pdf of two contiguous image frames (5387_5949 and 5387_5967) along an orbit crossing the marginal ice zone. The peaks in their image histograms overlap, but for a small transition from the swell-disturbed marginal ice zone to the outer ice edge. Mixtures of brash ice and small first-year ice floes, together with occasional multi-year ice floes characterized the northwestern, swell-rocked ice margin of the Weddell Sea at the time of SAR imaging [*Haas et al.*, 1992]. Air temperatures and windspeeds recorded on board R.V. *Polarstern* at the time of imaging were between -5 and 0° C, and 5 and 10 m s^{-1}, respectively. These ice mixtures have an extremely distinctive signature, with $\sigma°_{SAR}$ values ranging between -10 and -5 dB, with a mode at -7.3 dB. Rough surface scattering, occurring from sub-resolution floes, has a distinctively uniform texture and bears a close resemblance to wave-disturbed northern hemisphere MIZ ice signatures observed by airborne SAR and Scatterometer in the Labrador Sea [*Livingstone and Drinkwater*, 1991; *Drinkwater and Squire*, 1987]. The further into the marginal ice zone one traverses, the higher the probability of observing large undeformed first-year ice floes and new ice in leads. These pixels appear as a toe of darker material in the image 5387_5949 pdf, extending from -12 dB down to around -19 dB.

Similar signatures to the previous MIZ cases, in terms of backscatter amplitude, may be encountered in the central ice pack, but with the obvious distinction of individual ice floe units becoming visible. SAR image 5249_4941, shown in Figure 5 (along the R.V. *Polarstern* path), illustrates a typical mixture of perennial and seasonal ice. The mixture histogram combining these two distinctive end-members appears as a broad pdf in Figure 4. Shipborne observations, noted by *Haas et al.* (1992) as *Polarstern* traversed the region encompassed by this image (less than 24 hours after SAR imaging—see Table 1), indicated a closed ice cover (100% concentration) comprising level first-year ice punctuated by distinct ridges and rougher multiyear ice in fractions up to 30%. First-year ice was extremely rubbled and ridged, particularly around the perimeter of thick perennial ice floes. The broad pdf comprises two overlapping distributions without distinctive peaks, due to the continuum of states of the ice ranging from patches of smooth, relatively undeformed first-year ice, to old, deformed snow-covered floes. Individual samples identified in Figure 5 are shown later and used to illustrate the distribution of backscatter in small-scale sample areas.

The least deformed, or level first-year ice is observed in image 5377_5031 from the central Weddell Gyre region, and image 5449_5139 from the southwestern Weddell Sea (see Figure 3). Both yield distinctive pdf modes at between -12 and -13 dB. During early winter the former is a largely divergent region, in which vast expanses of undeformed white ice alternated with snow-covered stony fields along the transect of *Polarstern* [*Haas et al.*, 1992]. July air temperatures remained stable around -27° C and ice formation was immediate during divergent periods. The SAR image 5377_5031 indicates no open water, and an extremely low ridge density. In contrast, the example from the southern Weddell (image 5449_5139) contains a larger proportion of deformed seasonal ice (with higher ridging density), and a small fraction of grounded icebergs and perennial ice leaving the basin northwards [*Viehoff and Li*, 1995]. These higher backscatter components of the scene account for the higher pdf mode for image 5449_5139 in Figure 4, and the tail extending to values exceeding -5 dB. Based on the predominant drift direction, old ice and icebergs found in this region appears to have originated in the vicinity of icebergs A22 and A23 grounded to the north of the Filchner Ice Shelf. Large 5–10 km diameter high backscatter (-7 to -2 dB) multiyear ice floes are clearly conglomerates of old ice floes cemented together by a matrix of what appears to have been

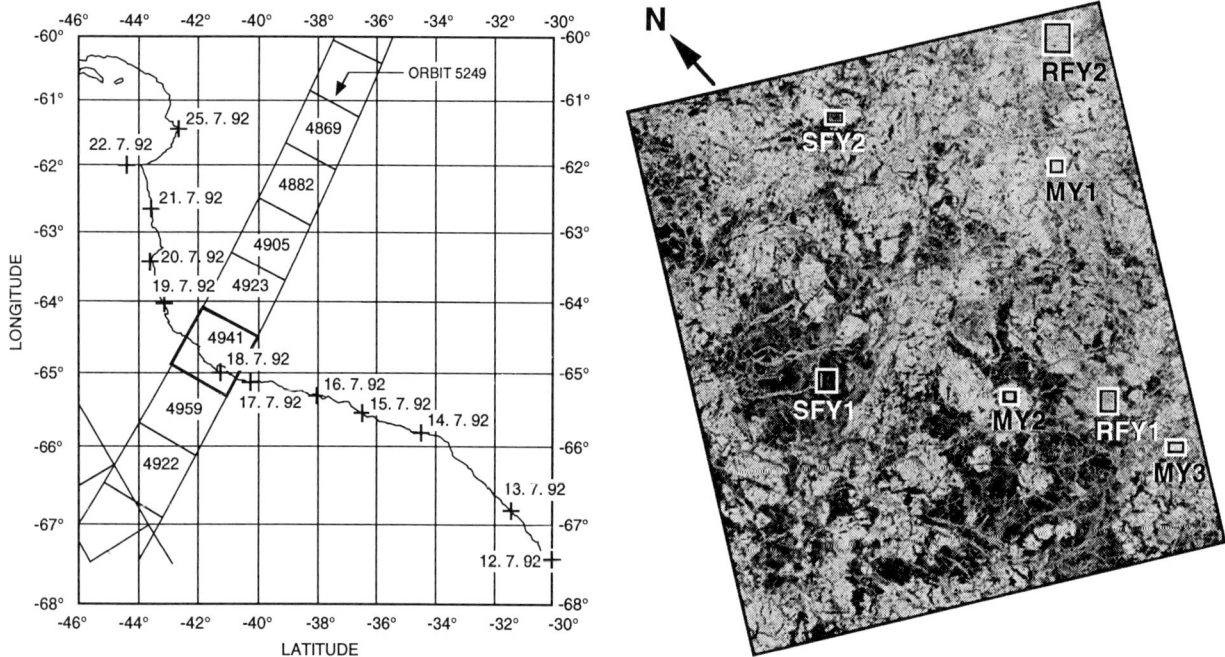

Figure 5. ERS-1 SAR image from orbit 5249, frame 4941, taken on 17 July, 1992 [©*ESA*, 1992] and located in Figure 3. Boxes highlight regions from which component backscatter distributions are plotted in Figure 6. The inset marks the location of the image with respect to the track location of the R.V. *Polarstern*.

snow-covered land-fast ice. These multiyear floes broke off from the fast ice and drifted northwestward. The iceberg chain, described by *Viehoff and Li* (1995), originated from one large iceberg which calved off the Filchner ice shelf in the late 1980's and subsequently grounded on the broad General Belgrano Bank, on the northern flank of shallow topography over the Berkner Seamount. Ice-motion data obtained from SAR, Scatterometer and buoys in the following section provide complementary evidence for this assertion, but further supporting evidence is provided in the form of tracks of deformed sea-ice created as the sea-ice drifts past the location of each of these icebergs (see *Viehoff and Li*, Figure 9). The ice cover had drifted consistently northwards, leaving linear tracers of deformed material extending 100 km or more in length in the direction of drift. At a typical drift speed of 16 cm s^{-1} north from the iceberg barrier, the length of each feature represents a period of around 1 week. Furthermore, as pointed out by *Viehoff and Li* (1995), the persistence of such bands (in the lee of the icebergs), in time as well as in space ($\geq$ 100 km), can be interpreted as an indication for a relatively homogeneous ice motion field in this location.

In summary, examples from images 5377_5031 and 5449_5139 indicate that the least deformed first-year ice, observed in large proportions in the Weddell Sea, comprises the lowermost peak in the trimodal global pdf (in the thick black histogram in Figure 4). Old ice and marginal ice on the other hand fill an overlapping range in backscatter to form the central peak in the global pdf. As expected, heavily ridged or deformed first-year ice fall squarely in the middle of this range, and comprise the main peak in image 5090_5121. The roughness of sea ice in the location of this SAR image is further described by laser altimeter flights made from *Polarstern* when mean ridge heights were typically of the order of 1.2 m, with a mean ridge spacing of 54 m [*Dierking*, 1995]. Therefore, as previously noted in *Drinkwater et al.* (1995a) for the central ice pack, linear mixing of the two end members (i) smooth undeformed ice and (ii) thick perennial ice accounts for a continuum of σ°_{SAR} values between -19 and -4 dB. Other than the overlap existing between MY ice and marginal ice zone signatures, the value of backscatter generally appears related to the amount of deformation and ridging density of the ice, and the ice thickness. Although backscatter values may be encountered in the range

TABLE 2. Summary of Backscatter Samples From ERS-1
SAR Image 5249_4941 in Figure 5.

Ice Type	Sample #	Mean/Median (dB)	Std. Deviation (dB)
Smooth First-Year	FYS 1	-16.09/-16.10	±0.99
	FYS 2	-15.46/-15.50	±1.05
Rough First-Year	FYR 1	-10.14/-10.30	±1.83
	FYR 2	-9.54/-9.50	±1.69
Multiyear Ice	MY 1	-5.72/-5.50	±1.88
	MY 2	-6.81/-6.50	±2.27
	MY 3	-5.92/-5.50	±2.03

20 to -25 dB, for newly growing areas of nilas and grey ice in leads or ice shelf (and coastal) polynya systems, large expanses of such low backscatter material are rarely encountered in the Weddell Sea.

3.2.3 Snapshot signature summary. Based upon surface measurement data, and its primary location with respect to the occurrence of perennial ice, the SAR image in Figure 5 provides the broadest range of ice ages, ice deformation styles and snow depths. The pdf for this SAR scene in Figure 4 confirms a broad peak comprising a variety of pixels spanning the range of possible white, first-year and multiyear ice σ°_{SAR} values. A number of samples are extracted from Figure 5 to generate Table 2. The values describe the signatures of the primary ice components, each of which is extracted on the basis of the scattering characteristics of extensive, homogeneous ice floe units. Smooth and rough first-year and multiyear ice mean σ°_{vv} values range from -16.1 dB to -5.7 dB. In total, a further 60 samples (each of 100 or more pixels) were collected in the same manner from the other shaded SAR images shown in Figure 3, to generate a summary of snapshot winter backscatter characteristics of these primary ice types. All except the southernmost of these SAR images were acquired in regions in which WWGS '92 surface samples and ice cores were collected [Drinkwater and Haas, 1994]. Figure 6 compares and contrasts the ranges and statistics of these data with matching summary statistics inferred from the mean field scatterometer σ°_{Ship} curves (i.e. shaded pdf's) in Figure 2. Additional comparison data are included for open water backscatter coefficients, and scattering from brash ice or pancakes in the MIZ.

Inevitably, the marginal ice zone provides one of the largest ranges of backscatter signatures, due to the influence of floe size and packing density, wave environment, and variable air temperature. Depending on the season and location, frazil ice growth and pancake formation may be favored. However, brash ice also provides extremely high σ°_{vv} values due to its similar characteristics in terms of sub-resolution floes and floe packing. A good analogy may be drawn between the Weddell and the Labrador Sea MIZ which was observed by SAR during the LIMEX experiments [Drinkwater, 1989; Livingstone and Drinkwater, 1991]. In Figure 6, SAR and EScat values overlap due to the broad range in scattering conditions described in section 3.1.

Field sampling of smooth first-year ice floes is inherently biased, since an icebreaking ship normally seeks the path of least resistance. Although generally true during WWGS '92, the route was largely based upon large-scale information provided by AVHRR satellite images, and did not always enable sampling of the smoothest first-year ice. The benefit of SAR images is that they enable the smoothest first-year ice forms to be easily found and selected. Samples shown in Figure 6 show that the measurement ranges plotted from the shipborne scatterometer and SAR do not overlap particularly well. Similarly, with multiyear ice floes, the ship navigated around heavily ridged ice. As a consequence, the multiyear ice sampled is at the least deformed extreme of such perennial ice floes. Haas et al. (1992) document the level nature of most of the multiyear ice floes sampled. SAR backscatter values in Figure 6 indicate that the field scatterometer measurements are biased to the lower end of the multiyear ice range.

In terms of the signatures from the ERS scatterometer, shipborne scatterometer measurements are consistently lower at 40° incidence than equivalent EScat image data extracted in the same locations. Spatial resolution and the averaging over a number of resolution cells influences mean EScat backscatter measurements. Intervening ice motion and resulting mixtures of backscatter elements causes a relatively higher σ°_{vv} than that exhibited by any individual ice component. In addition, the larger roughness elements dominate backscattering at higher incidence angles, yet shipborne scatterometer measurements cannot

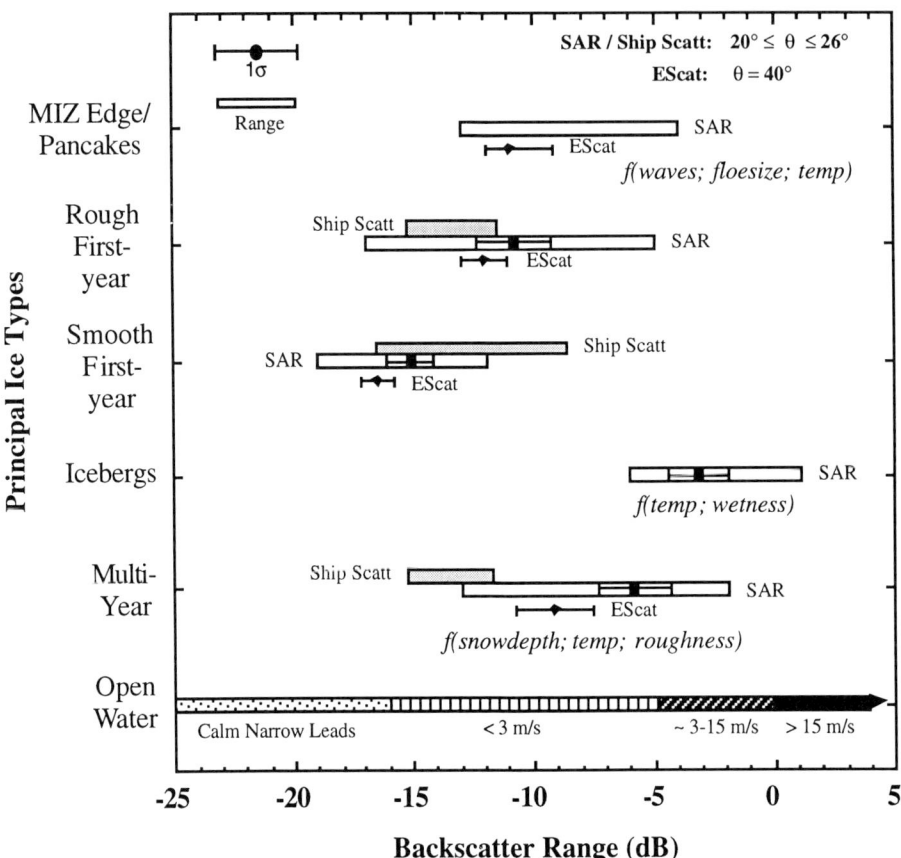

Figure 6. Summary plot of C-band Weddell Sea ice and ocean signatures indicating the ranges and/or standard deviation of SAR, shipborne, and EScat backscatter measurements of each broad ice category.

easily measure the signature of such nonuniform or deformed ice floe surfaces.

Open water signatures are generated in Figure 6 for comparison, using the CMOD4 algorithm [*Stoffelan et al.*, 1993]. As expected, the range of backscatter values can, under differing wind conditions, bracket the entire range of sea-ice signatures. Nevertheless, using image context and knowledge regarding the shape of leads within the ice pack, together with ice dynamics information from ice tracking, it is possible to distinguish open water areas within the pack. In any case, the probability of observing a lead within the Weddell Sea ice pack in winter is rare, and for the most part new leads freeze within a matter of hours after opening.

3.3 *Large Scale Characteristics of the Weddell Sea*

The large-scale variability of the mean ice conditions are successfully monitored using EScat enhanced resolution images [*Drinkwater et al.*, 1993a]. EScat images shown in this paper indicate the mean amplitude of the C-band vv-polarized backscattering coefficient (*A*) normalized to a 40° incidence angle (*i.e.* $\sigma°_{EScat}$). Plate 1 illustrates *A* images during (*a*) the austral summer ice minimum, during the first week of February 1992 (Days 32–37) and (*b*) during the austral winter ice edge advance (Days 210–216). Each image is masked in the open ocean using an ice-edge contour derived from the mean SSM/I-

TABLE 3. Summary of Winter C-band EScat Backscatter Ranges ($\theta = 40°$ Incidence) of the Main Categories of Sea Ice in the Weddell Sea.

Ice Type	Backscatter Range (dB)
Icebergs	$-6.0 \leq \sigma°_{EScat} \leq 0.0$
Multiyear/Pancakes	$-11.0 \leq \sigma°_{EScat} < -6.0$
Rough First-year	$-14.0 \leq \sigma°_{EScat} < -11.0$
Smooth First-year	$-20.0 \leq \sigma°_{EScat} < -14.0$
Nilas	$-32.0 \leq \sigma°_{EScat} < -20.0$

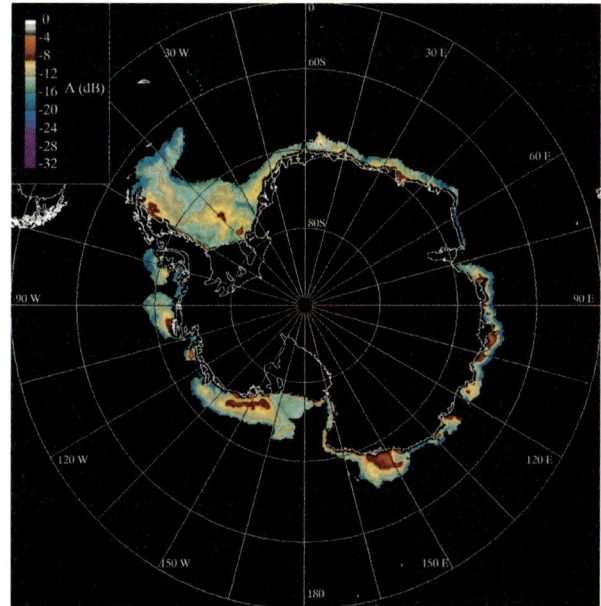

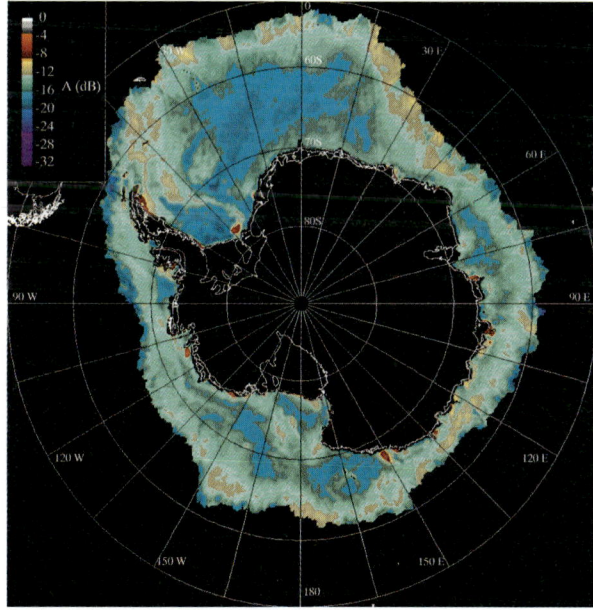

Plate 1. EScat A images (normalized to 40° incidence), for (a) February 26 (days 54-59); and (b) 31 July (days 210-215), 1992.

values, while the most deformed and thicker ice has the highest $\sigma°_{EScat}$ values.

3.3.1 Summer scatterometer images. Austral summer sea-ice characteristics and the distribution of perennial ice in the Weddell Sea in 1992 are represented by Plate 1a. Minimum ice extent typically occurs in February in the Weddell Sea. The 1991 to 1995 EScat time-series indicates that peak A values coincide with mid-austral-summer ablation [Drinkwater et al., 1995c], and that the timing of this peak varies interannually within the December to mid-March window. A values in some locations in Plate 1a exceed -5 dB, owing to predominantly rough surface scattering from deformed ice floes during damp snow-surface conditions. Furthermore, regions with the highest residual concentrations of rough perennial ice appear to match the pattern of A values exceeding -12 dB. Regional backscatter values in the central part of the perennial ice region uniformly exceed -16 dB, as do those of a tongue of material extending eastwards from the tip of the Antarctic peninsula at ~ 63°S, where northwards drifting ice becomes entrained in the Antarctic Circumpolar Current (ACC). A noticeable polynya type feature also becomes captured by rapid ice edge recession in the location 74°S 32°W, where low backscatter (< -20 dB) values are observed.

Three locations of the highest backscatter values correspond with (i) a cluster of grounded giant icebergs (fragments of A22 and A23) off Berkner Island (at 77°S 45°W), between the Ronne and Filchner ice shelves [Zibordi and Van Woert, 1993], (ii) a region east of the northern Larsen ice shelf (at 66°S 59°W), and (iii) a large residual patch of rough perennial ice between (i) and (ii) in the region 73°S 45°W. Three small areas of relatively lower backscatter and lower concentration deformed ice occur along the Riiser-Larsen (75°S 30°W), and the east and west end of the Ronne Ice shelf front (at 77°S 47°W and 74°S 59°W, respectively.

3.3.2 Winter scatterometer images. As the winter cooling progresses, and as less deformed, higher salinity seasonal ice cover grows rapidly to latitudes north of 60° S, A values decline. Plate 1b shows the mean ice cover in the last week of July, 1992. Samples of Weddell Sea $\sigma°_{EScat}$ values from the winter ice cover indicate a typical variation of C-band backscatter between -6 and -23 dB. The pattern of backscatter values is reflective of the structure, variability and dynamics of the sea-ice cover [Drinkwater et al., 1993a]. As the concentration of perennial ice is reduced by diffusion and northwards drift (including melting at the northern ice margin), lower backscatter material grows on the continental shelf region in the predominantly divergent area along the Ronne ice-

derived zero ice concentration isoline generated for the 6-day period of imaging [after NSIDC, 1995]. Normalized C-band backscatter images at this incidence angle largely reflect the surface roughness of the sea ice [Drinkwater, 1989; Drinkwater et al., 1995a; Drinkwater and Lytle, 1997]. As described in the previous sections, smoother, less deformed young ice results in the lowest backscatter

shelf front. Similar low backscatter ice may be observed in the center of the Weddell Gyre at 65°S 20°W.

The streamline of the northward limb of the Weddell Gyre is traced in Plate 1*b* by the western margin of a northwest to southeast extending band of low concentration perennial ice. This feature is expressed as a meandering band of higher backscatter (exceeding -12 dB) material which traces the isobaths of the eastern flank of the continental shelf break. This streamer carries old deformed ice away from the site of the grounded icebergs near the Filchner ice shelf. The climatological significance of this feature is further described later (see also Plate 4). Winter images from 1991 to the present day also indicate that these large icebergs played a significant role in preventing the influx of thick deformed ice into the southwestern Weddell Sea. As in the 1992 example, this periodically causes reduced concentrations of perennial ice in this region during summer and at the beginning of autumn freeze-up. This perennial-ice-starved region is consequently a high heat flux and vigorous ice growth location during early autumn freeze-up, as offshore winds and tidal currents carry the ice northwards away from the Ronne ice shelf front [*Drinkwater and Kreyscher*, unpublished data]. In this argument, slow northwards ice drift on the continental shelf is responsible for the export of ice. Supporting evidence for an ice factory on the continental shelf is provided by overlapping ERS SAR and SSM/I images [*Viehoff et al.*, 1994; *Drinkwater et al.*, 1994]. SSM/I data confirm the characteristic appearance of higher concentrations of first-year ice in this region north of the Ronne ice-shelf front, together with a typical northward spread in the extent of this feature throughout winter. A more rapid northwards transport of ice off the continental shelf (*i.e.* in the primary Gyre circulation) drives a shear feature at 68°S 55°W in which smoother first-year ice separates the slowly moving perennial ice on the continental shelf from the more rapid motion in the streamer of perennial ice originating to the south (described previously).

A further key feature of the winter ice cover is the bright fringe bordering the marginal ice zone. Examples of such high backscatter values were described in section 3.1.1, and correspond with zones of pancake ice formation at the pack ice edge [*Early and Long*, in press].

3.3.3 *Scatterometer time-series.* Time-series maps of the Southern Ocean are a primary application of the Scatterometer imaging technique. Due to restrictions on receiving station operation described in Section 2.3, SAR can retrieve only limited spatial and temporal coverage of dynamic phenomena observed in response to ocean and atmosphere forcing. The scale of coverage offered by ESCat is sufficient to monitor these processes and to place the 25 m resolution 100 km × 100 km SAR images in the context of synoptic or basin-scale ice conditions. The regional time-series of unmasked images shown in Figure 7 indicates the A value of radar backscatter during eight weekly periods, at 21-day intervals in 1992. The main characteristic of these relatively high incidence-angle (θ = 40°) images is that sea ice typically demonstrates smoothly-varying isotropic returns in direct contrast to the highly variable returns of open ocean equatorward of the sea-ice margin (indicated by bold white line drawn from the National Ice Center ice-edge charts). Variability in the backscatter from wind-generated waves is large on time-scales of several days, and the advantage of the EScat is that it views the same location from a number of incidence and azimuthal angles during the period of image integration. This gives a powerful method for separating highly anisotropic or azimuthally-dependent variable wind waves from relatively isotropic and azimuthally independent scattering from collections of sea-ice floes [*Early and Long*, in press].

In Figure 7, weekly A images are generated at 21-day intervals from the first week in February (Days 32–37) in (*a*) through the first weeks of July (Days 188–193) in (h). Heavy operation of the SAR mode of the AMI results in data 'drop-outs' (black diamond-shaped regions) in EScat images, particularly in the region around the O'Higgins Station (at 63° 19'S 57° 54'W) in (*b*) (*i.e.* during early testing of the SAR receiving capability). The 1992 minimum ice extent occurred at the end of February, with a subsequent rapid northeastwards advance during the onset of austral winter, under predominantly thermodynamic ice growth [*Massom*, 1992]. As the Weddell Sea becomes ice-bound, wind- and current-driven ice motion carries some of the large units of relatively high backscatter sea-ice away from their origins along the coastal and ice shelf margins. The bright patch of perennial ice originating in the southern central Weddell Sea in summer is stretched and dilated by shear and divergence during its northwards drift as winter progresses. Another rather characteristic area of smooth first-year ice in Figure 7c, grows and extends northwards along the continental shelf margin in a distinctive shear boundary, separating continental shelf ice from that driven northwards in the westward limb of the Weddell Gyre. This shear margin is expressed as a finger of dark material which appears in Figure 7g and extends and protrudes northwards (at 70°S 50°W) from the area of low backscatter material in the south west Weddell Sea basin. This shear margin eventually separates the distinctive perennial ice area on the shelf (along the eastern flank of the Antarctic peninsula), from that perennial ice originating in the central Gyre region.

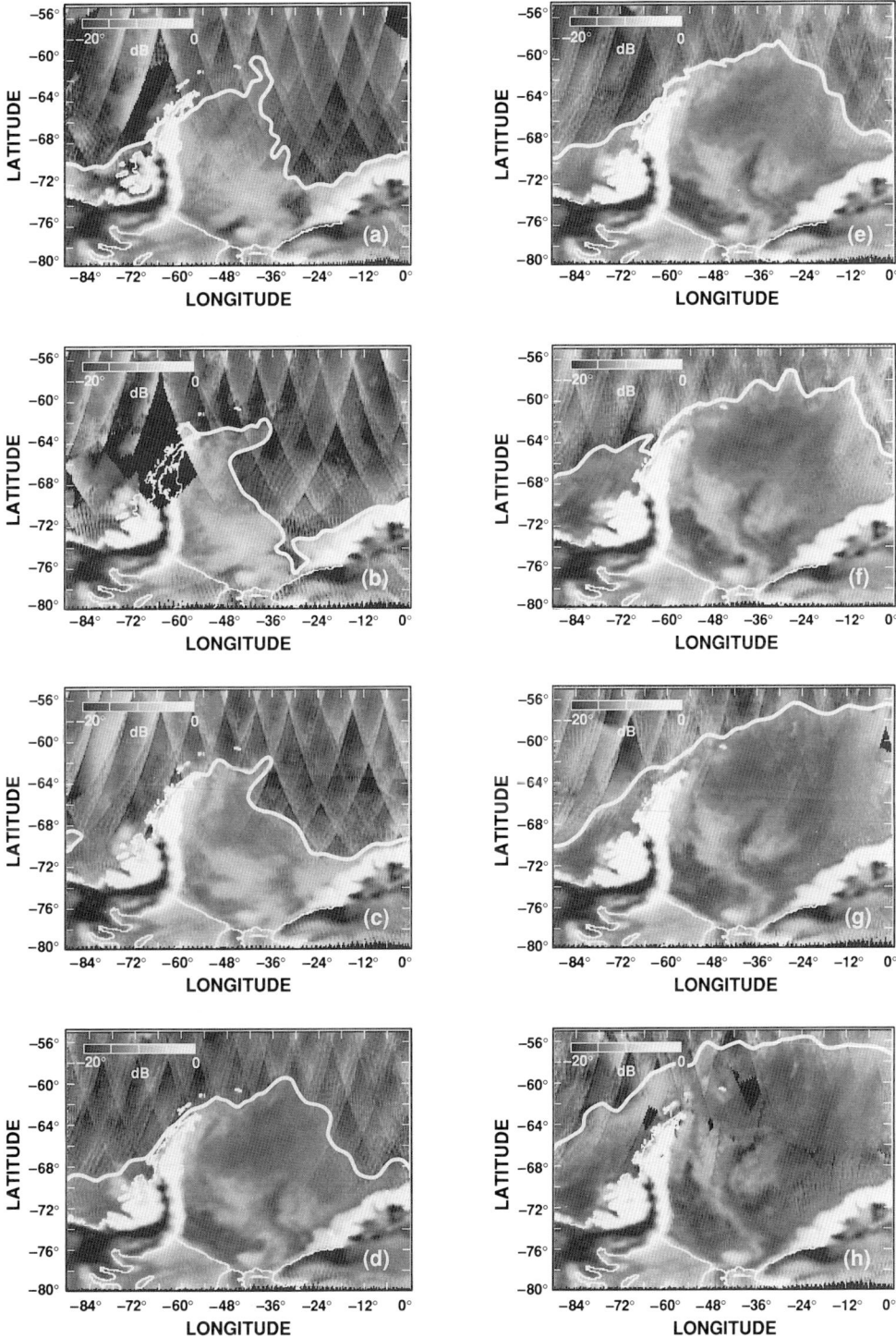

Figure 7. Weddell Sea time-series of EScat A images, illustrating evolving backscatter characteristics from minimum to maximum sea-ice extent in 1992. Day of Year periods illustrated are; (a) 32-37; (b) 53-58; (c) 74-79; (d) 104-109; (e) 125-130; (f) 146-151; (g) 167-172; (h) 188-193. Coastlines and ice shelves are indicated by a thin white line while the National Ice Center sea-ice margin is identified by a thick line. ERS Scatterometer data © ESA.

In Figure 7a, two summer polynya regions of relatively low ice concentration appear in the early series as low backscatter areas in the southern and southeastern Weddell Sea. The latter (at 74°S 30°W), is sufficiently close to the ice margin that it becomes captured in the sea-ice retreat by the end of February, 1992 in Figure 7b. Nevertheless, as the ice margin advances the relatively darker signature of these previously open water areas reappears and they become source regions for new-ice formation, as identified by their low backscatter. Undeformed, low backscatter ice is transported away from these locations as it becomes embedded and advected northwestwards in the motion of the Weddell Gyre circulation. As the ice is carried northwards, new seasonal ice is formed particularly along the front of the Ronne ice shelf. By early May (Figure 7e), a large expanse of low backscatter (dark) has appeared, confirmed in *Drinkwater et al.* (1993a) and *Viehoff et al.* (1994) using ERS–1 SAR images to be the northward expansion in extent of seasonal, undeformed ice produced off the ice-shelf front.

In addition to the stationary, grounded icebergs described in the previous section, large icebergs are also observed drifting in the pack ice. For instance, the tabular iceberg identified in SAR image 5387_5967 and noted in section 3.2.2, is clearly identifiable in Figure 7g and (h) as a high backscatter point drifting to the west of the South Orkney islands.

3.4 Seasonal Backscatter Variability

Time-series of ERS radar data demonstrate that backscatter signatures of a particular ice class vary considerably seasonally and interannually in the Weddell Sea. The primary distinction, therefore, between seasonal and perennial ice covers can only be made by separating data acquired during the winter ice maximum and austral summer minimum, or alternatively by continuously tracking perennial ice floes identified during the summer months, through time and space. During February and March of 1992, repeat swath SAR acquisitions were planned over Ice Station Weddell (ISW) [*Gordon et al.*, 1993]. In total, 35 calibrated images were acquired over the perennial ice zone. Previously described winter SAR data acquisitions were planned during WWGS '92 in locations where R.V. *Polarstern* was traveling across the Weddell Sea, and in a number of other instrumented locations. The result in 1992 is a total collection of more than 55 images (exceeding 3.5×10^9 pixels), of regions where sea-ice cover was either measured *in-situ* or for which independent sources of information are available. Two examples of seasonal backscatter variability are illustrated below. In the first, all ERS-1 SAR images in

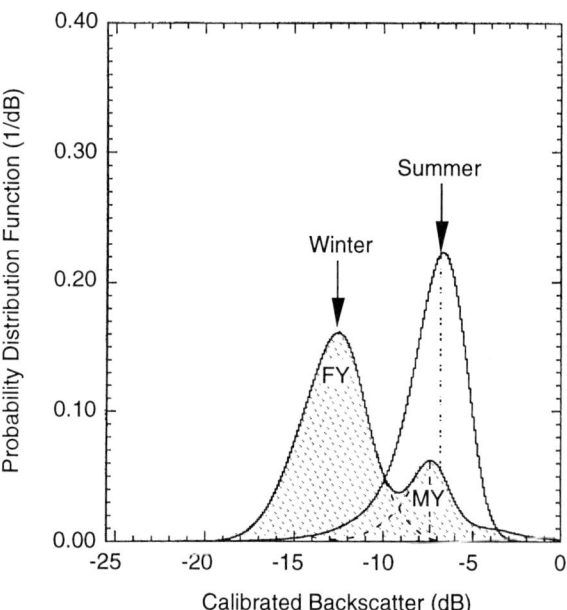

Figure 8. Seasonal SAR backscatter probability distribution functions from grouped winter and summer images. Bimodal peaks in the winter distribution comprise first-year or seasonal (FY), and multiyear (MY) or perennial ice.

winter and summer states are separated and grouped, and their backscatter statistics plotted in Figure 8. In the second, in Figure 9, a number of ice parcels were followed in time and space, to illustrate the progression in signature states [after *Drinkwater and Lytle*, 1997].

3.4.1 Summer and winter SAR image characteristics.
If we consider purely mid-winter and mid-summer conditions, then we may make a collection of all statistics of calibrated backscatter images on the basis of growth or melt seasons. Since the Weddell Sea ice cover is largely composed of seasonal ice, the residual ice cover at the end of summer typifies the various states of perennial ice: all ice surviving the summer melt is defined here as multiyear ice. All the grouped pixel values are combined into either summer and winter pdf's in Figure 8 to describe the differences between the summer and winter perennial or annual ice signatures, respectively. The summer pdf shows a mode around -6 dB, with a small tail to higher values comprising highly deformed ice and icebergs. Ice concentrations typically exceeded 95 % in the perennial ice pack and the contribution from windroughened open water is negligible [*Drinkwater and Lytle*, 1997]. At the lower end of the summer pdf is a longer tail extending to -15 dB and beyond. This includes level, undeformed multi-year ice, and snow-covered first-year ice thick enough to survive the summer melt, *i.e.* second-year ice.

floor of the SAR (~ -25 dB) includes small proportions of new and young ice forms, and recent results from Morris and Jeffries (*this volume*) in the Bellingshausen Sea indicate that typical undeformed new ice appears with a mean backscatter value of -23 ± 0.5 dB in small areas of coastal polynyas and flaw leads.

3.4.2 *The summer-autumn transition*. Studies focusing specifically on multiyear ice in summer and winter SAR images show that its backscatter coefficient changes seasonally, and the mode of the multiyear ice pdf in Figure 8 is observed to shift slightly from around –6 dB to -7 dB between warm and cold seasons. Surface measurements made at ISW [*Lytle and Ackley*, 1996] illustrate that perennial ice retains a deep snowcover on its surface. But a large proportion of the ISW camp floe and surrounding ice floes experienced summer flooding at the base of the snow. The following discussion illustrates how the freezing of this saturated basal snow layer may contribute to the -1 dB shift in the mode over the autumn freeze-up period.

Figure 9 illustrates the temporal change of the SAR backscatter signature of tracked perennial ice floes in the vicinity of ISW as they cooled during the summer-autumn transition period. As the cold wave (*i.e.* 0° C isotherm) propagates downwards through the snow, areas of flooded or saturated basal snow (slush) cool and refreeze. In response to this process, the permittivity of the newly formed snow-ice decreases significantly, thereby causing the C-band microwave backscatter values shown in Figure 9 to decrease [*Drinkwater and Lytle*, 1997]. This period of decreasing air temperatures at ISW (Figure 9*a*) is shown in parallel to thermistor profiles at a variety of sites with differing snow depth (Figure 9*b*). Thermistors indicate that protracted cold air temperatures after day 63 enable the snow layer to develop a strong thermal gradient. This cools the basal snow sufficiently quickly that the slush freezes around day 67.

Large numbers of individual multiyear ice floes were tracked in ERS-1 SAR images to derive the microwave backscatter trend shown in the upper trace of Figure 9*c*. For comparison, global SAR-image backscatter means and EScat $\sigma°_{vv}$ values, measured over the identical 100 × 100 km SAR-imaged areas, each report essentially the same decreasing trend. The largest decrease in $\sigma°_{vv}$ is observed before and after the freeze-up of the slush. Thus, the change from rough surface scattering from snow-covered slush to rough surface scattering from dry snow-covered perennial ice appears responsible for the observed shift in the multiyear ice peak in Figure 8.

3.4.3 *Melt ponding*. Additional effects of transformation at the surface of perennial ice floes occur as a

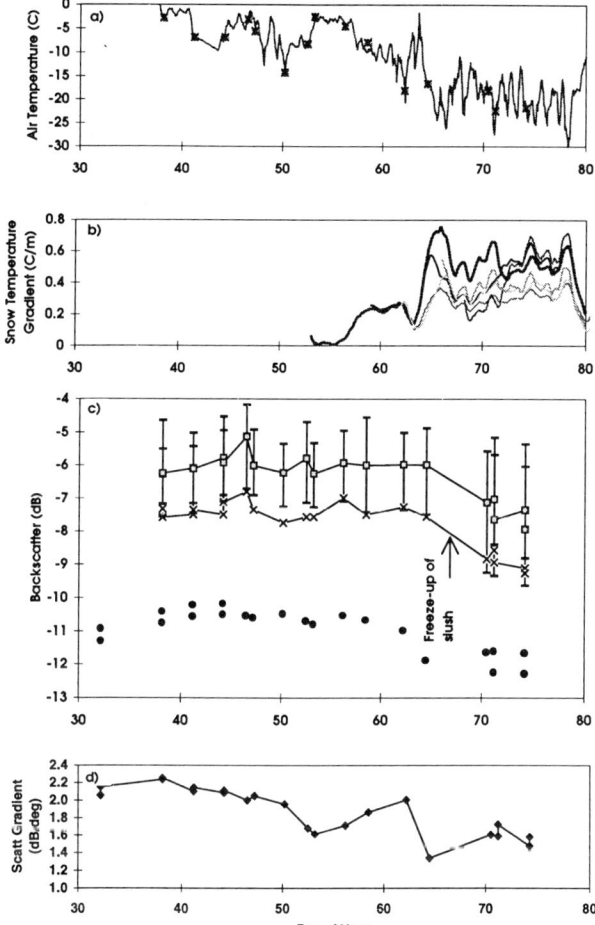

Figure 9. Time-series of (a) surface air temperatures from the Finnish Salargos buoy [id: #5908]; (b) snow-temperature gradients at various thermistor sites on snow-covered perennial ice floes near ISW; (c) mean backscatter (■), $\overline{\sigma}°_{MY}$ of multiyear floes (vertical bars indicate range of values), global SAR image mean (×), $\overline{\sigma}°_{SAR}$, and 40° incidence-angle-normalized backscatter coefficient (●), $\overline{\sigma}°_{EScat}$; and (d) EScat backscatter gradient (linear regressed) at 40° incidence, *M*.

In winter, the pdf becomes bimodal with the appearance of seasonal ice. As time progresses, advection of an increasing fraction of multiyear ice northwards out of the basin reduces the probability of occurrence of multiyear ice and produces the imbalanced bimodal mid-winter distribution shown in Figure 8. While a secondary peak remains, with a mode at around -7 dB, the main peak now occurs at around -12.5 dB and comprises more prevalent forms of level first-year ice. The low end of the pdf, at values of -15 dB and below, indicates the least deformed first-year ice. A winter tail extending down to the noise-

consequence of summer melt processes and periodic winter excursions in air temperature due to storm systems [*Massom et al.*, 1997]. Large excursions in air temperature have been observed to have considerable impact upon the microwave backscattering characteristics of sea ice, in response to the changing surface snow and ice characteristics [*Drinkwater et al.*, 1995a]. Melt ponds have commonly been observed near the ice margin during shipborne experiments [*Haas et al.*, 1992]. The appearance of classical Arctic-style melt ponds in the Weddell Sea (*i.e.* expressed at the surface as a consequence of melting rather than by the mechanism of saturation and slush development at the base of the snow) appears to be checked by the latitudinal limit of southward penetration of the zero-degree isotherm. Recent studies have recorded observations of surface ponding on perennial ice floes further to the south in the Weddell Sea. The appearance and expression of melt ponding in SAR images is discussed by *Low* (1995), and observations are confirmed by both snow and ice surface measurements and helicopter airborne photography.

4. WEDDELL SEA ICE DYNAMICS

Studies by *Viehoff and Li* (1995), *Vihma et al.* (1996) and *Kottmeier and Sellmann* (1996) have pieced Lagrangian buoy drift statistics together to reconstruct the seasonal and regional patterns of ice drift in the Western Weddell Sea in response to winds and currents. Recent work by *Drinkwater and Kottmeier* (1994) introduced ice-motion tracking from mesoscale SAR data as an alternative method for deriving kinematic measurements of the Weddell ice cover. In contrast to the Arctic ice motion studies using ERS SAR [*Kwok et al.*, 1990], the capability to track Antarctic ice using automated SAR motion tracking had not been demonstrated. The reason was because, in the Arctic, large fractions of multiyear ice provide high contrast targets which can be successfully tracked in time using radar images. Weddell Sea ice by comparison exhibits large expanses of low contrast, level first-year ice, and a lack of distinct features with which computer algorithms may successfully track the ice.

4.1 Tracking Ice Station Weddell

As previously described in section 2, a series of SAR image acquisitions was planned in conjunction with the 1992 drift of ISW. Plate 2 indicates a result from tracking snow-covered perennial ice floes (in the vicinity of ISW, marked by +), during the early drift phase of the experiment (February–March 1992). Two overlapping images acquired along 1-day spaced crossing orbits (during the so-

Plate 2. Three-day ice motion vectors superimposed onto the ERS-1 SAR image for 15 March 1992 [Image ©ESA, 1992]. The scale indicates the distance (in km) of the vector displacements and the mean u and v components of ice velocity are given at upper left.

called "Ice Phase" orbit of ERS-1) were correlated to produce the displacement vectors using the algorithm described by *Kwok et al.* (1990). Plate 2 shows the 5 km gridded ice drift displacement field superimposed on the ERS-1 SAR reference image (orbit 3058 frame 5103), acquired at 11:34 GMT on 15 February, 1992 at 71.59° S 53.04° W. Spots indicate the gridded starting positions of tracked features (*i.e.* vector tails), and the scaled vector arrows indicate the distance traveled in the 17h period separating the two images. Vectors imply strong cyclonic motion, and mean and standard deviations of the u and v velocity components are -10.43±2.5 and 5.95±2.7 cm s^{-1}, respectively. Partial derivatives of velocity over the entire tracked portion of the scene indicate a mean divergence of 0.15±0.63 % d^{-1}; a mean vorticity of -0.45±0.70 % d^{-1}; and mean shear of 2.01±1.02 % d^{-1}.

4.1.1 GPS validation of ice velocity. The period of consecutive ERS-1 SAR imaging of the ISW camp floe lasted between day 38 (7 Feb.) through day 75 (15 Mar), in 1992. Initial results, such as Plate 2, demonstrate that algorithms developed for winter Arctic sea-ice tracking can be used without serious problems in both cases of ISW summer perennial ice or relatively featureless smooth Weddell Sea winter first-year ice. In order to establish the success and accuracy of the tracking

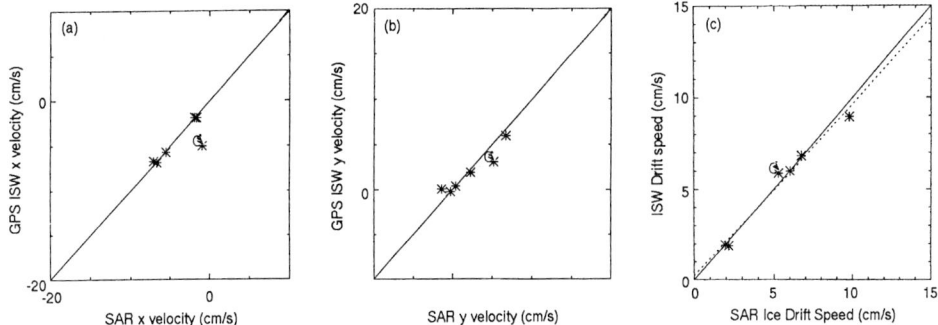

Figure 10. Comparison of SAR-tracked 3-day velocities with equivalent 3-day velocities computed from instantaneous GPS locations of Ice Station Weddell.

algorithms, positions from a fixed GPS receiver at the ISW camp are compared with SAR ice-tracked drift vectors. GPS positional fixes were not logged until after day 57, preventing accurate interpolation of latitude and longitude positions of the camp earlier than this date. The period of highest frequency GPS measurements overlaps with 6 consecutive SAR ice velocity image "pairs" (*i.e.* 7 consecutive images), allowing comparisons to be made between day 57 and 75. Figure 10 shows the correlation between mean velocity components derived from SAR (such as that in Plate 2) and those derived from instantaneous GPS ISW locations (interpolated to the exact SAR imaging times). SAR u and v velocity components are rotated into the local coordinate system to match calculated GPS velocities.

In terms of different mean ice drift speeds, Figure 10 indicates a sample ranging from 2 to 10 cm s^{-1}. Figure 10a shows the west-east (*i.e.* x or u component) and Figure 10b the south-north (*i.e.* y or v) components of velocity, while Figure 10c indicates the magnitude of the velocity. Ice drift is largely westward and northward during this period. Though a small sample, the plot shows a high degree of accuracy in deriving both the direction and magnitude of ice drift. Although spatial mean SAR velocities are used in the regression analysis, as opposed to values of the SAR-derived velocity interpolated to the ISW floe location, the result in Figure 10c indicates a correlation of 0.98 (illustrated by the dotted regression line). The rms error in SAR motion tracking is computed to be less than 0.5 cm s^{-1} for 3-day motion tracking. This translates into a relative error of ~100 m, or equivalently ~1 pixel after accounting for error variance generated by temporal interpolation of GPS locations and spatially averaged mean SAR velocity vectors.

The largest outlying vector in Figure 10 is identified throughout by a curly arrow, indicating that the motion field displays a large amount of rotation or vorticity. As Plate 2 graphically shows, the mean velocity components of such a field are not sufficiently representative of any single point in space for a good comparison to be made. Though selecting the nearest tracked vector slightly improves the correlation coefficient to 0.995, it is felt that spatial interpolation of velocity field to the GPS-estimated drift-camp location is not representative of the true inter-particle floe motion, and so further efforts were not made to try to improve upon this result.

4.1.2 Ice kinematics time-series. Results of processing all 3-day or 1-day repeat overlapping "pairs" into ice kinematics information are shown as a velocity time-series in Figure 11 together with surface wind speed and the GPS monitored drift of the ISW camp floe. In Figure 11a ISW measurements of the wind speed are shown as a dotted line. Prior to completion of the meteorological mast on day 57, windspeed was recorded hourly by the Finnish Salargos buoy to the east of the camp. An overlapping solid line indicates the 3-day running mean of the Salargos surface wind speed prior to day 57. Thereafter, the solid line is a 3-day running mean of the hourly data from the ISW meteorological station anemometer.

Figure 11b and c show the mean SAR instantaneous velocity components joined by a dotted trend line. Horizontal bars indicate the period separating the tracked image pair, and the vertical bar indicates one standard deviation (1σ) about the mean velocity vector (spatially averaged over a 100 × 100 km scene). An overlapping solid line indicates the 3-day smoothed velocity of ISW and begins upon installation of a fully functional GPS at ISW on day 57. Hollow diamond symbols represent the velocity computed from distances traveled between instantaneous GPS locational "fixes", between SAR imaging times. Where diamonds overlap mean SAR velocity mea-

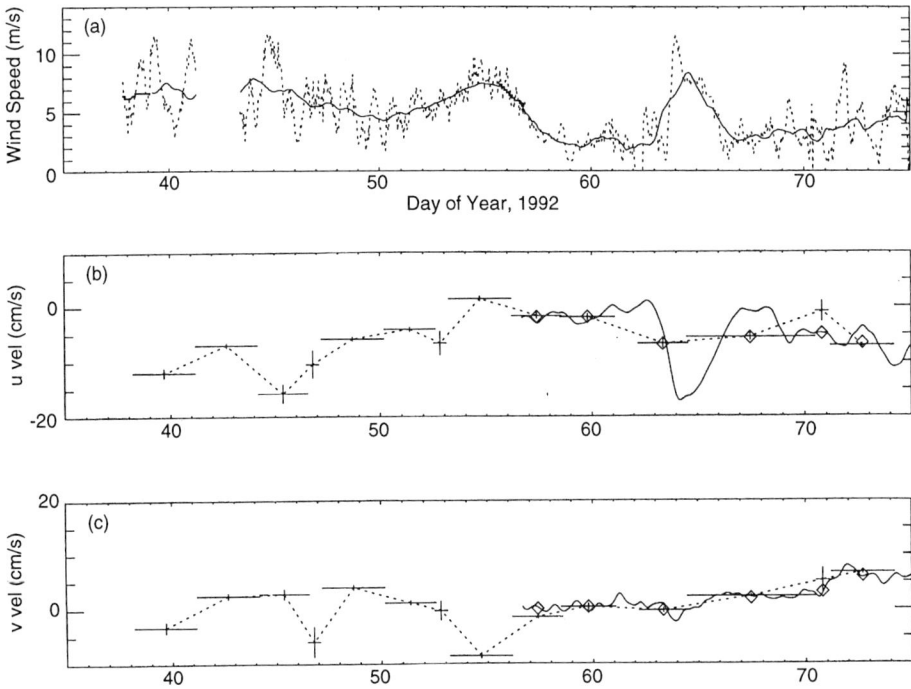

Figure 11. Time-series of ERS-1 SAR-tracked ISW ice motion. (a) shows measured surface windspeed (dotted) and 3-day smoothed windspeed (solid); (b) shows SAR (dotted) and GPS-derived (solid) u component of velocity (*i.e.* east is positive); (c) v component of velocity (north positive) from SAR (dotted) and GPS (solid—3-d running mean).

surements, the spatially averaged SAR velocity is representative of a single GPS velocity vector. In a spatial context, where the mesoscale velocity field does not represent translational motion, the ISW GPS velocity diamond does not overlap the mean SAR velocity. Thus, in the case of day 70, the discrepancy shown in Figure 11*b* and *c* is explained by strong cyclonic motion, and corresponds to the tracked example in Plate 2. A mean velocity of a spatial field of cyclonic or anticyclonic flow is not representative of any single vector within that field, and such means cannot be directly compared to the GPS ISW velocity. Nevertheless, smoothed 3-day velocity components respond to sustained bursts of wind shown in the solid line in Figure 11*a*. Additionally, though not in pure "free-drift", the SAR reported drift velocity components (dotted line in Figure 11*b* and *c*) clearly respond to the 3-day filtered wind forcing. The largest discrepancy between SAR velocity and 3-day smoothed ISW velocity occurs as a result of the wind burst from day 63 to 67, when ISW is pushed to the west at -15 cm s^{-1}. In the corresponding SAR image record the closest pair of image acquisitions was 6-days, and thus the minimum in u velocity is missed. A higher temporal resolution in SAR imaging is required to capture such deviations in ice motion.

Generally, examples such as Figure 11 indicate that the sea-ice drift field for the most part responds swiftly to the synoptic pressure-field-driven geostrophic winds. The temporal frequency of fluctuations in the field of ice velocity vectors closely match those in wind velocity. Temporal spacing of less than 3-d sampling is necessary if the small-scale response to synoptic storms is to be studied in more detail in an open basin such as the Weddell Sea. In addition, 1-day repeat orbits, plotted at day 46, 52, and 71 curiously capture more variance in ice velocity. Preliminary spectral studies using buoy data confirm this phenomena to be attributable to the aliasing of velocity components induced by tidal currents. Importantly, therefore, exact 3-day repeat orbits do not alias diurnal or semidiurnal (which are close to 12 hr. period) tidal motions, whereas non-multiples of 12 hours contain significant tidal variance in this location. Ultimately, a trade-off is necessary between resolution in temporal sampling and the sampling interval over which ice floes are tracked. The best solution appears to be 3-day SAR ice-motion tracking at regular one day intervals, but issues such as the SAR data volumes (for requisite coverage of the entire Weddell Sea) become paramount. Recent, limited ERS-1/2 Tandem phase (24-hour separated) images offer interesting possibilities for high-frequency revisit and 1-day ice-motion tracking, while future Radarsat ScanSAR image coverage is the only high-resolution SAR solution to tracking large areas.

4.2 Rheological Response

The ultimate scientific goal of employing automatically-generated SAR ice kinematics products is to develop a spatial and temporal picture of Weddell Sea ice drift and opening and closing of the ice in response to various components of the momentum balance. Plate 3 illustrates a product derived from the SAR ice velocity field in Plate 2. Ice deformation is defined in terms of the ratio of strain-rate invariants E_I and E_{II}, themselves calculated from the spatial derivatives of drift velocity on the 5-km grid [*Thorndike*, 1986]. The characteristic angle,

$$\theta = \tan^{-1}(E_{II}/E_I) \quad (1)$$

expresses the style of deformation, where a positive value of E_I represents divergence (negative = convergence) and E_{II} represents the rate of shearing. The result is a graphical illustration of whether the motion is accompanied by either predominantly divergence (i.e. $\theta \sim 0$), shear ($\theta \sim 90°$), or convergence ($\theta \sim 180°$). We note for the field of cyclonic motion recorded in Plate 2, that a large proportion of the deformation represents shear-style motion (indicated by dark blue and pink). A single purple 5 km grid cell located to the northwest of the ice camp (marked by +) illustrates a situation tending towards pure convergence. It begins as an area of small ice floes, with a large open water (i.e. lead fraction), which after 3 days has clearly closed. Such sea-ice deformation data can be used in validating ice rheologies presently employed in coupled ice-ocean-atmosphere models [*Stern et al.*, 1995]. More accurately parameterized coupled regional ocean-ice-atmosphere models together with SAR data will be used to evaluate the relationship between the changing basin-wide distribution of sea ice and the momentum flux [*Drinkwater et al.*, 1995b]. Accurate monitoring of Weddell Sea ice formation, drift, deformation and divergence are also of primary importance to estimating surface fluxes of heat, freshwater and salt [*Drinkwater et al.*, 1995c].

4.3 Synoptic Scale Motion Tracking and Ice Motion Climatologies

Updated tracking schemes are presently being tested with EScat images to derive large-scale motion fields at a reduced resolution. Plate 4 shows a preliminary example of the 1992 climatological Weddell Sea ice-drift. More than 100 3-day ice motion products were initially generated on a regular grid (~ 100 km) from "pairs" of EScat images at 3-day intervals throughout 1992. An unweighted, unfiltered average was then computed from these data. The colorized climatology shown in Plate 4 is the result of a spatial interpolation of the gridded, basin-wide ice-drift speed

$$|V| = \sqrt{v_x^2 + v_y^2} \quad (2)$$

in the region south of 64° S latitude, and in the 0–66° W meridional sector. White streamlines are computed which are locally tangential to the mean Eulerian velocity field, and the direction of flow along each line is annotated by arrowheads. Streamlines give both a spatial description of the basin-wide flow, and indicate regions of convergence and divergence. Importantly, though, locations where seasonal ice does not appear year-round (i.e. having fewer tracked vectors), have less reliable statistics. In Plate 4, this causes both the drift speed interpolator and streamline function to produce spurious results, notably in the northeast part of the Weddell Sea. These streamlines must be filtered in future versions of the algorithm.

Drift speeds in Plate 4 are multiplied by a factor of 10 km d^{-1} to highlight the regional differences. Together with the streamline characteristics, and information in the pre-

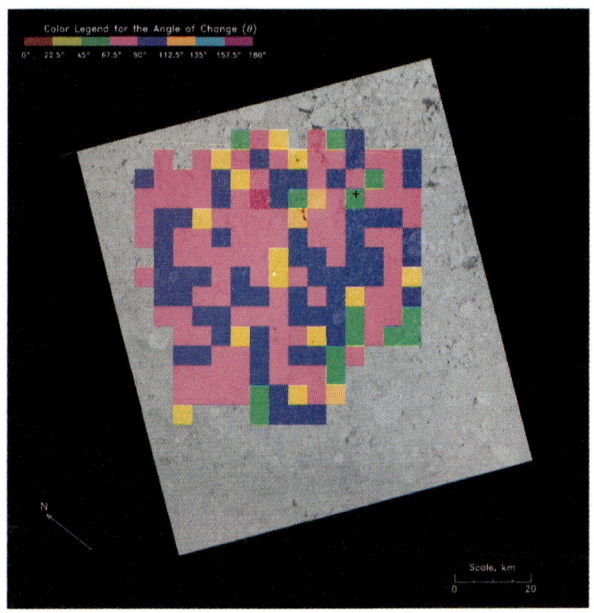

Plate 3. Color-coded display of deformation angle θ (i.e. tan^{-1}(E$_{II}$/E$_I$)) for the SAR ice-motion example in Plate 3.

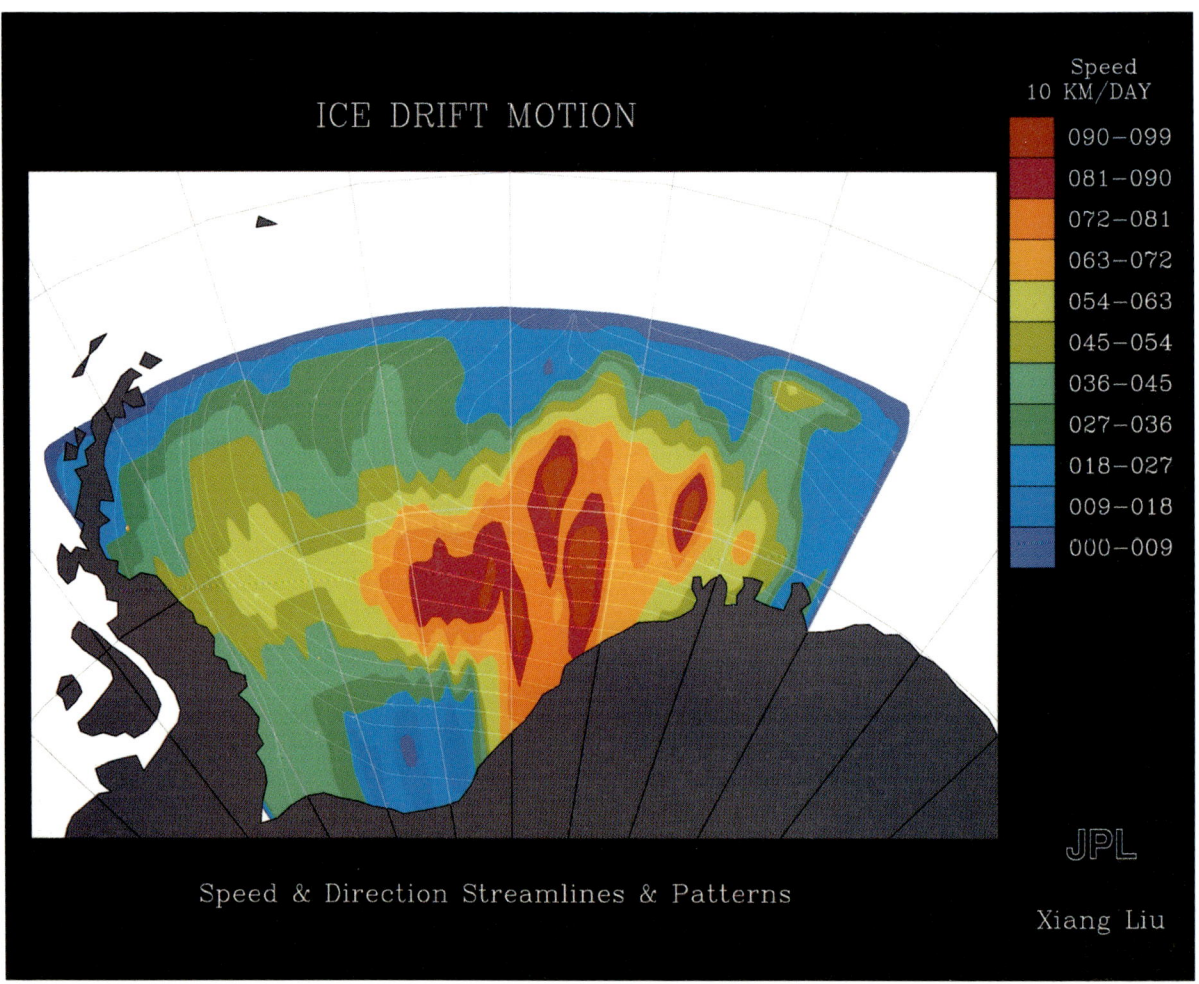

Plate 4. Preliminary Weddell Sea EScat ice-drift climatology constructed from an unweighted average of ice motion grids generated at 3-day intervals throughout 1992. Colors indicate the magnitude of the ice-drift velocity, and white lines trace the streamlines of drift.

vious section, it is evident that most sea ice originating in the southwestern part of the Weddell Sea follows a northward trajectory before turning and leaving the basin eastwards, or melting at the ice edge. Clearly, the large grounded icebergs discussed in section 3.3.1 also have a significant impact on drift in the southernmost region, and typical stagnant drift speeds of between 0 and 1.8 km d^{-1} (*i.e.* 0–2 cm s^{-1}) are observed. Most ice originating along the eastern coast of the Weddell Sea, drifts northwestwards before becoming constrained by the Antarctic peninsula to turn northwards. This is illustrated by the convergence zone in the western Weddell Sea around 73° S. The highest long-term mean drift speeds (exceeding 10 cm s^{-1}) occur in the eastern and central Weddell Sea, where the ice drifts primarily away from the coast.

It is recognized that significant work is required to validate and quantify errors in these satellite-tracked motion products. Nevertheless, details of the Weddell sea-ice circulation are clearly highlighted in climatologies displayed in the manner shown in Plate 4. Furthermore, changes in patterns observed in the backscatter images in section 3.1 are more easily explained once dynamic information such as this becomes available. Future comparisons will be made between individual EScat ice motion products and buoy drift trajectories, and other climatologies constructed from historical buoy drift datasets [*Kottmeier and Sellmann*, 1996]. Historical datasets such as the passive microwave record are also now being used in the same manner as these scatterometer data, in order to build up a long-term ice motion climatology.

5. CONCLUSIONS

Sea-ice conditions in the southwestern Weddell Sea are analyzed using available satellite and shipborne radar remote sensing techniques. In particular, the temporal and spatial variability in seasonal and perennial ice are identified in ERS SAR and Scatterometer data. Variability is described in terms of both the radar response to changing surface conditions and sea-ice dynamics. It is shown that large-scale ice patterns observed using C–band satellite radar progressively change as a function of the ice-growth season and the residual amount of perennial ice in the Weddell Sea. Resolution-dependent microwave signatures of individual components of the ice cover are identified, which may be used as markers in the microwave data record. Furthermore, the complementary nature of the high resolution SAR and Scatterometer data is exploited to highlight these characteristics in the context of both snapshot and time-series records.

Combinations of microwave data sources, together with ground based data, indicate that the southern Weddell Sea comprises a number of distinct sea-ice regimes. A number of large icebergs which have been grounded off Berkner Island in much the same location over the early part of this decade have markedly altered the dynamics and character of the sea ice, especially in the southern part of the basin. Being starved of deformed ice, this region is quite different from the remaining winter sea-ice conditions. To date, during the 1990's it appears to have been composed primarily of seasonal ice. Temporary polynyas forced by katabatic winds off the Ronne Ice Shelf in this location might be considered as the main ice formation source for the southwestern continental shelf region. The spatial separation of this and other regimes is correlated with the continental slope, indicating that the barotropic part of the Weddell Gyre ocean circulation, together with the synoptic pressure pattern, are dominant factors responsible for the long-term forcing of the ice cover in the remainder of the basin. Perennial ice, formed by compression around the grounded icebergs, meanders its way northwestwards before being ejected eastwards out of the basin. A streamer of this high backscatter material closely follows the climatological drift streamlines and represent an important diagnostic of the seasonal drift record over the preceding ice-growth season. Old, deformed ice, originating in the vicinity of the grounded icebergs, later becomes separated from perennial ice forming on the shelf along the Antarctic peninsula by a northwards, seasonally-propagating shear line at around 54°W. The location of this shear line is oriented approximately along the 2000-m isobath, and defines the margin of more slowly moving ice to the west. ERS data indicate that areas of increased compactness and ice deformation may be found in this northwestern part of the Weddell Sea, especially off the Larsen Ice Shelf. In contrast, the central part of the Gyre, defined from the ice-motion climatology as approximately 67°S 20°W, experiences largely shear and divergence and thus formation of largely undeformed, level first-year ice. Laser profiling from R.V. *Polarstern* during WWGS '92 confirms that the central Weddell Sea ice has a much lower ridging density than is encountered on both the westernmost or easternmost flanks of the basin [*Dierking*, 1995].

Despite some problems experienced in tracking the more dynamic ice margins during periods of maximum ice extent or pancake ice growth, ice-motion tracking using EScat images appears to be a promising addition to SAR ice-floe tracking. Large-scale ice tracking methods promise to fill in missing information about basin-wide drift conditions, especially during periods without buoy drift campaigns, or between SAR imaging overpasses. More detailed analysis of ice dynamics data must now be made in conjunction with time-series of ERS images to prove various hypotheses about the connection between deformed ice, high backscatter regions and convergent flow, or *vice-versa*, between level ice, low backscatter and divergent flow. Presently, motion data are being compared in a quantitative fashion with results of theoretical sea ice models of the Weddell Sea [e.g. *Drinkwater et al.*, 1995b] to improve our understanding of the ice dynamics contributing to observed patterns. Now that ice-tracking techniques have been successfully tested and applied on both SAR and Scatterometer images from the Weddell Sea, a detailed historical ice-motion database can be built up. The response of the sea ice to synoptic-scale variations in the momentum balance can be more effectively studied using these data. Similarly, information provided on the rheological and dynamical response of the ice cover will benefit the latest generation of Weddell Sea ice models.

In this paper, it is shown that although uncertainties remain in the interpretations of the C-band backscatter data, examples indicate that radar images obtained by satellites such as ERS and RADARSAT are contributing significantly towards broadening our understanding of sea-ice processes in Antarctica. Though limitations still exist on the reception of SAR data, Scatterometer images have been shown to provide an essential complement. The advantage of Scatterometer images, in contrast to SAR or 25 km resolution passive microwave images, is that they trace and highlight large-scale ice characteristics directly resulting from dynamical processes. The recent successful launch of the ADEOS satellite, which carries the Ku-band

NASA Scatterometer (NSCAT) instrument, will ensure future uninterrupted production of full-polar images at an even higher resolution. Various improvements can be made over the present EScat image processing, due to additional Doppler information provided in the NSCAT datastream. Backscatter cell location is more precisely known, and consequently NSCAT images will double the resolution presently achieved using ERS scatterometer data. Furthermore, the NSCAT instrument traces swaths on both sides of the spacecraft, effectively doubling the amount of coverage achieved on a daily basis. This implies that a higher temporal resolution may be derived from NSCAT for use in ice motion tracking. Future plans include implementing ice motion tracking over the entire north and south polar regions as NSCAT becomes operational.

Acknowledgments. Enhanced resolution ERS-1 scatterometer data were processed by David Long and David Early of Brigham Young University for this study, as part of the collaborative ESA-supported AO2.USA.119 project. SAR data are processed and supplied by the German Processing and Archiving Facility, and special thanks go to Jörg Gredel, Birgit Schättler and Klaus Reiniger. Ron Kwok and Shirley Pang are thanked for generating ice motion products from EScat images, and Peter Woiceshyn for his help in development of the streamline processing method. Timo Vihma is sincerely thanked for supplying us with the Finnish ISW buoy data, and Prasad Gogineni likewise for his loan of the Kansas scatterometer. I am grateful also for the data analysis support of Xiang Liu. MRD performed this work at the Jet Propulsion Laboratory, California Institute of Technology under contract to NASA, and the support of Robert H. Thomas of Code YSG, NASA.

REFERENCES

Augstein, E., N. Bagriantsev, and H.W. Schenke. (Eds.), The Expedition Antarktis VIII/1-2, 1989 with the Winter Weddell Gyre Study of the research vessels "Polarstern" and Akademik Fedorov, *Reports on Polar Research*, 84, Alfred-Wegener-Institut für Polar- und Meeresforschung, Bremerhaven, Germany, 134pp, 1991.

Bally, P., and K. Fellah, Evaluation of the accuracy of the backscattering coefficient measurement in SAR data products, *ESA/ESTEC Earth Sciences Division Technical Note*, July 1995.

Brennecke, W., Die Ozeanographischen arbeiten der deutschen Antarktischen Expedition 1911-1912, *Aus Archiven Deutsche Seewarte*, 34, 195, 1921.

Dierking, W., Laser profiling of the ice surface topography during the Winter Weddell Gyre Study 1992, *J. Geophys. Res.*, 100, C3, 4807-4820, 1995

Drinkwater, M.R., Satellite microwave radar observations of Antarctic sea ice, in *Recent Advances in the Analysis of SAR for Remote Sensing of the Polar Oceans*, edited by C. Tsatsoulis and R. Kwok, chapt. 8, Springer-Verlag, New York, in press, 1997.

Drinkwater, M.R. (Ed.), PIPOR Special Issue, *Int. J. Remote Sens.*, 16,17, 1995.

Drinkwater, M.R., LIMEX '87 Ice surface characteristics; implications for C-band SAR backscatter signatures, *IEEE Trans. Geosci. and Remote Sens.*, 27, 5, 501-513, 1989.

Drinkwater, M.R., and V.I. Lytle, ERS-1 SAR and field-observed characteristics of austral fall freeze-up in the Weddell Sea, Antarctica, *J. Geophys. Res.*, 102, C6, 12593-12608, 1997.

Drinkwater, M.R., and C. Kottmeier, Satellite microwave radar- and buoy-tracked ice motion in the Weddell Sea during WWGS '92, *Proc. IGARSS '94*, vol. 1, Pasadena, CA, Aug. 8-12, IEEE Cat. No. 94CH3378-7, 153-155, 1994.

Drinkwater, M. R., and Squire, V. A., C-band SAR observations of marginal ice zone rheology in the Labrador Sea, *IEEE Trans. Geosci. and Remote Sens.*, 27, 5, 522-534, 1989.

Drinkwater, M.R., and C. Haas, Snow, sea-ice and radar observations during ANT X/4: summary data report, *Berichte aus dem Fachbereich Physik*, 53, Alfred Wegener Institut für Polar- und Meeresforschung, Bremerhaven, Germany, 58 pp., 1994.

Drinkwater, M.R, R. Hosseinmostafa, and S.P. Gogineni, C-band backscatter measurements of winter sea ice in the Weddell Sea, Antarctica, *Int. J. Remote Sens.*, 16, 17, 3365-3389, 1995a.

Drinkwater, M.R., H. Fischer, M. Kreyscher, and M. Harder, Comparison of seasonal sea-ice model results with satellite microwave data in the Weddell Sea, *Proc. IGARSS '95*, Vol. 1, Florence, Italy, 10-14 July, IEEE Cat. No. 95CH35770, 357-359, 1995b.

Drinkwater, M.R., D.G. Long, and D.S. Early, Comparison of variations in sea-ice formation in the Weddell Sea with seasonal bottom-water outflow data, *Proc. IGARSS '95*, Vol. 1, Florence, Italy, 10-14 July, IEEE Cat. No. 95CH35770, 402-404, 1995c.

Drinkwater, M.R., D.S. Early and D.G. Long, ERS-1 investigations of Southern Ocean sea-ice geophysics using combined scatterometer and SAR images, *Proc. IGARSS '94*, Vol. 1, 165-168, Pasadena, CA, Aug. 8-12, IEEE Cat. No. 94CH3378-7, 1994.

Drinkwater, M.R., D.G. Long, and D.S. Early, Enhanced resolution scatterometer imaging of Southern Ocean sea ice, *ESA Journal*, 17, 307-322, 1993a.

Drinkwater, M.R., Hosseinmostafa, R., and W. Dierking, C-band microwave backscatter of sea ice in the Weddell Sea during the winter of 1992, *Proc. IGARSS '93*, Tokyo, Japan, Aug. 18-21, Vol. 2, 446-448, 1993b.

Early, D.S., and D.G. Long, Azimuth modulation of C-band scatterometer $\sigma°$ over Southern Ocean sea ice, *IEEE Trans. Geosci. and Remote Sens.*, in press, 1996.

Battrick, B. (Ed.), A Programme for International Polar Oceans Research, (PIPOR), ESA SP-1074, European Space Agency, Paris 1985.

ESA Earthnet, ERS-1 System, *ESA Special Publication*, SP-1146, ESA Publications Division, ESTEC, Noordwijk, The Netherlands, 87 pp., 1992.

Fahrbach, E., G. Rohardt, M. Schröder, and V. Strass, Transport and structure of the Weddell Gyre, *Ann. Geophys.*, 12, 840-855, 1994.

Fischer, H., Comparison of an optimised dynamic-thermodynamic sea-ice model with observations in the Weddell Sea, *Reports on Polar Research*, 166, Alfred-Wegener-Institut für Polar- und Meeresforschung, Bremerhaven, Germany, 130pp, 1995.

Gloersen, P., W.J. Campbell, D.J. Cavalieri, J.C. Comiso, C.L. Parkinson, and H.J. Zwally, Arctic and Antarctic sea ice, 1978-1987: satellite passive-microwave observations and analysis. *NASA SP-511*, Washington DC, National Aeronautics and Space Administration, 1992.

Gohin, F., and A. Cavanié, A first try at identification of sea ice using the three beam scatterometer of ERS-1, *Int. J. Remote Sens.*, 15, 1221-1228, 1994.

Gohin, F., Some active and passive microwave signatures of Antarctic sea ice from mid-winter to spring 1991, *Int. J. Remote Sens.*, 16, 11, 2031-2054, 1995.

Haas, C., T. Viehoff, and H. Eicken, Sea ice conditions during the Winter Weddell Gyre Study 1992 ANT X/4 with R/V *Polarstern*: shipboard observations and AVHRR imagery, *AWI Berichte aus dem Fachbereich Physik*, 34, Alfred Wegener Institut für Polar- and Meeresforschung, Dec. 1992.

Haykin, S., E.O. Lewis, R.K. Raney, and J.R. Rossiter, *Remote sensing of sea ice and icebergs*, Wiley, New York, 686pp., 1994.

Hosseinmostafa, A.R., V.I. Lytle, K.C. Jezek, S.P. Gogineni, S.F. Ackley, and R.K. Moore, Comparison of radar back-scatter from Antarctic and Arctic sea ice, *Electromag. Wave Applications*, 9, 421-438, 1995.

Ice Station Weddell Group, Weddell Sea exploration from Ice Station, *EOS*, 74, 11, 121 and 124-126, 1993.

Kottmeier, C., and L. Sellmann, Atmospheric and oceanic forcing of Weddell Sea ice motion, *J. Geophys. Res.*, 101, C9, 20809-20824, 1996.

Kwok, R., J.C. Curlander, R. McConnell, and S.S. Pang, An ice-motion tracking system at the Alaska SAR facility, *IEEE J. Oceanic Eng.*, OE-15, 1, 44-54, 1990.

Laur, H., P. Bally, P.J. Meadows, J.I. Sanchez, B. Schaettler, and E. Lopinto, Derivation of the backscattering coefficient σ^o in ESA ERS SAR PRI products, *ESA Pub. ES-TN-RS-PM-HL09*, ESA/ESRIN Issue 2, Rev. 2, 41pp., 1996.

Lemke, P. (Ed.), The Expedition ANTARKTIS X/4 of R.V. "Polarstern" in 1992, *Reports on Polar Research*, 140, Alfred-Wegener-Institut für Polar- und Meeresforschung, Bremerhaven, Germany, 90pp, 1994.

Livingstone, C.E., and M.R. Drinkwater, Springtime C-band SAR backscatter signatures of Labrador Sea marginal ice: measurements vs modelling predictions, *IEEE Trans. Geosci. and Remote Sens.*, 29, 1, 29-41, 1991.

Long, D.G., P.J. Hardin, and P.T. Whiting, Resolution enhancement of spaceborne scatterometer data. *IEEE Trans. Geosci. and Remote Sens.*, 31, 3, 700-715, 1993.

Low, D., The Validation of ERS-1 Summer SAR data for Antarctic summer sea ice, M.Sc. Thesis, University of Cambridge, Cambridge, 1995.

Lytle, V.I., and S. Ackley, Heat flux through sea ice in the western Weddell Sea: convective and conductive transfer processes, *J. Geophys. Res.*, 101, C4, 8853-8868, 1996.

Lytle, V.I., K.C. Jezek, S.P. Gogineni, and A.R. Hosseinmostafa, Field observations of microwave backscatter from Weddell Sea ice, *Int. J. Remote. Sens.*, 17, 1, 167-180, 1996.

Martinson, D.G., and C. Wamser, Ice drift and momentum exchange in the winter Antarctic pack ice, *J. Geophys. Res.*, 95, 1741-1755, 1990.

Massom, R.A., Observing the advection of sea ice in the Weddell Sea using buoy and satellite passive microwave data, *J. Geophys. Res.*, 97, C10, 15559-15572, 1992.

Massom, R., M.R. Drinkwater and C. Haas, Winter snowcover on sea ice in the Weddell Sea, *J. Geophys. Res.*, 102 (C1), 1101-1117, 1997.

Morris, K. and M.O. Jeffries, Sea ice characteristics and seasonal variability of ERS-1 SAR backscatter in the Bellingshausen Sea, *Antarctic Res Ser.*, this volume.

Muench, R.D., and A.L. Gordon, Circulation and transport of water along the western Weddell Sea, *J. Geophys. Res.*, 100, C9, 18503-18515, 1995.

NSIDC, *DMSP SSM/I Brightness temperature and sea ice concentration grids for the polar regions on CD-ROM*, User's Guide. National Snow and Ice Data Center Special Report, 1, Boulder, CO, USA, 1995.

Orsi, A.H., W.D. Nowlin, Jr., and T. Whitworth III, On the circulation and stratification of the Weddell Gyre, *Deep Sea Res.*, 40, 169-203, 1993.

Schnack-Schiel, S. (Ed.), The winter-expedition of R.V. "Polarstern" to the Antarctic (Ant V/1-3). *Reports on Polar Research*, 39, Alfred-Wegener-Institut für Polar- und Meeresforschung, D-27568, Germany, 259pp, 1987.

Stern, H.L., D.A. Rothrock, and R. Kwok, Open water production in Arctic sea ice: satellite measurements and model parameterizations, *J. Geophys. Res.*, 100, C10, 20601-20612, 1995.

Stoffelan, A., and D.L.T. Anderson, ERS-1 scatterometer data characteristics and wind retrieval skill, Space at the Service of Our Environment, *Proc. 1st ERS-1 Symp.*, 4-6 Nov. 1992, Cannes, France, ESA SP-359, Vol. 1, 41-47, 1993.

Thorndike, A.S., Kinematics of Sea Ice, In *The Geophysics of Sea Ice*, (Ed.) N. Untersteiner, NATO ASI Series B, Physics, Plenum Press, New York, 489-549, 1986.

Tucker, W.B., D.K. Perovich, A.J. Gow, W.F. Weeks, and M.R. Drinkwater, Physical properties of sea ice relevant to remote sensing, In *Microwave Remote Sensing of Sea Ice*, (Ed.) F.D. Carsey, American Geophysical Union, Geophysical Monograph 28, Chapt. 2, 9-28, 1992.

Viehoff, T., and A. Li, Iceberg observations and estimation of submarine ridges in the western Weddell Sea, *Int. J. Remote Sens.*, 16, 17, 3391-3408, 1995.

Viehoff, T., A. Li, C. Oelke, and H. Rebhan, Characteristics of winter sea-ice conditions in the southern Weddell Sea in 1992 as derived from multi-sensor observations. *Proc. IGARSS '94*, Pasadena, California, USA, Aug. 8-12, Vol. 1, 150-152, 1994.

Vihma, T., J. Launiainen, and J. Uotila, Weddell Sea ice drift: kinematics and wind forcing, *J. Geophys. Res.*, 100, C9, 18503-18515, 1995.

Zibordi, G., and M.L. Van Woert, Antarctic sea ice mapping using the AVHRR, *Remote Sensing of the Environment*, 45, 155-163, 1993.

M. Drinkwater, Earth and Space Sciences Division, Jet Propulsion Laboratory, California Institute of Technology, 4800 Oak Grove Drive, Pasadena, CA 91109

(Received September 16, 1996;
accepted April 25, 1997)

SEA ICE CHARACTERISTICS AND SEASONAL VARIABILITY OF ERS–1 SAR BACKSCATTER IN THE BELLINGSHAUSEN SEA

Kim Morris, Martin O. Jeffries and Shusun Li

Geophysical Institute, University of Alaska Fairbanks, Alaska

Radiometrically–corrected synthetic aperture radar (SAR) data acquired by the European ERS–1 satellite have been used to document the backscatter ($\sigma°$) variability of sea ice in the Bellingshausen Sea in winter (August and September, 1993) and summer (January, February and March, 1994). In Antarctica, such data were previously only available for the Weddell Sea. The backscatter signatures of first–year and multiyear ice in the Bellingshausen Sea are similar to those in the Weddell Sea and probably reflect a similarity in surface properties, particularly ice surface roughness and the occurrence of seawater flooding at the snow/ice interface. These factors contribute to higher winter first–year ice $\sigma°$ values than those of arctic first–year ice. Antarctic winter multiyear ice $\sigma°$ values are similar to those of the arctic winter multiyear ice. On the other hand, antarctic summer multiyear ice $\sigma°$ values are higher than those of arctic summer multiyear ice, probably due to some snow metamorphosis and increased ice surface flooding. The greatest contrast in antarctic sea ice backscatter signatures is between new/young ice and the older, thicker ice types. That a similar contrast occurs in arctic $\sigma°$ values reflects the fact that, at these early stages of ice development, the new/young ice surfaces are subject to similar influences in both polar regions. At later stages of ice development such phenomena as seawater flooding and a perennial snow cover affect antarctic, but not arctic, sea ice. The backscatter signals of many antarctic ice types do not vary greatly during the course of winter and summer; consequently, backscatter anomalies are very distinctive when they do occur. Two such anomalies occurred in summer; they were a consequence of the introduction of warm, marine air over the perennial pack ice by an intense cyclone which resulted in a uniform backscatter reduction of 5 dB.

1. INTRODUCTION

Satellite passive microwave data show that antarctic sea ice extent varies seasonally from a minimum of slightly less than 4×10^6 km^2 to a maximum of roughly 19×10^6 km^2 [*Gloersen et al.*, 1992]. The passive microwave data have provided a rich vein for studies of the spatial and temporal variability of the sea ice in relation to climate and environmental variability [*Gloersen and Campbell*, 1988, 1991; *Jacobs and Comiso*, 1993; *Parkinson*, 1992, 1994; *Gloersen*, 1995]. In addition, these data have been used to investigate smaller scale process studies such as polynya and pack ice variability and oceanographic effects [*Carsey*, 1980; *Cavalieri and Martin*, 1985; *Zwally et al.*, 1985; *Jacobs and Comiso*, 1989], and atmospheric forcing of ice motion and ice concentration variability [*Massom*, 1992].

In contrast to the frequent use of passive microwave data for antarctic sea ice studies, there have been far fewer applications of active microwave remote sensing. Until recently, this has been primarily a consequence of the lack of data, from either airborne or spaceborne sources, and it contrasts starkly with the frequent use of active microwave sensors for studies of arctic sea ice processes since the early 1970s [*Carsey*, 1992; *Haykin et al.*, 1994]. Active microwave remote sensing has a distinct advantage over passive microwave and other remote sensing instruments for sea ice studies because of its high spatial resolution in addition to its all–weather, all–season capability to obtain earth surface data. The potential of active microwave remote sensing of antarctic sea ice was demonstrated by the SIR–B experiment, in which a small amount of data was acquired from the northern parts of the Weddell Sea ice pack with a synthetic aperture radar (SAR) aboard the Space Shuttle [*Carsey et al.*, 1986; *Martin et al.*, 1987]. Real aperture radar data of antarctic sea ice acquired by instruments operated by the former Soviet Union is illustrated by *Comiso et al.* [1992].

The lack of opportunity for acquiring active microwave data for antarctic sea ice studies changed dramatically in July 1991 when the first European Remote Sensing Satellite (ERS–1) was successfully launched by the European Space Agency (ESA). ERS–1 carries an active microwave instrument (AMI) that acts as a SAR when operated in image mode. Since 1991, ERS–1 SAR data have been acquired primarily over the Weddell Sea and used to investigate sea ice motion, detection of the autumn freeze–up, and the factors that affect backscatter from the ice [*Drinkwater and Kottmeier*, 1994; *Drinkwater et al.*, 1994, 1995]. Since the launch of ERS–1, additional SAR instruments have become operational: the Japanese National Space Development Agency (NASDA) has been operating JERS–1 since February 1992; ESA and the Canadian Space Agency (CSA) launched ERS–2 and RADAR-SAT respectively in April and November 1995. There is now considerable potential for active microwave remote sensing of antarctic sea ice.

Since the 1970s, the identification of the factors that affect backscatter from sea ice in the Arctic has been the subject of considerable research. This work was designed to determine the optimum SAR sensor characteristics (e.g., frequency, polarization) for sea ice studies and which sea ice properties and processes can be detected by SAR. For detailed discussions of these arctic investigations, the reader is referred to the many papers in *Carsey* [1992] and *Haykin et al.* [1994]. While the arctic research has established the fundamentals of microwave scattering from sea ice, there remains a need for investigations of the factors that affect backscatter from antarctic sea ice and, thus, the interpretation of sea ice signatures in antarctic SAR images. This is made necessary by some fundamental differences in arctic and antarctic sea ice properties and processes that are relevant to the understanding of remotely sensed data [*Gow and Tucker*, 1990; *Tucker et al.*, 1992].

Two aspects of antarctic sea ice properties and processes in particular are of considerable interest with regard to their possible effects on active microwave remote sensing; both involve the snow cover on the ice surface. First, in summer, it is uncommon for the snow on antarctic floes to melt away completely and puddle the ice surface with melt ponds [*Andreas and Ackley*, 1982]. Extensive snow melt and puddling occurs each summer in the Arctic, where it plays a key role in changing the near–surface ice properties and enables the differentiation of multiyear ice from younger ice types during subsequent winters using both passive and active microwave sensors [*Comiso*, 1986; *Kwok et al.*, 1992]. Second, it is common for large areas of the surface of antarctic floes to be flooded with seawater, thereby soaking the base of the snow cover [*Wadhams et al.*, 1987; *Worby et al.*, 1994], a phenomenon uncommon in the Arctic. The survival of the snow cover and resultant increase in depth through time, and the flooding of the base of the snow cover, effectively mask the ice surface and make it more difficult to identify antarctic multiyear ice using passive microwave data [*Comiso et al.*, 1992]. This might also be the case with SAR, as suggested by field investigations using ship–mounted radar scatterometers which indicate that thick snow and seawater flooding significantly affect active microwave sea ice signatures [*Lytle et al.*, 1990, 1996; *Hosseinmostafa et al.*, 1995; *Jeffries et al.*, 1995c].

Most of the investigations of sea ice using ERS–1 SAR and surface radar scatterometers have been undertaken in the Weddell Sea. Likewise, there have been many studies of sea ice properties and processes in this region. The considerable focus on Weddell Sea ice has raised the question of just how representative the data acquired there might be of the circum–antarctic sea ice cover. The short answer to that question is that sea ice properties and processes in other parts of the Southern Ocean are broadly similar to those in the Weddell Sea [*Jeffries and Weeks*, 1992; *Jeffries et al.*, 1994a, 1995b; *Allison and Worby*, 1994]. However, in view of the limited number of investigations of antarctic sea ice properties and processes in relation to radar backscatter, it remains reasonable to question how representative Weddell Sea active microwave backscatter and SAR image signatures might be of other Southern Ocean sea ice zones.

With the last question in mind, we have investigated the radar backscatter variability of sea ice in the Bellingshausen Sea [*Jeffries et al.*, 1995a; *Morris and Jeffries*, 1995]. This is an interesting region from the point of view of sea ice, climate variability and remote sensing. There is evidence that the sea ice cover is responding to increasing air temperatures on the west side of the Antarctic Peninsula [*Jacobs and Comiso*, 1993]. This sensitivity was manifested as a record decrease in summer ice extent from mid–1988 through early 1991, which might have been hastened by higher than normal ice temperatures and more extensive seawater flooding of the ice surface [*Jacobs and Comiso*, 1993]. As noted above, flooding might be responsible for masking the multiyear ice signature in passive microwave data, and for affecting radar backscatter. On the other hand, at other times in the Bellingshausen Sea, the sea ice passive microwave signature has been observed to more closely resemble the arctic multiyear sea ice signature than other regions of the Southern Ocean ice cover [*Zwally and Gloersen*, 1977; *Comiso et al.*, 1992].

The factors that affect the microwave remote sensing signatures of sea ice in the Bellingshausen Sea remain to be fully understood. In the case of active microwave remote sensing, one approach to improving this situation is to begin with a basic survey of the spatial and temporal variability of backscatter. This paper documents the results of just such a survey of ERS–1 SAR backscatter variability in winter and summer in the Bellingshausen Sea between the ice edge and the continent. The discussion focuses primarily on a comparison of the Bellingshausen Sea ice backscatter data with Weddell Sea and Arctic Ocean sea ice

TABLE 1. ERS–1 SAR orbits from which sea ice backscatter statistics were derived for this study.

Map ID	Orbit No.–Direction[a]	Date of Acquisition	Time of Aquisition[b] (hhmm:ss)	Number of SAR images
WINTER 1993				
A	10976 – D	Aug. 21	1345:32	4
B	11000 – A	Aug. 23	0607:42	12
C	11019 – D	Aug. 24	1349:01	13
D	11086 – A	Aug. 29	0619:27	9
E	11091 – D	Aug. 29	1432:08	11
F	11134 – D	Sept. 1	1437:53	11
G	11158 – A	Sept. 3	0702:20	11
H	11172 – A	Sept. 4	0630:58	8
I	11177 – D	Sept. 4	1443:39	10
J	11201 – A	Sept. 6	0708:05	10
K	11263 – D	Sept. 10	1456:55	5
SUMMER 1994				
L	13177 – A	Jan. 22	0714:40	5
	13220 – A	Jan. 25	0714:39	5
	13263 – A	Jan. 28	0714:39	5
	13306 – A	Jan. 31	0714:38	4
	13349 – A	Feb. 3	0714:39	4
	13392 – A	Feb. 6	0714:39	5
	13478 – A	Feb. 12	0714:39	5
	13607 – A	Feb. 21	0714:36	5
	13650 – A	Feb. 24	0714:38	5
	13693 – A	Feb. 27	0714:38	5
	13736 – A	Mar. 2	0714:39	4
	13779 – A	Mar. 5	07:14:39	4
	13822 – A	Mar. 8	0714:39	5
	13865 – A	Mar. 11	0714:39	5
	13908 – A	Mar. 14	0714:39	5
	13951 – A	Mar. 17	0714:39	4
	13994 – A	Mar. 20	0714:39	4

[a] A – ascending orbit, D – descending orbit
[b] UT start time of acquisition

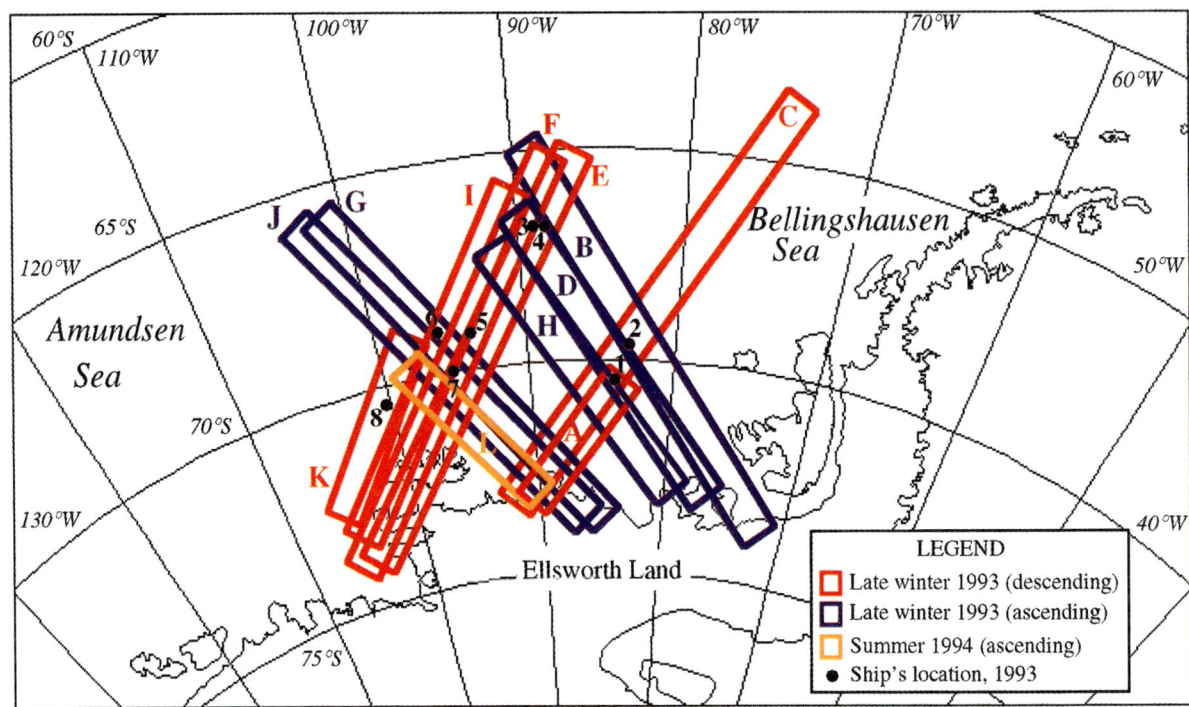

Plate 1. Map showing the locations of the ground tracks of ERS-1 SAR orbits analyzed for this study in the Bellingshausen Sea. The orbit details are summarized in Table 1. The solid circles show the locations of the R.V. *Nathaniel B. Palmer* concurrent with these satellite overpasses (see Table 2).

backscatter data, and considers some of the snow and ice properties and processes that might account for the backscatter variability.

2. STUDY AREA AND METHODS

2.1. Study Area and SAR Data Acquisition

The Bellingshausen Sea is located between the Antarctic Peninsula and 103°W, the longitude of Cape Flying Fish in westernmost Ellsworth Land (Plate 1). During August and September 1993, the R. V. *Nathaniel B. Palmer* operated in this area supporting late winter field research on the physical characteristics and processes of the first-year sea ice and its snow cover. See *Jeffries* [1993] for a map of the cruise track. The ship penetrated no further south than 70.33°S, 96.18°W where it encountered the northern margin of the perennial pack ice which was characterized by thick floes with a deep snow cover that severely impeded progress.

From August 21 to September 10, the German Antarctic Receiving Station (GARS) located at the Chilean station Base O'Higgins on the Antarctic Peninsula was open for the reception of ERS-1 SAR data. A total of 104 SAR images from 11 different orbits (Plate 1, Table 1) from this time period were subsequently ordered for analysis. There is some geographic overlap between some of the selected orbits, e.g., orbits E, F and I, but none of the orbits completely overlaps another. Only two orbits do not extend from the ice edge to the continent or nearshore islands (orbits 10976 and 11263). This data set offers a number of "snapshots" of the late winter sea ice within the general area of the Bellingshausen Sea. The ship's location, when it fell within the ground swath of the ERS-1 satellite as SAR data were acquired at GARS, are listed in Table 2.

GARS was again open to receive SAR data at the beginning of 1994. From this data set, a total of 79 SAR images from 17 orbits (Plate 1, Table 1) acquired between January 22 and March 20, 1994 were ordered for analysis. All of these orbits, acquired during the 3 day repeat cycle or Ice Phase of the satellite, cover the same area (L in Plate 1). With this exact repeat orbit data set it is possible to follow the same ice floes and features for several days and sometimes weeks.

2.2. SAR Data Characteristics and Analysis

The ERS–1 SAR operates at C–band (5.3 GHz ± 0.2 MHz, 53 mm wavelength) with a mid-swath incidence angle of 23°. All of the data used in this study were SAR Precision Images (PRI) processed from the raw signal data at the German Processing and Archiving Facility. A precision image is a multi–look, ground range, digital image with a nominal geographical coverage of 100 km by 100 km, a pixel size of 12.5 m by 12.5 m (16–bits per pixel) and a nominal ground resolution of <33 m. Post–processing of the images was performed at the Geophysical Institute, University of Alaska Fairbanks. This included radiometrically calibrating the images using the ESA algorithm [*Laur*, 1992] and reducing the data to an 8–bit, 25 m by 25 m pixel size by merging four pixels into one (approximately 40 m spatial resolution). The backscatter coefficient calculated for a given pixel from these images has an approximate error of ±2.0 dB at a 90% confidence level (M. Drinkwater, pers. comm., 1996). However, this error is greatly reduced when pixels are combined into a sample. For example, a sample consisting of 6 pixels (the smallest sample in this study) has an error of ±0.8dB. Since most of the samples are large (see below), the error for any given sample is small.

All of the images were analyzed using MacSigma_0, a software package designed at the Jet Propulsion Laboratory to compute statistics from digital images. Zones of homogeneous tone and texture were sampled and backscatter ($\sigma°$) statistics (mean, standard deviation and distribution [histogram]) were calculated. Individual sample sizes ranged from as few as 6 pixels to as many as 772,000 pixels. The number of samples per image ranged from 5 to 55, depending on the complexity of the sea ice cover. Samples of similar tone, texture and location in the pack were grouped in an unsupervised classification scheme and "global" statistics were calculated for each class of ice.

As can be seen in Plate 1 and Table 1, all of the summer ERS–1 SAR data were acquired over exactly the same location at exactly the same time of day (0714 UT). As a consequence, there are no diurnal effects in the sample backscatter statistics for this data set. The winter ERS–1 SAR data acquisitions fall into two general groups (Plate 1 and Table 1). The ascending orbits were acquired between 0605 and 0710 UT, while the descending orbits were acquired, on average, 7 hours and 45 minutes later in the day. When taken as a whole, the sample $\sigma°$ values from the ascending and descending orbits are not statistically different. (See Morris and Jeffries, 1995 for a detailed floe–to–floe comparison of August 29 data.) Comparisons of sample $\sigma°$ values between the two groups for individual ice types show that those designated Type W1 and W2 (see Section 4.1.3) are the most dissimilar (Figure 1). Since most of the sample means lie within 1 standard deviation of the overall mean for each ice type for these least similar data, we decided it was possible to group the ascending and descending $\sigma°$ values into one data set without introducing significant errors.

TABLE 2. Location of ship during concurrent satellite overpasses.

Location Numbers [a]	Position	In Orbit - Date
1	70.5°S 83.8°W	10976 – Aug. 21
2	69.7°S 83.4°W	11019 – Aug. 24
3	66.8°S 89.0°W	11086 – Aug. 29
4	66.8°S 88.6°W	11091 – Aug. 29
5	69.2°S 94.2°W	11134 – Sept. 1
6	69.1°S 96.4°W	11158 – Sept. 3
7	70.1°S 95.8°W	11201 – Sept. 6
8	70.5°S 101.0°W	11263 – Sept. 10

[a] The locations numbers correspond to those in Plate 1.

2.3. Late Winter Field Observations and Measurements

During the August and September 1993 cruise of the R. V. *Nathaniel B. Palmer*, several different kinds of observations and measurements were made. While the ship was underway in the ice, hourly observations were made of the snow and ice thickness, ice concentration and areal extent and height of ridges using a standardized system [*Allison et al.*, 1993; *Allison and Worby*, 1994]. Each day the ship was stopped to allow different ice floes to be investigated. Measurements included snow and ice thickness and freeboard along 100 m transects [*Worby et al.*, 1994, 1996], sea ice temperature, salinity and crystal structure as determined from ice cores [*Jeffries et al.*, 1994b; 1997] and snow temperature, salinity, and general stratigraphy as determined from snow pits [*Jeffries et al.*, 1994c]. In addition to determining the salinity of snow samples from each snow pit, a visual assessment of the wetness of the snow was made. The designations ranged from completely saturated (slush), through wetted (snow has a gray appearance) to dry. In general, the slush had the highest salinities, while the wetted snow had lower salinities and the dry snow was non-saline (<0.5‰).

218 ANTARCTIC SEA ICE: PHYSICAL PROCESSES, INTERACTIONS AND VARIABILITY

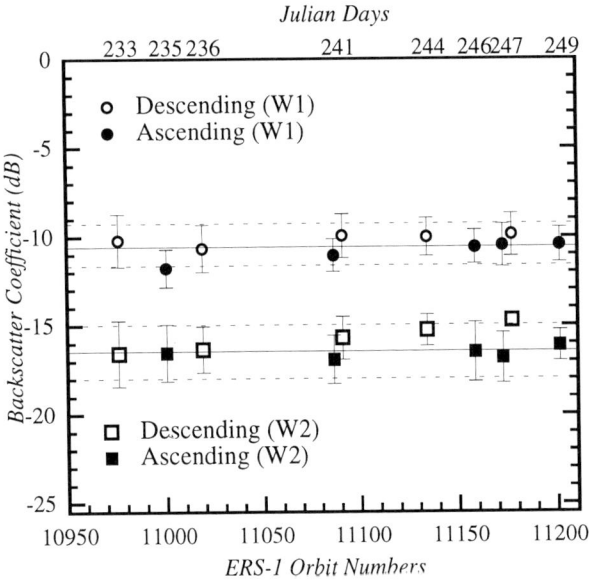

Fig. 1. Comparison of mean σ° values for ascending and descending passes for two winter perennial ice types (See Section 4.1.3 for ice type descriptions). The solid lines are the overall mean σ° values for each ice type and the dashed lines represent one standard deviation from those means (W1=-10.4±1.3 dB, W2=-16.5±1.5 dB).

3. GENERAL DESCRIPTION OF SNOW AND ICE CHARACTERISTICS

3.1. Winter 1993

The hourly observations of ice concentration, areal extent of ridging and ridge height are presented as a function of latitude in Figure 2. Ice concentration was high throughout the study area, which is consistent with the high ice concentrations subsequently observed in the SAR images. Deformation features (ridges and rafts) made up 20–30% of the total area of floes and the deformation features were typically 0.5 m to 1 m high. Deformation features of this height protrude above the snow surface, but below it are smaller–scale roughness features that may be masked by the snow. This method of snow and ice observations [*Allison et al.*, 1993; *Allison and Worby*, 1994] does not include deformation features with an average height less than 0.5 m. As a consequence, the more common, small–scale, surface deformation features that result from rafting are not included in the areal deformation estimates. The typical small–scale roughness of the ice surface can be gauged from the floe illustrated in Plate 2 and its snow and ice thickness profile shown in Figure 3. The floe comprised rafted cakes that had subsequently consolidated creating a rough surface of upturned edges. For the entire cruise, the mean floe thickness was 0.90±0.6 m with snow depths of 0.23±0.2 m; in general, snow and ice thicknesses increased with distance from the ice edge [*Worby et al.*, 1994, 1996].

Although no seawater flooding of the ice surface was observed along the transect on Floe 235 (Figure 3), flooding was quite common throughout the study area; negative freeboard values were recorded at 200 (18%) of the total of 1,113 holes that were drilled along transects on 30 different floes [*Worby et al.*, 1994, 1996; *Jeffries et al.*, 1995c]. A total of 118 snow pits were investigated on 50 different floes and a layer of slush comprising snow and seawater was observed at the base

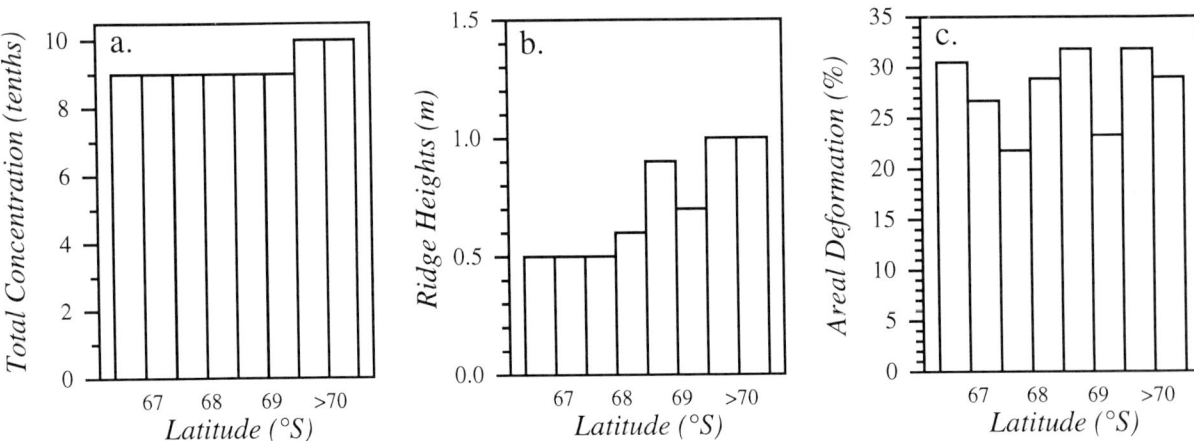

Fig. 2. Total ice concentration (*a*), ridge height (*b*) and areal extent of deformation (*c*) as a function of latitude determined from hourly ice observations made while the R. V. *Nathaniel B. Palmer* was underway in the Bellingshausen Sea in August and September, 1993. Each bar in the graphs represents an area averaged value, and the values in (*c*) are weighted according to the ice concentration.

of 35 pits [*Jeffries et al.*, 1994c]. Wetted but not saturated snow was often observed immediately above the slush layers, and also immediately above ice surfaces that were not flooded. In both cases, upward migration of brine from the slush or the ice surface by a wicking process leads to increased wetness and salinity. Most of the dry snow had salinity values <0.5‰, whereas values >20‰ were commonly observed in the slush layers (Figure 4).

Even in winter, the snow cover on the antarctic sea ice is very icy due to the presence of ice lenses, ice layers, percolation columns and melt–grain clusters [*Sturm et al.*, 1995]. Ice layers were common in the snow cover throughout the study area, with two to three such layers often being observed in a single snow pit [*Jeffries et al.*, 1994c]. These layers were generally 2–4 mm thick and extended horizontally for distances of metres. The ice layers are primarily melt and condensation features that form during periods of warm, moist marine air circulation over the pack ice [*Jeffries et al.*, 1994c; *Sturm et al.*, 1995]. Ice layers in the late winter snow cover have since been observed throughout the pack ice of the Pacific sector of the Southern Ocean [*Sturm et al.*, this volume].

During the period for which the winter ERS–1 SAR imagery was acquired (August 21 to September 10), the mean air temperature was -9.8±7.6 °C with a minimum temperature of -25.2 °C. The average wind speed was 9.6±4.2 m s^{-1} with gusts up to 22.0 m s^{-1}. Wind direction was variable but primarily from SW to N with only three episodes of easterly winds.

3.2. Summer 1994

Sea ice research was performed from R. V. *Polarstern* in the Bellingshausen and Amundsen Seas from mid–January to March 1994 [*Haas and Viehoff*, 1994]. In general, the sea ice in the Bellingshausen Sea was 1.0–1.2 m thick; the mean ridge height was 1.06 m (0.8 m cut-off height) with a spacing of 207 m (C. Haas, pers. comm.). Two sets of sampling sites from this cruise are of particular interest as they are close to the summer SAR image footprint [*Haas and Viehoff*, 1994]. Data collection on 13, 14, 17 and

Plate 2. Photographs taken from the ships's bridge (*a*) and on the ice (*b*) illustrate the areal extent of small scale deformation and the height of the deformation features due to extensive rafting of a floe located at 69.73°S, 83.02°W in the central Bellingshausen Sea.

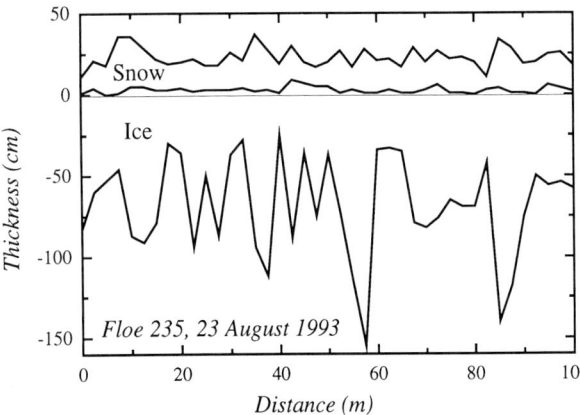

Fig. 3. Snow and ice thickness, and top and bottom roughness profiles for the floe illustrated in Plate 2. The thin, horizontal line at 0 m represents sea level. The low top surface relief compared to the very rough bottom topography is typical of heavily rafted floes.

18 February was centered near 70.33°S, 87.5°W: the ice concentrations ranged from 3 to 9.9 tenths (20 of 37 observations >8/10); ice thickness was 0.4–3.0 m; snow depths were 0.05–0.60 m; and floe sizes ranged from <1 m to >1000 m. The mean air temperature was -1.9±1.4 °C (min. -8.3 °C), average wind speed was 7.8±3.7 m s^{-1} with gusts up to 19.5 m s^{-1} and direction was 98.4±59.5° (NE to SSE). The data acquired from 22–24 February, located in the general vicinity of 71.5°S, 105.25°W, revealed high ice concentrations (9.5-10/10), ice thicknesses of 0.8–4 m, snow depths of 0.05–0.40 m and floe sizes from <1 m to 1000 m. During this second period, the mean air temperature was -3.9±1.7 °C (min. -6.9 °C), average wind speed was 11.6±2.2 m s^{-1} with gusts up to 17.5 m s^{-1} and direction was 100±19.1° (E to ESE).

During February 1992, 12 ice cores were obtained in the Bellingshausen Sea in the vicinity of Thurston Island, just west of the summer 1994 ERS–1 SAR footprint [*Jeffries et al.*, 1994a]. The mean length of these ice cores was 1.86±1.2 m (range: 0.77 to 4.6 m) with an average snow depth of 0.34±0.15 m (range: 0.16 to 0.70 m). These values are similar to those obtained in February 1994 and, taken together, they probably represent the general sea ice and snow conditions in the summer perennial pack ice in the Bellingshausen Sea.

4. BACKSCATTER VARIABILITY

4.1. Winter Sea Ice

4.1.1. General description of the winter pack ice. The antarctic sea ice pack reaches its maximum extent between August and October. By mid–August 1993, when the R. V. *Nathaniel B. Palmer* entered the Bellingshausen Sea ice pack, the ice cover was well established and almost at its full extent (Plate 3). The National Ice Center (NIC) ice charts show that at that time, with the exception of a 2° band of lower ice concentration along the ice edge and some lower ice concentrations zones along the coastline, the ice concentration in the perennial and annual ice zones was typically 9/10 to 10/10. This agrees well with our own observations (Figure 2a) and with the SAR image mosaic (Figure 5). The antarctic sea ice pack is reduced to its minimum extent during late February or early March. According to the NIC ice charts, the minimum ice extent in summer 1993, i.e., the summer immediately prior to this study, occurred in early March when the ice edge was located generally between 70° and 72°S (Plate 3). The ice charts show that the ice concentration at this time was variable in the perennial ice zone, ranging from 1–3/10 to 9/10, with primarily 9–10/10 in the area of interest (i. e., in the vicinity of Farwell Island).

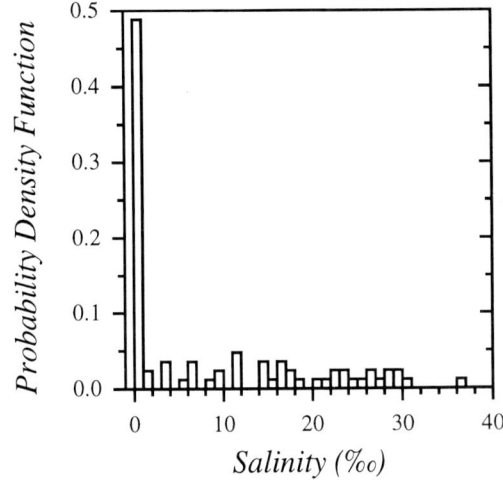

Fig. 4. Probability density function of snow salinity values in the late winter 1993 pack ice.

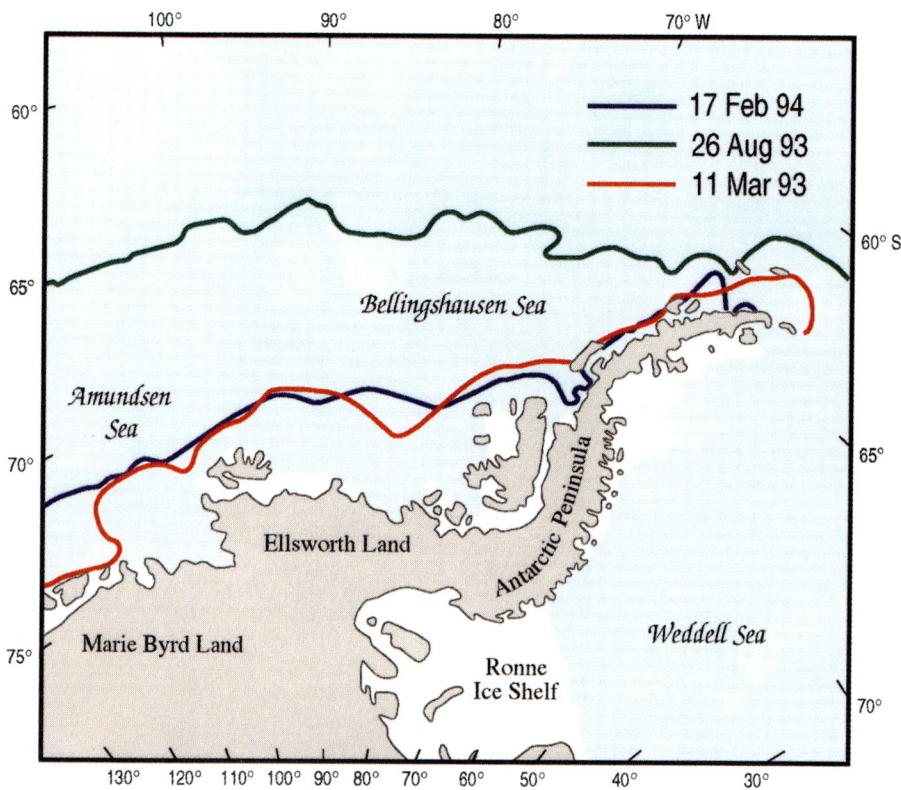

Plate 3. Map of the Bellingshausen and east Amundsen seas showing the minimum and maximum extent of the pack ice in 1993 and the minimum extent of the pack ice in 1994 as determined from U.S. National Ice Center antarctic sea ice charts.

The SAR mosaic (Figure 5) illustrates the general sea ice conditions in the area during the late winter cruise (see Table 1 for time of data acquisition). The ship was in the general vicinity when these data were acquired (Plate 1 and Table 2). Shipboard weather instruments indicate that 38 hours before the satellite overpass the air temperature was -11.6 °C with 12.9 m s^{-1} winds from the southwest. Over the next 31 hours, the mean air temperature was -16.5±1.6 °C with westerly winds at an average speed of 9.15±3.4 m s^{-1}. Seven hours before the overpass, the winds became northerly with mean speeds of 5.5±1.3 m s^{-1} and mean air temperatures of -12.2±1.6 °C.

On the basis of sea ice features and variations in tone and texture, three zones are identified: the coastal zone, the perennial ice zone and the annual ice zone. The boundary between the annual and perennial ice coincides closely with the previous summer's ice edge (Plate 3 and Figure 5). Overall, gray tones increase in variability and, thus, contrast from the ice edge to the continent. The annual and perennial pack ice have quite different textures, with the former having more curvilinear lead features that contrast with the greater linearity of leads in the perennial ice zone. Tone and texture variations and the identification and unsupervised classification of ice types for analysis of their backscatter are described in more detail below.

4.1.2. Winter coastal ice. The coastal region lies within approximately 50 to 60 km of the coast and is dominated by flaw leads and polynyas. This is an area of often intense atmosphere–ice–ocean interaction where offshore winds force the ice seaward creating leads and polynyas which become sites for further ice production. An example of this offshore ice movement can be seen in Figure 5: the distinctive shapes of some of the ice

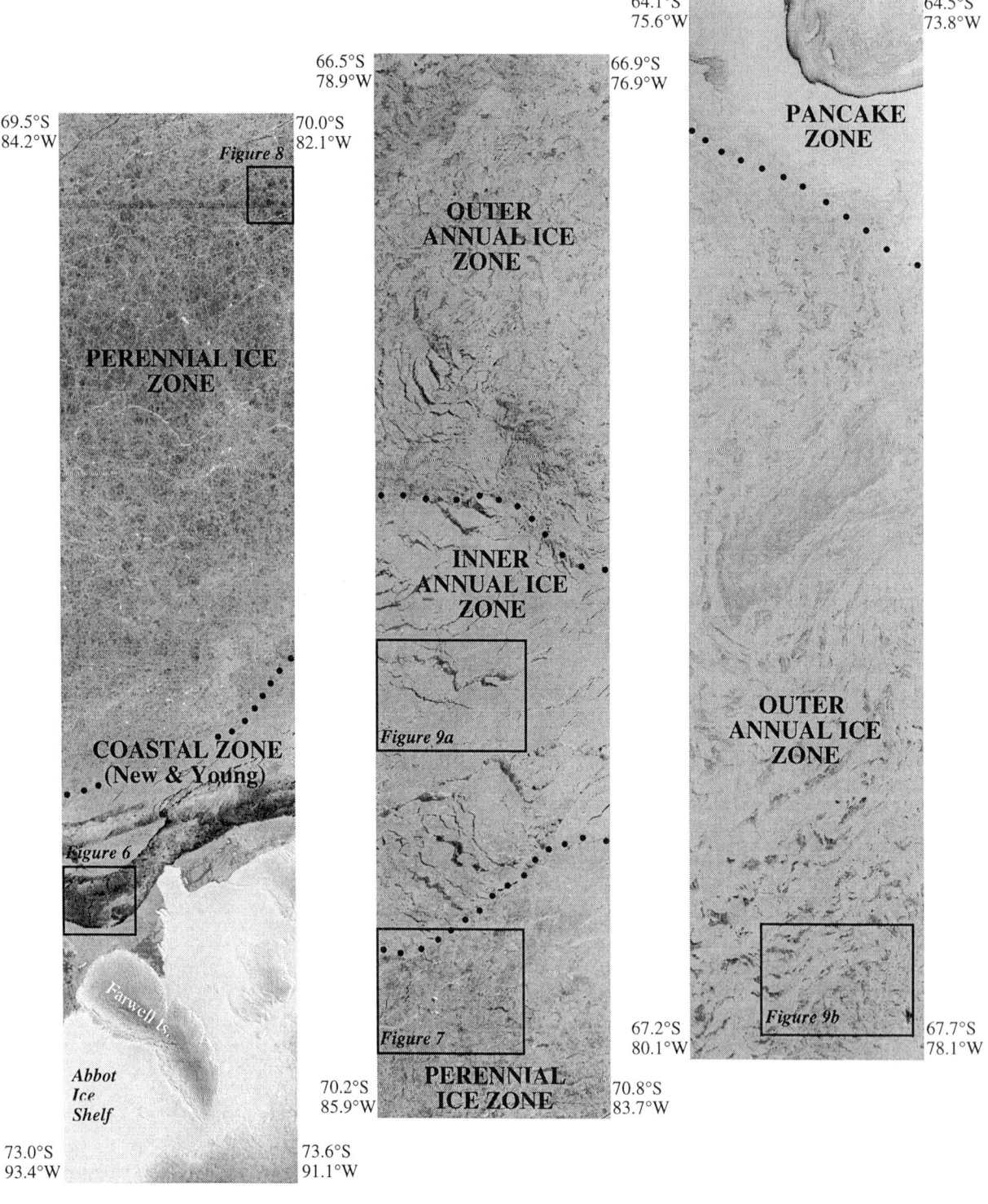

Fig. 5. Mosaic of ERS-1 SAR orbit 11019, acquired on August 24, 1993 illustrating typical sea ice conditions. The three main ice regimes are identified and their boundaries identified by dotted lines: annual pack (subdivided into inner and outer); perennial pack and coastal zone. The bright tones at the northern margin of the outer annual pack probably delineate a zone of pancake ice. Each strip of images has surface dimensions of approximately 100 km by 445 km. The locations of the five subscenes presented in Figures 6, 7, 8 and 9 are identified by rectangles.

floes allow their original, shoreward location to be identified. The ice forming the northern and southern boundaries of the lead is most likely thin first–year ice and, until the upper portion broke away, was probably landfast. The landfast ice on the north side of Farwell Island (Figures 5, 13 and 14), which was only becoming established during the late winter period, remained in place for the rest of the winter and survived the following summer. Summer landfast ice $\sigma°$ values have been derived and will be described in a later section.

A subscene of the lead/polynya near Farwell Island (Figure 6) illustrates some of the tonal and textural details. For the purpose of determining backscatter statistics, we identify two ice types. First, there are extensive areas of very dark tone with a near uniform texture broken only by some faint and localized areas of stronger backscatter. The latter probably results from the presence of some widely distributed rafting features in an otherwise extensive area of sheet ice such as nilas. Second, there are extensive areas of brighter tone with a more variable texture. In a dynamic area such as this, thin ice can be quickly and extensively deformed [*Eicken and Lange*, 1989], and the resultant increase in surface roughness leads to increased backscatter. It is also possible that these might be areas of new and young ice with an extensive frost flower cover, which is known to significantly increase backscatter [*Onstott*, 1992; *Hallikainen and Winebrenner*, 1992; *Jeffries et al.*, 1995a]. However, frost flowers usually do not form under strong (>5 m s^{-1}) wind conditions [K. Steffen, pers. comm. cited in *Martin et al.*, 1996]. Thus, if we assume that the strong winds recorded in the area to the north prior to the imagery being acquired also prevailed in the coastal zone, frost flower formation is unlikely. In summary, the areas of dark tone are probably undeformed nilas, and the areas of brighter tone are probably moderately to strongly deformed new and young ice.

4.1.3. Winter perennial pack ice. The perennial ice zone lies south of approximately 70°S in Figure 5. In view of the location of the late summer 1993 ice edge and the ice concentration observed in the NIC ice charts, it is reasonable to assume that this zone comprises a large amount of multiyear ice. The perennial ice zone has a complex appearance with tones varying between very bright and very dark (Figure 5). A subscene illustrates some of this variation in more detail (Figure 7). The very bright, angular features are icebergs, which in this case have dimensions ranging from 0.2 km by 0.15 km to 1.8 km by 1.0 km.

The elongated features with the very dark tone are leads. Given the relatively high wind speeds (>5 m s^{-1}) in the area, their low backscatter is probably due to the presence of smooth, relatively thin, new ice rather than calm water. The bright elongated features are also leads. Their high backscatter might be a consequence of several factors. These may be leads covered with rough, deformed thin ice or they may be open water zones whose bright tone can be attributed to wind roughening. The high wind speeds in the area preclude the formation of frost flowers [*Martin et al.*, 1996] which could also produce high $\sigma°$ values on these linear features. Because of the great tonal variability and small size of leads within very small areas of the pack ice, lead ice backscatter statistics were not computed. Instead, backscatter statistics for new and young ice were determined from coastal leads/polynyas where we believe there is less ambiguity about the history and nature of the ice.

For the purpose of determining backscatter statistics for the perennial ice zone, three main ice types (W1, W2, and W3; where W signifies winter) have been identified according to their tone and form. Type W1 ice is the dominant ice type and is characterized by mid–range gray tones. It has no characteristic form but rather provides the "background" from which the other two ice

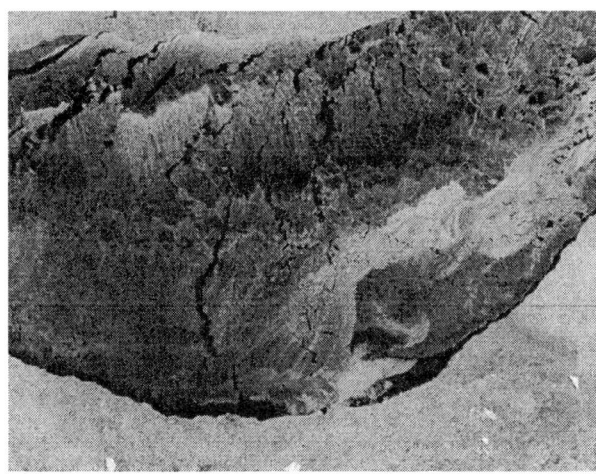

Fig. 6. A mosaic of subscenes from Frames 5121 and 5139 (Orbit 11019) shows some details of the backscatter variability from new and young ice types in a coastal flaw lead or polynya that is typical of this zone. The subscene has surface dimensions of 25.5 km by 33.5 km.

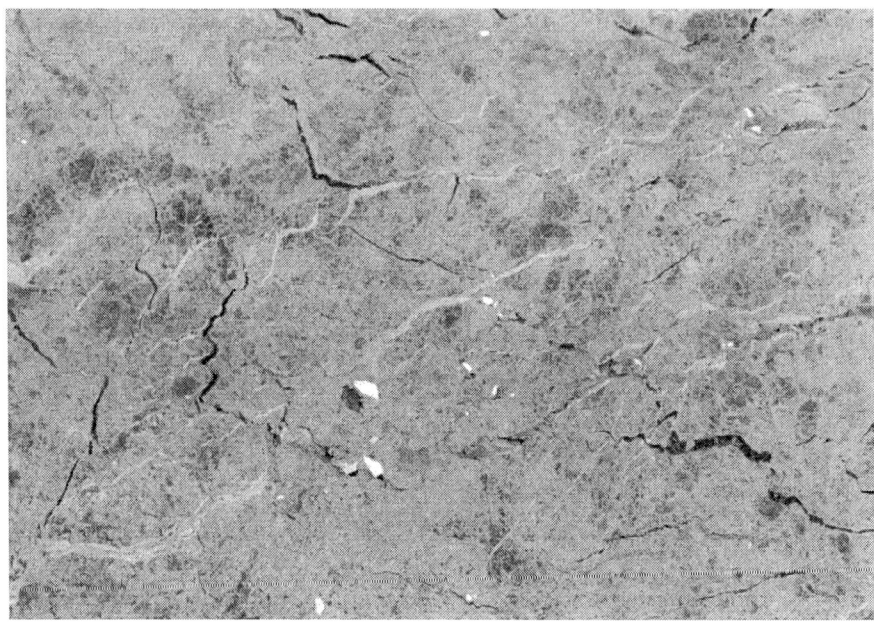

Fig. 7. A subscene of the perennial pack ice (Orbit 11019, Frame 5067) comprising ice that had survived the previous summer and ice that had formed during winter 1993. The subscene has surface dimensions of 42.5 km by 62.0 km.

types are distinguished. Type W2 ice (Figure 8) appears darker than Type W1 ice and takes the form of rounded, floe–like structures that are found throughout the perennial pack ice, with localized clustering. Type W3 ice is brighter and occurs in much smaller quantities than Types W1 and W2. Type W3 ice is as bright as some of the bright leads but unlike those long, thin linear features, it occurs as small, floe–like features. Backscatter statistics were also determined for icebergs in this zone.

4.1.4. Winter annual pack ice. The annual pack ice comprising primarily first–year ice can be divided into two zones: the inner and outer annual pack ice. Taken together, these ice zones extend from the consolidated ice edge to, on average, 70°S (Figure 5). Apart from the brightest tones in pancake ice fields near the ice edge, the inner annual pack ice is not significantly different from the outer annual pack ice in terms of gray tones; rather, they differ in general appearance and texture, which are shown in detail in two subscenes (Figure 9). The outer pack ice has a more broken and chaotic appearance than the inner annual pack ice, and a texture of curvi–linear features that suggest ice circulation driven by the atmosphere and/or ocean. The inner annual pack ice, which is spatially less extensive than the outer annual pack ice, has a more massive and consolidated appearance. This difference can be attributed to the age, thickness and degree of consolidation of the ice as it relates to the northward expansion of the ice cover in autumn and winter. The ice located in the inner annual pack ice adjacent to the northern edge of the perennial ice must be older than that in the outer annual pack ice. Outer annual ice located near the inner annual ice pack boundary may be months older than the outer annual ice near the ice edge. Consequently, the inner annual ice pack has had more time to become well consolidated into larger, more massive, floes than the outer pack ice. The hourly ice observation program aboard the R.V. *Nathaniel B. Palmer* in late winter 1993 confirms that ice thickness and floe size increase with increasing latitude in the annual pack ice zone [*Worby et al.*, 1994, 1996].

4.1.5 Swell in winter annual pack ice. In the SAR imagery acquired between 1 and 10 September, a wave pattern was clearly visible in the annual pack ice (Figure 10). During this time, the ice edge was located at approximately 66°S (Plate 3). The wave pattern was seen in the SAR images as far south as 68.5°S to 69.5°S between 89°W and 100°W, i.e., some 275 km to 400 km into the annual pack ice. The occurrence of the wave pattern in the SAR images corresponds to our observations at this time aboard the R.V. *Nathaniel B.*

Palmer when a significant swell was visible in the pack ice.

Points 5, 6 and 7 in Plate 1 correspond to the locations (Table 2) where the ship was in the swell as ERS–1 passed over the area. A swell was first detected on August 28 aboard the ship at 66.72°S, 88.88°W. On the same day, working on a 73 cm thick floe with 16 cm of snow, the heaving of the ice was noticeable underfoot and the movement was visible to the naked eye. Late on August 29, when the swell was estimated to have a wavelength of approximately 400 m and a height of 2–3 m, it was large enough to appear on the ship's radar screen and it broke up the consolidated ice cover into small floes, typically 10–20 m across (Figure 11). The above estimate of the swell's wavelength is close to that estimated from the SAR data (Figure 10).

Subsequently, the swell diminished briefly, but by the afternoon of September 1 it was once again visible on the radar screen at 69.15°S, 94.18°W. After September 1, the swell was no longer apparent aboard the ship or when we were on the ice, but the consequences of the swell continued to be seen in the form of the broken up ice cover comprising small floes until late on September 5 when thicker, consolidated floes were encountered at 70.05°S, 95.80°W, some 440 km from the ice edge. A similar phenomenon has been reported in the Weddell Sea where the wave field was damped down sufficiently to permit pancakes to freeze together fully only at distances almost 300 km south of the ice edge [*Wadhams et al.*, 1987; *Wadhams*, 1991].

Similar swell (wave) features have been seen in SAR images of arctic sea ice. *Wadhams and Holt* [1991] used L–band SAR from the Seasat satellite to study and model the propagation of waves in frazil/pancake ice in the Chukchi Sea. The Labrador Extreme Waves Experiment (LEWEX '87) and Labrador Ice Margin Experiment (LIMEX '87) and LIMEX '89 used airborne C–band SAR to study and model the penetration of ocean waves into the marginal ice zone east of Newfoundland [*Vachon et al.*, 1988; *Liu et al.*, 1991a, 1991b].

4.1.6 *Time series and probability density functions of winter ice.* Time series of mean $\sigma°$ values for each of the ice types identified in the three sea ice zones are presented in Figure 12. The $\sigma°$ values do not vary greatly through time and those fluctuations that do occur probably can be attributed to the fact that each value represents the average backscatter from multiple samples throughout a large study area.

In the coastal and perennial ice zones, the time series show that the backscatter signature for each ice type within each zone is statistically significantly different from the other ice types (Figure 12). This is clearly apparent in the probability density functions constructed from the sample means that summarize the backscatter for each ice type (Plate 4). There is some overlap in the tails of each distribution, but the peaks are clearly well separated from each other. The overall mean and standard deviation values of backscatter for the coastal and perennial ice types are summarized in Table 3.

The PDF for the calm water/undeformed new ice in the coastal zone is symmetrical about the mean and has a very narrow range of low values near the noise floor of the SAR system (Plate 4a and Table 3). The narrow range of values reflects the nearly uniform texture and limited deformation that is evident in the SAR images (Figures 5 and 6). In contrast, the PDF for the deformed new and young ice has a very broad base (Plate 4a), which is a consequence of the variation in the degree of deformation and surface characteristics. In the perennial ice zone, the Type W1 and Type W3 ice $\sigma°$ distributions are essentially symmetrical about the mean (Plate 4b).

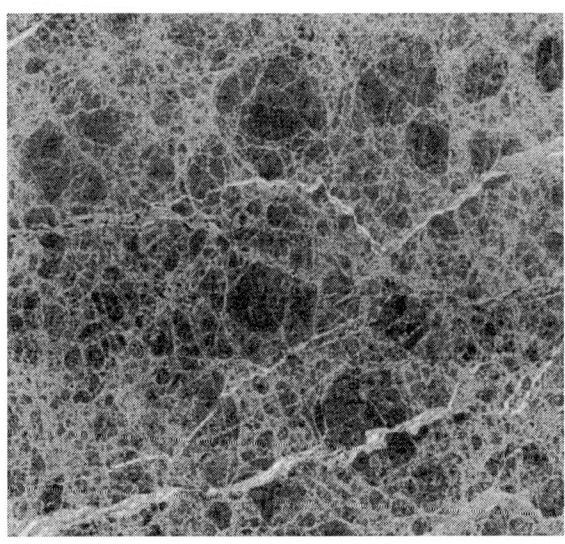

Fig. 8. A subscene of Frame 5085, Orbit 11019 showing Type W2 ice with its characteristic dark tone and rounded form in the Perennial Ice Zone. The brighter, elongated features are most likely leads covered with deformed ice or wind roughened water. The subscene has surface dimensions of 19.8 km by 21.5 km.

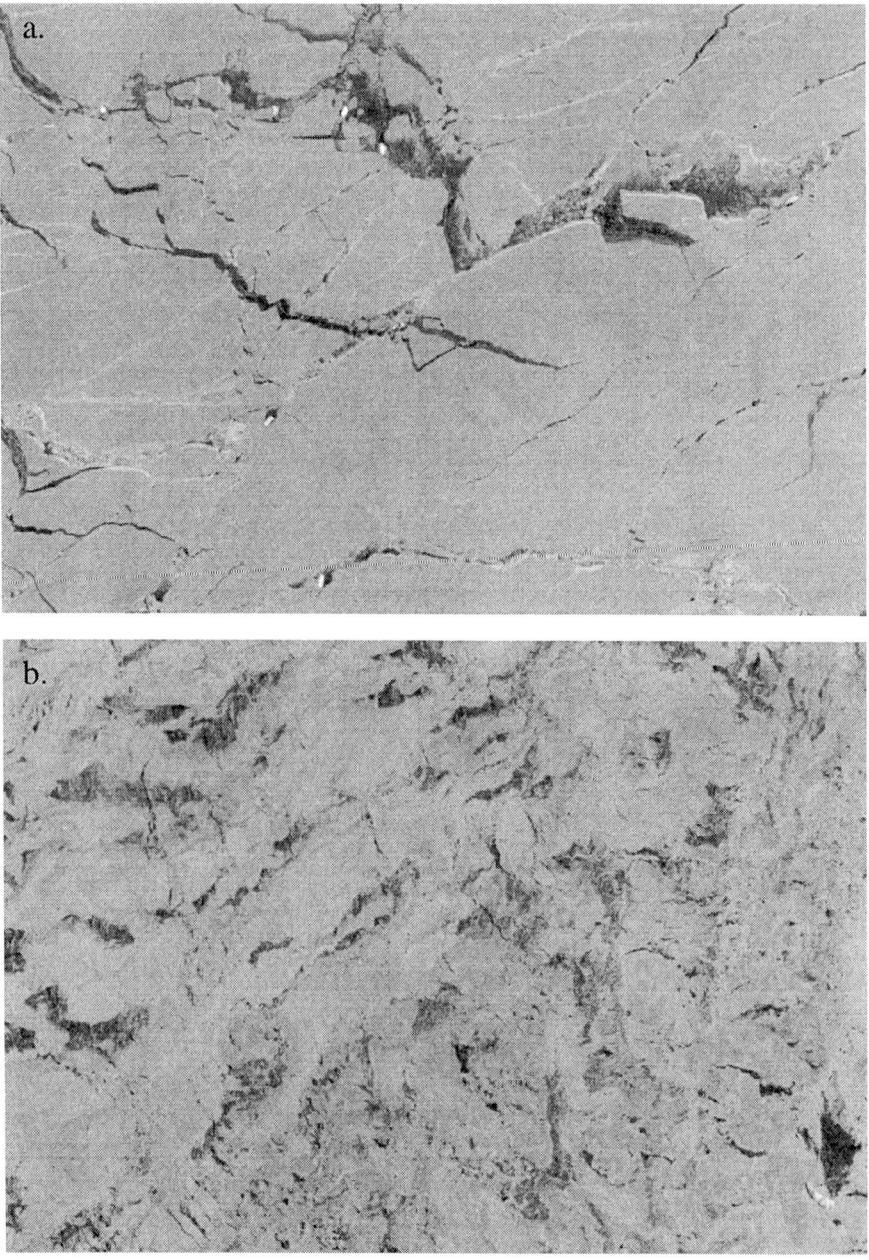

Fig. 9. Subscenes of (a) the Inner Annual Pack Ice (Orbit 11019, Frame 5049) and (b) the Outer Annual Pack Ice (Orbit 11019, Frame 4995). Each subscene has surface dimensions of 42.5 km by 62.0 km. The Inner Annual Pack Ice is the region of the oldest annual ice and the pack ice here is made up of large, well consolidated floes. On average, this zone extends as far south as 70°S. The small, very bright, angular targets are icebergs. The dark, elongated features are probably leads with a smooth, new ice cover, while the brighter, elongated features are probably covered with thin, deformed ice or wind roughened open water. Our shipborne observations show that in the Outer Annual Pack Ice the ice cover was less consolidated than that in the Inner Annual Pack Ice and extensive areas were made up of smaller floes comprising consolidated pancakes and cakes. The areas of darkest tone are probably thin, undeformed new ice.

On the other hand, the PDF for the Type W2 ice shows peaks on either side of the mean. The significance of this is unclear; it might be a consequence of under-sampling the central portion of this $\sigma°$ range or this curve may contain two different, but closely related ice types (see PDF from summer landfast ice in Plate 5a). It is noteworthy that the Type W2 ice distribution in the perennial ice zone falls between each of the coastal ice distributions, while the distributions for the other two perennial ice types (W1 and W3) occur to the right of the coastal ice types. The iceberg $\sigma°$ values are very high and near the noise ceiling of the SAR system.

In contrast to the different backscatter signatures for the coastal and perennial ice types, within the annual ice there is little difference between the outer pack ice, inner pack ice and swell-dominated ice types (Figure 12, Plate 4, and Table 3). There is no statistically significant difference between the backscatter distribution of the unperturbed outer annual pack ice and the swell-dominated outer annual pack ice. The difference between the two appears to be limited to the dramatically different texture of these two zones (Figures 9b and 10). It is noteworthy that the annual ice $\sigma°$ distributions, particularly that for the inner annual pack ice, are very similar to that of the Type W1 perennial ice. The overall mean and standard deviation values of backscatter for the annual ice types are summarized in Table 3.

4.2 Summer Sea Ice

4.2.1. General description of the summer pack ice. Two SAR image mosaics of the perennial ice acquired in late February and early March 1994 are presented in Figure 13 (see Table 1 for time of data acquisition). The ice edge is clearly visible and the pack ice is consolidated with almost 10/10 concentration. A similar ice edge location and ice concentration values (9/10) are seen in the NIC ice charts of this time period. As was indicated in Section 3.2, air temperatures over the perennial pack ice were generally below freezing and wind speeds were quite high during February. According to *Haas and Viehoff* [1994], the snow cover remained intact although isolated areas appeared grey (wet) from above. Fresh/recent snow was also seen on some floes. Grease and nilas ice were forming in open water zones.

The February 24 mosaic (Figure 13a) is representative of the general sea ice conditions and backscatter variability seen in the imagery acquired during the previous month. In contrast to the winter perennial sea ice, the summer perennial sea ice tones appear quite uniform with only minor variability, and little in the

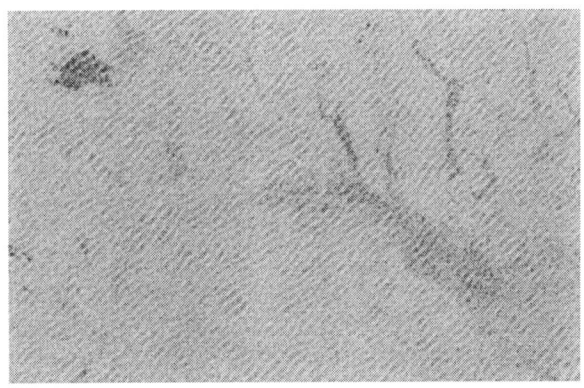

Fig. 10. A subscene (Orbit 11134, Frame 5031) showing a swell running through the Outer Annual Pack Ice on September 1, 1993. The wavelength of the swell is roughly 500m. From the R. V. *Nathaniel B. Palmer*, we estimated the wavelength to be 400 m. The subscene has surface dimensions of 16.3 km by 25.9 km.

way of large-scale floe structure is evident. The March 5 mosaic (Figure 13b) has a quite different appearance and illustrates one of only two late summer episodes when there was a dramatic change to a darker tone throughout the ice pack. During these backscatter anomalies some ice features that previously were not easily discernible become visible. This is particularly true of the outer part of the pack ice where large rounded floe features, with dimensions of as much as 15 km by 10 km, become visible, in contrast to the inner part of the pack ice where backscatter also decreases but the textural changes are more subtle. This reduction in backscatter also occurred on the continental ice margin (Figure 13b). For the purpose of determining summer backscatter variability, the ice pack was divided into coastal and perennial pack ice zones.

4.2.2. Summer coastal ice. The polynya in front of Farwell Island (Figures 13 and 14) was the sole coastal zone location from which $\sigma°$ values for the new and young ice types were derived. A landfast ice category, the ice that began to form in late winter 1993 adjacent to Farwell Island (Figure 14), is included in this zone. The SAR image time series (Figure 14) illustrates the dynamic nature of this location where the ice is moved around and out of the polynya resulting in changing textural characteristics. The size and shape of the polynya is determined in part by the tongue of landfast ice that is anchored at its seaward edge by a swarm of grounded icebergs.

Fig. 11. View aft from the starboard bridge wing of the R.V. *Nathaniel B. Palmer* shows small floes created from the formerly large/massive floes by a swell propagating through the pack ice in the western Bellingshausen Sea during late August and early September, 1993.

At times, large areas of the polynya had a nearly uniform bright tone (Figure 13a), which we attribute to wind–roughening of the water surface and a resultant increase in backscatter. On many other occasions, for example those illustrated in Figure 14, there was greater textural variability in the polynya and ice features could be easily recognized for sampling for backscatter determination. As in the winter coastal ice zone, we sampled the dark areas with quite uniform texture, i.e., new ice, and the brighter areas with greater textural variability that we believe are deformed new and young ice.

4.2.3. *Summer perennial ice.* While the SAR image mosaics such as that in Figure 13a give the impression of minimal backscatter variability in the perennial ice pack, an examination of the details of the ice in SAR subscenes reveals greater complexity (Figure 15). Based on the analysis of such subscenes, three ice types (S1, S2 and S3; where S signifies summer) have been identified and their backscatter determined. Type S1 ice, the dominant ice type, is composed of mid–range gray tones and has no characteristic form. Instead it is the "background" from which the other two ice types are distinguished. Type S2 ice is darker than Type S1 and primarily takes the form of small, rounded, floe–like structures. Type S3 ice is brighter than Type S1 and Type S2 ice and appears as small floe–like structures of variable definition. The very brightest, well–defined, angular targets are icebergs.

4.2.4. *Time series and probability density functions of summer ice.* Time series of mean $\sigma°$ values for each of the ice types in the coastal and perennial ice zones from late January to late March are presented in Figure 16. In general, there is little variability in the perennial and landfast ice $\sigma°$ values through time, but they appear to decline beginning on Day 70 (March 11), particularly from the Type S2 ice. The new and young ice type values are more variable than as those of the older ice types.

In a previous section, it was noted that there were two occasions in late summer when the tone darkened and the texture increased in the SAR image mosaics (Figure 13). These two events, on February 27 (Julian Day 58) and March 5 (Julian Day 64), are evident as sharp decreases in the backscatter time series, particularly for the perennial, landfast and iceberg categories (Figure 16). These events are illustrated in more detail in the polynya image time series (Figure 14) and a time series of subscenes of the perennial ice (Figure 17). In the polynya time series, the backscatter decrease is particularly evident on the fast ice, but is also clearly visible on the continental ice. In the perennial sea ice, there are significant changes in floe definition, with the best definition and appearance of floes occurring on March 5 (Figures 13b and 17).

Probability density functions (PDFs) for each of the ice types in the two sea ice zones are presented in Plate 5. The means and standard deviation values are listed in Table 4. Data acquired on February 27 and March 5 were excluded from the calculation of the overall means and standard deviations and probability density functions as they were considered to represent anomalous conditions. The PDF for the new ice is symmetrical about the mean and has a very narrow range of possible values. This is very similar to the winter new ice (Plate 4a). The PDF for the deformed new and young ice is more complex than that of the new ice (Plate 5a); in the broadest terms it might be described as bimodal and it covers a similar range of values as its winter equivalent (Plate 4a). The $\sigma°$ values for the landfast ice were derived from the same five sites in each image. The landfast ice PDF has two distinct peaks (Plate 5a) due to the fact that it comprises two different ages of ice as a consequence of the pattern of ice development.

The mean $\sigma°$ values for the perennial ice types are distinct from each other (Table 4). Despite differences in the height and width of the curves, all three PDFs are roughly symmetrical about their means (Plate 5b). The extent of overlap between the tails of the distributions, particularly between Type S1 and Type S3 ice, demonstrates a greater ambiguity in the nature of the radar backscatter from summer sea ice. Regardless of this ambiguity, the backscatter signatures of the three summer perennial ice types are higher than those observed in the winter perennial ice zone. As in the winter, the icebergs have the strongest backscatter, but the summer backscatter is slightly lower than the winter backscatter.

5. DISCUSSION

5.1. Winter Annual Ice

The backscatter from the annual ice in the Bellinghausen Sea is roughly 4 dB and 7 dB higher than the ice types identified by *Kwok and Cunningham* [1994] as deformed and undeformed first–year ice in the Beaufort Sea, Arctic Ocean, and almost identical to that of deformed and undeformed Beaufort Sea multiyear ice. The backscatter signatures of the Bellinghausen annual ice are almost identical to those of rough first–year ice in the Weddell Sea [*Drinkwater et al.*, 1995]. The high $\sigma°$ values are consistent with our observations of widespread deformation and rough ice throughout the annual ice cover. Rough ice surfaces resulting from ridging and rafting cause strong scattering from the angular, tilted surfaces [*Onstott*, 1992; *Drinkwater et al.*, 1995; *Jeffries et al.*, 1995a]. While it is certain that the extensive deformation features observed in the pack ice contribute to the high backscatter, it is unlikely that they alone account for the backscatter being greater than

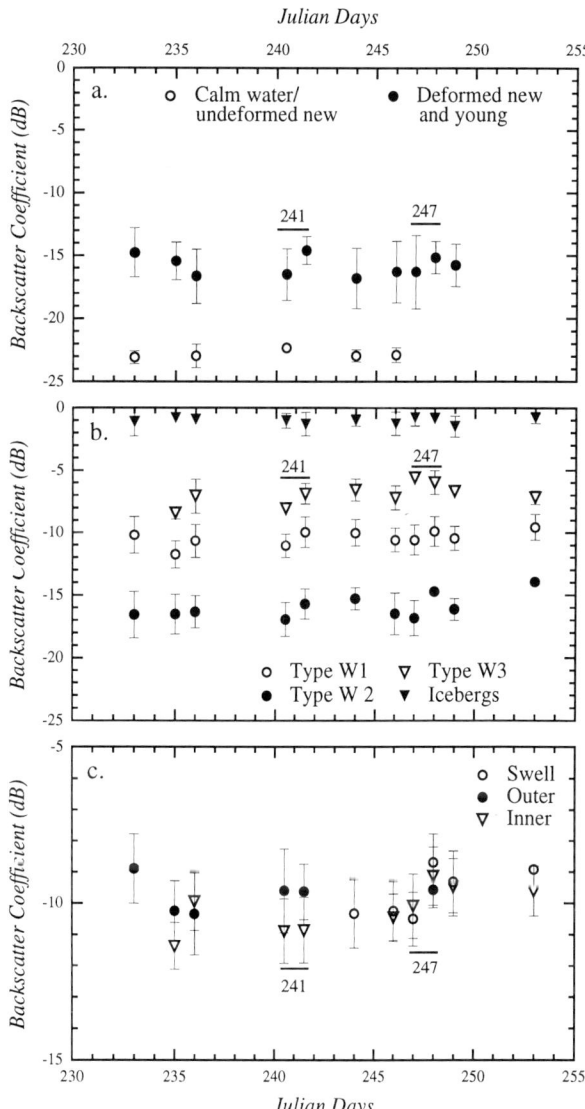

Fig. 12. Time series of backscatter coefficients (mean ± 1 standard deviation) for the different ice types identified in (a) the coastal zone, (b) the perennial pack ice, and (c) the annual pack ice in winter in the Bellinghausen Sea. The horizontal lines marked 241 and 247 denote pairs of values for the same ice type from ascending and descending passes on the same day that have been offset to avoid overlay and crowding of the symbols.

that of ice of similar age in the Arctic. An additional factor in determining the backscatter is seawater flooding of the ice surface.

Field investigations with radar scatterometers show that seawater flooding and the resultant slush and brine

almost identical to those derived for the annual pack ice from the ERS-1 SAR images [*Jeffries et al.*, 1995a]. Laboratory and numerical modelling studies of the effects of flooding on backscatter indicate that the high dielectric contrast between the dry snow and the slush, sometimes compounded by a rough surface created by brine wicking upward into the snow cover, leads to strong backscatter [*Lytle et al.*, 1993, 1996; *Fung et al.*, 1994].

Despite the age gradient between the ice edge and the southern boundary of the annual ice, there is very little difference in σ° values between the outer and inner annual pack ice. This probably reflects the fact that the surface roughness is largely independent of the age of the ice and flooding may be present, to some degree, on any ice type. Even in its earliest stages of development, when pancake formation and consolidation is common [*Lange et al.*, 1989], antarctic sea ice is rough due to the upturned edges of the pancakes. These upturned edges are a source of strong backscatter, which can be enhanced by wave washing of the ice surface [*Drinkwater et al.*, 1995]. We have observed that flooding is common as pancakes are rafted, with the latter further increasing the surface roughness. This roughness is preserved as the pancakes consolidate into cakes and ultimately large, older floes where, together with further flooding, they will contribute significantly to the backscatter.

5.2. Winter Perennial Ice

5.2.1. Type W1 ice. According to the NIC ice charts, at the end of summer 1993, the perennial ice zone had an extensive, high concentration ice cover. These same charts show that these high ice concentrations persisted through to late winter; thus, we conclude that the 1993 winter perennial ice zone contained a large amount of multiyear ice. Since the Type W1 ice is the most common in the winter perennial ice zone, it is probable that it represents multiyear ice.

One of the criteria for visual identification of multiyear floes as part of our hourly ice observation program aboard the R. V. *Nathaniel B. Palmer* was the presence of an obviously deeper snow cover than occurs on the first-year ice. This is the result of the survival of the snow cover in summer and further snow accumulation as the multiyear ice ages. *Lytle et al.* [1996] used a similar criterion for identifying multiyear floes in the Weddell Sea. The amount of flooding on the few multiyear floes that we have investigated in winter in

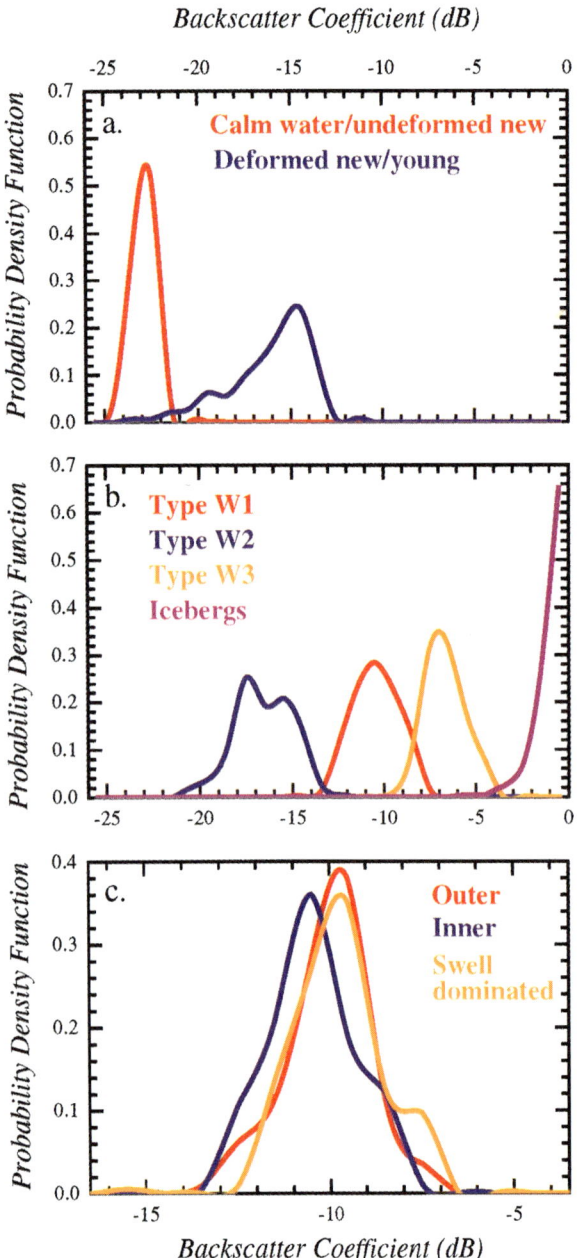

Plate 4. Probability density functions of σ° values created from the sample means of the different winter ice types identified in (*a*) the coastal zone, (*b*) the perennial pack ice, and (*c*) the annual pack ice.

wicking into the overlying snow significantly increase the backscatter [*Lytle et al.*, 1990, 1996; *Hosseinmostafa et al.*, 1995]. Backscatter values obtained with a scatterometer mounted aboard the R.V. *Nathaniel B. Palmer* in late winter 1993 were

TABLE 3. Winter ice mean and standard deviation backscatter ($\sigma°$) values.

Ice type	Mean (dB)	Standard deviation (dB)	Sample size (floes)
Coastal			
Calm water/ undeformed new	-22.9	0.6	73
Deformed new/young	-16.1	2.2	144
Perennial			
Type W1	-10.4	1.3	578
Type W2	-16.5	1.5	202
Type W3	-6.7	1.1	102
Annual			
Inner	-10.5	1.2	197
Outer	-10.0	1.2	113
Swell-dominated	-9.8	1.2	196
Icebergs	-1.0	0.7	270

the Bellingshausen Sea and elsewhere in the Pacific sector of the Southern Ocean has been as great as that on the many first-year floes that we have investigated. The flooded surfaces are a source of considerable backscatter. The Type W1 multiyear ice backscatter signature is 3–4 dB lower than that in the Weddell Sea in winter 1992 (M. Drinkwater, pers. comm., 1995). This might be due to some combination of more extensive flooding and rougher ice in the Weddell Sea at that time compared to the Bellingshausen Sea one year later.

The backscatter signatures of the Type W1 multiyear ice are similar to those of deformed and undeformed multiyear ice in the Beaufort Sea, which have PDF peaks at roughly -11 to -10 dB [*Kwok and Cunningham*, 1994]. However, the processes respons-ible for the arctic multiyear ice backscatter signatures are quite different from those for the antarctic ice. The complete melting of the snow cover and the resultant puddling and melt pond formation on arctic floes in summer leads to desalination of the upper portion of the floe due to freshwater flushing; this results in significant volume scattering and increased returns to the radar in winter [*Tucker et al.*, 1992; *Hallikainen and Winebrenner*, 1992]. The snow cover on antarctic sea ice remains largely intact during the summer and there is no ice surface manifestation of melting (melt ponds). Rather, the snow pack undergoes some internal melting during periods of near to above freezing temperatures producing ice layers, percolation columns and, in some cases, superimposed ice. Saline slush layers at the ice surface persist throughout the summer so the upper layer of the antarctic multiyear ice does not undergo extensive freshwater flushing. Thus, the $\sigma°$ values from this ice are primarily from the snow/ice or snow/slush interface and are not the consequence of volume scattering from the bubbly upper layer of the ice, as is the case for multiyear ice in the Arctic [*Hosseinmostafa et al.*, 1995].

5.2.2. *Type W2 ice*. The smaller size and more rounded appearance of the Type W2 ice floes initially suggests that they might be old floes that have been worn down as they grind around in the pack ice. However, if they are old floes, it would be reasonable to expect them to have much higher backscatter for the reasons discussed in the previous section. Consequently, we believe the Type W2 ice is first-year ice at an early stage of development, since the backscatter PDF is intermediate between that of the calm water/undeformed nilas and deformed new/young ice in the coastal zone. These backscatter characteristics suggest that the Type W2 ice is young, relatively thin first-year ice that has not been as extensively deformed as similar ice in the more dynamic coastal zone. Radar backscatter data acquired from the R.V. *Nathaniel B. Palmer* in the annual pack ice zone in late winter 1993 showed that 30 cm thick undeformed sheet ice, for example, can have $\sigma°$ values as low as -20 to -25 dB [*Jeffries et al.*, 1995a]. These values fall within the left tail of the Type W2 ice backscatter distribution (Plate 4b). The occurrence of extensive areas of Type W2 young ice at all times during the study period suggests that there may be considerable new and young ice growth, and, thus, significant heat and salt fluxes, in leads that open frequently in the perennial ice cover on the continental shelf.

5.2.3. *Type W3 ice*. Type W3 ice occurs in much smaller amounts than the other winter perennial ice types. Its high backscatter signature suggests that it is either completely flooded and/or extremely deformed. Thus, it might represent either occasional rubble fields or remnants of the very oldest multiyear ice.

5.3. Summer Perennial Ice

5.3.1. *Type S1, S2 and S3 ice*. The Type S1 ice is the most common of the summer perennial ice types and as such it probably represents a combination of multiyear and second-year ice that have been part of the perennial pack ice since the previous summer, and thick first-year ice that formed during the previous winter. The backscatter signature of the S1 ice is

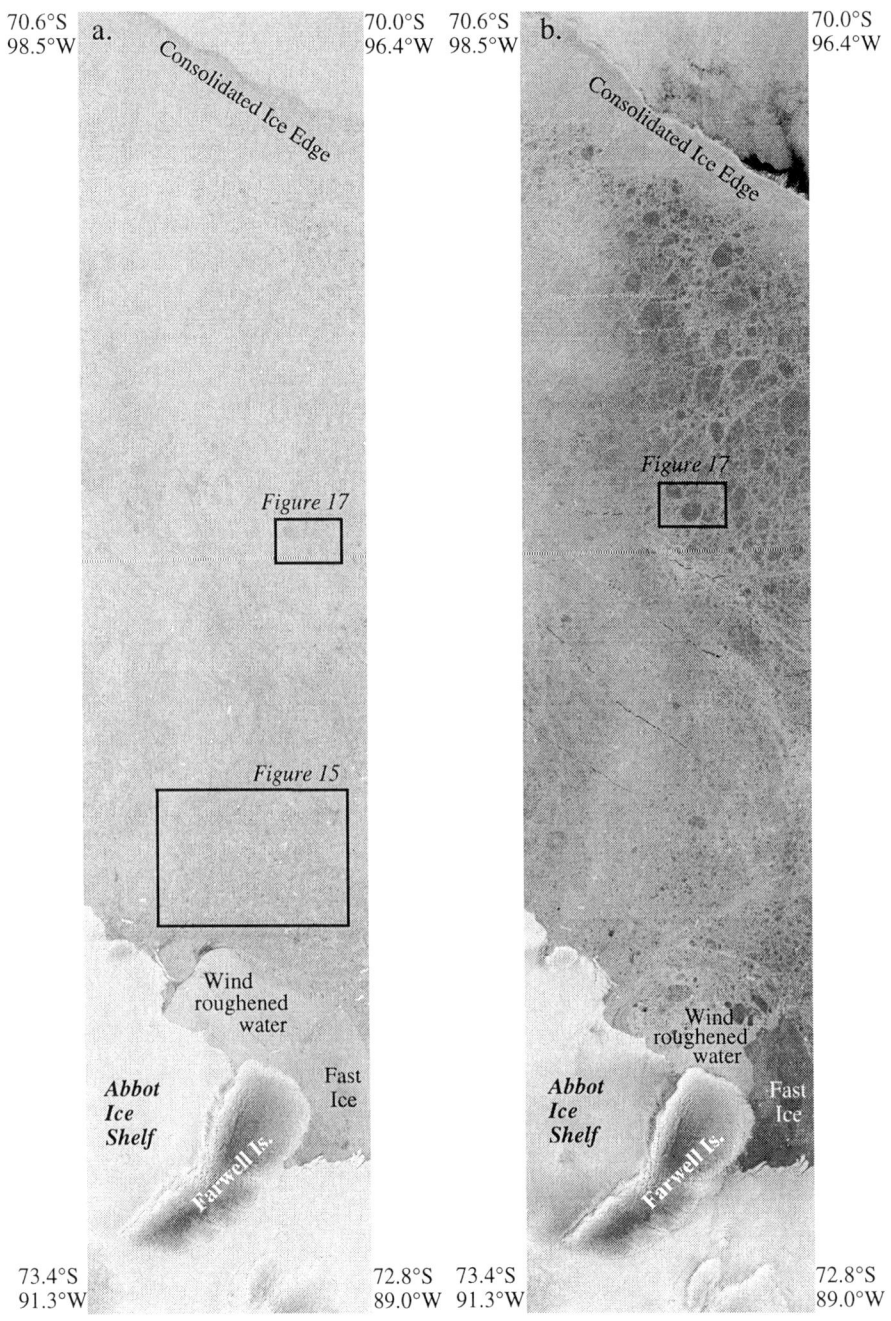

Fig. 13. Mosaics of scenes of the summertime Bellingshausen Sea perennial pack ice on February 24 (*a* – Orbit 13650) and March 5 (*b* – Orbit 13779). Each image swath has surface dimensions of approximately 100 km by 440 km. Note the difference in the backscatter intensity and variability between the two data sets which were acquired at the same time of day. The large rectangle in the February 24 mosaic identifies the location of a subscene presented as Figure 15. The small rectangle in each mosaic identifies the location of subscenes presented in Figure 17. The subscenes are of the same parcel of sea ice which has drifted to the west between February 24 and March 5.

similar to that of second–year (multiyear) ice in summer 1992 in the Weddell Sea [*Drinkwater et al.*, 1994]. During August and September 1993, approximately 18% of the annual sea ice floes sampled had negative freeboards. We speculate that a similar or greater proportion of the winter perennial ice was flooded. In the Weddell Sea, comparable values (8–40%) for negative freeboards and extent of seawater flooding have been measured in the winter pack ice [*Ackley et al.*, 1990; *Eicken et al.*, 1995], and even higher values have been estimated in late summer/early fall [*Lytle and Ackley*, 1996]. This increase in surface flooding from winter to summer is probably brought about by melting at the base of the floes which thins the ice and effectively increases the snow load. It is reasonable to assume that a similar increase in the areal extent of seawater flooding of the snow/ice interface takes place in the summer perennial ice zone in the Bellingshausen Sea with an accompanying increase in area averaged $\sigma°$ values. This argument assumes that the wetness of the late winter (August and September) and mid–to–late summer (February and March) snow covers are not sufficiently different to drastically alter the penetration depth of the radar. It may be that low relative humidity [*Andreas and Ackley*, 1982], high wind speeds and generally below–freezing temperatures [*Haas and Viehoff*, 1994] prevent the elevation of the snow wetness to such high levels except for short periods of time. During the late summer in the Weddell Sea, *Drinkwater and Lytle* [in press] found that 50% of the sea ice was covered with a 0.05–0.30 m slush layer overlain by 0.20–0.50 m dry snow. Two very short periods of decreased $\sigma°$ values were found in the summer SAR imagery and are linked with warm, moist conditions and, hence, increased snow wetness (see Section 5.3.2).

The winter snow cover characteristics on annual sea ice have been examined in the Bellingshausen, Amundsen and Ross seas [*Sturm et al.*, 1995, this volume]. The degree of iciness (percentage of total snow cover consisting of ice lenses, icy layers, percolation columns, and melt–grain clusters) for the southernmost stations remains fairly constant at 2 to 10%; iciness increases both towards the ice edge and as the season progresses. These melt–freeze features are a consequence of the alternation of warm marine cyclones, often accompanied by rain or above–freezing temperatures, with prolonged periods of cold weather. It is reasonable to assume that these features are more numerous and extensive in the snow cover on the summer perennial sea ice as it continues to undergoes melt–freeze cycles

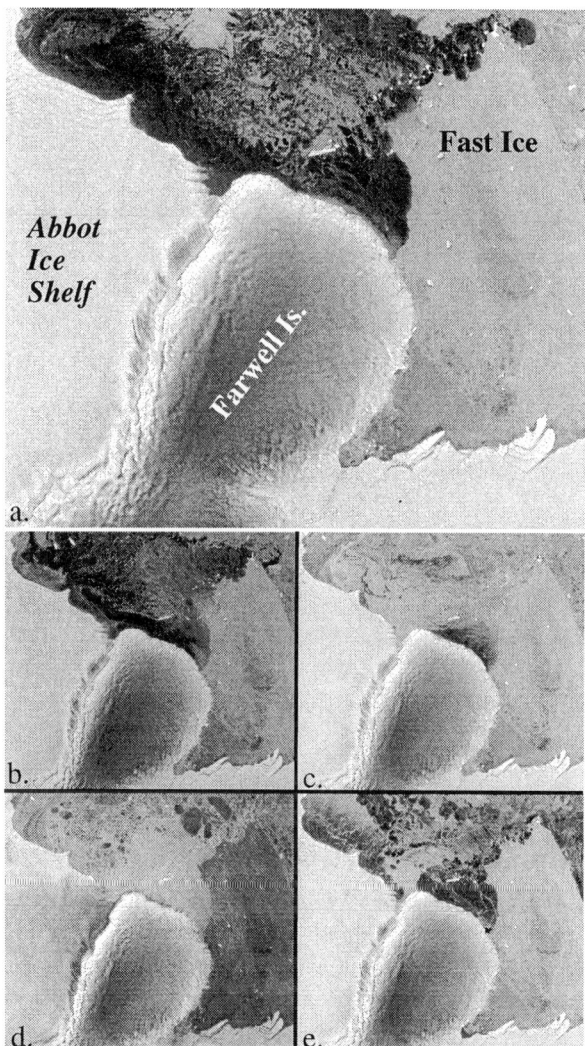

Fig. 14. Time series of subscenes of a polynya adjacent to Farwell Island in austral summer 1994 illustrating the dynamic nature of the ice conditions and backscatter variability of new and young ice in this coastal zone. The dates for each subscene are (*a*) February 21, (*b*) February 27, (*c*) March 2, (*d*) March 5, and (*e*) March 11. Each subscene has surface dimensions of 70.7 km by 60.9 km.

throughout the summer. On the Greenland Ice Sheet, ice lenses and percolation columns in the percolation zone have proven to be effective microwave backscatterers at C–band due to the roughness of their upper surfaces [*Rignot et al.*, 1993; *Fahnestock et al.*, 1993; *Zabel et al.*, 1995]. It is possible that the melt–freeze features within the antarctic snow cover contribute to the overall high $\sigma°$ values of the sea ice throughout the year but

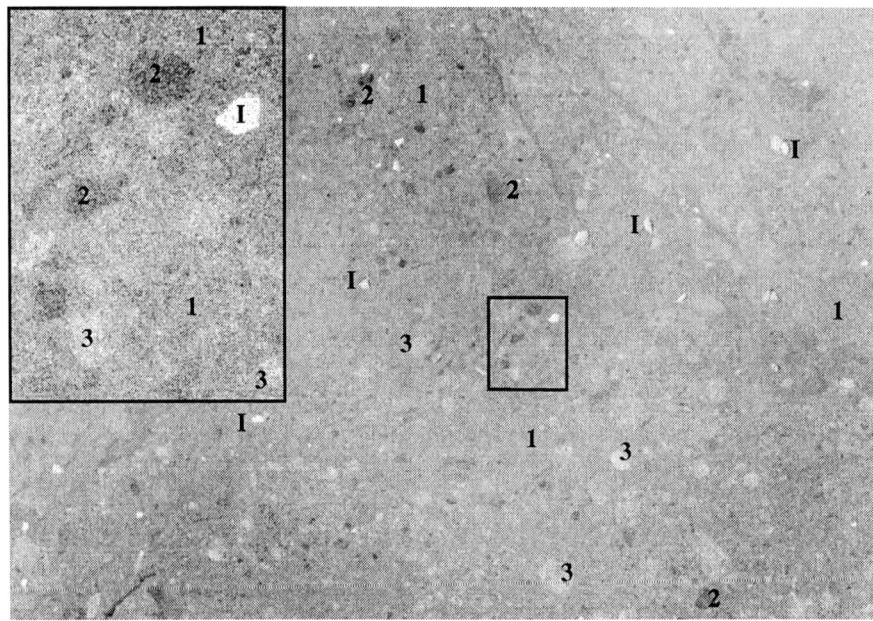

Fig. 15. A subscene of the summer ice pack (Orbit 13650, Frame 5679) shows some details of ice floes and backscatter variability. Numbers 1, 2 and 3 denote Ice Types S1, S2, and S3 as discussed in the text. Icebergs are identified by I. The subscene has surface dimensions of 42.5 km by 62.0 km. The insert in the upper left corner is an enlargement of the small boxed area, with surface dimensions of 6.7 km by 4.9 km, and shows additional details of the ice.

especially in summer when they are more numerous and extensive. The observation of superimposed ice at the surface of floes by the end of late summer 1992 in the western Bellingshausen Sea (Jeffries et al., 1997), suggests that, on some occasions, snow melt reaches an advanced stage with liquid water percolating down to the snow/ice interface. This superimposed layer is rough on the centimeter length scale and so may cause significant microwave scattering [*Onstott and Gogineni*, 1985]. The same explanations for increased summer $\sigma°$ values can also be applied to the Type S2 and S3 ice types. Of these three ice types, Type 2 ice is probably the thinnest to survive through the summer.

An increase in $\sigma°$ values from winter to summer in the antarctic sea ice is the opposite of what occurs in the Arctic. During the melt onset on multiyear arctic sea ice, $\sigma°$ values decrease initially by 6–9 dB from their winter values, as a consequence of warmer temperatures and the appearance of liquid water in the overlying snow cover as it melts away. They then stabilize, later in the summer, at values 4–5 dB below winter levels once the snow is completely removed [*Winebrenner et al.*, 1994].

A backscatter PDF for arctic multiyear ice is presented in Plate 5c. This PDF is constructed from two sets of sample backscatter means. The first set is taken from floes located in the Chukchi Sea (75°N, 147–156°W) and extends from June 29 to July 27, 1992 (mean backscatter -14.30±1.1 dB, n=68). The second set of floes was located in the vicinity of Meighen Island (81°N, 99–114°W) and extends from July 14 to August 26, 1992 (mean backscatter -13.96±1.6 dB, n=153). Since there is no statistical difference between these two data sets, they were pooled (overall mean -14.06±1.5 dB, n=221). These data represent $\sigma°$ values taken after the snow cover had melted away completely and the $\sigma°$ values had stabilized. They compare well with the -9 to -13 dB late summer (August and early September) $\sigma°$ values for thick ice in the Chukchi and Beaufort Seas in 1991 [*Winebrenner et al.*, 1996].

5.3.2. Backscatter anomalies. Apart from two notable anomalies, there is little variation in the summer $\sigma°$ values of the three perennial sea ice types for the days for which there are data. That the $\sigma°$ values quickly recovered to values similar to those prior to each anomaly indicates that the effect was not permanent. Furthermore, the anomalies included the icebergs and the continental ice (Figures 13 and 14), suggesting that the cause of the backscatter decrease is related to snow cover changes rather than sea ice changes. The backscatter decrease on each occasion and for each ice type is 3–5 dB. A backscatter anomaly of this magnitude has been observed in radar scatterometer data in the winter in the Weddell Sea when air temperatures increased rapidly from -20 °C to 0 °C during the passage of a warm front [*Drinkwater et al.*, 1995]. *Winebrenner et al.* [1994] observed similar "perturbations" in the $\sigma°$ values of arctic sea ice at the beginning of spring when temperatures rose to near or just above 0 °C for 2–3 days, causing a decrease in $\sigma°$ values, and then temperatures decreased again causing a corresponding increase in the $\sigma°$ values to previous levels. We attribute the summer backscatter anomalies to a similar cause.

Between 2 and 5 March 1993, an intense low pressure system developed roughly 200 km NNW of the study area causing a strong northerly flow over the perennial pack ice in the SAR images (Figure 17). Under these circumstances, warm, moist air was brought into the perennial ice zone and onto the continental ice. A warm front over the area of interest is clearly marked on the Australian Bureau of Meteorology Southern Hemisphere surface chart for March 4, 1200 UT (not shown). The air temperatures over the perennial ice were probably below freezing during much of February and March (see Section 3.2) and the introduction of this marine air represented a temporary warming, perhaps to above freezing temperatures, in this area. This probably caused an increase in the water content of the snow and a resultant decrease in scattering [*Stiles and Ulaby*, 1980]. Subsequently, the low pressure system weakened and a cold front moved through the study area. This cooled the snow cover and reduced its water content through freezing, thereby reducing the signal losses and allowing the backscatter to recover to its previous values. The February 27 backscatter anomaly (Figure 17) is not as strong as the later anomaly, perhaps because the westerly flow conditions that prevailed at the time did not introduce as much warm, marine air as the northerly air flow that occurred on March 5, or because the February 27 data represent a partial recovery from the maximum effects of the passing weather system.

5.3.3. End of season backscatter decline.

The backscatter anomalies affecting the sea ice and glacier ice were quite brief and the backscatter recovered to its former value once cooler conditions returned. At this time of year, late summer, even cooler conditions

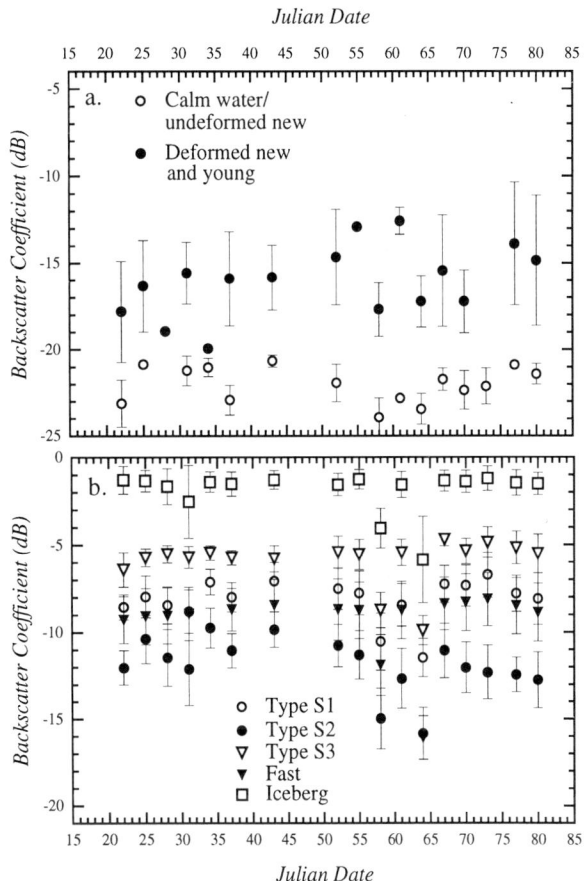

Fig. 16. Time series of backscatter coefficients (mean ± 1 standard deviation) for the different summer ice types identified in (*a*) the coastal zone and (*b*) the perennial pack ice.

are imminent as the seasons change. In the Weddell Sea, it has been observed that the onset of freeze–up in autumn results in a decrease in backscatter from the sea ice [*Drinkwater et al.*, 1994]. The decline in sea ice backscatter beginning on about Day 70 (March 11) in the Bellingshausen Sea may also represent the onset of freeze–up. This decline might be a consequence of the freezing of slush at the base of the snow cover and resultant signal loss due to high brine volumes in the newly formed saline ice [*Drinkwater and Lytle*, in press]. This seems quite plausible in view of the fact that the iceberg backscatter, which is unlikely to be affected by saline ice, does not decline at this time.

5.4. Coastal ice

Backscatter variability in the coastal zone was examined not only because this is a distinctive zone

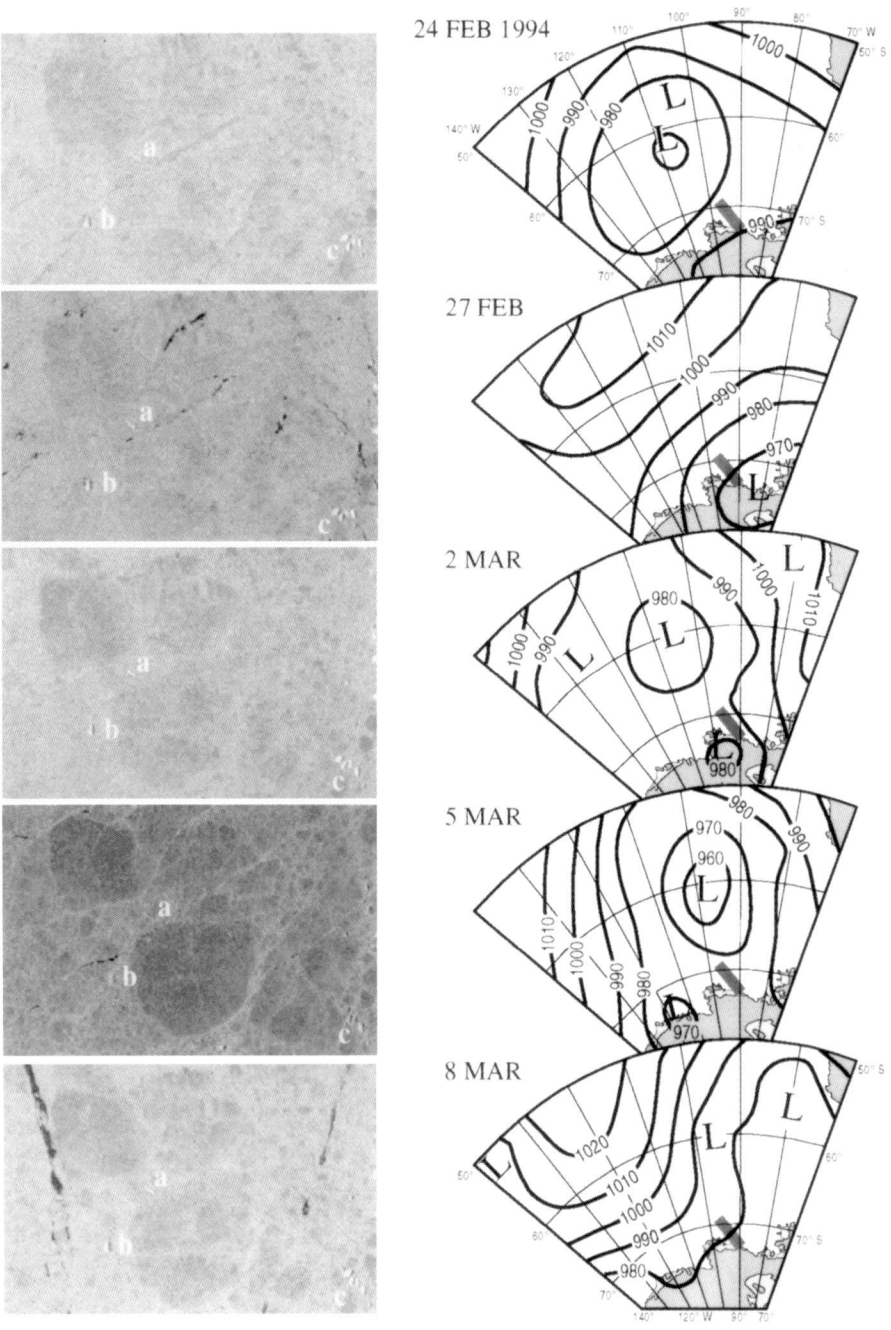

Fig. 17. A series of subscenes of the perennial pack ice illustrating some of the details of the backscatter changes in late February and early March when there were two brief episodes (February 27 and March 5) when backscatter decreased sharply and floe definition improved. Each subscene contains the same pieces of ice, including three icebergs (a, b and c) which are identified to aid in feature recognition. Each subscene has surface dimensions of 13.7 km by 21.4 km. The meteorological maps are excerpted from the Australian Bureau of Meteorology Southern Hemisphere charts (surface analysis). The shaded rectangle on each meteorological map represents the location and areal extent of the summer ERS-1 SAR orbits (see also Plate 1 and Figure 13).

within the sea ice cover, but also because it offered opportunities for sampling new and young ice types. Such ice types occur in leads in the pack ice, but because of the size of the leads they are more difficult to sample than the large coastal polynyas. Furthermore, little other than new and young ice occur in the flaw leads and polynyas at the coast; thus, ice type identification is more reliable. *Kwok and Cunningham* [1994] described this problem in their study of arctic winter sea ice backscatter variability in which they derived backscatter statistics for a generic "lead ice" comprising the many different ice types and surface characteristics. The backscatter signatures of the Bellingshausen Sea coastal ice types are probably an accurate representation of ice in leads in the perennial and annual ice.

In summer, apart from occasional periods of very strong winds when the flaw leads and polynyas are characterized by uniform high backscatter due to wind roughening of the water, it is clear that there is ice growth in the coastal zone. There might be melting in the snow cover further offshore where the ice is subject to a considerable marine influence, but close to shore the continental influence will be greater and temperatures are likely to be sufficiently low for ice formation. When taken as a whole, the backscatter signatures of the summer coastal ice types are not statistically different from those in winter. The differences in the sample means that do exist, such as that between the summer and winter deformed new and young ice, are probably more a function of the dynamic nature of this environment and resultant deformation and effects on surface roughness than they are due to other environmental influences such as temperature effects on snow and ice properties. The apparently limited influence of possible environmental variability, other than the winds which influence ice motion and deformation, is evident in the low backscatter variability of the landfast ice.

The peak of the backscatter PDF for arctic lead ice is -17 dB [*Kwok and Cunningham*, 1994]. The antarctic winter calm water/undeformed new ice and deformed new/young ice categories bracket the arctic lead ice backscatter distribution. Regardless of polar region, bare, undeformed new ice will have very low $\sigma°$ values caused by specular reflection away from the satellite coupled with signal absorption due to high liquid brine volumes. Backscatter values from young ice (<30 cm) increase as the ice either becomes deformed (rafting and ridging) or thickens thermodynamically and cools, resulting in some brine loss [*Hallikainen and*

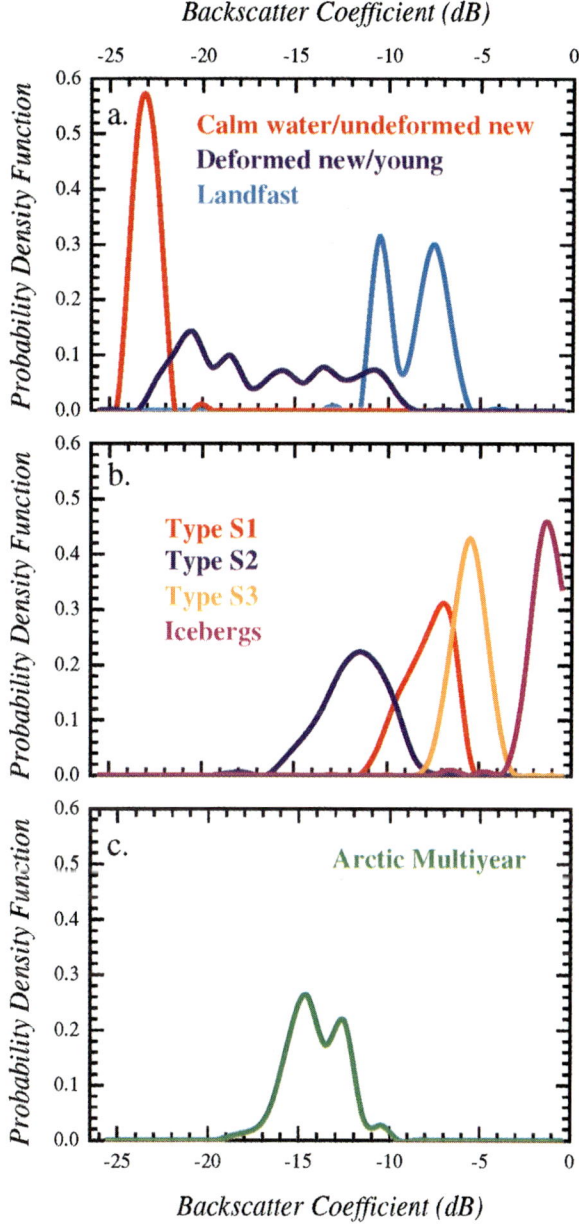

Plate 5. Probability density functions of $\sigma°$ values created from the sample means of the different summer ice types identified in (*a*) the coastal zone and (*b*) the perennial pack ice. Anomalous $\sigma°$ values from February 27 and March 5 are excluded from the PDFs. A probability density function of $\sigma°$ values from arctic multiyear ice in late summer 1992 is shown in (*c*).

Winebrenner, 1992]. Once a snow cover becomes established on the ice, the main source of backscatter may be from the formation of slush at the snow/ice

TABLE 4. Summer ice mean and standard deviation backscatter (σ°) values.

Ice type	Mean (dB)[a]	Standard deviation (dB)	Sample size (floes)
Coastal			
Calm water/ undeformed new	-23.2	0.5	27
Deformed new/young	-16.9	3.9	141
Landfast	-8.7	1.4	70
Perennial			
Type S1	-7.8	1.2	297
Type S2	-11.7	1.7	237
Type S3	-5.5	0.8	86
Icebergs	-1.5	0.9	303

[a] Anomalous backscatter values from February 27 and March 5 are excluded from these statistics.

interface [*Hallikainen and Winebrenner*, 1992] or it may continue to be from ice surface roughness if the snow remains cold and dry and is, therefore, transparent to the radar [*Rott and Mätzler*, 1987]. The similarity between the arctic and antarctic backscatter signatures at these early stages of ice development arises because the ice surfaces are subject to similar influences, unlike the later stages of ice development when phenomena such as flooding and a perennial snow cover affect the antarctic, but not the arctic, sea ice.

6. SUMMARY AND CONCLUSION

The seasonal variability of spaceborne radar backscatter from sea ice in the Bellingshausen Sea has been described for the first time. The data have been compared with those available for Weddell Sea ice and arctic sea ice, and the results have been explained on the basis of the current understanding of differences between arctic and antarctic sea ice and snow cover properties and processes and their role in affecting backscatter. The winter and summer backscatter signatures for the perennial and annual sea ice in the Bellingshausen Sea are summarized in Figure 18. The main findings and conclusions of this study are:

1. First-year ice in the Bellingshausen Sea has a backscatter signature similar to that of first-year ice in the Weddell Sea. Both are higher than that of winter first-year ice in the Arctic Ocean, probably due to the combined effects of surface roughness and seawater flooding of the ice surface and brine wicking into the overlying snow and ice lenses and percolation columns within the snow cover. Flooding is much less common in the Arctic Ocean.

2. In the Bellingshausen Sea, the winter backscatter signatures for multiyear and thick first-year sea ice are similar. There is also no obvious difference between multiyear and thick annual ice backscatter signals in the Weddell Sea. The similarity between the backscatter signatures of these two different ages of ice is probably due to the strong influence of ice surface roughness and flooding of the snow/ice interface of the floes on both ice types. This contrasts with the winter Arctic Ocean data in which multiyear ice is easily distinguishable from thick first-year ice types. Although it appears to be difficult to differentiate between thicker first-year and multiyear ice in the Antarctic, the fact that it is possible to identify new and young ice, i. e. thinner first-year ice types, (see point 5 below), means that SAR may be applied to the study of the ice thickness distribution and the heat and mass balance of the pack ice in this region.

3. Summer multiyear ice in the Bellingshausen Sea has a backscatter signature similar to that in the Weddell Sea and these are 2–5 dB higher than the winter backscatter signatures. This is probably due to the increase in the areal extent of slush at the snow/ice interface in the summer which would produce higher σ° values. The backscatter values may be enhanced by

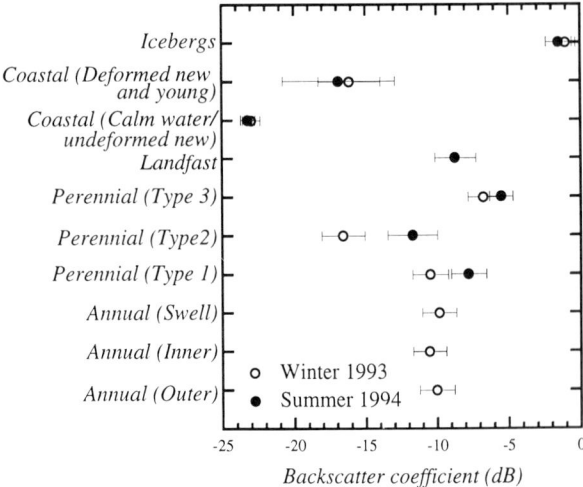

Fig. 18. Summary of backscatter statistics (mean ± 1 standard deviation) for each of the ice types identified in the winter and summer pack ice in the Bellingshausen Sea. Anomalous σ° values from February 27 and March 5 are excluded from the summer values.

scattering from the rough surfaces of numerous and extensive ice lenses and percolation columns within the summer snow cover. Backscatter from the antarctic summer multiyear ice is much higher than that in the Arctic. In the Arctic, much of the snow melts completely and puddles on the ice surface. In late summer, these bare and puddled ice surfaces have lower $\sigma°$ values (4–5 dB) than their winter snow-covered counterparts.

4. The antarctic summer backscatter signatures of all the perennial ice types are higher than the winter signatures and they do not vary greatly throughout the summer unless they are affected by some extreme event. In this study, a significant decrease in backscatter associated with an intense cyclone and influx of warm, marine air over the perennial pack ice to the edge of the continent indicates that SAR data could be used for monitoring atmospheric processes and variability.

5. The backscatter signatures of new and young ice in antarctic polynyas, and by implication those in leads in the pack ice, are similar in winter and summer and quite different from the older, thicker ice types. Consequently, SAR data have considerable potential for use in studies of the heat and salt fluxes associated with ice growth on polynyas and leads, as is currently the case in the Arctic.

6. Icebergs are easily detected in SAR images due to their very high backscatter signature. SAR could be used for iceberg censuses and tracking, which would provide important information related to the mass balance of the antarctic ice sheet, ocean currents and, where icebergs ground on the sea–floor, bathymetry.

7. Ocean swells penetrate into the pack ice for distances exceeding 400 km from the ice edge. Penetration of waves into the pack ice has also been observed in the Weddell Sea and in the marginal ice zones of the Arctic.

8. The results presented in this paper represent only one summer and one winter in the Bellingshausen Sea. Nevertheless, there are many similarities with Weddell sea ice backscatter signatures. Ideally, a baseline data set of sea ice backscatter signatures in all regions of the Southern Ocean in all seasons over the course of two to three years is required in order to identify any regionally unique signatures and the processes responsible for them. In combination with field validation campaigns, this would contribute to a better understanding of the sea ice properties and processes that affect radar backscatter and how SAR can be best applied to remote sensing studies of air–ice–ocean interactions.

Acknowledgments. This work was supported by the National Science Foundation grant OPP–9117721. We thank Shusun Li and Chuah Teong Sek for their assistance in radiometrically calibrating and post-processing the ERS–1 SAR images. Our understanding of SAR backscatter and antarctic sea ice properties and processes has benefited from numerous discussions with Mark Drinkwater, Shusun Li, Vicki Lytle, Rob Massom and Willy Weeks. All of the SAR scenes are copyrighted to ESA and were acquired through ESA ERS–1 Project US1–1c. Klaus Reiniger of the German Processing and Archiving Facility provided information on the late winter 1993 SAR image data acquisition plan for GARS and supplied invaluable paper copies of the digital SAR images that were used in this study. Christian Haas and Gert Koenig-Langlo made the meteorological data from the 1994 cruise of R. V. *Polarstern* in the Bellingshausen and Amundsen Seas available to us via the Alfred Wegener Institute World Wide Web homepage. Willy Weeks provided the photograph in Figure 11. Hugh Hutchinson and Terry Hart, Australian Bureau of Meteorology, helped us to obtain the Southern Hemisphere charts used in Figure 17. Ken Schwartz made available the arctic summer multiyear ice backscatter data in Plate 5c. Edison Chouest Offshore and Antarctic Support Associates personnel contributed to the success of the late winter 1993 field investigation. Stan Jacobs had editorial responsibility for this paper.

REFERENCES

Ackley, S. F., M. A. Lange and P. Wadhams, Snow cover effects on antarctic sea ice thickness, *Sea Ice Properties and Processes*, edited by S. F. Ackley and W. F. Weeks, CRREL Monogr. 90-1, 16–21, 1990.

Allison, I., R. E. Brandt and S. G. Warren, East Antarctic sea ice: Albedo, thickness distribution and snow cover, *J. Geophys. Res., 98*, 12417–12429, 1993.

Allison, I. and A. P. Worby, Seasonal changes in sea ice characteristics off east Antarctica, *Ann. Glaciol., 20*, 195–201, 1994.

Andreas, E. L. and S. F. Ackley, On the differences in ablation seasons of Arctic and Antarctic sea ice, *J. Atmos. Sci., 39*, 440–447, 1982.

Carsey, F. D., Microwave observations of the Weddell Polynya, *Mon. Weather Rev., 108*, 2032–2044, 1980.

Carsey, F. D. (editor), *Microwave Remote Sensing of Sea Ice*, Geophysical Monograph 68, 462 pp., AGU, Washington, D.C., 1992.

Carsey, F. D., B. Holt, S. Martin, L. McNutt, D. A. Rothrock, V. A. Squire and W. F. Weeks, Weddell–Scotia Sea marginal ice zone observations from space, *J. Geophys. Res., 91*, 3920–3924, 1986.

Cavalieri, D. J. and S. Martin, A passive microwave study of polynyas along the Antarctic Wilkes Land coast, in *Oceanology of the Antarctic Continental Shelf*, edited by S. S. Jacobs, Antarc. Res. Ser., vol. 43, pp. 227–252, AGU, Washington, D.C., 1985.

Comiso, J. C., Characteristics of Arctic winter sea ice from satellite multispectral microwave observations, *J. Geophys. Res., 91*, 975–994, 1986.

Comiso, J. C., T. C. Grenfell, M. A. Lange, A. W. Lohanick, R. K. Moore and P. Wadhams, Microwave remote sensing of the Southern Ocean sea ice cover, in *Microwave Remote Sensing of Sea Ice*, edited by F. D. Carsey, Geophysical Monograph 68, pp. 243–259, AGU, Washington, D.C., 1992.

Drinkwater, M. R. and C. Kottmeier, Satellite microwave radar– and buoy–tracked ice motion in the Weddell Sea during WWGS'92, in *Proceedings of the 1994 International Geoscience and Remote Sensing Symposium (IGARSS'94)*, Catalog No. 94CH3378-7, pp. 153–155, Inst. of Electr. and Electron. Eng., New York, 1994.

Drinkwater, M. R., D. S. Early and D. G. Long, ERS–1 investigations of Southern Ocean geophysics using combined scatterometer and SAR images, *Proceedings of the 1994 International Geoscience and Remote Sensing Symposium (IGARSS'94)*, Catalog No. 94CH3378-7, pp. 165–167, Inst. of Electr. and Electron. Eng., New York, 1994.

Drinkwater, M. R., R. Hosseinmostafa and S. P. Gogineni, C–band backscatter measurements of winter sea ice in the Weddell Sea, *Int. J. Rem. Sens., 16*, 3365–3389, 1995.

Drinkwater, M. R. and V. I. Lytle, ERS–1 SAR and field–observed characteristics of austral fall freeze-up in the Weddell Sea, Antarctica, *J. Geophys. Res.*, in press.

Eicken, H. and M. A. Lange, Development and properties of sea ice in the coastal regime of the southeastern Weddell Sea, *J. Geophys. Res., 94*, 8193–8206, 1989.

Eicken, H., H. Fischer and P. Lemke, Effects of the snow cover on antarctic sea ice and potential modulation of its response to climate change, *Ann. Glaciol., 21*, 369–376, 1995.

Fahnestock, M., R. Bindschadler, R. Kwok and K. Jezek, Greenland ice sheet surface properties and ice dynamics from ERS-1 SAR imagery, *Science, 262*, 1530–1534, 1993.

Fung, A. K., S. Tjuatja, S. Beaven, S. P. Gogineni, K. Jezek, A. J. Gow and D. K. Perovich, Modeling interpretation of scattering from snow–covered ice, in *Proceedings of the 1994 International Geoscience and Remote Sensing Symposium (IGARSS'94)*, Catalog No. 94CH3378-7, pp. 617–619, Inst. of Electr. and Electron. Eng., New York, 1994.

Gloersen, P., Modulation of hemispheric sea–ice cover by ENSO events, *Nature, 373*, 503–506, 1995.

Gloersen, P. and W. J. Campbell, Variations in the Arctic, Antarctic and Global sea ice covers during 1978-1987 as observed with the Nimbus 7 Scanning Multichannel Microwave Radiometer, *J. Geophys. Res., 93*, 10666–10674, 1988.

Gloersen, P. and W. J. Campbell, Recent variations in Arctic and Antarctic sea–ice covers, *Nature, 352*, 33–35, 1991.

Gloersen, P., W. J. Campbell, D. J. Cavalieri, J. C. Comiso, C. L. Parkinson and H. J. Zwally, *Arctic and Antarctic Sea Ice, 1978-1987: Satellite–Passive Microwave Observations and Analysis*, NASA Spec. Publ., SP-511, 290pp., 1992.

Gow, A. J. and W. B. Tucker, III, Sea Ice in the Polar Regions, in *Polar Oceanography, Part A, Physical Science*, edited by W. O. Smith, Jr., pp. 47–122, Academic Press, San Diego, 1990.

Haas, C. and T. Viehoff, Sea ice conditions in Bellingshausen/Amundsen Sea: Shipboard observations and satellite imagery during ANT XI/3, *Berichte aus dem Fachbereich Physik*, Alfred Wegner Institute für Polar und Meeresforschung, Report 51, 55pp., 1994.

Hallikainen, M. and D. P. Winebrenner, The physical basis for sea ice remote sensing, in *Microwave Remote Sensing of Sea Ice, Geophys. Monogr. Ser.*, vol. 68, edited by F. D. Carsey, pp. 29–46, AGU, Washington, D.C., 1992.

Haykin, S., E. O. Lewis, R. K. Rainey and J. R. Rossiter (editors), *Remote Sensing of Sea Ice and Icebergs*, 686 pp., John Wiley & Sons, Inc., New York, 1994

Hosseinmostafa, A. R., V. I. Lytle, K. C. Jezek, S. P. Gogineni, S. F. Ackley and R. K. Moore, Comparison of radar backscatter from antarctic and arctic sea ice, *J. Electromag. Waves and Appl., 9*, 421–438, 1995.

Jacobs, S. S. and J. C. Comiso, Sea ice and oceanic processes on the Ross Sea continental shelf, *J. Geophys. Res., 94*, 18195–18211, 1989.

Jacobs, S. S. and J. C. Comiso, A recent sea–ice retreat west of the Antarctic Peninsula, *Geophys. Res. Lett., 20*, 1171–1174, 1993.

Jeffries, M. O. and W. F. Weeks, Structural characteristics and development of sea ice in the western Ross Sea, *Antarct. Sci., 5*, 63–75, 1992.

Jeffries, M. O., R/V Nathaniel B. Palmer Cruise NBP 93-5: Sea ice physics and biology in the Bellingshausen and Amundsen Seas, August and September 1993, *Antarct. J. of the U.S., 29(1)*, 7–8, 1994.

Jeffries, M. O., R. A. Shaw, K. Morris, A. L. Veazey and H. R. Krouse, Crystal structure, stable isotopes $\delta^{18}O$ and development of sea ice in the Ross, Amundsen and Bellingshausen Seas, Antarctica, *J. Geophys. Res., 99*, 985–995, 1994a.

Jeffries, M. O., K. Morris, A. P. Worby and W. F. Weeks, Late winter sea ice properties and growth processes in the Bellingshausen and Amundsen Seas, *Antarct. J. of the U.S.*, 29, 11-13, 1994b.

Jeffries, M. O., K. Morris, A. P. Worby and W. F. Weeks, Late winter characteristics of the seasonal snow cover on sea ice floes in the Bellingshausen and Amundsen Seas, *Antarct. J. of the U.S.*, 29, 9-10, 1994c.

Jeffries, M. O., T. S. Chuah and K. Morris, C-band radar backscatter from Antarctic first-year sea ice: I *in situ* scatterometer measurements, *Antarct. J. of the U.S.*, 30(1-4), 24-26, 1995a.

Jeffries, M. O., S. Cushing and M. Porter, Sea ice development in the Ross, Amundsen and Bellingshausen seas revealed by analysis of ice cores in late winter 1993 and 1994, *Antarct. J. of the U.S.*, 30(1-4), 16-18, 1995b.

Jeffries, M. O., R. Jaña, S. Li and S. McCullars, Sea ice and snow thickness distributions in late winter 1993 and 1994 in the Ross, Amundsen and Bellingshausen seas, *Antarct. J. of the U.S.*, 30(1-4), 18-21, 1995c.

Jeffries, M. O., A. P. Worby, K. Morris and W. F. Weeks, Seasonal variations in the properties and structural composition of sea ice and snow cover in the Bellingshausen and Amundsen seas, Antarctica. *J. of Glaciol.*, 43 (143), 1997, in press.

Kwok, R. and G. F. Cunningham, Backscatter characteristics of the winter ice cover in the Beaufort Sea, *J. Geophys. Res.*, 99, 7787-7802, 1994.

Kwok, R., E. Rignot, B. Holt and R. G. Onstott, Identification of sea ice type in spaceborne SAR data, *J. Geophys. Res.*, 97, 2391-2402, 1992.

Lange, M. A., S. F. Ackley, P. Wadhams, G. S. Dieckmann and H. Eicken, Development of sea ice in the Weddell Sea, *Ann. Glaciol.*, 12, 92-96, 1989.

Laur, H., ERS-1 SAR calibration: Derivation of backscattering coefficient $\sigma°$ in ERS-1.SAR.PRI products, Issue 1, Rev. 0, 16 pp., European Space Agency, ESA Publication, 1992.

Liu, A. K., B. Holt and P. W. Vachon, Wave propagation in the marginal ice zone: Model predictions and comparisons with buoy and synthetic aperture radar data, *J. Geophys. Res.*, 96, 4605-4621, 1991a.

Liu, A. K., P. W. Vachon and C. Y. Peng, Observations of wave refraction at an ice edge by synthetic aperture radar, *J. Geophys. Res.*, 96, 4803-4808, 1991b.

Lytle, V. I., K. C. Jezek, S. Gogineni, R. K. Moore and S. F. Ackley, Radar backscatter measurements during the Winter Weddell Gyre Study, *Antarct. J. of the U.S.*, 25, 123-125, 1990.

Lytle, V. I., K. C. Jezek, A. R. Hosseinmostafa and S. Gogineni, Laboratory backscatter measurements over urea ice with a snow cover at Ku band, *IEEE Trans. Geosci. Rem. Sens.*, 31, 1009-1017, 1993.

Lytle, V. I., K. C. Jezek, S. Gogineni and A. R. Hosseinmostafa, Field observations of microwave backscatter from Weddell Sea ice, *Int. J. Remote Sensing*, 17, 167-180, 1996.

Lytle, V. I. and S. F. Ackley, Heat flux through sea ice in the western Weddell Sea: Convective and conductive transfer processes, *J. Geophys. Res.*, 101, 8853-8868, 1996.

Martin, S., B. Holt, D. J. Cavalieri and V. Squire, Shuttle Imaging Radar B (SIR-B) Weddell Sea ice observations: a comparison of SIR-B and Scanning Multichannel Microwave Radiometer sea ice concentrations, *J. Geophys. Res.*, 92, 7173-7179, 1987.

Martin, S., Y. Yu and R. Drucker, The temperature dependence of frost flower growth on laboratory sea ice and the effect of the flowers on infrared observations of the surface, *J. Geophys. Res.*, 101, 12111-12125, 1996.

Massom, R. A., Observing the advection of sea ice in the Weddell Sea using buoy and satellite passive microwave data, *J. Geophys. Res.*, 97, 15559-15572, 1992.

Morris, K. and M. O. Jeffries, C-band radar backscatter from Antarctic first-year sea ice: II ERS-1 SAR measurements, *Antarct. J. of the U.S.*, 30(1-4), 26-28, 1995.

Onstott, R. G., SAR and scatterometer signaures of sea ice, in *Microwave Remote Sensing of Sea Ice, Geophys. Monogr. Ser.*, vol. 68, edited by F. D. Carsey, pp. 73-104, AGU, Washington, D.C., 1992.

Onstott, R. G. and S. P. Gogineni, Active microwave measurements of Arctic sea ice under summer conditions, *J. Geophys. Res.*, 90, 5035-5044, 1985.

Parkinson, C. L., Interannual variability of monthly Southern Ocean sea ice distributions, *J. Geophys. Res.*, 97, 5349-5363, 1992.

Parkinson, C. L., Spatial patterns in the length of the sea ice season in the Southern Ocean, 1979-1986, *J. Geophys. Res.*, 99, 16327-16339, 1994.

Rignot, E. J., S. J. Ostro, J. J. van Zyl and K. C. Jezek, Unusual radar echo from the Greenland Ice Sheet, *Science*, 261, 1710-1713, 1993.

Rott, H. and C. Mätzler, Possibilities and limits of synthetic aperture radar for snow and glacier surveying, *Ann. Glaciol.*, 9, 195-199, 1987.

Stiles, W. H. and F. T. Ulaby, The active and passive microwave response to snow parameters: 1. Wetness, *J. Geophys. Res.*, 85, 1037-1044, 1980.

Sturm, M., K. Morris and R. Massom, A description of the snow cover on the winter sea ice of the Amundsen and Ross seas, *Antarc. J. of the U.S.*, 30(1-4), 21-24, 1995.

Sturm, M., K. Morris and R. Massom, The winter snow cover of the West Anarctic Pack Ice: Its spatial and temporal variability, this volume.

Tucker, W. B., III, D. K. Perovich, A. J. Gow, W. F. Weeks and M. R. Drinkwater, Physical properties of sea ice relevant to remote sensing, in *Microwave Remote Sensing of Sea Ice*, edited by F. D. Carsey, Geophysical

Monograph 68, pp. 9–28, AGU, Washington, D.C., 1992.

Vachon, P. W., R. B. Olsen, C. E. Livingstone and N. E. Freeman, Airborne SAR imagery of ocean suface waves obtained during LEWEX: Some initial results, *IEEE Trans. Geosci. Rem. Sens.*, 26, 548–561, 1988.

Wadhams, P., M. A. Lange and S. F. Ackley, The ice thickness distribution across the Atlantic Sector of the Antarctic Ocean in midwinter, *J. Geophys. Res.*, 92, 14535–14552, 1987.

Wadhams, P., Variations in sea ice thickness in the polar regions, in *Proceedings of the International Conference on the Role of the Polar Regions in Global Change, June 1990*, edited by G. Weller, C. Wilson and B. Severin, pp. 4–13, Geophysical Institute, University of Alaska Fairbanks, 1991.

Wadhams, P. and B. Holt, Waves in frazil and pancake ice and their detection in Seasat synthetic aperature radar imagery, *J. Geophys. Res.*, 96, C5, 8835–8852, 1991.

Winebrenner, D. P., E. D. Nelson, R. Colony and R. D. West, Observation of melt onset on multiyear Arctic sea ice using ERS-1 synthetic aperture radar, *J. Geophys. Res.*, 99, 22425–22441, 1994.

Winebrenner, D. P., B. Holt, and E. D. Nelson, Observations of autumn freeze-up in the Beaufort and Chukchi Seas using ERS-1 synthetic aperture radar, *J. Geophys. Res.*, 101, 16401–16419, 1996.

Worby, A. P., W. F. Weeks, M. O. Jeffries, K. Morris and R. Jaña, Late winter sea ice and snow thickness distributions in the Bellingshausen and Amundsen Seas, *Antarct. J. of the U. S.*, 29, 13–15, 1994.

Worby, A. P., M. O. Jeffries, W. F. Weeks, K. Morris and R. Jaña, The thickness distribution of sea ice and snow cover during late winter in the Bellingshausen and Amundsen Seas, Antarctica, *J. Geophys. Res.*, 101, 28441–28455, 1996.

Zabel, I. H. H., K. C. Jezek, P. A. Baggeroer and S. P. Gogineni, Ground-based radar observation of snow stratigraphy and melt processes in the percolation facies of the Greenland ice sheet, *Ann. Glaciol.*, 21, 40–44, 1995.

Zwally, H. J. and P. Gloersen, Passive microwave images of the polar regions and research applications, *Polar Rec.*, 18, 431–450, 1977.

Zwally, H. J., J. C. Comiso and A. L. Gordon, Antarctic offshore leads and polynyas and oceanographic effects, in *Oceanology of the Antarctic Continental Shelf*, edited by S. S. Jacobs, Antarc. Res. Ser., vol. 43, pp. 203–226, AGU, Washington, D.C., 1985.

K. Morris, M. O. Jeffries and Shusun Li, Geophysical Institute, University of Alaska Fairbanks, 903 Koyukuk Drive, P. O. Box 757320, Fairbanks, AK 99775-7320.

(Received March 28, 1996
accepted September 22, 1996)

ANTARCTIC OCEAN-ICE INTERACTION: IMPLICATIONS FROM OCEAN BULK PROPERTY DISTRIBUTIONS IN THE WEDDELL GYRE

Douglas G. Martinson and Richard A. Iannuzzi

Lamont-Doherty Earth Observatory and Department of Earth and Environmental Sciences, Columbia University Palisades, NY

The sea ice distribution in the Antarctic polar oceans is intimately tied to the underlying ocean structure, which controls the oceans' vertical heat flux and stability. The former determines the rate at which ice grows for a given air-sea heat flux, while the latter limits the amount of sea ice that can grow locally before overturning the water column. These relationships have been described through a set of scalings, allowing us to estimate, through examination of the vertical distributions of ocean temperature and salinity: (1) the maximum amount of *in situ* ice growth in any one location, (2) the ratio of ice melt to ice growth, (3) the amount of ice that has melted in any particular summer location, (4) the ocean winter-averaged heat flux. Climatological maps of these quantities are presented for the Weddell gyre region and general results described. Results include: (1) the sea ice cover throughout the seasonal sea ice region is typically 0.6 m thick or less by the spring melt period, though it is thinner than 0.3 m in some regions near the gyre core; (2) the ocean-ice system manages to liberate heat from the deep water at an average winter rate of 25-35 W m^{-2} throughout the gyre, regardless of the large scale stratification and dynamic setting which reflect different processes by which the heat makes its way to the surface from the deep waters; (3) strong mixing due to the passage of intense polar lows may serve to reduce the bulk stability of the water column by as much as 75%; (4) most of the bulk stability of the water column is attributed to the enthalpy content of the thermocline, not by direct reduction in ice growth by a strong diffusive heat flux; and (5) positive perturbations (i.e., excess ice growth) in the annual *in situ* ice growth of ≥80% are required to overturn the water column throughout much of the Weddell gyre. The bulk parameters presented here involve vertically-integrated property distributions, and as such they provide constraints or limitations on the ocean-ice system behavior over seasonal time scales. Consequently, they imply a mean seasonal evolution which may be considerably different from the actual time-dependent behavior.

1. INTRODUCTION

Numerous studies have suggested that the polar oceans play an important role in global climate over a broad range of time scales [e.g., *Washington and Meehl*, 1984; *Hanson et al.*, 1984; *Schlesinger and Mitchell*, 1985; *Meehl and Washington*, 1990; *Manabe et al.*, 1991; *Ledley*, 1991; *Imbrie et al.*, 1992; *Wadhams*, 1994; *Rind et al.*, 1995]. In the marginally stable Antarctic polar oceans, the sea ice distribution, ocean heat flux and ocean stability represent three fundamentally important components of this coupled polar-climate system. The sea ice distribution modulates climate through its insulating effect, high albedo and freshwater transport. The latter influences ocean stability and ventilation of deep waters. The ocean heat flux directly influences the rates of ice growth and deep water ventilation. The stability controls the likelihood of a system mode change [e.g., *Gordon*, 1991] from its current semi-stable mode, which supports a seasonal sea ice cover with coastal deep-water formation, to an unstable mode with open ocean deep water formation and the inability to support a sea ice cover. The Weddell polynya was the signature of the most conspicuous example of the unstable mode on a regional scale [*Gordon*, 1978; *Killworth*, 1979; *Martinson et al.*, 1981; *Parkinson*, 1983; *Motoi et al.*, 1987; *Walin*, 1993].

Given the climatic relevance of the sea ice distribution, ocean heat flux and ocean stability, the purpose of this paper is to present a set of parameters, derived from easily observed features of the water column, that impose controls or limitations on these 3 characteristics. The ability to do this reflects the fact that the ocean-atmosphere-ice (OAI) system is so highly interactive that a change in one part of the system influences all other parts. Since the ocean structure has the longest integrated-property memory within the seasonal OAI system, it can be examined anytime within a year and still provide the relevant information [e.g., *Gordon*, 1981]. Ocean properties thus provide an ideal component for use in establishing such parameters. Also, since the sea ice spatial distribution is the most easily observed polar characteristic (from space), the parameters are presented so as to provide information regarding that component of the sea ice distribution which is most difficult to obtain: sea ice thickness.

Specifically, through examination of the vertical distributions of ocean temperature (T) and salinity (S), we estimate: (1) sea ice growth and thickness constraints, (2) winter mean contributions of the ocean heat flux, and (3) susceptibility to deep convection (i.e., overturning the water column, ice elimination and mode change). Time-averaged (climatology) spatial distributions of the parameter values within the Weddell gyre are then provided to demonstrate their usefulness and interpretation, though the concepts apply to anywhere within the sea ice fields. The benefits of such parameters lie in their ability to easily encapsulate fundamental seasonally-averaged characteristics of the OAI system and their sensitivities, and present them in a manner amenable for evaluation of their spatial and temporal variations. They are not appropriate for estimating detailed time-dependent behavior, which may deviate significantly at any one time from the predicted mean evolution presented here.

2. CONCEPTS

To estimate the OAI system parameters of interest based on critical features of the upper water column requires knowledge of the system's external parameter dependence. This was developed for the winter season by *Martinson* [1990], in which a linear system was reduced to a set of simple scaling relationships that captured most ($\geq$83%) of the variance of the full linear model (the implications and deficiencies of a linear treatment are evaluated below). These relationships provide prognostic estimates of ice thickness, mixed layer entrainment (destabilization), and mixed layer salinity as a function of the external parameters of the system: mixed layer and pycnocline thicknesses; T and S gradients (∇T, ∇S) through the pycnocline; the external surface forcing (heat and freshwater); and, diffusive mixing across the pycnocline. The latter is represented, for lack of a better parameterization, by a turbulent diffusivity coefficient. The scalings and diagnostic relationships revealing their inter-dependencies, establish the foundation from which parameters relating the ocean features to ice thickness, ocean heat flux and bulk stability are defined here.

A fundamental, though non-limiting, assumption in the use of these scalings and related parameters is that the OAI system is predominantly controlled by vertical processes. This assumption should be fairly reasonable throughout most of the Weddell gyre, as evidenced by relatively weak (in the mean) lateral property gradients [e.g., *Gordon and Molinelli*, 1982; *Bagriantsev et al.*, 1989], away from the gyre boundaries and the predominant topographic feature, Maud Rise at 1° E and 67° S. Where the assumption is violated, that is, where lateral fluxes contribute significantly to the local evolution of the surface water column, the scalings, and integrated (bulk) parameters presented here will still represent valid approximations if spatially averaged over scales comparable to those representing the range of influence of the lateral fluxes over the length of a season (the integration time of the parameters). This averaging is introduced when producing the climatological maps of Section 4.

Where spatial averaging does not properly accommodate lateral fluxes is in the vicinity of the continental margins where new water masses are being formed, and thus their properties are not accounted for in the integrated ocean profiles. Since these new water masses typically exit the surface layer by flowing along the continental margins [e.g., *Foldvik and Gammelsrød*, 1988; *Gordon et al.*, 1993], the regions for which the parameters are invalid should mainly be constrained to the region of the continental shelf. The depth contour delimiting this region, at approximately 1000 m depth, is outlined in the parameter maps presented later, and provides an approximate southern limit to the parameter validity.

Other basic assumptions in this linear system are that the upper ocean property profiles must display the general shapes shown in Figure 1, and the surface forcing is assumed to be smooth — that is, constant or slowly varying within a season relative to the time constants of the OAI system. Significant departures from the general profile shapes are unacceptable; in such cases, the profiles are excluded from the analysis as discussed in Section 3. Violation of the smooth forcing is more difficult to anticipate, but attempts to do so are considered later in this section.

Linear Winter System: Primary Parameters

Winter salt deficit (SD_w) and thermal barrier (TB_w). In essence, much of the oceanic control on the ice thickness, ocean bulk stability and heat flux, revolves around the distribution of heat and salt within the mixed layer and pycnocline. Specifically, as defined in *Martinson* [1990], the relationship between the winter "thermal barrier" (TB_w) and "salt deficit" (SD_w), both described below, dictates a considerable amount of the system evolution and control. The term "winter", as used here, is that period of the year for which the T and S profiles display the general form of Figure 1a. That is, the surface layer does not contain a seasonal pycnocline; the surface mixed layer is more or less in direct contact with the permanent pycnocline.

The total winter salt deficit, SD_w^T is the stabilizing

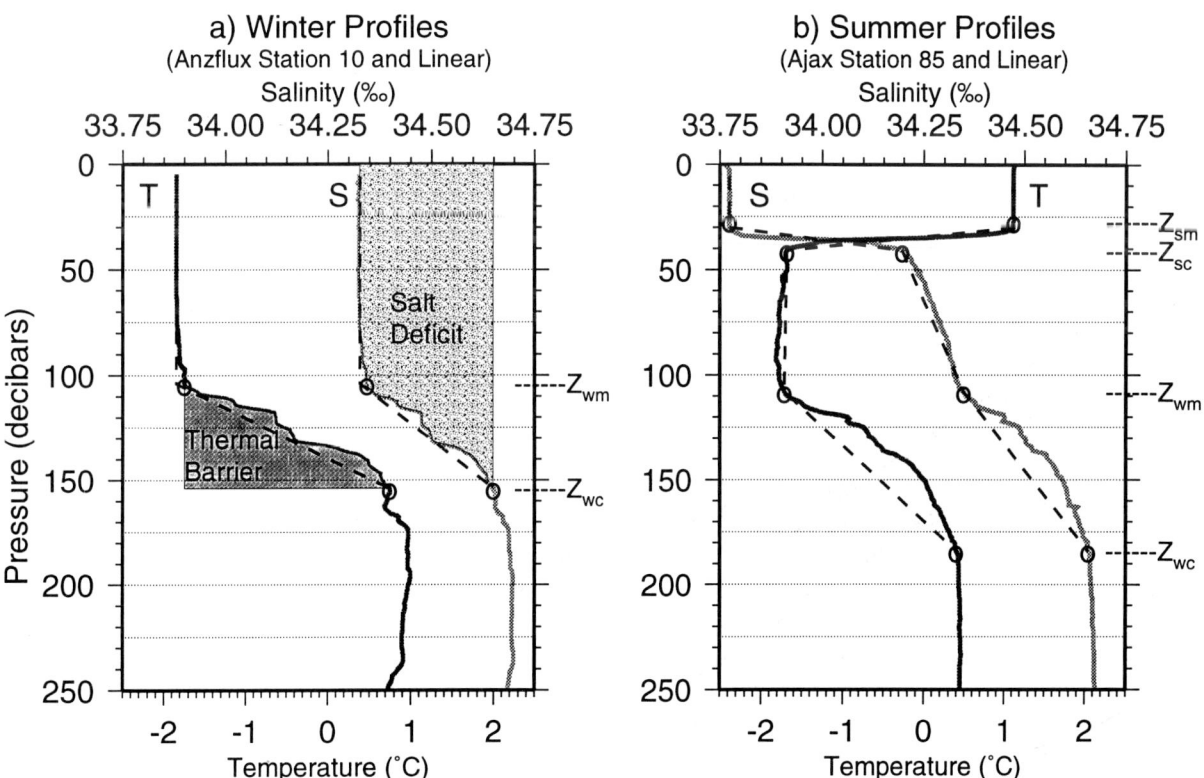

Figure 1: Ideal winter (a) and summer (b) profiles of T and S. Dashed line shows ideal profile shapes; shading indicates the area integrated to produce the Thermal Barrier and Salt Deficit for winter profiles. Key integration depths referred to in the text are also indicated.

freshwater content of the winter mixed layer, relative to the salinity near the base of the pycnocline, that must be eliminated in order to destabilize the surface layer, overturning the water column and driving deep convection. Some of this salt is supplied by the deep ocean, via turbulent diffusion and entrainment. An estimate of this ocean contribution can be removed leaving the corrected winter salt deficit, SD_w (hereafter referred to simply as winter salt deficit), that must be eliminated predominantly by salt rejection driven by ice growth.

TB_w is the sensible heat (enthalpy in excess of the freezing point, T_f) available in the thermocline that must be vented by erosion of the pycnocline during destabilization. As it is vented, that is, mixed into the surface layer, this enthalpy effectively stabilizes the water column by melting ice or, equivalently, by preventing ice growth which would otherwise destabilize through salt rejection. Therefore, as the SD_w is reduced by salinization during ice growth, static instability drives entrainment that gradually vents the TB_w, freshening the surface layer and restabilizing it to some degree. The TB_w thus provides a negative feedback to the ice-growth-driven destabilization process, though it is also vented independently of ice growth through mechanical mixing events driven by intense storms [*McPhee et al.*, 1996].

Both of these quantities, SD_w and TB_w, can be normalized into equivalent units of buoyancy, though for practical purposes their equivalencies in terms of effective ice thickness per unit area is more useful, especially since ice growth and decay is the principal source of buoyancy forcing during the winter season. Therefore, SD_w is the thickness of *in situ* ice growth required to reject enough salt to destabilize the surface layer; TB_w is the thickness of ice that could be melted by completely venting the thermocline of its sensible heat. Though the units of ice thickness are convenient, in an absolute sense they ultimately reflect a heat loss or gain, since that drives the thermodynamic ice growth and directly reflects the TB_w venting. This latter equivalency is preferable for some studies, and is simply proportional to the ice thickness units, but it is not employed here.

For real profiles, TB_w and SD_w are computed by vertical integration. For "ideal" profiles, i.e., geometrically perfect shapes shown superimposed on the real profiles of Figure 1, the integrals reduce to simple geometric relationships:

$$TB_w = [(T_{wp} - T_f)h_{wp}/2](\rho_w c_w/\rho_i L_i) \quad (1)$$

and

$$SD_w^T = (S_{wp} - S_{wm})(h_{wm} + h_{wp}/2)\sigma^{-1} \quad (2a)$$

$$SD_w = SD_w^T - SE \quad (2b)$$

where: T_{wp} and S_{wp} are T and S at z_{wc}, the critical depth near the base of the winter (permanent) pycnocline, below which additional entrainment occurs via cooling alone (i.e., no additional salinization is required for further destabilization once entrainment mixes down to this depth); h_{wp} is the thickness of the winter pycnocline to the critical depth; ρ_w and ρ_i are the densities of water and sea ice, respectively; c_w is the specific heat of seawater; L_i is the latent heat of fusion of sea ice; S_{wm} is the salinity of the winter mixed layer; h_{wm} is the thickness of the winter mixed layer; and σ (= 30‰) is a convenient means of converting a unit of ice to an ocean mixed layer salinity increase. SE is the net contribution of salt from other non-ice sources, such as eddy diffusion across the base of the mixed layer and freshwater input at the surface due to snow entering through leads. Winter snow input is estimated from *Martinson* [1990] as ~10^{-9} ‰ m s^{-1} through ~5% leads [*Wadhams et al.*, 1987]. This value is negligible in terms of equivalent ice thickness and thus ignored for winter conditions. Therefore, SE is predominantly the diffusive input of salt across the mixed layer base. This contribution is estimated below with the ocean heat flux.

The value of the ice-salinity conversion factor, σ, represents the salinization of a 100 m thick mixed layer at 35‰ salinity due to the extraction of a unit volume of water, in the form of ice, at 5‰ salinity. Though not an exact parameterization of the true salinization effect, it provides a close approximation which does not warrant better analytic treatment since the precise salinity of ice at initial formation, and its subsequent brine drainage, is itself variable and thus represents a small uncertainty in SD_w^T which is ignored here, though its potential (minor) influence is discussed in Section 4.

Because the water grows significantly warmer with depth through the pycnocline, the thermal contribution to density ($\alpha \partial \rho/\partial T$; where α is the thermal expansion coefficient) becomes increasingly important with depth. Consequently, z_{wc} may be significantly above the pycnocline base (as defined by maximum $|\nabla^2 \rho|$) if the ratio $\alpha \nabla T/\beta \nabla S$ is small, where β is the haline contraction coefficient. The influence of thermobaricity can reduce z_{wc} further (e.g., *Garwood et al.*, 1994; *Akitomo et al.*, 1995) though it is not considered here.

In isolation, SD_w indicates the overall degree of stability in the water column associated with the surface freshwater cap, while TB_w indicates the potential to resist overturn due to the heat storage in the thermocline. In various combinations these fundamental parameters provide the basis for several more quantities of interest.

Bulk stability. The most notable combination of SD_w and TB_w occurs in the form of net surface water, or bulk, stability (Σ), where $\Sigma = TB_w + SD_w$. This measure of stability indicates the amount of *in situ* ice growth (or its heat loss equivalent) sufficient to overturn the water column and drive deep convection, ignoring storm influences or thermobaric effects, both discussed later.

Bulk stability is the sum of both SD_w and TB_w because the destabilization induced by growing an amount of ice equivalent to SD_w will completely eliminate the thermocline, melting or preventing the growth of an amount of ice equivalent to TB_w. Thus, an additional amount of ice equal to TB_w must then grow in order to overcome the freshwater introduced by the melt (or its effective freshening by the prevention of ice growth). This value is an upper limit since storms can effectively reduce its influence on the ice budget, as discussed later.

Diffusive heat flux. The ability of the water column to resist destabilization by ice growth lies in the ocean heat flux; that is, the transfer of heat from the warm deep water into the mixed layer (by entrainment), reduces ice growth. Turbulent diffusion accomplishes this directly by mixing the warm deep water upward, which continually decreases the density stratification at the base of the mixed layer, making it possible to mix this weakly stratified warmer water into the mixed layer without a change in surface mixing intensity. Therefore, on average, turbulent diffusion directly effects a heat flux into the mixed layer, without the need to explicitly account for the entrainment process (i.e., the background mixing in the mixed layer is sufficient).

The direct contribution of ocean sensible heat via turbulent diffusion across the pycnocline (F_{DT}), preventing a destabilizing ice growth, is given as $F_{DT} \approx \rho_w c_w K_z \nabla T$, where all quantities here (and in following definitions) are for seasonally-averaged values unless otherwise noted; K_z is an eddy diffusivity coefficient. Results of the recent ANZFLUX experiment [*McPhee et al.*, 1996] reveals that K_z changes significantly with the intensity of the surface stress forcing [*Stanton*, personal communication]. Thus, it is difficult to assign a single constant value for K_z. On the other hand, property distributions [e.g., *Gordon and Huber*, 1990; *Schlosser et al.*, 1987] and simple modeling studies [e.g., *Martinson*, 1990] suggest that a seasonally-averaged value, $K_z \sim 0.6 \times 10^{-4}$, is required to achieve the observed seasonal heat balance. This value is large [e.g., *Gregg*, 1988, *Ledwell*, 1993], but apparently reflects the tremendous episodic increase in turbulent diffusion during the frequent storm events, as suggested by the ANZFLUX experiment. It may also reflect the weak stratification of the Antarctic pycnocline. Regardless, this is approximately the average winter value required to achieve the observed average seasonal heat flux estimated in a number of previous studies, referenced above. The impact of using a single, invariant value for K_z is discussed in Section 4 when discussing the results. Note that the dependency of K_z on the surface stress suggests that during summer, when the permanent pycnocline is isolated from this stress by the seasonal pycnocline, that the average value of K_z will be small, which is consistent with the fact that the remnant winter mixed layer survives relatively intact through the summer months.

As an ice-melt, or ice-growth-inhibitor, potential, this turbulent-diffusive heat flux can be provided in terms of equivalent ice thickness: $\Theta_{DT} = F_{DT} \Delta t / \rho_i L_i$, where Δt is the time period (~5 months) over which the ocean is in its winter configuration (i.e., with the deep winter mixed layer present). This provides an estimate of the thickness of ice that is prohibited from growing, or is actively melted, by this heat flux component.

Entrainment heat flux. Whereas the turbulent diffusive flux directly resists destabilization by reducing ice growth by an amount Θ_{DT}, any net ice growth that is realized must salinate the mixed layer, driving static instability and a density adjustment through free convection. The free convection, or entrainment, erodes the thermocline, releasing heat stored in the thermal barrier, venting TB_w as the negative feedback mechanism previously discussed.

In the context of the water column's ability to resist overturning, this entrainment heat flux is most effectively presented as a *TB_w efficiency ratio*: $\gamma_{TB} \approx TB_w / \Sigma$. This ratio provides an indication of the overall fraction of bulk stability attributed to the thermal barrier, or negative feedback. Where the value is large (approaching 1), the bulk stability is dominated by the large enthalpy content of the thermocline; where it is small, the surface freshwater content dominates. This differentiates between subsurface versus surface stabilization. The latter reflects stability largely attributed to ice drift which controls the surface freshwater balance, while the former reflects stability due to enthalpy contained within the thermocline. Both sources are linked through the larger scale surface stress forcing.

As a heat flux potential, γ_{TB} represents the percent of the air-sea heat flux provided initially as latent heat of fusion (i.e., the net air-sea heat flux minus F_{DT}) that will ultimately be realized as an ocean sensible heat flux. For example, if $\gamma_{TB} = 0.2$, $F_{DT} = 20$ W m^{-2} and the average air-sea heat flux, $F_a = 35$ W m^{-2}, then 15 W m^{-2} must initially be provided to the atmosphere as latent heat of fusion, F_{LT}, or $F_{LT} = F_a - F_{DT}$. This grows enough ice to drive an entrainment heat flux (i.e., venting of TB_w), $F_{ET} = \gamma_{TB} F_{LT} = 0.2 \times 15 = 3$ W m^{-2} of sensible heat from the TB_w; or, given as an equivalent ice-growth-inhibitor potential: $\Theta_{ET} = F_{ET} \Delta t / \rho_i L_i$. Therefore, the net ocean sensible heat flux is the diffusive plus entrainment fluxes, or $F_T = F_{DT} + F_{ET} = 23$ W m^{-2}. This net value can be given as a total ice-growth-inhibitor: $\Theta_T = F_T \Delta t / \rho_i L_i$. Alternatively, 20% ($\gamma_{TB} = 0.2$) of every watt of heat released to the atmosphere as latent heat of fusion (ice growth) is ultimately converted, through the negative feedback, to a sensible heat flux. Thus, γ_{TB} indicates the efficiency of the negative feedback mechanism in extracting additional ocean sensible heat (TB_w-ventilation) from the subsurface waters that is otherwise not directly accessible to the surface. In a seasonal bulk analysis, this ratio is arguably more important than the average entrainment heat flux, as it gives a direct indication of the relative importance of the thermal barrier in maintaining the water column stability, and thus some indication of the mechanistic controls, and relative sensitivities, of the system.

Linear Winter System: Additional Diagnostic Parameters

Ice melt to growth ratio. Another measure of the efficiency of the negative feedback is given by the ratio $\gamma_\Theta = TB_w / SD_w$. This ratio indicates the effectiveness of the feedback mechanism in melting or inhibiting ice growth. For each unit of ice growth (destabilization), γ_Θ-units of ice melt (stabilization) are introduced by the venting of the TB_w. Thus γ_Θ is the ratio of *in situ* melt to growth. If $\gamma_\Theta > 1$, for each unit of ice grown, more than one unit of ice melts, resulting in a significant reduction in the net growth rate.

For example, if $\gamma_\Theta = 9$, and one unit of ice grows per day, then one day's growth vents enough of the TB_w to prohibit ice formation for the next 9 days, or it melts 9 units of ice immediately, requiring the next 9 days to regrow the melted ice and overcome the 9 units of stabilizing meltwater. In a 10-day period, net ice growth occurs for only 1 day so a growth efficiency is defined as $\gamma_{SD} = (\gamma_\Theta + 1)^{-1} = SD_w / \Sigma$. This is the salt deficit equivalent to γ_{TB}. It controls the *effective* latent heat flux required to make a net reduction in the stability. The effective latent heat flux, $<F_{LT}^{eff}> = \gamma_{SD} <F_{LT}>$, gives the latent heat flux resulting in the net increase in ice after compensating for ice melt due to TB_w venting. So, $<F_{ET}> + <F_{LT}^{eff}> = (\gamma_{SD} + \gamma_{TB}) <F_{LT}> = <F_{LT}>$, the original (gross) latent heat flux, of which some fraction, γ_{TB}, is converted to a sensible heat flux via venting of the TB_w, melting ice, while the complement goes toward the actual net increase in ice thickness. Also, $\gamma_\Theta = \gamma_{TB} / \gamma_{SD}$.

Ocean heat flux distribution. The ocean heat flux, as partitioned here, is realized through the eddy-diffusive flux, F_{DT}, and entrainment-driven flux, F_{ET} [for a more complete discussion of the ocean heat flux dependencies, relative contributions and sensitivities to diffusion, upwelling and free/forced convective entrainment on the ocean heat flux, see *Martinson*, 1990, 1993]. The diffusive flux draws heat directly from the deep water, a near infinite reservoir, while the entrainment flux taps the finite reservoir stored within the thermocline. This separation is artificial since entrainment is required to incorporate all warmer water into the mixed layer, and the ultimate source of the enthalpy comes from the deep water in each case. However, it represents a rather natural separation reflecting the processes and time scales of the different mechanisms by which heat is transferred to the surface. Upwelling and other gyre-scale dynamics also influence this categorization, though these are treated through their influence on the mixed layer depth and pycnocline characteristics, which control the bulk parameterizations. In any case, the ratio of these two seasonally-averaged fluxes, or their ice-melt equivalencies, $\gamma_T = \Theta_{ET} / \Theta_{DT}$, provides an indication of the immediate source of the ocean heat.

For small values of γ_T, the ocean heat flux is dominated by the diffusive flux; for large values the entrainment flux dominates. The primary control on this ratio is the thickness of the thermocline, since, for a given T_{max} at the base of the thermocline (i.e., for a given deep water T), this thickness controls both ∇T (dominating F_{DT}) and TB_w (dominating F_{ET}). The explicit (bulk) covariation between F_{DT} and F_{ET} can be determined analytically, but the result is algebraically tedious and not particularly insightful. In essence, small ratios of γ_T reveal areas where deep water enthalpy is most effectively vented directly across a thin thermocline that provides little thermal storage itself. Large values reflect a deep water that is more effectively buffered by a thick thermocline that undermines diffusion but which stores considerable enthalpy that is vented via entrainment.

The thin pycnocline reflects: (1) a stronger upwelling, effectively forcing the deep water closer to the surface layer, and/or (2) a greater mean surface stress, or greater frequency/intensity of storms, that increase the depth of the surface layer, effectively forcing it closer to the deeper water. Both scenarios allow a more direct interaction between the deep water and surface layer. Conversely, the thick pycnocline indicates the opposite, and the interaction between the surface and deep waters must be effected through an intermediate process, entrainment. However, either situation allows the atmosphere access to the deeper ocean sensible heat it ultimately requires.

The relative heat flux contributions can also be given as $\Gamma_{ET} = \Theta_{ET}/(\Theta_{ET} + \Theta_{DT})$ and $\Gamma_{DT} = \Theta_{DT}/(\Theta_{ET} + \Theta_{DT})$. These ratios provide the fraction of the total heat flux (or ice-growth inhibition) contributed by the entrainment-driven, or diffusive components, respectively.

Salt flux distribution. An eddy diffusive flux for salt, F_{DS}, is estimated by: $F_{DS} \approx K_z \nabla S$. Like the eddy diffusive heat flux, it too can be given in terms of equivalent units of ice, $\Theta_{DS} = F_{DS}\Delta t/\sigma$, representing the ice growth required to contribute this degree of salinization. Also, a ratio relates the sources of salt, $\gamma_S = SD_w/SE$, where $SE \approx \Theta_{DS} + \Theta_{ES}$, and Θ_{DS} is the diffusive contribution as defined previously and Θ_{ES} is the contribution of excess salt as the halocline is mixed into the surface layer via entrainment. This ratio indicates how much of the destabilization, measured as the elimination of the total salt deficit, SD_w^T, is contributed by deep ocean salt sources relative to that which must come from surface freshwater extraction (ice growth). Large values of γ_S indicate that the predominant source of salt must be forced by heat loss driving ice growth.

Since Θ_{ES} is driven by a latent heat flux through the negative feedback mechanism, ice growth is ultimately responsible for reducing SD_w^T by both SD_w (salinization by ice) and Θ_{ES} (direct consequence of the salinization). Thus, the salt ratio can be defined in terms of the forcing, rather than sources, of salt: $\gamma_{Si} = (SD_w + \Theta_{ES})/\Theta_{DS}$, where γ_{Si} is a measure of the salt deficit reduction by latent heat loss relative to that by ocean diffusion.

The relative salt contributions can also be given as: $\Gamma_{ST} = SD_w/SD_w^T$, $\Gamma_{SS} = SE/SD_w^T$ and $\Gamma_{Si} = (SD_w + \Theta_{ES})/SD_w^T$. These provide the fraction of the total salt deficit that must be contributed by ice growth, ocean salt sources, and latent heat loss, respectively.

Linear Summer System: Primary Parameters

In addition to the winter parameters, the presence of the seasonal pycnocline in the summer provides an opportunity for additional quantities involving the surface layer that is warm and fresh, representing the spring meltwater and summer warming. In these cases the permanent pycnocline features are preserved at depth, though slightly diffused, still allowing computation of the winter parameters just described as well.

Summer salt deficit and thermal barrier. As with winter, the freshwater and thermal content associated with the seasonal mixed layer and pycnocline can be classified in terms of a summer salt deficit, SD_s, and thermal barrier, TB_s. The SD_s in this case is the freshwater content of the summer surface layer relative to the salinity of the remnant winter mixed layer below the seasonal pycnocline. The TB_s, unlike TB_w, is predominantly contained within the surface mixed layer, since temperature decreases through the seasonal thermocline reflecting a diminishing enthalpy content with depth. Consequently, most of this heat is vented via direct exchange with the atmosphere and its primary role is not as a negative feedback inhibiting ice growth, but rather as a thermal buffer that must be eliminated before ice can grow at all. Thus, its main influence is on the seasonality of the ice, and only through that influence can it affect ice thickness (some of the heat is indeed vented by ice growth destroying the seasonal thermocline, but this is a relatively minor fraction). Consequently, it is not sensible to present TB_s in terms of equivalent ice thickness. It is given as enthalpy relative to freezing, though SD_s is still given as equivalent ice thickness.

Consistent with the winter parameters, TB_s and SD_s are computed by vertical integration for real profiles. For ideal summer profiles (Figure 1b), the integrals again reduce to simple geometric relationships:

$$TB_s = [(T_{sm} - T_{min})(h_{sm} + h_{sp}/2) \\ + (T_{min} - T_f)(z_{sc} + h_{wr})]\rho_w c_w \quad (3)$$

and

$$SD_s = (S_{wr} - S_{sm})(h_{sm} + h_{sp}/2)\sigma^{-1} \quad (4)$$

where: T_{sm} and S_{sm} are T and S within the summer mixed layer; h_{sm} is the thickness of the summer mixed layer; h_{sp} is the thickness of the seasonal pycnocline; T_{min} is the minimum temperature, which lies within the remnant winter mixed layer between the seasonal and permanent pycnoclines; z_{sc} is the depth at the base of the seasonal pycnocline (and top of the remnant winter mixed layer); h_{wr} is the thickness of the remnant winter mixed layer; and S_{wr} is the average S of the remnant winter mixed layer.

The second term on the right-hand-side of (3) is broken into two components: the enthalpy content to the base of the seasonal pycnocline, and then through the remnant winter mixed layer. The former is predominantly vented prior to ice growth, whereas the latter, a relatively small amount, is vented by mixing during removal of the seasonal halocline with the initial ice growth. This latter component does not contribute to the seasonality of the ice cover, but does introduce a slight temporary reduction in initial growth rate during the fall.

SD_s does not have a correction term, corresponding to SE in (2b), since elimination of SD_s during fall ice growth is too fast for diffusion to influence it. The fall ice growth is rapid because: (1) it does not begin until TB_s is vented, eliminating the seasonal thermocline and thus any potential summer negative feedback comparable to that in the winter; (2) the seasonal halocline isolates the surface layer from the deeper ocean heat flux associated with the permanent pycnocline, so there is no inhibiting ocean heat flux, either diffusive or via the deeper negative feedback, save the minor amount $(T_{min}-T_f)h_{wf}\rho_w c_w$; and (3) there is no insulating ice cover to restrict the air-sea heat flux initially. Consequently, the ice grows very rapidly until SD_s is eliminated (along with the seasonal halocline/pycnocline), after which the winter mixed layer is fully developed and in direct contact with the permanent pycnocline, making available the diffusive and entrainment heat fluxes and greatly inhibiting further (i.e., winter) ice growth.

SD_s is dominated by the ice and snow melt from the previous winter as well as runoff, summer precipitation-evaporation and diffusion across the seasonal halocline. Contributions from the latter three are presumably quite small [e.g., *Jacobs et al.*, 1992; *Martinson et al.*, 1981; *Martinson*, 1990] so SD_s is predominantly a meltwater signal. Regardless of the source of freshwater, SD_s represents the thickness of the fall ice growth that will grow rapidly until the winter mixed layer is developed initiating the ocean heat flux influence.

TB_s is of questionable value without specific information regarding the fall air-sea heat flux. For an approximate fall regional heat flux curve, TB_s can be presented as the amount of time until initial ice growth following the onset of fall cooling. However, the fall air-sea heat flux can show tremendous variability in the absence of an insulating ice cover, introducing unacceptably large errors relative to the signal. Alternatively, since satellite coverage reveals the time of initial ice growth, TB_s, in conjunction with the climatological average of when fall cooling begins, allows an estimate of the average fall air-sea heat flux and thus its spatial distribution. However, here too, given the dramatic increase in heat loss later in the fall and potential errors in initiation of fall cooling, such estimates may also be of questionable value.

Linear Summer System: Additional Diagnostic Parameters

Critical interannual ice growth perturbation. The total amount of stabilizing freshwater contained above the permanent pycnocline that must be eliminated, via ice growth, in order to destabilize the water column is given by the sum of the summer salt deficit and winter stability, $\Sigma_T = SD_s + \Sigma$. The fraction of this realized through the fall ice growth is $\gamma_{ms} = SD_s/\Sigma_T$. Since SD_s is predominantly a measure of the spring ice and snow melt, this ratio provides an indication of what fraction of the net stabilizing surface freshwater content is mobile each year through actual ice *in situ* growth/decay, ice divergence/convergence, and snow accumulation. For fractions approaching 1, a relatively small change in ice growth/decay, divergence/convergence, or winter snow accumulation, relative to the seasonal average, may lead to destabilization.

For example, if $\gamma_{ms} = 0.9$, then 90% of the destabilization required to overturn the water column is achieved in a typical year. Conversely, the fraction $\Gamma_{ms} = 1/\gamma_{ms}$ indicates the size perturbation in annual ice thickness relative to the climatological mean required to destabilize the water column. For the above example, a perturbation of ~11% would be sufficient to destabilize the (climatological) water column. Therefore, this parameter can identify regions most susceptible to overturn given interannual variability. In fact, given an estimate of interannual variability at any given location, one can estimate the likelihood of achieving a critical perturbation sufficient to destabilize the water column and induce a mode change. This interpretation is only approximate however, since divergence and ice drift assures that ice does not melt where it forms and thus the actual *in situ* melt may not be indicative of the *in situ* ice growth in that same location.

Nonlinear System

The linear summer and winter descriptions above suffer from several weaknesses. The most conspicuous ones are associated with the assumptions of general profile shapes and a steady surface stress forcing. The former influences the manner in which the parameters are computed as well as the uncertainties associated with their bulk (ideal-geometry) calculation, though this calculation represents a convenience, and is not a computational restriction. This is treated in the next section. The second is the most obvious of a more general problem associated with ignoring nonlinearities. The specific influence of a variable surface stress forcing can be estimated to some extent through heuristic arguments. Its most impressive impact, as evident during the 1994 winter ANZFLUX experiment, is the extensive entrainment driven by turbulent mixing during the passage of frequent and intense storms [McPhee et al., 1996]. These vent enormous amounts of TB_w without a corresponding reduction in SD_w due to ice growth. This particular response introduces nonlinearities in the form of discontinuities attributed to the transition from ice growth, or destabilization periods, to melting, or stabilization periods. Fortunately, for time-integrated quantities such as the bulk parameters, this type of nonlinearity effects a minimal seasonal impact and thus can be reasonably accounted for (though it may introduce other nonlinear influences or feedbacks, particularly through the covariation of external parameters and forcing that have not been anticipated or treated here).

The decoupling of TB_w and SD_w via storm-induced mixing events may, in the limit, result in the complete elimination of the thermocline and venting of TB_w, leaving SD_w essentially unchanged. SD_w is unchanged because the freshwater content of both the mixed layer and halocline is included in its calculation, thus the entrainment of the halocline into the mixed layer, while influencing the vertical distribution of the freshwater, does not influence the net amount of freshwater. Because TB_w and SD_w are intimately coupled under a smooth forcing, where salinization reduces SD_w while simultaneously venting TB_w, the time-dependent behavior of the variable forcing (storm) scenario will be quite different from that of the smooth forcing. However, when integrated to the limits of stability, both scenarios ultimately require the same amount of net *in situ* ice growth to eliminate SD_w.

Even though SD_w is not altered by storm events, the bulk stability, Σ, of the water column can be. The storms erode the pycnocline, mixing the TB_w enthalpy into the mixed layer. This changes the nature of TB_w venting from a negative feedback as described for the linear winter, to a direct venting, comparable to that in summer (which is independent of SD_s). Given a large enough mixing event, the vented TB_w may have the potential to melt more ice than present and/or significantly reduce the ice concentration. In either case, the venting of TB_w to the atmosphere through a reduced ice concentration is considerably more efficient than that expected during the smooth forcing scenario. Therefore, the storms may effectively reduce TB_w as an ice-growth-inhibitor equivalent. If the entire TB_w is vented in one storm, a limiting, albeit unlikely, scenario, its influence on the freshwater balance will be restricted to being less than or equal to the local ice thickness, h_i. That is, it cannot put more freshwater into the surface layer than it can melt (h_i).

In regions where the pycnocline stratification is weak enough to allow a significant or complete turbulent erosion of the thermocline during intense storms, the bulk stability may be reduced from $\Sigma = TB_w + SD_w$ to $\Sigma_e \approx h_i + SD_w$, the latter being the effective bulk stability. Throughout much of the seasonal ice region $h_i \sim 0.6$ m [Wadhams et al., 1987; Ackley et al., 1990; Eicken and Lange, 1989]. So, wherever TB_w is substantially larger than h_i, storms can introduce a substantial decrease in ocean stability.

For the other winter parameters, the decoupling of TB_w and SD_w alters the nature of the time-dependent behavior relative to the linear case, but should not significantly alter the overall seasonal interpretation of the parameter. The veracity of this statement is to a large part dependent upon the averaging used to compute the external parameters and their uncertainty. For example, the value used for ∇T when computing F_{DT} must represent a temporal mean accounting for differences preceding and following storms, and the variability about the mean must be incorporated into the uncertainties of the ultimate parameter estimate.

In summer, storms may initiate winter conditions before ice growth eliminates the seasonal halocline, thus venting TB_s but leaving a fresher winter mixed layer than otherwise expected from the steady forcing scenario. Therefore SD_s, as an estimate of fall ice growth, is an upper limit which conveys the potential for rapid ice growth contribution. However, because the fall ice growth is so rapid, the storms must occur within a relatively short window of time (during the fall growth, or just before it) in order to alter this particular estimate.

We are not sure of the degree or nature of additional nonlinearities that may be associated with storm events, or the variable surface stress forcing in general, so additional refinements may still be required. For example, there may be a strong covariance between storms, pycnocline doming (influencing mixed layer depth among other things), ice divergence and effective salinization distribution, K_z and lead area through increased surface divergence, driving nonlinearities or feedbacks that are not presently accounted for and that ultimately drive the net seasonal response from that expected for the smoothly forced linear system. A potentially more important impact of such storm induced effects however, is their influence on the upper ocean structure that controls the external parameters (e.g., mixed layer depth and pycnocline thickness), and thus controls the linear bulk parameters directly. From this perspective, even the linear analysis presented here may account for a significant influence of persistent storm tracks.

3. METHODS

Data Base

Antarctic CTD station data collected between the Antarctic Peninsula and 20° E, during 28 cruises conducted over the last 25 years were used in this study. These data are readily available from the NOAA Ocean Data Center, the Alfred-Wegner-Institut für Polar- und Meeresforschung, the Nemo Oceanographic data server at Scripps Institution of Oceanography, and from *Heywood and King* [1996]; a complete listing (as well as a postscript version of this paper) is provided in the dgm home page at http://www.ldeo.columbia.edu:80/~dgm/ (or link through the Lamont-Doherty Physical Oceanography Web site: http://www.ldeo.columbia.edu).

Of the initial 2016 CTD station profiles available from the 28 cruises, 306 were rejected immediately because they resided outside of the polar gyre or were incomplete (large data gaps or missing variables). The remaining 1710 stations were then processed, as described below, and inspected at several stages to cull severely corrupted data or those whose shape deviated significantly from the ideal shapes of Figure 1, preventing identification of the integration limits. This hand culling eliminated another 287 stations leaving a total of 1423 stations used to construct the climatologies. The surviving 1423 hydrographic station locations used in the analysis are identified via small white dots in each of the parameter maps of the next section. The winter parameters are computed using all 1423 stations (as described previously), but only 715 stations are available for the summer calculations, thus these latter results are rather sparse and can only demonstrate the concepts and describe the broadest sense of the features.

Typical temperature and salinity profiles, are shown in Figure 1 superimposed on the ideal profiles to give an indication of how they look relative to the ideal shapes, though, as described below, it is important to realize that the parameter calculations are based on the true profile shapes, not the ideal shapes; the latter are derived strictly to facilitate the error analysis (which takes into account deviations of the true shape from the ideal ones), and to allow quick estimation of the parameters from profiles without performing a full integration.

Data Processing

Smoothing. Most of the T and S profiles were recorded at 1 db intervals. Those recorded at lower vertical resolution were interpolated to 1 m resolution so that the same vertical smoothing function could be applied to all profiles, though this diminished the effective level of noise reduction in the more coarsely sampled profiles. The data were smoothed with a 19-point vertical median filter. A rather extensive set of tests suggested that the bulk parameter estimates were extremely robust to the actual degree of smoothing and type of smoother applied. Consequently, the filter width used here was determined experimentally and found to provide the minimal filter necessary to qualitatively smooth the noise from the profiles while not overly destroying the critical features within it.

Parameter calculations. To avoid uncertainties associated with deviations of the true profiles from the ideal, the parameters are calculated by vertically integrating heat, salt and buoyancy through the upper water column, over the appropriate limits. In order to partition the different sources of heat and salt into their natural physical constituent components (e.g., differentiating the deep TB_w from the shallower TB_s), integration limits are defined at several critical features within the upper ocean profiles. The critical features are labeled in Figure 1. Most of these are identified based on their physical interpretation and thus uniquely identifiable during the integration itself. For example, the lower integration limit is typically the depth at which no additional salinization is required to drive overturn of the water column, z_{wc}. In other words, once the mixed layer deepens to this point by the elimination of the SD_w, cooling the mixed layer back toward the freezing point is sufficient to drive additional convection, so that deep overturn is essentially assured.

Similar arguments apply to all other points in the water column except for the interface between the winter mixed layer and permanent pycnocline, z_{wm}. This feature is the critical limit from which most summer integrations end and most winter ones begin. Identification of z_{wm} is done through use of a penalty function which seeks the maximum curvature of a normalized smoothed salinity profile, with the minimum deviation from the mixed layer salinity. That is, it rewards high curvature, but recognizes that some deeper feature within the pycnocline such as a step, intrusion or other abrupt feature may actually possess the global curvature maximum. Thus, it penalizes for deviating from the mixed layer salinity, which proceeds quite rapidly with depth in the pycnocline. The normalized salinity profile, $S^*(z)$, is given by:

$$S^*(z) = S(z) - \frac{1}{z}\int_0^z S(z)dz \qquad (5)$$

This quantity is the salinity perturbation from the mixed layer salinity, given mixing to any depth z.

Once the integration limits are picked, the fundamental parameters can be computed according to the geometric relationships provided in (1) - (4), or via the full depth integration. The difference between these two estimates is a measure of how much the profiles deviate, nonsymmetrically, from the ideal profiles. This difference is relatively unimportant when full profiles are available, since the depth integrated values used here are accurate and easily calculated. However, the more simple geometric calculations are important for assessing the sensitivities of the parameters to changes in the external parameters (e.g., by computing the derivatives with respect to the external parameter of interest). They are also good for quick assessments of profiles, and ultimately for model-based computations where the deviation from ideal should be minimal in most cases and the depth integration too computationally-intensive to compute regularly.

While there is considerable variability across the gyre, typical values of the integration points, defined in Figure 1, are as follows: $z_{wm} = 117 \pm 46$ m, $z_{wc} = 205 \pm 94$ m, $z_{sm} = 22 \pm 22$ m, $z_{sc} = 57 \pm 41$ m. Source code for the depth-picks and integrations to produce the various parameters presented here can be obtained from the web site (dgm home page) listed previously.

Uncertainties

Several types of uncertainties are expected in the parameter values: (1) methodological error (random and bias), reflecting the ability of the penalty function to capture the desired physical characteristic, (2) analytical error, reflecting the precision with which the critical features in the profiles can actually be located, (3) temporal variability, (4) temporal bias, arising from the time of season when the profiles were acquired, and (5) sampling errors.

The methodological and analytical errors are dominated by the uncertainty in identifying the mixed layer-pycnocline interface depth, z_{wm}, which is difficult due to smearing by turbulent diffusion, though entrainment tends to drive the interface back toward the ideal profile shape in Figure 1. The other critical features (integration limits) are identified via physically unique criteria as previously described and are consistently identified within a smoothed profile to the resolution at which the profile is sampled; typically one or two meters for the data available here.

Methodological error (random error and bias). The methodological error reflects the ability of the penalty function to pick that z_{wm} which is representative of the physical characteristic sought; in this case, the boundary between the mixed layer and pycnocline. This error manifests itself in two stages. The first involves the variables used in the penalty function and its functional form. The second involves the ability of the optimal penalty function to identify the interface, given irregular profile shapes and smearing by diffusion. These are addressed by generating a set of control profiles with a known interface depth that have then been subjected to varying degrees of diffusive smoothing (applied as a cascading filter, and spanning a range comparable to that present in the data set). This testing involved a variety of penalty functions and variables (e.g., T, S and ρ profiles, and various hybrid combinations), which led to our ultimate choice of penalty function described above.

Further testing revealed that the diffusive smearing of the mixed layer-pycnocline interface introduces a bias in the pick, with a precision about the biased-pick of better than ±2 m (the methodological random error). The bias itself, ε, is as large as 10 m shallower than the true z_{wm}, though the typical bias appears to be 2-3 m. It also shows a functional dependence on the local curvature ($\partial^2 S/\partial z^2$) at z_{wm} and ∇S through the pycnocline (that is, the angle at which the pycnocline intersects the mixed layer). The steeper the pycnocline, the stronger the bias. The functional relationship of the bias to the curvature and halocline slope was determined by two-dimensional regression, of the form:

$$\varepsilon = a_1 + a_2 \log(\nabla S^{*2}) + a_3 \log\left(\left.\frac{\partial^2 S^*}{\partial z^2}\right|_{z_{wm}}\right) \qquad (6)$$

with $a_1 = -5.47$, $a_2 = 16.57$ and $a_3 = -10.26$. This correction captures ~54% of the bias variance.

Despite what appears to be a rather large bias in the actual location of z_{wm}, its influence is minor in the actual parameter values. This is because the integration in the vicinity of this interface (and shallower, in the direction of the bias) is over a nearly vertical (no-property change) slope, and thus, even though it covers ≤10 m depth, its net contribution to the total property integral is trivial; for example, it introduces errors of ~3-4% in SD_w in the representative cruises for typical bias (2-3 m), and <11% error for the infrequent but worst case bias. Consequently, because of the weak influence and our mediocre success with (6), we did not apply the bias correction before processing the data.

Analytical error. The analytic error, i.e., the ability to precisely pick a unique z_{wm} depth, is related to the curvature and noise level in the profile at z_{wm} (ignoring the second component of the penalty function that simply prevents the picking of a feature with stronger curvature elsewhere). Conceptually, the stronger the curvature relative to the level of noise in the profile, the more uniquely a maximum can be identified. The weaker the curvature relative to the noise level, the larger the uncertainty in identifying the maximum. This error is estimated by $[\sigma_S/(\partial^2 S^*/\partial z^2)]^{1/2}$, where σ_S is the typical sample standard deviation in the S^* profiles (in the vicinity of the interface) from which the pick is being made.

For the data used here, the analytical error is typically quite small, ~0.65 m — comparable to the resolution of the data itself (based on analysis of two cruises thought to be representative of the entire data base). This states that the profiles are smooth enough to allow a clear determination of the point of maximum curvature, so this error can be safely ignored as the resolution itself subsumes it.

Temporal variability. While the above errors are generally small and manageable, the major source of uncertainty is associated with vertical migrations of the water column, possibly in response to surface forcing variations. These temporal migrations are difficult to separate from spatial variability, but can be estimated from the rather extensive buoy data archives [*Sellmann and Kottmeier*, 1996; *Kottmeier et al.*, 1997]. Examination of the longest duration temperature-depth sections along drift tracks suggest average changes in mixed layer and pycnocline thicknesses (h_{wm} and h_{wp}, respectively) of, $\sigma_{h_{wm}} \sim 15$ m, and $\sigma_{h_{wp}} \sim 12$ m, for the winter months. The excursions appear to be slightly smaller for summer months. For both seasons, the intra-seasonal covariance between h_{wm} and h_{wp}, cov$[h_{wm},h_{wp}]$ < 5%.

The influence of $\sigma_{h_{wm}}$ and $\sigma_{h_{wp}}$ on errors in SD_w and TB_w is estimated via the expectance operator. Specifically, since TB_w and SD_w are both linear in h_{wm} and h_{wp}, the variance of TB_w scales linearly with $\sigma_{h_{wp}}^2$, the scaling factor given by $[(T_{wp}-T_f)\rho_w c_w/2\rho_i L_i]^2$; the variance of SD_w scales linearly with the sum of $\sigma_{h_{wm}}^2$, $\sigma_{h_{wp}}^2/4$ and cov$[h_{wm},h_{wp}]$ (<5%, so neglected here), the scaling factor given by $[(S_{wp}-S_{wm})/\sigma]^2$. So, $\sigma_{h_{wp}} \sim 12$ m and $\sigma_{h_{wm}} \sim 15$ m contribute to temporal uncertainties in TB_w and SD_w of $\sigma_{TB_w}^t \sim 0.05$ m and $\sigma_{SD_w}^t \sim 0.02$ m, respectively, for typical T and S differences across the pycnocline near the gyre center.

For climatologies, in the densely sampled regions, we typically average 3 to 5 data points, reducing $\sigma_{TB_w}^t$ and $\sigma_{SD_w}^t$ by half (though the reduction is not completely realized since some spatial variability is introduced during climatological averaging within spatial bins). For parameters involving differences or ratios, this error is again inflated by the operation, which approximately doubles it back to the original, unaveraged size. In either case, the temporal variations introduce errors of 5-10% at most of the station locations ($\sigma_{SD_w}^t/SD_w$ is constant for any particular $h_{wm} \pm \sigma_{h_{wm}}$ and $h_{wp} \pm \sigma_{h_{wp}}$, thus $\sigma_{SD_w}^t$ can be given as a percentage of SD_w; likewise for TB_w and $\sigma_{TB_w}^t$). More importantly, the temporal errors are more than an order of magnitude smaller than most of the spatial variability displayed in the parameter maps below, so they preserve a very good spatial signal-to-noise ratio of ~20 db.

The temporal variability subsumes the smaller analytical and methodological uncertainties discussed above in all but the limiting cases.

Temporal bias. This bias arises because the parameters, which represent seasonal limits, are determined from observational profiles that were not necessarily acquired at the start of the season. Therefore, some of the seasonal evolution has already occurred, but the parameters have not been corrected for this. SD_w for example, represents the amount of freshwater in the surface layer that must be removed by ice growth in order to overturn the water column. If the profile from which this quantity is calculated was taken in mid-winter, some fraction of SD_w will have already been eliminated by ice growth that does not appear in the calculation. Therefore, there is a bias associated with each parameter estimate that is proportional to the length of time that has elapsed since the start of the relevant season.

This temporal bias influences the interpretation of the data in all situations, not just those involving the bulk parameters presented here, so an estimate of its impact is necessitated for all analyses. In order to properly correct for this bias we need estimates of winter ice growth and entrainment rates, and the time of onset of the various seasons from either the data or models. At present, this information is nominal for both the data and models. Comparing ice thickness changes through time from data suggests that the ice undergoes a minimal growth of ~1.6x10^{-3} m day^{-1} [*Wadhams et al.*,

1987] in winter, while simple model estimates suggest ~1.3×10^{-3} m day^{-1} [*Martinson*, 1990] (these are ~25 and 20 cm per 5 months of winter, respectively). For every month that passes following the onset of winter conditions, SD$_w$ is thus decreased by ~0.04-0.05 m. For entrainment rates, the modeling [*Martinson*, 1990] suggests mixed layer deepening of 0.03-0.12 m day^{-1}, the smaller values in regions where TB$_w$ is smallest, so the bias approximately scales with TB$_w$. TB$_w$ varies proportionally to mixed layer deepening at about 2% of the change in depth; that is, for every month that passes following the onset of winter, the above estimates suggest that TB$_w$ decreases by ~0.02-0.07 m. Therefore, data acquired in late winter may introduce a bias as large as 30% or so, while data from early winter introduces a bias of <5%.

At present, we do not correct for this bias given the questionable quality of the model and limited data estimates. Consequently, the values may under-estimate some of the quantities they purport to represent. Data acquired exceptionally late in the seasons were not included in the analysis to minimize their particularly large impact on the bias. The remaining data from different times within a season should average out the bias toward mid-season values, typically around 10-30%. Furthermore, the bias is expected to be spatially homogeneous and thus should not significantly reduce the spatial signal-to-noise ratio in the parameter maps below.

Sampling Error

Sampling errors are assumed to be responsible for "bad" profiles — those that display grossly uncharacteristic shapes (relative to surrounding stations) or contain exceptional data values. The majority of these profiles were identified and eliminated prior to processing as described previously. However, a second attempt at eliminating bad profiles was made upon completion of the processing by examining those profiles responsible for introducing any exceptional features in the parameter spatial distributions (e.g., local minima or maxima). Only one isolated feature from the parameter maps was eliminated as a result of this particular quality control measure.

4. RESULTS

Climatology Maps

The various parameters discussed above have been computed for each station using the real profiles. Using the GMT gridding program [*Wessel and Smith*, 1991; *Smith and Wessell*, 1990], station values for individual cruises are then interpolated spatially onto a tight, 0.25° (latitude) x 0.5° (longitude) grid using a taut spline that minimizes overshoot across abrupt gradients and is constrained to minimize overshoot at the boundaries, which reduces the already small interpolation error at the edge of grid cells. Parameter values within this grid are then averaged through time onto a broader 0.5° (latitude) x 1.0° (longitude) grid to produce a climatological (time-averaged) spatial distribution for each. The climatology grid size approximates the typical spatial decorrelation lengths within the eastern region, where spatial variability is highest, so the bin size should be a reasonable estimate throughout the remainder of the gyre region.

The spatial averaging also accounts for the influence of lateral fluxes in the vertically-integrated bulk parameter values, thus extending the spatial range of the parameter maps. North of the polar front, however, these fluxes begin to dominate and the vertical distribution of ocean properties is such that the underlying assumptions of the analysis become questionable and should be ignored. The position of the polar front, according to *Orsi et al.* [1995] is indicated in Figure 2 and on each of the parameter maps. The parameter values are computed for regions north of this front, but the values frequently lie well outside the standard range and are not included on the color scale of the maps; instead they appear as white color ("off scale"). As seen in the maps, this white region often appears just north of the polar front, and thus the parameters themselves seem to nicely delimit the natural boundary of the polar gyre.

Also, in the vicinity of the perennial ice in the western Weddell Sea, indicated in Figure 2 by the February ice extent, the parameter interpretations become equivocal since the seasonal conditions assumed elsewhere in the gyre are not applicable in the perennial region. Therefore, while some of the parameters still have a physical interpretation of interest, they may not be consistent with the broader interpretation presented for the rest of the gyre. The discussion here is thus limited to the broad gyre-scale implications and basic concepts. More detailed discussion regarding particular parameters or their spatial and temporal variability will be given elsewhere.

In order to relate the spatial distributions to the gross features of the Weddell gyre region, and delimit the regions discussed above, Figure 2 presents the bottom topography, position of summer and winter ice extents, and polar front.

Winter thermal barrier. The winter thermal barrier is presented in Plate 1. The TB$_w$ is seen to clearly re-

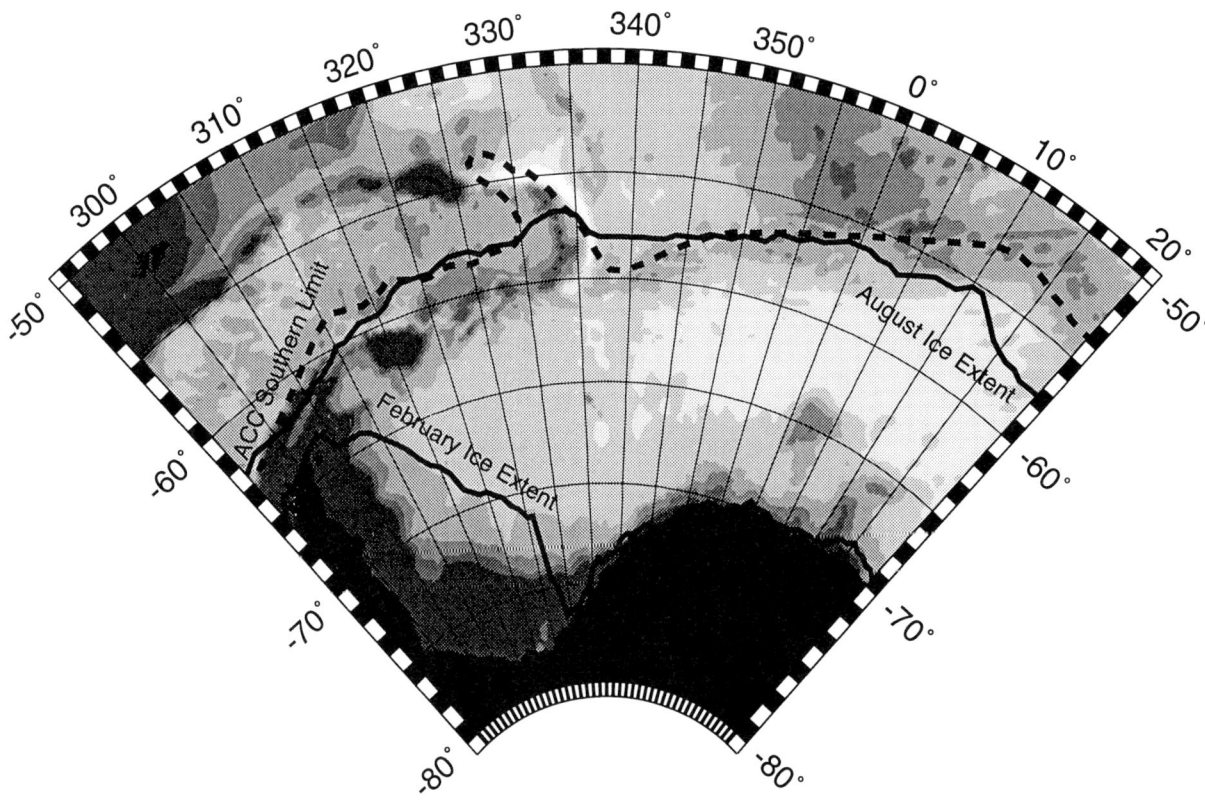

Figure 2: General physical setting and characteristics of the Weddell gyre region. Bottom topography is shaded at 1000 m intervals; the winter (August) and summer (February) ice extents are indicated by solid lines (the latter represents the extent of the perennial ice cover in the region); the approximate location of the southern limit of the Antarctic Circumpolar Current, as defined by *Orsi et al.* [1995] gives the approximate location of the northern limit of the polar gyre.

flect the gyre geometry, with increasing TB_w near the gyre margins where reduced upwelling allows a broader thermocline that contains more stored enthalpy. TB_w depends predominantly on the thickness of the pycnocline (Plate 2) with ~48% of its variance attributed to this specific water column characteristic. As seen in the gyre's zonally elongated core (~66° S), stronger upwelling leads to a thinner thermocline that stores only enough heat to melt ~0.5 m of ice (i.e., a relatively weak thermal buffer), whereas TB_w is an order of magnitude larger at the margins. Note that at the northern margins, even if lateral (ageostrophic) fluxes begin to play a dominant role in the property balance, additional spatial averaging would accommodate these fluxes. As clearly evident from the figure, such additional averaging, while smearing the zonal TB_w gradient somewhat, will not eliminate this overall rapid increase in its value at the northern margins of the polar gyre.

Winter salt deficit. The total amount of salt required to eliminate the winter surface freshwater content, SD_w^T, is presented in Plate 3. It shows a distribution somewhat similar to TB_w, i.e., reflecting the gyre geometry, though the relationship is not quite as clear. As seen, the surface freshwater content throughout the vast majority of the central Weddell gyre is less than 0.5 m of equivalent ice growth. Toward the northern extreme of the gyre, particularly in the east where the Antarctic Circumpolar Deep Waters enter the Weddell gyre, SD_w^T increases by a factor of two or three. This may reflect the northward and eastward drift of the sea ice and thus a convergence of ice melt in those regions.

An estimate of that portion of the salt deficit which is eliminated by non-ice related sources of salt, SE, is shown in Plate 4. The entrainment-driven salt flux is relatively small given the predominant role of salinity on density, so $\Theta_{DS} \gg \Theta_{ES}$ and $SE \approx \Theta_{DS}$ in all but a

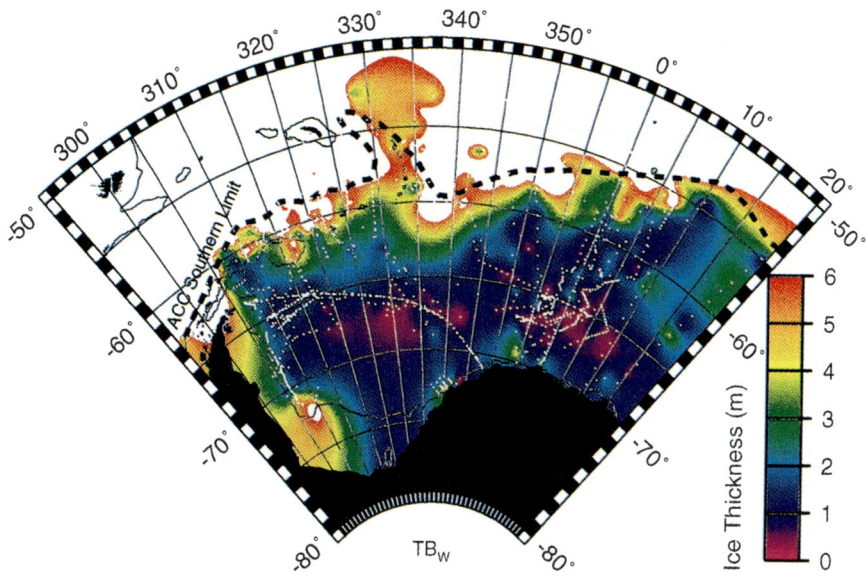

Plate 1: Winter Thermal Barrier (TB$_w$) in units of equivalent units of ice thickness (i.e., how much ice can be melted by enthalpy content of the thermocline). White areas exceed the standard range of the parameter within the polar gyre (they are "off scale"); 1000 m depth contour is given to indicate approximate location of shelf-slope break; dashed line shows approximate northern limit of polar gyre.

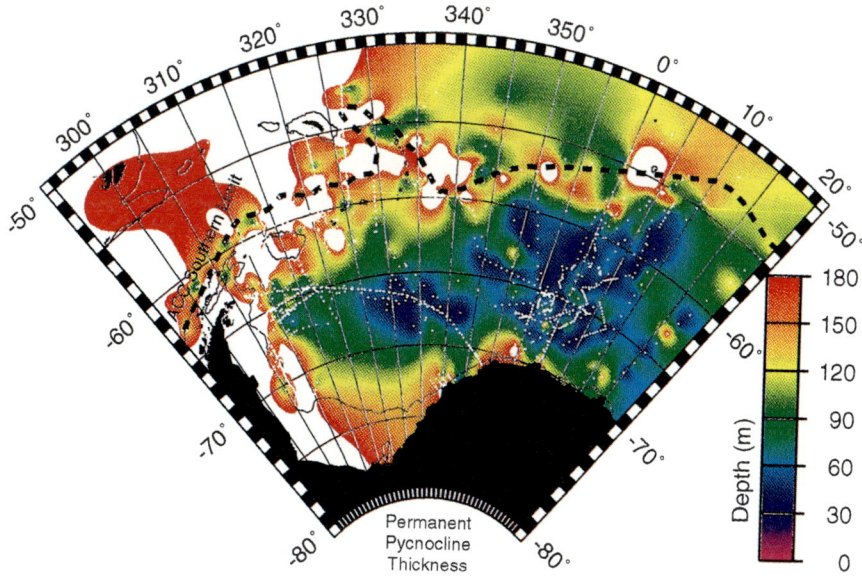

Plate 2: Permanent pycnocline thickness. Contours and white areas as in Plate 1.

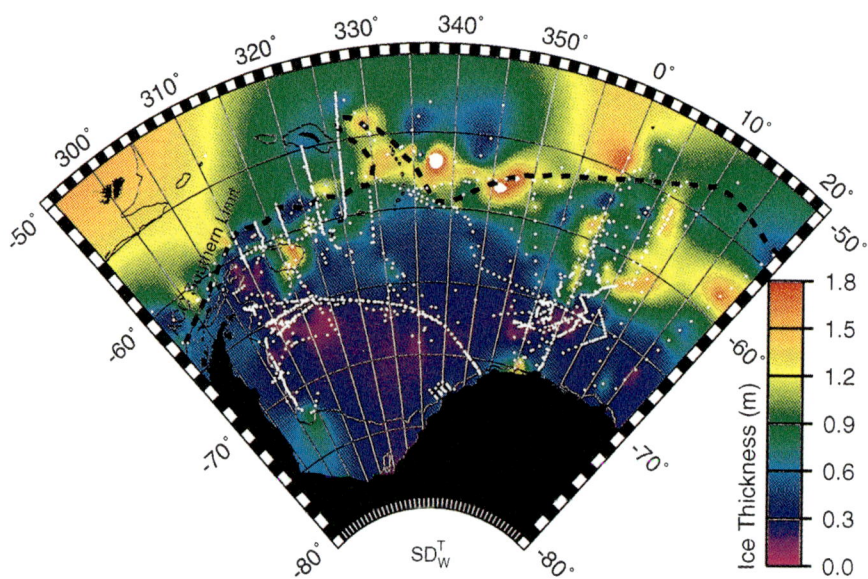

Plate 3: Total Winter Salt Deficit (SD_w^T), given in units of equivalent ice thickness (i.e., how much ice must grow to inject enough salt into the surface ocean to eliminate stabilizing freshwater layer). Contours and white areas as in Plate 1.

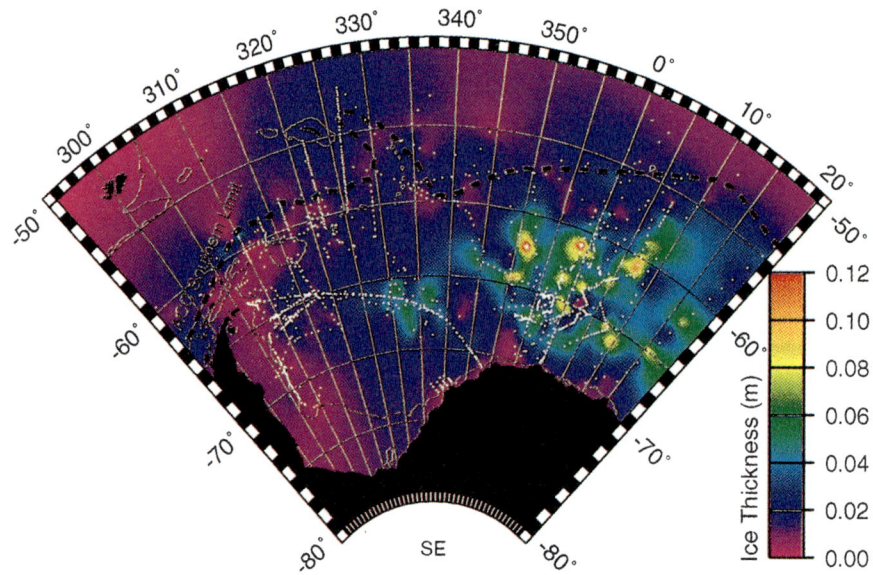

Plate 4: Contribution of "non-ice" sources of salt to winter mixed layer (SE) in units of equivalent ice thickness (i.e., how salinization is contributed to upper ocean over winter due to sources of salt other than ice growth). This is predominantly an indication of diffusion across the pycnocline. Contours and white areas as in Plate 1.

few locations where the haloclime is exceptionally weak due to abnormally large thickness. Therefore, Plate 4 is an approximation of the diffusive salt flux. Since this diffusion is proportional to the pycnocline thickness, the largest values coincide with the thinnest pycnocline, as shown in Plate 2.

Applying the SE correction to the total salt deficit, SD_w^T, gives the corrected winter salt deficit, SD_w (Plate 5). Because SE is small, typically <0.05 m, its influence on the pattern of SD_w^T is minimal so SD_w shows a similar pattern. This is not true near Maud Rise (65° S, 0° E) however, where SE represents a relatively large contribution to a relatively small SD_w^T resulting in the near elimination of SD_w^T. In other words, in that area there is a minimal stabilizing freshwater cap since the freshwater present can be almost eliminated by oceanic processes alone in the absence of an ice growth salinization contribution. Bulk stability here is most likely maintained by the stabilizing influence of the strong diffusive heat flux (shown below) and/or the potential influence of lateral processes in this rather spatially heterogeneous region. Given the latter, the bulk parameters still provide the desired spatially-averaged limitations and constraints when integrated over a slightly broader spatial area so that the full upper water column budget is properly accounted for.

SD_w indicates the maximum net thickness of *in situ* ice growth that can be realized in winter since any more ice growth rejects enough salt to overturn the water column. The gross ice thickness (more accurately, the heat loss in units of equivalent ice thickness) is equal to SD_w + TB_w, but venting of TB_w melts (or inhibits from growing) an amount equivalent to TB_w whose meltwater must then be overcome by growth of an amount equivalent to TB_w again. Thus, SD_w is ultimately the maximum *net* amount of winter *in situ* ice growth. The maximum amount of annual ice growth is the fall plus winter ice growth, or SD_s + SD_w.

Throughout much of the central gyre region the amount of ice growth required to eliminate the freshwater storage in the winter mixed layer is fairly small, typically ≤0.30 m (as previously stated, this is in addition to the fall ice growth, SD_s).

Bulk stability. The bulk stability, Σ, is shown in Plate 6. Consistent with the comments above, the least stable portion of the gyre lies along the zonal core where approximately 1 m or less of total winter ice growth, or its heat loss equivalent, overturns the water column. The minimum value occurs near the Greenwich Meridian and is equivalent to ~0.2 m. However, at this location, near Maud Rise, the lateral processes may be significant contributors to the OAI interaction and compromise the vertical bulk stability value suggested here [*Gordon and Huber*, 1984; *Bersch et al.*, 1992].

Near the gyre rim the bulk stability is approximately 5-9 m of *in situ* ice growth. There, the ability to resist overturn is formidable despite the weak pycnocline, relative to mid- and low-latitude profiles, and relative to the typical magnitude of the forcing, which is sufficient to grow approximately 3 m of ice in the absence of the ocean heat flux [*Martinson*, 1993].

The potential influence of storms on the bulk stability, that is, $\Sigma_e \approx h_i + SD_w$, is grossly approximated by assuming $h_i \sim 0.6$ m as discussed previously. This is a simple scaling of SD_w and is presented by the lower-left color bar in Plate 5. As seen, while the pattern is fairly similar to that of bulk stability (Plate 6), the Σ_e values are considerably reduced in those regions where TB_w is large (along the gyre margins). Storm-induced bulk stability reduction is less in the central gyre region where TB_w contributed less to the bulk stability initially.

While the absolute reduction in Σ is relatively small throughout the central gyre relative to the reduction at the gyre margins, bulk stability is still reduced by ~40% by storms (as seen below, TB_w represents a considerable fraction of bulk stability even in its weakest locations). This parameter needs to be supplemented by one estimating the amount of ventilation expected per typical storm, and one estimating the magnitude of storm required to completely ventilate TB_w, in order to determine which areas are most susceptible to realizing the full bulk stability reduction by storms.

Plate 7 shows that fraction of bulk stability that is attributable to TB_w, $\gamma_{TB} = TB_w/\Sigma$. This clearly reveals that bulk stability throughout most of the area is due to TB_w, that is, to the deep ocean heat, not the surface freshwater layer. In fact, TB_w accounts for 70-90% of the bulk stability throughout most of the region. Its influence is weakest near the Greenwich Meridian, where it still accounts for almost 50% of the bulk stability, and in a few regions where it gets as low as 30% where the thermocline is exceedingly sharp so its enthalpy content is quite small.

The effectiveness of TB_w in maintaining bulk stability is given by the ice melt to growth ratio, $\gamma_\Theta = TB_w/SD_w$ (Plate 8). As seen, the gyre is dominated by values of this ratio greater than 1. In these regions TB_w is sufficient to significantly dampen the winter ice growth to an effective growth rate of $(\gamma_\Theta+1)^{-1}$. Thus where the values are large, a significant amount of time

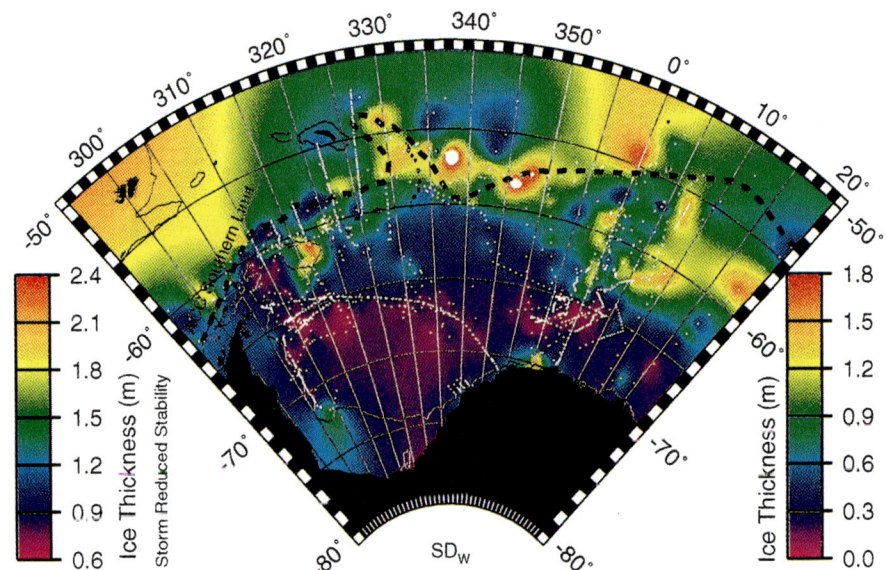

Plate 5: Winter Salt Deficit (SD_w), given in units of equivalent ice thickness (i.e., how much ice must grow to inject enough salt into the surface ocean to eliminate stabilizing freshwater layer after allowing for salt contributions by SE in Plate 4). Color scale on left indicates stability (see Plate 6) after accounting for potential influence of storms. Contours and white areas as in Plate 1.

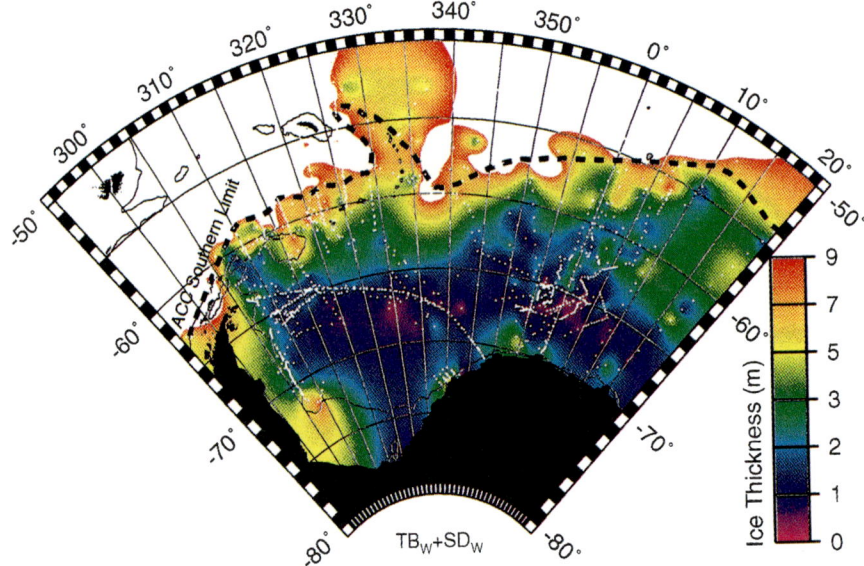

Plate 6: Stability (TB_w+SD_w), given in units of equivalent ice thickness (i.e., how much ice must grow to inject enough salt into the surface ocean to overcome both stabilizing freshwater layer and thermocline heat content, destabilizing water column, driving deep ocean convection and eliminating ice cover). Contours and white areas as in Plate 1.

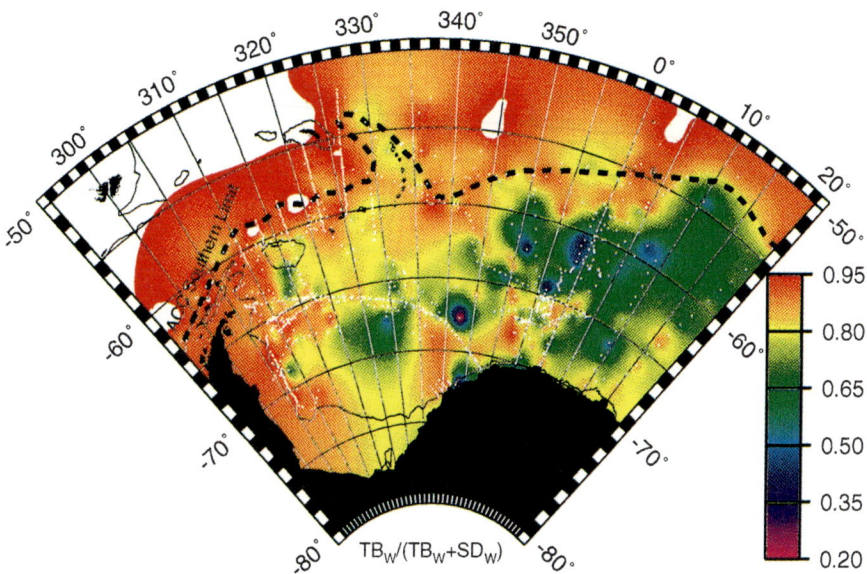

Plate 7: Winter Thermal Barrier fraction of Stability (TB$_w$/Σ). Indicates fraction of stability attributed to heat content of thermocline as opposed to freshwater content of surface layer (the latter related to ice growth/melt patterns). Contours and white areas as in Plate 1.

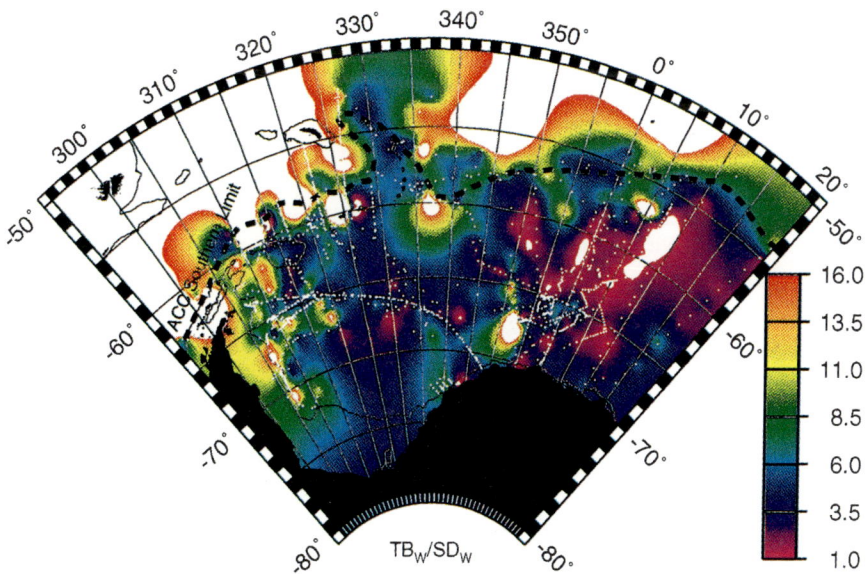

Plate 8: Ice Melt to Growth Ratio (TB$_w$/SD$_w$). Indicates how many units of ice are melted (by venting of ocean heat driven by ice growth salinization) for every unit of ice grown. Contours and white areas as in Plate 1, though in this figure, white areas also indicate regions in the parameter value less than the minimum value presented on the color bar.

will be spent under melt conditions with a minimal amount of ice growth.

Only in those regions where the surface freshwater content dominated the bulk stability, can the water column support efficient winter ice growth with respect to the negative feedback. These regions approximately parallel the primary storm tracks which may serve to keep the thermocline thin while venting the TB_w more effectively, as previously discussed. This may be a strong indication of the storm influence on the potential bulk stability.

Also, the general east-west trend, with smaller values in the east, reveals more feedback in the west. That initially seems counter-intuitive since the west is the cold regime [*Bagriantsev et al.*, 1989] where the deep waters are ~1° cooler than those to the east, and thus the deeper waters are apparently already vented, implying less resistance to ice growth. However, because the thermocline is thicker to the west, the cooler deep water is overcompensated by a thicker thermocline, storing more warm water closer to the surface and making it more accessible through a weaker stratification. In other words, the ocean can vent more heat per unit of ice grown, so the destabilization is more effective in tapping this stored enthalpy, even though the deeper water is cooler in an absolute sense.

Ocean heat flux. The parameters discussed so far give an indication of the influence of freshwater versus thermal storage in bulk stability and ice growth limitations. The final set of winter parameters provide an indication of the ocean heat flux, which more explicitly reveals the implications of the east-west increase in γ_Θ.

Plate 9 shows the winter-average eddy diffusive heat flux, F_{DT}, and its ice thickness equivalent, Θ_{DT}. Since this flux is proportional to ∇T and choice of K_z, the relative values, or spatial patterns, are more robust than the absolute heat flux values provided. From that perspective, Plate 9 reveals that the diffusive heat flux is highest in the eastern gyre where TB_w contributes relatively little to the net stability (Plate 7). This area is where the thermocline is thinnest, likely due to stronger upwelling and/or the more regular passage of intense polar lows. Both factors compress the thermocline which keeps TB_w small and ∇T large, so $F_{DT} \sim 15$ W m^{-2}. This is sufficient to prevent almost a meter of ice growth over the course of a 5-month winter growth season. Near the gyre margins, the thicker thermocline dominates, resulting in an order of magnitude reduction in diffusion.

Some fraction of the amplitude of the spatial pattern in F_{DT} may reflect the use of a spatially invariant K_z in its computation. Since surface stress influences the value of K_z and drives upwelling, which controls the characteristics of ∇T to some extent, it is possible that K_z and ∇T covary. A linear covariation would lead to an enhancement or attenuation of the spatial amplitude shown here, though it would not alter the general shape of the pattern. Thus the order of magnitude change in F_{DT} from the center of the gyre to its margins may in fact be larger or smaller depending on the degree and nature of any covariation between K_z and ∇T, but the pattern itself should be relatively robust.

Plate 10 shows the estimated entrainment heat flux, F_{ET}, and its ice thickness equivalent, Θ_{ET} assuming an average 35 W m^{-2} heat loss over the gyre. The pattern shown for F_{ET} is strongly anti-correlated with the diffusive heat flux. F_{ET} reflects the accessibility of the TB_w, which depends on both the ratio of $\nabla T/\nabla S$, as well as the thickness of the mixed layer. The close anti-correlation to the diffusive flux indicates that the mixed layer depth is fairly similar throughout the region and the dominant control on $\nabla T/\nabla S$ is the thickness of the pycnocline as described above.

The entrainment heat flux varies over the gyre by almost a factor of five, and it contributes enough heat to melt or inhibit from growing 0.4 to 1.7 m of ice (smaller values in the east and larger values in the west). Analogous to the situation with the diffusive heat flux, the entrainment heat flux may show some alteration of its amplitude if one allows for a spatially varying air-sea surface heat flux, here specified as a spatially invariant 35 W m^{-2}.

Despite strong spatial gradients in the entrainment and diffusive heat fluxes, realizing an order of magnitude difference in the diffusive flux for example, and reflecting gyre-scale processes such as upwelling, the total heat flux, $F_T = F_{DT} + F_{ET}$, shows a remarkably uniform value ($\pm 30\%$ change) throughout the gyre, of between 25-35 W m^{-2} (Plate 11). That is, even though the diffusive flux dominates in the gyre core, and the deep water is much warmer to the east, approximately 75% of a 35 W m^{-2} air-sea heat flux is ultimately provided in the form of ocean sensible heat. That is, this winter-average flux is realized either via direct ventilation, or via more indirect means in which the heat is slowly stored in an intermediate buffer, the thermocline, which is more easily eroded via surface-induced mixing, either by storms or free convection. If ∇T and K_z covary, it is possible that some spatial inhomogeneity may emerge in the total heat flux value, with the emerging pattern more similar to the diffusive heat flux spatial pattern.

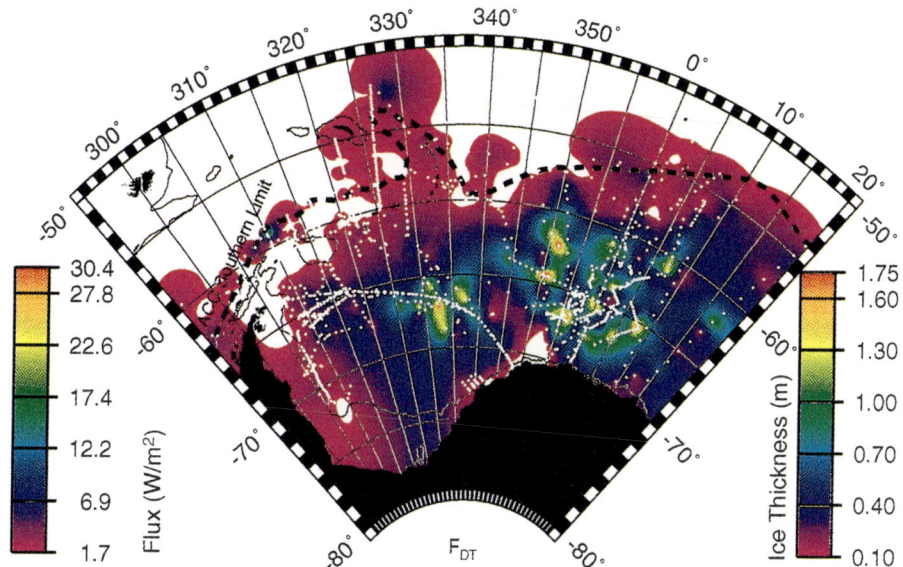

Plate 9: Average winter diffusive heat flux in units of W m^{-2} and in units of ice melt over the course of a 5 month winter. This value is proportional to the thermal gradient through the pycnocline, so the spatial pattern is more robust than the absolute numbers. Contours and white areas as in Plate 8.

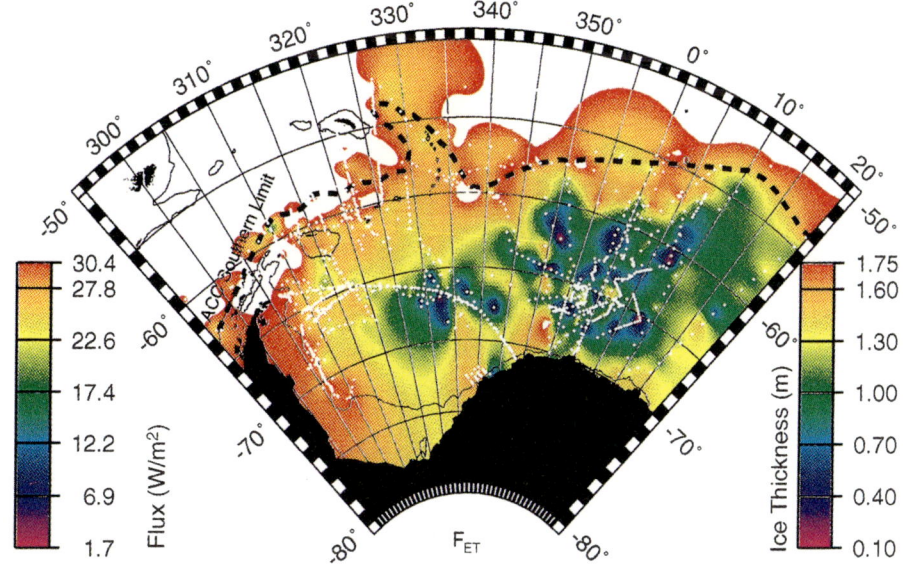

Plate 10: Average winter entrainment heat flux (units as for Plate 9). Estimate of ocean heat flux driven by entrainment in response to salinization during ice growth assuming a 35 W m^{-2} air-sea heat flux. As with diffusive heat flux, spatial patterns are likely to be more robust than absolute values. Contours and white areas as in Plate 1.

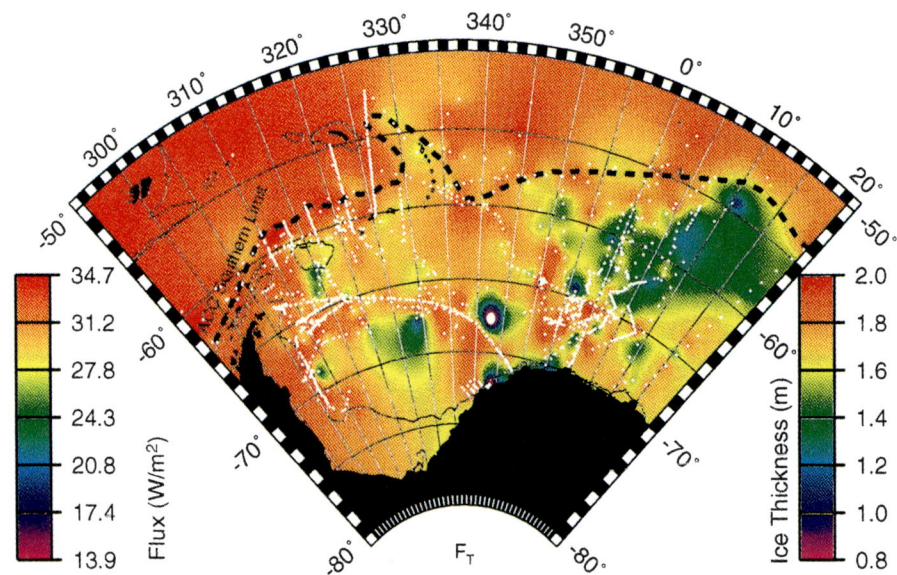

Plate 11: Total average winter ocean heat flux (units as for Plate 9). Sum of diffusive and entrainment heat fluxes. Note significant reduction in spatial variability relative to that of the two component fluxes (Plates 10 and 11). Contours and white areas as in Plate 8.

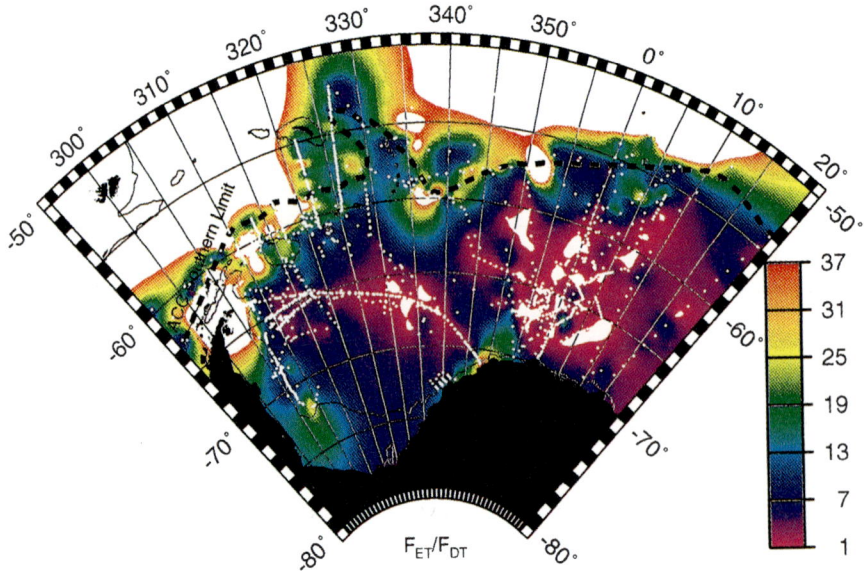

Plate 12: Ratio of average winter entrainment heat flux to average winter diffusive heat flux (units as for Plate 9). Indicates the mechanism by which heat is vented to surface. Where entrainment heat flux dominates (ratio > 1), the heat flux is predominantly driven by negative feedback in which ice growth drives entrainment and associated heat flux by salinization. Contours and white areas as in Plate 8.

Finally, Plate 12 shows the spatial patterns of the ratio between the entrainment and diffusive heat fluxes, $\gamma_T = \Theta_{ET}/\Theta_{DT}$. As seen, the entrainment heat flux is considerably larger than the diffusive heat flux in all but the core regions. Therefore, the storage of heat within the thermocline is a significantly more efficient way of venting heat from the system than simple diffusion which vents the deep water directly, but apparently far less efficiently.

The pattern here reinforces the concept that, where the pycnocline is relatively thick and the diffusive heat flux weak, ocean heat is predominantly supplied by an easily eroded pycnocline (whose enthalpy is replenished later via diffusion). In areas where upwelling brings the deep waters close to the surface (or storms bring the surface waters closer to the deep waters), exposing the deep water almost directly to the atmospheric interaction, the thermocline is stronger, and more resistant to erosion and heat release via entrainment, but it provides a considerably higher diffusive flux to accomplish a similar magnitude venting. Furthermore, where the diffusive flux is larger, ice growth is reduced so the entrainment heat flux is further inhibited by weaker salinization-induced destabilization.

Summer salt deficit. Plate 13 shows the summer salt deficit, SD_S, which varies from 0.2 - 1 m of ice. In the west, and north, the signal is predominately one of ice/snow melt, not growth because of the perennial ice cover. Also, given the relatively sparse summer data set, the values contain considerable uncertainty since we do not have enough samples to adequately average and make all of the necessary corrections, particularly removal of the temporal bias. However, the map does show a general reduction in SD_s from west to east, reflecting a thinner fall ice growth cover to the east. That is, the ice cover will be thinner in the east at the time when the winter conditions set in and the negative feedback mechanism becomes active, limiting further winter ice growth.

Critical interannual ice growth perturbation. Plate 14 shows the size of the perturbation in annual ice thickness required to destabilize the water column, $\Gamma_{ms} = 1/\gamma_{ms} = (SD_s + \Sigma_e)/SD_s$ (the relationship breaks down for reasons already discussed in the perennial ice fields to the west as indicated in Figure 2). In the eastern Weddell region, a couple of locations achieve ratios as low as 1.8. This indicates that the ice growth would have to exceed the annual climatological average by 80% in order to overturn the system. Unfortunately, the sparse summer coverage and temporal statistics are insufficient to provide decent spatial coverage, and to evaluate the likelihood of obtaining a perturbation of this magnitude during any one particular year. Once we obtain enough data to determine interannual ice thickness variance, we can estimate the likelihood of destabilization for any particular location. At present, the data can only demonstrate the concept and provide an indication of the approximate size of critical perturbations required in a few isolated locations.

Temporal Variability

The temporal variability in the various parameter values provides the variance, time-scales of variability and longer-term trends. These reveal tendencies for change, the magnitude of variability (allowing estimates of the likelihood of exceeding critical stability values as discussed above) and climate drift. Unfortunately, the current data base does not allow meaningful interannual comparisons since there is inadequate multi-year sampling with close enough spatial proximity, as dictated by local decorrelation lengths, to differentiate spatial from temporal variability.

For example, the area around Maud Rise has been sampled a number of times over the last couple of decades, but the actual overlap of stations within the local decorrelation lengths is quite small. Plate 15 shows bulk stability, Σ, for seven different years in this locale: 1977, 1981, 1984, 1986, 1989, 1992 and 1994. The parameters are interpolated between stations spanning gaps as large as 3° longitude, which is approximately three times the spatial decorrelation length. This large spread is necessary to convey some sense of the parameter distribution for comparison of one year to the next.

Focusing on the Greenwich Meridian at 65° S provides some sense of an increase in bulk stability in 1977 (immediately following restabilization of the area after termination of the Weddell polynya, *Zwally and Gloersen*, 1977). The values increase from ~0.7 m to ~2 m in 1984-1992, then decrease to ~0.5 m in 1994. Presently, it is difficult to distinguish whether this reflects a systematic change in bulk stability of the region, or minor spatial shifts in the presence of the extremely high local lateral gradients.

5. DISCUSSION AND CONCLUSIONS

The bulk parameters presented here are designed to encapsulate the physical essence of much of the ocean-ice interaction within the Antarctic polar oceans. In particu-

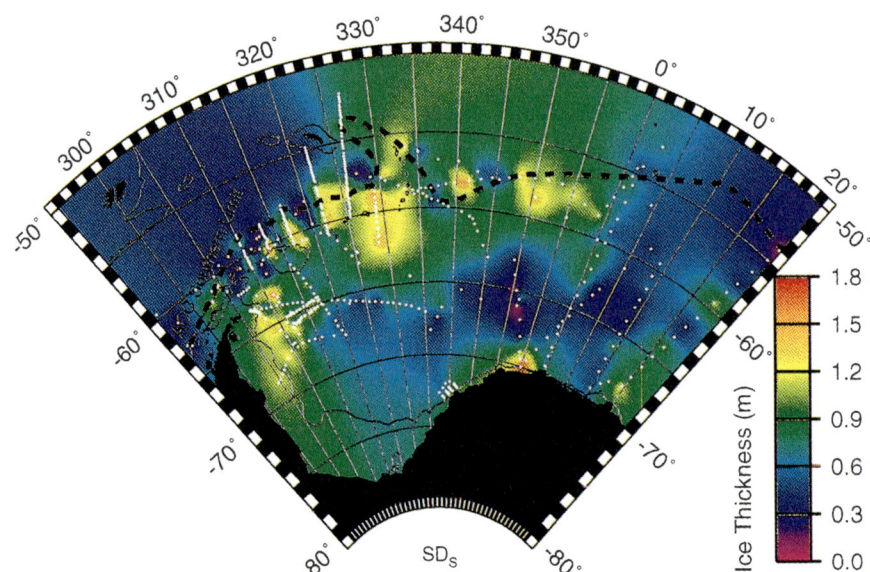

Plate 13: Summer Salt Deficit (SD_s). Similar to Plate 3, but for summer surface layer. Indicates amount of freshwater contained in summer surface layer, predominantly an indication of ice meltwater from previous winter, and an indication of how much ice will grow rapidly in fall before winter conditions are achieved and winter heat fluxes reduce ice growth rate. Contours and white areas as in Plate 1.

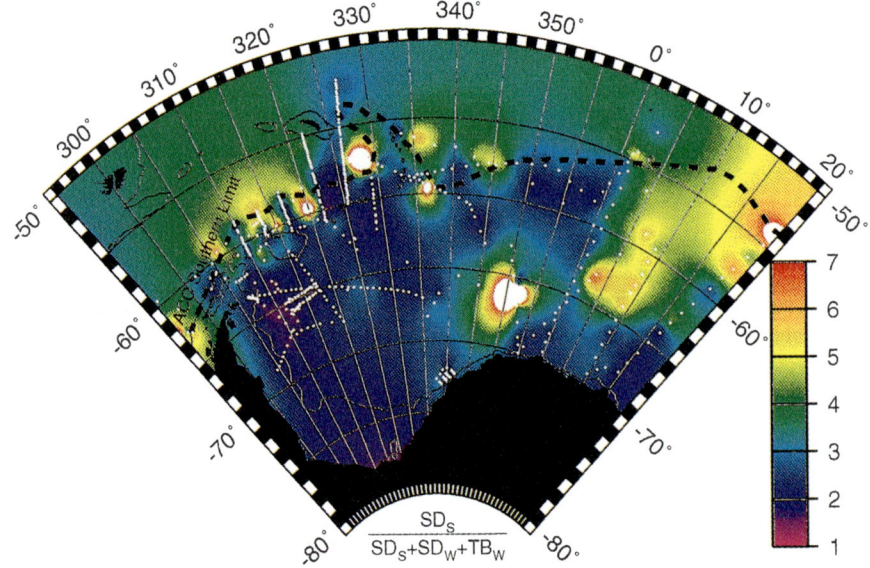

Plate 14: Critical Ice Growth Perturbation. This gives the amount of ice growth that is cycled each year through ice growth relative to the total amount of ice growth required to destabilize the water column. The fraction indicates how much of an interannual perturbation in annual average ice growth is required to destabilize the water column. Interpretation does not hold for the perennial ice regions in the western Weddell Sea. Contours and white areas as in Plate 1.

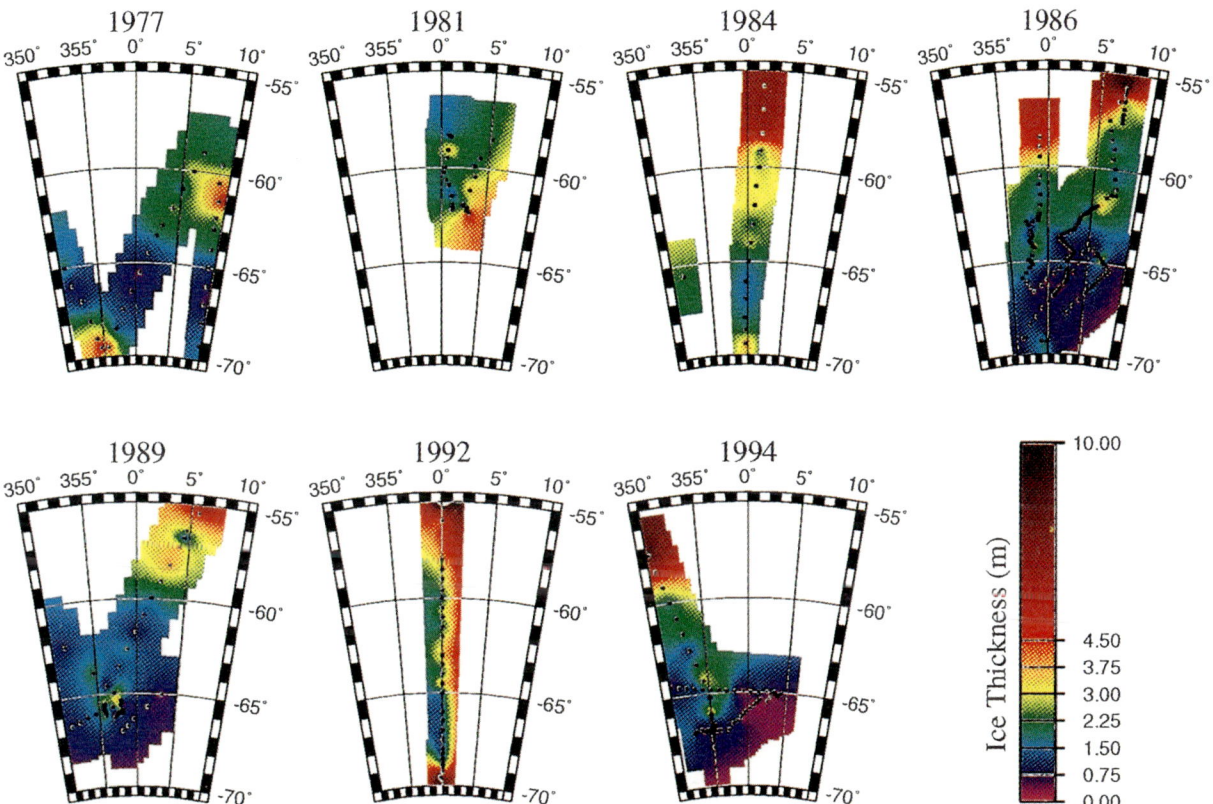

Plate 15: Stability (as in Plate 6) for 7 different years near the Greenwich Meridian. Decorrelation lengths are only about one-third of the color swath widths, so comparisons are difficult except in areas where repeat stations have been obtained.

lar, they provide insights and constraints on the system's ability to grow ice, the rates and limits of ice growth, and the influence of ice growth/melt on the ocean stability and heat flux. The absolute values of the parameters vary within relatively large (≤30%) intrinsic uncertainties, many owing to a lack of sufficient data, but their relative distributions show a good spatial signal to local noise ratio (~20 db). As such, the climatological maps of the parameters provide some intriguing relationships and patterns within the Weddell gyre where these general relationships have been applied. Specifically:

(1) The upper ocean freshwater content, controlled predominantly by the overlying ice dynamics and thermodynamics, and the upper ocean heat distribution, controlled predominantly by the large scale gyre dynamics and deep water characteristics, dictate the maximum amount of *in situ* ice growth and growth rates throughout the Weddell gyre region. Regarding thickness (growth rates are considered in point 4 below), it is seen from the summer salt deficit (Plate 13), that the fall ice growth, that is, the thickness of ice that grows rapidly in fall while eliminating the seasonal pycnocline prior to the onset of winter conditions, is typically 0.6 m or less. It is thinner than 0.3 m in some regions near the gyre core. This 0.3-0.6 m of ice can grow rapidly each fall since the seasonal pycnocline buffers the surface from the warm ocean deep water. Once the summer pycnocline is eliminated by this growth, the freshwater and heat content of the upper ocean during winter limits the ability of the water column to supporting only 0.5-1.0 m of additional winter ice growth throughout much of the gyre core (bulk stability; Plate 6). There are large regions, however, that can support another 1-3 m of growth before destabilizing the water column. If this maximum growth is exceeded, as has happened in the past as evidenced by the Weddell polynya, the water column will overturn and the resulting mode change cannot support an ice cover until the water column is eventually restabilized by a significant influx of freshwater at the surface.

(2) One of the most interesting results is that on regional scales, the ocean-ice system manages to vent the deep water at an average winter rate of 25-35 W m^{-2} throughout the gyre (Plate 11), regardless of the large scale stratification and dynamic setting. That is, despite the fact that the turbulent diffusive ocean heat flux varies by over an order of magnitude throughout the gyre (Plate 9), and the ocean entrainment heat flux varies by just under an order of magnitude (Plate 10), their sum, representing the net ocean sensible heat flux, only varies by ~30% across the gyre.

This predominantly indicates that where the pycnocline is relatively thick and the diffusive heat flux correspondingly weak, ice growth, unencumbered by a strong ocean heat flux, drives static instability due to salinization of the water column, which in turn drives an entrainment heat flux by eroding the weak pycnocline (whose enthalpy is replenished later via diffusion). The erosion may also be accomplished by storm-induced mixing (not associated with ice growth).

Alternatively, where the deep waters are close to the surface due to upwelling or the surface waters are close to the deep waters by storm mixing, the thermocline is stronger and more resistant to erosion and heat release via entrainment, but it provides a considerably higher diffusive flux to accomplish a similar magnitude venting. Furthermore, where the diffusive flux is larger, ice growth is reduced so the entrainment heat flux is further inhibited by weaker salinization-induced destabilization.

The diffusive and entrainment fluxes have complementary dependencies, so the system is ultimately successful in extracting the 25-35 W m^{-2} from the deep water regardless of which process dominates. Some of this spatial homogeneity in total ocean heat flux may be the result of using a spatially-invariant eddy diffusivity coefficient for estimating the diffusive heat flux, and assuming a spatially-invariant air-sea heat flux when computing the entrainment heat flux.

(3) Intense mixing due to the passage of intense polar lows may serve to reduce the bulk stability of the water column by as much as 75% (Plates 6 and 7) by venting the stabilizing heat contained within the thermocline independently of free convection driven by salinization during ice growth. This decouples the ice growth destabilization from the entrainment ocean heat flux. Also, storms may contribute to the particularly thin pycnocline in the eastern portion of the gyre, which enhances the ocean's diffusive heat flux, but reduces the ability of the ocean to resist ice growth through the negative feedback mechanism in which the ocean heat flux is increased by entrainment driven by salinization during ice growth.

(4) The large enthalpy content of the thermocline throughout most of the Weddell gyre region effectively reduces the ability to grow ice by a factor of 2-6 (see Plate 8). That is, the ice grows at a rate 2-6 times slower than expected by only considering the surface heat loss and ocean diffusive heat flux. Also, in the regions where the ice melt to growth rate ratio exceeds one, especially where it is considerably higher than one, we might expect long periods of significant melting. This basal melting may lead to negative ice freeboard given the weight of the snow on the ice, and thus these

regions where the ice melt to growth ratio exceeds one may correspond with regions in which ice flooding by seawater is most prevalent.

(5) Most of the bulk stability of the water column (given as the maximum amount of winter ice growth) is attributed to the enthalpy content of the thermocline (Plate 7), not by direct reduction in ice growth by a strong diffusive heat flux. That is, the majority of the ocean heat flux appears to originate from either entrainment driven by storms, or entrainment driven by ice growth. In both cases, the entrainment releases the enthalpy stored within the thermocline, which then acts to melt existing ice or to inhibit additional ice growth. This form of stabilization involves a more active ice growth-melt cycling since ice growth drives the ocean heat flux which drives ice melt, etc. In regions dominated by a diffusive heat flux, the ocean heat simply reduces the rate of ice growth and the entrainment heat flux is relatively minor because of the strong pycnocline.

The entrainment heat flux, when driven by ice growth, is the mechanism by which ocean sensible heat is vented to the atmosphere even when the surface layer is initially at the freezing point and thus can only give up heat in the form of latent heat of fusion. This latent heat loss must generate ice growth which drives entrainment, releasing sensible heat stored within the permanent thermocline. The results here suggest that the entrainment heat flux dominates the total ocean heat flux.

(6) Perturbations in the annual *in situ* ice growth of ≥80% are required to destabilize the water column throughout much of the Weddell gyre where summer data are available. However, these estimates are based on a small summer sample size. The likelihood of perturbations of such size in any one particular year must be estimated from more extensive multi-year sampling.

Finally, the bulk parameters presented here involve vertically-integrated property distributions, and, as such, they provide constraints or limitations on the ocean-ice system behavior over the appropriately averaged time scale — in this case, seasonal time scales. Consequently, they imply a mean seasonal evolution which may be considerably different from the actual time-dependent behavior. Also, they must still be diagnosed against complete models and modified to include any relevant nonlinear physics influencing the mean behavior. Some attempt was made to estimate the important influence of storms in this analysis. In general, the parameters serve to demonstrate the extent to which fairly fundamental characteristics of the OAI system may be extracted from simple-to-observe features of the water column. Additional temporal coverage is required to ultimately determine the distributions, allowing assessment of the likelihood of destabilization in the system and significant change in the ocean-ice behavior.

Because the parameters discussed here represent physically meaningful combinations of the water column features, these, or other such combinations, may represent more physically meaningful (and sensitive) diagnostics for model-data comparison than profile shapes or individual property values.

Acknowledgments. Special thanks to Karen Heywood, Eberhard Fahrbach and Michael Schroeder for making their CTD data readily available for this study, and to Christoph Kottmeier for making his buoy data available. This work was supported by National Science Foundation research grant OPP93-17231. Lamont-Doherty Earth Observatory contribution number 5690.

REFERENCES

Ackley, S.F., M. Lange and P. Wadhams, Snow cover effects on Antarctic sea ice thickness. In, *Sea Ice Properties and Processes*: Proc. of the W.F. Weeks Symposium, S.F. Ackley and W.F. Weeks, eds., CRREL Monograph 90-1, 16-21, 1990.

Akitomo, K., T. Awaji and N. Imasato, Open-ocean deep convection in the Weddell Sea: two-dimensional numerical experiments with a nonhydrostatic model, *Deep-Sea Res.*, 42, 53-73, 1995.

Bagriantsev, N.V., A.L. Gordon, and B.A. Huber, Weddell Gyre: Temperature maximum stratum, *J. Geophys. Res.*, 94, 8331-8334, 1989.

Bersch, M., G.A. Becker, H. Frey, K.P. Koltermann, Topographic effects of Maud Rise on the stratification and circulation of the Weddell Gyre, *Deep Sea Res.*, 303-331, 1992.

Eicken, H., and M.A. Lange, Development and properties of sea ice in the coastal regime of the southeastern Weddell Sea, *J. Geophys. Res.*, 94, 8193-8206, 1989.

Foldvik, A. and T. Gammelsrod, Notes on Southern Ocean hydrography, sea-ice and bottom water formation, *Palaeogeogr., Palaeoclimatol., Palaeoecol.*, 67, 3-17, 1988.

Garwood, R.W., Jr., S.M. Isakari and P.C. Gallacher, Thermobaric convection, in *The Polar Oceans and Their Role in Shaping the Global Environment*, Johannessen, Muench and Overland (eds.), Geophys. Monogr. 85, 199-209, 1994.

Gordon, A.L., Deep Antarctic convection west of Maud Rise, *J. Phys. Oceanogr.*, 8, 600-612, 1978.

Gordon, A.L., Seasonality of Southern Ocean sea ice, *J. Geophys. Res.*, 85, 4193-4197, 1981.

Gordon, A.L., Two stable modes of Southern Ocean winter stratification, in *Deep Convection and Deep Water Formation in the Oceans*, eds. Chu and Gascard, Elsevier Science Publishers, 17-35, 1991.

Gordon, A.L. and B.A. Huber, Thermohaline stratification below the Southern Ocean sea ice, *J. Geophys. Res.*, 89, 641-648, 1984.

Gordon, A.L., and B.A. Huber, Southern Ocean winter mixed layer, *J. Geophys. Res.*, 95, 11655-11672, 1990.

Gordon, A.L., B.A. Huber, H.H. Hellmer and A. Ffield, Deep and bottom water of the Weddell Sea's western rim, *Science*, 262, 95-97, 1993.

Gordon, A.L. and E.J. Molinelli, Southern Ocean Atlas: Thermohaline and Chemical Distributions and the Atlas Data Set, International Decade of Ocean Exploration, Columbia University Press, 1982.

Gregg, M.C., The dependence of turbulent dissipation on stratification in a diffusively stable thermocline, *J. Geophys. Res.*, 93, 12381-12392, 1988.

Hansen, J., A. Lacis, D. Rind, G. Russell, P. Stone, I. Fung, R. Ruedy, and J.Lerner, Climate sensitivity: analysis of feedback mechanisms. *Climate Processes and Climate Sensitivity*, Geophys. Monogr. 29, Maurice Ewing Volume 5, American Geophysical Union, 130-163, 1984.

Heywood, K.J. and B.A. King, WOCE section A23 Cruise Report, Series No. 1, Southampton Oceanography Center, University of East Anglia, Norwich, 75 pp, 1996.

Imbrie, J., E.A. Boyle, S.C. Clemens, A. Duffy, W.R. Howard, G. Kukla, J. Kutzbach, D.G. Martinson, A. McIntyre, A.C. Mix, B. Molfino, J.J. Morley, L.C. Peterson, N.G. Pisias, W.L. Prell, M.E. Raymo, N.J. Shackleton, and J.R. Toggweiller, On the structure and origin of major glaciation cycles. 1. Linear responses to Milankovitch forcing. *Paleoceanography*, 7,6, 701-738, 1992.

Jacobs, S.S., H.H. Hellmer, C.S.M. Doake, A. Jenkins and R.M. Frolich, Melting of ice shelves and the mass balance of Antarctica, *J. Glaciology*, 38, 1992.

Killworth, P.D., On "chimney" formations in the ocean, *J. Phys. Oceanogr.*, 9, 531-554, 1979.

Kottmeier, C., S. Ackley, E. Andreas, D. Crane, H. Hoeber, J. King, J. Launiainen, D. Limbert, D. Martinson, R. Roth, L. Sellmann, P. Wadhams and T. Vihma, Wind, temperature and ice motion statistics in the Weddell Sea, World Climate Research Programme, International Programme for Antarctic Buoys, WMO/TD-No.797, 1997.

Ledley, T.S., The climatic response to meridional sea-ice transport, *J. Climate*, 4, 147-163, 1991.

Ledwell, J.R., A.J. Watson and C.S. Law, Evidence for slow mixing across the pycnocline from an open-ocean tracer-release experiment, *Nature*, 364, 701-703, 1993.

Manabe, S., R. J. Stouffer, M.J. Spelman and K. Bryan, Transient responses of a coupled ocean-atmosphere model to gradual changes of atmospheric CO_2. Part I: Annual mean response. *J. of Climate*, 4, 785-818, 1991.

Martinson, D.G., P.D. Killworth, A.L. Gordon, A convective model for the Weddell polynya, *J. Phys. Oceanogr.*, 11, 466-488, 1981.

Martinson, D.G., Evolution of the Southern Ocean winter mixed layer and sea ice; open ocean deepwater formation and ventilation, *J. Geophys. Res.*, 95, 11641-11654, 1990.

Martinson, D.G., Ocean heat and seasonal sea ice thickness in the Southern Ocean, NATO ASI Series, Vol. I, Ice in the Climate System, ed. W. Richard Peltier, Springer-Verlag, 1993.

McPhee, M.G., S.F. Ackley, P. Guest, B.A. Huber, D.G. Martinson, J.H. Morison, R.D. Muench, L. Padman, T.P. Stanton, The Antarctic Zone Flux Experiment, *Bull. Amer. Met. Soc.*, 77, 1221-1232, 1996.

Meehl, G.A. and W.M. Washington, CO_2 climate sensitivity and snow-sea-ice albedo parameterization in an atmospheric GCM coupled to a mixed-layer ocean model, *Climate Change*, 16, 283-306, 1990.

Motoi, T., N. Ono and M. Wakatsuchi, A mechanism for the formation of the Weddell polynya in 1974, *Am. Meteorol. Soc.*, 17, 2241-2247, 1987.

Orsi, A.H., T.I. Whitworth and W.D.J. Nowlin, On the meridional extent and fronts of the Antarctic Circumpolar Current, *Deep Sea. Res.*, 42, 641-673, 1995.

Parkinson, C.L., On the development and cause of the Weddell polynya in a sea ice simulation, *J. Phys. Oceanogr.*, 13, 501-511, 1983.

Rind, D., R. Healy, C. Parkinson and D. Martinson, The role of sea ice in $2XCO_2$ climate model sensitivity. Part I: The total influence of sea ice thickness and extent, *Amer. Meteorol. Soc.*, 1-15, 1995.

Schlesinger, M., and J. Mitchell, Model projections of the equilibrium climatic response to increased carbon dioxide, in *The Potential Climatic Effects of Increasing Carbon Dioxide*, DOE/ER-0237, 81-148, U.S. Dept. of Energy, 1985.

Schlosser, P., W. Roether and G. Rohardt, ^{3}He balance of the upper layers of the northwestern Weddell Sea, *Deep-Sea Res.*, 34, 365-377, 1987.

Sellmann, L., and Ch. Kottmeier, Sea ice buoys 1991 - 1995 data documentation, Alfred-Wegener-Institut für Polar- und Meeresforschung, *Berichte aus dem Fachbereich Physik*, 1996.

Smith, W.H.F. and P. Wessel, Gridding with continuous curvature splines in tension, *Geophys.*, 55, 293-305, 1990.

Wadhams, P., M.A. Lange and S.F. Ackley, The ice thickness distribution across the Atlantic sector of the Antarctic Ocean in midwinter, *J. Geophys. Res.*, 92, 14535-14552, 1987.

Wadhams, P., Sea ice thickness changes and their relation to climate, In: *The Polar Oceans and Their Role in Shaping the Global Environment*, eds., Johannessen, Muench and Overland, Geophys. Monogr., 337-361, 1994.

Walin, G., On the formation of ice on deep weakly stratified water, *Tellus*, 45A, 143-157, 1993.

Washingon, W.M. and G.A. Meehl, Seasonal cycle experiment on the climate sensitivity due to a doubling of CO_2 with an atmospheric general circulation model coupled to a simple mixed-layer ocean model, *J. Geophys. Res.*, 89, 9475-9503, 1984.

Wessel, P. and W.H.F. Smith, Free software helps map and display data, *EOS Trans. Amer. Geophys. U.*, 72, 441, 445-446, 1991.

Zwally, H.J. and Gloersen, P., Passive microwave images of the polar regions and research application, *Polar Rec.*, 18, 431-450, 1977.

_____

Douglas G. Martinson and Richard A. Iannuzzi, Lamont-Doherty Earth Observatory, and Department of Earth and Environmental Sciences, Columbia University, Palisades, NY 10964

(Received October 29, 1996;
accepted August 14, 1997.)

ICE FORMATION IN COASTAL POLYNYAS IN THE WEDDELL SEA AND THEIR IMPACT ON OCEANIC SALINITY

Thorsten Markus[1]

Institute of Environmental Physics, University of Bremen, Germany

Christoph Kottmeier[2] and Eberhard Fahrbach

Alfred-Wegener-Institute for Polar and Marine Research, Bremerhaven, Germany

The intense ice production and resultant brine rejection in coastal polynyas is widely recognized. Quantitative analyses need continuous information on polynya sizes and meteorological conditions. We apply a special method using satellite passive microwave data to obtain polynya areas in the Weddell Sea and the ECMWF atmospheric model analyses as inputs for heat flux calculations for the years 1992 through 1994. The passive microwave method allows us to estimate polynya areas continuously along the coast which are not resolved in operational ice concentration analyses or have large errors due to land contamination. The results show that the coastal polynyas occupy only a relatively small area of the ice covered part of the Weddell Sea, about 0.2%, during the winter months when the ice production is highest. However, in spite of the small areal fraction of the polynyas, they produce between 2.5% and 5% of the total Weddell Sea ice volume. The fixed position of the polynyas at the coast, together with relatively slow mean currents on the shelves allow the accumulation of salt in coastal waters. To derive time series of the salinity increase in the coastal current due to brine release in the polynyas, we prescribe a clockwise circulation along the coastline. The mean drift of a water parcel along that trajectory is determined by adjusting the salt gain to the observed salinity, 34.06 psu, in the coastal belt in the eastern Weddell Sea and 34.49 psu off the northern tip of the Antarctic Peninsula. The results indicate an effective drift velocity of the coastal current of 2 to 5 km d^{-1}.

1. INTRODUCTION

A polynya is an oceanic area which remains either partially or totally ice free at times and under meteorological conditions when the ocean is expected to be ice-covered [*Smith et al.*, 1990]. This situation generally can be attributed to one of two mechanisms. (1) Ice formed within the area is continually removed by the physical action of winds or currents on the ice. The oceanic heat lost to the atmosphere is balanced in this case by the latent heat of fusion of the ice which continually forms. (2) Ice formation is prevented by the influx of oceanic heat into the local region. These two mechanisms are not mutually exclusive, and in many cases both contribute to the maintenance of polynyas [*Jacobs and Comiso*, 1989]. *Gordon and Comiso* [1988] distin-

[1] Now at NASA Goddard Space Flight Center, Greenbelt, Maryland
[2] Also at Institute for Environmental Physics, University of Bremen, Germany

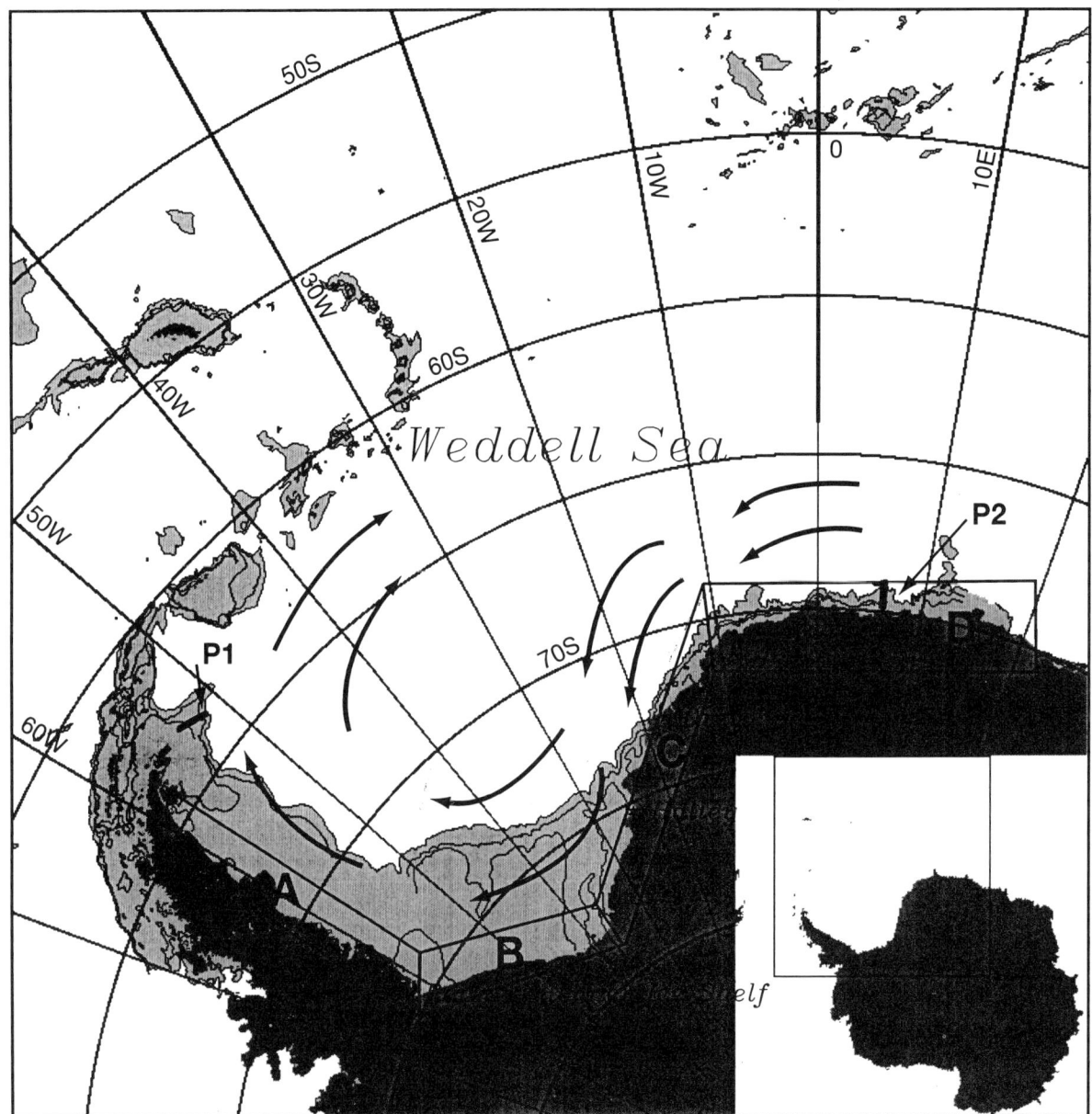

Fig.1. The area of study. The Antarctic continent, including ice shelves, is dark gray. The continental shelf, from the General Bathymetric Chart of the Oceans (GEBCO) Digital Atlas published by the British Oceanographic Data Centre on CD-ROM, is light gray. Lines indicate the 2000, 1000, and 500 m depth contours. The arrows indicate the mean direction of the Weddell Gyre and Coastal Current. The boxes, labeled A–D, indicate four regions of the Weddell Sea coast which are distinguished to study regional differences. The thicker bars, P1 and P2, show the locations of in-situ temperature and salinity measurements presented in Figure 2.

guish polynyas according to their location, either as coastal polynyas or open ocean polynyas. Along the Antarctic coast, coastal polynyas are mostly wind-driven, since over the continental shelf the entire water column is near the freezing point [*Zwally et al.,* 1985] and thus no heat from below can be supplied. Therefore, Antarctic coastal polynyas are sites of high ice production and are responsible for a large amount of the Antarctic sea ice [*Gordon and Comiso*, 1988]. The presence of open water in the pack ice can increase the ocean-atmosphere heat flux by a factor of 10 to 100, depending on atmospheric conditions [*Maykut*, 1978]. The high ice production rate, about 10 m yr^{-1} [*Cavalieri and Martin*, 1985; *Zwally et al.*, 1985], in coastal polynyas induces intensive brine release from sea ice into the water column. This flux of salt is an important condition for the formation of Antarctic Bottom Water, 70 % of which contains water masses that obtained their characteristics in the Weddell Sea [*Carmack*, 1977]. In comparison, ice formation rates in the central Weddell Sea are about 1 m yr^{-1} [*Wadhams et al.*, 1987], (Strass and Fahrbach, pers. comm.), which, under normal conditions, is not enough to destabilize the water column. We, therefore, concentrate our study on the coastal region of the Weddell Sea from 10°E to the tip of the Antarctic Peninsula (Figure 1).

In order to quantify the role of Antarctic coastal polynyas in determining sea ice coverage and salinity distribution variability, it is necessary to continuously measure their extent as well as estimate the energy exchange over the polynyas. *Kottmeier and Engelbart* [1992] studied the forcing mechanisms for Antarctic coastal polynyas and the thermodynamic effects by means of an air-sea interaction experiment at the eastern coast of the Weddell Sea. Because of their large spatial and temporal variability, as coastal polynyas form most rapidly with offshore ice motion, it is necessary to measure their sizes on a daily basis. Here, we use satellite passive microwave data from the DMSP special sensor microwave imager (SSM/I) to monitor polynya recurrence, persistence, and variability. Unlike measurements with satellite high-resolution visible and infrared sensors, these observations are not hampered by frequent cloud cover and the absence of solar illumination during the polar night. However, due to the long wavelength, passive microwave sensors have a relatively coarse resolution of about 1000 km^2. Therefore, operational ice concentration algorithms [e.g. *Cavalieri et al.*, 1984; *Comiso*, 1995] cannot unambiguously distinguish coastal polynyas from leads further offshore, since both might lie inside one antenna footprint. Also, they may overestimate open water area because of land contamination in coastal pixels. A method specially developed for the detection of coastal polynyas from passive microwave data [*Markus and Burns*, 1995] is applied in this study. With this method, estimates of polynya areas are derived daily, allowing the opening and closing of individual polynyas along the coastline of the Weddell Sea to be monitored.

The energy exchange with the atmosphere and the resultant ice production is calculated with a simple thermodynamic model. The contribution of ice production in coastal polynyas to the total Weddell Sea ice volume is investigated by comparison with ice concentration analyses. The effect on the ocean due to brine release is calculated as the integrated salt flux into the water column. Previous studies of coastal polynyas [*Cavalieri and Martin*, 1985; *Zwally et al.*, 1985; *Martin and Cavalieri*, 1989] neglect oceanic currents in their salt budget calculations. There, the salt released from polynya ice production increases the salinity of the water column without being advected. In this study, we include the ocean current along the continental shelf around the Weddell Sea. The movement of water parcels most strongly affects the distribution of water properties [*Carmack*, 1986]. The combination of local salt release with the motion of the water parcels along the coast allows us to simulate the salt increase of a coastal flow in the southern Weddell Sea which is suggested by the large-scale salinity distribution with significantly higher salinities in the western Weddell Sea than in the Indian Ocean sector [e.g. *Carmack*, 1977]. This is also shown by in-situ hydrographic measurements from R/V Polarstern which show mean surface layer salinities of 34.06 psu in the eastern Weddell Sea at 6°E (Figure 2a (top)) and 34.49 psu in the western Weddell Sea (bottom). With this study, we hope to give a better feel for the importance of coastal polynyas in the Weddell Sea on sea ice coverage and salt budgets and whether there are regional differences.

2. CALCULATION OF HEAT FLUX AND ICE PRODUCTION IN POLYNYAS

The surface energy balance in Antarctic coastal polynyas can be written as

$$H_s + H_e + Q_{lu} - Q_{ld} - Q_s = H_{tot} = \varrho_i L_f F \quad (1)$$

where H_s and H_e are the sensible and latent heat fluxes respectively, Q_{lu} and Q_{ld} are the upward and downward longwave radiative fluxes, and Q_s is the net solar radiative flux. The atmospheric energy

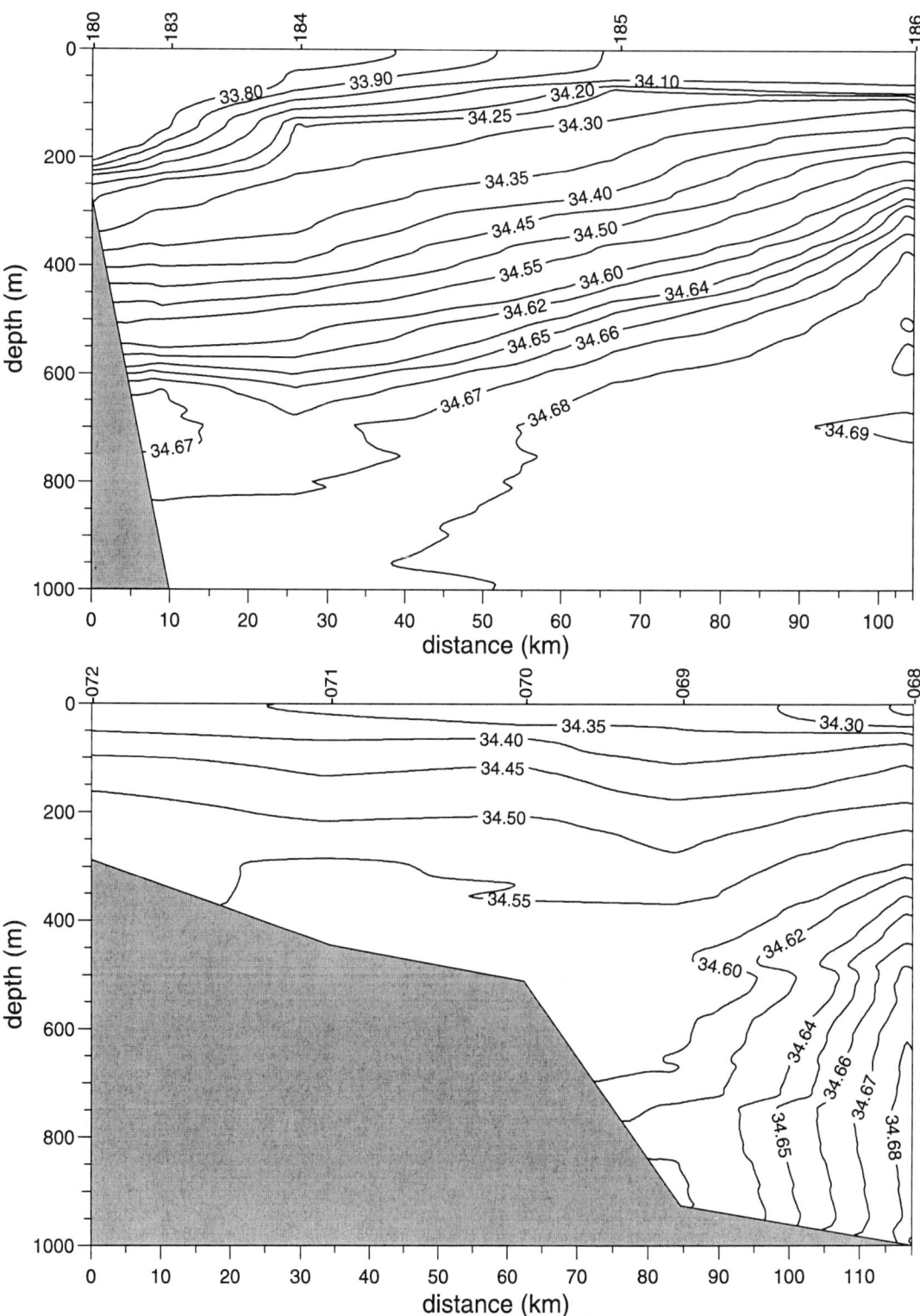

Fig. 2a. Salinity (a) and potential temperature (b) transects perpendicular to the depth contours across the coastal current in the eastern Weddell Sea from 69°57'S, 6°19'E to 69°01'S, 6°12'E (top) and in the western Weddell Sea from 63°37'S, 51°31'W to 63°13'S, 53°42'W (bottom) carried out by 'R/V Polarstern' in February 1991 and January 1993. The locations are indicated in Figure 1 as P1 and P2, respectively. Numbers at the top of each plot indicate CTD measurements.

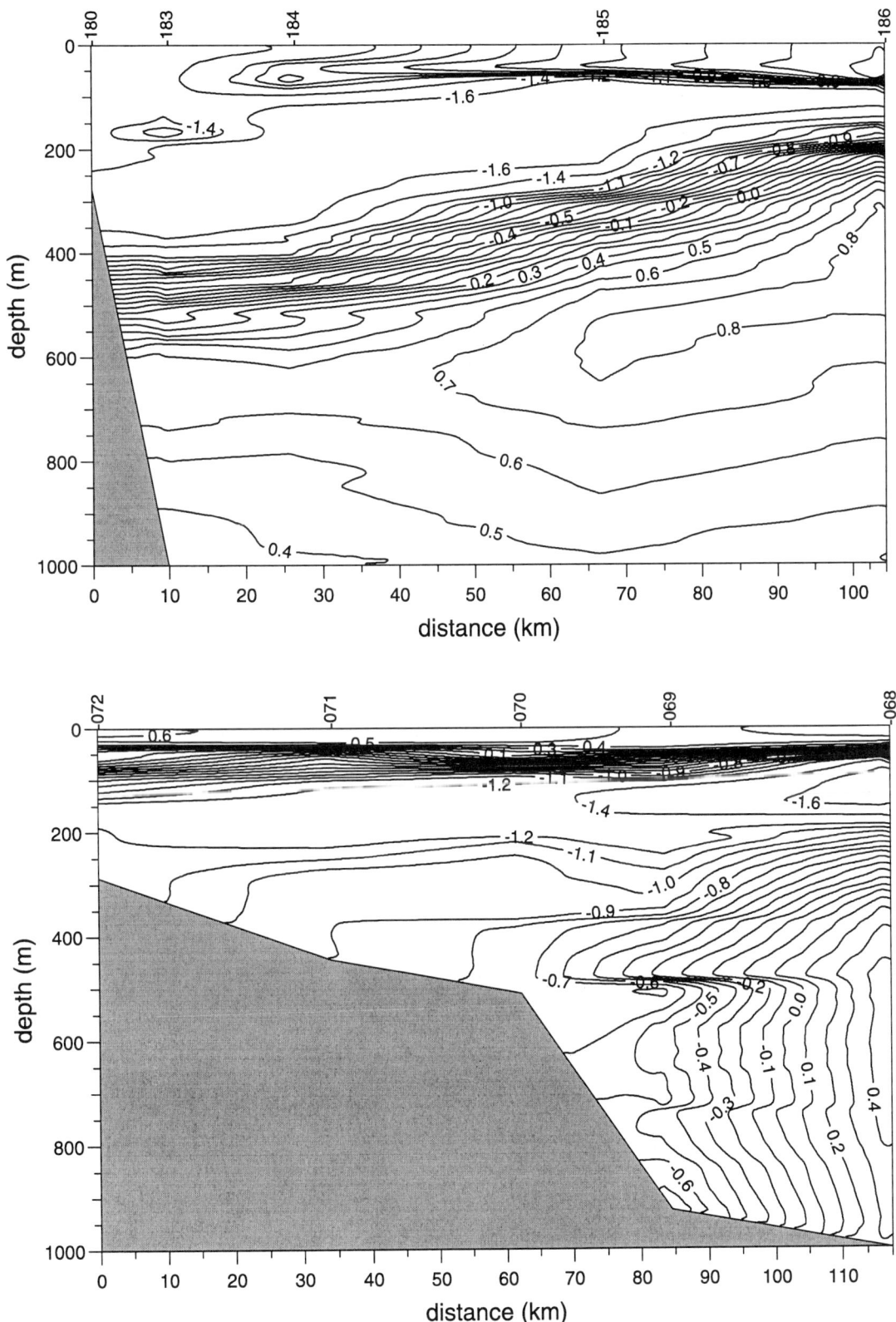

Fig. 2b. See Figure 2a.

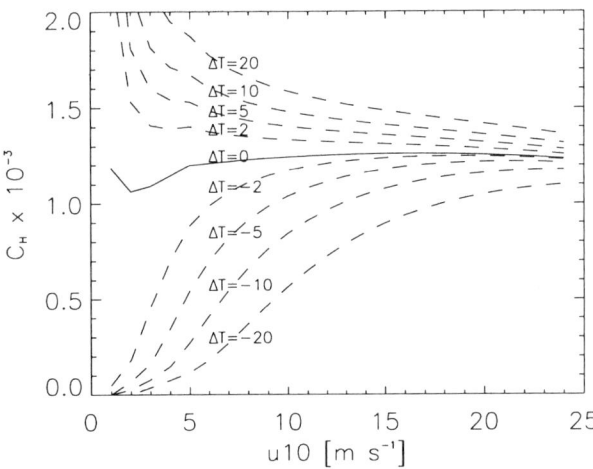

Fig. 3. Sensible heat transfer coefficients. C_H, for different wind speeds at 10 m height (u10) and temperature differences after Kondo [1975] (Note that $\Delta T = T_s - T_a$).

fluxes (terms on the left side of (1), summarized as H_{tot}) are balanced by $\varrho_i L_f F$ which describes the heat flux due to the release of latent heat of fusion during ice production. The density of young saline ice ϱ_i is 0.95×10^3 kg m^{-3}, $L_f = 3.34 \times 10^5$ J kg^{-1} is the latent heat of freezing salt water [e.g. Pease, 1987], and F is the ice production rate in m s^{-1}.

The fluxes H_s and H_e are calculated from standard bulk aerodynamic formulae according to

$$H_s = \varrho_a C_H c_p V_a (T_s - T_a) \quad (2)$$

and

$$H_e = \varrho_a C_E L V_a (q_s - q_a) \quad (3)$$

where $\varrho_a = 1.30$ kg m^{-3} is the density of air, $c_p = 1004$ J K^{-1}kg^{-1} the specific heat of air, V_a the wind speed at the reference height of 10 m, T_s the surface temperature and T_a the air temperature at 2 m height, and $L = 2.5 \times 10^{-6}$ J kg^{-1} is the latent heat of evaporation. The mixing ratio at T_a is q_a, and q_s is the saturated mixing ratio at T_s. The sensible heat and water vapor transfer coefficients (C_H and C_E) are calculated to account for wind and stability effects [Kondo, 1975]. Field studies have shown that, for wind speeds between 2 and 12 m s^{-1} and temperature differences from 0 K to 20 K, the transfer coefficients vary by a factor of 2 [e.g. Smith et al., 1983]. The scheme of Kondo [1975] is based on the Monin-Obukhov similarity and accounts for the existence of a viscous sub-layer above water surfaces for small Reynolds roughness numbers. The relationships for the transfer coefficients are based on empirical approximations to the iterative solutions for neutral and diabatic (stable and unstable) conditions given in the appendix of Kondo [1975]. Heat transfer studies for polynyas and leads from the Arctic Ice Dynamics Joint Experiment (AIDJEX) and other experimental programs, summarized by Smith et al. [1990], reveal fetch dependent intensified vertical heat exchange for lead widths of up to about 30 m. The wind fetch dependent effects are not important for areally averaged fluxes of the larger polynyas considered here. The heat transfer coefficients for neutral stability and wide polynyas are close to those from the method of Kondo [1975]. A stability dependence of C_H similar to that for open oceans has been found for polynyas [Smith et al. 1990]. The dependence of C_H on wind speed and temperature difference (ΔT) is presented in Figure 3. For positive ΔT (flux from ocean to atmosphere) and moderate wind speeds, C_H varies from 1.2×10^{-3} to 1.9×10^{-3}. When the turbulent heat flux is directed downwards, the variation is even higher. The dependence of turbulent heat fluxes on ΔT and wind speed u is nonlinear in contrast to the assumption of a constant C_H (Figure 4).

The longwave radiation flux emitted from the surface (Q_{lu}) is given by the Stefan Boltzmann law

$$Q_{lu} = \sigma \varepsilon_w T_w^4 \quad (4)$$

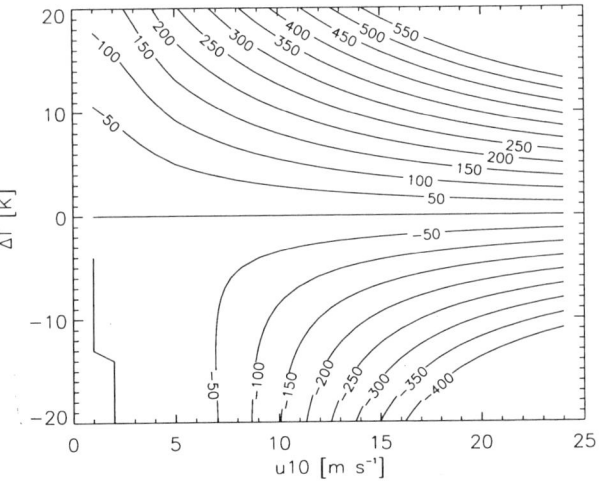

Fig. 4. Turbulent heat fluxes in W m-2 for different combinations of ΔT and 10-m wind speed, u10, with transfer coefficients, calculated after Kondo [1975].

where $\sigma = 5.67 \cdot 10^{-8}$ W m^{-2}K^{-4} is the Boltzmann constant, $\varepsilon_w = 0.98$ the emissivity of water [e.g. Maykut, 1986], and T_w is the surface water temperature.

The downward longwave radiation (Q_{ld}) is also approximated by the Stefan Boltzmann law [Pease, 1987] with T_a replacing T_w. The effective emissivity of air ε_a is calculated after Maykut and Church [1973]

$$\varepsilon_a = 0.7829(1 + 0.2232 \times CL^{2.75}) \qquad (5)$$

where CL is the fractional cloud cover. The parameterization is in close agreement with the results of König-Langlo and Augstein [1994] based on data from two polar stations (Koldewey in the Arctic and Georg-von-Neumayer in the Antarctic).

According to Maykut [1986], when taking clouds and the albedo into account, the effective solar radiation (Q_s) can be written

$$Q_s = (1-a)\kappa Q_0 \qquad (6)$$

where a is the surface albedo (0.08 for open water [Maykut, 1982]) and κ is the cloud correction term after Laevastu [1960]

$$\kappa = 1 - 0.6 \times CL^3. \qquad (7)$$

The solar radiation incident at the surface in the absence of clouds Q_0 is taken from Zillman [1972] which is commonly used for polar conditions [Parkinson and Washington, 1979; Maykut, 1986; Cavalieri and Martin, 1994]

$$Q_0 = \frac{S_0 \cos^2 z}{1.085 \cos z + (2.7 + \cos z)v_p \times 10^{-3} + 0.1} \qquad (8)$$

where $S_0 = 1353$ W m^{-2} is the solar constant, v_p is the vapor pressure and z is the solar zenith angle which depends on latitude, day of the year and solar time. Therefore, $\cos z$ is given by

$$\cos z = \sin \phi \sin \delta + \cos \phi \cos \delta \cos h \qquad (9)$$

where ϕ is the latitude. The solar inclination angle (δ) depends on the day of the year (d) (integer number from 1 to 365) according to

$$\delta = 23.44° \cos(172 - d) \qquad (10)$$

and varies between +23.44° on June 21 and −23.44° on December 22.

The solar hour angle (h) is given by

$$h = (12 - t_s) \times 15 \qquad (11)$$

where t_s is the solar time. The solar hour angle is 0° at noon local time and changes at a rate of 15° per hour. The vapor pressure v_p

$$v_p = v_{\text{sat}} \times RH \qquad (12)$$

is calculated from the relative humidity RH and from the saturation vapor pressure v_{sat} (in hPa), which depends only on T_a (in °C) and is derived from an empirical relationship

$$v_{\text{sat}} = 6.11 \times 10^{\alpha T_a/(\beta + T_a)} \qquad (13)$$

where $\alpha = 7.5$ and $\beta = 237.3$ are empirical constants. Arithmetically, Q_0 becomes negative for times with no sunlight. During these times we set Q_0 to zero.

In winter, the complete water column over the continental shelf is assumed to be at the freezing point of −1.8°C. The lower parts of the water column remain at the freezing point throughout the year, but in summer, the temperature of the surface layer is higher mainly because of solar heating and the low albedo of water. The temporal change of the surface layer temperature ΔT_s is calculated after Parkinson and Washington [1979] by

$$\Delta T_s = \frac{\Delta t \times H_{\text{tot}}}{d_{\text{mix}} \times c_w} \qquad (14)$$

where Δt is the time step, H_{tot} the surface heat flux, and $c_w = 4.19$MJ m^{-3} is the volumetric heat capacity of water. Ship-borne measurements show that the depth of this warm summer surface layer d_{mix} is around 150 m (Figure 2 b). On the western transect, the warm summer layer is well developed and reaches to a depth of 150 m (Figure 2 b, bottom). The coastal part of the survey in the eastern Weddell Sea occurred shortly after a heavy storm and is consequently relatively well mixed. Therefore, only a weak temperature maximum indicates the depth of the summer layer (Figure 2 b, top).

3. DATA AND METHODOLOGY

3.1. Meteorological Data

In this study, we use the uninitialized global atmospheric analyses of the European Center for Medium

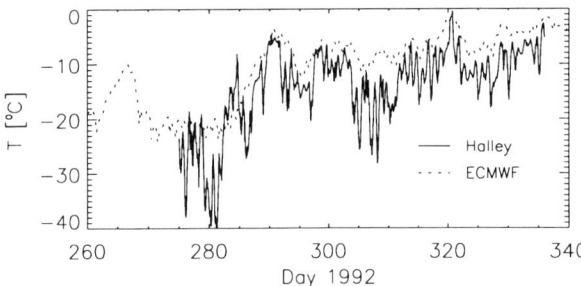

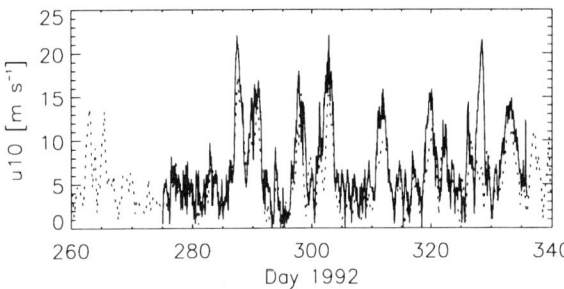

Fig. 5. Measured air temperature (a) and 10-m wind speed (b) from Halley statioin (solid line) and analyzed ECMWF data (dotted line) at the closest grid point for a 2-month period in 1992.

Range Weather Forecasts (ECMWF), which specify the 2-m temperatures and 10-m wind speeds, as atmospheric forcing data. The relative humidity is set to a constant value of 80%. The uncertainty because of this constant value is discussed in the following section.

The ECMWF analyses include the data from manned and automatic stations by the ECMWF-assimilation procedure [ECMWF, 1992]. The spatial resolution of the ECMWF-data is 1.125 degrees both for latitude and longitude and the time resolution is 6 hours. In 1992, the regular observational network of 12 wintering stations around the periphery of the Weddell Sea [Scientific Committee on Antarctic Research, 1992] was extended by 5 automatic weather stations and up to 12 drifting buoys. Radiosonde data from the majority of the manned stations and the surface data provided a relatively dense data coverage for the Weddell Sea, which improved the quality of the analysis.

The accuracy of the ECMWF analyses can only be estimated by comparison with local observations. The wind analyses and measurements agree well, when measured wind data are inserted into the Global Telecommunication System (GTS) and thus are included in the analyses. This is illustrated by the comparison between ECMWF 10m-wind and 2m-temperature data at the ocean grid point next to the British station Halley and the actual measurements at Halley, which were transmitted to the GTS and contributed to the analyses (Figure 5). The average temperature and average wind speed at Halley are 3.5 K lower and is 2.5 m s^{-1} higher respectively than the ECMWF analysis. A similar wind speed bias is found for the Neumayer station and can be explained by the ECMWF assimilation scheme, which does not aim for perfect agreement with local wind data. The temperatures biases mainly result from cold periods, when ECMWF-data are up to 10 K warmer than observed temperatures. This effect can be only explained by rejection or very low weights of temperature data in the assimilation scheme. From in-situ measurements in Antarctic coastal polynyas [Engelbart, 1992; Kottmeier and Engelbart, 1992], we conclude that the observed biases between Halley and ECMWF data tend to make the ECMWF data even more realistic than data from stations on ice shelves for estimating the heat transfer of coastal polynyas. The simultaneous measurements showed that the temperature of the cold air advected from the ice shelf increases by about 5 K over the polynya, while the mean wind speed decreases by about 2 m s^{-1}.

Comparisons with buoy observations by Kottmeier and Sellmann [1996] show that the horizontal pressure gradients reflected by the ECMWF grid point data generally tend to smooth the actual pressure gradient, even when pressure data are assimilated. The estimated standard errors in geostrophic wind components from ECMWF analyses are 3 m s^{-1} in regions where buoy data go into the GTS, and 7 m

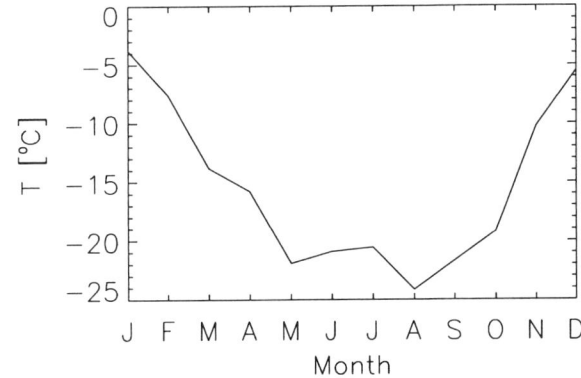

Fig. 6. Monthly mean 2-m air temperature for the eastern coast of the Weddell Sea near Halley station.

s^{-1} without any local data. Due to the manned and automatic weather stations along the Weddell Sea coast wind speeds are expected to be correct to ± 3.5 m s^{-1} for the polynya regions in general.

The cloud cover information is derived from the International Satellite Cloud Climatology Project (ISCCP) with a spatial resolution of 2.5° [*Rossow and Schiffer*, 1991]. The monthly averaged ISCCP-C2 data for the period from July 1983 through June 1991 provide the cloud input data, which are used as climatological information. Using the ISCCP-C2 data for a climatology with dependence on the latitude and season only (similar to the climatology of *van Loon* [1972]), standard deviations are about 20% for the cloud cover fraction. Nevertheless, the average values are in reasonable agreement with the *van Loon* values. The mean cloud cover decreases from about 75% in summer and 80% in winter at 65°S to 30% in summer and 10% in winter at 80°S. The high standard deviations result from the zonally uneven distribution of land and sea ice which strongly influences the cloud coverage. The standard deviations for the individual grid cells lie between 2% and 7% for the eight-year period. *Schweiger and Key* [1994] compared an ISCCP C-2 climatology in the Arctic with the cloud atlas from *Warren et al.* [1988]; during winter there is very good agreement, whereas in summer the ISCCP data are about 20% (absolute) lower than the atlas data. No validation of the ISCCP data is available for the Antarctic region.

3.2. Sensitivities

The sensitivity of the surface heat flux (H_{tot}) to different meteorological inputs is studied in order to estimate the influence of errors within these quantities on the results. For that purpose, variations in H_{tot} due to variations of the standard input values by a range of deviations are calculated for each month. Except for the air temperature, the individual parameters are set to be constant, so that seasonal variation results from air temperature and solar radiation. The mean air temperatures at the coast of the Eastern Weddell Sea, obtained from Halley station, show a significant seasonal cycle, causing large temperature differences between open water at the freezing temperature and the air in winter months (Figure 6). The standard values for the calculations are listed in Table 1. The uncertainties in relative humidity do not have any noticeable effect on H_{tot} (Figure 7a). The influence of cloud coverage varies seasonally with the solar radiation (Figure 7b). During

TABLE 1. Standard values in the sensitivity study.

Parameter	Value	
Cloud coverage	80	%
Rel. humidity	80	%
Wind speed	5.0	m s^{-1}
Air pressure	1×10^5	Pa
GMT	12	h
Latitude	70	°S

the winter period with no solar radiation, cloud coverage is not a critical parameter; changes are about ± 25 W m^{-2}. During summer on the other hand, uncertainties in cloud coverage of 20% can result in errors of ±200 W m^{-2}. The dependence on wind speed is larger in winter than in summer (Figure 7c). The wind speed directly influences the sensible heat flux. In summer, when radiative processes are dominant, the importance of the wind is small. During winter, the turbulent heat fluxes are much more important, since large temperature differences arise between the surface and the air. Therefore, errors in wind speed of 2 m s^{-1} can cause changes in total heat flux of ±60 W m^{-2}. Uncertainties in air temperature (Figure 7d) have similar importance in all seasons because the air temperature influences the sensible heat flux as well as the downward longwave radiation. Air temperature biases of 4 K can change the total heat flux by 50 W m^{-2} throughout the year.

3.3. Polynya Area Estimates

The polynya areas are derived from SSM/I data with an iterative method described in detail by *Markus and Burns* [1995] (henceforth referred as PSSM (Polynya Signature Simulation Method)). The SSM/I data are daily averaged brightness temperatures mapped to a polar stereographic projection available on CD-ROM from the National Snow and Ice Data Center (NSIDC) in Boulder, Colorado. The PSSM uses 85 and 37 GHz data (both horizontal and vertical polarizations) successively in order to take full advantage of the higher resolution at 85 GHz (about 15 km; 37 GHz: ≈30 km; 19 GHz: ≈60 km) while compensating for its sensitivity to atmospheric effects with the 37 GHz data. Figure 8 gives an overview of the algorithm.

The polarization ratio (*PR*) at both 37 and 85 GHz is calculated from the measured SSM/I data with land masked out. The maximum *PR* value of the ocean region (indicating open water) minus an arbitrary value Δ*PR* gives a threshold for a first guess

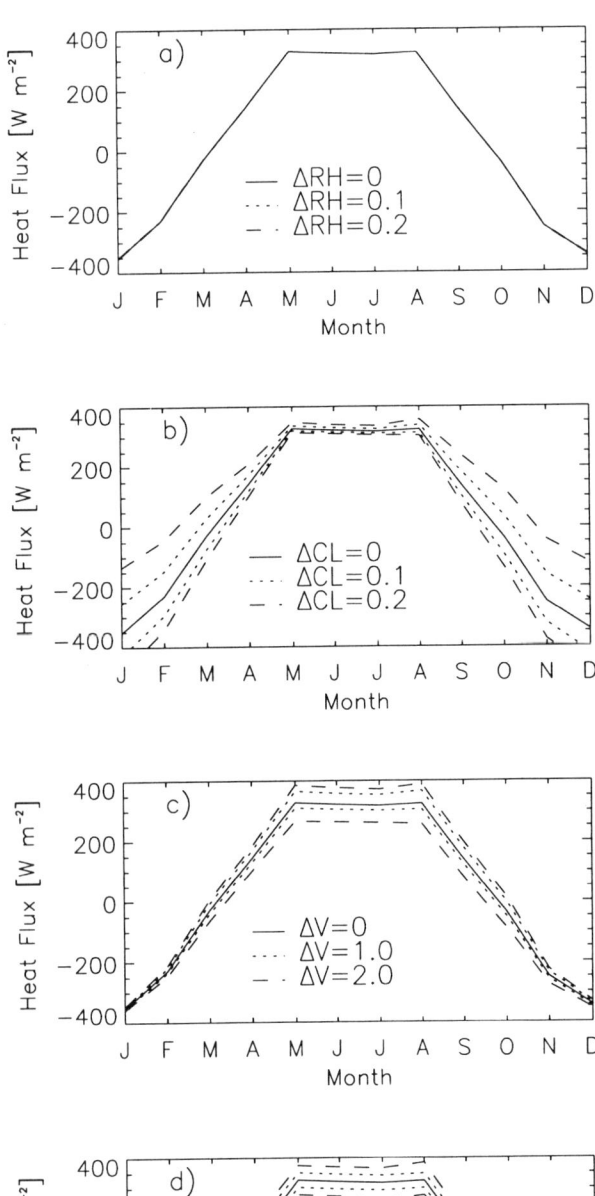

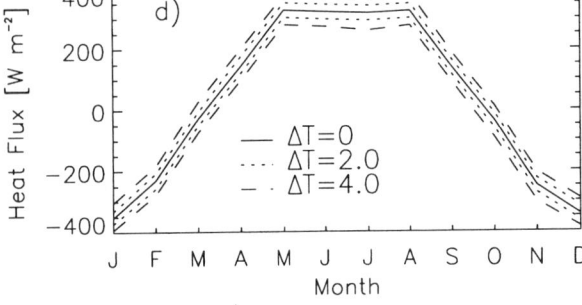

Fig. 7. Sensitivities of heat flux to different input uncertainties: (a) total heat flux for different relative humidity; (b) total heat flux for different cloud coverages; (c) total heat flux for different wind speeds; (d) total heat flux for different air temperatures.

of open water area. All pixels with a *PR* higher than this threshold are classified as open water in the first iteration step. The remaining ocean pixels are classified as sea ice. This results in a three-class image (land, sea ice, and open water). Two images of brightness temperature are generated by assigning average brightness temperatures for each class as measured by the SSM/I 37 GHz vertical and horizontal polarization channels. This leads to synthetic brightness temperature distributions for 37 GHz with a resolution of the 85 GHz channels. Convolution with the respective 37 GHz antenna patterns (including the sensor's sample spacing) results in simulated 37 GHz images and thus simulated *PR*s for 37 GHz. Measured and simulated *PR*s are compared by calculating the correlation coefficient and the mean difference of the two images. This classification, simulation, and comparison process is repeated with reduction of ΔPR (decreasing open water area) until maximum agreement is reached. The 85 GHz data are additionally interpolated into a 5 km × 5 km grid. This, of course, does not provide additional information, but does allow finer changes in open water area for each iteration step. Simulations of cloud coverage have shown that clouds lead to an underestimation of polynya areas of 10%, which is about the order of magnitude of which clouds influence the open water brightness temperature at 37 GHz with respect to sea ice. This means that the sensitivity to weather effects in the algorithm is determined by the 37 GHz channels. The maximum ice thickness which is included in the identified polynya area is about 0.06 m. Because the adjacent land is included in the simulation process explicitly, land contamination effects are accounted for.

Figure 9 shows a short time series of the coastal polynyas in the southern Weddell Sea in November 1992 (days 313, 314, 316, 320). The images in the left column are the results from the PSSM and in the right column are coincident images using the visible channel of the operational linescan system (OLS), a sensor also aboard the DMSP satellites. Although the polynyas are rather small, comparison between the two columns demonstrates the very good qualitative agreement between PSSM and OLS. Additionally, it shows that the coastal polynya conditions can change completely in a few days, confirming the need for continuous daily information.

4. DATA ANALYSIS AND RESULTS

The coast is divided into four regions in order to study regional differences. The basis for this division is the change in coastline orientation. These are the coast of the Antarctic Peninsula (western), the coast along the Filchner-Ronne Ice Shelf (southern), the south-eastern (from the Ronne-Filchner Ice Shelf at 35°W to 8°W) and the eastern coast (from 8°W to 20°E). The regions are indicated in Figure 1 (labeled A–D). Ice production starts when the surface temperature has reached the freezing point. Figure 10 shows the calculated surface layer temperatures for the year 1992 using (14) at different locations along the Weddell Sea coast indicated in Figure 1. The relatively high summer temperatures of 1.2°C in the northwestern Weddell Sea agree with measurements from Polarstern cruises [Fahrbach et al., 1994a].

4.1. Polynya Heat Exchange

Based on the meteorological input parameters and the polynya size estimates, all components of the surface energy balance are calculated with the thermodynamic ice growth model at a temporal resolution of 6 hours. The resulting monthly averages of the turbulent fluxes $(H_s + H_e)$ and the net radiative fluxes $(Q_{lu} - Q_{ld} - Q_s)$ show pronounced seasonal cycles in all polynya regions (Figure 11). Energy losses by turbulent fluxes occur throughout the year, with maximum values in winter months (between 200 and 300 W m^{-2}). The total atmospheric energy loss from polynyas in winter consists of about 60% from turbulent fluxes and 40% from radiative fluxes. The surface loses energy by net upward radiative flux during the period from February/March to September. The duration of the period with net upward energy flux varies regionally. In the southern region along the Filchner-Ronne Ice Shelf it lasts about 20 days longer than along the more northerly coasts. Due to the relatively warm surface, the energy loss by radiative fluxes is greater than 100 W m^{-2} in midwinter. A mean cloud cover of about 50% in winter is the limiting factor for the radiative loss of energy from the surface. Another factor limiting very large turbulent fluxes is the rapid decrease of surface temperature when new ice forms in the polynya.

Along the southern and south-eastern coast, the maxima of mean monthly total atmospheric fluxes are 450 W m^{-2} and 430 W m^{-2}, respectively, whereas it is only 340 W m^{-2} at the western coast. The regional differences result primarily from different turbulent heat fluxes. They are due to warmer air temperatures at the northern part of the Antarctic Peninsula and to the effects of cold air outbreaks off the Filchner-Ronne Ice Shelf and of katabatic winds

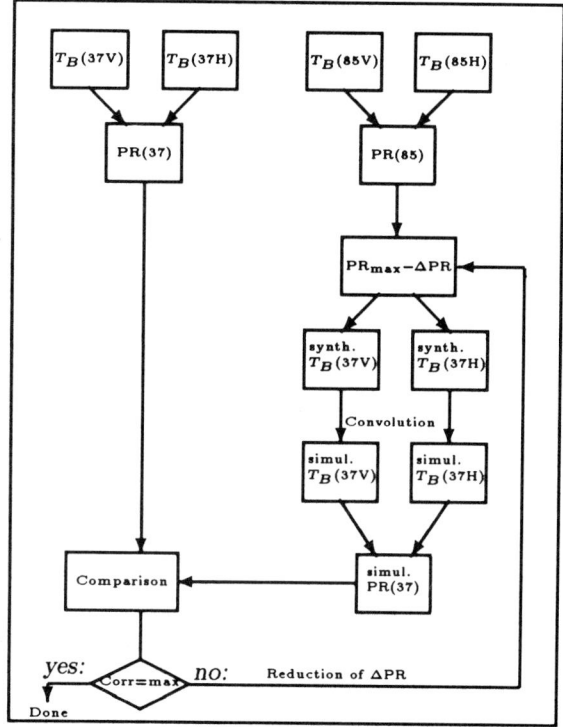

Fig. 8. Flow chart of the polynya detection algorithm (PSSM), after Markus and Burns [1995].

in the other regions. Warm cyclones regularly reach the eastern coast of the Antarctic Peninsula [Scherdtfeger, 1984] and restrict the maximum mean monthly energy losses to 370 W m^{-2}. The annual means are listed in Table 2.

4.2. Polynya Areas

The polynya areas are calculated for the years 1992 through November 1994. The mean autumn-winter (May through October) polynya areas for the four regions of the Weddell Sea coast are presented in Table 3. Except for the southern coast (Filchner-Ronne Ice Shelf) there is no significant interannual variation. The data from the western and the south-eastern coasts show opposite trends for that short period. During the summer and spring months (January through April, and November/December), the ice-free ocean reaches the Antarctic continent in certain areas, so that the definition of a polynya is no longer valid for this period. The open water area, as measured by the PSSM, is limited by the pixels investigated. This is about 50 km offshore. No polynyas with larger widths occur during periods of freezing.

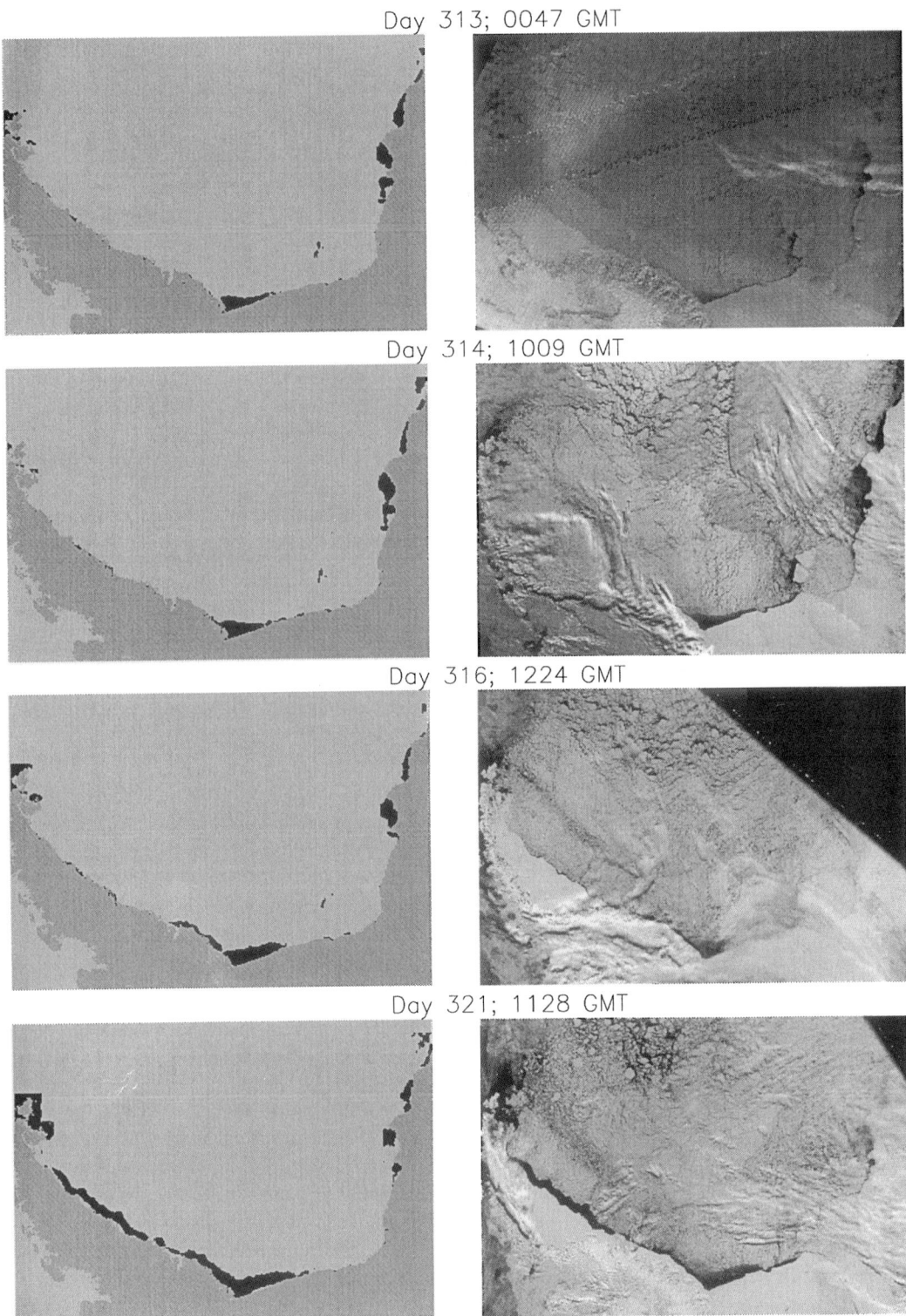

Fig. 9. Results from the PSSM (left) at the southern Weddell Sea coast in comparison with high resolution visible OLS data for November 1992 (right). The days are indicated in the PSSM images; OLS overflight time is also indicated in the respective image. The resolution is 5 km, and the projection is the same as in Figure 1.

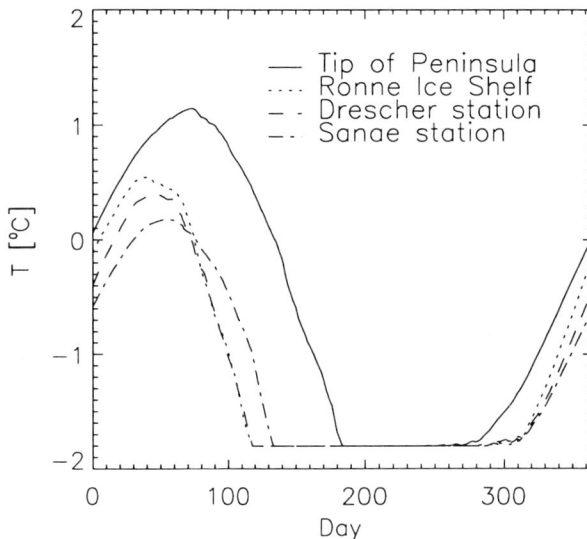

Fig. 10. Surface water temperatures for 1992 at different locations along the Weddell Sea coast derived from heat flux calculations.

Polynya widths are calculated by dividing the total polynya area of a region by the respective coastline length (Figure 12). The mean winter polynya widths are 1.2 km for the western coast, 2.5 km for the southern coast, 1.5 km for the south-eastern coast, and 2.4 km for the eastern coast. It should be noted that polynyas always cover only part of the coastline. Therefore the mean winter polynya width of around 1 km is significantly smaller than the typical polynya width, which is in the order of 10 km. The maximum width at the southern coast agrees with large coastal polynyas that are often observed at the western part of the Ronne Ice Shelf [*Gammelsroed et al.*, 1994], where huge grounded tabular icebergs in front of Berkner Island prevent the ice pack from drifting westwards [*Markus*, 1996]. Off the Antarctic Peninsula, the mean ice motion prevents the formation of large polynyas. Also, only off the Antarctic Peninsula is a significant amount of sea ice present during summer [see e.g. *Gloersen et al.*, 1992]. These factors result in the minimum polynya size as well as the smallest seasonal variation.

4.3. Ice Production and Impact on Sea Ice Coverage

The impact of the ice produced in polynyas on the total sea ice coverage in the Weddell Sea depends on the net heat flux (which determines the ice production rate) and the polynya area, both of which vary spatially and temporally as demonstrated in the previous section. The average total ice production in the polynyas is found to be between 11 and 13 m yr^{-1}. These values are significantly lower (ca. 5 m yr^{-1}) along the western coast and higher along the southern coast (between 12 m yr^{-1} (1993, 1994) and 17 m yr^{-1} (1992)). For ice-covered pixels, the ice production rate is assumed to be zero. As we assume that thin ice is piled up immediately by the offshore wind at the polynya boundary to a significant ice thickness, further growth is highly reduced.

The heat flux calculations assume that the whole water column under a polynya is at the freezing point. This is only valid over the shallow continental shelf. Over the deep ocean, the heat flux from the Warm Deep Water below the mixed layer, with temperatures of more than 0.5°C, would compensate partially for the heat loss to the atmosphere and consequently suppress the ice formation. The extent of the continental shelf is very variable along the Weddell Sea coast (Figure 1). Whereas in the eastern Weddell Sea, a polynya may exceed the shelf width, the shelf areas off the Antarctic Peninsula and the Filchner-Ronne Ice Shelf normally extend much further offshore than the polynyas. We neglect all polynya pixels which lie over the deep ocean in our calculations. For the south-eastern and eastern coast, the total ice production is reduced by neglecting the deep ocean pixels, so that ice production is greatest along the southern coast (Table 4). Nevertheless, along the western coast the total amount of ice produced in polynyas is half as much as that in the other regions. The total polynya ice production per month for the years 1992 through 1994 is shown in Figure 13.

In order to the relate this to the overall Weddell Sea ice coverage, the calculated ice production is compared with ice extents from large-scale ice concentration analyses using the NASA-Team algorithm [*Cavalieri*, 1992]. All data points with an ice concentration higher than 15% are summed [*Zwally et al.*, 1983] and multiplied by the pixel size (625 km^2). The monthly mean ice extent for the years 1992 through 1994 is also presented in Figure 13. Each year, the minimum extent occurs in the second half of February, with an area between 1.4×10^6 and 1.7×10^6 km^2 for all three years. The maximum is at the end of July with an area between 7×10^6 and 8×10^6 km^2. Ice production in the coastal polynyas starts in April and achieves its maximum one or two months before the maximum pack ice extent. The polynya ice production, however, continues after the pack ice has

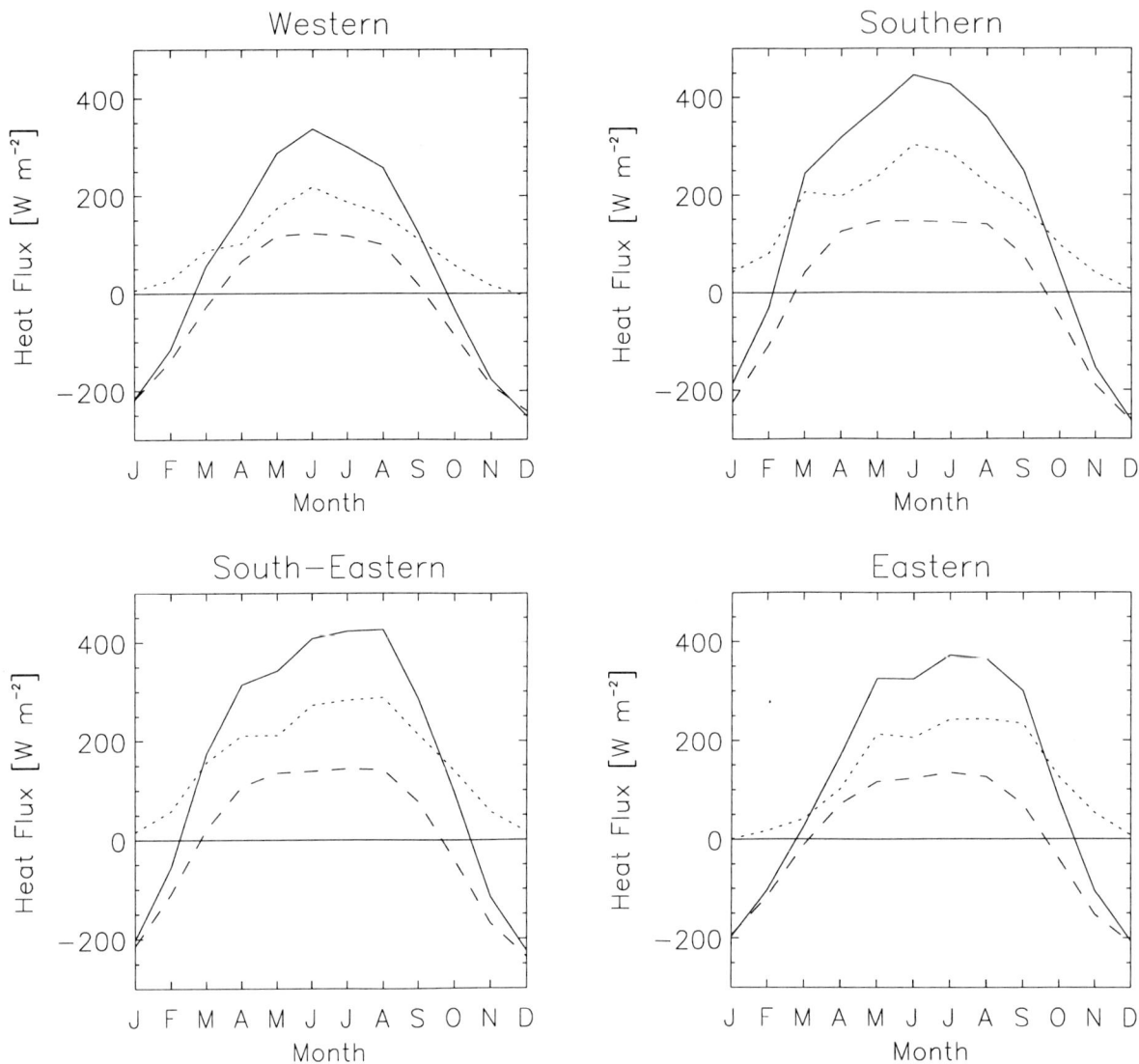

Fig. 11. Total heat flux (solid line), turbulent heat flux (dotted line) and radiative heat flux (dashed line) for the four parts of the Weddell Sea coast during the year 1992.

reached its maximum extent and therefore retards the sea ice decay. This is in agreement with calculations with a dynamic-thermodynamic sea ice model [Lemke et al., 1990], which shows intense melting in September in the marginal ice zone, whereas the ice production is still positive in the coastal polynyas. This difference results primarily from air temperatures still well below freezing in September at the Weddell Sea coast, while temperatures are higher in the northern marginal ice zone.

To determine the contribution of ice volume produced in the polynyas to the overall production in the Weddell Sea, the ice extent area is converted to equivalent ice volume by multiplication with an average ice thickness. In-situ measurements gave an average ice thickness of 1 m for the Weddell Sea [Lange and Eicken, 1991], (Strass and Fahrbach, pers. communication). The mean monthly ice volume for the period of ice extent increase, from March to August, is shown in Figure 14a. In 1993 the volume is 0.5×10^{12} m^3 smaller throughout the year compared to 1992 and 1994. The cumulative polynya ice production reaches up to 130×10^9 m^3 by the end of July (Figure 14b). During this period the Weddell Sea sea ice changed its volume by 4.5×10^{12} m^3, so that about 2.6% of the total sea ice volume results from coastal polynyas. The areal fraction of coastal polynyas with respect to ice extent decreases from between 0.7%

TABLE 2. Annual mean heat fluxes in coastal polynyas for the different parts of the Weddell Sea coast.

Flux[a]	Western	Southern	S-E[b]	Eastern
Total	61	153	155	113
Turbulent	95	158	160	124
Radiative	−30	−1	−1	−7

[a] Heat flux in [W m^{-2}].
[b] S-E stands for South-Eastern.

and 1.8% in April to only 0.2% during June and July (Figure 14c). Since polynya ice production is highest during the latter period, the ice volume produced per month in polynyas amounts to 3% - 4% of the monthly change in Weddell Sea ice volume (Figure 14d). As these results depend linearly on the assumed average ice thickness, if ice thickness is reduced by 50%, i.e. ice thickness of 0.5 m, the importance of coastal polynyas would double. Then, about 5% of the Weddell Sea ice volume would result from ice production in coastal polynyas. Although this represents a small contribution to the ice budget, the fixed location of polynyas at the coastline results in a significant contribution to the water mass formation.

4.4. Impact on Water Mass Formation

The brine rejection associated with new ice production is also determined by the ice production rate and polynya area. According to *Gow et al.* [1990], the salt released in a polynya per day is

$$S_F = \varrho_i F_i A_p (s_w - s_i) \quad (15)$$

where S_F is the salt released in kg, F_i is the ice production rate in m d^{-1}, A_p the polynya area, s_w the water salinity, and s_i the salinity of frazil ice, which, after *Martin and Kaufmann* [1981], is

$$s_i = 0.31 s_w \quad (16)$$

The salt mixes with the underlying water column and increases the salinity s_w, which feeds back into eq.15 for the next day. We take 400 m as an average depth of the shelf areas in this paper.

The cyclonic flow of the Weddell Gyre moves the water from the eastern Weddell Sea westward along the coast towards the Antarctic Peninsula. The shelfwater enters the eastern boundary of our area of study with a relatively low salinity. The average salinity of the shelf water layer at 10°E, obtained from hydrographic measurements from R/V Polarstern, is 34.06 psu (Figure 2a, top). This continuous freshwater input at the eastern coast prevents the water from increasing its salinity and becoming dense enough to form Bottom Water [*Fahrbach et al.*, 1994b]. Consequently, the increase in salinity due to polynya brine rejection is calculated for water parcels moving along the coast with the aid of a simple model prescribing the coastal current. The final salinity at the north-western Weddell Sea is observed to be 34.49 psu (Figure 2a, bottom).

The mean drift velocity in the core of the Antarctic Coastal Current (u_{drift}) over the continental shelf is between 10 and 20 km d^{-1} [*Fahrbach et al.*, 1992]. Measurements with buoys show a drift of about 5 km d^{-1} [*Kottmeier and Sellmann*, 1996]. There are various reasons why the velocities obtained from the buoys are lower than from the moored current meters: first, the measurements did not occur at the same time; second, the moorings and the buoys were not at the same locations; and third, and most important, the buoys represent a random spatial sampling whereas the moored instruments were located in the current core. It has to be noted that the buoys provide measurements from the near-surface currents whereas the current meters were approximately 50 m below the surface. However, as the polynya is mostly over the shelf in the lateral boundary layer, and as we have to take a spatial and temporal mean value to describe the motion of the water from pixel to pixel along the coastline, we have to assume significantly smaller velocities. Therefore, we run the model with drift velocities between 2 and 10 km d^{-1}.

We assume that the width of the drift belt along the coast is 50 km, to include all polynyas and also the effect of horizontal mixing. This width corresponds to the extent of the low salinity shelf water layer as derived from shipborne salinity measurements. According to the resolution of the derived polynya areas, this current band is divided into 10 sub-bands of 5 km width parallel to the coast. The salinity increase along each band of a parcel with January, April, July, and October start times and $u_{\text{drift}} = 2$ and 5 km d^{-1} is shown in Figure 15. The

TABLE 3. Mean autumn and winter (May through October) polynya areas for the four regions and for the years 1992, 1993, and 1994 in km^2.

Year	Western	Southern	S-E[a]	Eastern
1992	4243	3214	2960	4643
1993	4521	5233	2633	4530
1994	5283	2896	2203	4864

[a] S-E stands for South-Eastern.

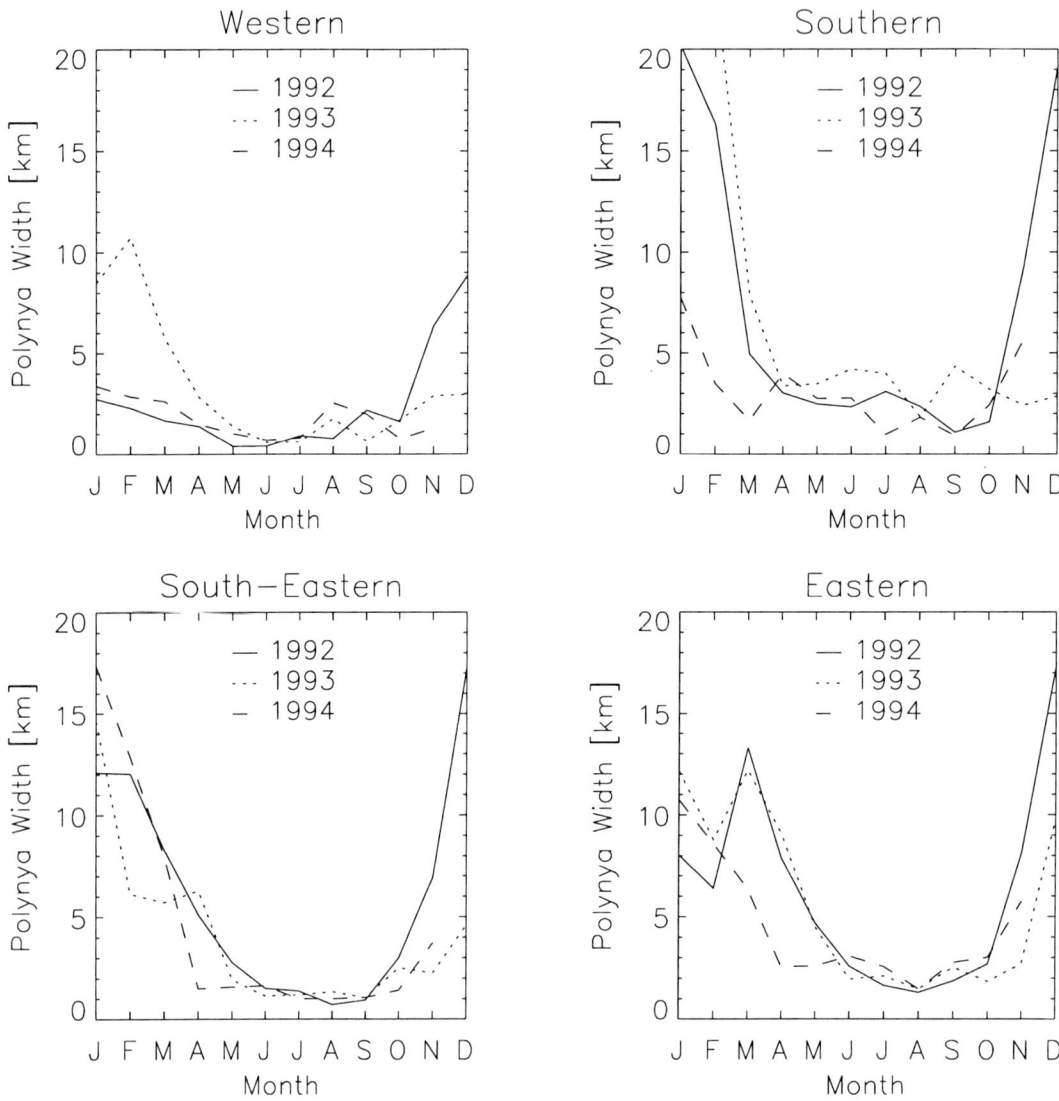

Fig. 12. Monthly mean polynya widths for the four regions.

length of the Weddell Sea coast is about 6500 km. Starting at 10°E, Kapp Norvegia is located at about 2000 km, and the Filchner-Ronne Ice Shelf between 3600 km and 4800 km. Therefore, assuming a drift velocity, u_{drift}, of 2 km d^{-1}, a water parcel takes up to 9 years to reach the northern tip of the Antarctic Peninsula. The parcels with different distances from the coast achieve different salinity increases due to different residence times in the polynyas. The line with highest salinity increase is the one directly along the coast (0-5 km), because this parcel is most frequently in the coastal polynya. No increase in salinity results from the outermost flow band (45-50 km), since polynya widths greater than 50 km did not occur during our period of study. The final salinity of the water parcel in the north-western Weddell Sea depends primarily on the number of winters it has spent on its way along the coast. As the ice production is highest at the coast of the Filchner-Ronne Ice Shelf, the water parcels spending more winters at this location experience the highest salinity increase. The salinity of the 50 km wide current band in the northwestern Weddell Sea amounts to 34.17 - 34.21 psu for a mean flow of 5 km d^{-1} and 34.37- 34.45 psu for a mean flow of 2 km d^{-1}. While the increase obtained with 5 km d^{-1} is too small, the

TABLE 4. Total annual ice production with and without deep ocean pixels (in $10^6\,m^3$).

Region	all pixels	no deep ocean pixels
Western	37010	37010
Southern	87229	87229
South-Eastern	84261	69981
Eastern	97493	82853

one with 2 km d^{-1} is close to the observations. If we take into account that the ocean gains freshwater in the polynya from precipitation, and near the coast from melting ice shelves and icebergs [*Fahrbach et al.*, 1994c], the mean drift of 2 km d^{-1} is more realistic as it provides enough salt gain to compensate for the freshwater gain. However, a detailed estimate of the salt budget of the coastal belt would require an estimate of the exchange between this layer and the open ocean. This would include salt gain from Warm Deep Water penetrating on the shelf [*Fahrbach et al.*, 1992] and salt loss or gain according to the difference between the shelf water and the open ocean Winter Water or Surface Water. Present data are not sufficient to carry out such an estimate.

5. CONCLUSIONS

In this study we have analyzed coastal polynyas in the Weddell Sea for the years 1992 through 1994. With satellite passive microwave data and meteorological information, polynya areas and ice production

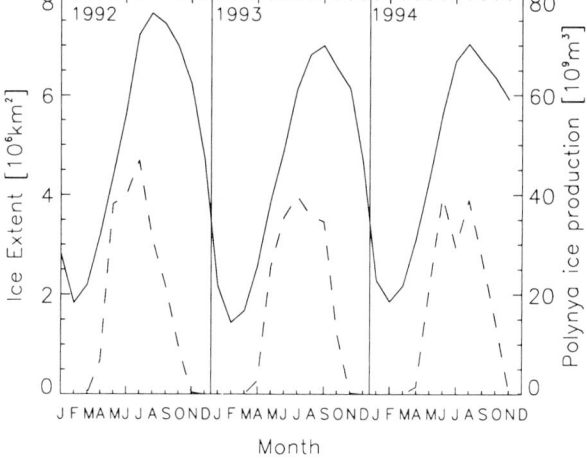

Fig. 13. Weddell Sea ice extent (solid line) and monthly polynya ice production (dashed line) for the years 1992, 1993, and 1994.

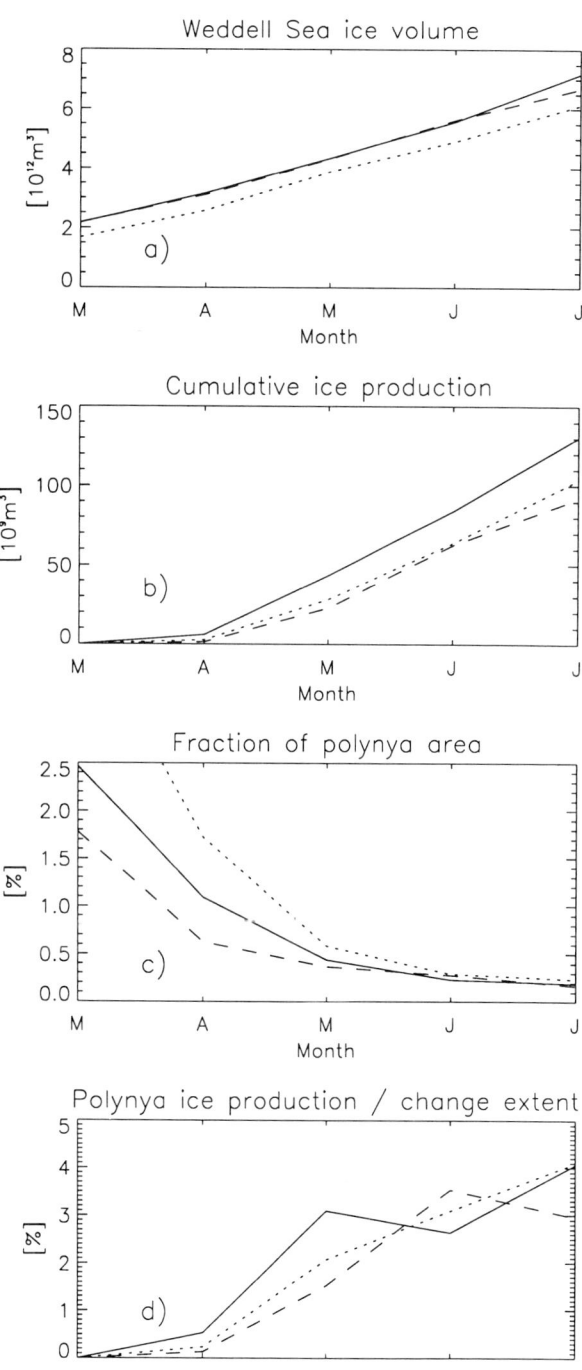

Fig. 14. Polynya ice production compared with Weddell Sea ice volume: (a) Mean monthly ice volumes; (b) cumulative ice production in coastal polynyas; (c) percentage of polynya areas expressed as a fraction of Weddell Sea ice extent; (d) ratio of polynya ice production to change in Weddell Sea ice extent for each month. Solid line: 1992; dotted line: 1993; dashed line: 1994.

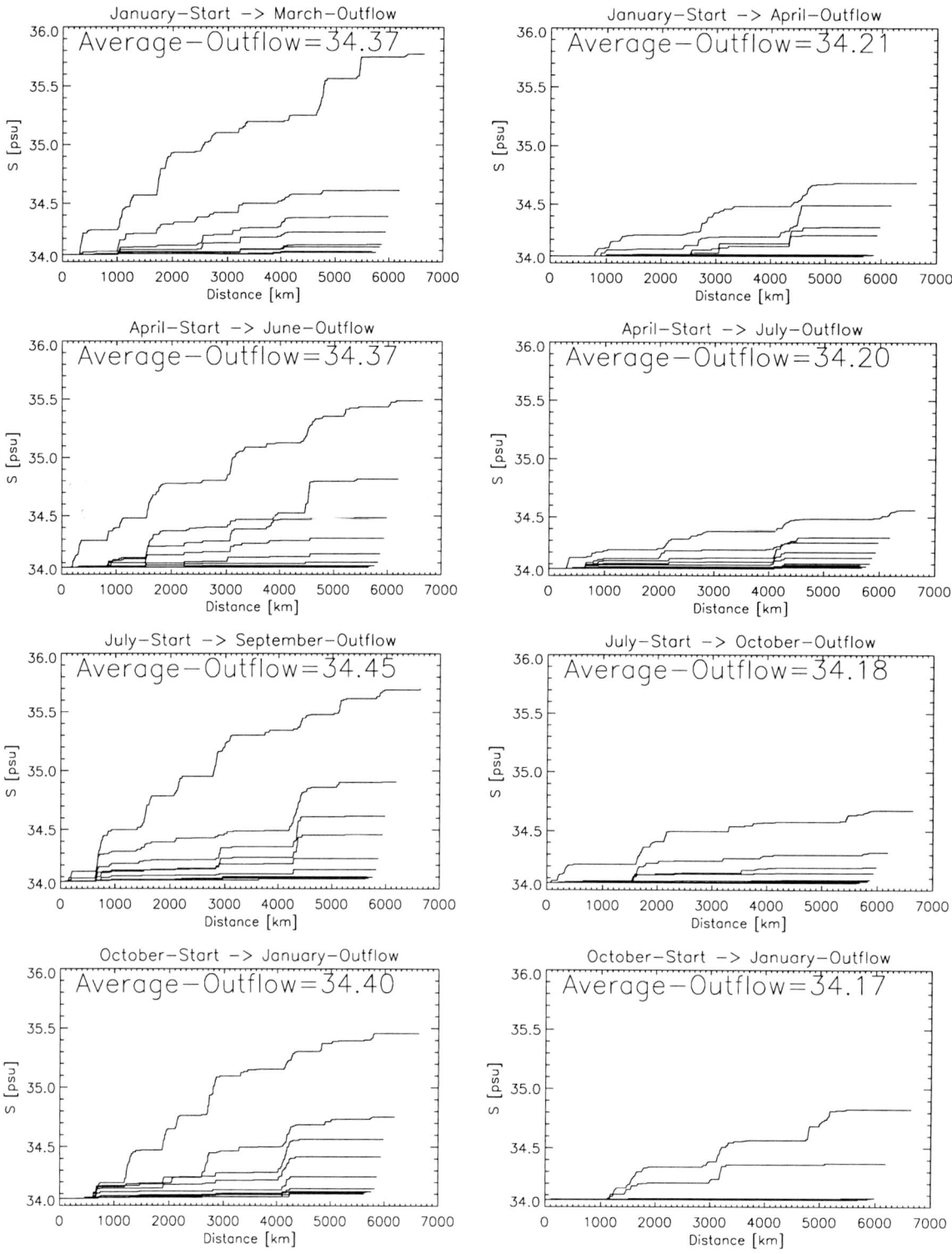

Fig. 15. Increase of salinity with drift position for January, April, July, and October start for drift speeds of 2 km d^{-1} (left) and 5 km d^{-1} (right). The single lines represent the different drift bands with a width of 5 km. The lines with greatest salinity increase are the bands directly at the coast (0-5 km). The second greatest is 5-10 km etc.

were calculated on a daily basis. Because of the large heat flux to the atmosphere, with winter monthly averages of about 400 W m^{-2} and single-day maxima of almost 1000 W m^{-2}, there is intense ice production amounting to about 12 m yr^{-1}, which is ten times the average ice production over the central Weddell Sea. The ice produced in coastal polynyas contributes between 2.5% and 5% to the total ice volume formed in the Weddell Sea. This is rather small bearing in mind that coastal polynyas are often considered to be "ice factories". The reason for this is their small areal fraction, especially in winter when polynya ice production is highest, so that the contribution to the total ice budget is not significant. Even if one assumes a permanent undetected open water band of 1 km width along the entire Weddell Sea coast, the contribution does not increase significantly. However, as polynyas occur at fixed geographical locations with relatively long water mass residence times, they significantly affect water mass formation in the Weddell Sea.

The impact on the ocean due to brine release during ice production was studied with a simple model of water drift over the continental shelf. A mean drift velocity of 2 km d^{-1} gives the most reasonable results. In spite of the simple assumption regarding the oceanic drift and mixing, the calculated increase of salinity from the eastern to the northwestern Weddell Sea is consistent with measured data. As there is a continuous fresh water input due to melting under the ice shelves or precipitation, our calculations give only upper limits on the salinity increase which can be produced. However, these values might help to obtain better estimates of these fresh water fluxes.

Acknowledgments. The authors are grateful to John King at British Antarctic Survey for the meteorological data of Halley station, and to Bob Whritner at Scripps Institution of Oceanography for the OLS data. Furthermore, we appreciate the very constructive comments of Rob Massom and the other two reviewers as well as the editorial and grammatical comments of Martin Jeffries. This is AWI publication number 1205.

REFERENCES

Carmack, E.C., Water characteristics of the southern ocean south of the polar front, in *A Voyage of Discovery, George Deacon 70th Anniversary Volume*, edited by M. Angel, pp. 15-37, Pergamon, New York, 1977.

Carmack, E.C., Circulation and mixing in ice-covered waters, in *The Geophysics of Sea Ice*, edited by N. Untersteiner, NATO ASI Series, vol. 146, 641-712, 1986.

Cavalieri, D.J., Sea ice algorithm, in *NASA Sea Ice Validation Program for the Defense Meteorological Satellite Program Special Sensor Microwave Imager: Final Report*, edited by D.J. Cavalieri, pp. 25-31, NASA Technical Memorandum 104559, 1992.

Cavalieri, D.J., and S. Martin, A passive microwave study of polynyas along the Antarctic Wilkes Land coast, in *Oceanology of the Antarctic Continental Shelf, Antarct. Res. Ser.*, vol. 43, edited by S. Jacobs, pp. 227-252, AGU, Washington, D.C., 1985.

Cavalieri, D.J., and S. Martin, The contribution of Alaskan, Siberian, and Canadian coastal polynyas to the cold halocline layer of the Arctic Ocean, *J. Geophys. Res.*, 99, 18,343-18,362, 1994.

Cavalieri, D.J., P. Gloersen, and W.J. Campbell, Determination of sea ice parameters with the NIMBUS 7 scanning multichannel microwave radiometer, *J. Geophys. Res.*, 89, 5355-5369, 1984.

Comiso, J.C., SSM/I sea ice concentrations using the bootstrap algorithm, *NASA Reference Publication, 1380*, 49 pp., 1995.

ECMWF, The description of the ECMWF/WRCP Level III-A Global Atmospheric Data Archive. *European Center For Medium-Range Weather Forecasts*, 1992.

Engelbart, D., Thermodynamik und Dynamik von Küstenpolynyen im Weddell-Meer, *Berichte d. Instituts für Meteorologie und Klimatologie der Universität Hannover*, 132 pp., Hannover, 1992.

Fahrbach, E., G. Rohardt, and G. Krause, The Antarctic Coastal Current in the southeastern Weddell Sea, *Polar Biology*, 12, 171-182, 1992.

Fahrbach, E., et al., Cruise Report ANT X/7 in: The Expeditions ANTARKTIS X/6-8 of the Research Vessel "Polarstern" in 1992/93, in *Ber. Polarforsch.*, 135, edited by Bathmann, U., V. Smetacek H. de Baar, E. Fahrbach and G. Krause, 127-197, 1994a.

Fahrbach, E., R.G. Peterson, G. Rohardt, P. Schlosser, and R. Bayer, Suppression of bottom water formation in the southeastern Weddell Sea, *Deep Sea Res.*, 41, 389-411, 1994b.

Fahrbach, E., G. Rohardt, M. Schroder, and V. Strass, Transport and structure of the Weddell Gyre, *Ann. Geophysicae*, 12, 840-855, 1994c.

Gammelsroed, T., A. Foldvik, O.A. Noest, Oe Skaseth, L.G. Anderson, E. Fogelqvist, K. Olsson, T. Tanhva, E.P. Jones, and S. Oesterhus, Distribution of water masses on the continental shelf in the Southern Weddell Sea, *The Polar Oceans and Their Role in Shaping the Global Environment, Geophysical Monograph, 84*, 159-176, AGU, 1994.

Gloersen, P., W.J. Campbell, D.J. Cavalieri, J.C. Comiso, C.L. Parkinson, and H.J. Zwally, Arctic and Antarctic sea ice, 1978-1987: Satellite passive microwave observations and analysis, *NASA SP-511*, 290 pp., Washington, D.C., 1992.

Gordon, A.L. and J.C. Comiso, Polynyas in the Southern Ocean, *Sci. Am., 256,*, 90-97, 1988.

Gow, A.J, D.A. Meese, D.K. Perovich, and W.B. Tucker III, The Anatomy of a Freezing Lead, *J. Geophys. Res., 95*, 18,221-18,232, 1990.

Jacobs, S.S. and J.C. Comiso, Sea ice processes on the Ross Sea continental shelf, *J. Geophys. Res., 94*, 18,195-18,211, 1989.

Kondo, J., Air-sea bulk transfer coefficients in diabatic conditions, *Boundary Layer Meteorol., 9*, 91-112, 1975.

König-Langlo, G. and Augstein, E., Parameterization of the downward longwave radiation at the Earth's surface in polar regions, *Met. Zeitschrift, Neue Folge, 6*, 343-347, 1994.

Kottmeier, C., and D. Engelbart, Generation and atmospheric heat exchange of coastal polynyas in the Weddell Sea, *Boundary Layer Meteorol., 60*, 207-234, 1992.

Kottmeier, C., and L. Sellmann, Atmospheric and oceanic forcing of Weddell Sea ice motion, *J. Geophys. Res., 101*, 20,809-20,824, 1996.

Laevastu, T., Factors affecting the temperature of the surface layer of the sea, *Commentat. Phys. Math., 25*, 1-36, 1960.

Lange, M.A. and H. Eicken, Sea-ice thickness distribution in the northwestern Weddell Sea, *J. Geophys. Res., 96*, 4821-4837, 1991.

Lemke, P., W.B. Owens, and W.D. Hibler III, A coupled sea ice-mixed layer-pycnocline model for the Weddell Sea, *J. Geophys. Res., 95*, 9513-9525, 1990.

Markus, T., The effect of the grounded tabular icebergs in front of Berkner Island on the Weddell Sea ice drift as seen from satellite passive microwave sensors, *IGARSS Proceedings*, 1791-1793, 1996.

Markus T. and B.A. Burns, A method to estimate subpixel scale coastal polynyas with satellite passive microwave data, *J. Geophys. Res., 100*, 4473-4487, 1995.

Martin, S. and P. Kaufmann, A field and laboratory study of wave damping by grease ice, *J. Glaciol., 27*, 283-314, 1981.

Martin, S. and D.J. Cavalieri, Contributions of the Siberian Shelf polynyas to the Arctic Ocean intermediate and deep water, *J. Geophys. Res., 94,*, 12,725-12,738, 1989.

Maykut, G.A., Energy exchange over young sea ice in the central Arctic, *J. Geophys. Res., 83*, 3646-3658, 1978.

Maykut, G.A., Large-scale heat exchange and ice production in the Central Arctic, *J. Geophys. Res., 87*, 7971-7984, 1982.

Maykut, G.A., The surface heat and mass balance, in *The Geophysics of Sea Ice*, edited by N. Untersteiner, NATO ASI Series, vol. 146, 395-464, 1986.

Maykut, G.A., and P.E. Church, Radiation climate of Barrow, Alaska, 1962-1966, *J. Appl. Meteorol., 12*, 620-628, 1973.

Parkinson, C.L., and W.M. Washington, A large-scale numerical model of sea ice, *J. Geophys. Res., 84*, 311-337, 1979.

Pease, C.H., The size of wind-driven coastal polynyas, *J. Geophys. Res., 92*, 7049-7059, 1987.

Rossow, W.B. and R.A. Schiffer, ISCCP cloud data products, *Bull. Amer. Meteor. Soc., 72*, 2-20, 1991.

Scientific Committee on Antarctic Research, Stations of SCAR nations operating in the Antarctic, winter 1992, *SCAR Bulletin*, No.107, 1992.

Schweiger, A.J. and J.R. Key, Arctic ocean radiative fluxes and cloud forcing estimated from the ISCCP C2 cloud data set, 1983–1990, *J. Appl. Meteor., 33*, 948-963, 1994.

Schwerdtfeger, W., Weather and climate of the Antarctic, *Developments in Antarctic Science, 15*, 261 pp., Elsevier, Amsterdam, 1984.

Smith, S.D., R.J. Anderson, G. Den Hartog, D.R. Topham, and R.G. Perkin, An investigation of a polynya in the Canadian Archipelago 2, structure of turbulence and sensible heat flux, *J. Geophys. Res., 88*, 2900-2910, 1983.

Smith, S.D., R.D. Muench, and C.H. Pease, Polynyas and leads: An overview of physical processes and environment, *J. Geophys. Res., 95*, 9461-9479, 1990.

Van Loon, H., Cloudiness and precipitation in the southern hemisphere, Met. Monographs, 13, edited by C.W. Newton, Met. of the Southern Hemisphere, Amer. Met. Soc., Boston, Mass., 1972.

Wadhams, P., M. Lange, and S.F. Ackley, The ice thickness distribution across the Atlantic sector of the Antarctic Ocean in midwinter, *J. Geophys. Res., 92*, 14,535-14,552, 1987.

Warren, S.G., C.J. Hahn, J. London, R.M. Chervin, and R. Jenne, Global distribution of total cloud cover and cloud type amounts over the oceans, *NCAR, Technical Note TN-317+STR*, NCAR, Boulder, Colorado, 1988.

Zillman, J.W., A study of some aspects of the radiation and heat budgets of the southern hemisphere oceans, *Meteorol. Studies, 26, Bureau of Meteorol.*, Dept. of Interior, Canberra, Australia, 1972.

Zwally, H.J., J.C. Comiso, W.J. Campbell, F.D. Carsey, and P. Gloersen, Antarctic Sea Ice, 1973-1976: Satellite Passive-Microwave Observations, *NASA SP-459*, 1983.

Zwally, H.J., J.C. Comiso, and A.L. Gordon, Antarctic offshore leads and polynyas and oceanographic effects, in *Oceanology of the Antarctic Continental Shelf, Antarct. Res. Ser.*, vol. 43, edited by S. Jacobs, pp. 227-252, AGU, Washington, D.C.. 1985.

E. Fahrbach and C. Kottmeier, Alfred-Wegener-Institut für Polar- und Meeresforschung, Postfach 120161, D-27515 Bremerhaven, Germany.

T. Markus, Code 971, NASA Goddard Space Flight Center, Greenbelt, MD 20771.

(Received July 9, 1996; accepted February 3, 1997.)

INTERANNUAL VARIABILITY IN SUMMER SEA ICE MINIMUM, COASTAL POLYNYAS AND BOTTOM WATER FORMATION IN THE WEDDELL SEA

Josefino C. Comiso

Laboratory for Hydrospheric Processes, NASA Goddard Space Flight Center

Arnold L. Gordon

Lamont-Doherty Earth Observatory of Columbia University

As sea ice formation plays a key role in driving Antarctic Bottom Water (AABW) production through brine rejection and the associated production of High Salinity freezing point Shelf Water (HSSW), a possible link between interannual variability of the ice cover and changes in AABW production is expected. Analysis of satellite data indicates that the summer minimum extents (and area) in the Weddell Sea were highly variable from 1979 through 1995, but the correlations of summer ice characteristics to those of subsequent winter are generally weak. We observed, instead, that unusually low summer minimum extents which occurred in 1981, 1985, 1988, and 1993 were preceded by high winter maximum extents (with effects in 1985 not as large as other years). During years of high winter maximum extents, the meridional winds were observed to be relatively strong and coastal polynya areas were greater than average. Wind effects appear to be an important factor in causing the large variability observed in the ice cover while the effects of air temperature variations are not apparent. Temporal variability of AABW in the western Weddell Sea from 1963 through 1993 also is observed, with the bottom water salinity being significantly lower in 1992 and 1993 than those in earlier decades. A strong direct connection of the ice data with AABW is not apparent from available data but our results suggest the intriguing possibility that AABW formation is affected by the passage of the Antarctic Circumpolar Wave.

1. INTRODUCTION

Around the periphery of Antarctica, the Weddell Sea has the most extensive sea ice cover both in summer and in winter. The sector that extends from 60°W to 20°E [after *Zwally et al.*, 1983b] has a sea ice areal extent that usually ranges from a summer minimum of about 0.9×10^6 km^2 to a winter maximum of about 7.9×10^6 km^2. The Weddell Sea has been cited as one of the major sources of bottom water for the world oceans [*Gordon*, 1991]. Recent studies [*Gordon et al.*, 1993; *Muench and Gordon*, 1995; *Fahrbach et al.*, 1995] indicate that in the western rim of the Weddell Sea, a 300 to 500 m thick bottom layer of very cold water (the average potential temperature of the lower 400 m of the water column is -0.8°C) is transported north-ward at a rate of 5 to 6×10^6 m^3 s^{-1}. Salinity stratification within this cold benthic layer suggests that a variety of processes may feed this large outflow of the concentrated Antarctic Bottom Water [AABW; *Gordon et al.*, 1993], which is referred to as the Weddell Sea Bottom Water (WSBW).

Passive microwave remote sensing from satellites, uninterrupted by the presence of clouds, provides an extraordinary time series of the Southern Ocean sea ice cover from 1973 to the present. As the sea ice is so closely linked with ocean-atmosphere coupling due to brine release during ice production, and is a causal factor in ocean overturning related to AABW formation, its long-term behavior is an important indicator of climate change. While the existing Weddell Sea oceanographic data is too sparse to construct a meaningful time series, sea ice variability may provide a proxy indicator of changes in ocean stratification, including the production of AABW.

In this paper, we first characterize the spatial extent and interannual variability of the Weddell Sea summer ice cover that reflects initial conditioning at the onset of the growth season. Relationships of the summer extent with subsequent and previous winter extents are then studied. Coastal polynyas are investigated in the context of ice production and the variability of winter extents. We assess the ice production contribution in both the western section of the Weddell Sea where the shelf is wide, and the eastern section where the shelf is narrow. Wind, surface air temperature, and hydrographic data are analyzed and relationships to the sea ice cover are explored.

2. THE WEDDELL SEA SUMMER ICE COVER

The summer sea ice cover in the Weddell Sea is generally located in the western region adjacent to the Antarctic Peninsula [*Zwally et al.*, 1983b; *Gloersen et al.*, 1992]. The presence of near-year-round ice cover keeps the surface of the ocean in summer colder than in other regions as the high albedo of ice keeps the ocean from absorbing solar energy. Additionally, summer meltwater from sea ice and the neighboring ice shelves contributes to a cold surface layer. The advection characteristics of the ice cover in autumn through winter, as inferred from buoy data [*Ackley and Holt*, 1984; *Massom*, 1992; *Kottmeier and Sellmann*, 1996], indicate that the region is a major source of sea ice and high salinity shelf water (HSSW), the key component of ice shelf water (ISW) and AABW.

The ice concentration maps used in our analysis were derived from Special Sensor Microwave Imager (SSM/I) data using the Bootstrap technique as described in *Comiso* [1995]. Changes in surface emissivity during the summer cause errors in the retrieval but the latter is minimized by using a set of reference brightness temperatures based on results of comparative analysis of SAR and SSM/I data [*Comiso and Ackley*, 1994]. Although an ocean masking has been applied on the daily images, some residuals in the open ocean remained and had to be removed manually using an interactive computer technique.

The summer minimum in 1995 (Plate 1a), which occurred on February 15, extended as far east as 20°W with a large fraction to the west of the 45°W. While a large part of the immediate coastal region in the eastern Weddell Sea was free of ice (east of 30°W), the normal situation, the 1995 summer ice cover appears different when compared with previous years [e.g. *Zwally et al.*, 1983b; *Gloersen et al.*, 1992] due to the presence of an extensive ice-covered area between 30°W and 45°W over the deep ocean. The average of the summer ice concentration maps in the Weddell Sea during summer minima from 1979 through 1995 (Plate 1b) depicts an ice cover significantly different from the single year coverage in 1995 (Plate 1c). The yellows, oranges, and reds represent positive changes (more ice in 1995 relative to the mean) in the ice cover, while the grays, greens, and blues represent negative changes (less ice). The difference image shows more quantitatively the large positive change in extent that occurred in the eastern region in 1995. Some negative changes are notable in the southern region where the coastal polynya occurred in 1995 and in the northern region adjacent to the tip of the Antarctic Peninsula, demonstrating the extent of variability in the summer ice cover that may be expected.

The interannual variability is indeed large for the years 1979 through 1994 (Plate 2). The summer ice cover is always concentrated in the western region adjacent to the Antarctic Peninsula, but the size, shape, and general location change considerably from one year to the next. During some years (e.g., 1980), a narrow tongue of ice extends from the tip of the Antarctic Peninsula towards the east. In other years (e.g., 1988, 1989), the coastal regions are basically free of ice from the east to as far west as the Filchner Ice Shelf (35°W), while in other years (e.g., 1991), the coastline is covered up to the Riiser-Larsenisen Ice Shelf (15°W). There are also years (e.g., 1993) when the ice cover is extremely

---

Plate 1. (*a*) Color-coded ice concentration map of the Weddell Sea ice cover during summer ice minimum in 1995; (*b*) Average ice concentration map of summer minima from 1979 through 1995; and (*c*) Difference between the average map and the 1995 summer ice minimum map.

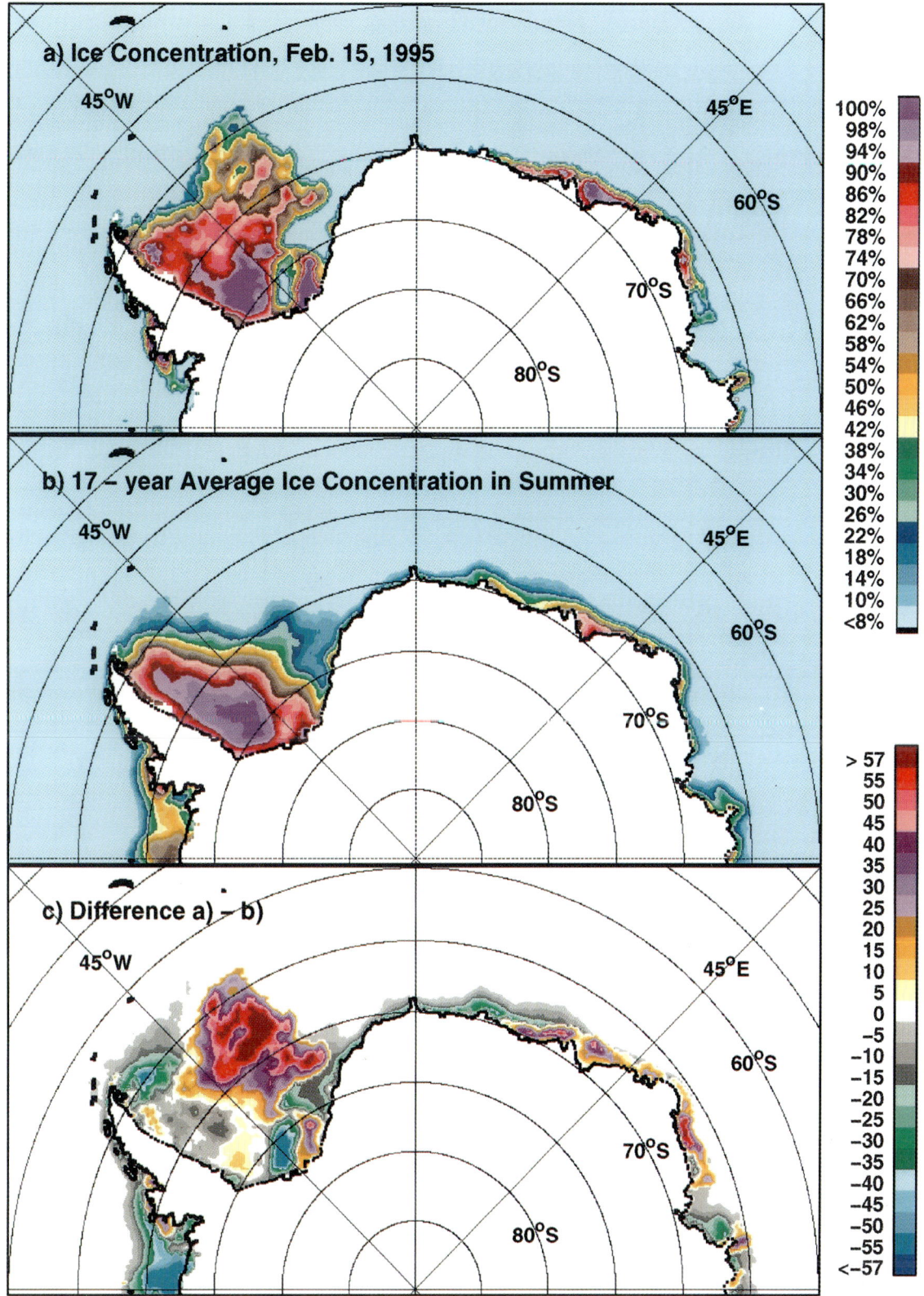

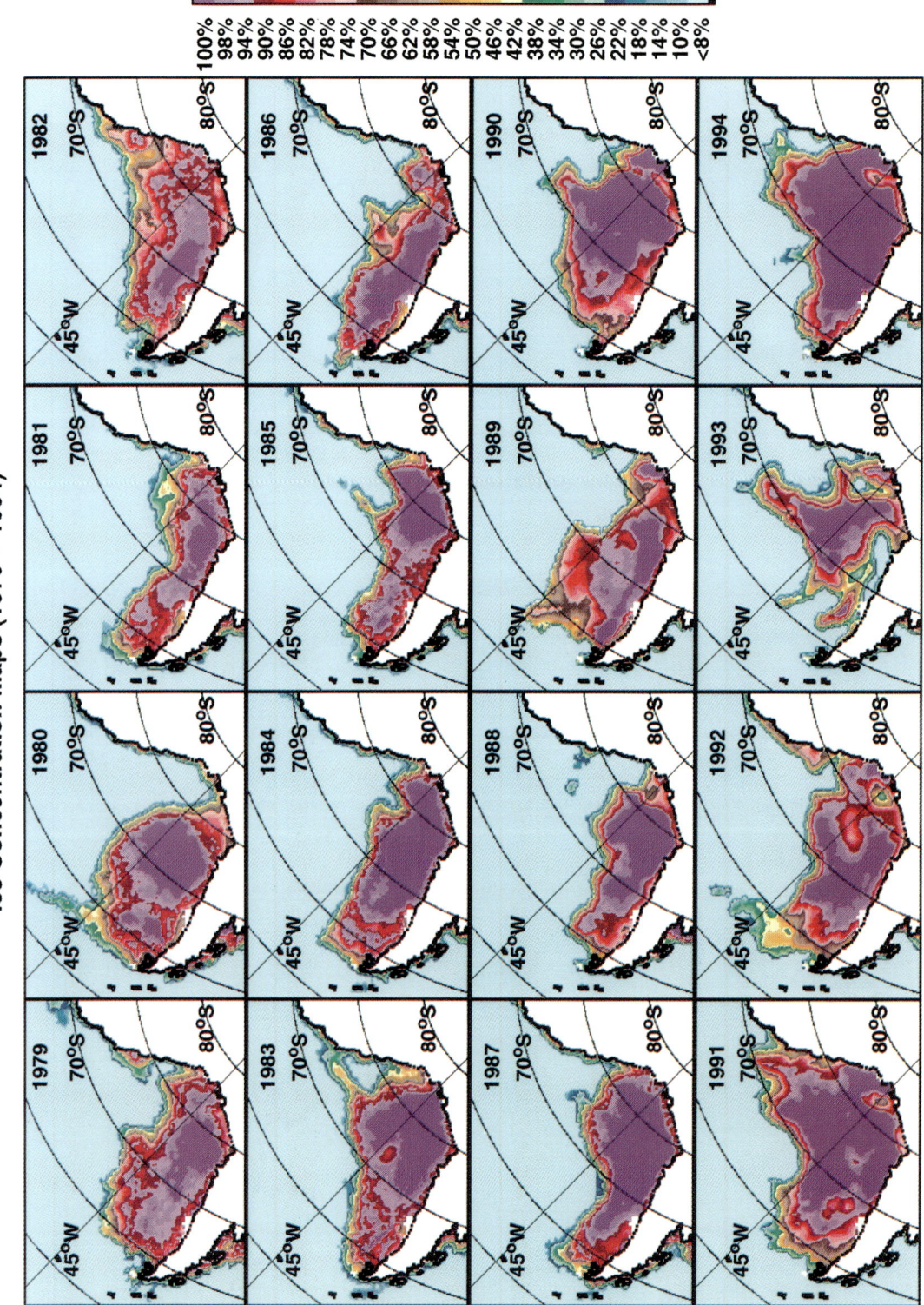

Plate 2. Color-coded ice concentration maps of the Weddell Sea ice cover during summer minimum for each year from 1979 through 1994. The Julian date of the occurrence of the minimum for each year is given in Table 1.

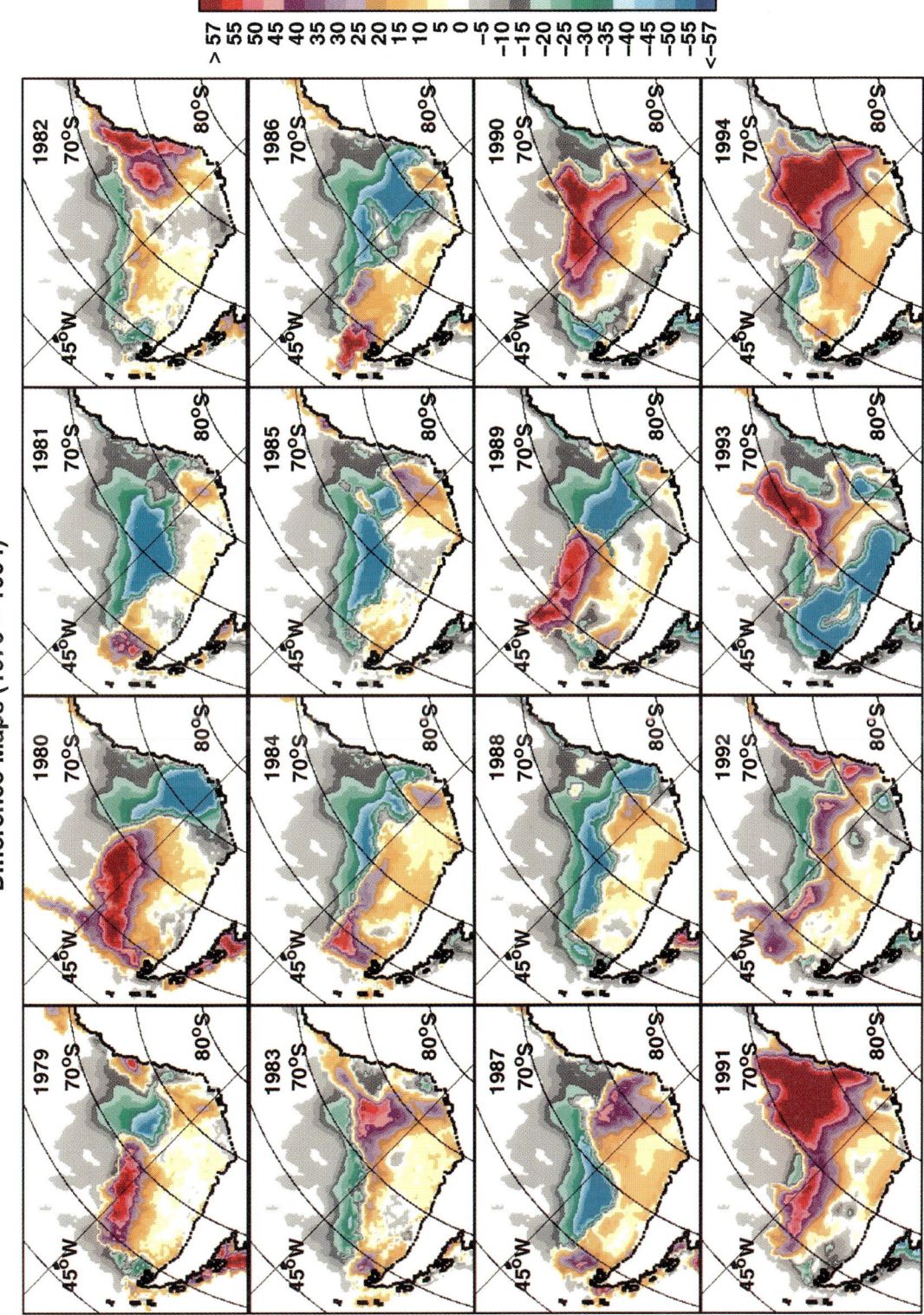

Plate 3. Color-coded summer sea ice minimum anomaly maps in the Weddell Sea from 1979 through 1994. Each map represents the difference between the ice concentration map of each year and the average ice concentration map shown in Plate 1b.

Table 1. Ice extents during summer minima and winter maxima, lengths of summer and winter, lengths of growth and decay, and expansion rate in winter. The numbers in parenthesis are the Julian dates (JD). Low values in summer and high values in winter are in bold type.

Year	Minima Summer 10^6km^2 (JD)	Maxima Winter 10^6km^2 (JD)	Length Summer(days) <2M km^2 (JD)	Length Winter(days) >6M km^2 (JD)	Length Growth (days)	Length Decay (days)	Ave. Exp. Rate
1979	1.33(48)	6.86(276)	59(23-83)	125(189-313)	228	147	0.0242
1980	1.37(57)	**7.89(267)**	77(22-98)	157(180-336)	210	166	0.0310
1981	**0.94(67)**	7.39(263)	104(4-107)	129(190-318)	196	169	0.0329
1982	1.33(66)	6.63(272)	87(9-95)	83(217-299)	206	143	0.0257
1983	1.25(49)	6.75(261)	71(18-88)	83(234-316)	212	153	0.0259
1984	1.14(48)	**6.90(258)**	79(15-93)	131(187-317)	210	158	0.0274
1985	**0.93(50)**	6.48(234)	87(12-99)	113(209-321)	184	189	0.0302
1986	1.08(57)	6.15(235)	91(16-106)	79(230-308)	178	183	0.0284
1987	1.20(52)	**7.37(243)**	89(15-103)	124(207-330)	191	181	0.0323
1988	**0.93(58)**	7.21(252)	99(10-108)	139(196-334)	194	179	0.0324
1989	1.11(65)	6.82(246)	91(11-101)	92(213-304)	181	169	0.0315
1990	1.38(49)	6.19(264)	73(13-95)	58(217-274)	215	159	0.0224
1991	1.58(57)	7.26(271)	54(26-79)	133(198-330)	214	152	0.0265
1992	1.37(57)	**7.55(220)**	58(24-81)	138(181-318)	163	191	0.0379
1993	**1.17(45)**	6.82(263)	80(13-92)	110(203-312)	218	151	0.0259
1994	1.48(48)	6.97(241)	59(19-77)	131(186-316)	193	171	0.0284
1995	1.67(46)	6.97(231)	53(30-82)		185		0.0286

different, with eastward projections in the north and the formation of large polynyas in the south.

The deviations from the average (Plate 3) provide a quantitative measure of how typical the ice cover has been for each year up to 1994. Summer ice minimum extent was close to average in 1983, and 1987. Relatively low values of ice area occurred in 1981, 1985, 1986, 1988, 1989 and 1993, while relatively high values occurred in 1979, 1980, 1982, 1990, 1991, 1992, and 1994. The historical record from 1979 through 1995 indicates that the highest value among the minimum extents occurred in 1995. Overall, the summer ice extent has been very variable, with those in the 1990s being more variable than in other years.

3. INTERANNUAL CHANGES IN ICE EXTENT AND ACTUAL ICE AREA

Ice extent is defined as the ocean surface area that is partly or completely covered by sea ice and is quantified by taking the sum of the areas of each pixel with ice concentrations greater than 15%. Actual ice area, on the other hand, is defined as the surface area occupied by sea ice and is derived from the sum of the product of the ice concentration and the area of each pixel with concentrations greater than 15%. The precision of our estimates of ice extent and actual ice area is of the order of 2% based on statistical and sensitivity analyses [*Zwally et al.*, 1983b; *Gloersen et al.*, 1992]. However, based on available validation data sets, the absolute error is larger and estimated to be about 3% or greater. The use of different sensors (i.e., SMMR from 1979 to 1987 and SSM/I from 1987 to 1995) introduces some uncertainties in the time series because of different antenna patterns and effects of sidelobes. However, the difference in extent during the sensor overlap period from mid-July through mid-August 1987 was very small.

Minimum ice extents and actual ice areas in the Weddell Sea sector from 1979 through 1995 during summer are presented in the first columns of Tables 1 and 2, respectively. Maximum ice extents and actual ice areas for the subsequent winter of each year are given in the second columns of the two tables. In this study, the western limit of the Weddell sector has been modified from 60°W to 62°W to include all sea ice to the east of the Antarctic Peninsula.

It is apparent that there are large seasonal and inter-

Table 2. Actual ice areas during summer minima and winter maxima, lengths of summer and winter, lengths of growth and decay, and ice expansion rate in winter. The numbers in parenthesis are the Julian dates (JD). Low values in summer and high values in winter are in bold type.

Year	Minima Summer 10^6km² (JD)	Maxima Winter 10^6km² (JD)	Length Summer(days) <1.5M km² (JD)	Length Winter(days) >5M km² (JD)	Length Growth (days)	Length Decay (days)	Ave. Exp. Rate
1979	1.01(48)	6.24(272)	58(20- 77)	163(172-334)	224	141	0.0233
1980	0.99(47)	**6.91(269)**	75(22- 96)	161(171-331)	222	160	0.0267
1981	**0.67(63)**	6.69(263)	105(2-106)	150(177-326)	200	141	0.0287
1982	0.93(38)	5.81(272)	83(5- 87)	118(190-307)	234	161	0.0218
1983	1.01(67)	6.12(259)	67(18- 84)	123(193-315)	192	165	0.0266
1984	0.90(58)	**6.18(260)**	77(11- 87)	139(184-322)	202	156	0.0261
1985	**0.72(50)**	5.96(263)	91(7- 97)	149(178-328)	188	160	0.0278
1986	0.80(57)	5.72(263)	93(2-104)	109(203-311)	206	155	0.0239
1987	0.96(52)	**6.52(252)**	89(13-101)	136(186-321)	200	171	0.0278
1988	**0.71(57)**	6.51(255)	103(1-104)	156(183-338)	198	162	0.0293
1989	0.86(51)	6.29(247)	91(10-100)	128(191-318)	196	167	0.0277
1990	1.12(48)	5.68(249)	78(14- 91)	110(198-307)	200	172	0.0228
1991	1.35(55)	6.58(274)	33(31- 69)	139(190-328)	219	147	0.0239
1992	1.01(55)	**6.45(220)**	53(22- 74)	147(173-319)	165	194	0.0330
1993	**0.75(48)**	6.15(263)	87(4-90)	138(187-324)	215	151	0.0251
1994	1.25(48)	6.29(242)	41(25- 65)	154(175-328)	194	171	0.0260
1995	1.11(47)	6.31(232)	55(27-82)		185		0.0281

annual variabilities in both extent and actual area. The ice extent is shown to have summer minimum values ranging from 0.9 x 10^6 km² to 1.9 x 10^6 km², while the winter maximum ranges from 6.1 x 10^6 km² to 7.9 x 10^6 km² (Table 1). High peak values during winter occurred in 1980, 1984, 1987, and 1992. These high values are shown to be followed by low summer values in 1981, 1985, 1988, and 1993, respectively. Also, low peak winter values in 1982, 1990 and 1993 are followed by relatively high summer minima in 1983, 1991 and 1994. This phenomenon was observed as a general pattern by Zwally et al. [1983a] and will be addressed in more detail in later sections. The yearly and seasonal characteristics of the actual ice areas (Table 2) appear to be similar to those in the ice extent plots. There are small differences in the relative location of peaks in the actual area data when compared with those of the ice extent (e.g., slightly higher winter maximum in 1988 than in 1987), but generally the features are similar.

To evaluate the possible direct influence of the summer ice cover, a regression analysis was done between the minimum extents and other parameters associated with ice growth. Among these parameters are lengths of summer and winter, defined (for the purpose of making comparisons) as the number of days in which the areal extents are <2 x 10^6 km² (summer) and >6 x 10^6 km², (winter), respectively. The respective values for summer and winter lengths are given in columns 3 and 4 of Table 1. Finally, the number of days from summer minimum to winter maximum and from winter maximum to summer minimum for each year are given in columns 5 and 6, respectively, while the areal expansion rates (defined as the change in ice extent/area from summer minimum to winter maximum divided by the number of days between minimum and maximum) are given in column 7.

The correlations of the summer ice minimum extent to the various variables are generally weak. The ice minimum is best correlated with the length of summer, the correlation coefficient being -0.88. This is expected since an early (or long) summer would allow for more solar radiation and warming of the ocean (due to lower albedo) leading to further melting of sea ice. The other parameters that correlate somewhat with minimum extent are the net change in extent and the areal expansion rate of the subsequent winter, with correlation coefficients of -0.42 and -0.30, respectively. These coefficients imply that 18% and 9% of the variations in the net change in extent and expansion rate, respectively, are accounted for by differences in summer minimum extents. However, the negative sign is opposite to that expected from the conditioning effect: that an expansive summer ice causes the enhancement of ice growth and ice production in the subsequent winter. Similar regression analysis was done using the ice area and the results are very similar to those of ice extent. The lack of a strong linear correlation between summer minima and the variables in subsequent winter is an indication that there are other factors that affect the relationships, as discussed in the following sections.

4. ATMOSPHERIC AND OCEANOGRAPHIC CONNECTIONS

One of the interesting results of the ice extent analysis is that high maximum extents (or ice area) in winter are followed by low minimum extents (or ice area) during the subsequent summer (in bold, Tables 1 and 2). This phenomenon occurred four times (1980, 1984, 1987, and 1992) during the 17-year study period, with the effect in 1984 not as big as those in the other three years. These high winter maximum values coincide approximately with the extended sea ice edge reported by *White and Peterson* [1996] for the Weddell region, which they connected to the eastward propagating Antarctic Circumpolar Wave (ACW) observed in a variety of ocean and atmospheric parameters. *White and Peterson* [1996] reported a more northerly sea ice edge passing by 60°W to 30°W in 1979-1980, 1983 (weak), 1987-1988 (strong), and 1991. Their study period ended in 1991 but two years are usually affected and the last occurrence likely extended through 1992. Also, the anomalously low summer minima immediately follow the winter maxima and are also likely influenced by the passage of the ACW.

4.1. Air Temperature and Ice Extent

Surface air temperature has been shown to be well correlated with sea ice extent [*Weatherly et al.*, 1991; *Jacka and Budd*, 1991; *Jacobs and Comiso*, 1993; *Smith et al.*, 1996]. A similar relationship is not apparent in the Weddell Sea region from averages of surface temperature data between 65°S and 78°S and from 1979 to 1995 (Figure 1a). The data used are from three different sources: (a) Temperature-Humidity Infrared Radiometer (THIR) satellite data (1979-1985) as described in *Comiso* [1994]; (b) European Center for Medium Range Weather Forecasts (ECMWF) data (1985-1995); and (c) the Australian Bureau of Meteorology (ABM) data set. The ECMWF temperature record is further subdivided into ECMWF-1 data (1985-

1992), derived from climatological data, and ECMWF-2 data (1992-1995) derived using a prognostic equation for modeling surface air temperature. A summary of ECMWF analysis updates is given by *Trenberth et al.* [1989]. The change in technique on August 17, 1992 caused an apparent discontinuity or bias in the temperature record. The apparent difference is minimized by subtracting 5K (based on the difference of the averages between the two data sets) from each of the ECMWF-2 data points. We shall assume that the precision of data from each set, which appears consistent and coherent, is good enough to provide useful information

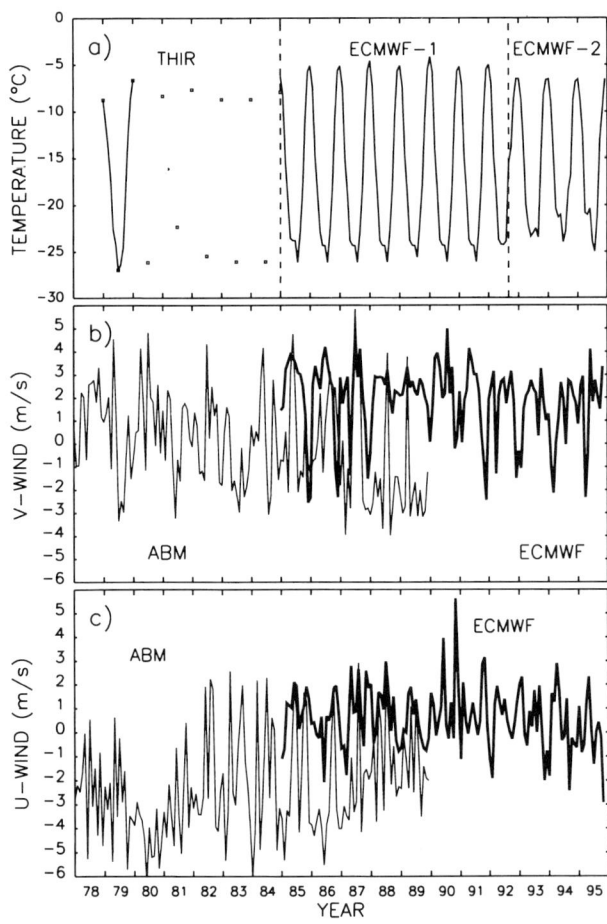

Fig. 1. (*a*) Monthly average surface temperatures in the Weddell Sea Sector for >65°S from 1979 through 1985 using satellite Nimbus-7 THIR data and ECMWF data; (*b*) Monthly meridional (V) winds in the Weddell Sea Sector from 1985 through 1995 using ABM and ECMWF data; and (*c*) Monthly zonal (U) wind in the Weddell Sea Sector from 1985 through 1995 using ABM and ECMWF data. The averages were over the area 62°W to 20°E and 65°S to the continent. Each tic mark along the abscissa represents one year of data starting with January.

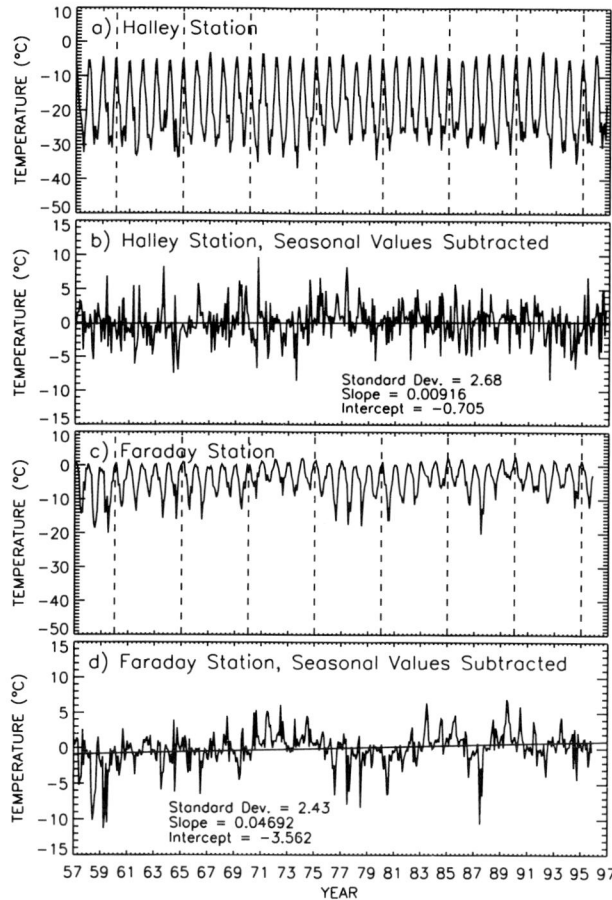

Fig. 2. Surface temperature air measurements (3 m from ground) from the meteorological stations in (a) Halley Station; (b) Halley Station with seasonal averages subtracted; (c) Faraday Station; and (d) Faraday Station with seasonal averages subtracted.

about the temporal variability of the surface air temperature.

The plot in Figure 1a shows that cold temperatures cannot explain the relatively high ice extent in 1980, since the THIR surface temperature record indicates warmer winter temperatures in 1980 than 1979 when the ice extent was considerably lower. Also, the ice extent peak values in 1987 and 1992 do not correspond to unusually cold winters in the ECMWF-1, and ECMWF-2 data. Similarly, the low extents in the summer of 1981, 1988, and 1993 do not appear to correspond to exceptionally warm periods.

The effect of temperature was further examined using meteorological station data at the Halley station (75.88°S, 26.07°W; Figures 2a and 2b) for the period 1957 through 1996. We analyze the entire data set, including those before 1978 to examine possible long term cyclical patterns that may be present in the temperature data. The Halley air temperature record reveals colder winter temperatures during some years (e.g., 1961, 1964, 1968, 1970, 1973, 1979, 1984, 1989, 1992, and 1994) than other years. Except for 1992, however, these years (from 1979) do not match those of the peaks in ice extent during winter. Some peaks in temperature (e.g., 1959, 1967, 1971, 1977, 1979, 1984, 1987, and 1992) also do not exactly match the dips in ice extent (Table 1). Meteorological station data at Cape Adams (75.01°S, 62.53°W), Butler Island (72.20°S, 60.34°W) and Larsen Ice Shelf (66.97°S, 60.55°W) from 1986 through 1995 were also examined and, again, the peaks (or troughs) in the temperature data do not show direct relationships with the peaks (or troughs) in the ice extent and ice area data.

To gain insight into the relationship of monthly temperature (as observed at the Halley Station) with ice extent, the same set of data is plotted twice in Plate 4 to show: (a) seasonal effects, indicated by different colors for each month (Plate 4a), and (b) interannual variability, indicated by different symbols and colors for each year (Plate 4b). A hysteresis loop is evident, showing a relatively strong but non-linear relationship of the two variables during the growth season but a much weaker one during the decay season. Ice extent is shown to have the highest values during the months of July, August, September, and October. During these months, the temperature ranges from -10°C through -37°C, which is about the same temperature range for the ice cover during the growth period. Also, during the decay period (i.e., December), the ice extent varied substantially from one year to another, although the temperatures were nearly the same (at close to melting temperatures). These phenomena may partly explain why the ice extent in this region is not well correlated with temperature. The scatter plots also show, from a different perspective, the relative location of the previously mentioned low summer extents in 1981, 1985, 1988, and 1993 and high winter extents in 1980, 1984, 1987, and 1992. The data for 1980 and 1990 are in closed symbols to illustrate the modulation (when matched with summers of 1981 and 1991) of the ice extent distribution as indicated earlier. Note that in 1992 when the surface temperature was coldest, the ice extent was not as high as in 1980 when the extent was highest.

The surface temperature data at the Faraday station (Figure 2c) were also analyzed because, although the station is located on the western side of the Antarctic Peninsula (65.25°S, 64.27°W), it is near the northern

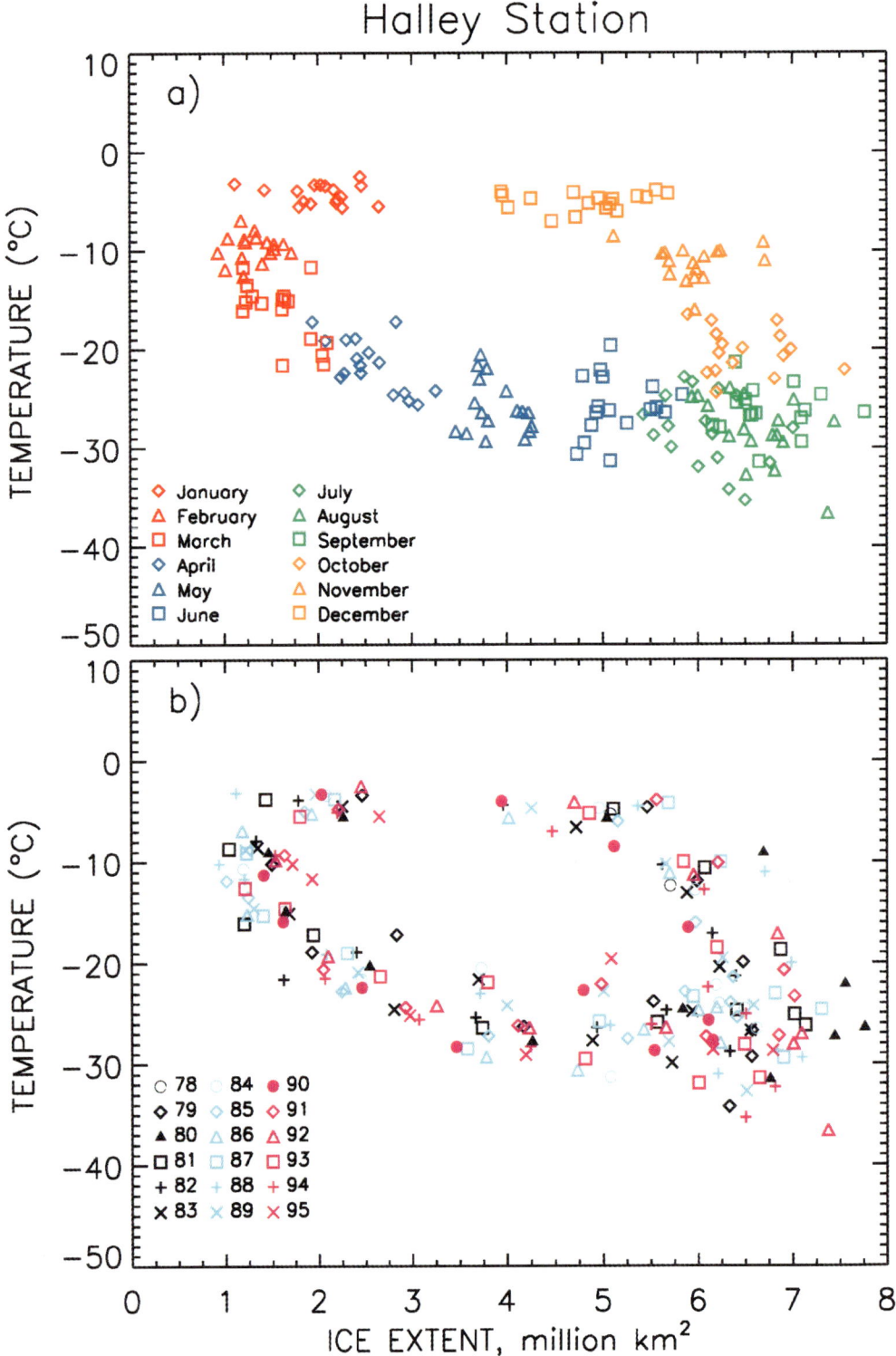

Plate 4. Surface temperature at Halley Station versus ice extent for each month from November 1978 through December 1995 with (a) different color symbols representing different months and (b) different symbol and color for each year (1978-1995).

Table 3. Average concentrations (weighted) during summer minima and winter maxima and polynya areas in Zones-, Zone-2, and Zone-3 in the Weddell Sea. Values in parentheses were derived using ice concentrations (IC-2) from the alternate (Arctic) algorithm.

Year	<C> Summer	<C> Winter	Winter Polynya Area(10^3km^2) Zone-1	Zone-2	Zone-3	Salinization (‰) Zone-1	Zone-2	Zone-3	Weighted
1979	0.759	0.910	20.8(13.4)	11.4(7.8)	39.5(31.4)	0.210	0.304	0.245	0.241
1980	0.723	0.876	24.4(11.4)	12.3(9.7)	39.9(36.6)	0.246	0.328	0.238	0.253
1981	0.713	0.905	17.5(11.6)	11.2(8.5)	35.3(29.4)	0.177	0.300	0.213	0.212
1982	0.699	0.876	19.6(15.6)	11.3(9.2)	32.7(27.3)	0.198	0.301	0.208	0.217
1983	0.808	0.907	18.4(12.4)	10.7(7.9)	39.3(33.8)	0.186	0.287	0.235	0.224
1984	0.789	0.896	18.2(13.9)	10.2(8.2)	33.2(27.6)	0.184	0.274	0.216	0.212
1985	0.774	0.920	25.5(17.7)	10.8(8.8)	33.6(28.9)	0.257	0.290	0.211	0.238
1986	0.741	0.930	18.9(13.1)	11.2(8.7)	33.3(27.3)	0.191	0.299	0.213	0.217
1987	0.800	0.885	17.4(9.4)	13.0(9.9)	36.2(28.8)	0.176	0.346	0.220	0.221
1988	0.763	0.903	19.5(7.9)	12.1(8.8)	40.2(32.7)	0.197	0.323	0.257	0.244
1989	0.775	0.922	19.4(12.7)	11.0(9.1)	37.4(28.9)	0.196	0.295	0.217	0.220
1990	0.811	0.917	20.7(7.8)	10.4(8.1)	33.6(26.1)	0.218	0.278	0.215	0.221
1991	0.854	0.906	24.8(11.2)	12.6(10.1)	30.8(25.1)	0.250	0.336	0.197	0.235
1992	0.737	0.854	21.5(9.8)	15.2(11.5)	40.5(25.2)	0.217	0.406	0.253	0.261
1993	0.641	0.902	21.5(10.3)	15.3(12.5)	34.0(23.7)	0.217	0.409	0.217	0.243
1994	0.844	0.902	20.9(10.1)	14.3(11.7)	32.1(24.1)	0.211	0.382	0.207	0.232
1995	0.665	0.905	32.0(7.6)	14.4(12.0)	35.6(26.3)	0.323	0.385	0.230	0.232

end of the peninsula. The surface air temperatures from this station may thus reflect those at the eastern side of the peninsula, especially near the ice edges. The correlation of these temperatures with the ice extents in the Weddell Sea is, however, poor. This may be partly because the average air temperatures in the Weddell Sea region are generally colder than those in the Bellingshausen Sea, with mean values in the former well below the threshold temperature for melting. Mean values of -18.4° are observed at Halley Bay and -5.5°C at the Faraday Station [King, 1991]. Also, after subtracting seasonal averages, the temperature plots shown in Figures 2b and 2d also indicate slopes of 0.009 and 0.047 for Halley and Faraday Stations, respectively. A significant warming trend observed at Faraday is not reflected in the Halley station data. Increasing air temperatures eventually affect the ice extent and the differences in magnitude and slope may explain why the decreasing ice extent observed in the Bellingshausen Sea region from 1979 through 1994 by *Jacobs and Comiso* [1997] is not observed in the Weddell Sea.

4.2. Wind Effects and Ice Extent

To investigate the effect of wind, we first examine interannual changes in the interior of the ice cover. The weighted averages of ice concentration in the Weddell Sea sector during summer minima and winter maxima for each year are given in columns 1 and 2 of Table 3. The weighted values, <C>, are calculated from

$$<C> = \Sigma_i C_i A_i / \Sigma_i A_i \qquad (1)$$

where C_i is the ice concentration in the data grid element, i, with area A_i, and the summation, Σ_i is taken for all data elements with ice concentration greater than 15% in the study area. The average concentration ranges from about 90% during the winter to about 70% during the summer. Note that in the winter of 1980, 1984 1987, and 1992, the weighted average concentrations were relatively lower than in other years except 1982 when the average values are similarly low. This suggests more divergence and hence larger extents during these years than other years. The data also show relatively low average concentrations in early summer (i.e., January) of 1981 and 1993, when the summer minima were unusually low, suggesting the possible influence of wind on the ice cover in summer.

The short term effect of wind on ice concentration was studied using concurrent daily wind and ice concentration data, from the Halley station and SSM/I, respectively, in a study area adjacent to the station. The results show good negative correlations between wind and ice concentrations during the time periods when the fluctuation in size of the coastal polynya areas was significant. The correlation coefficients for total and meridional winds versus ice concentrations are -0.63 and -0.46, respectively.

Monthly meridional (V) and zonal (U) winds from 1979 through 1995 (Figures 2b and 2c, respectively) are used to investigate long connections between wind and ice extent. We present data from 1979 through 1989, compiled by the Australian Bureau of Meteorology, and those from 1985 through 1995, from ECMWF. Again, we treat these two data sets separately because of possible biases which are apparent during overlap periods (1985-1989). However, many of the data points show similar temporal variabilities during the time of overlap. In 1980 when winter ice extent was relatively high, the meridional wind was also high and northerly while the zonal wind was high and easterly. The strong wind caused divergence that likely led to more extensive ice cover than other years.

Monthly averages of wind data may not provide full information on wind effects because of large fluctuations during each month, and the short-term effects of some of the strong winds may have been averaged out. Short-term effects, however, are difficult to evaluate in the context of long-term changes because so many complex processes are involved. To gain additional insights into long-term variability, we analyzed one-year running averages of the meridional and zonal components of the wind in two study areas. Analysis of the components provides a means to sort out wind advection effects, especially to the north or to the west. The two study areas are (a) the western region (between 60°W to 40°W and called A1), where much of the summer ice cover is located; and (b) the eastern region between 40°W to 20°E (called A2), which is primarily in the seasonal ice region. Only wind data between 65°S to 72°S are analyzed in both areas.

One-year running averages of meridional and zonal components of the wind from 1978 through 1995 are shown in Figure 3, using the same set of data used in Figure 1. The data points in Figure 3 are relatively low compared to those in Figure 1 because each represents an average of monthly data over an entire year. While differences in the meridional wind distribution are apparent during the overlap period (Figure 3a), except for 1988, the values from the two sources (ABM and ECMWF) show a similar variability in A1 during this period. In A2, the values from the two sources during the overlap period agree very well (Figure 3b). Overall, the plots show that the one-year running averages of meridional winds provide distinct interannual variations

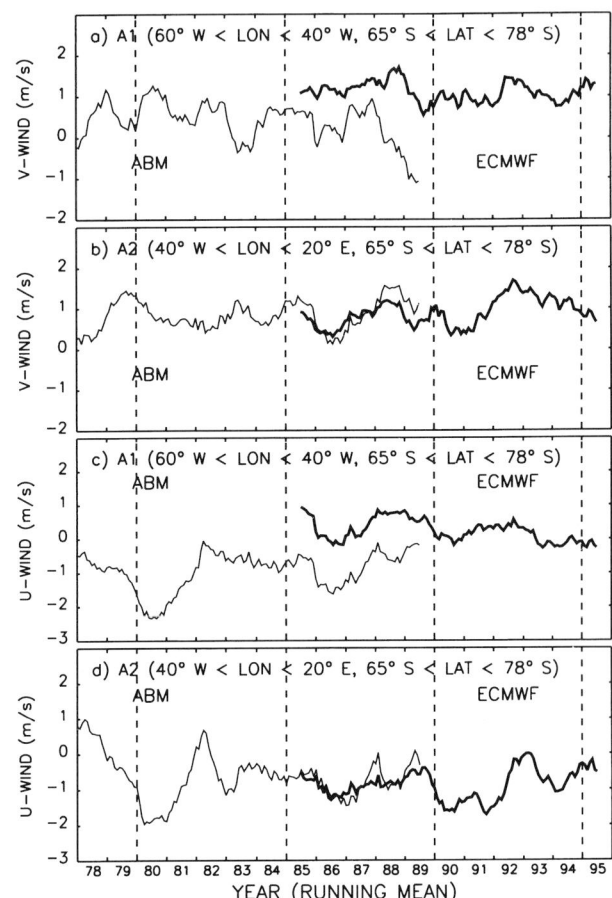

Fig. 3. Yearly-running averages of monthly meridional (a and b) and zonal (c and d) winds over sea ice in A1(65°S to 78°S, 60°W to 40°W) and A2(65°S to 78°S, 40°W to 20°E) study areas.

and other information that are not apparent from the monthly plots (Figure 1). In A1, ABM winds have peak values in 1979, 1980, 1982, 1984, and 1987, while the ECMWF wind data show peak values in 1988 and 1992, and 1995. In A2, enhancements in the ABM meridional winds are apparent in 1979, 1983, 1985, and 1988, while enhancements in the ECMWF winds are evident in 1988, 1989, and 1992. For completeness, the zonal components of the wind are also shown. Except for a few exceptions, the U-wind is generally westerly, and where it is easterly, as with ECMWF winds, there appears to be a bias with respect to the ABM values. The U-wind appears to be most prominent in 1980, as cited earlier, when the wind is relatively strong and negative in both A1 and A2 study areas and when ice extent was most extensive.

The years of occurrence of peak values in the winter ice extents (e.g., 1980, 1984, 1987, and 1992) are generally consistently close to those of peak values in the meridional winds. The relatively lower average ice concentrations during these periods (Table 3) are also consistent with wind effects. The smaller enhancement in the ice extent in 1984 compared to the other years may be associated with relatively lower wind speed during the year. The correlation of wind with ice extent is slightly better in A2 than in A1. This will be discussed in the context of coastal polynya areas in the next section.

We present the following hypotheses to explain the phenomenon of a "low summer minimum following a large winter maximum ice cover":

1. The Antarctic Circumpolar Wave (ACW) leads to greater winter ice coverage every 4 to 5 years [*White and Peterson, 1996*]. Increased brine release associated with more ice production causes the formation of more saline surface water, reducing the strength of the pycnocline and allowing greater exchange with the warmer deep water. This additional heat further reduces the ice cover during the summer minimum.

2. Exceptionally strong southerly winds result in higher ice velocities, pushing the ice edge further north. Increased lead and coastal polynya areas produce a larger percentage of the thinner ice than normal. This would make the ice cover more vulnerable to melting, leading to a less extensive summer ice cover.

3. Wind advection effects (regional wind stress curl) on the ocean may cause a spin up of the Weddell Gyre. An increase in wind speed during the year makes the gyre move faster while a decrease produces a slower gyre. In winter, a faster Weddell Gyre (stronger regional curl of the wind stress) moves more ice out of the cold southern region (where it "leaves room" for increased ice production) of the western boundary region, moving the winter ice edge further north than usual. By summer, the perennial ice field of the western Weddell Sea has been flushed out of the region by the stronger Weddell Gyre, into the melting zones to the north. A faster Weddell Gyre may thus produce greater differences between winter and summer sea ice cover.

A periodicity of approximately four years is apparent in the data, and matches the period of the Antarctic Circumpolar Wave [*White and Peterson, 1996*], lending support to Hypothesis 1. The observed wind effects also support Hypothesis 2 but it is not known which of Hypothesis 1 or 2 is more important. Wind effects could also lead to a vigorous Weddell Gyre (Hypothesis 3). It is apparent that the key factor is the influence of wind and that the three hypotheses are not independent of each other.

5. ROLE OF LATENT HEAT POLYNYAS IN ICE PRODUCTION

To understand the relationship of the ice cover to AABW, the relative influence of polynyas to total ice production should be evaluated. Two basic types of polynyas, each of which has different roles in ice/ocean/atmosphere coupling, have been defined [e.g., *Gordon and Comiso*, 1988; *Comiso and Gordon, 1996*]. One type is the "latent heat polynya" induced by wind removal of newly formed sea ice. Latent heat polynyas usually form along coasts or obstructions (islands and big icebergs) and have been called coastal polynyas. The other type is the "sensible heat polynya", which is maintained by the upwelling of relatively warm deep water to the sea surface. Depending on size and surface conditions, the latter may be either convective or non-convective. In this study, we focus on the role of the latent heat polynyas because they are directly linked to the production of HSSW and ultimately to bottom water formation.

Coastal polynyas around Antarctica have been referred to as "ice factories" as the constant removal of newly formed ice by the wind results in the continuous formation of sea ice. During the slacking of the wind, the polynya may be covered by thin ice, but this would be removed with the next wind burst, once more exposing the ocean to the atmosphere. It has been inferred indirectly from sea ice core studies that a large fraction of the Weddell Sea ice pack may originate from sea ice formed in coastal polynyas [*Eicken and Lange*, 1989]. Examples of latent heat polynyas in the Antarctic coastal region have been cited and their net effects studied [*Zwally et al.*, 1985; *Cavalieri and Martin*, 1985; *Jacobs and Comiso*, 1989; *Darby, et al.* 1995; *Kottmeier and Sellmann*, 1996]. Estimates of salinization from polynyas in the Weddell Sea are also presented by *Markus et al.* [this volume] for the 1992 to 1994 period.

5.1. Variability in the Areas of Coastal Polynyas

Typical coastal polynyas in winter have widths of the order of less than one to several kilometers. The best resolution available from current satellite passive microwave systems is 12.5 km from the 85 GHz channels. Generally, even these channels cannot resolve most of the coastal polynya features and therefore the latter are represented in the data as reduced ice concentrations. Because of the high sensitivity to weather and other atmospheric effects as well as the high variability of the emissivity of the ice surface at this frequency [*Comiso et al.*, 1989], the 85 GHz channels have not been used to infer global sea ice concentration. When the conditions are good, however, they may provide better spatial details than other SSM/I channels [*Markus and Burns*, 1995]. Since satellite 85 GHz data are currently available only with the SSM/I sensor and for the period July 1987 to 1988 and from December 1992 to the present, they are presently not suitable for long-term studies. Our study makes use of ice concentrations derived from the 37 and 18 GHz (or 19 GHz) channels [*Comiso*, 1995]

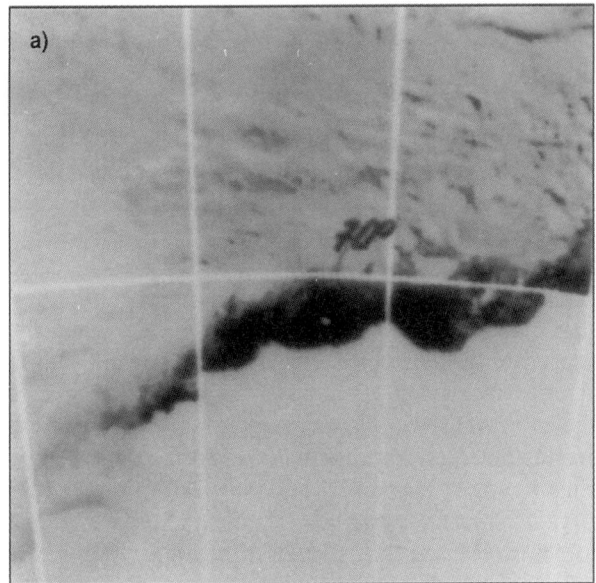

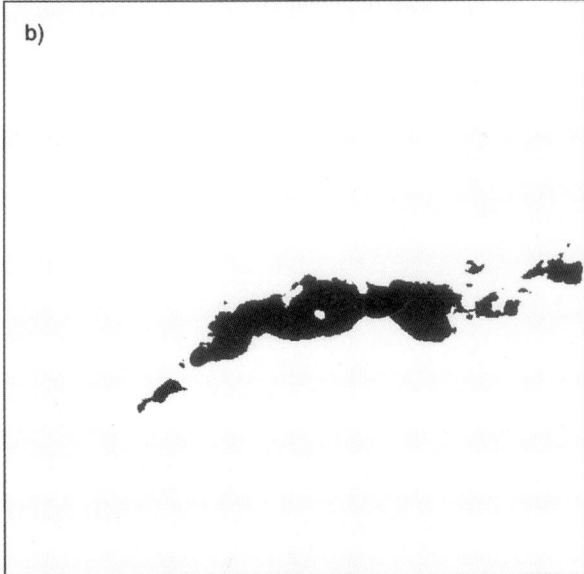

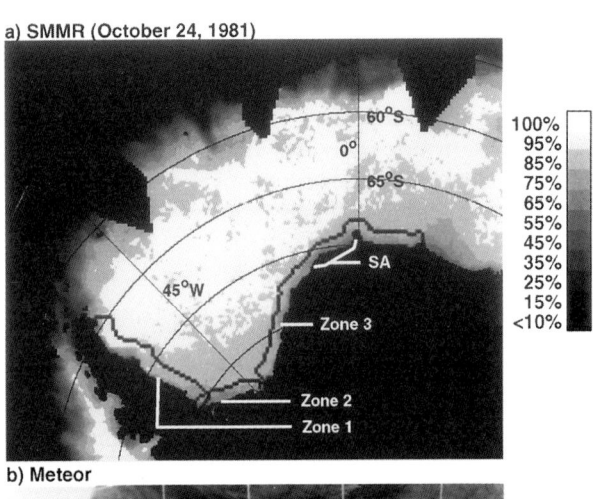

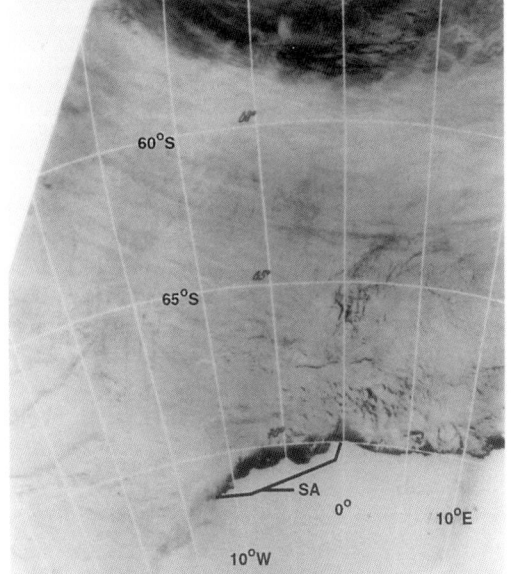

Fig. 5. (*a*) An enlarged visible channel image of Figure 4*b*, including SA, and (*b*) a binary image of ice and water using a thresholding technique on the image data in (*a*).

Fig. 4. Ice cover characteristics on October 24, 1981 as depicted by (*a*) a coded ice concentration map using SSM/I data and (*b*) a visible channel image from the Russian Satellite Meteor. Regions of Zone-1 through Zone-3 are indicated as well as a polynya study area (SA).

which provide consistent polynya area information for the period from 1979 to the present.

The overall error in our estimates of polynya area is difficult to quantify because of the lack of validation data sets in the region during winter. The relatively low emissivity of new ice compared to thick ice and lower surface ice temperatures than average in these regions (especially for young ice), causes overestimates in open water areas. However, polynyas are known to be covered predominantly by new and young ice [*Martin et al.*, 1992]. In polynya heat flux studies, the errors due to new and young ice are partly compensated because fluxes from new and young ice are known to be an order of magnitude lower than those of thick ice [*Maykut*, 1978; *Lytle and Ackley*, 1996]. A crude assessment of the accuracy of the area estimates during spring has been made through a comparative analysis of SSM/I ice concentration data (Figure 4a) and Russian Meteor satellite data in the visible channel (Figure 4b). The two data sets show similar features in the coastal regions, especially in the study area (SA) between 0° and 15°W. A thresholding algorithm based on the large contrast of the albedo of thick ice with snow cover and open water was applied to the visible channel image (magnified in Figure 5a), and the result is shown as a binary map in Figure 5b. The estimate of open water from SSM/I was 12,700 km² while the area indicated as open water in the binary map is 12,345 km². The thresholding technique may classify some forms of new ice (i.e., those below the threshold) as open water since new ice has an albedo closer to that of open water than thick ice [*Perovich, et*

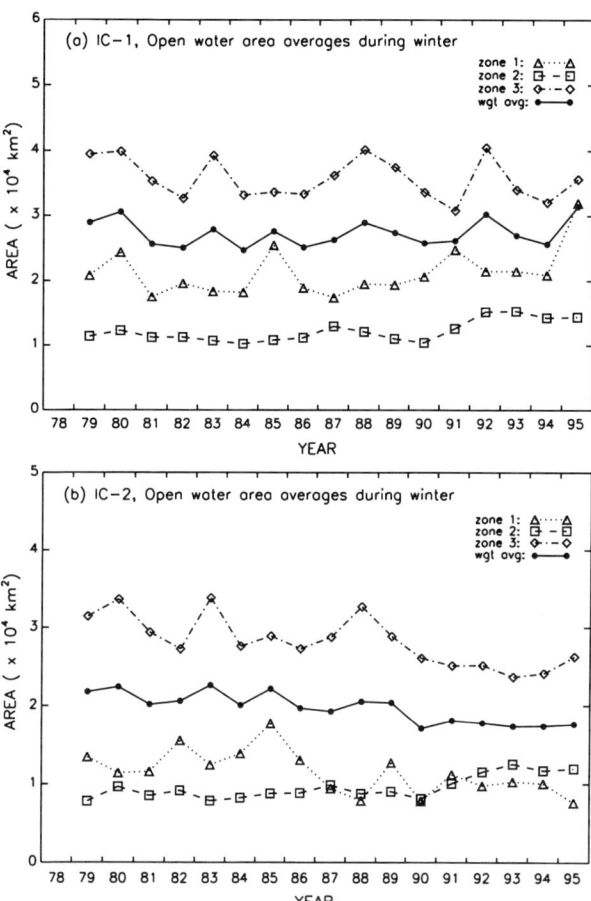

Fig. 7. Averages of daily coastal polynya open water areas during each winter from 1979 through 1995 using ice concentration values from (a) the standard Bootstrap Algorithm for the Antarctic (IC-1), and (b) the modified Bootstrap algorithm for the Arctic (IC-2). The number of days in the average depends on the length of winter which varied from year to year.

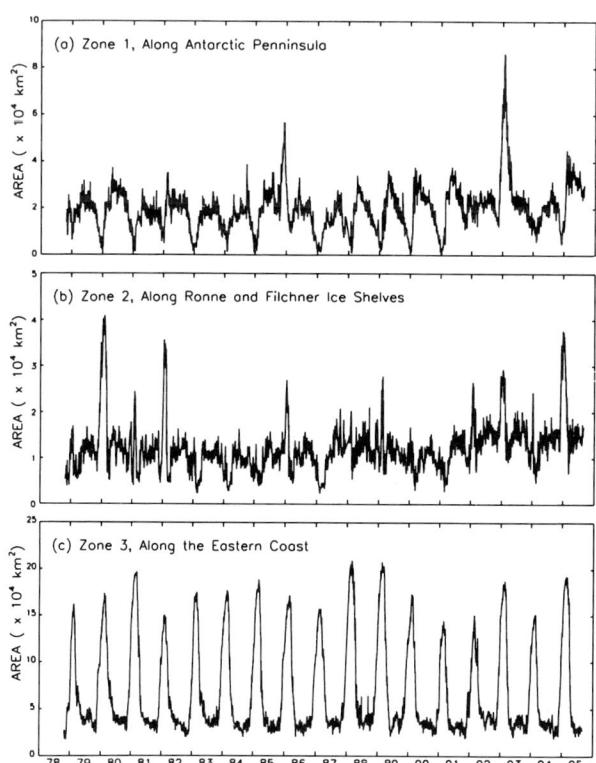

Fig. 6. Weddell Sea coastal polynya open water areas derived from SSM/I daily average data in (a) Zone-1, (b) Zone-2, and (c) Zone-3 from 1979 through 1995.

al., 1986; *Allison et al.*, 1993]. The good agreement in the estimates (within 2.9%) indicates that the polynya areas inferred from SSM/I data reflect quantities associated with a real surface feature such as those in Figure 4a (labeled as SA) that may be partly frozen. Although the spatial details of the polynya as observed from the visible image is not reproduced in the SSM/I data because of the much coarser resolution of the latter, the inferred magnitude of the polynya size from the two systems are in reasonable agreement.

Estimates of daily open water areas along the coastline of the Weddell Sea study region are shown in Figure 6. The coastline is divided into three study areas: Zone-1, along the Antarctic Peninsula (64°S to 75°S);

Zone-2, along the Ronne/Filchner Ice Shelves (37°W to 60°W); and Zone-3, along the eastern coastline of the Weddell Sea (20°E to 37°W). The outlines of the three zones are indicated in Figure 4a. The width of the study areas is approximately the same and about 100 km (4 pixels) from the continental margin. The total areas of Zone-1, Zone-2, and Zone-3 are 16.06×10^4, 6.07×10^4, and 22.6×10^4 km^2, respectively.

The seasonal variability of open water areas in Zone-1 is shown to be quite large (Figure 7a). Apart from a few exceptions, open water areas are unexpectedly higher during winter than in summer. During spring and summer, the areas of open water in this zone are reduced to almost zero (consistent with *Markus et al.* [this volume]), suggesting that, during this time period, ice is advected towards the Antarctic Peninsula, probably due to the influence of wind (as observed in the 1988 and 1989 data) and the Weddell Gyre. The exceptions (when open water area are more substantial) are in the springs of 1985, 1992, and to some degree 1991. The pronounced peak starting in the spring of 1992 through the summer of 1993 was due to a large coastal polynya (see Plate 2). The presence of this polynya in summer is among the factors that led to the low minimum in 1993 which was preceded by a high winter maximum as pointed out earlier. The unusually dispersed features also suggest strong wind effects as described earlier. The occurrence of the peak value during the spring of 1985 coincided with the absence of ice near the eastern tip of the Antarctic Peninsula which may be associated with wind effects or possible intrusion of warm water masses from the north. During some years (e.g., 1982, 1983, 1992), the size of the coastal polynyas is essentially sustained during much of the growth period, while in other years (e.g., 1988, 1989, 1990) the size is enhanced in early winter and gradually decreases towards the summer. Persistently strong winds could cause the former while a decreasing wind intensity and a heavy ice cover could cause the latter. Short term (e.g., daily) variations are also apparent, likely caused by short term oscillations of wind effects.

The variability of the coastal polynya areas in Zone-2 (Figure 6b), is similar to that in Zone-1 but the summer values are not as consistently close to zero in this zone as in the latter, and there are more years when the coastal areas have substantial open water during the summer. It is also apparent that Zone-2 is more active (i.e., higher open water area per kilometer of coastline) during winter than Zone-1. In 1986, large icebergs about 12,000 km^2 in total surface area calved from the Filchner Ice Shelf (near 45°W) and were grounded near the coastline. The presence of the icebergs apparently caused a higher open water area than during previous years, except in 1990. A bias (estimated to be about 2,500 km^2) caused by the presence of the icebergs, which have lower emissivities than those of sea, has been removed. During the summer, the open water areas are shown to be relatively high in 1980, 1981, 1982, 1986, 1989, 1992, 1993 and 1995. This is partly due to the formation of polynyas and partly due to the seasonal retreat of the ice cover (see Plates 1 and 2).

The coastal polynya area distribution for Zone-3 (Figure 6c) is different from those in the other zones in that the coastal region is largely a part of the seasonal ice region. Some uncertainties are introduced in the retrieval of ice concentrations near coastlines because of instrumental (e.g., antenna pattern and side lobes) effects and contamination of the pixels by continental contributions. The yearly peak values were relatively low in 1982, 1991, and 1992, which are the years when much of the coastline was occupied by ice during the summer. Large interannual variations in winter are also apparent, and the total areas are slightly higher than those in Zone-1.

To evaluate the interannual variability of coastal polynya areas, averages of these open water areas during winter (arbitrarily defined as the period from April through November when the study areas were fully ice covered) were calculated for each zone and results are shown in Figure 7a. The numerical values are also given in Table 3. The percentages of open water in winter in Zone-1, Zone-2, and Zone-3 are 12%, 20%, and 16%, respectively, indicating that Zone-2 has the most polynyas per unit area among the three. In Zone-1, polynya areas are relatively enhanced in 1980, 1985, and 1992. In Zone-2, the polynya areas are slightly higher than average in 1980 and 1987, but significant increases in area occurred in the 1992, 1993 and 1994 period. In Zone-3, distinct peaks in 1980, 1983, 1988, and 1992 are apparent. These years coincide approximately with years of enhanced ice extent and meridional winds as discussed previously. The coherence is even better when the net impact from all three zones are taken into account as indicated by the weighted averages of the polynya areas shown as bold lines in Figure 7a.

Our estimates of polynya areas are in general agreement with those of *Markus et al.* [this volume] during the summer, when validation data are available. However, we found large discrepancies between our estimates and those of *Markus et al.* [this volume] in the winter period. The difference is attributed to the use of different sets of SSM/I channels, as indicated earlier, which have different sensitivities to new ice and atmospheric effects. To gain insights into the difference, ice concentrations were derived using the Arctic version of the Bootstrap algorithm [*Comiso*, 1995] in which the vertically and horizontally polarized channels at 37 GHz (called the HV37 set) are utilized for highly consolidated ice areas. The use of this set minimizes surface temperature and ice sheet contamination effects where they are applied. *Markus et al.* [this volume] also uses the 37 GHz channel in combination with the 85 GHz channels for deriving ice concentration.

The results of the use of ice concentrations (IC-2) from the Arctic algorithm are plotted in Figure 7b and values are given (in parenthesis) in Table 3. These results show considerably lower polynya areas than the previous ones. The average percentage differences between the two techniques are 42%, 22% and 20%, in Zone-1, Zone-2, and Zone-3, respectively. The discrepancies are believed to be mainly due to the presence of new and young ice, since the difference is caused by the use of the HV37 set which is more sensitive to these types of ice surfaces. Assuming that this is the case, the results from the two techniques can be used to assess the effect of having new and young ice cover instead of open water. Since heat and salinity fluxes are very different for new and young ice surfaces compared to those of thick ice, ability to detect them as part of a polynya feature is an important consideration. The percentage difference can thus be used as a guide in readjusting the estimates presented in Table 3. The use of the HV37 also minimizes the bias due to contamination from the continental ice sheets. However, overall, the distributions from the two results are similar except for the years after 1991. This may mean more thin ice production in the region in recent years than previously.

The polynya areas derived from the Arctic algorithm are closer to those of *Markus et al.* [this volume] but still significantly higher. The use of the 85 GHz data has its advantages due to higher resolution but has not been validated during adverse weather conditions. The NOAA-AVHRR images in Figure 8 illustrate that polynya sizes can be relatively large (and comparable to our results) in the western Weddell region. The images also show substantial atmospheric effects in the region. Such effects are usually opaque to the 85 GHz signal and can cause large errors in ice concentration estimates when 85 GHz data are used. The strength of our technique is the temporal consistency in the estimates because we use the 19 GHz data (in combination with a 37 GHz data), which are less vulnerable to atmospheric and surface effects. A weakness is the coarser resolution of the 19

Fig. 8. NOAA-AVHRR images of the western Weddell region on (*a*) November 20, 1988, (*b*) November 27, 1988, and (*c*) November 9, 1989. Polynya features in Zone-1 and Zone-2 are apparent but sometimes they are hidden by clouds that can adversely affect retrievals from the 85 GHz data.

GHz data, which may cause some contamination of the polynya area by the continental ice sheet, and hence a bias, as pointed out in *Zwally et al.* [1985]. The Arctic algorithm reduces the bias and may provide a better estimate of actual open water area but a fraction of new and young ice is classified as thick ice. Further studies are needed to establish the accuracy of the various techniques, especially during winter.

5.2. Oceanographic Significance of the Coastal Polynyas

Coastal polynyas are formed as wind blowing off Antarctica sweeps clear the ice from adjacent ocean [e.g., *Smith et al. 1990*]. They facilitate the production of high salinity shelf water [HSSW, *Foster and Carmack*, 1976a,b; *Foldvik et al.*, 1985], through production of vast amounts of sea ice [*Zwally et al.*, 1985]. The total ice production within a coastal polynya can be quite high, perhaps tens of meters per winter season, as newly formed ice within the polynya is continuously removed by the wind exposing the ocean to the very cold atmosphere allowing continued ice production. In the Weddell Sea, HSSW is exported to the deep ocean within the convective plumes over the continental slope to feed the saline variety of the Antarctic Bottom Water observed in the western rim of the Weddell Sea [*Gordon et al.*, 1993]; a similar process occurs in the Ross Sea [*Jacobs et al.*, 1985]. Additionally, HSSW migrates below the ice shelves, where melting glacial ice emerges as slightly fresher Ice Shelf Water (ISW) with temperatures well below the one atmospheric pressure freezing point [*Carmack and Foster*, 1975; *Foldvik et al.*, 1985]. The fresher ISW flows into the deep ocean to feed a low salinity variety of the Antarctic Bottom Water in both the Weddell and Ross Seas [*Gordon et al.*, 1993; *Jacobs et al., 1985*]. The partitioning of HSSW that flows directly off the shelf versus that which passes through the ice shelf contact is not known. One might expect that an increase in HSSW directly adjacent to the shelf ice barrier would ultimately contribute to increased Ice Shelf Water production.

The premise relating coastal polynyas to Antarctic Bottom Water production is: coastal polynyas result in more ice production, enhancing the production of HSSW, with commensurate increases in slope plumes of HSSW and of ISW leading to the formation of more Antarctic Bottom Water. Enhancement of shelf salinity depends on the existence of a broad continental shelf, allowing for accumulation of HSSW. In Zone-3, where the brine release may be considerable, bottom water formation appears to be suppressed [*Fahrbach et al.*, 1994], because of the narrow width of the continental shelf. Zones-1 and Zone-2 may be more effective in leading to bottom water production.

Zwally et al. [1985] estimated the added ice production stemming from Antarctic coastal polynyas as 0.10 to 0.17 m/day with an increase in shelf water salinity of 0.2 to 0.4 $^o/_{oo}$, making coastal polynyas the major contributor to the creation of HSSW. Although it is just a crude estimate, we use the *Zwally et al.* relationship to determine the salinity increase within the Weddell Sea coastal polynyas as a simple and consistent indicator of the interannual changes in HSSW production that is most likely linked to coastal polynya processes. More sophisticated techniques require the use of accurate and temporally consistent wind and temperature data which, as indicated earlier, are available only in a few stations. The *Zwally et al.* [1985] equation, which relates increasing shelf water salinity to polynya ice production is

$$\Delta S = s T A_r R_i / 0.1 h, \qquad (2)$$

where ΔS is the salinity increase resulting from ice formation in the coastal polynya; s is the salt rejection in g cm^{-2} for each meter of sea ice formed (a value of 2.5 is used); 0.1 in the denominator is a scaling factor to yield ΔS values in parts per thousand; T is the duration in days of the winter period, taken as 240 days for Zone-1 and Zone-2 and varied from 180 to 216 days in Zone-3; A_r is the area ratio of open water to total area of the shelf (in our case, the study area is used as an approximation since the dilution factor is not known); R_i is the rate of ice formation within the polynya in m d^{-1}; h is the water column thickness (a 518 m average is used for the glacially depressed Weddell Sea Shelf). Using the open water polynya areas over the continental shelf (Figure 7a) the associated boost to shelf water salinity is calculated (Table 3). The R_i value used is 0.14 m d^{-1}, the mean of the estimated range of *Zwally et al.* [1985]. As we are mainly interested only in the interannual variations of shelf water production, the exact value of R_i is not critical, though we assume it to be a constant. The derived ΔS values are high compared to previous values [e.g., *Zwally et al.*, 1985] because the areas of each zone, instead of the larger, total shelf area, is used in the denominator of A_r.

The 1979 through 1995 mean salinity increase for the coastal zones of the Weddell Sea is 0.232 $^o/_{oo}$ with a standard deviation of 0.021 $^o/_{oo}$. With the Arctic algorithm, the corresponding value is 0.157 $^o/_{oo}$. Relatively high values are encountered in 1980, 1988, 1992, and 1995, which represent the years following a low summer minimum extent, which, as discussed above, follow the high winter maximum occurrence associated with passage of the Antarctic Circumpolar Wave [ACW, *White and Peterson,* 1996].

The bottom water connection may be best associated with Zones-1 and -2, as Zone-3 is mostly over a narrow shelf, prohibiting the accumulation of HSSW. Zone-1 shows maximum production of saline shelf water in 1980, 1985, and 1991 while Zone-2 shows high values in 1980, 1988, and for the years 1991 to 1995. An intriguing possibility exists that the passage of the ACW may be related to increased bottom water formation.

Instead of causing an increase in shelf water salinity, enhanced sea ice production within the coastal polynyas may lead to increased production of HSSW of constant salinity. Assuming that the entrainment ratio of brine to resident surface water of the winter mixed layer remains the same, increased sea ice production is expected to be linearly related to HSSW volume production of constant salinity. Hence, the interannual variations in HSSW production rate would be directly proportional to polynya areal extent, similar to the salinity increase listed in Table 3.

6. RELATIONSHIPS TO THE ANTARCTIC BOTTOM WATER

Among the key results of the Ice Station Weddell project [*Gordon et al.*, 1993] is the identification of two types of Weddell Sea Bottom Water (WSBW) forming in the western Weddell Sea: a salty denser type that draws water directly from the HSSW; and a slightly fresher, less dense water type that is drawn from Ice Shelf Water. The two types of WSBW are found within the lower 300 meters of the water column. At many sites, both WSBW types occur in the same benthic layer. In the northwestern Weddell Sea, the benthic layer has mixed upward into the water column removing the complex salinity stratification.

The T/S relationship of all station pairs in 1° latitude by 1° longitude bins in the western Weddell Sea (Plate 5) indicates that in 1992 and 1993, the WSBW is colder and fresher than that of previous years. During the 1960s and 1970s, the -0.8° to -1.0° C bottom water is markedly more saline than that observed in 1992-1993. All but one station in the western rim of the Weddell Sea are from 1992 and 1993. That station (black circle) represents the 1960s period; it has a bottom T/S that is clearly more saline, and warmer than that observed in 1992/93 (red circle). In the northwestern corner of the Weddell Sea the data array representing a wider range of years, also shows the 1992/93 bottom water (red triangles) as being fresher and colder than that observed in previous years (e.g., black triangles). While the time series is far from complete, it strongly suggests that WSBW observed in 1992 and 1993 is both fresher and colder than that observed in earlier years. The movement towards fresher and colder water is indicative of an increased Ice Shelf Water contribution.

The relationship of increased coastal polynya area and associated greater HSSW production to increased ISW escape from the continental margin is not known and numerical model research is encouraged. One possibility that should be investigated is that increased HSSW production is expected to occur during more extensive coastal polynya periods. This may lead to a thicker layer of HSSW which would in turn lift the ISW layer to a shallower depth, providing less impeded overflow of the outer shelf topography. However, a factor that must be considered is that the HSSW and ISW water types that feed WSBW have residence times measured in years. *Mensch et al.* [in press] estimate the residence time of HSSW in the Weddell Sea as 5 years, and that the ISW is renewed from the HSSW on a time scale of 14 years. A direct year-by-year connection of ice extent and coastal polynyas with the variability of WSBW may not be obvious or expected, as lower frequency thermohaline events may determine the availability of the shelf water for bottom water formation. Other processes such as wind induced shelf water circulation changes may influence higher frequency changes in the escape of shelf water to the deep ocean. For example, wind changes may control the spreading of HSSW, determining the percentage of HSSW that flows directly into the deep sea versus that component which spreads below the glacial ice to form ISW. From the continuity equation, an increase of the HSSW percentage flowing under the shelf ice would be expected to force increased escape of ISW.

Although no imaging passive microwave satellite data were available before 1973, some visible channel data indicate that the areal extent of Antarctic sea ice was higher in the early 1970s com-pared with later years [*Zwally et al.*, 1983a]. This is compatible with a saltier and colder bottom water (Plate 5) for these years and as observed by *Foster and Carmack* [1976b]. However, *Foster and Middleton* (1979) also indicated that the year to year variability in the bottom water of the Weddell Sea can be significant. Needless to say, improved monitoring of the Weddell thermohaline stratification and the sea ice cover is required to gain a meaningful statistical relationship between sea ice and WSBW.

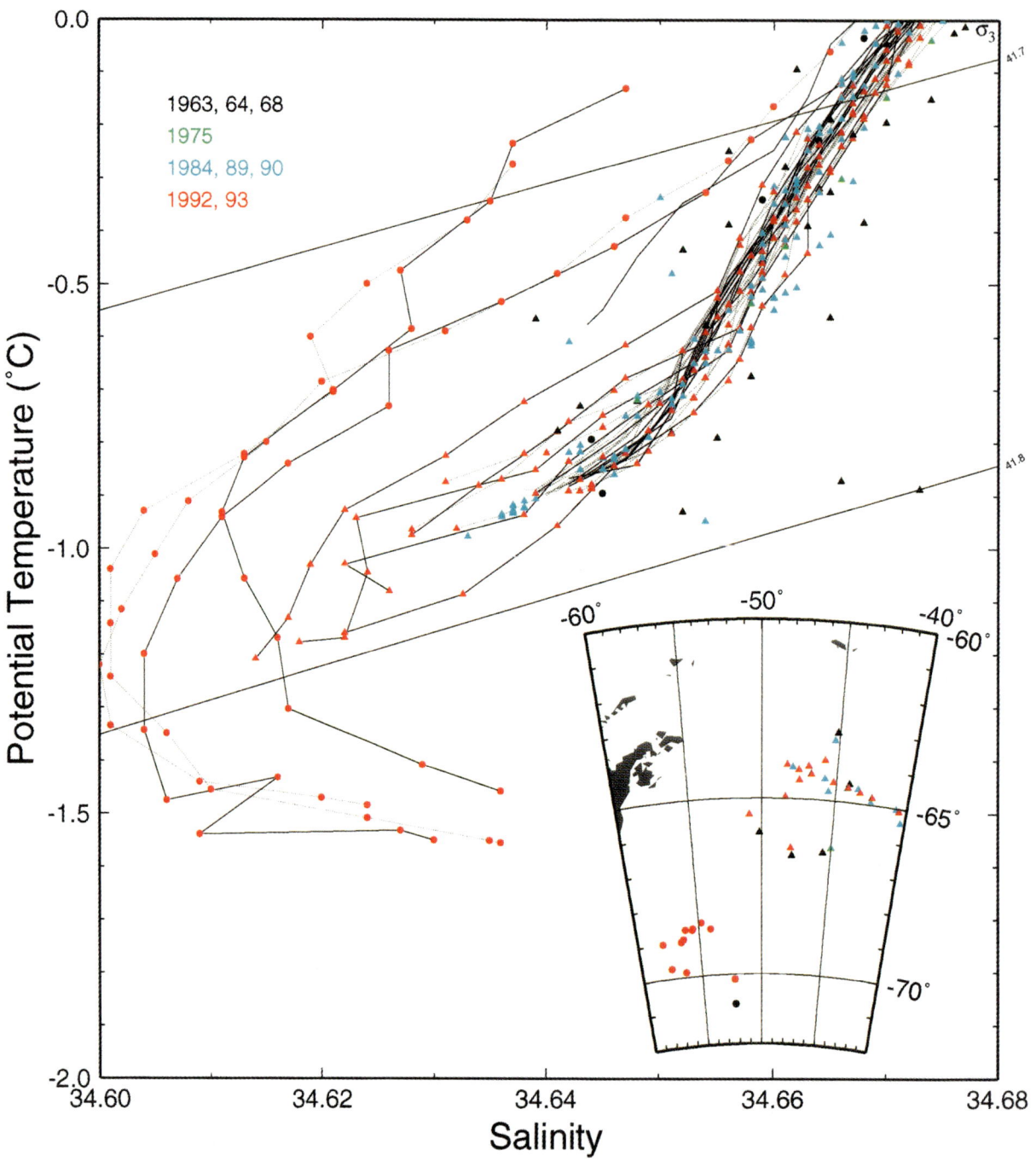

Plate 5. The potential temperature-salinity relationship for the lower water column in the Weddell Sea derived using hydrographic stations from different years within 1° latitude by 1° longitude bins. The water column progressively gets colder as the sea floor is approached. For a station to qualify for this figure, the deepest T/S value has to be less than 50 m off the sea floor and there must be at least one other station within the same bin from a different year. When there was more than one station from a particular year, only one was used, so as not to over represent any specific year in the T/S scatter. Symbols, shown in the insert, are keyed to specific year intervals represented in the figure. The stations used are shown in the map insert, using the same symbol notation. Solid lines and dotted lines represent data from 1992 and 1993 stations, respectively. The circles represent the western rim of the Weddell Sea while the triangles represent the northwestern corner. Both of these areas are along the primary pathway for the escape of newly formed bottom water from the Weddell Sea (*Gordon et al., 1993; Muench and Gordon, 1995*).

7. SUMMARY AND CONCLUSIONS

The characteristics and interannual variability of the sea ice cover in the Weddell Sea from 1979 through 1995 have been examined. Our results indicate a highly variable summer sea ice cover confined mainly in the western Weddell Sea but covering different areas of the region during different years. The summer ice extent has been shown to vary by as much as 6.5×10^5 km^2 during the 17 year period with summer ice in the 1990s significantly more extensive than in earlier years. However, summer ice minimum extent does not correlate well with winter parameters and is highly correlated only with the length of summer, with a correlation coefficient of -0.88, confirming the effect of increased solar heating when summer occurs early. It is also weakly correlated with net ice production, and ice expansion rate, with correlation coefficients of -0.42, and -0.30, respectively. The negative signs for these coefficients are unexpected, since more extensive summer minima were expected to cause a conditioning in the ocean that would enhance ice production and ice areal expansion rate.

Our analysis shows that unusually high ice extents during the winter are followed by unusually low extents during the summer. These optimum winter events, which happened in 1980, 1984, 1987, and 1992, are shown to be correlated with meridional winds, which are coherent with the observed passages of the Antarctic Circumpolar Wave [White and Peterson, 1996]. Good correlation of the ice cover with available surface temperature data is not as apparent in the Weddell as in the Bellingshausen Sea. A hysteresis loop is observed in the relationships of surface temperature and ice extent, which may partly explain the lack of strong correlation. Also, the average temperature in the western Weddell Sea is a lot lower than that of the Bellingshausen Sea [King, 1991].

The areal extent of coastal polynyas in the Weddell sector also has been analyzed in three zones and our results indicate a good correlation of polynya areas with ice extent and meridional winds. The coherence of the occurrence of peak values of areas of coastal polynyas and ice extent confirms that polynyas play a strong role in the production of sea ice and hence of cold, saline, dense water. The polynyas in the eastern Weddell Sea is comparable in size with those in the western region and are slightly better correlated with wind and ice extent. While one might expect production of HSSW in the eastern region, the narrow shelf prohibits accumulation of HSSW and thus suppresses the formation of dense bottom water [Fahrbach et al., 1994].

Large interannual variability of WSBW from 1963 through 1993 has been observed from available hydrographic data. We suggest, but cannot prove with the limited hydrographic data set, that the WSBW variability is related to the large interannual variability in the sea ice cover. The fresher, colder WSBW of 1992 and 1993 than in earlier years (e.g., 1970s) suggests a changing ratio of HSSW to Ice Shelf Water-derived WSBW varieties. Quantitative analysis of the sea ice - WSBW link is complicated by other factors such as changing shelf water circulation in response to wind variability, though this factor is also expected to be reflected in sea ice distribution, shelf water mass residence times, and associated glacial ice melting.

Acknowledgments. We gratefully acknowledge Larry Stock of Caelum and Rob Gersten of Hughes STX for programming support in the analysis of wind and satellite data. Members of the Goddard sea ice group (D.J. Cavalieri, P. Gloersen, C. Parkinson, J. Zwally, and one of the authors [JCC]) were responsible for producing the consistent time series of SMMR and SSM/I data used in this paper. The critical and valuable comments by Stan Jacobs and four anonymous reviewers are very much appreciated. We also thank Rob Massom and Per Gloersen for providing very useful suggestions. This research was supported by the Cryosphere Program at NASA Headquarters and by NSF Grant OPP 93-13700 to Lamont-Earth Observatory of Columbia University. Lamont-Doherty Earth Observatory contribution number 5710.

REFERENCES:

Ackley, S. F., and E. T. Holt, Sea ice data buoys in the Weddell Sea, *CRREL REP. 84-11*, 18 pp., 1984.

Allison, I., R. E. Brandt and S. G. Warren, East Antarctic sea ice: albedo, thickness distribution, and snow cover, *J. Geophys. Res.*, 98(C7), 12,417-12,429, 1993.

Carmack, E. C. and T. D. Foster, Circulation and distribution of oceanographic properties near the Filchner Ice Shelf, *Deep-Sea Res.*, 22, 77-90, 1975.

Cavalieri, D. J. and S. Martin, A passive microwave study of polynyas along the Antarctic Wilkes Land coast, *Antarc. Res. Ser.*, 43, 227-252, 1985.

Comiso, J. C., Surface Temperatures in the Polar Regions using Nimbus-7 THIR, *J. Geophys. Res.*, 99(C3), 5181-5200, 1994.

Comiso, J. C., SSM/I Concentrations using the Bootstrap Algorithm, *NASA Ref. Publication- 1380*, 40pp, 1995.

Comiso, J. C., and S. Ackley, Antarctic sea ice passive microwave signatures during summer and autumn, *IEEE IGARSS'94 Digest*, Vol. 1, 143-146, 1994.

Comiso, J. C., and A. L. Gordon, The Cosmonaut Polynya in the Southern Ocean: Structure and Variability, *J. Geophys. Res.*, 101(C8), 18,297-18313, 1996.

Comiso, J. C., T. C. Grenfell, D. L. Bell, M. A. Lange, and S. F. Ackley, Passive Microwave In Situ Observations of Winter Weddell Sea Ice, *J. Geophy. Res., 9(C8)*, 10891-10905, 1989.

Darby, M. S., A. J. Willmott, and T. A. Somerville, On the influence of coastline orientation on the steady state width of a latent heat polynya, *J. Geophys. Res., 100(C7)*, 13,525-13,633, 1995.

Eicken, H., and M. A. Lange, Development and properties of sea ice in the coastal regime of the southeastern Weddell Sea, *J. Geophys. Res., 9(C6)*, 8193-8206, 1989.

Fahrbach, E., R. G. Peterson, G. Rohardt, P. Schlosser, and R. Bayer, Suppression of bottom water formation in the southeastern Weddell Sea, *Deep-Sea Res., 41(2)*, 389-411, 1994.

Fahrbach, E., G. Rohardt, N. Scheele, M. Schroder, V. Strasss, and A. Wisotski, Formation and discharge of deep and bottom water in the northwestern Weddell Sea, *J. Marine Res., 53*, 515-538, 1995.

Foldvik, A., T. Gammelsrod, and T. Torreesen, Circulation and water masses on the southern Weddell Sea, *Antarct. Res. Ser., 43*, 5-20, 1985.

Foster, T. D, and E. C. Carmack, Frontal zone mixing and Antarctic bottom water formation in the southern Weddell Sea, *Deep-Sea Res., 23*, 301-317, 1976a.

Foster, T. D., and E. C. Carmack, Temperature and salinity structure in the Weddell Sea, *J. Phys. Oceanog., 6*, 36-44, 1976b.

Foster, T. D., and J. H. Middleton, Variability in the bottom water in the Weddell Sea, *Deep-Sea Res., 26A*, 743-763, 1979.

Gloersen, P., W. Campbell, D. Cavalieri, J. Comiso, C. Parkinson, H. J. Zwally, Arctic and Antarctic Sea Ice, 1978-1987: Satellite Passive Microwave Observations and Analysis, *NASA Spec. Publ. 511*, 1992.

Gordon, A. L., Two stable modes of Southern Ocean winter stratification, in *Deep Convection and Water Mass Formation in the Ocean*, edited by J. Gascard, and P. Chu, Elsevier Publisher, pp. 17-35, 1991.

Gordon, A. L., and J. C. Comiso, Polynyas in the Southern Ocean, *Scientific American, 258*, 90-97, 1988.

Gordon, A. L., B. A. Huber, H. H. Hellmer, and A. Ffield, Deep and bottom water of the Weddell Sea's western rim, *Science, 262*, 95-97, 1993.

Jacka, T. H., and W. F. Budd, Detection of sea ice and temperature changes in the Antarctic and Southern Ocean, in *International Conference on the Role of the Polar Regions in Global Change*, Geophysical Institute and Center for Global change and Arctic System Research, University of Alaska Fairbanks, 1, 63-70, 1991.

Jacobs, S. S., and J. C. Comiso, Satellite passive microwave sea ice observations and oceanic processes in the Ross Sea, Antarctica, *J. Geophys. Res., 94(C12)*, 18195-18211, 1989.

Jacobs, S. S, and J.C. Comiso, A recent sea-ice retreat west of the Antarctic Peninsula, *Geophys Res. Letter, 20(12)*, 1171-1174, 1993.

Jacobs, S. S., and J.C. Comiso, A climate anomaly in the Amundsen and Bellingshausen Seas, *J. Climate, 10(4)*, 697-709, 1997.

Jacobs, S. S., R. Fairbanks and Y. Horibe, Origin and evolution of water masses near the Antarctic continental margin: Evidence from $H_2^{18}O/H_2^{16}O$ ratio in seawater, *Antarctic Research Series, 43*, 59-85, 1985.

King, J. C., Recent climate variability in the Antarctic Peninsula, *Proceedings of the 5th Conference Climate Variations*, Denver, Oct. 14-18, 1991, Boston: American Meteorological Soc., pp. 354-357, 1991.

Kottmeier, C., and L. Sellmann, Atmospheric and oceanic forcing of the Weddell Sea ice motion, *J. Geophys. Res., 101(C9)*, 10,809-20,824, 1996.

Lytle, V. I., and S. F. Ackley, Heat flux through sea ice in the Western Weddell Sea: Convective and conductive transfer processes, *J. Geophys. Res., 101(C4)*, 8853-8868, 1996.

Markus, T., and B. Burns, A method to estimate subpixel-scale coastal polynyas with satellite passive microwave data, *J. Geophys. Res., 100(C3)*, 4473-4488, 1995.

Markus, T., C., Kottmeier, and E. Fahrbach, Ice formation in coastal polynyas in the Weddell Sea and their impact on oceanic salinity, *Antarct. Res. Ser. (this volume)*.

Martin, S., K. Steffen, J. Comiso, D. Cavalieri, M. R. Drinkwater, and B. Holt, Microwave remote sensing of polynyas, edited by F. Carsey, *AGU Geophysical Monograph Series, 68*, 303-311, 1992.

Massom, R. A., Observing the advection of sea ice in the Weddell Sea using buoy and satellite passive microwave data, *J. Geophys. Res., 97(C10)*, 15,559-15,572, 1992.

Maykut, G. A., Energy exchange over young sea ice in the central Arctic, *J. Geophys. Res., 83(C7)*, 3646-3658, 1978.

Mensch, M., R. Bayer, J. Bullister, P. Schlosser, R. Weiss, The distribution of Trituim and CFCs in the Weddell Sea during the mid 1980's, *Progress in Oceanography* (In press).

Muench, R. D., and A. L. Gordon, Circulation and transport of water along the western Weddell Sea margin, *J. Geophys. Res., 100(C9)*, 18503-18515, 1995.

Perovich, D. K., G. A. Maykut, and T. C. Grenfell, Optical properties of ice and snow in the polar oceans, 1: Observations, *Proceedings of SPIE: Ocean Optics VIII, 637*, 232-241, 1986.

Smith, S., R. Muench, and C. Pease, Polynyas and leads: An overview of physical processes and environment, *J. Geophy. Res., 95(C6)*, 9461-9479, 1990.

Smith, R. C., S. E. Stammerjohn, and K. S. Baker, Surface air temperature variations in the western Antarctic Peninsula region, *Antarct. Res. Ser., 70*, 81-104, 1996.

Trenberth, K. E., J. E. Olson, and W. G. Large, A global ocean wind stress climatology based on ECMWF Analyses, NCAR/TN-338, National Center for Atmospheric Research, Boulder Colorado, 93 pp., 1989.

Weatherly, J. W., J. E. Walsh, and H. J. Zwally, Antarctic sea ice variations and seasonal air temperature relationships, *J. Geophys. Res., 96(C8)*, 15,119-15,130, 1991.

White, W. B. and R. G. Peterson, An Antarctic circumpolar wave in surface pressure, wind, temperature, and sea-ice extent, *Nature, 380,* 699-702, 1996.

Zwally, H. J., J. C. Comiso, and A. Gordon, Antarctic offshore leads and polynyas and oceanographic effects, *Antarc. Res. Ser., 43,* 203-226, 1985.

Zwally, H. J., J. C. Comiso, and C. L. Parkinson, Variability of Antarctic sea ice and changes in carbon dioxide, *Science, 220,* 1005-1012, 1983a.

Zwally, H. J., J. C. Comiso, C. L. Parkinson, W. J. Campbell, F. D. Carsey, and P. Gloersen, Antarctic Sea Ice 1973-1976 from Satellite Passive Microwave Observations, *NASA Spec. Publ. 459,* 1983b.

J. C. Comiso, Laboratory for Hydrospheric Processes, Code 971, NASA/Goddard Space Flight Center, Greenbelt, MD 20771.

A. L. Gordon, Lamont-Doherty Earth Observatory of Columbia University, Route 9A, Palisades, NY 10964-8000.

(Received November 15, 1996; accepted August 5, 1997)

MESOSCALE ICE FEATURES IN THE SUMMER MARGINAL ICE ZONE OFF EAST QUEEN MAUD LAND OBSERVED IN NOAA AVHRR IMAGERY

Yasushi Fukamachi, Kay I. Ohshima, and Takayuki Ishikawa[1]

Institute of Low Temperature Science, Hokkaido University, Sapporo, Japan

Mesoscale ice features off East Queen Maud Land in Antarctica are examined using NOAA advanced very high resolution radiometer imagery during summer between 1987 and 1990. Characteristic mesoscale features are found at two locations. Ice tongues and eddies are identified in the marginal ice zone and wavelike patterns are identified on the offshore edge of a polynya located between regions of pack ice and landfast ice. The growth of these features is not correlated with the wind field. Patterns of the ice tongues and eddies are associated with those of cold water offshore. This suggests that ocean currents advect both ice and cold water to generate these patterns. The wavelike patterns at the polynya edge appear at various locations that are not correlated with bottom topography in the area.

1. INTRODUCTION

Mesoscale ice features, such as meanders, tongues, and eddies, in the marginal ice zone (MIZ) have been observed from remote sensing data. They are reported in the Fram Strait and Greenland Sea [*Johannessen et al.*, 1983; *Johannessen et al.*, 1987; and others], the Labrador Sea [*LeBlond*, 1982], and the Sea of Okhotsk [*Ginsberg and Fedorov*, 1989; *Wakatsuchi and Ohshima*, 1990]. Fram Strait and the Greenland Sea in particular have been studied intensively since the Marginal Ice Zone Experiment (MIZEX) in summer of 1983 and 1984. Most of the mesoscale ice features are observed during the melting season. *Johannessen et al.* [1987] identified the following oceanic processes as possible generation mechanisms of the observed features in the MIZ of Fram Strait; barotropic and baroclinic instabilities of background currents; topographic steering and trapping; and, the interaction between open ocean eddies and the ice edge.

There are few studies on mesoscale ice features in the Antarctic MIZ. From NOAA advanced very high resolution radiometer (AVHRR) imagery received at Syowa Station, *Yamanouchi and Seko* [1992] reported ice tongues and eddies in the MIZ off East Queen Maud Land. Figure 1 is an AVHRR visible-band image showing these phenomena on January 21, 1988. (Figure 2 is a map of the area shown in Figure 1.) There are many ice tongues extending offshore in the MIZ located north of the landfast ice region. The ice tongues and eddies are observed during summer when sea ice melts and decreases in concentration. Mesoscale ice features are also found at the offshore edge of a flaw polynya observed off Prince Olav Coast (east of Syowa Station). The large areal extent of open water during summer in this area was previously noted by *Zwally et al.* [1985]. This polynya is seen in Figure 1 at the boundary between the pack ice and landfast ice regions, which roughly coincides with a continental shelf break 70-80 km offshore (see Figure 2). *Ishikawa et al.* [1996] showed that this polynya appeared along this boundary when offshore winds prevailed. When the area of this polynya becomes larger during summer, its offshore edge occasionally exhibits wavelike patterns (see Figures 7 and 8).

In this paper, we examine mesoscale ice features in the MIZ and at the polynya edge off East Queen Maud Land using NOAA AVHRR data. Our goal is to propose possible generation mechanisms of these features. Section 2 describes satellite, wind, and oceanographic data used in this study. In section 3, the satellite data and their relation to the wind data are analyzed. Finally, summary and discussion are given in section 4.

2. DATA

In this study, we examine satellite and wind data obtained at Syowa Station, and oceanographic data collected in the MIZ off Syowa Station. All these data are obtained by the Japanese Antarctic Research Expeditions. NOAA AVHRR imagery was received daily at Syowa Station from

---
[1]Now at Toshiba Co., Shimoishigami 1385, Ohtawara, Tochigi, 224 Japan.

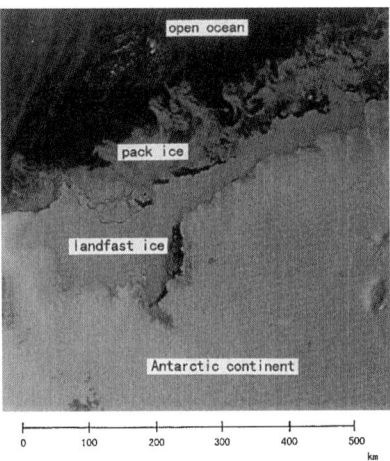

Fig. 1. An AVHRR visible-band image on January 21, 1988 obtained at Syowa Station (39°35'E, 69°00'S). A map showing the areal coverage of this image is given in Figure 2. Lighter tones correspond to higher albedo values, denoting land ice, sea ice, and clouds, and darker tones correspond to lower albedo values, denoting open ocean. Ice tongues extending offshore are seen in the MIZ.

1980 to 1991. In this study, 1.1 km-resolution images in the area shown in Figure 2 during summer between 1987 and 1990 are examined. Based on the difference in albedo values of ice and water, visible images of channel 2 are used to detect ice-edge locations. Infrared images of channel 4 are used to examine patterns of sea surface temperature in ice-free regions. Surface winds are measured at Syowa Station twice a day. In addition to these winds, the winds of European Centre for Medium-Range Weather Forecasts (ECMWF) are examined. There are few oceanographic data available in the MIZ of the study area even during summer. Figure 3 shows vertical profiles of temperature, salinity, and σ_t in the MIZ early in February 1966 and 1974. Due to ice melting, salinity values are significantly lower near the surface than deeper in the water column and there is a well-defined mixed layer in both cases. Based on these profiles, the thickness of the upper layer and the density difference between the upper layer and deep ocean are estimated to be 30 m and 1.08×10^{-3} g/cm^3, respectively. Using the Coriolis parameter at 68°S, these values give the internal Rossby radius of 4.2 km.

3. DATA ANALYSIS

3.1. Mesoscale Ice Features in the MIZ

The ice tongues and eddies were observed every summer from 1987 to 1990. In 1987 and 1988, especially, they were seen several times. Figure 4 shows three time series of ice-edge patterns detected from the AVHRR images. Polynya regions are shaded. Figure 4a displays ice-edge locations from January 1 to 8, 1987. Small-scale bumps, seen at the ice edge on January 1, evolve into ice tongues extending offshore in the next four days. (Four prominent tongues are labeled a_1 to a_4 in the figure.) Note that tongue a_1 looks like a dipole pair on January 4 and tongue a_2 looks like an anticyclonic eddy on January 2. These tongues disappear and the ice edge becomes straight by January 8. Figures 4b and 4c show ice-edge locations from January 25 to 29, 1987 and from January 6 to 10, 1988, respectively. (A visible-band image on January 9, 1988 is shown in the top panel of Figure 6.) As in Figure 4a, several ice tongues develop in these periods. (Four prominent tongues are labeled b_1 and b_2 in Figure 4b, and c_1 and c_2 in Figure 4c.) In Figure 4b, polynyas located south of tongues b_1 and b_2 become elongated as these tongues extend offshore. Note that tongue c_1 looks like a cyclonic eddy on January 9. The tongues are 10-30 km wide and located with a wide range of spacings of 20-90 km. Based on these time series, offshore extending and zonal propagation speeds for labeled tongues are estimated as shown in Table 1. Negative values in the zonal propagation speed indicate that these tongues move westward. Note that westward propagation speeds are lower than offshore extending speeds except for tongue a_3.

Figure 5 shows time series of surface winds at Syowa Station for January 1987 and 1988. These winds are regarded as synoptic ones in the study area after comparison with the ECMWF winds. For both years, the prevailing winds are northeasterlies. Wind speeds are generally lower than 10 m/s. They are higher than 15 m/s only on January

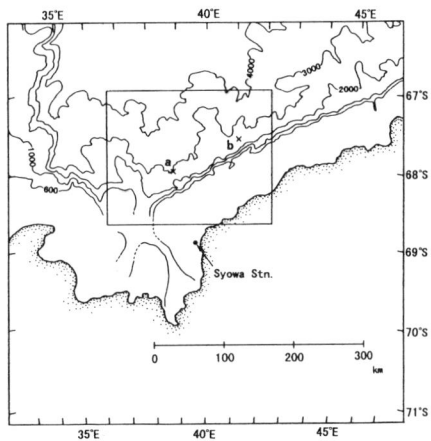

Fig. 2. The area of satellite images analyzed in this study. Bathymetry of the coastal ocean is also shown. Crosses labeled a and b indicate locations of hydrographic surveys displayed in Figure 3. The rectangular area denotes the region of satellite images shown in Figure 6.

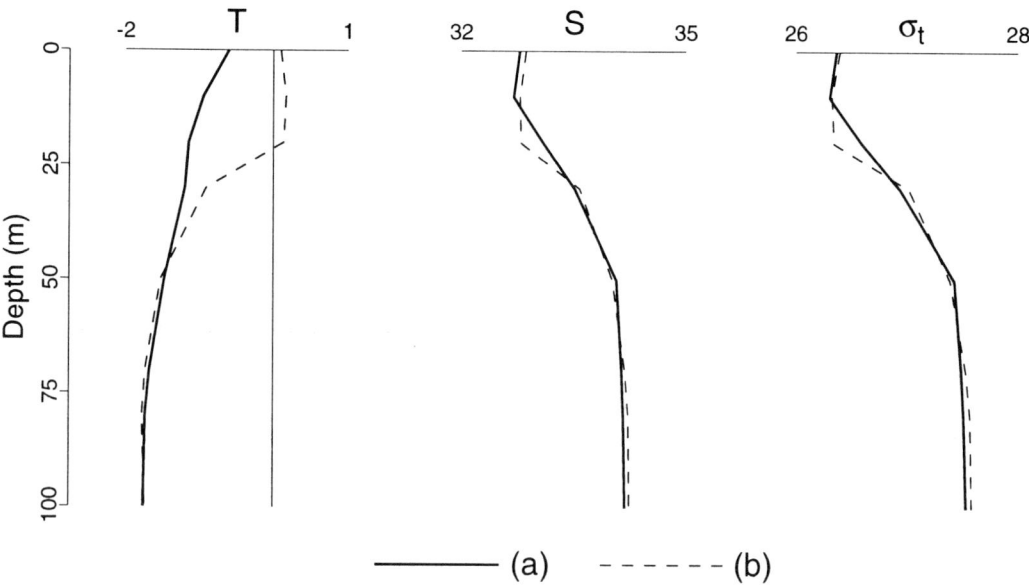

Fig. 3. Vertical profiles of (left) temperature, (middle) salinity, and (right) σ_t in the upper ocean of the MIZ early in February. Profiles (a) and (b) are taken at (38°50'E, 68°07'S) on February 2, 1966 and at (41°06'E, 67°42'S) on February 9, 1974, respectively. These locations are shown in Figure 2. There is a thin mixed layer due to ice melting.

6 and 7, 1987 and January 3 and 4, 1988. An examination of the ECMWF winds reveals that these high winds are associated with low pressure systems approaching the study area. During the time series shown in Figure 4a, the ice tongues and eddies grow when winds are weak and their directions are random from January 1 to 5. The tongues and eddies disappear by January 8 after northeasterly winds exceed 10 m/s on January 6 and 7. For the time series displayed in Figure 4b, the ice tongues continue to extend offshore even after the easterlies become stronger than 10 m/s on January 27 and 28. For the time series shown in Figure 4c, the tongues grow continuously throughout the period when wind speeds are constantly lower than 5 m/s. Before January 6, the mesoscale ice features, which are present until the end of December, disappear after northeasterlies exceed 15 cm/s on January 3 and 4.

The relation between the time series of ice-edge patterns and winds reveals that the development of ice tongues and eddies occurs as long as the onshore component of winds is not strong enough to push the sea ice toward the shore. This suggests that wind forcing is not an essential generation mechanism of the observed ice tongues and eddies in this area.

Figure 6 displays both visible and infrared images on January 9, 1988. In the visible-band image, the ice-edge pattern is characterized by two tongues (labeled c_1 and c_2 in Figure 4c) and several small-scale bumps to the east. This ice-edge pattern is also recognized as an isotherm of 270.5 K in the infrared-band image. In addition, there is another isotherm of 272.0 K (indicated by an arrow) off the ice edge showing the presence of cold water. The presence of this cold water can be explained by local ice melting near the ice edge and advection of cold water from the ice-covered region. As seen in Figure 6, patterns of some tongues and bumps are correlated well with those of sea surface temperature away from the ice edge. Similar features are seen occasionally, including in the visible and infrared images for January 21, 1988 (the visible-band image shown in Figure 1). This suggests that ocean currents play an important role in generating the patterns of both ice edge and sea surface temperature. It is likely that these currents are confined to the upper water column because a thin mixed layer is well developed, as shown in Figure 3.

3.2. Mesoscale Ice Features at the Polynya Edge

When a flaw polynya off Prince Olav Coast becomes larger during summer, its offshore edge facing the region of pack ice occasionally evolves into wavelike patterns. Figure 7 shows a time series of polynya areas detected in the AVHRR images from December 8 to 16, 1990. Note that its onshore edge is the margin of landfast ice. As the polynya extends, the wavelike patterns exhibit a series of anticyclonic eddies. The average wavelength of the patterns is about 70 km.

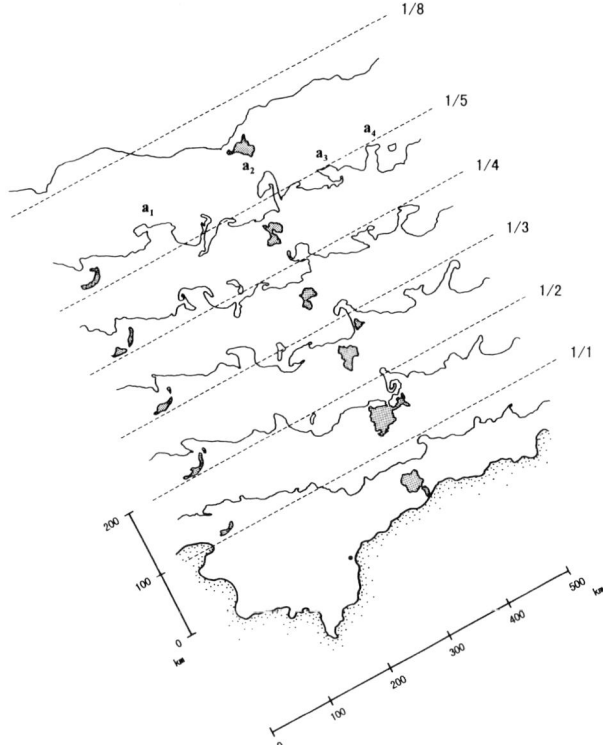

Fig. 4a. The evolution of ice-edge patterns detected from the AVHRR imagery for January 1-8, 1987. The first pattern is drawn at the actual location and subsequent patterns are shown offshore in sequence. Dashed lines are datums located roughly 100 km off Prince Olav Coast. The location of Syowa Station is indicated by a solid circle. The growth and decay of ice tongues and eddies are seen in this time series.

This time series does not show significant zonal propagation of the patterns. Similar patterns were clearly identified on two other occasions from 1987 to 1990. Characteristics of the three wavelike patterns are summarized in Table 2. Average wind speeds during the three periods are also listed. Like the example shown in Figure 7, the other two cases show negligible zonal propagation speeds. Average wavelengths range from 50 to 90 km. Average wind speeds are moderate to light during all these periods. Predominant winds are mostly northeasterly, except for the period from December 22 to 26, 1989 when wind directions are highly random. It seems that there is no apparent relationship between the growth of the wavelike patterns and winds. This suggests that wind forcing is not an essential generation mechanism of the observed wavelike patterns.

Since the mesoscale features at the polynya edge appear near the shelf break, we examine whether these patterns correspond with those of bottom topography. Figure 8 displays polynya areas for the three cases listed in Table 2, together with the 1000-m and 2000-m isobaths. Note that locations of crests and troughs of the wavelike patterns vary in these three cases. In addition, these patterns do not agree well with those of local bottom topography as indicated by 1000-m and 2000-m isobaths. This suggests that bottom topography does not play an important role in generating the wavelike patterns at the polynya edge.

4. SUMMARY AND DISCUSSION

In this study, we examine mesoscale ice features, such as those in Figure 1, in the summer MIZ off East Queen Maud Land using NOAA AVHRR imagery received at Syowa Station. During summer, the upper ocean in the MIZ is characterized by a thin mixed layer caused by ice melting (Figure 3). The growth and decay of ice tongues and eddies in the MIZ are observed from the AVHRR images. The ice tongues are separated by distances of 20-90 km along the ice edge (Figure 4). They extend offshore when the onshore component of winds is weak, and disappear after periods of strong onshore winds (Figure 5). Comparison between visible and infrared images reveals that characteristic patterns of the ice edge, such as tongues and eddies, correspond with those of cold water off the ice edge (Figure 6). These facts suggest that ocean currents, not winds, play an important role in generating the ice tongues and eddies. The offshore edge of the polynya, which exists at the boundary between the pack ice and landfast ice regions, occasionally exhibits wavelike patterns during summer (Figure 7). These patterns have wavelengths of 50-90 km and develop under a variety of wind conditions. Comparison

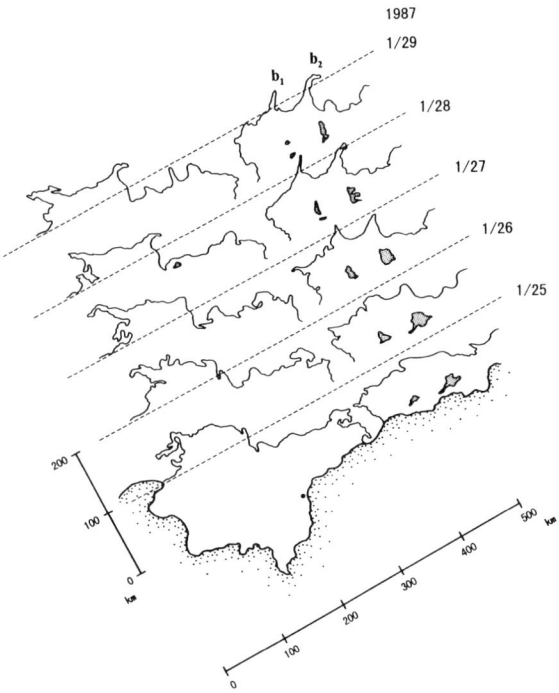

Fig. 4b. Similar to Figure 4a, except for January 25-29, 1987.

of wavelike patterns for different instances shows that their crests and troughs do not have any preferred locations. In addition, these patterns are not associated with bottom topography in the area (Figure 8). These facts suggest that the wavelike patterns are generated by a mechanism other than winds and topographic effects.

This study suggests the importance of ocean currents for the generation of the mesoscale features in the MIZ of the study area. *Johannessen et al.* [1987] pointed out four oceanic processes (described earlier in the introduction), which could result in mesoscale features observed in the MIZ of the Fram Strait. It is not possible, however, to identify what kind of process is important to generate such ocean currents in the study area from observational data used here. Nevertheless, thermodynamic processes are presumably important because the mesoscale ice features appear during the melting season. There is a possibility that instabilities of a density front caused by ice melting may be the source of these features. The same instabilities may occur both in the MIZ and at the polynya edge because both locations are boundaries between the ice-covered region with melting ice and the ice-free region. There is also a possibility that instabilities of the Antarctic Coastal Current may be the source of the mesoscale ice features. There are, however, no observational data of this current in the study area.

There have been several numerical studies that focus on mesoscale ice features in the MIZ. *Smith and Bird* [1991] applied an ice-ocean coupled model, consisting of a dynamic ice model and a two-layer ocean model, in the East Greenland Current MIZ. They showed that the interaction of open ocean eddies with a jet in the MIZ and the eddy-topography interaction can cause mesoscale ice features. *Johannessen et al.* [1995] used a two-layer ocean model to investigate the stability of the East Greenland Current. They concluded that mixed barotropic/baroclinic instability of this current is responsible for generating mesoscale ice features. These models, however, did not include a density front across the MIZ, which can be important in generating mesoscale ice features during the melting season. In the study of *Qiu et al.* [1988], a density front caused by the freshwater runoff was considered in an oceanic general circulation model. They showed that coastal currents associated with the density front become baroclinically unstable. The growing instabilities carry light water offshore and heavy water onshore to produce mesoscale wave patterns in the upper ocean (their Figure 3). We

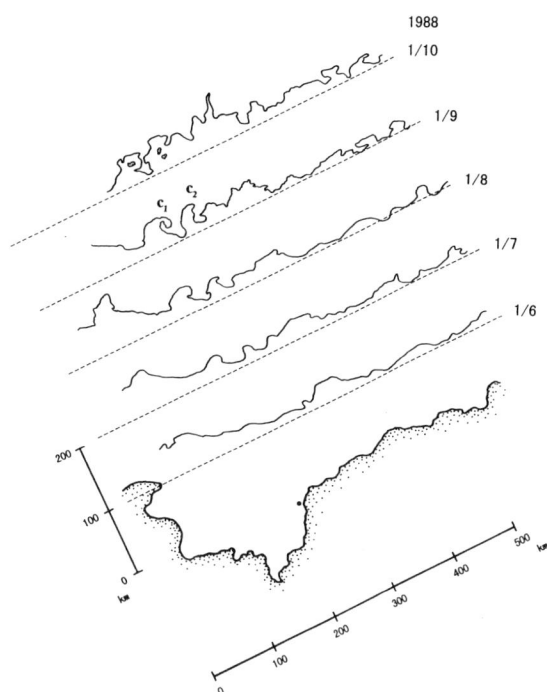

Fig. 4c. Similar to Figure 4a, except for January 6-10, 1988.

are currently developing an ice-ocean coupled model that basically repeats the study of *Qiu et al.* [1988], but which uses melting ice as the source of freshwater. The coupled system consists of a dynamic-thermodynamic ice model and a thermodynamic, reduced-gravity ocean model, in which a density front can be considered. In preliminary runs of the model, the application of the surface heat flux to a model MIZ leads to the formation of a density front at the ice edge because ice melting reduces salinity in the ice-covered region. The ice-edge currents associated with this density front become baroclinically unstable. The energy source of this instability is the available potential energy associated with horizontal density gradients [*McCreary et al.*, 1991; *Fukamachi et al.*, 1995]. The resulting meandering ocean currents advect cold and fresh water along with ice from the ice-covered region to the ice-free region. This ice-ocean circulation is similar to that inferred from visible and infrared images shown in Figure 6. Although this similarity may indicate that our model includes the essential factors governing the circulation in the summer MIZ, further experiments are necessary to draw any firm conclusions.

TABLE 1. Characteristics of Tongues

Ice tongue	a_1	a_2	a_3	a_4	b_1	b_2	c_1	c_2
Offshore extending speed [km/day]	5.3	6.6	3.9	3.3	7.8	10.3	6.3	8.2
Zonal propagation speed [km/day]	-4.3	-5.3	-4.0	-0.7	-0.3	-1.7	-3.0	-3.0

322 ANTARCTIC SEA ICE: PHYSICAL PROCESSES, INTERACTIONS AND VARIABILITY

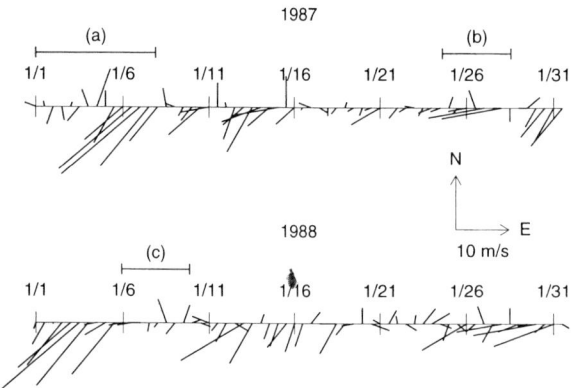

Fig. 5. Stick diagrams showing surface-wind vectors at Syowa Station during (top) January 1987 and (bottom) January 1988. The sticks point in the direction to which the wind is blowing. The stick length represents the wind speed (a 10 m/s scale is shown at right). Bars labeled (a), (b), and (c) show periods of the time series shown in Figure 4. Northeasterly winds are common.

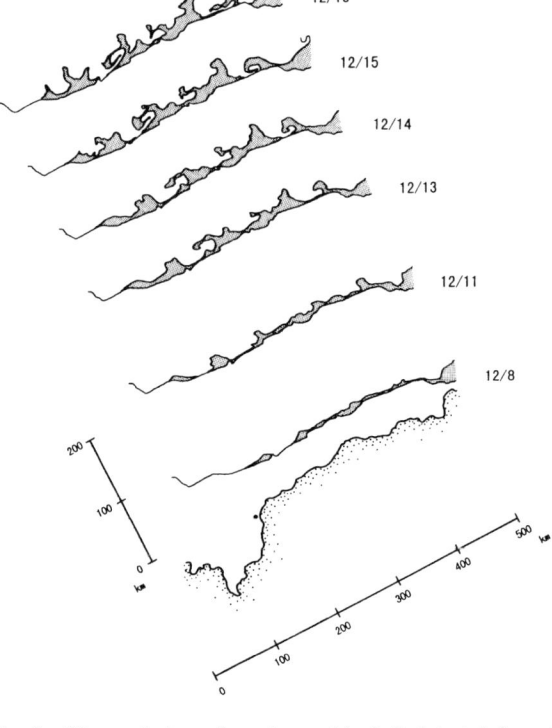

Fig. 7. The evolution of a polynya (shaded) detected from the AVHRR images from December 8 to 16, 1990. A polynya on December 8 is drawn at the actual location and its subsequent development is shown in the sequence going further offshore. The offshore edge of the polynya evolves into wavelike patterns as its area becomes larger.

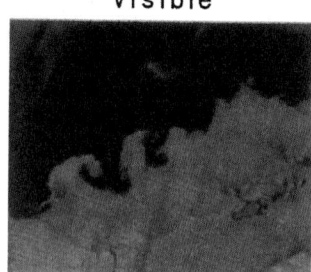

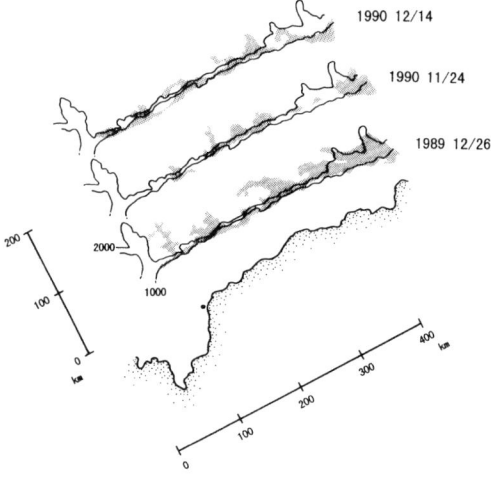

Fig. 6. (Top) Visible and (bottom) infrared AVHRR images on January 9, 1988 of the rectangular area in Figure 2. In the infrared-band image, lighter tones indicate colder temperatures and darker tones indicate warmer temperatures. Patterns of the ice edge in the visible-band image correspond well with those of sea surface temperature off the ice edge (indicated by an arrow) in the infrared-band image.

Fig. 8. Polynyas (shaded) for the three time periods listed in Table 2. Contours are the 1000-m and 2000-m isobaths. A polynya on December 26, 1989 is drawn at the actual location and the subsequent polynyas are shifted offshore in sequence. Wavelike patterns seen at the offshore edge of these polynyas are not located at the same places in the three cases and are not associated with local bottom topography.

TABLE 2. Characteristics of Wavelike Patterns at Polynya edge

Period	12/22-26, 1989	11/21-25, 1990	12/11-16, 1990
Wavelength [km]	90	50	70
Number of waves	4	3	5
Average wind speed [m/s]	3.0	5.9	7.4

The accumulation of more oceanographic data is essential for identifying what kind of oceanic process is responsible for the generation of the mesoscale ice features and validating numerical model results. For example, we need to carry out detailed hydrographic surveys across the ice edge to examine upper-ocean structure and mooring experiments to monitor the Antarctic Coastal Current.

Acknowledgments. This work was supported by a Grant-in-Aid (07740380) for Encouragement of Young Scientists form the Ministry of Education, Science, Sports, and Culture, Japan. We thank Shoichi Kizu, Naohiko Hirasawa, Sei-ichi Saitoh, Shuki Ushio, and Shoshiro Minobe for their help in processing satellite imagery. Discussions with Takashi Yamanouchi and Masaaki Wakatsuchi were helpful. The comments of Martin Jeffries and two anonymous reviewers were also helpful. We also thank Noriaki Kimura for his drafting.

REFERENCES

Fukamachi, Y., J.P. McCreary, and J.A. Proehl, Instability of density fronts in layer and continuously stratified models, *J. Geophys. Res., 100*, 2559-2577, 1995.

Ginsburg, A.I., and K.N. Fedorov, On the multitude of forms of coherent motions in marginal ice zones (MIZ), in *Mesoscale/Synoptic Coherent Structures in Geophysical Turbulence*, edited by J.C.J. Nihoul and B.M. Jamart, pp. 25-39, Elsevier, 1989.

Ishikawa, T., J. Ukita, K.I. Ohshima, M. Wakatsuchi, T. Yamanouchi, and N. Ono, Coastal polynyas off East Queen Maud Land observed from NOAA AVHRR data, *J. Oceanogr., 52*, 389-398, 1996.

Johannessen, J.A., O.M. Johannessen, E. Svendsen, R. Shuchman, T. Manley, W.J. Campbell, E.G. Josberger, S. Sandven, J.C. Gascard, T. Olaussen. K. Davidson, and J. Van Leer, Mesoscale eddies in the Fram Strait marginal ice zone during the 1983 and 1984 Marginal Ice Zone Experiments, *J. Geophys. Res., 92*, 6754-6772, 1987.

Johannessen, O.M., J.A. Johannessen, J. Morrison, B.A. Farrelly, and E.A.S. Svendsen, Oceanographic conditions in the Marginal Ice Zone north of Svalbard in early fall 1979 with an emphasis on mesoscale processes, *J. Geophys. Res., 88*, 2755-2769, 1983.

Johannessen, O.M., S. Sandven, W.P. Budgell, J.A. Johannessen, and R.A. Shuchman, Observation and simulation of ice tongues and vortex pairs in the marginal ice zone, in *The Polar Oceans and Their Role in Shaping the Global Environment*, edited by R.D. Muench and J.E. Overland, pp. 109-136, American Geophysical Union, 1995.

LeBlond, P.H., Satellite observations of Labrador Current undulations, *Atmos.-Ocean, 20*, 129-142, 1982.

McCreary, J.P., Y. Fukamachi, and P.K. Kundu, A numerical investigation of jets and eddies near an eastern ocean boundary, *J. Geophys. Res., 96*, 2515-2534, 1991.

Qiu, B., N. Imasato, and T. Awaji, Baroclinic instability of buoyancy-driven coastal density currents, *J. Geophys. Res., 93*, 5037-5050, 1988.

Smith, D.C., IV, and A.A. Bird, The interaction of an ocean eddy with an ice edge ocean jet in a marginal ice zone, *J. Geophys. Res., 96*, 1675-1689, 1991.

Wakatsuchi, M., and K.I. Ohshima, Observations of ice-ocean eddy streets in the Sea of Okhotsk off the Hokkaido Coast using radar images, *J. Phys. Oceanogr., 20*, 585-594, 1990.

Yamanouchi, T., and K. Seko, *Antarctica from NOAA Satellites, -Clouds, Ice and Snow-*, National Institute of Polar Research, Tokyo, 91pp., 1992.

Zwally, H.J., J.C. Comiso, and A.L. Gordon, Antarctic offshore leads and polynyas and oceanographic effects, in *Oceanology of the Antarctic Continental Shelf, Antarc. Res. Ser.*, Vol. 43, edited by S.S. Jacobs, pp. 203-226, American Geophysical Union, 1985.

Y. Fukamachi, T. Ishikawa, and K. I. Ohshima, Institute of Low Temperature Science, Hokkaido University, Sapporo, 060 Japan.

(Received July 8, 1996;
accepted February 11, 1997.)

WAVE DAMPING IN COMPACT PANCAKE ICE FIELDS DUE TO INTERACTIONS BETWEEN PANCAKES

Hayley H. Shen

Department of Civil and Environmental Engineering, Clarkson University, Potsdam, New York

Vernon A. Squire

Department of Mathematics and Statistics, University of Otago, Dunedin, New Zealand

A new wave damping model is proposed that invokes ice floe collisions arising from the differential drift of individual floes to explain the attenuation of ocean waves as they pass through a field of pancake ice. The dissipated energy is found to be a function of the restitution coefficient of the pancakes and their impact velocities. For small dissipation, where the velocity of energy transport is unaffected by the damping, the spatial attenuation of the wave energy follows a power law rather than an exponential function. Although the value of the exponent is different, this is in agreement with the steady-state creep model of *Wadhams* [1973] to which it offers a physical basis. Results from the model are compared with a data set collected during the 1986 Winter Weddell Sea Project.

INTRODUCTION

Over three decades of scientific observations and centuries-old anecdotal knowledge from seafarers and the Eskimo establish that when ocean surface waves enter an area of sea ice they are altered by the ice floes they encounter. Moreover, the ice itself may be altered by the waves; floes may be broken up into smaller pieces or have their edges mechanically abraded, they may raft over one another, or they may melt faster. A most noticeable effect is a gradual loss of energy from the penetrating wave field, which causes the shape of the incoming wave spectrum to change as it travels through the ice because the level of damping experienced is a function of the wave period. Insofar as waves are concerned, sea ice in its various forms may be viewed as a low-pass filter with a gain that depends on the ice morphology and, in particular, on the floe size and thickness distributions and concentration.

Three major processes are involved in wave propagation under an ice cover: attenuation, scattering, and refraction. Attenuation results from the energy dissipation mechanisms in the wave-ice interaction. Scattering results from the reflection of wave energy from ice floe edges. Refraction is a macroscale phenomenon caused by the difference between the dispersion relations for the open water wave and the wave under an ice cover. While scattering does not consume energy, it reduces the forward propagating energy and thus contributes to wave decay. In this study we will concentrate on the attenuation phenomenon.

Many experiments in different ice fields have been conducted to measure wave attenuation; the reader is referred to *Squire et al.* [1995] for an up-to-date review and an explanation of results. Here, to remain brief, we cite only the key works [*Robin*, 1963; *Squire and Moore*, 1980; *Wadhams*, 1972, 1973, 1975, 1978; *Wadhams et al.*, 1986, 1988; *Liu et al.*, 1991a, 1992]. It is generally supposed that several processes to be discussed below act to attenuate the ocean waves, and it is highly probable that the relative intensity of these processes is dependent on the character of the sea ice through which the waves are traveling.

One of these processes, namely, absorption arising from inelastic collisions between ice floes, has been omitted in studies to date. It is the purpose of the current paper to remedy this in the context of pancake ice, where it is believed to be a dominant mechanism. Few field data have been published on wave propagation through pancake ice, e.g., *Squire* [1989], *Wadhams and Holt* [1991], yet, particularly in the Southern Ocean, this ice type may occupy over a million square kilometers at certain times of the year [*Wadhams et al.*, 1987].

Pancake ice floes are strikingly circular and uniform in size. The common belief at present is that frazil crystals produced in supercooled conditions are swept together by wave motions to form these circular floes. They typically have raised edges due to their soft consistency and the bumping and pumping action that occurs between adjacent pancakes. The size of these floes is of the order of 1 m in diameter with an aspect ratio of roughly 10 to 1. Examples of pancake ice fields are given later in the paper.

This paper applies modern granular flow theory to floe interactions occurring in pancake ice fields typical of the winter Southern Ocean. Instead of the conventional heuristic approach, it will be based on a known physical phenomenon that has been neglected hitherto, namely, collisions between ice floes.

We begin by surveying and estimating all possible mechanisms that lead to a reduction in forward-going wave energy and then introduce a model that accounts for floe collisions. It will be shown that floe collisions are a major energy dissipator, especially in pancake ice fields.

ATTENUATIVE PROCESSES

The total rate of energy loss per unit area, $\dot{\epsilon}$, for a swell or sea penetrating and traveling through a sea ice field is composed of four parts as follows:

$$\dot{\epsilon} = \dot{\epsilon}_1 + \dot{\epsilon}_2 + \dot{\epsilon}_3 + \dot{\epsilon}_4 \quad (1)$$

In (1) $\dot{\epsilon}_1$ denotes the absorption due to collisions and other interactions between ice floes, $\dot{\epsilon}_2$ denotes absorption due to hysteresis as floes deform on the passing wave field, $\dot{\epsilon}_3$ denotes absorption in the water column from processes such as wave breaking and viscous losses in the turbulent boundary layer, and $\dot{\epsilon}_4$ represents the redistribution of wave energy due to scattering by the floes. The relative balance of the terms in (1) is affected by wave conditions and the type of ice cover. For example, in an open ice field of low concentration in which the floes have diameters of the same order of magnitude as the wavelengths in the incoming sea, significant bending and scattering will occur. While the bending may not influence wave decay directly, it certainly affects the scattering of waves substantially [*Meylan and Squire*, 1996].

Further, one would expect hydrodynamic absorption to be significant, particularly when the seas are intense enough to break over the floes, but collisions between floes to be rare. In a compact ice field of the same type, $\dot{\epsilon}_1$ will increase, $\dot{\epsilon}_2$ will remain of similar magnitude, while $\dot{\epsilon}_3$ will probably decrease slightly. Hereinafter we focus our discussion on pancake ice as opposed to larger floes, as we wish to model the vast pancake ice fields of winter Antarctica. The words 'floe' and 'pancake' will therefore be used interchangeably. We also assume deep water.

Energy Loss by Collision

We will establish later in the paper that the quantity $\dot{\epsilon}_1$ is the most significant term in (1) within a compact pancake ice field.

Hysteresis Loss by Ice Cake Flexure

The work of *Fox and Squire* [1991, 1994] and *Meylan and Squire* [1994] allows an estimate to be made of the bending of individual floes making up the pancake ice field. Weddell Sea pancake ice floes, found to be about 1-m diameter and 0.3-m thick near the ice edge, would not change a deep ocean wave typically found in the marginal ice zone [*Meylan and Squire*, 1994]. Individual floes are so small relative to the wavelength that they hardly bend at all. We compute a surface strain of 3×10^{-4} for waves of amplitude $a = 1$ m using *Meylan and Squire* [1994].

To find the approximate rate of energy absorption per unit area, $\dot{\epsilon}_2$, we need to know the molecular dynamic viscosity, μ, of the pancake ice, a quantity that has not been measured in the field. Indeed, measurements of the viscosity of sheet sea ice generally are rare, as ice is usually assumed to be more complicated than a simple Newtonian fluid. Because we are only interested in an order of magnitude estimate, one way of finding μ is to assume that the pancakes behave collectively as a densely packed, highly viscous surface layer. Then, after *Phillips* [1977] and *Gaster* [1962], the spatial attenuation rate α will be given by $\alpha^2 = \mu\omega^7/(2\rho_i g^4)$, where g is the acceleration due to gravity, ω is the circular wave frequency, and ρ_i is the ice density. This expression assumes that the deep water group velocity $d\omega/d\kappa = \omega/2\kappa = g/2\omega$, where κ is the wave number, is unaffected by the ice layer. For 10-s waves we find $\mu \approx 4 \times 10^8 \alpha^2$ Pa s.

No measurements of α exist that account only for hysteresis loss in the pancakes. However, an upper bound on α will be sufficient, as we shall find that the power density dissipation $\dot{\epsilon}_2$ is negligible even then. We choose a value for $\alpha \approx 10^{-4}$ as appropriate for

10-s waves, from measurements in the Weddell Sea taken during the austral winter. Then $\mu \approx 4\,\text{Pa\,s} = 40\,\text{P}$.

A second estimate for μ is reported by *Tabata* [1959], who studied the internal friction of rectangular bars of sea ice at various temperatures up to within a degree of the melting point. A very strong dependence on temperature was found, which is significant to the present study because we would expect pancake ice to exist at high homologous temperatures. This being so we use Tabata's measurement of $\mu \approx 3 \times 10^8\,\text{P}$, taken at his highest recorded temperature of $-3°\,\text{C}$, but recognize that this value will probably grossly overestimate the true value of μ for pancake ice, which we would expect to be closer to its melting point.

In a viscous fluid the rate of working against viscous forces is made up of a viscous energy flux and a positive quantity $2\mu(\dot{e}_{ij})^2$ that represents the rate of energy dissipation per unit volume by molecular viscosity, where $\dot{e}_{ij}$ denotes the strain rate tensor. The quantity we require, the power density dissipation $\dot{\epsilon}_2$, is found by integrating through the thickness of the pancake. When this is done we find that $\dot{\epsilon}_2 \approx 10^{-9}\,\text{W\,m}^{-2}$ from our first analysis and $\dot{\epsilon}_2 \approx 10^{-2}\,\text{W\,m}^{-2}$ based on Tabata's value of viscosity. While these estimates are many orders of magnitude apart because the viscosity of pancake ice is unknown, they are both upper bounds, and we shall discover that they are dwarfed by the dissipation due to collisions.

Shorter period waves or waves of greater amplitude can, in principle, lead to larger strains. But, because $a/d \ll 1$ is required for wave stability when the wavelength becomes close to the diameter d of the pancake, the energy absorption directly due to hysteresis is always negligible for pancake ice.

Water Column Losses

Ice floes create a solid boundary on top of the water column through which the wave travels. When floes are dispersed, each floe acts like a solid body around which a wake is generated due to the relative motion of the floe and the water surrounding it. When floes are concentrated, such as in the case of the compact pancake ice that we are considering here, the gaps between adjacent pancake ice floes are too small to allow wake development. Instead, the bottom surfaces of the pancakes, together with their uneven edges and any stacking up due to rafting, present a rough boundary to the water particles beneath. Shear flow across this rough boundary creates turbulent eddies and they in turn dissipate through viscous effects. This mechanism is described in *Liu and Mollo-Christensen* [1988], where it is modeled by an eddy viscosity. However, if wave attenuation is due solely to this dissipation scenario, the range of values for eddy viscosity embraces three orders of magnitude when fitted to field data [*Liu et al.*, 1991a].

A different approach is adopted here. We will apply two independent methods to estimate $\dot{\epsilon}_3$ directly.

The first method begins with the definition that the turbulent energy dissipation per unit volume of fluid is $\rho \nu u^2/D^2$, where ρ is the fluid density, ν is the eddy viscosity, u is the characteristic relative velocity between the water particles and the pancake ice floe, and D is the boundary layer thickness. We use field data reported by *Hunkins* [1966], measuring the drift velocity of an ice floe in the central Arctic, to continue this estimate. There, u and D were found to be of the order of $0.1\,\text{m\,s}^{-1}$ and $10\,\text{m}$ respectively. The eddy viscosity from the drift data was $24\,\text{St}$. From these data, the rate of energy dissipation per unit volume is in the order of $10^{-4}\,\text{W\,m}^{-3}$. When integrated through the boundary layer thickness, the turbulent power density dissipation $\dot{\epsilon}_3$ is approximately $10^{-3}\,\text{W\,m}^{-2}$. Because the turbulence level in the central Arctic is likely to be lower than that at the ice edge, a more plausible value is $\dot{\epsilon}_3 \approx 10^{-2}\,\text{W\,m}^{-2}$, based on $\nu = 160\,\text{St}$ for the Weddell Sea [*Brennecke*, 1921].

The second method is more closely related to the actual physical processes. It applies to a collection of individual pancakes as will be discussed below. Consider a single ice floe oscillating in a wave field. The forces acting on the floe due to the water particles arise from the viscous drag and inviscid effects. The former is modeled as $\rho C_w A(u-v)^2$ and the latter is $m C_m \partial(u-v)/\partial t$, in which C_w is the drag coefficient, C_m is the added mass coefficient that accounts for the inertia effect of the inviscid flow surrounding the oscillating pancake, ρ is the sea water density, m is the mass of the floe, A is the surface area of a single floe, and u and v are the floe and water velocities respectively. These two forces act like resistance to the floe motion, so the rate of energy that is consumed to maintain the floe motion is therefore

$$\left(\rho C_w A(u-v)^2 + m C_m \frac{\partial(u-v)}{\partial t}\right)(u-v)$$

The power density dissipation is the above multiplied by the number of pancakes per unit area. Consider the case where the ice concentration is nearly 100%, the remaining parameters are estimated as follows: $u - v$ is order a/T, where T represents the wave period; and the acceleration $\partial(u-v)/\partial t$ is of order

a/T^2. Substituting these into the power density dissipation we have $(\rho a C_w + \rho_i h C_m) A a^2 / T^3$, where h is the thickness of the pancake. Suppose $a = 1$ m, $T = 10$ s, $A = 1$ m^2, $h = 0.3$ m. A nominal value of $C_w = 0.05$ and $C_m = 0.3$ may be used. These values are in the order of those found with plastic and real ice floes in the laboratory [*Frankenstein*, 1996]. With these values the power density dissipation $\dot{\epsilon}_3 \approx 0.1$ W m^{-2}. For short periods $\dot{\epsilon}_3$ will be larger and for longer periods it will be smaller (5 s: 1 W m^{-2}; 15 s: 10^{-2} W m^{-2}).

Consequently, both estimates of the power density dissipation arising from losses in the water column are about or less than 10^{-1} W m^{-2} for 10 s waves.

Scattering

Each pancake scatters a portion of the wave energy incident upon it, thereby reducing the forward-going energy. Dissipative processes will attenuate the scattered energy. The most sophisticated models that include scattering are due to *Masson and LeBlond* [1989] and *Meylan and Fox* [1996] based on an array of rigid [*Masson*, 1991] or flexible [*Meylan*, 1995; *Meylan and Squire*, 1996] floating circular disks. Each model assumes an ice field of low concentration, which is not the case being considered here. Accordingly, they cannot be applied directly.

In two-dimensions, if floes of diameter d occupy a proportion c of the sea surface area, and unit wave energy per unit area is incident on the ice edge, then, after n rows of identical ice cakes with transmission coefficient t, the forward-going energy is reduced to $E_f = t^n$ when the backward-going energy is assumed to be entirely dissipated (simple scattering). This elementary model can be applied at higher concentrations by including multiple scattering [*Wadhams*, 1986]. It is sufficient here to give an order of magnitude estimate of $\dot{\epsilon}_4$ employing *Meylan and Squire* [1994] to provide an estimate of t and the energy reflection coefficient r.

Denote the ice concentration by c and floe diameter by d. The approximate spatial energy attenuation coefficient $\alpha = -cr/d$ for simple scattering or $\alpha < -cr/d$ for multiple scattering [*Wadhams*, 1986] is used to write down an expression for the rate of energy loss by scattering via $\partial E / \partial x = \alpha E$. The energy density in deep water is $E = \frac{1}{2} \rho g a^2$. Then, if the deep water group velocity is unaffected by the ice, the power density dissipation $\dot{\epsilon}_4 \approx \alpha \rho g^2 a^2 / 4\omega$. For a 1-m amplitude, 10-s wave impinging on 1×0.3-m^3 pancakes, we find, using the numerical code of *Meylan and Squire* [1994], that $r \approx 10^{-10}$. Consequently even though the multiplying factor is large, we estimate that for multiple scattering $\dot{\epsilon}_4 \approx 10^{-6}$ W m^{-2}.

At 5 and 15-s period, estimates for $\dot{\epsilon}_4$ are 10^{-4} and 10^{-8} W m^{-2} respectively. Thus, over the range of wave conditions that are commonly found in the marginal ice zone, the attenuation of forward propagating wave energy due to reflection is negligible.

OTHER PROCESSES

In addition to the decrease in energy, an ocean wave field traveling through sea ice is known to change in other ways. On entry at the ice margin, for example, waves are believed to refract. Consequently, as they penetrate further into the ice their directional structure is observed to broaden in a manner which depends on wave period. The argument for refraction, which is very difficult to observe unequivocally, rests especially on synthetic aperture radar data reported by *Liu et al.* [1991b] and on casual observation. It is also nourished by theoretical models based on the thin elastic plate that predict a length change as waves cross the edge, and, concomitantly for oblique incidence, refraction. The change in wavelength arises because of the rigidity of the ice cover: similar to the wave propagation in a spring, the greater the spring constant, the longer the wavelength. *Masson and LeBlond* [1989] and *Meylan and Fox* [1996] show that scattering causes the change in the directional spread of wave energy as the waves move through the ice cover.

In pancake ice, where the average floe size is many times smaller than the typical wavelength present for waves of any reasonable amplitude, scattering cannot be a significant factor in modifying the wave field, as shown in the analysis given in the previous section. Likewise, refraction is likely to occur only at wavelengths comparable with the floe diameter, associated with a change in the dispersive nature of the waves as they cross the ice edge. The dispersion of typical open ocean waves from about 5 to 20-s period is unlikely to be affected by a field made up of 1×0.3-m^3 pancakes. Ocean waves of shorter period would disperse differently within the ice, however, and substantial absorption could affect their velocity of energy transport [*Brillouin*, 1960]. In this scenario, refraction would occur due to the change in phase speed on entry to the pancake ice mosaic, leading to the concept of pseudo-rigidity.

THE MODEL

We now begin the modeling of the collision-induced wave energy dissipation, i.e., the term $\dot{\epsilon}_1$ in (1). Conservation of energy applied across a control volume within the pancake ice field suggests

$$\frac{\partial}{\partial x}(EU) = -\dot{\epsilon} \quad (2)$$

where E is the energy density of the wave, U is its velocity of transport, $\dot{\epsilon}$ is the energy dissipated per unit time per unit surface area due to ice floe collisions, assuming other dissipative processes are small in comparison, and x is the position coordinate which increases with penetration. The quantity EU is called the energy flux. Because the quantity $\dot{\epsilon}$ depends on the collision frequency per pancake f, the energy lost during each collision e, and the number of pancakes present per unit area, i.e., the concentration c,

$$\frac{\partial}{\partial x}(EU) = -fec \quad (3)$$

This idea is illustrated in Figure 1 where the rate of energy loss per unit area is due to floe collisions alone. The energy flux between the upstream face and the downstream face of the control volume must be balanced with this energy loss rate. The quantities f and e are themselves functions of a, ω, c, and the floe properties, which we denote by $\mathcal{F}$, i.e.,

$$f = f(a, \omega, c, \mathcal{F}) \quad (4a)$$
$$e = e(a, \omega, c, \mathcal{F}) \quad (4b)$$

The determination of f and e is difficult, and this is where granular flow theory is employed to model the pancake ice field, as described in greater detail below.

Granular Flow Theory

In this section we describe how the functions f and e and their product are found. Floe collisions were first included in ice dynamics modeling by *Shen et al.* [1986]. The collision frequency of ice floes in a fragmented ice cover was derived based on the energy input from the mean deformation field and the dissipation from energy loss in floe collisions. This theory was later extended to obtain the constitutive relation for a fragmented ice cover [*Shen et al.*, 1987]. In these studies, wave effects were absent. *Shen and Ackley* [1991] adopted the theory proposed by *Rumer et al.* [1979], in which the equation of motion for a single ice floe in a wave field was given. By considering the impact between neighboring floes, this theory was extended to show that waves could induce and maintain steady state floe collisions.

The equation of motion for a single ice floe in a wave field is [*Rumer et al.*, 1979]

$$m(1 + C_m)\frac{d^2x}{dt^2} = F_g + F_w \quad (5)$$

where the gravitational force on the floe is modified by the slope of the water surface due to the passing wave as follows:

$$F_g = -\frac{1}{2}m\left(g + \frac{\partial^2 \zeta}{\partial t^2}\right)\sin\left(2\tan^{-1}\frac{\partial \zeta}{\partial x}\right) \quad (6)$$

and the drag force between water and the floe is

$$F_w = \rho C_w A \left|u_o - \frac{dx}{dt}\right|\left(u_o - \frac{dx}{dt}\right) \quad (7)$$

The yet undefined quantities in the above are the water surface elevation, ζ, and the fluid velocity in the x direction at the water surface, u_o. The water surface elevation in a monochromatic wave field is determined by the wave amplitude a as

$$\zeta = a\sin(\kappa x - \omega t) \quad (8)$$

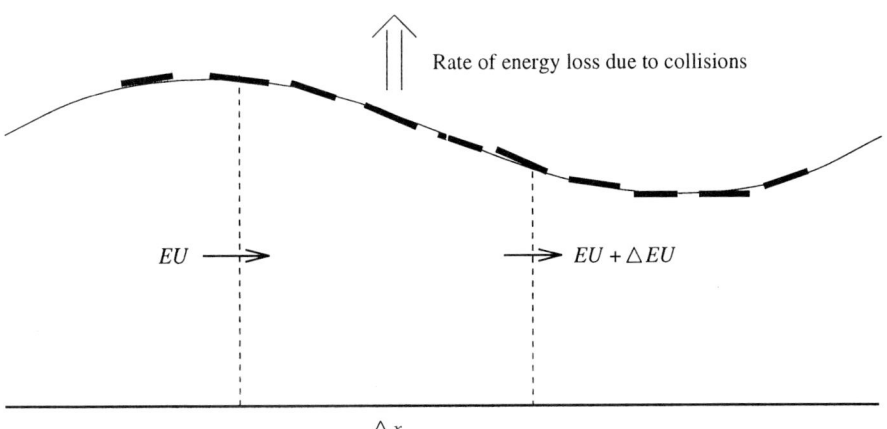

Figure 1. Schematic, illustrating energy balance Equation (3).

and [Stoker, 1957]

$$u_o = \omega\zeta \qquad (9)$$

Because we adopt the deep water, small amplitude wave theory

$$\omega = \sqrt{g\kappa} \qquad (10)$$

and (6) simplifies to

$$F_g = -mg\frac{\partial \zeta}{\partial x} \qquad (11)$$

When a collision between ice floes occurs, an additional force F_c must be added to (5). This force is modeled as a spring-dashpot pair [Shen and Ackley, 1991]. Accordingly

$$F_c = k\delta x + \eta \delta \dot{x} \qquad (12)$$

where δx and $\delta \dot{x}$ are the relative center-to-center position and velocity of two colliding cakes, k is the spring constant, and η is the dashpot coefficient.

Equation (5) does not include wind stress, which will be present in most field situations. Over homogeneous pancake ice, we believe that wind stress acts as a constant body force that moves all floes in the same direction with equal intensity. Consequently, it does not contribute substantially to interactions between floes, which are due only to their relative motion. But, because wave-ice and ice-ice collision interactions are highly nonlinear processes, the wind stress effect warrants future modeling effort.

The collision of ice floes in a randomly distributed pancake ice field is a stochastic process and solutions of the above equations cannot be found analytically. Instead, computer simulation is adopted to determine the right hand side of (3). In the computer simulation, a number of uniform pancakes are randomly positioned in a preset finite domain. Since this is a one-dimensional theory, motion will only be simulated in the x direction for all floes and the wave. Periodic boundary conditions are employed so that floes moving out of the domain will re-enter from the opposite side with identical kinematic conditions. The collision events are recorded and the energy dissipation rate at each time step, if a collision event is detected, is calculated as $\eta\delta\dot{x}^2$. The total energy dissipation in the entire simulation duration after steady state is reached is then found by accumulating the energy loss at each step. Upon dividing by the steady state simulation duration and the domain length, the energy dissipation rate per unit length is found. The time step is set so that results are independent of the resolution. It is found to be a function of the spring constant used in the collision model. For the current case, 10^{-3} was tested to be sufficiently small.

Because the spring constant is modeled to measure the force and compression relation, it is related to the Young's modulus E in the following way

$$\begin{aligned} k &= \frac{\text{force}}{\text{compression}} \\ &= \frac{\text{stress} \times \text{area of contact}}{\text{strain} \times \text{floe length}} \\ &= \frac{E \times \text{area of contact}}{\text{floe length}} \\ &= Eh \end{aligned} \qquad (13)$$

The above is valid for frontal impacts between rectangular blocks with length and width d and thickness h. For general geometry, the relation is very complicated and thus the above can only be viewed as an approximate parameterization for the circular floes used herein. The dashpot coefficient is related to the restitution coefficient e by [Babić et al., 1990]

$$\frac{\eta}{\sqrt{mk}} = \frac{-\ln(e)}{(\pi^2 + \ln^2 e)^{\frac{1}{2}}} \qquad (14)$$

The parameters of this model are summarized as follows. Those related to wave characteristics: a, κ or ω; those related to the pancake: d, h, E, e; those related to the hydrodynamic interactions: C_m, C_w; and those related to the distribution of pancakes: concentration c and domain size. A constant specific gravity of 0.9 is assumed and the domain size is also fixed at 1000 m. The other fixed parameters are $c = 0.9$, $d = 1$ m, $h = 0.3$ m, $E = 10$ MPa, $e = 0.2$, $C_m = 0.3$, and $C_w = 0.05$. The last four are in the range of typical values measured in the laboratory of urea ice and polypropylene floes [Frankenstein, 1996]. Pancake ice floes are very soft, especially in the newly formed stage. Their Young's moduli are expected to be much lower than that for polycrystalline ice. An LVDT test was made to determine the average Young's modulus in a laboratory grown ice sheet that had the consistency of a pancake ice floe. Its value was found to be of the order given above. Although the consistency and surface roughness of laboratory pancake ice is very similar to that found in the field, and hence the added mass and drag coefficients should be similar in each case, field measurement is necessary to validate this.

Before presenting the results of the above derivation, a formal analysis may be made to estimate the functional form of the attenuation. When the processes leading to attenuation of the waves as they travel through the pancake ice field are such that en-

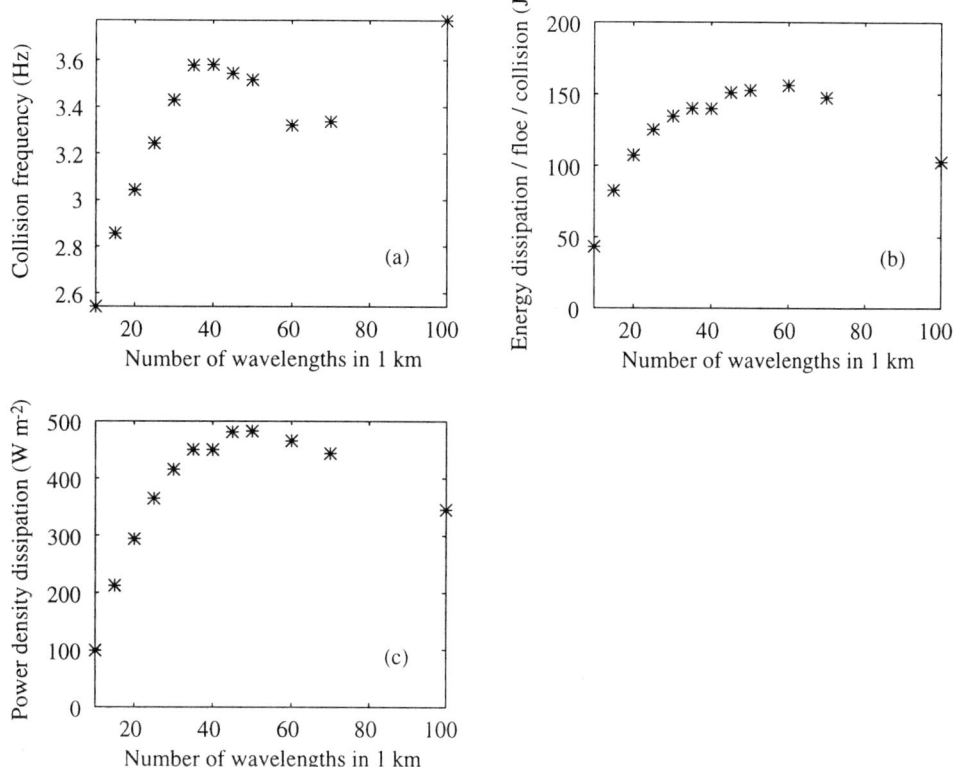

Figure 2. (a) The collision frequency, (b) the energy dissipation per pancake per collision, and (c) the power density dissipation. ($c = 0.9$, $d = 1$ m, $h = 0.3$ m, $E = 10$ MPa, $e = 0.2$, $C_m = 0.3$, $C_w = 0.05$ and $a = 2.3$ m.)

ergy dissipation is small, the velocity of energy transport and the group velocity are equal. We propose also that U doesn't change significantly on entry into the ice or subsequently but remains approximately at its open sea value. This is reasonable as long as the length of the wave does not approach the order of magnitude of the dimensions of the pancake, typically chosen to be 1×0.3 m^3. The behavior of the waves then resembles electromagnetic wave propagation through a slightly lossy ionosphere, as analysed by *Budden* [1988]. Consequently, equation (3) becomes

$$\frac{\partial a^2}{\partial x} = -\frac{4\omega}{\rho g^2} fec \qquad (15)$$

To a first guess we might take $f \propto a$ and $e \propto a^2$. This is because collision events must increase with floe velocity, which increases with wave amplitude, and the energy loss in each collision is proportional to the kinetic energy, or the square of the velocity of the floes. Then (15) becomes simply

$$\frac{\partial a^2}{\partial x} \propto -a^3 \quad \text{or} \quad \frac{\partial a}{\partial x} \propto -a^2 = -Sa^2 \qquad (16)$$

with solution $a = (Sx + 1/a_0)^{-1}$, where a_0 is the amplitude at the ice edge. The hyperbolic decay predicted by this simple approximate formula differs from the usual exponential decay, but still remains an acceptable fit to the field experimental results [*Wadhams*, 1973].

When granular flow theory is used to find f and e, it will be shown in the following section that (16) is modified slightly to become

$$\frac{\partial a^2}{\partial x} \propto -a^{n+1} \quad \text{or} \quad \frac{\partial a}{\partial x} \propto -a^n = -Sa^n \qquad (17)$$

where the power n depends on the floe conditions. Equation (17) is actually equivalent to that derived by *Wadhams* [1973] based on steady state creep argu-

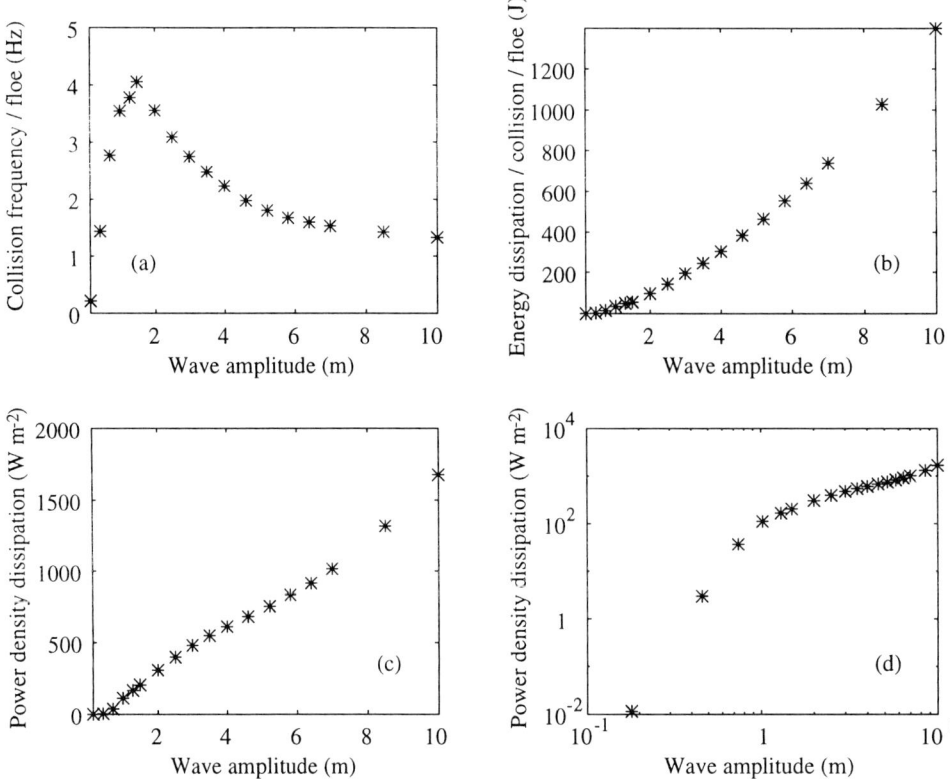

Figure 3. (a) The collision frequency, (b) the energy dissipation per pancake per collision, and (c, d) the power density dissipation, each plotted against the wave amplitude. Wave period is 5 s. ($c = 0.9$, $d = 1$ m, $h = 0.3$ m, $E = 10$ MPa, $e = 0.2$, $C_m = 0.3$, $C_w = 0.05$ and $a = 2.3$ m.)

ments, though the n here is derived by totally different methods. The general solution to equation (17) is

$$a^{n-1} = \frac{1}{(n-1)Sx + a_0^{(1-n)}} \quad (18)$$

The link between the current collision model and Wadhams' theory is most satisfying, as it provides a physical basis for the steady-state creep model of wave decay in an ice field composed of discrete floes. The creep exponent is strongly dependent on ice conditions, however, and can take on values that are different from those that typically describe the flow of polycrystalline ice, namely, $n \approx 3$.

The assumption that the group velocity and the velocity of energy transport at each frequency are the same, and do not change on entry to the pancake ice field, ceases to hold when the absorption is large. When this occurs the manner in which the waves disperse is affected significantly by the dissipation, and the group velocity and velocity of energy transport are no longer the same [*Brillouin*, 1960]. This situation would occur for very short period waves of length comparable to the dimensions of the pancakes present in the ice field.

RESULTS

Because the theoretical framework is actually quite straightforward, the majority of the results presented herein concern either the functions shown in (4), which are not simple, or predictions by the complete theory in relation to the modification of waves entering and travelling through pancake ice.

Attenuation Versus Wave Number

The result of power density dissipation in the wave direction due to collisions between pancake floes is summarized in Figure 2, where the collision frequency, the energy dissipation per collision per floe, and the power density dissipation are presented respectively against the number of waves per kilometer, i.e., the normalized wave number $\hat{\kappa} = 1000\kappa/2\pi$.

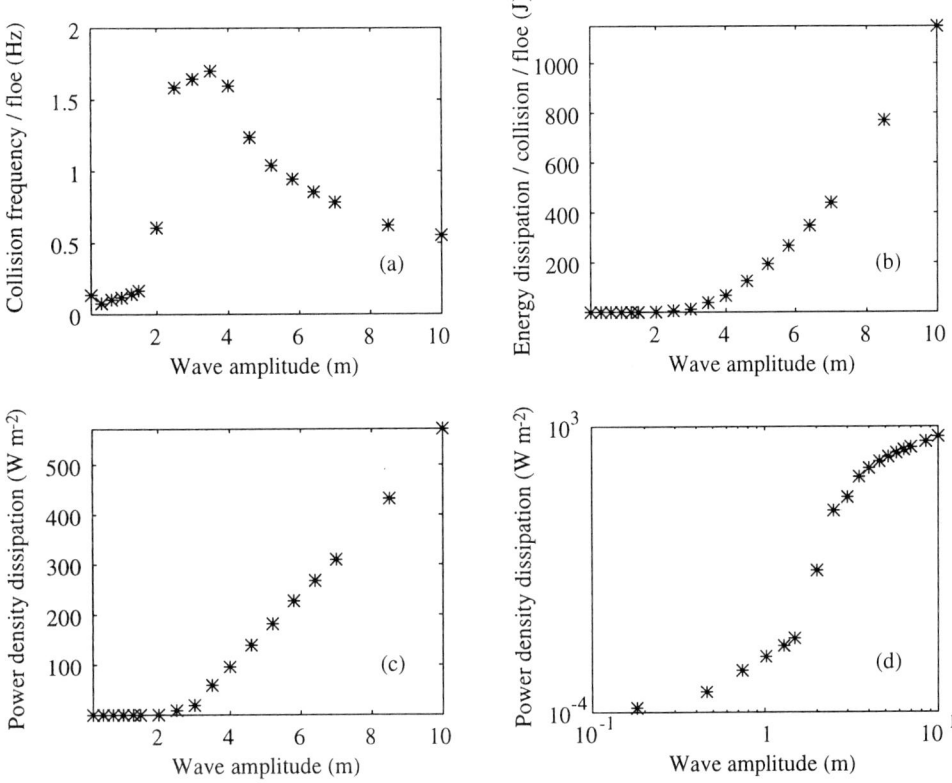

Figure 4. As Figure 3 for 10.3 s period waves.

The wave amplitude in these figures is fixed at 2.3 m. These results will depend quantitatively on the input parameters but the trend remains the same. A detailed parametric study awaits further investigation. An interesting observation can be made in the three energy plots making up Figure 2b, c, d. All three curves peak first then drop off. It is intuitive that for the same wave amplitude the oscillatory motion of the floes should increase with wave number, i.e., frequency. Therefore, it is expected that the collision frequency, the energy dissipation per collision activity, and the power density dissipation increase with the wave number at first. It is however surprising to know that this trend can reverse as the wave number increases beyond a critical value. Looking closely at the floe trajectories near the critical wave number, it is seen that floes become more and more synchronized with the wave motion. At some wave number that depends on the conditions of the waves and ice, pancakes move harmonically in the wave field in such a way that before they approach an adjacent floe the wave motion forces them to change their directions. Hence, collisions reduce drastically. This phenomenon may be specific to the monochromatic wave assumption. Although we speculate that it should also be present for a wave field that is narrow in spectral form, data from simulations are needed to confirm this. The collision frequency plot in Figure 2 shows a rise again after the initial fall. A closer look at floe trajectories shows that when floes are moving in near synchronized motion, a collision is often followed by rapid recurrences with weak impact. This explains the drop in the resulting energy dissipation despite a rise in collision frequency.

Concentrating on Figure 2c, the power density dissipation is precisely the absolute value of the right hand side of (3). This drop off of power density implies a decrease in wave attenuation; the so-called "rollover" phenomenon seen in field data [*Wadhams et al.*, 1988]. An eddy viscosity model was proposed by *Liu and Mollo-Christensen* [1988] that successfully captured this phenomenon with properly chosen eddy viscosities [*Liu et al.*, 1991a], but the eddy viscosities chosen to fit data ranged from 4 to 3450 St. Estimates of eddy viscosity from field studies are 160 St [*Brennecke*, 1921] and 24 St [*Hunkins*,

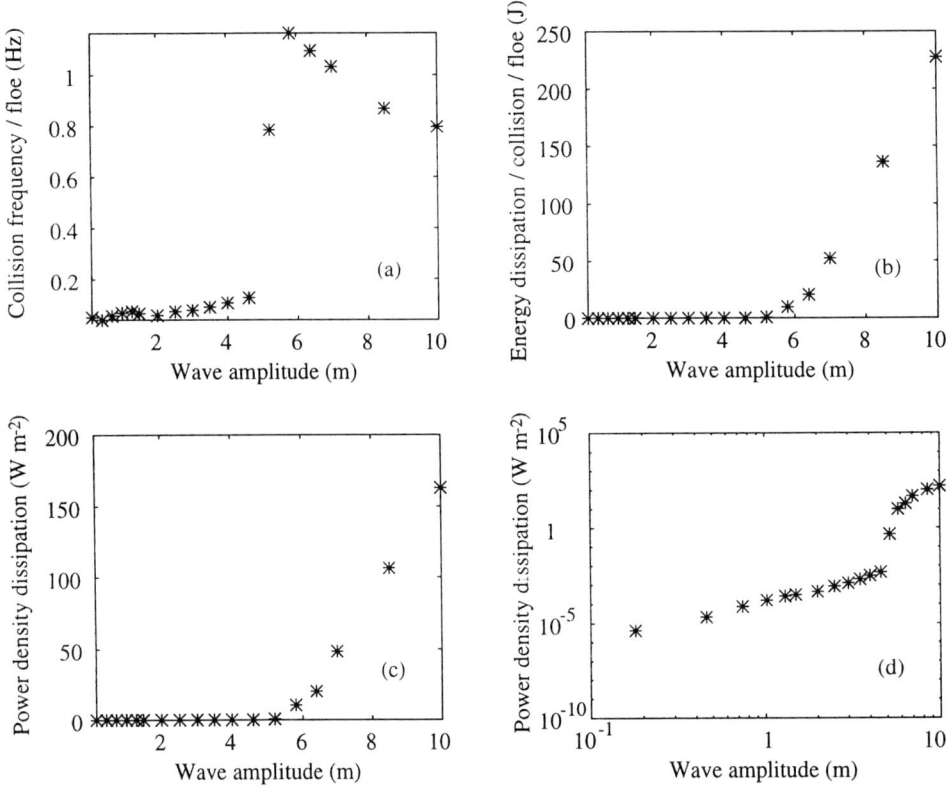

Figure 5. As Figure 3 for 14.6 s period waves.

1966]. Whether the present model can capture the rollover phenomenon with more fundamental parameters such as the wave conditions and the floe characteristics remains to be seen.

Power Law Determination

The collision frequency per ice floe, the associated energy dissipation per collision, and the power density dissipation are presented against wave amplitude in Figures 3–5 for wave periods of 5, 10.3 and 14.6 s respectively. These wave periods are chosen so that the number of waves in the computer simulation domain of 1000 m is an integer (25, 6, and 3 respectively). This integer requirement is due to the periodic boundary condition imposed on the simulation. As expected, when wave energy increases the energy dissipation per collision and the power density dissipation increase, but the collision frequency has a peak in each of the cases run. This phenomenon corresponds to the synchronization of floe motion as described before.

The log-log plots of Figures 3d, 4d, and 5d have been combined into Figure 6, where the structure of each curve is more easily seen. All curves increase monotonically. The points of both the 10.3 and 14.6-s curves, denoted by × and ∘, respectively, increase in a roughly linear manner at first from the lowest amplitude plotted, namely, 0.18 m. Power dissipation then begins to change rapidly at a specific wave amplitude that is different for each period. Finally, at a larger amplitude, the curves become less steep again, with a gradient similar to that at low amplitudes. The results for 5-s waves, represented by +, appear at first sight to be slightly different. However, the points trace out a curve that actually resembles the other curves shifted to the left. At very large amplitudes all curves tend asymptotically to one another, though this situation becomes physically meaningless as waves of such amplitude cannot exist. (A 5-s wave becomes unstable at 2.8 m, a 10-s wave at 11.1 m, and a 15-s wave at 24.9 m.) A picture emerges then of an amplitude dependence that is qualitatively similar at each wave period, namely, a sheared, s-shaped curve

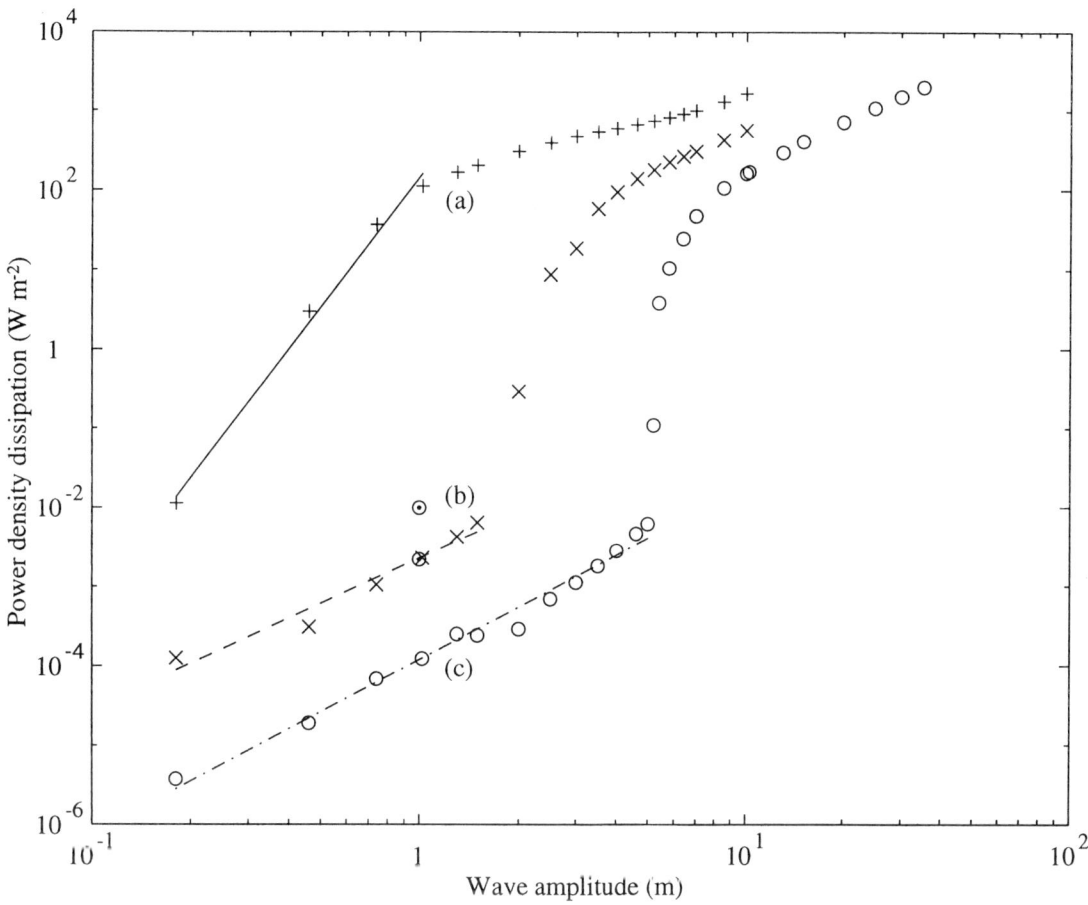

Figure 6. Curves (d) of Figures 3, 4, and 5 replotted with a partial straight line fit. The fit in each case has an approximate slope and intercept of (a) 5.40, $10^{2.16}$, (b) 1.89, $10^{-2.64}$, (c) 2.14, $10^{-3.87}$ for 5, 10.3, and 14.6-s waves, respectively. The two extra ⊙ points corresponding to $C_w = 0.02$ (highest dissipation) and 0.08 (indistinguishable from the $C_w = 0.05$ point) on the 10-s curve are included to illustrate sensitivity to the drag coefficient. See text for more details.

that is positioned and sloped according to period. It is clear that (17), which suggests a linear dependence between $\log a$ and the logarithm of the dissipation, cannot be used over the entire range of physically possible amplitudes.

Returning briefly to Figures 3b, 4b, and 5b, it is clear that the energy dissipated during each pancake ice collision increases steadily with amplitude, possibly following a simple power law. The unusual structure of the curves in Figure 6 is caused by differences in phase between the floes and the wave, which can prevent ice cakes from colliding and thereby cause changes in collision activity. For a real sea made up of waves of different periods with random phases, the s-shaped curves would be expected to be smoothed out to some degree.

Within prescribed ranges of amplitudes we may fit a power law piecewise to the points of Figure 6. We do this anticipating the range of amplitudes that will be encountered in the next section, where actual data will be considered. This results in the linear fits shown, where (a) representing 5 s has a slope of 5.40 and an intercept of $10^{2.16}$, (b) representing 10.3 s has a slope of 1.89 and an intercept of $10^{-2.64}$, and (c) representing 14.6 s has a slope of 2.14 and an intercept of $10^{-3.87}$. Accordingly, the power law parameter n that appears in (17) and (18) is $n = 4.4$ for 5 s, $n = 0.89$ for 10 s, and $n = 1.14$ for 15 s. The actual values of these power law coefficients will vary with ice properties, concentration, and floe size. Power law attenuation levels off more rapidly than the usual exponential decay, and consequently greater penetra-

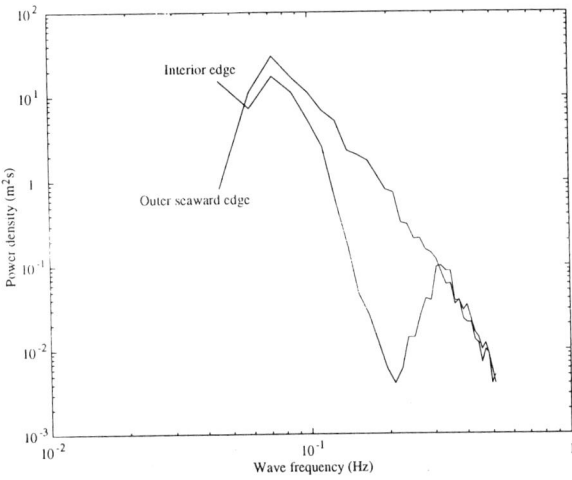

Figure 7. Energy density spectra for wave time series on the seaward and inward sides of a band of pancake ice of width 4 km. (After *Squire* [1989] with modification.)

tion of wave energy is achieved. In Figure 6, for the 10-s wave case at 1-m amplitude, two extra points are calculated to give an idea of the sensitivity to C_w, which is believed to be an important parameter. The upper one corresponds to $C_w = 0.02$ and the lower one corresponds to $C_w = 0.08$. The latter case is indistinguishable from the nominal value of 0.05 chosen for demonstrating the model. The two extremes of 0.02 and 0.08 were obtained, respectively, from a smooth plastic floe and a rough ice floe produced by slushy ice crystals. Likewise, two extreme values of $C_m = 0.1$ and 0.5 give power density dissipations of 0.48×10^{-2} and $0.15 \times 10^{-2}\,\mathrm{W\,m^{-2}}$ respectively. Since both are of the same order of magnitude as that obtained from $C_m = 0.3$ used for the cases in Figure 6, it is likely that the added mass coefficient will not affect the order of magnitude of the collisional power density dissipation. It is worthwhile stressing that, since the parameter values for a field case are unknown, both model sensitivity analysis and field measurements should be carried out and these are planned for future work.

Comparison with Data

Figure 7 shows two energy spectra computed from a data set obtained during a field experiment in the Weddell Sea [*Squire*, 1989]. The data were collected

Figure 8. Photograph of a U.K. Institute of Oceanographic Sciences pitch-roll buoy located in pancake ice of approximate diameter 1 m and thickness 0.3 m.

Figure 9. Aerial photograph of the band of pancake ice in the Weddell Sea, showing the R. V. Polarstern in the distance. Long period waves can be seen propagating through the band.

using a U.K. Institute of Oceanographic Sciences pitch-roll buoy (Figure 8) either side of a highly compact, 4-km band of pancake ice made up of floes of average dimension $1 \times 0.3\,\text{m}^3$ (Figures 9 and 10). Waves were incident on the band, propagating in a direction approximately parallel to a bearing of 060° from the west. Thus, the effective width of the band was 8 km. Each 4096-s record was sampled at 2 Hz, and the spectra were smoothed to 56 degrees of freedom. Because the physical parameters for the simulations done earlier were chosen to represent the conditions during data collection, we may apply these model predictions directly.

To compare the data with the theoretical results presented earlier, we consider three frequency bands of 5 contiguous power density values centered at 5, 10 and 15 s, i.e., close to the theoretical results of 5, 10.3 and 14.6 s. Within each bandwidth we compute the energy density and hence, the amplitude, for the measurements made each side of the band of pancake ice. By this means the mean amplitude of waves at 5, 10 and 15 s can be found before and after passing through the band, and 95% confidence intervals can be established. The comparison is shown in Figure 11.

It is clear that the collision model cannot account for all of the dissipation that is occurring as the waves traverse the pancake ice band. For 5-s waves (Figure 11a), the power dissipated by collisions from the collision model at the observed amplitude amounts to about $3\,\text{W}\,\text{m}^{-2}$, which, when the measured amplitude is accommodated, is of similar order but somewhat larger than the energy lost in the water column ($\sim 0.5\,\text{W}\,\text{m}^{-2}$). In contrast, the effect of collisions at longer wave periods is less than that due to water column losses. At 10 s (Figure 11b), it is about $10^{-2}\,\text{W}\,\text{m}^{-2}$ compared to $2 \times 10^{-1}\,\text{W}\,\text{m}^{-2}$, and at 15 s (Figure 11c) it is $4 \times 10^{-4}\,\text{W}\,\text{m}^{-2}$ compared to $2 \times 10^{-2}\,\text{W}\,\text{m}^{-2}$. We conclude that collisions have their greatest impact at short wave periods, where they may be significantly more efficient at removing the energy from an incoming sea than other processes. For the size, thickness and concentration of pancake ice taken, collisions between pancakes become less important when the wave period exceeds about 10 s.

Figure 10. Closeup photograph of the band, showing its composition of pancakes in brash ice, the latter presumably generated by collisions between ice cakes.

SUMMARY AND CONCLUSIONS

Granular flow theory has been used to model a pancake ice field composed of interacting pancake ice floes in differential motion, with the purpose of devising a new, physically plausible theoretical description of ocean wave propagation in the winter Weddell Sea. The advantage of this approach is that the ice conditions, namely, concentration, floe size, and thickness, etc., are a functional part of the theory, so that the difficult step of translating sea ice morphology into mathematics is avoided. Moreover, the current model is totally based on fundamental physics, rather than being a perceived parameterization of what is happening in the ice or an empirical fit to field data.

We summarize our results and conclusions as follows.

1) The four energy dissipation mechanisms and their order of magnitude dependence on the wavelength and wave amplitude are:

$$\dot{\epsilon}_1 = a^{n+1}\left[k_1 - k_2(L-L_0)^2\right]$$

where the functional dependence on wavelength L is proposed to fit the rollover seen in Figure 2 and k_1, k_2 are related to the wave and ice properties,

$$\dot{\epsilon}_2 = \mu g h^3 a^2 L^{-5}$$
$$\dot{\epsilon}_3 = \rho C_w g^{3/2} a^3 L^{-3/2}$$
$$\dot{\epsilon}_4 = \frac{cr}{d}\rho g^{3/2} a^2 L^{1/2}$$

2) We have found for compact pancake ice fields that two dissipative processes compete to dominate the removal of energy from incoming ocean waves, namely, water column losses due to turbulence, $\dot{\epsilon}_3$, and absorption arising from collisions between adjacent pancakes, $\dot{\epsilon}_1$. Which one of the two is the major energy dissipator depends on the wave and floe properties. At shorter wavelengths $\dot{\epsilon}_1$ quickly wins because of its rapid increase with wave amplitude ($a^{5.4}$). At very long wavelengths $\dot{\epsilon}_4$ can become dominant since it is the only process that increases with wavelength. A schematic diagram of the behavior of these four energy dissipation mechanisms is given in Figure 12 for the 10.3-s wave case. The

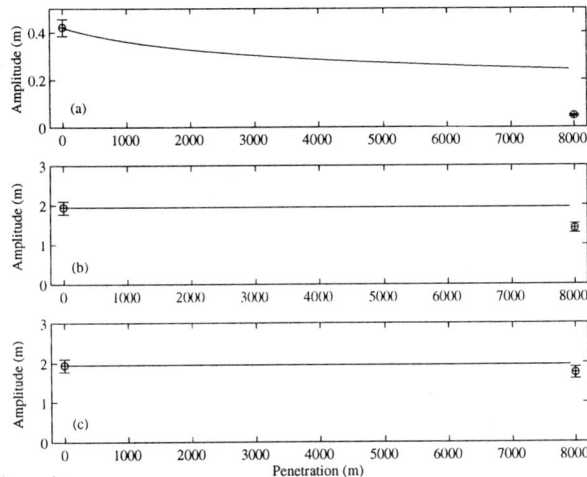

Figure 11. Predicted wave decay due to collisions compared with the Weddell Sea pancake ice band data for waves of period centered at (a) 5 s, (b) 10 s, and (c) 15 s.

intercept of each log-log plot (on the vertical axis passing through the 1-m wave amplitude point) is estimated in the following way. For $\dot{\epsilon}_1$ it is $10^{-2.64}$ from the mathematical modeling. For $\dot{\epsilon}_2$, $\dot{\epsilon}_3$ and $\dot{\epsilon}_4$ we used the order of magnitude estimates, each in SI units, $\mu = 10\text{--}10^7$, $g = 10$, $h = 10^{-1}$, $L = 10^2$, $\rho = 10^3$, $C_w = 10^{-2}$, $r = 10^{-10}$, $d = 1$. Each intercept will change as the wavelength is altered: for $\dot{\epsilon}_1$ it moves downward as L increases; for $\dot{\epsilon}_2$ it goes down as L^{-5}; for $\dot{\epsilon}_3$ it goes down as $L^{-3/2}$; and for $\dot{\epsilon}_4$ it goes up as $L^{1/2}$. The intercepts will also affect the relative magnitude of the dissipation. Out of the four, the only one that will change its slope under different wave conditions is $\dot{\epsilon}_1$. Summarizing, neither hysteresis loss due to bending nor scattering play a significant role for pancake ice, though these mechanisms will be important for other classes of ice cover. The relative balance between $\dot{\epsilon}_1$ and $\dot{\epsilon}_3$ changes with the wave period, with collisions being most influential at shorter periods and turbulence in the water column increasing in significance as the wave becomes longer.

3) The rollover seen consistently in ice field wave attenuation observations [*Wadhams et al.*, 1988] is predicted by the current theory. It is shown in Figure 2c that with increasing wave number the dissipation rises up to a maximum, and then decreases again. While this phenomenon has been modeled before, this is the first time that rollover has been produced from basic physical laws that govern the motion and interaction of ice floes.

4) Although an increase in wave amplitude always results in an increase in dissipation, the relationship is not simple. Relatively small changes in wave amplitude can cause considerable variation in power dissipation, depending on the wave period and on the initial amplitude. For example, whereas an increase in amplitude from 1 to 1.5 m leads to a doubling of the dissipation at 10.3 s period, an increase in the same ratio from 4.6 m causes the dissipation to increase by a factor 10^4. This highly nonlinear behavior is attributed to phase differences between the waves and pancake ice floes that control whether or not a collision occurs between adjacent floes. While this effect is exacerbated by the use of monochromatic waves in the model, and we would expect it to be smeared out when broad ocean wave spectra were considered instead, narrow spectra could give rise to a similar phenomenon.

5) The theory in this paper has been applied only to a highly concentrated field of pancake ice of a particular geometry, principally because of the importance of this type of ice canopy to the Southern Ocean and the scarcity of published research on the topic. However, explicit collisions between ice floes have also been omitted in all models that deal with

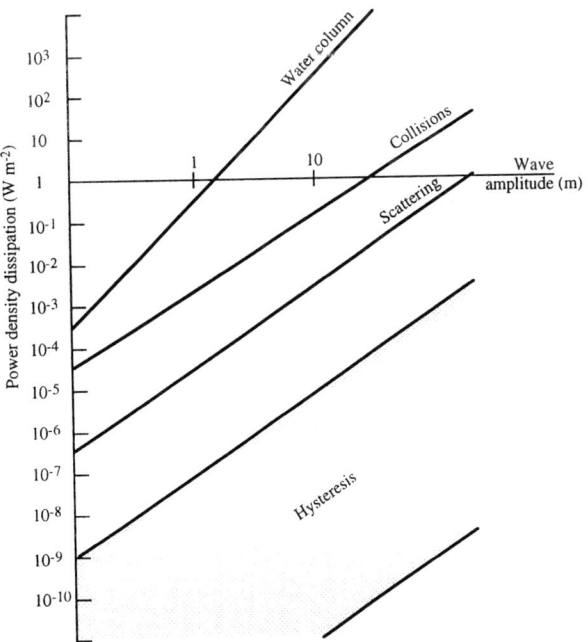

Figure 12. Schematic showing behavior of various energy dissipation mechanisms with amplitude when period is 10.3 s. The hysteresis band represents the uncertainty in μ.

wave propagation through sea ice generally. We have shown that they can be important, and that their effect is dependent on period. For larger ice floes, such as those at the edge of the Greenland Sea marginal ice zone, we would expect collision-induced wave energy dissipation to be significant at higher periods than those studied here, where the ratio of wavelength to floe diameter is still small. The heavily pommeled brash zone at the extreme edge of the ice cover may be evidence of this.

6) Provided that the model presented here is correct, then wave energy penetration into a fragmented ice cover is more persistent than previously thought. A rapid decay near the ice edge is common from both the exponential law and the power law, hence short distance wave measurement cannot distinguish between the two. However, over a long distance, the power law results predict higher residual wave energy. There may be interesting implications of this to oceanic mixing under ice covers.

Acknowledgments. We are grateful to the U. S. National Science Foundation for Grant #OPP-9219165, the N. Z. Foundation for Research, Science and Technology, the Royal Society of New Zealand, the University of Otago, and Clarkson University for their continued financial support. The paper was begun while V. A. S. was a guest of Clarkson University, and completed while H. H. S. was visiting the University of Otago.

REFERENCES

Babić, M., H. H. Shen, and H. T. Shen, The stress tensor in granular shear flows of uniform, deformable disks at high solids concentrations, *J. Fluid Mech., 219,* 81, 1990.

Brennecke, B., Die ozeanographischen arbeiten der Deutschen Antarktischen expedition 1911–1912, *Arch. Dtsch. Seewarte, 39,* 206, 1921.

Brillouin, L., *Wave Propagation and Group Velocity,* Academic Press, New York, 154 pp., 1960.

Budden, K. G., *The Propagation of Radio Waves,* Cambridge University Press, New York, 669 pp., 1988.

Fox, C., and V. A. Squire, Strain in shore fast ice due to incoming ocean waves and swell, *J. Geophys. Res., 96,* 4531, 1991.

Fox, C., and V. A. Squire, On the oblique reflexion and transmission of ocean waves from shore fast sea ice, *Philos. Trans. R. Soc. London A, 347,* 185, 1994.

Frankenstein, S., *The Effect of Waves on Pancake Ice,* Ph. D. dissertation, Clarkson University, New York, 1996.

Gaster, M., A note on the relation between temporally-increasing and spatially-increasing disturbances in hydrodynamical stability, *J. Fluid Mech., 14,* 222, 1962.

Hunkins, K., Ekman drift current in the Arctic Ocean, *Deep Sea Res., 13,* 607, 1966.

Liu, A. K., and E. Mollo-Christensen, Wave Propagation in a solid ice pack, *J. Phys. Oceanogr., 18,* 1702, 1988.

Liu, A. K., B. Holt, and P. W. Vachon, Wave Propagation in the marginal ice zone: model predictions and comparisons with buoy and SAR data, *J. Geophys. Res., 96,* 4605, 1991a.

Liu, A. K., P. W. Vachon, and C. Y. Peng, Observation of wave refraction at an ice edge by SAR, *J. Geophys. Res., 96,* 4803, 1991b.

Liu, A. K., P. W. Vachon, C. Y. Peng, and A. S. Bhogal, Wave attenuation in the marginal ice zone during LIMEX, *Atmos. Ocean, 30,* 192, 1992.

Masson, D., Wave-induced drift force in the marginal ice zone, *J. Phys. Oceanog., 21,* 3, 1991.

Masson, D., and P. H. LeBlond, Spectral evolution of wind-generated surface gravity waves in a dispersed ice field, *J. Fluid Mech., 202,* 43, 1989.

Meylan, M., The motion of a floating flexible disk under wave action, Proceedings 5th International Offshore and Polar Engineering Conference, 3, International Society of Offshore and Polar Engineers, Golden, Colorado, 450, 1995.

Meylan, M., and C. Fox, A model for the propagation of waves through the MIZ from a single floe solution, Proceedings 6th International Offshore and Polar Engineering Conference, 3, International Society of Offshore and Polar Engineers, Golden, Colorado, 321, 1996.

Meylan, M., and V. A. Squire, The response of ice floes to ocean waves, *J. Geophys. Res., 99,* 891, 1994.

Meylan, M., and V. A. Squire, Response of a circular ice floe to ocean waves, *J. Geophys. Res., 101,* 8869, 1996.

Phillips, O. M., *The Dynamics of the Upper Ocean,* 2nd ed., Cambridge University Press, New York, 1977.

Robin, G. de Q., Wave propagation through fields of pack ice, *Phil. Trans. R. Soc. A, 25,* 5313, 1963.

Rumer, R. R., R. Crissman, and A. Wake, Ice transport in Great Lakes, *Water Resources and Environmental Engineering Res. Rep. 79-3,* pp. 275, State Univ. N.Y., 1979.

Shen, H. H., and S. F. Ackley, A one-dimensional model for wave-induced ice-floe collisions, *Ann. Glaciol, 15,* 87, 1991.

Shen, H. H., W. D. Hibler III, and M. Leppäranta, On applying granular flow theory to a deforming broken ice field, *Acta Mech., 63,* 143, 1986

Shen, H. H., W. D. Hibler III, and M. Leppäranta, The role of floe collisions in sea ice rheology *J. Geophys. Res., 92,* 7085, 1987

Squire, V. A., Super-critical reflection of ocean waves; a new factor in ice edge dynamics, *Ann. Glaciol, 12,* 157, 1989.

Squire, V. A., and S. C. Moore, Direct measurement of the attenuation of ocean waves by pack ice, *Nature, Lond., 283,* 365, 1980.

Squire, V. A., J. P. Dugan, P. Wadhams, P. J. Rottier, and A. K. Liu, Of ocean waves and sea ice, *Annu. Rev. Fluid Mech., 27,* 115, 1995.

Stoker, J. J., *Water Waves: The Mathematical Theory with Applications,* Wiley-Interscience, New York, 1957.

Tabata, T., Studies on the mechanical properties of sea ice IV, *Low Temp. Sci., A, 18,* 131, 1959.

Wadhams, P., Measurement of wave attenuation in pack

ice by inverted echo sounding, In *Sea Ice,* ed. T. Karlsson, Nat. Res. Counc., Reykjavik, 255, 1972.

Wadhams, P., Attenuation of swell by sea ice, *J. Geophys. Res., 78,* 3552, 1973.

Wadhams, P., Airborne laser profiling of swell in an open ice field, *J. Geophys. Res., 80,* 4520, 1975.

Wadhams, P., Wave decay in the marginal ice zone measured from a submarine, *Deep Sea Res., 25,* 23, 1978.

Wadhams, P., The seasonal ice zone, In *The Geophysics of Sea Ice,* ed. N. Untersteiner, Plenum, N.Y., 825, 1986.

Wadhams, P., and B. Holt, Waves in frazil and pancake ice and their detection in Seasat synthetic aperture radar imagery, *J. Geophys. Res., 96,* 8835, 1991.

Wadhams, P., M. A. Lange, and S. F. Ackley, The ice thickness distribution across the Atlantic sector of the Antarctic Ocean in midwinter, *J. Geophys. Res., 92,* 14535, 1987.

Wadhams, P., V. A. Squire, J. A. Ewing, and R. W. Pascal, The effect of the marginal ice zone on the directional wave spectrum of the ocean, *J. Phys. Oceanog., 16,* 358, 1986.

Wadhams, P., V. A. Squire, D. J. Goodman, A. M. Cowan, and S. C. Moore, The attenuation of ocean waves in the marginal ice zone, *J. Geophys. Res., 93,* 6799, 1988.

H. H. Shen, Department of Civil and Environmental Engineering, Clarkson University, Potsdam, N.Y. 13699-5710, U.S.A. (email: hhshen@sun.soe.clarkson.edu)

V. A. Squire, Department of Mathematics and Statistics, University of Otago, P.O. Box 56, Dunedin, New Zealand. (email: vsquire@maths.otago.ac.nz)

(Received July 17, 1996;
accepted January 9, 1997.)

SOME FEATURES OF THE GROWTH, STRUCTURE AND METAMORPHISM OF EAST ANTARCTIC LANDFAST SEA ICE

V. I. Fedotov, N. V. Cherepanov, and K. P. Tyshko

The State Research Center of the Russian Federation - Arctic and Antarctic Research Institute, St.-Petersburg, Russia

Landfast ice is one of the important elements of the Antarctic region. Its growth and expansion depend on geographical peculiarities of the continental nearshore regions and varied meteorological and ice-hydrological processes. According to the results of many years of observations in East Antarctica, the width of landfast ice zones is controlled by underwater relief and continental shelf depth. The crystal structure of the fast ice during its growth is nearly always layered and may consist of combinations of congelation, frazil, shuga, infiltration and platelet ice. Infiltration ice (snow ice) increases multiyear ice thickness as a consequence of snow accumulation and ice formation on the upper ice surface. Characteristic layered structures appear in the landfast ice in spring; these are caused by metamorphism processes. On the basis of pronounced variations of ice crystal structure, and physical and mechanical properties in this period, the different and changing stages of the physical state of the ice can be distinguished.

INTRODUCTION

The beginning of the systematic study of sea ice in Antarctica by Russian investigators dates back to the first expedition aboard the ship Ob in 1955-56. In order to supply the first polar stations that were placed on the continental coast, unloading vessels, construction of ice roads and spring-summer transport were undertaken on the landfast ice. It was for this reason that systematic observations of landfast ice formation, development and extent in East Antarctica were initiated.

The fact that the fast ice is a product of multiple interactions of the ocean and atmosphere demanded complex investigations. Year–round studies of snow and ice surface morphology, ice crystal structure, its physical and mechanical properties, and also their spatial and temporal variability were carried out at the polar stations and occasionally from vessels and aircraft. The methods of these investigations were much the same as those in the Central Arctic. It should be mentioned here that the application of ring drills [*Cherepenov and Sokolov*, 1969] with diameters of 110, 220, and 310 mm for taking ice cores through the entire thickness of the sea ice cover noticeably increased the quantity and quality of the field-work. In the 1980s and 1990s, satellite data for ice investigations also became very important [*Zwally et al.*, 1983; *Rey and McCormick Rey*, 1988; *Kirillov and Paramanov*, 1991].

The results of Soviet/Russian sea ice investigations in Antarctica have been published systematically in periodicals such as the Information Bulletin of the Soviet Antarctic Expedition (SAE), Proceedings of the SAE, Proceedings of the Arctic and Antarctic Research Institute (AARI), and Problems of the Arctic and Antarctic. Several monographs have been devoted to the results of sea ice studies in East Antarctica [*Gordienko et al.*, 1970; *Buynitskiy*, 1973; *Kozlovsky et al.*, 1977]. Using data from these and other publications, we present in this paper a short review of the most common and typical processes that affect the development and extent of landfast sea ice and its characteristics in East Antarctica between 0° and 160°E (Figure 1).

GENERAL FEATURES OF ANTARCTIC FAST ICE

The formation, growth and expansion of the landfast ice in Antarctica depend on the environmental conditions in the nearshore regions. It doesn't form an unbroken ice ring around the Antarctic even under the most severe en-

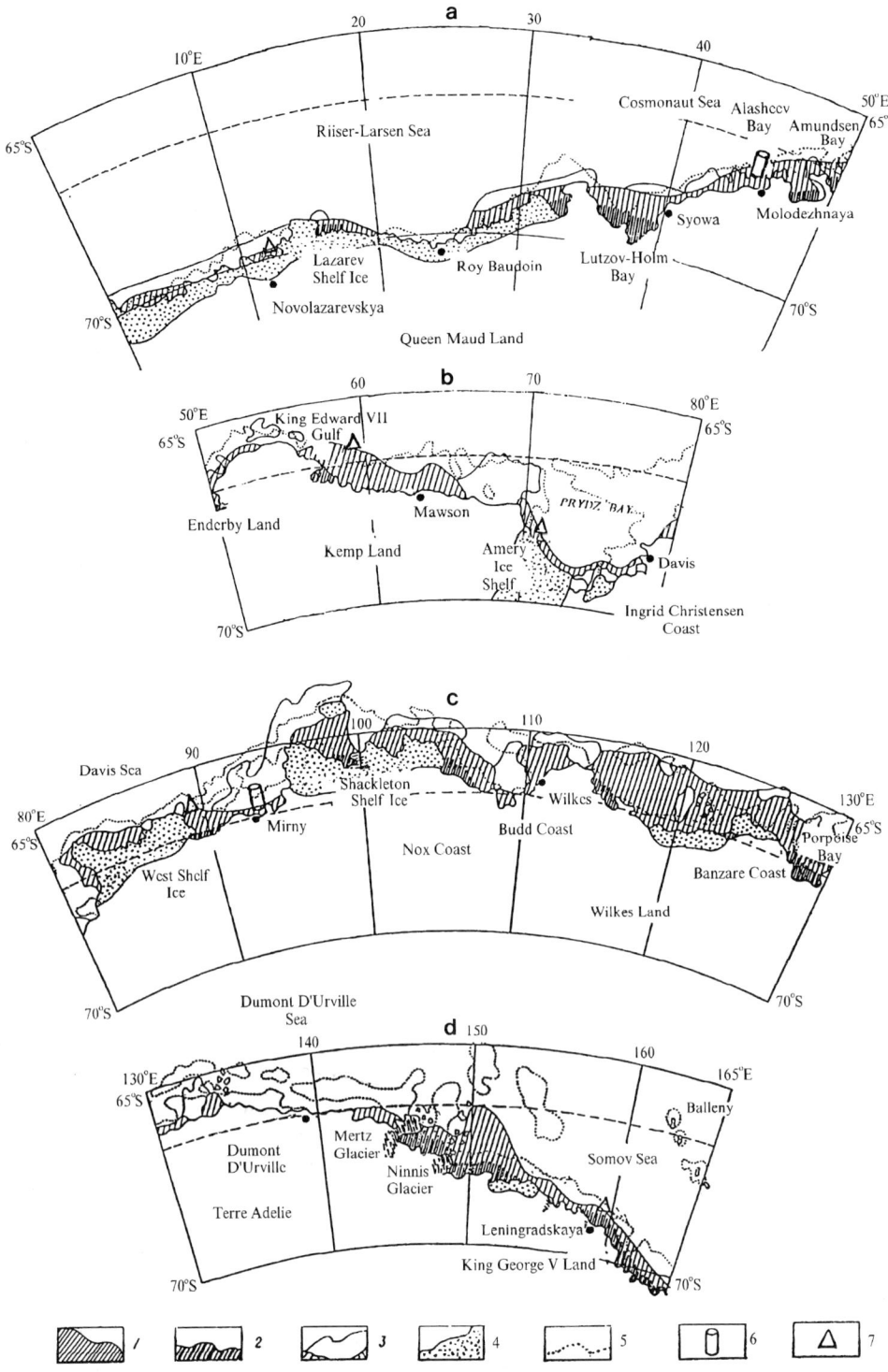

Fig. 1. Maximum extent of the landfast ice in East Antarctica in November [*Kozlovsky et al.*, 1977] and regions of long term ice investigations. 1: first-year fast ice; 2: multiyear fast ice; 3: near-barrier and flaw polynyas; 4: ice shelf; 5: 500 m isobath; 6: regions of long term ice investigations near Russian and other bases; 7: regions of occasional ice investigations.

vironmental conditions. Its average width ranges from 45 to 65 km and the maximum extent of the fast ice is observed along the Sabrina Coast (Wilkes Land, 120°E), where it is up to 270 km wide (Figure 1c). November is the month of the maximum extent of the fast ice around the Antarctic continent when it usually occupies an area near 550 000 km^2. The extent of the fast ice in East Antarctica in November is presented in Figure 1 [*Kozlovsky et al.*, 1977].

According to some investigators [*Kozlovsky et al.*, 1977], the maximum area that the landfast ice may occupy, taking into account the area of the continental shelf, is about 800 000 km^2. Comparing the area of the shelf with the area occupied by the fast ice one can conclude that the expansion of the fast ice is affected by the shelf depths. For example, near ice shelves limited by large water depths, such as Riiser-Larsen ice shelf (depths 200-600 m and in some places even 2000 m), and the Ross and Amery ice shelves (depths 500-800 m), the fast ice does not extend as far as it does near the West or Shackleton ice shelves (Figure 1c), where water depths are not more than 100-200 m.

Underwater relief affects the extent of the landfast ice mainly through the icebergs that drift into shallow nearshore regions, where they ground on the bottom and form ice islands that actively participate in the formation and retention of the ice cover. Usually the landfast ice consists of first-year ice. Second-year and multiyear fast ice occur predominantly in the most calm parts of embayments. For example, the total area occupied by multiyear fast ice in the Eastern Antarctic does not exceed 18 000 km^2 [*Kozlovsky et al.*, 1977].

In most regions, the landfast ice in the Antarctic forms in autumn and at the end of its growth can reach a thickness of 2 m. Considering its morphological features, one can note that usually it has a level upper surface. Hummocked parts of the fast ice are observed at its edges and also in regions with many icebergs, whose motion deforms the thin and fragile ice.

The surface of the fast ice in winter is affected by intensive snow accumulation and wind erosion. In summer, in the nearshore snowless zone, the surface of the fast ice becomes rough as a result of melting. Considering the formation of snow cover on the fast ice, three zones can be distinguished: a nearshore zone of snow accumulation; a snowless strip with a width of 0.5-2.0 km, where snow is removed by continental winds of constant direction; and, the most distant, seaward snow-covered zone where the snow blown off the continent is redeposited and precipitation from cyclones accumulates. The average snow thickness in this seaward zone ranges from 0.6 to 1.0 m, depending on location. Maximum snow accumulation is observed at a distance of 6-15 km from shore in a zone where continental winds diminish. Icebergs and growlers in the fast ice promote snow accumulation on their lee sides where drifts form. These subsequently become an additional barrier, making the process of snow accumulation more intensive especially when the wind direction changes.

One of the typical features of the fast ice surface is the presence of numerous cracks. Also, some parts have an undulating surface with so-called "compression waves" formed under pressure of the movement of glacier tongues. Depending on the mechanism of their formation, the following types of cracks can be distinguished: tidal, thermal, wave, and those formed under static or dynamic pressure. The appearance of cracks is usually related to changes of sea level, shearing of ice at the shift of glacial tongues, icebergs and growlers, intensive accumulation of snow, and waves formed during the calving or upset of icebergs.

FAST ICE STRUCTURE

The dynamic ice-ocean regime in nearshore regions, often with strong storms and intensive snowfall, is one of the main reasons for multilayer structure formation, where the layers of congelation, congelation/frazil, frazil/shuga, platelet and infiltration ice occur in different combinations [*Vtjurin*, 1959; *Fedotov*, 1970; *Dubrovin and Savatyugin*, 1970; *Cherepanov and Kozlovsky*, 1973b; *Kozlovsky and Cherepanov*, 1973; *Crocker and Wadhams*, 1989; *Jeffries and Weeks*, 1991; *Jeffries et al.*, 1993; *Veazey et al.*, 1994; *Gow et al.*, this volume]. This multiple layering can be seen in Figure 2, a schematic structural profile of the fast ice near Mirny station (Figure 1c) on the shore of the Davis Sea [*Cherepanov and Kozlovsky*, 1972].

In many places, congelation ice of type C2 [*Cherepanov*, 1974; *Cherepanov et al.*, in press] is the main component of the fast ice. This type of ice has an aggregated fibrous (columnar) structure with horizontal orientation of C-axes in crystals that can form only in calm conditions. In some regions, such as near Mirny station (Figure 1c) and Alasheyev Bay, Cosmonaut Sea (Figure 1a), oriented structures with alignments of crystals have been observed. Such alignments appear under the effects of surface water currents with near-constant direction [*Cherepanov*, 1971; *Weeks and Gow*, 1978; *Strakhov*, 1987, 1989; *Jeffries et al.*, 1993; *Veazey et al.*, 1994; *Gow et al.*, this volume] and have both scientific and practical significance because they explain the pronounced anisotropic behavior of the physical properties of sea ice over large areas.

In Antarctic fast ice, the formation of two types of ice are closely connected with blowing snow and snowfall.

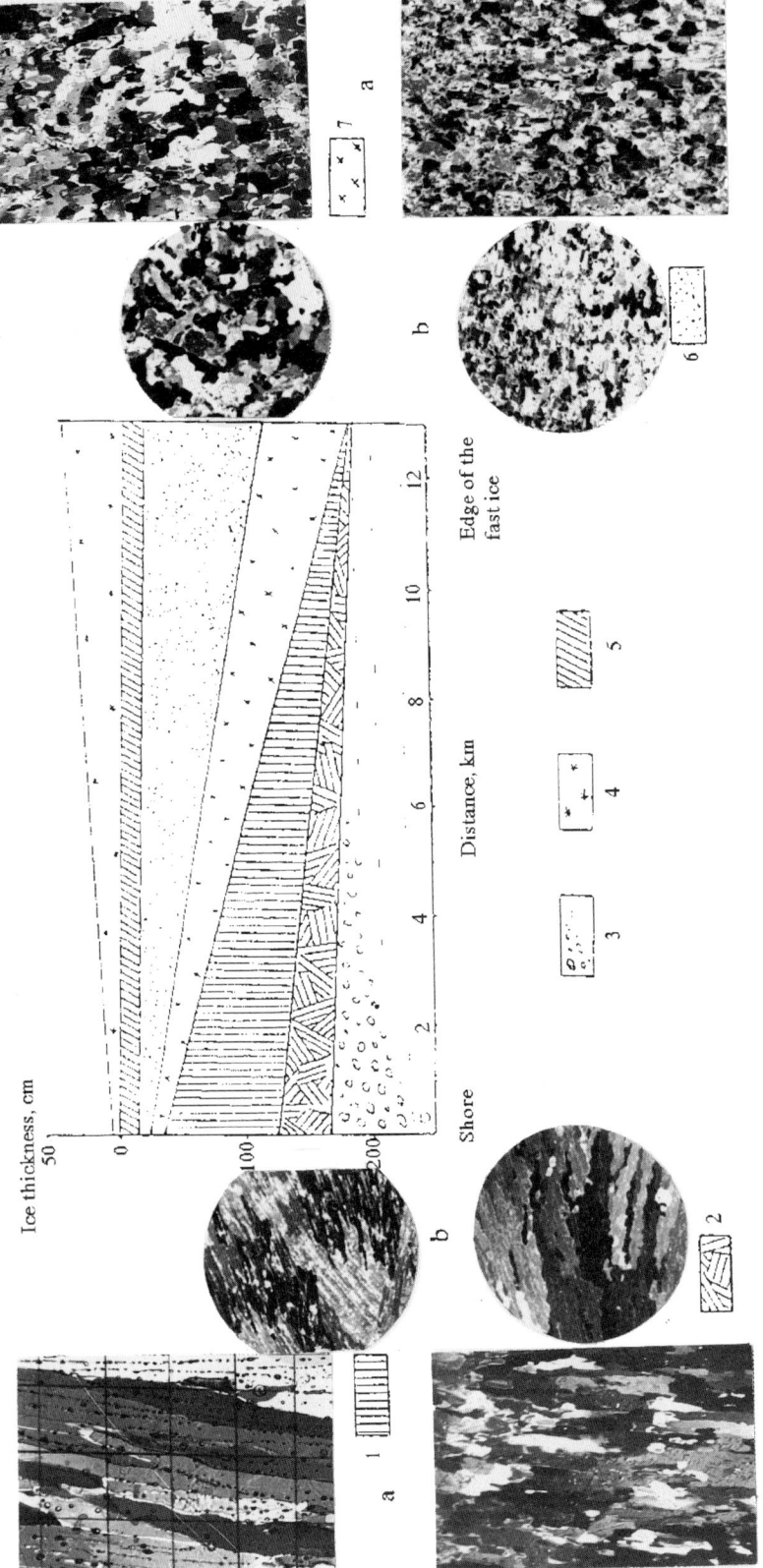

Fig. 2. Typical structural profile of the fast ice near Mirny station (Figure 1b), January 1971 [*Cherepanov and Kozlovsky*, 1972], with (a) vertical and (b) horizontal thin sections of several structural ice types [*Cherepanov*, 1974; *Cherepanov et al.*, in press]. 1: congelation ice (type C2); 2: spring congelation/frazil ice (type C4); 3: large ice plates; 4: snow; 5: recrystallized ice; 6: infiltration (snow) ice (type C9); 7: frazil and shuga layer formed in autumn-winter period (type C8). Photographs of ice types C2, C4, C8 and C9 are presented. The 10 mm grid in the upper left section of congelation ice represents the scale for all the ice sections.

Frazil/shuga type C8 [*Cherepanov*, 1974; *Cherepanov et al.*, in press] (Figure 2) forms in the autumn-winter period when snow and frazil crystals freeze together in the surface water layer. In Antarctic waters it is encountered everywhere, both in the pack ice and fast ice. As a consequence of an initial high concentration of snow and frazil crystals in sea water, the structure of this ice type is fine grained. The distribution of air and brine inclusions is irregular and also depends on the initial concentration of crystals and the velocity of sea ice growth. Small spherical bubbles and capillary pores (1-2 mm) are the most predominant inclusion to develop. Crystal sizes are in the 1-4 mm range and exhibit isometric shapes. Often there is some development of hypidiomorphic crystal shapes. The C-axis orientation is random and the physical properties are isotropic.

The second type of ice that is related to snow, type C9 (Figure 2), forms by the infiltration of water into the lower layers of the snow cover. This can occur as a result of the isostatic depression of the ice surface under the weight of the snow and the associated formation of cracks, which facilitate the transfer of sea water to the upper ice surface. Snow ice differs significantly in its structure, composition and physical properties from congelation and congelation/frazil ice, and even from frazil/shuga, in that it has a higher porosity, smaller grain size and a random orientation of the C-axes (Figure 3, upper right fabric diagram).

Like type C8, the C9 infiltration ice (snow ice) is also widespread in the fast ice of Antarctic, especially in the zones of significant snow accumulation [*Barkov and Ivanov*, 1967]. A schematic structural section of the fast ice in the Cosmonaut Sea is presented in Figure 3. Here, the upper layer of the ice cover consists of ice type C9. Infiltration ice in Antarctic waters is often encountered in first-year fast ice and always constitutes the upper layer of the multiyear fast ice [*Fedotov*, 1976; *Kozlovsky*, 1976] (Figure 4). The congelation ice fabric diagrams in Figure 3 also illustrate the initial girdle fabric (center right diagram) of randomly oriented crystals with horizontal C-axes, which can subsequently become strongly aligned (Figure 3, lower right diagram) due to the water current effects on crystal growth and orientation at the bottom of the ice [*Cherepanov*, 1971; *Weeks and Gow*, 1978].

In the autumn-winter period, during highly dynamic conditions, the crystals of frazil ice and shuga formed in polynyas and flaw leads are transported for large distances under the fast ice by surface currents. In this case, the concentration of crystals is not high and their sizes are larger than those of ice types C8 and C9. All these factors result in the formation of medium-grained and non-uniform crystal structure of type C7 (Figure 4) [*Cherepanov and Kozlovsky*, 1972; 1973a; *Cherepanov*, 1974; *Cherepanov et al.*, in press; *Kozlovsky and Cherepanov*, 1977]. The thickness of such ice layers depends on the environmental conditions, including the duration of dynamic conditions, and the distance from the polynyas and flaw leads. The larger this distance, the smaller the thickness of the C7 ice layer (Figure 2). In many cases, when the number of granular isometric crystals is small (for instance, in regions remote from the source of such crystals) an ice cover primarily composed of congelation/frazil ice of type C4 will prevail (Figure 2). This ice type usually contains two types of typical crystals: fibres (columns), as in C2 ice, and isometric crystals with diameters of 3-10 mm that are incorporated into the C2 ice and are located randomly or, in same cases, form agglomerations. These are located between the columnar crystals and occupy less than 25% of the cross sectional area. The C-axis orientation of the fibrous crystals is horizontal and that of the incorporated crystals vertical or close to vertical.

Small crystals of frazil ice form also in spring, and even in summer, in connection with the inflow of melting fresh water from ice shelves. This results in the formation of a spring congelation/frazil ice layer, with its maximum thickness near the shore, and then a gradual decrease in thickness towards the fast ice edge (Figure 2). In the period of intensive ice melting in summer, large platelet crystals, with diameters of 10-12 cm also appear under the fast ice (Figure 2). Their origin is likely related to the inflow of fresh water melted from the lower layers of the tongues of ice shelves [*Cherepanov and Kozlovsky*, 1972]. Western investigators suggest that this mechanism is related to the flow of low-density sea water masses from beneath ice shelves [*Dieckmann et al.*, 1986; *Jeffries et al.*, 1993; *Gow et al.*, this volume]. In both cases, fresh meltwater or low-density sea water rising from depths of several hundred meters is constantly in the supercooled state [*Foldvik and Kvinge*, 1974]. Taking into account the growth velocity of ice crystals at small values of supercooling, both mechanisms of formation seem probable.

THERMOMETAMORPHIC PROCESSES

In spring, when the ice thickness reaches its maximum value, clearly expressed changes in its crystal structure and physical and mechanical properties occur under the influence of thermometamorphic processes. These processes result in surface and internal melting of the ice, recrystallization, increases in brine discharge and ice porosity, and also in the development of a destruction layer and a layer of secondary recrystallized ice (Figure 4) [*Nazintzev and Fedotov*, 1984]. By the term "destruction layer" we mean an ice layer with noticeably different struc-

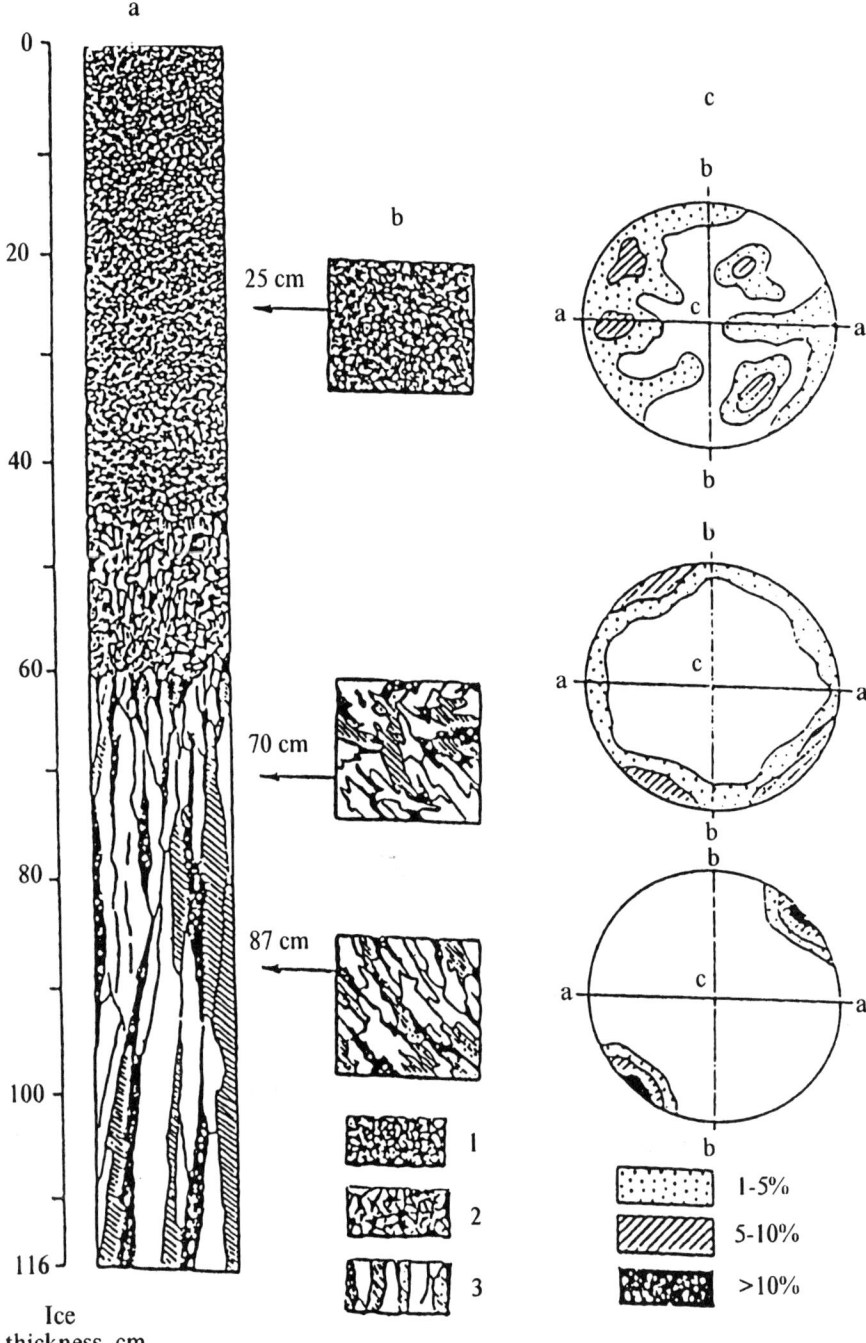

Fig. 3. Sketches of (a) vertical structure and (b) horizontal structure observed in thin sections, and (c) fabric diagrams of first-year fast ice in the Alasheyev Bay, Cosmonaut Sea (Figure 1a), December, 1976 [*Kozlovsky et al.*, 1977]. 1: snow ice/infiltration ice of type C9; 2: frazil ice of type C7; 3: congelation ice of type C2.

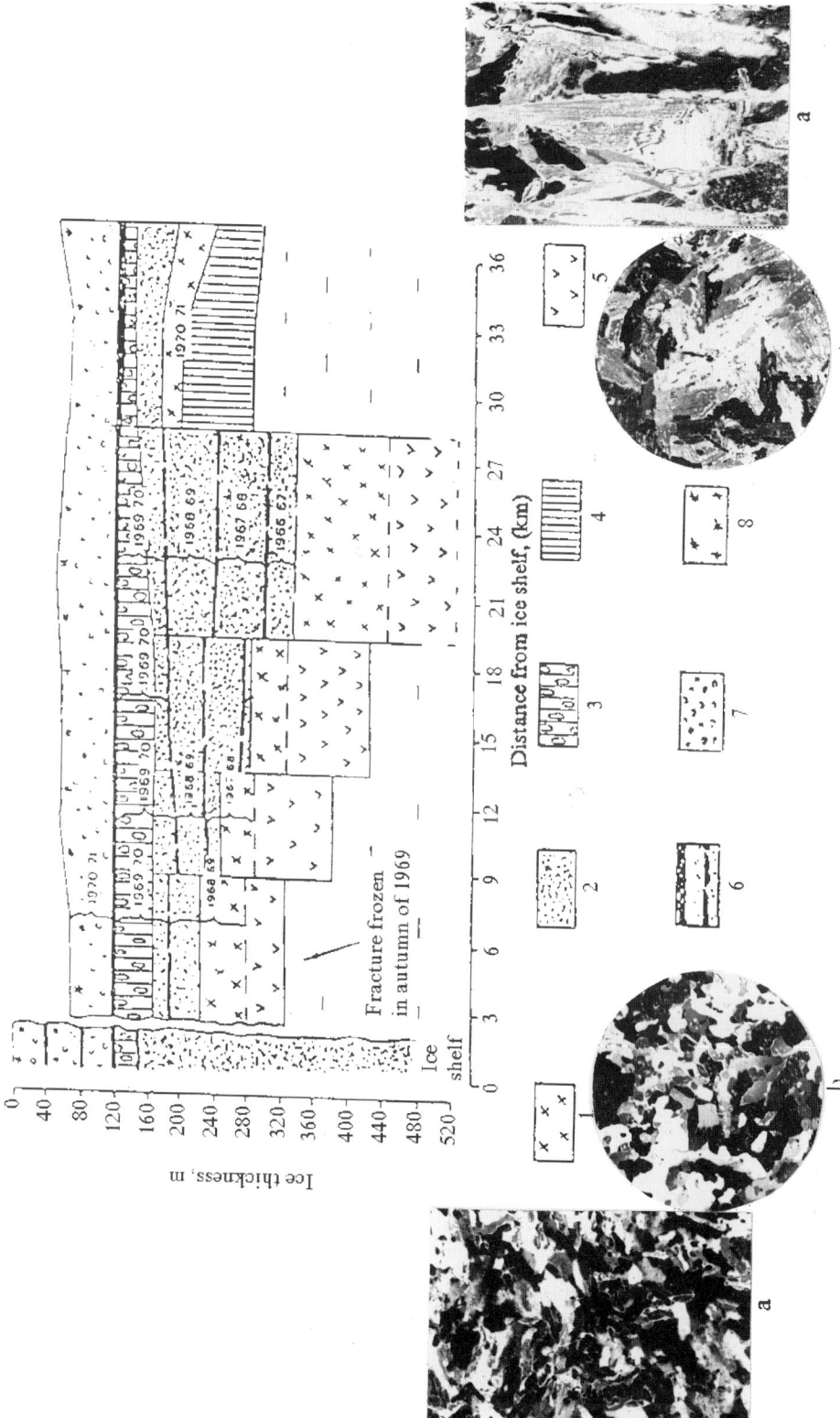

Fig. 4. Structural profile of multiyear fast ice near Leningradskya station (Figure 1d) in February 1971 [*Kozlovsky and Cherepanov*, 1972] with (a) vertical and (b) horizontal thin sections of several structural ice types [*Cherepanov*, 1974; *Cherepanov et al.*, in press]. 1: frazil ice (symbol and photos of type C9); 2: snow ice/infiltration ice (type C7); 3: destruction layer (water, poorly connected ice crystals); 4: congelation ice (type C2); 5: platelet ice (symbol and photos of type C6); 6: layers of recrystallized ice; 7: firn; 8: snow. The thin section photographs have the same scale as those in Figure 2.

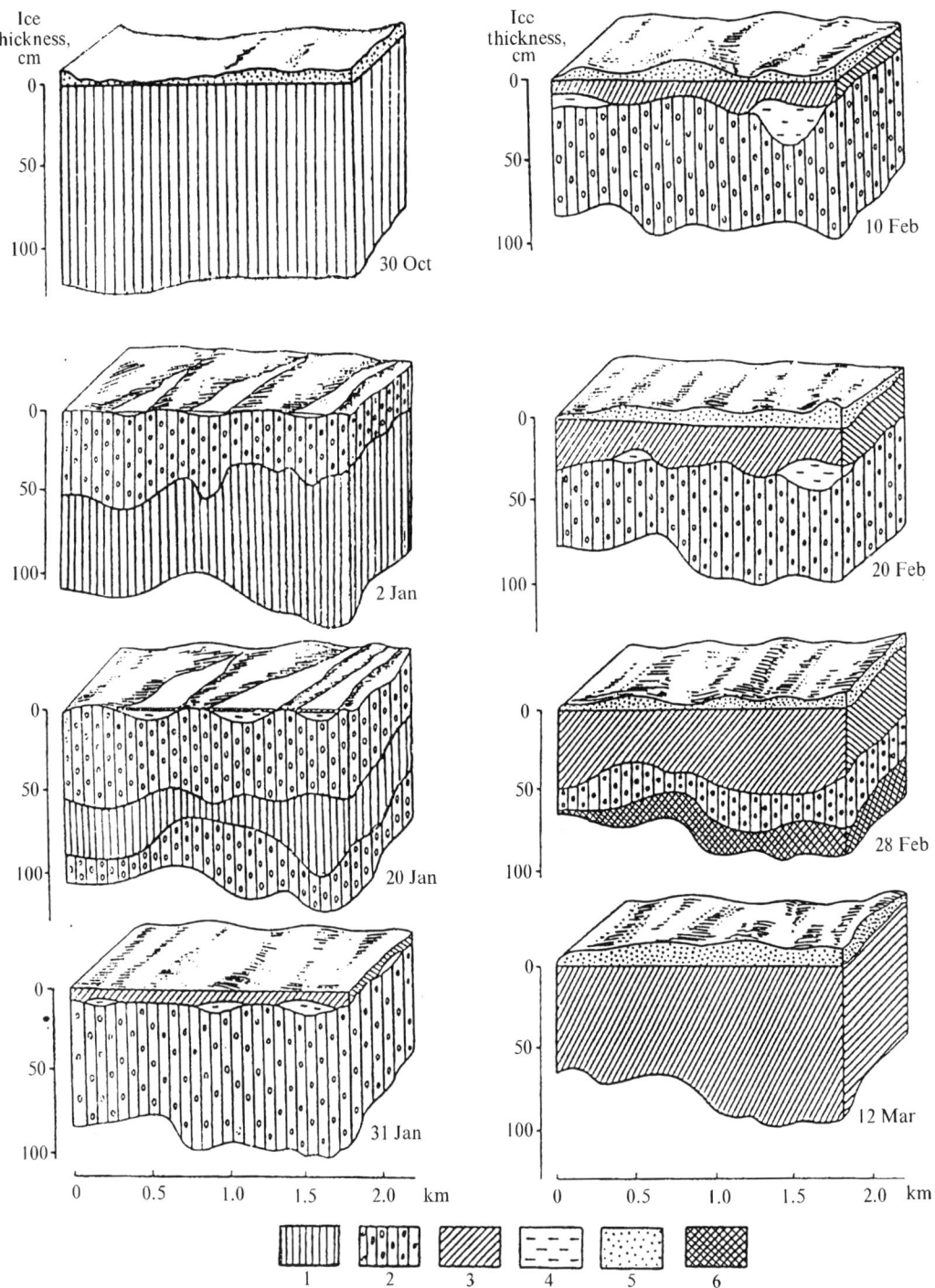

Fig. 5. Development of a seasonally layered structure in the ice cover of the Antarctic coastal zone as observed along a 2 km profile orientated from the shore to the edge of the fast ice in Alasheyev Bay, Cosmonaut Sea (Figure 1a), in the period October 1975 to March 1976 [*Kornilov et al.*, 1987]. 1: ice with original winter structure; 2: layer of destruction ice; 3: recrystallized ice; 4: water layer; 5: layer of newly fallen snow; 6: layer of sea water infiltration.

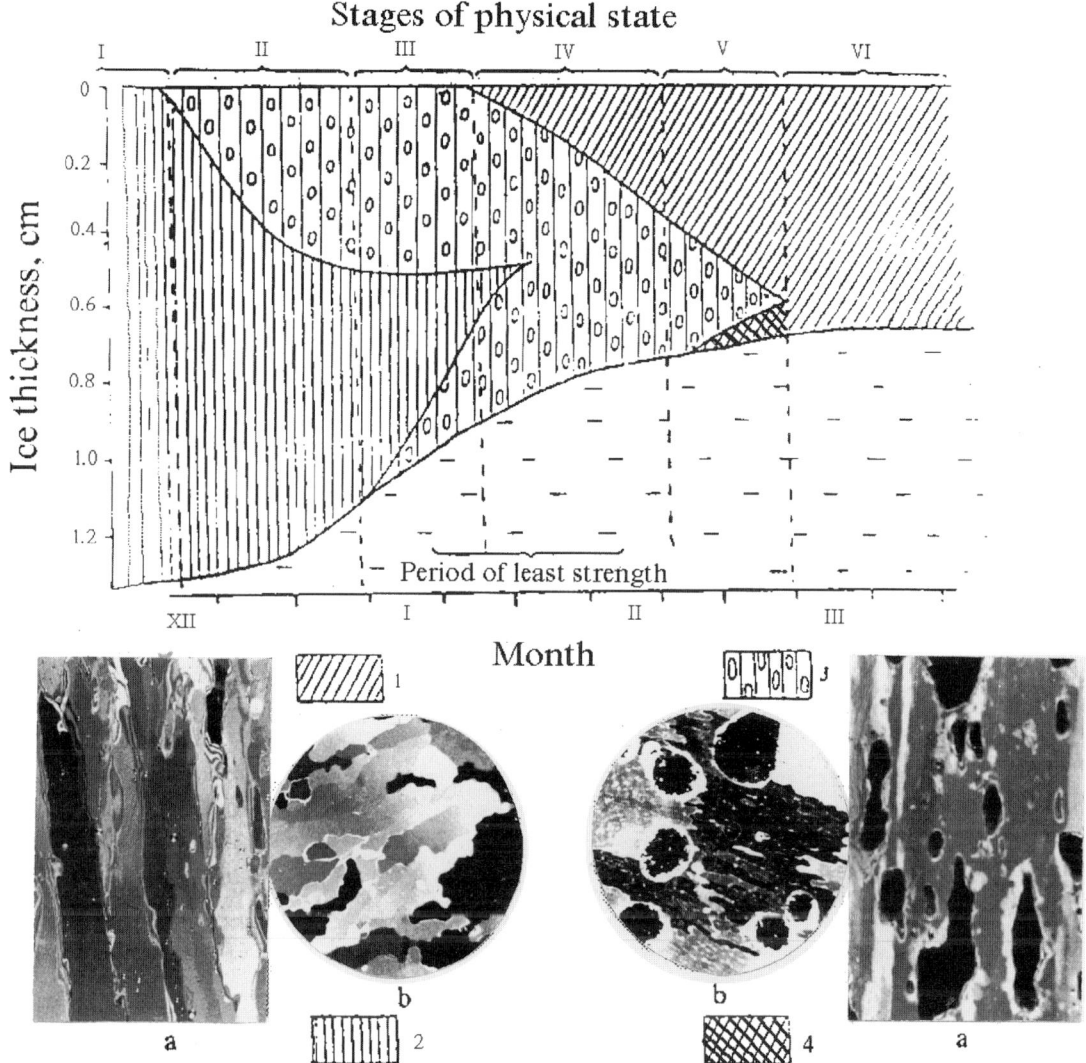

Fig. 6. Development of a seasonally layered structure in fast ice at Alasheyev Bay, Cosmonaut Sea (Figure 1a) in combination with the stages of its physical state, [*Nazintzev and Fedotov*, 1984], plus (a) vertical and (b) horizontal thin sections of the metamorphic ice types. 1: recrystallized ice; 2: ice with original winter structure; 3: layer of destruction ice; 4: layer of sea water infiltration. See Table 1 for a description of the stages of physical state.

tural and textural characteristics compared to those of the original winter ice.

The changes listed above are due to the effects of short-wave radiation and internal melting. The character of these processes is determined by climatic conditions and the radiation regime during the spring-summer period [*Fedotov*, 1971a, 1971b]. The intensity of the changes also depends on the thickness of the snow cover, the relief of the ice surface and the type of crystal structure in the upper layer of the ice sheet. For instance, during radiative heating, a thick snow cover and a fine-grained upper ice layer (ice types C8 and C9) absorb the larger part of the incoming short-wave radiation and, as a result, protect the lower ice layers from heating, internal melting and rapid loss of strength. The greater strength of such ice is confirmed by the results of annual investigations of the fast ice where optimal ice routes for heavy transport and sites for unloading vessels are selected [*Kozlovsky et al.*, 1977]. Mineral and organic inclusions in ice also can play an important role in its spring transformation. For example, due to the presence of brown diatomic alga in the ice cover, the absorption of short-wave radiation increases and heat-

Table 1. Main stages in the physical state of first-year ice located in near-shore regions of the East Antarctic during the spring-summer-autumn period (prevailing types C2, C4, and C6).

Stages	Description	Inclusions	Average total ice salinity, ‰	Flexural strength of ice in % of its winter value
I	Destruction of the upper ice layer by radiation. Slight melting of the upper ice surface.	Development of brine drainage channels in the upper ice layer.	3.5	70-80
II	Intensive surface melting. Development of puddles and a surface destruction layer.	Integration of drainage channels and appearance of vertical pores through the complete ice thickness.	2.0	40-50
III	Intensive melting of both the upper and lower ice surfaces with the associated development of a destruction layer.	Further development of vertical pores with diameters of 5-7 mm.	1.0	20-30
IV	Beginning of the development of a recrystallized layer near the ice surface with simultaneous intensive development of a destruction layer under the influence of "hotbed" effects and melting of lower ice surface. Layers of very loose ice crystals impregnated with water are observed within ice sheet.	The vertical pores become larger and integrate as a result of eroding their walls.	0.3	5-10
V	Development of recrystallized ice in the upper layer of the ice cover. Intensive infiltration of sea water into the lower ice layer. Melting and decreases in ice thickness cease.	Secondary inclusions develop in the recrystallized ice.	0.3-2.0	10-30

ing of the ice and its destruction noticeably accelerate [*Buynitskiy*, 1965; *Zeebe et al.*, 1996].

The intensity of the surface melting and of internal metamorphic transformations in the ice cover also depends upon the effects of diurnal fluctuations in air temperature. During periods of daylight, solar radiation is absorbed by the entire ice thickness, while at night, longwave radiation only is emitted from the upper ice surface. The quantitative difference between these values result in a "hotbed" effect that intensively heats the internal ice layer where the destruction layer develops (Figure 4). Meanwhile, at the upper ice surface, where maximum diurnal temperature fluctuations occur, recrystallized ice forms as fresh meltwater fills and freezes in the voids of the rotten ice. Thus, in autumn, in the fast ice of the Antarctic, a recrystallization layer that strengthens the ice cover usually forms simultaneously with a destruction layer that weakens it. Characteristically, these seasonally layered structures develop rapidly [*Kozlovsky et al.*, 1977]. Figure 5 illustrates the seasonal transformations of structure and decrease of the thickness of the fast ice in Alasheyev Bay in the Cosmonaut Sea (Figure 1a).

The role of recrystallized ice also is very important for the distinction of the annual ice layers in the vertical structure of multiyear fast ice. Forming beneath the infiltration ice, recrystallized ice layers indicate the boundaries of annual layers and allow the determination of the age of the multiyear fast ice (Figure 4).

The successive character of the thermometamorphic transformations of the snow and ice cover in the spring-summer period, leading to the development of summer seasonal layers and to considerable variations in the physical properties of the ice, allows one to distinguish different stages in the physical state of an ice cover (Table 1).

Under the term "stages of physical state" we refer to the main successive temporal changes of the structure and the physical properties of the ice under the effect of the thermometamorphic processes. These, in turn, depend on hydrometeorological conditions during the melt period and the period during the autumn when the ice cover gains strength. The development of a seasonal layered structure in the ice cover, in combination with the distinguishing features of the different stages, is shown in Figure 6. Here it is necessary to mention that we considered the development of a seasonal layered structure and corresponding stages of physical state only for bare fast ice in nearshore regions. In the zones of snow accumulation, and also in the regions occupied by multiyear fast ice, the spring-summer metamorphic transformations may have a different character.

SUMMARY AND CONCLUSION

In conclusion, we would like to note that currently the regional special features of the Antarctic ice cover and the effects of thermometamorphic transformations on ice characteristics in the spring-summer period are not studied sufficiently. These data are necessary for different purposes. For example, to correctly distinguish the different stages in the physical state of the ice, a good knowledge not only of all the temporal transformations of ice structure and physical properties is required but also the regional peculiarities of each ocean area must be taken into account. Such temporal divisions of the physical state of the ice may have practical significance because they allow one to know more precisely the structure and physical properties of the ice at any time during the period when the ice strength decreases in the spring, during the summer destruction of the ice, and during the autumn when the ice strengthens. This is very important for planning and implementation of spring-summer transport on the fast ice and unloading of vessels at the ice edge. Such information may also be very useful in preparing maps of the ice cover of the polar seas based on prevailing ice types, crystal structures and physical and mechanical properties.

Acknowledgements. We thank M. O. Jeffries, I. Allison, and two anonymous reviewers for useful, constructive comments.

REFERENCES

Barkov, N. I., and V. B. Ivanov, Snow accumulation on the ice slope and sea ice in the area of Mirny in 1960 (in Russian), *Sov. Antarct. Exped. Inf. Bull.*, *63*, 17-21, 1967

Buynitskiy, V. Kh., The influence of diatoms upon the structure and strength of sea ice (in Russian), *Proc. Sov. Antarct. Exped.*, *44*, 83-88, 1965.

Buynitskiy, V. Kh., Sea ice and icebergs in the Antarctic (in Russian), *Izdatelstvo LGU*, 1-254, 1973.

Cherepanov, N. V., and F. D. Sokolov, New ice ring drill (in Russian), *Probl. Arct. Antarct.*, *32*, 122-125, 1969.

Cherepanov, N. V., Spatial arrangement of sea ice crystal structure (in Russian), *Probl. Arct. Antarct.*, *38*, 137-140, 1971.

Cherepanov, N. V., and A. M. Kozlovsky, Underwater ice in the coastal waters of Antarctica (in Russian), *Sov. Antarct. Exped. Inf. Bull.*, *84*, 61-65, 1972.

Cherepanov, N. V., and A. M. Kozlovsky, Autumn formation of underwater ice in the Lazarev Ice Shelf area (in Russian), *Sov. Antarct. Exped. Inf. Bull.*, *86*, 38-44, 1973a.

Cherepanov, N. V., and A. M. Kozlovsky, Classification of Antarctic sea ice by the conditions of its formation (in Russian), *Probl. Arct. Antarct.*, *42*, 61-73, 1973b.

Cherepanov, N. V., Classification of ice of natural water bodies, in *IEEE Ocean'74*, 197-201, 1974.

Cherepanov, N. V., V. I. Fedotov, and K. P. Tyshko, The Structure of Sea Ice, in *Sea Ice Structure and Properties*, Handbook on the Analysis and Forecasting of Sea Ice, edited by W. F. Weeks and Yu. P. Doronin, (Section 1, Committee for Marine Meteorology, World Meteorological Organization), in press.

Crocker, G. B., and P. Wadhams, Modelling Antarctic fast-ice growth, *J. Glaciol.*, *35*(119), 3-8, 1989.

Diekmann, G. S., G. Rohardt, H. Hellmer, and J. Kipfstuhl, The occurrence of ice platelets at 250m depth near the Filchner Ice Shelf and its significance for sea ice biology, *Deep Sea Res., Part A*, *33*(2), 141-148, 1986.

Dubrovin, L. I., and L. M. Savatyugin, On the fast ice dynamics in the Mirny region (in Russian), *Proc. Sov. Antarct. Exped.*, *53*, 208-215, 1970.

Fedotov, V. I., Some features of the ice regime and physical properties of sea ice in Alasheyev Bay (February-December, 1966) (in Russian), *Proc. Sov. Antarct. Exped.*, *53*, 216-228, 1970.

Fedotov, V. I., Investigations of Antarctic landfast ice (in Russian), *Proc. Arct. Antarct. Res. Inst.*, *300*, 110-120, 1971a.

Fedotov, V. I., Destruction of Antarctic landfast ice under the effect of short-wave radiation (in Russian), *Proc. Arct. Antarct. Res. Inst.*, *300*, 128-136, 1971b.

Fedotov, V. I., Comparison of landfast ice in Arctic and Antarctic (in Russian), *Proc. Arct. Antarct. Res. Inst.*, *331*, 172-178, 1976.

Foldvik, A., and T. Kvinge, Conditional instability of sea water at the freezing point, *Deep-Sea Res.*, *21*(3), 169-174, 1974.

Gordienko, P. A., V. I. Fedotov, and V. I. Shilnikov, Ice cover in nearshore regions of East Antarctica (in Russian), *Proc. Sov. Antarct. Exped.*, *11*, 1-117, 1960.

Gow, A. J., W. F. Weeks, S. F. Ackley, and J. W. Govoni, Physical and structural properties of land-fast sea ice in McMurdo Sound, Antarctica, *Ant. Res. Ser.*, this volume.

Jeffries, M. O., and W. F. Weeks, Fast-ice properties and structure in McMurdo Sound, *Antarct. J. of the U.S.*, *26*(5), 94-95, 1991.

Jeffries, M. O., W. F. Weeks, R. Shaw, and K. Morris, Structural characteristics of congelation and platelet ice and their role in the development of Antarctic land-fast sea ice, *J. Glaciol.*, *39*(132), 223-238, 1993.

Kirillov, V. A., and A. I. Paramonov, Identification of sea ice types and native rocks in Antarctica using IR-images, *Proc. Arct. Antarct. Res. Inst.*, *421*, 105-115, 1991.

Kornilov, N. A., Ye. B. Leont'ev, and V. I. Fedotov, Ice conditions, physical state of ice cover and some characteristics of fast ice in the Alasheev Bay (in Russian), *Sov. Antarct. Exped. Inf. Bull.*, *109*, 18-28, 1987.

Kozlovsky, A. M., Some peculiarities and causes of multi-year sea ice formation in the Cosmonaut Sea (in Russian), *Sov. Antarct. Exped. Inf. Bull.*, *92*, 40-46, 1976.

Kozlovsky, A. M., and N. V. Cherepanov, Ice studies (in Russian), *Proc. Soviet Antarct. Exped.*, *61*, 92-138, 1973.

Kozlovsky, A. M., and N. V. Cherepanov, The effect of coastal morphological features on the formation of underwater ice in Antarctic seas, *CRREL Draft Translation 654*, 1-7, 1977.

Kozlovsky, A. M., Yu. L. Nazintsev, V. I. Fedotov, and N. V. Cherepanov, Fast ice of the Eastern Antarctic (in Russian), *Proc. Sov. Antarct. Exped.*, *63*, 1-129, 1977.

Nazintzev, Yu. L., and V. I. Fedotov, Thermal metamorphism of sea ice in the Antarctic (in Russian), *Sov. Antarct. Exped. Inf. Bull.*, *105*, 23-29, 1984.

Rey, J. K., and M. J. McCormick-Rey, *The Fauna of Polar Regions* (in Russian), Leningrad, Gidrometeoizdat, 1-247, 1988.

Strakhov, M. V., The effects of surface currents on alignments of ice crystal structure near polar station Mirny (in Russian), *Proc. Sov. Antarct. Exped.*, *82*, 119-124, 1987.

Strakhov, M. V., The alignments of ice crystal structure in Antarctic seas (in Russian), *Sov. Antarct. Exped. Inf. Bull.*, *111*, 67-70, 1989.

Veazey, A. L., M. O. Jeffries, and K. Morris, Small-scale variability of physical properties and structural characteristics of Antarctic fast ice, *Ann. Glaciol.*, *20*, 61-66, 1994.

Vtjurin, B. I., Structure of the first-year fast ice in East Antarctica, *Sov. Antarct. Exped. Inf. Bull.*, *4*, 55-60, 1959.

Weeks, W. F., and A. J. Gow, Preferred crystal orientation in the fast ice along the margins of the Arctic Ocean, *J. Geophys. Res.*, *83*(C10), 5105-5121, 1978.

Zeebe, R. E., H. Eicken, D. H. Robinson, D. Wolf–Glashow and G. S. Dieckmann, Modeling the heating and melting of sea ice through light absorption by microalgae, *J. Geophys. Res.*, *101*(C1), 1163-1181, 1996.

Zwally, H. J., J. C. Comiso, C. L. Parkinson, W. J. Campbell, F. D. Carsey, and P. Gloersen, *Antarctic Sea Ice, 1973-1976: Satellite Passive-Microwave Observations*, Washington, DC, National Aeronautics and Space Administration. (NASA SP-459), 1983.

N. V. Cherepanov, V. I. Fedotov, and K. P. Tyshko, State Scientific Center of the Russian Federation - Arctic and Antarctic Research Institute, 38 Bering Street, St. Peterburg, 199397, Russian Federation.

(Received July 8, 1996;
accepted March 27, 1997)

PHYSICAL AND STRUCTURAL PROPERTIES OF LAND-FAST SEA ICE IN MCMURDO SOUND, ANTARCTICA

A.J. Gow, S.F. Ackley and J.W. Govoni

U.S. Army Cold Regions Research and Engineering Laboratory (CRREL), Hanover, New Hampshire

W.F. Weeks

Geophysical Institute, University of Alaska, Fairbanks, Alaska

The physical properties of land-fast sea ice in McMurdo Sound were investigated in cores drilled to the bottom of the ice at 27 widely separated sites. Three major ice types were identified, including an upper transition layer, representing 15% of the total ice thickness, that consisted mainly of ice formed during the earliest stages of growth of congelation ice. Most of the underlying ice consisted of columnar congelation ice exhibiting aligned c-axes horizontal fabrics which transitioned into platelet ice forming the base of the ice sheet. Columnar-congelation ice comprised on average, 72% (1.45 m) of the total ice sheet thickness or 83% (1.68 m) if transition congelation ice is included. Platelet ice averaged 13% or about 0.26 m of ice. Bulk salinities averaged 6 $^o/_{oo}$. C-axes alignment directions corresponded generally with known winter water circulation patterns in McMurdo Sound. However, currents measured beneath the ice during growth of the platelet ice phase often deviated from those inferred from c-axes alignments in columnar-congelation ice formed weeks or months earlier. These observations imply near-surface current circulation changes, possibly related to the onset of growth of the sub-ice platelet layer. The date of its initial appearance can vary from year to year and its thickness can vary appreciably with respect to location and from year to year. Platelet ice appears to form by direct attachment to the bottom of the ice sheet and its morphological characteristics are consistent with formation from adiabatically supercooled water originating from beneath the Ross Ice Shelf.

INTRODUCTION

An examination of the physical and structural properties of land-fast sea ice in McMurdo Sound was conducted in October and November 1980. This work, performed in conjunction with studies of pack ice in the western Weddell Sea, constituted the final phase of a 3-year field program devoted to investigations of the dynamics, thermodynamics and structural properties of Antarctic sea ice. The overall scope of these investigations, including some results, is reported in Ackley [1979, 1981], Ackley et al. [1980] and Gow et al. [1981]. A detailed report of the western Weddell Sea pack ice studies was published by Gow et al. [1987] in which the major finding was the discovery of frazil ice in amounts not previously observed in Antarctic sea ice or in sea ice of comparable age and thickness in the Arctic. Frazil ice is derived from small ice crystals that are nucleated within the water column, as opposed to congelation ice, which forms by freezing of seawater directly onto the bottom of an existing sea-ice sheet. Whereas congelation ice dominates the structure of Arctic sea ice [Weeks and Ackley, 1982; Gow and Tucker, 1991], frazil ice was found by *Gow et al.* [1987] to be the dominant component of sea-ice structure in the western Weddell Sea, averaging 72% of the thickness of 13 multiyear floes and 37% of the 49 first-year floes examined. These findings were subsequently confirmed by *Lange et al.* [1989] for the central and eastern Weddell Sea. This preponderance of frazil ice clearly reflected the frequency and importance of a turbulent water column in the production of ice in the Weddell Sea.

Large plate crystals, several centimeters long, were also observed forming at the bottom of the thickest floes in the Weddell Sea. According to *Gow et al.* [1987], these plate crystals appear to be structurally similar to the sub-ice platelet layer crystals that form at the base of sea-ice sheets at coastal locations in Antarctica. Two cases in point are McMurdo Sound, where the formation of a sub-ice platelet layer was first documented by *Paige* [1966], and the coast near Mirny, East Antarctica, where *Serikov* [1963] observed plate crystals, which he called "underwater ice," accreting to the bottom of the sea ice sheet. However, Paige's observations were limited to just a few sites in the immediate vicinity of McMurdo Station (mainly in regard to sea ice runways and freight hauling routes) that are not necessarily representative of the relatively large expanse of land-fast ice in McMurdo Sound.

A major objective of our 1980 project was to examine the overall structural and physical properties of the land-fast sea ice in McMurdo Sound and, in particular, to determine the thickness variations and extent of platelet ice growth. Preliminary results of observations made in October and November 1980 are reported in *Gow et al.* [1981, 1982]. The purpose of this paper is to present a more detailed account of these initial observations. These studies included salinity profile measurements and characterization of the stratigraphic and compositional properties, in addition to documentation of the texture and c-axis fabrics.

FIELD SETTING AND SITE LOCATIONS

McMurdo Sound is a large embayment occupying the part of the Ross Sea that is bounded by Ross Island to the east, the coast of Victoria Land to the west and the McMurdo Ice Shelf to the south. *Paige* [1966] has characterized the climate of McMurdo Sound as one with "low temperature, extreme temperature fluctuations, frequent high winds and drifting snow." The lowest temperatures occur in July and August, sometimes falling below –50°C; the highest temperatures occur during December and January and occasionally reach +5°C. Factors controlling the extent, thickness and duration of the annual sea-ice cover in McMurdo Sound include the temperature regime, frequency and severity of storms, and oceanic circulation. Typically, ice growth begins in late March or April and continues through to December, by which time ice thicknesses approach 2 m or more, with the thickest ice occurring on the western side of the sound, where it may remain bay-fast for several years (e.g., 1973-1979; *Leventer and Dunbar*, 1988). On the Ross Island (eastern) side of the sound seasonal breakouts generally take place in late January, February or March [*Heine*, 1963; *Paige*; 1971; *Kovacs et al.* 1982]. These breakouts often involve calving-off of parts of the McMurdo Ice Shelf bordering the southernmost extent of sea ice in McMurdo Sound.

Thinning of the ice, occurring during December and January, can be attributed primarily to bottom melting by warm ocean currents. Only minimal melting occurs at the top surface of the ice sheet. The position of the ice front, as observed towards the end of the seasonal ice growth, can vary considerably from year to year as indicated in Figure 1, showing the northern limits of land-fast ice (10/10 coverage) for 1 November 1977, 19 October 1979 and 30 October 1980, respectively.

The locations of ice sampled in October and November 1980 are indicated in Figure 2. An enlargement of the Cape Armitage area, located near site 1 and containing sites 1 to 5, is shown in Figure 3. The geographic locations of all sites were established from multiple compass bearings to prominent terrain features in McMurdo Sound. An additional sampling site (28) was located outside of McMurdo Sound in Lewis Bay near the northern shore of Ross Island. This, and sites in the immediate vicinity of McMurdo Station (1 through 5) and sites 23, 24, 25, and 26, were visited independently of an over-ice traverse that began at site 6 and ended at site 21. It was during this traverse that a major breakout of the Koettlitz Ice Tongue on the western shore of McMurdo Sound was observed [*Gow and Govoni*, 1994]. This break-out likely occurred at the same time that 7-year old sea ice north of the Koettlitz Ice Tongue disintegrated during early 1980 [*Leventer and Dunbar*, 1988]. This same break-out probably coincided also with the loss of large sections of the McMurdo Ice Shelf between 6 February and 22 March 1980 [*Kovacs et al.*, 1982]. It would appear from an examination of old records [*Heine*, 1963; *Paige*, 1971], that this particular break-out was one of the largest ever recorded in McMurdo Sound.

SAMPLING

Sampling was conducted during October and November 1980 and was restricted to land-fast ice that had formed since early April, following one of the largest break-outs observed during the past 40 years. This breakout eliminated virtually all existing sea ice in McMurdo Sound, including several generations of multiyear ice in the western part of the sound. Only at one location (site 18) near the front of the Koettlitz Ice Tongue was a multiyear floe, embedded in first-year ice, observed and sampled. Sampling was accomplished by sawing blocks with a chainsaw and by coring. The general procedure at each site was to

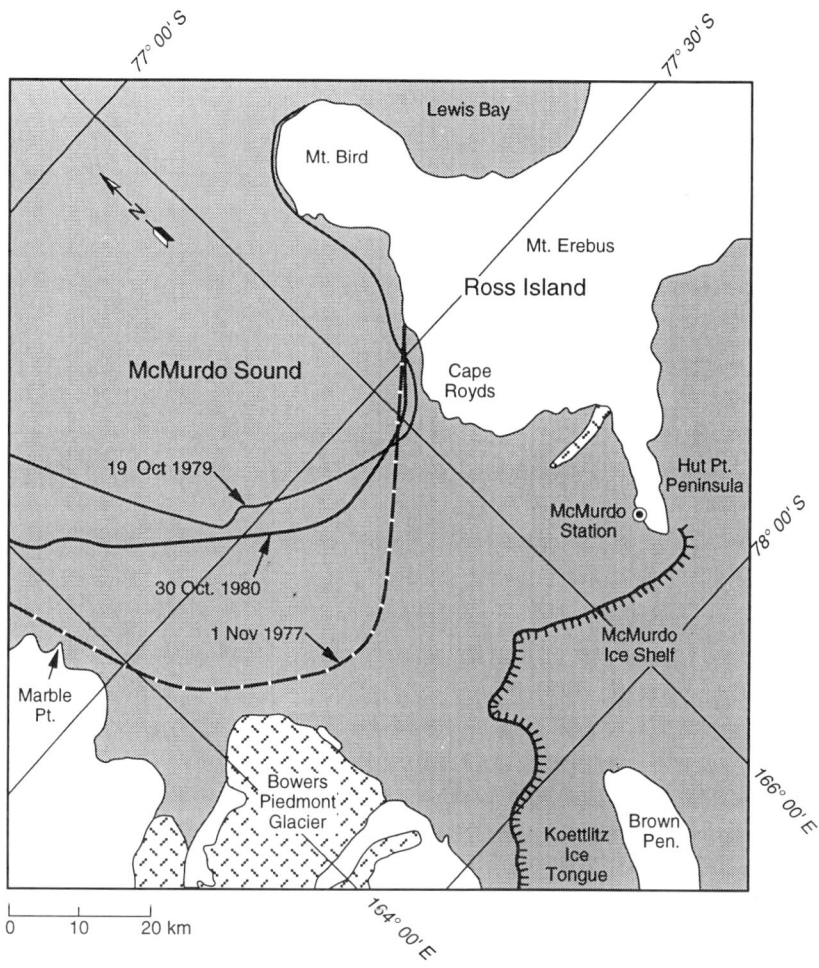

Fig. 1. Northern limits of land-fast sea ice in McMurdo Sound as observed on 1 November 1977, 19 October 1979 and 30 October 1980. Map data obtained through courtesy of U.S. Navy/NOAA National Ice Center, Suitland, Maryland, USA.

place a compass on the sea ice surface (previously cleared of snow), align the direction of magnetic north with a meter stick and mark the position with an ice chisel. This was followed by chainsawing a block of ice 0.6 m on a side and 0.6 m deep. The chisel and a pair of tongs were then used to pry loose the azimuthally oriented block and extract it. This block was used mainly to determine gross features of the structure, via an examination of vertical thick sections that were 10–20 mm thick, and to determine, by visual inspection, any alignment of the crystallographic c-axes at the base of the block. Next, a 78-mm-diameter CRREL coring auger was used to drill an azimuthally oriented structure core through the full thickness of sea ice. The procedure for orienting structure cores was the same as that used to orient ice blocks. Cores were usually retrieved in two or three pieces. It was a relatively easy matter to match core breaks, thereby ensuring that azimuthal orientation was maintained along the entire length of a core. At only one site was it necessary to drill a second structure core. A separate core was then drilled and processed on site for salinity measurements. A total of 28 sites was sampled, including a first-year site (28) outside McMurdo Sound, one site (18) located on multiyear ice and site 27 located in glacial ice on the McMurdo Ice Shelf.

Some processing of samples was performed in facilities at McMurdo Station. This included the preparation and provisional examination of thick and thin sections for structure, and the melting and measurement of salinity samples. However, much of the structural analysis was performed on samples returned to CRREL.

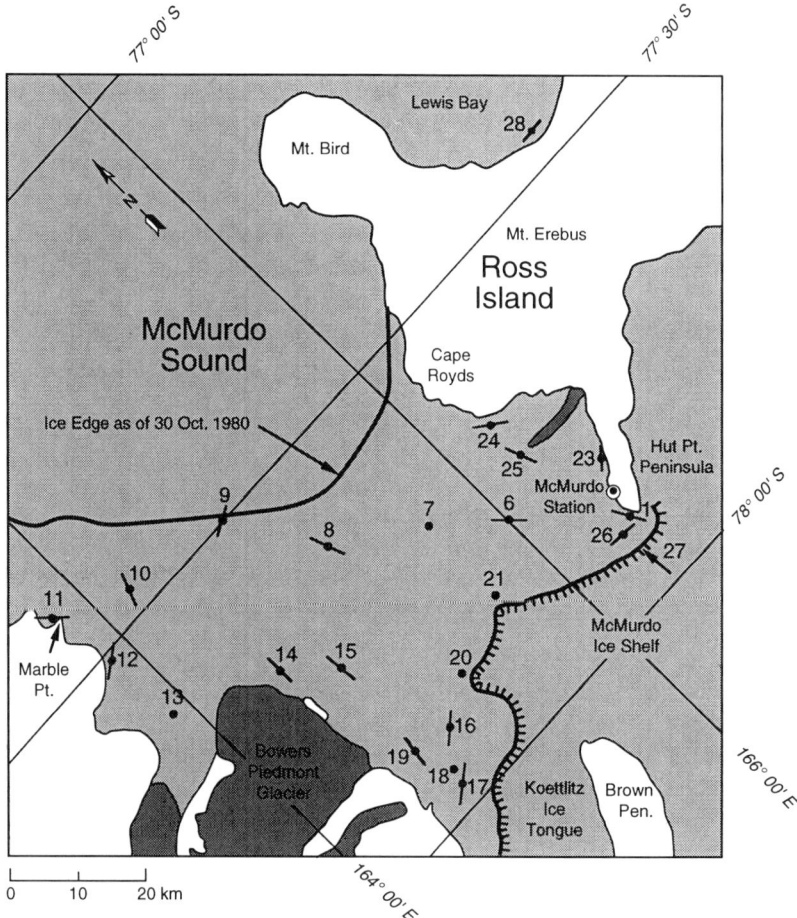

Fig. 2. Map of McMurdo Sound, showing locations of ice-sampling sites represented by solid circles. Lines passing through the solid circles indicate the c-axis alignment directions as determined from horizontal thin section studies. One site, also exhibiting a strong alignment of the c-axes, was No. 28, located in Lewis Bay outside McMurdo Sound. Solid circles lacking lines indicate sites at which c-axes alignments are either weakly developed or absent (sites 7, 13, 20 and 21). C-axis alignments were not measured at site 18 (multiyear ice) or at site 27 (shelf ice).

ANALYTICAL TECHNIQUES

Salinities

Cores used for salinity were divided into 10-cm-long sections on site, placed in plastic containers and returned to McMurdo Station for melting and analysis. Measurements were made with a temperature compensating Beckman Solubridge having an estimated measurement precision of $\pm 0.2\ °/_{00}$.

Crystalline Structures

Initial processing of cores and thick slices obtained from blocks included documentation of visible stratigraphic features, such as banding, structural discontinuities, brine drainage and algae incorporation. The structure of the McMurdo Sound sea ice was characterized through detailed examination of thin sections prepared according to techniques described by *Weeks and Gow* [1978, 1979]. A 5-mm-thick slice is sawed from the core, frozen to a glass plate, and thinned further to about 1 mm on a bandsaw. A microtome is then used to reduce the final thickness to about 0.5 mm. At this thickness, grain boundaries, brine drainage channels, and the brine pocket/ice platelet substructure of individual crystals are clearly revealed when the section is viewed between crossed polarizers.

Both vertical and horizontal thin sections were prepared in this manner. A representative selection of thin sections was then photographed between crossed polarizers to document major ice types and to identify individual crystals for c-axis fabric measurements. In addition, all structure

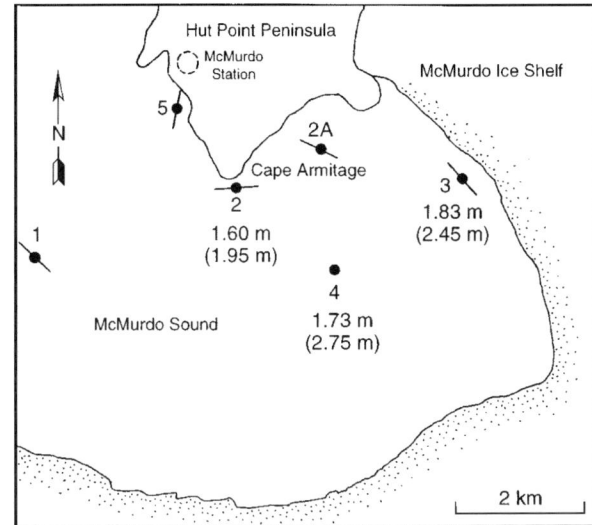

Fig. 3. Enlargement of the Cape Armitage, area situated near site 1 in Figure 2, showing locations of sites 1 to 5 and their c-axis alignment characteristics. No significant alignment of c-axes was observed at site 4. Also shown are ice thicknesses measured in 1980 relative to those, in parentheses, made at approximately the same locations in 1966 by *Paige* (1967).

cores were sliced vertically on a bandsaw and the slices examined on a light table to determine crystal structure variations over the full thickness of ice at each site. From the patterns revealed, a structure profile was then constructed for cores from all sites. These profiles were used to estimate percentage contributions of each ice type present at each of the 26 sites. These data, along with snow depth, ice thickness, bulk salinity and algae occurrence for each site are presented in Table 1. C-axis fabric measurements, performed on azimuthally-oriented, horizontal thin sections, entailed the use of a four-axis universal stage for determining the spatial distribution of the crystallographic c-axes in the sea ice crystal aggregate. In order to evaluate the precise nature of the fabrics, the c-axis measurements were transformed to a point scatter plot on an equal-area stereographic projection. The general procedure for determining c-axis fabrics in ice is given in *Langway* [1958]. Its application to sea ice is described in *Weeks and Gow* [1978, 1980].

RESULTS AND DISCUSSION

Surficial Characteristics and Ice Thickness

The depth of the snow cover varied according to the proximity of the sea ice to the McMurdo Ice Shelf, ranging from 0.5–0.6 m near the edge of the shelf (sites 1 to 4) to only trace amounts in the remote parts of McMurdo Sound, for example, the wind-swept area north of the Koettlitz Ice Tongue (sites 15 to 19 in Table 1). The surface of the ice beneath the snow cover was generally flat, except at sites 9 to 13 where the top 0.20 m of ice consisted in places of refrozen fragments of brash ice. This brash ice is thought to have originated sometime in June 1980 when the then 0.20-m-thick ice sheet in this part of McMurdo Sound was fragmented during a storm.

Ice thickness varied appreciably from site to site. As indicated in Table 1, it was generally thinnest where the snow was thickest, for example, in the immediate vicinity of Cape Armitage at sites 2, 2A, 3 and 4 (Figure 3), where appreciable flooding of the ice was also observed. Conversely, the ice was usually thickest at sites where the snow cover was sparse or absent. Overall, the ice thicknesses ranged from a low of 1.60 m at site 2 to a high of 2.49 m at site 26. The mean thickness for the 25 first-year sites in McMurdo Sound is 2.02 m. The thickness data were based on measurements of both the structural and salinity cores that were drilled within a meter or so of one another at each site. As indicated in Table 1, no significant differences in thickness between the two cores were observed at 14 of the 25 sites. At 8 of the 11 remaining first-year sites, the difference did not exceed 0.05 m. Only at sites 9, 13 and 19 were significant differences observed, 0.07 m at site 9, 0.12 m at site 13 and 0.09 m at site 19. Most of these thickness changes can be attributed to variations in the thickness of platelet ice growing onto the bottom of the ice sheets. This occurs during the latter stages of ice sheet growth in McMurdo Sound and is discussed in detail in later sections of this paper. At the single location outside McMurdo Sound (site 28), the ice thickness measured 1.89 m.

Ice thicknesses reported here are probably less than the maximum values which are usually attained by late November–early December. However, an examination of more recent measurements of ice thickness [*Leventer et al.*, 1987; *Jeffries et al.*, 1993] indicate that our 1980 thicknesses are probably within 0.05–0.10 m of the full season's growth. This slight increase in thickness would give a mean ice sheet thickness for 1980 of ≈2.10 m, in generally good agreement with recently published data. Morris and Jeffries [1992] examined McMurdo runway ice thickness data for 1989-91 and reported maximum ice thicknesses of at least 2.20/2.25 m. Significantly thicker ice was measured earlier by *Paige* [1967] in the small embayment between Cape Armitage and the McMurdo Ice Shelf, as depicted in Figure 3. Three of our 1980 sites were located close to a number of sites where Paige made his 1966 measurements. As indicated in Figure 3, Paige's late October measurements show appreciably thicker ice than we observed at the end of October 1980. At a number of additional locations in the same general area Paige measured ice in excess of 3 m thick. An examination of the 1966 winter temperatures in McMurdo Sound showed them to be appreciably lower than those measured in 1980. Paige also measured thicker ice during the 1965 winter, possibly indicating lower winter temperatures prevailing at the time that Paige made his

TABLE 1. Summary of Major Physical and Structural Properties of First-Year Sea Ice in McMurdo Sound, Antarctica, October–November 1980

Site	Snow Depth (m)	Ice Thickness (m)	Bulk Salinity (°/$_{oo}$)	Ice Type* T%	C%	P%
1	0.20	2.11	5.7	5	89^{60}†	6
2	0.56	1.60	5.6	14	75^{43}	11
2A	0.58	1.62	4.7	14	74^{43}	12
3	0.37	1.83	5.4	7	85^{59}	8
4	0.32	1.73	5.1	10	66^{0}	24
5	0.08	2.00	5.6	9	85^{52}	6
6	0.03	1.93–1.96	6.4	24	67^{19}	9A**
7	0.03	2.05	5.9	25	61^{0}	14A
8	0.04	2.12–2.15	6.1	13	73^{44}	14A
9	0.03	2.11–2.18	6.0	18	67^{40}	15A
10	0.03	2.08–2.11	6.3	13	70^{66}	17A
11	0.11	1.83	5.2	11	75^{85}	14A
12	0.02	2.14	nd	9	80^{92}	11A
13	0.26	2.12–2.24	6.1	39	52^{0}	9A
14	0.03	1.73–1.79	6.3	18	73^{52}	9
15	0.03	2.08–2.11	7.0	11	70^{49}	19A
16	0.03	2.41	6.0	11	80^{44}	9
17	<0.01	2.24	5.9	12	76^{55}	12A
18	nd	5.13	4.2	Multiyear ice core		
19	<0.01	2.05–2.14	6.2	10	81^{68}	9
20	0.05	2.12	5.8	33	54^{0}	13A
21	<0.01	1.98–2.00	7.1	35	39^{0}	26A
23	0.02	1.84	nd	19	65^{33}	16A
24	0.14	1.84–1.88	6.7	10	73^{75}	17A
25	<0.01	2.11–2.16	6.5	10	71^{59}	19A
26	0.03	2.49	6.3	10	86^{24}	4A
27	nd	nd	<0.1	Glacial ice shelf core		
28	<0.01	1.89	5.8	11	80^{85}	2A
28	<0.01	1.89	5.8	11	80^{85}	2A
Average Values	0.12	2.02	6.0	15	72	13

*With regard to ice type categories, T denotes transition ice, C denotes columnar-congelation ice and P denotes platelet ice.

†At this and other sites, the superscript value is the estimated percentage of columnar-congelation ice exhibiting aligned c-axes.

**At this and other sites, A indicates the presence of visible algae in bottom platelet ice.

measurements than those made since 1980. Though an examination of the detailed meteorological records at McMurdo Station is beyond the scope of this paper, the question of past variations in ice thickness relative to the meteorological record, could probably be answered from an examination of ice thickness measurements made during the past 40 years in conjunction with the preparation of the annual sea ice runway in McMurdo Sound. The value of such a data base for analyzing past variations in sea ice thicknesses in McMurdo Sound has already been emphasized by *Morris and Jeffries* [1992].

Salinity

Salinity-depth profiles, based on measurements of 0.10-m-long core segments, are presented in Figure 4. Cores from 22 first-year ice sites were analyzed, including a core

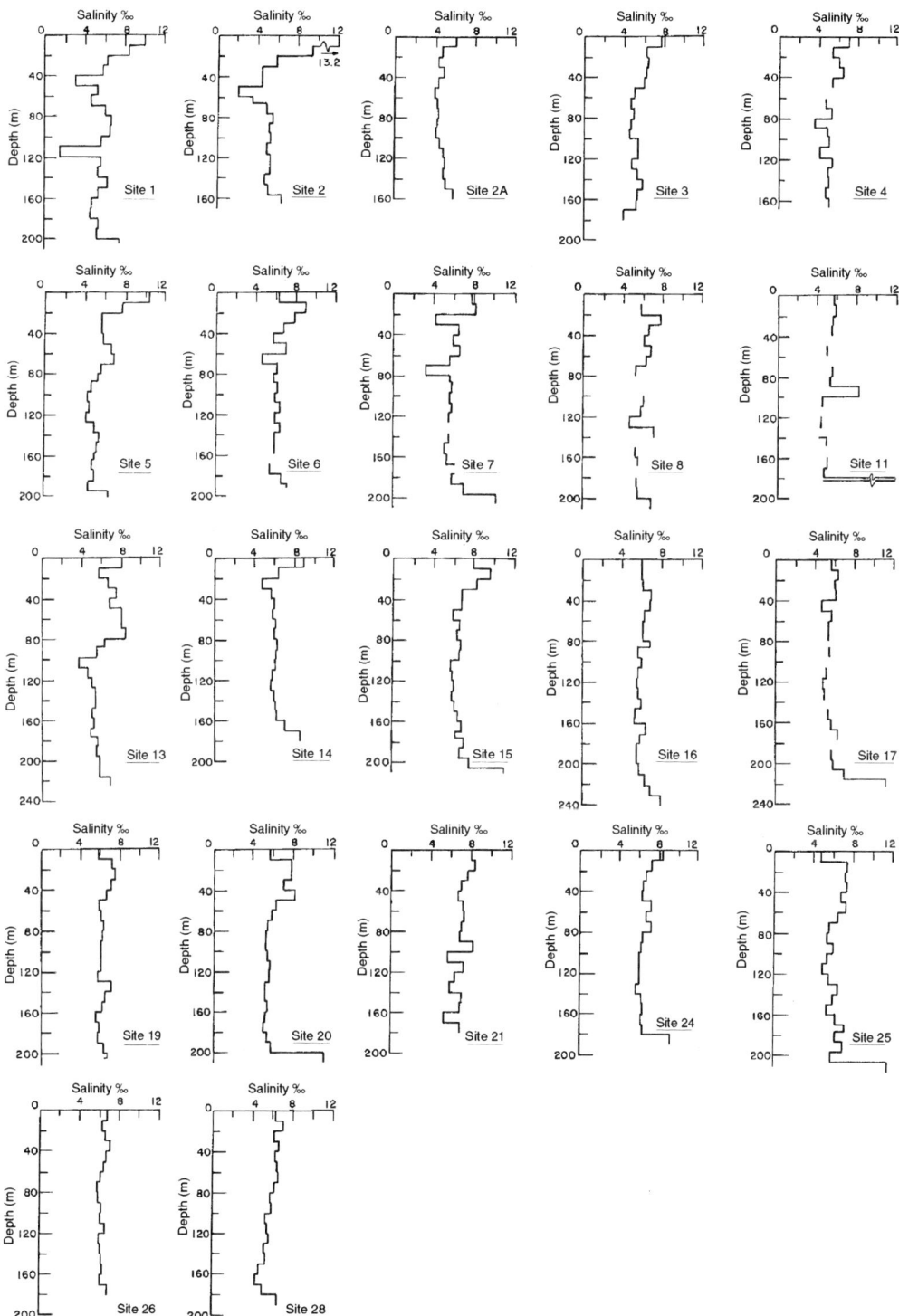

Fig. 4. Salinity-depth profiles of McMurdo Sound sea ice sites shown in Figure 2.

from site 28 located in Lewis Bay outside McMurdo Sound (see Figure 2). A number of profiles show C-shaped characteristics, arising from the appreciably higher salinities measured at the top and bottom of the ice column. However, most profiles lack the C-shaped features. In the case of C-shaped profiles, enhanced salinities at the top of a core can generally be attributed to growth conditions favoring sea water retention, for example, incorporation in the intergranular spaces of frazil ice formed during the initial stages of ice growth. In the case of enhanced salinities at the bottom of the ice these can be attributed largely to the relative effectiveness of residual brine retention, by capillarity, between the platelet crystals composing the bottom layers of ice. In these cases the C-shaped profile is only retained if samples of bottom ice are processed and containerized immediately after a core is retrieved; if not, then brine drainage will occur, effectively diminishing or even eliminating this feature of the salinity profile. However, for the greater part of the salinity profiles at all sites in McMurdo Sound, substantially uniform salinities of around 6 $^o/_{oo}$ are observed. This result is also in accord with the observations of *Nakawo and Sinha* [1981] who demonstrated for an entire winter's ice growth at Eclipse Sound, near Pond Inlet, Baffin Island, Canada, that salinities essentially stabilized at around 6.0 $^o/_{oo}$ after about 40 cm of ice growth. According to *Gow et al.* [1990] the establishment of quasi-stable salinity profiles in Arctic first-year ice can be attributed to a near-constant rate of salt entrainment associated with the progressive sealing off of brine pockets in the interdendrite spaces of freshly frozen columnar-congelation ice. This is especially interesting in light of the fact that the basal ice in McMurdo Sound consists of so-called platelet ice, not columnar-congelation ice. Yet, despite significant structural differences between these two ice types (discussed in a later section), the effect on the salinity distribution appears to be minimal. Platelet ice forms by the congelation of water in the spaces between individual platelet crystals. Salt entrainment incurred during freezing is apparently sufficient to maintain the bulk salinity at 6 $^o/_{oo}$.

Bulk (mean) salinities of first-year ice in McMurdo Sound ranged from 4.7 to 7.1 $^o/_{oo}$ with a mean value of 6.0 $^o/_{oo}$. No systematic variation of the mean salinity with either location or thickness of sea ice in McMurdo Sound is apparent. However, salinities measured in October–November 1980, are substantially higher than those measured by *Jeffries and Weeks* [1991] when they revisited many of the same geographically located sites in January 1991. Similarly, diminished salinities were also observed by *Veazey* [1994] at several multiple coring sites in McMurdo Sound in January 1992. *Jeffries and Weeks* [1991] measured salinities in the range 3.0 to 5.4 $^o/_{oo}$ that yielded a mean salinity of 4.2 $^o/_{oo}$ for the 15 first-year sites they sampled. They also observed a significant change to a roughly reverse S-shaped salinity profile, characterized by diminished salinities of the top of the ice sheet and higher salinities at the bottom, which they attributed to the effects of desalination: the change in the average bulk salinity from 6.0 $^o/_{oo}$ measured in October–November 1980 to 4.2 $^o/_{oo}$ in January 1991 corresponds to a brine loss of about 30% in a little more than 2 months, assuming the two data sets, taken 11 years apart accurately reflect winter/summer conditions. Such a change would then simply reflect the extent of thermally driven desalination occurring under summertime conditions in McMurdo Sound. It is likely that thermally controlled gravity drainage of brine is primarily responsible for the desalination of sea ice that *Jeffries and Weeks* [1991] observed in January 1991. They did not observe surface melting which is a major cause of summertime desalination of sea ice of comparable age and thickness in the Arctic, where the process is generally referred to as "flushing" [*Untersteiner*, 1968]. Compared to simple gravity drainage, "flushing" is likely to be of minimal importance in Antarctica where surface water and melt pond formation on sea ice are either rare or nonexistent.

The functional relationship between the bulk salinity of sea ice and its thickness was first examined by *Cox and Weeks* [1974], who obtained best-fit salinity trend curves for Arctic sea ice based on data for both cold first-year ice sampled during the winter growth season and for warm multiyear sea ice sampled during or at the end of the melt season. Figure 5 shows bulk salinities, obtained during October–November 1980, together with McMurdo Sound sea ice data from *Hendrickson and Rowland* [1965], *Veazey* [1994] and *Jeffries and Weeks* [1991], plotted on the Cox-Weeks salinity trend line diagram. These trend lines, designated the Arctic Cold Ice Trend (ACIT) and the Arctic Warm Ice Trend (AWIT), show both the first-year and multiyear ice salinities in McMurdo Sound to be appreciably higher than their Arctic counterparts. The average bulk salinity of all the 1980 profiles is 6.0 $^o/_{oo}$ compared to the 4.5–5 $^o/_{oo}$ for ice of comparable age and thickness in the Arctic. Even the warm first-year ice salinities plot close to the ACIT line and the difference between Arctic and Antarctic multiyear ice is even more pronounced. At the one multiyear site sampled in 1980, the salinity (not plotted on Figure 5) averaged 4.2 $^o/_{oo}$ which is nearly double the value reported by Cox and Weeks [1974] for Arctic multiyear ice of the same thickness (5 m). These data further reinforce the view that both first-year and multiyear sea ice in the Antarctic are much less prone to brine loss than their Arctic counterparts, mainly because of the absence of top surface melt, which, in the Arctic, promotes much more efficient desalination.

Ice Composition and Type—General Description

Banded ice. Horizontal bands of cloudy ice, 10 to 20 mm thick, interspersed with layers of clearer ice, are a common feature of McMurdo Sound sea ice. Such banding has been extensively studied by *Paige* [1966], who concluded that the origin of banding is linked closely to variations in the velocity of ice sheet growth resulting from

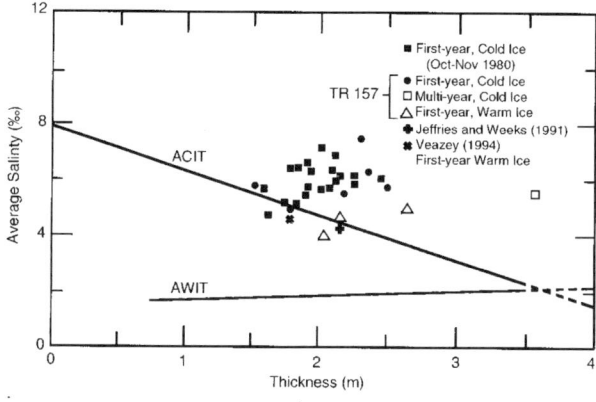

Fig. 5. Bulk salinities versus depth for McMurdo Sound sea ice plotted on the Cox-Weeks salinity trend line diagram derived from Arctic sea ice data sources by *Cox and Weeks* (1974). TR 157 data are from *Hendrickson and Rowland* (1965). The trend lines ACIT and AWIT denote Arctic Cold Ice Trend and Arctic Warm Ice Trend respectively.

surface temperature fluctuations. The actual nature of the bands has been attributed variously to zones of localized increased porosity (enhanced bubbliness), small-scale crystal orientation changes or reversals, and brine pocket concentration effects [*Bennington*, 1963], all of which would reflect fluctuations in growth velocity. Though readily observed in cores and blocks of sea ice, they became virtually indiscernible in thin sections of the order of 1 mm thick or less, thereby preventing determination of the precise nature of their formation. This is also true of widespread banding observed elsewhere in Antarctic sea ice, for example, by *Gow et al.* [1987] in Weddell Sea pack ice. Banded ice is also of equally widespread occurrence in Arctic sea ice [*Bennington*, 1963].

Our experience is that banding is only observed in columnar sea ice, so its appearance can be immediately identified with sea ice that is congelation in character, that is, ice that is formed by direct freezing of seawater to the underside of an ice sheet. The frequency of occurrence of banding in McMurdo Sound sea ice varied from core to core, almost certainly reflecting variations in growth conditions at the various locations. As many as 30 individual bands were counted in some cores, with most, however, occurring in the top meter. The disappearance of banding in the deeper ice can probably be attributed to the replacement of congelation ice by platelet ice, the growth of which is not conducive to the formation of bands. Another factor possibly contributing to the decrease in banding is the fact that as the ice becomes thicker there is a decrease in the change in the growth velocity resulting from a given change in the ice surface temperatures.

Texturally, the McMurdo Sound sea ice consisted predominantly of *columnar-congelation* ice, overlain by *transition ice* and, underlain by *platelet ice*. Structural-compositional observations of the 26 first-year ice sites examined are diagrammed in Figure 6.

Transition ice. As defined here, transition ice includes all granular ice at the top of a core plus that section of columnar-congelation ice growth leading to textures that display substantially horizontal c-axis orientations. The granular component can consist of either frazil and snow ice, or both. They are distinguished mainly on the basis of bubble content and grain size, bubbles generally being much more abundant and the grain size significantly larger in snow ice than frazil ice. Appreciable snow ice was encountered at sites 2 and 2A located near Cape Armitage [see Figure 3]. At both of these sites the snow load was sufficient to depress the freeboard below sea level. At other sites snow ice formation was minimal to nonexistent.

Frazil ice *per se* constituted generally less than 3% of the total ice sheet thickness in McMurdo Sound except at site 13, where the top 0.75–0.80 m consisted of frazil ice and brash ice derived from the freezing in of blocks of ice from a storm-fragmented ice sheet. This is in striking contrast to the situation in the Weddell Sea pack ice where frazil ice (as distinct from platelet ice) is the dominant ice type [*Gow et al.*, 1987; *Lange et al.*, 1989; *Eicken and Lange*, 1989]. Large amounts of frazil ice have also been observed in the Ross Sea and in the Indian Ocean sector of the Southern Ocean [*Jeffries and Weeks*, 1992; *Allison and Qian*, 1985; *Jacka et al.*, 1988]. Frazil ice formation in these seas is primarily the product of a turbulent water column, a condition that occurs much less commonly in the more protected land-fast ice environment of McMurdo Sound.

Transition ice also included the earliest stages of growth of columnar-congelation ice, leading to textures with substantially horizontal c-axis orientations. Ice exhibiting this orientation sometimes extended into the zone, designated here as columnar-congelation ice, prior to developing significant c-axis alignments. The columnar-congelation ice component of the transition zone represented as much as 12% of the total ice thickness at some sites in McMurdo Sound. Only five sites contained transition ice (including the granular ice component) in excess of 20% of the total ice sheet thickness; only four sites contained less than 10%. Its average contribution to the 1980 McMurdo Sound sea ice sheet was 15%.

Columnar-congelation ice. This ice type, which is the dominant component of McMurdo Sound sea ice, is formed by direct freezing of seawater to the underside of an existing ice sheet. It is characterized in vertical sections by elongated columnar crystals, which in horizontal cross section exhibit a characteristic substructure of ice plates sandwiched between layers of brine pockets. This substructure results from the dendritic nature of ice growth that occurs in response to the buildup of brine at the ice/water interface. Excess brine that cannot be expelled from the interface is systematically incorporated into the spaces between

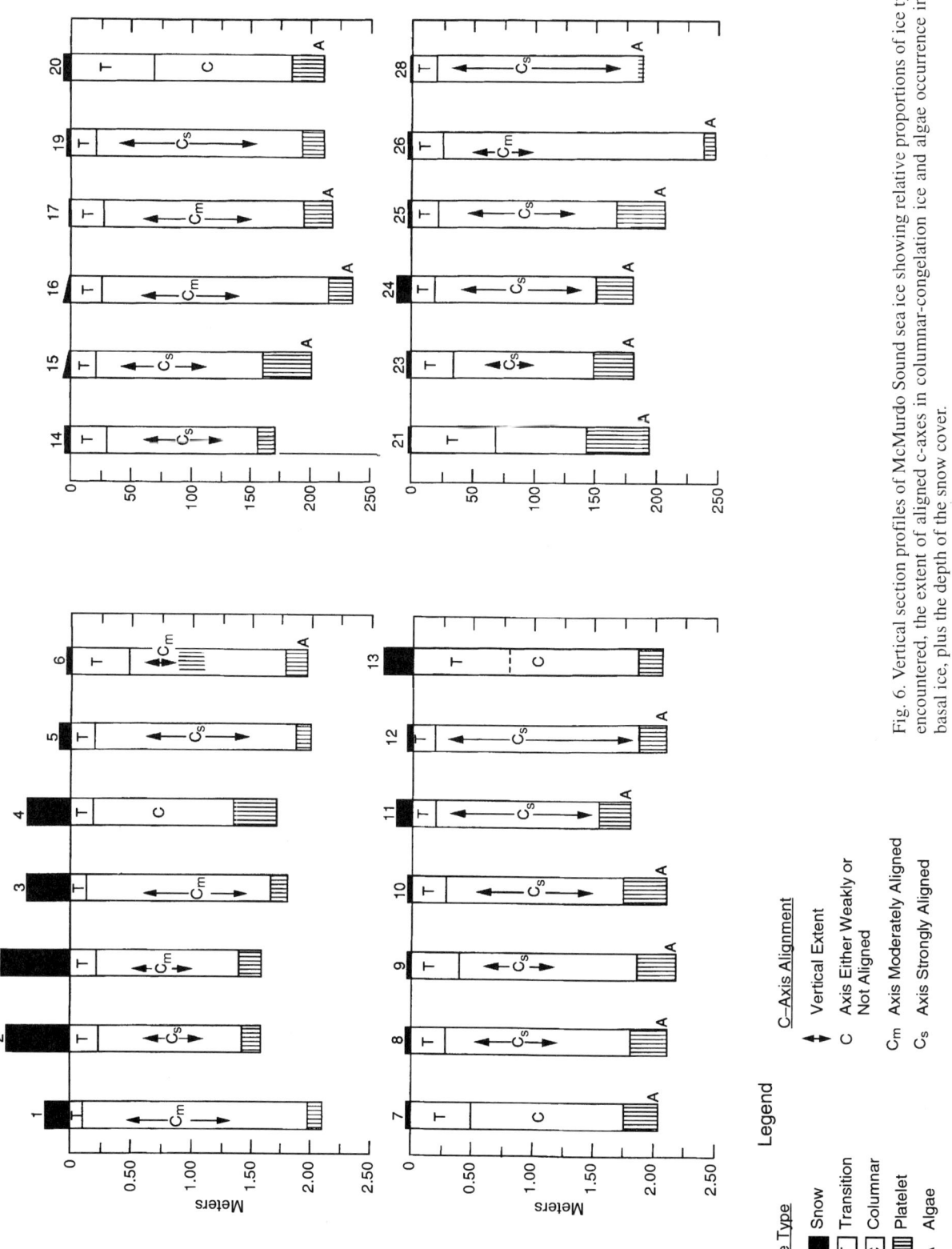

Fig. 6. Vertical section profiles of McMurdo Sound sea ice showing relative proportions of ice types encountered, the extent of aligned c-axes in columnar-congelation ice and algae occurrence in the basal ice, plus the depth of the snow cover.

dendrites. Actual incorporation of brine into the ice structure occurs by pinching off of the interdendrite spaces, leading to the formation of individual pockets of brine. In this process, described in detail by *Weeks and Ackley* [1982], the dendrites remain substantially free of brine and, with continued growth, constitute the ice plates between which layers of brine pockets become sandwiched. Individual crystals of sea ice may consist of 20 or more such plates and their intervening brine layers. Crystals themselves may extend vertically for as much as 100 mm; in cross section they may measure 10 mm or more.

At all first-year sites, except one (site 21), ice designated here as columnar congelation exceeded 50% of the total ice sheet thickness; at 12 of the 26 sites it exceeded 75% (Table 1) and its average contribution was 72%. If the columnar-congelation ice component of the transition ice is included then the total columnar-congelation ice would increase to about 83% of the average ice sheet thickness in McMurdo Sound. This is significantly larger than the 62.3 % reported by *Jeffries et al.* [1993] for the 1991 ice sheet in McMurdo Sound.

As noted above the earliest stages of congelation ice growth, leading to the formation of columnar crystals with substantially horizontally oriented c-axes, are included in the transition ice type. With continued growth, the c-axes generally begin to become aligned within the horizontal plane. Any such alignments or preferred orientation of the c-axes constitutes what is generally referred to as a c-axis fabric. The onset of an aligned c-axis fabric varied from site to site, but it generally occurred soon after the end of the transition stage. However, fully developed alignments (strong, moderate or otherwise) did not usually occur until some distance into the columnar-congelation ice zone. Also, aligned c-axis fabrics did not, in any instance, persist to the bottom of the ice sheet in McMurdo Sound. This is in striking contrast to the Arctic, where aligned c-axes, once developed, generally continue to the bottom of a mature winter ice sheet. However, in the case of McMurdo Sound sea ice, c-axis alignments at most sites terminated or weakened prior to the onset of growth of the platelet ice that characterized the final stages of ice sheet formation in McMurdo Sound. In a number of cases, weakening of the c-axis fabric occurred some distance above the transition to platelet ice. It appeared from an inspection of vertical thin sections that some comingling of congelation and platelet ice also occurred prior to full establishment of the platelet ice that underlies all previously formed columnar-congelation ice in McMurdo Sound. This comingled ice would appear to correspond to ice that *Jeffries et al.* [1993] designated as congelation/platelet ice. According to *Jeffries et al.* [1993] this ice constituted 9.3% of the total ice sheet thickness in McMurdo Sound in 1991. Horizontal thin sections invariably showed continued growth of congelation ice in the spaces between the platelets. Congelation ice in this particular situation was readily identified on the basis of the unique ice plate/brine layer substructure of its crystals, which those of platelet ice lack. A more detailed discussion of the nature of c-axis alignments in columnar-congelation ice McMurdo Sound and their relationship to measured and/or inferred directions of the current at the ice/water interface is given in a later section of this paper.

Platelet ice. Platelet ice is ubiquitous in McMurdo Sound, appearing as a major ice type in the final phase of sea ice growth at all locations sampled in October–November 1980. Platelet ice formation in McMurdo Sound was initially reported by *Wright and Priestley* [1922], but as an ice type, distinct from columnar-congelation ice, it was first documented by *Paige* [1966] and subsequently reported by *Gow et al.* [1981, 1982] *Lewis and Perkin* [1985], *Crocker and Wadhams* [1989] *Jeffries et al.* [1993] and *Veazey* [1994]. Its texture is distinctive, consisting mainly of disoriented, bladed, dendrite-like crystals up to 100 mm long and 50 mm wide. In cross-section platelet ice exhibits a criss-cross, even radiating arrangement of crystals, interspersed with crystals of conventional congelation ice.

As discussed in a later section, most platelets most likely originate as loose attachments growing directly to the underside of the existing ice sheet; it is only with the growth of congelation ice in the spaces between the platelets that consolidation to platelet ice occurs. Overall, the textural characteristics of platelet ice, as we observed them, agreed completely with descriptions originally reported in *Paige* [1966] and with the more detailed description of this ice type given by *Jeffries et al.* [1993].

Platelet ice underlying the 1980 sea ice sheet in McMurdo Sound varied in thickness from 10 to 50 cm, representing 5 to 25% of the total ice thickness and averaging 13%, equivalent to about 0.26 m of platelet ice growth. This amount of platelet ice growth is similar to that reported by *Paige* [1966]. However, it is much less than more recent observations in McMurdo Sound would indicate. *Lewis and Perkin* [1985], for example, reported a 40% contribution of platelet ice to the total ice thickness. *Jeffries et al.* [1993] measured platelet ice thicknesses ranging from 9.5 to 56.0% at 15 sites. Their mean value of 28.3% of the total ice thickness is more than double the platelet ice percentages we observed in 1980. However, our average platelet ice percentage (13%) is similar to the 11% reported for sea ice forming in the coastal polynya region of the southeastern Weddell Sea [*Eicken and Lange*, 1989].

Simple calculations of ice growth, based on freezing-degree-day records for 1980, assuming an initial freeze-up date of mid-March, indicate that platelet ice growth probably started during late August or early September. At locations with less than 5 cm of snow cover this would coincide with ice thicknesses of the order 1.75–2.00 m having formed prior to the onset of platelet ice growth. At the time of our observations we can only assume that the platelet ice layer still appeared to be in a state of active growth as was indicated by the fact that platelet crystal growth was also observed occurring on weighted polypropylene lines suspended beneath the ice sheet. We estimate that an additional 0.05–0.10 m of platelet ice growth would occur before maximum ice sheet thicknesses were reached in early to mid-December. There appears to be no systematic variation in platelet ice thickness with location in McMurdo Sound. Only minimal growth of platelet ice was observed at the one site (28) located outside McMurdo Sound. This would indicate that conditions for platelet ice formation at this site were much less favorable than in McMurdo Sound. This conclusion is also supported by the observations of *Jeffries and Weeks* [1992] who found platelet ice to be absent at locations north of McMurdo Sound, unless the sampled ice was collected in the vicinity of an ice shelf.

Only at a single location (site 6) did we observe platelet ice sandwiched between columnar-congelation ice. The precise origin of this layer of platelet ice, located between 0.87 m and 1.08 m depth in the ice sheet (Figure 6), is not known. The columnar-congelation ice directly above the platelet layer at site 6 exhibited a moderately strong c-axis alignment, but no significant alignment was observed in the columnar-congelation ice below it. This would indicate a significant change in the circulation pattern of water beneath the ice, one in which currents were no longer favorable to the development of aligned c-axes once the ice sheet had reached a thickness of 1.08 m. The significance of this and the issue of the directional control of the current on c-axis alignment at the growing ice/water interface are discussed in a later section.

At most locations (17 of 26 sites diagrammed in Figure 6) the bottom 0.02 to 0.10 m of platelet ice also contained brown algae. As indicated in Table 1, the presence of visible algae is strongly correlated with snow cover thickness, with those locations in which the snow layer exceeds 0.20 m apparently proving unfavorable to the growth and incorporation of algae in the basal ice. These observations accord with the view that the photosynthesis process, essential to the growth of algae in the basal ice, is greatly hindered or prevented altogether by the high light extinction properties of a modestly thick snow layer on top of the ice sheet [*Sullivan et al.*, 1983; *Sullivan et al.*, 1985; *Palmisano et al.*, 1985].

Details of c-axis fabrics. Herein, the term fabric is restricted to orientated features in the ice that are directly linked to a preferred orientation of the crystallographic c-axes. Since the crystallographic c-axis corresponds to the optic axis it is a relatively simple matter to measure the spatial orientation of the c-axes on a four-axis universal stage [*Langway,* 1958]. An additional fabric element, dimensional elongation of crystals in the horizontal plane, often develops in conjunction with a preferred c-axis orientation in sea ice.

It has been known for some time [see *Weeks and Ackley,* 1982] that in columnar-congelation ice growth a dominant horizontal c-axis crystal orientation develops rapidly after the formation of the initial skim ice. Horizontal disposition of the c-axes in sea ice is, in itself, a strong fabric that can be referred to as a girdle fabric. However, what was not well known was that as growth continued the c-axes could become aligned (directionally oriented) within the horizontal plane. The transformation to a directionally aligned fabric constitutes an even stronger fabric than the girdle. A few isolated observations, mainly on sea ice incorporated into ice islands [*Cherepanov*, 1964; *Smith*, 1964] and a single measurement on first-year ice by *Peyton* [1966], indicated the existence of strong azimuthal alignments of the c-axes, but the significance of this work was not fully appreciated by the sea ice community. Apart from later work by *Cherepanov* [1971] in the Kara Sea, the regional extent of such alignments and the probable explanation of their origin did not become apparent until the work of *Weeks and Gow* [1978, 1980], off the north coast of Alaska, which indicated that the direction of c-axis alignments in columnar-congelation sea ice was controlled by the direction of the average, long-term current acting at the ice/water interface. A similar situation became very much apparent in a re-examination of Cherepanov's [1971] results from the Kara Sea. Although Cherepanov did not include water current data in his diagram of c-axis alignments, when such information obtained from U.S. Navy sources was included, remarkable agreement between the c-axis alignment and current directions was observed. This result was even more interesting in light of the fact that *Cherepanov* [1971] actually considered current action as a possible mechanism for aligning the c-axes, but rejected it in favor of a mechanism involving the direct influence of the earth's magnetic field on the crystallization process. Such a coupling is now considered highly unlikely. Indirect evidence of a strong directional alignment of c-axes, based on studies of the electrical properties of sea ice, was also obtained by *Campbell and Orange* [1974] and subsequently confirmed by *Kohnen* [1976] and *Kovacs and Morey* [1978]. The relationship between the direction of the c-axis alignment and the direction of the current at the growing ice interface now appears to be an established fact, based not only on the results of numerous field studies, but also on the laboratory experiments of *Langhorne* [1983], *Langhorne and Robinson* [1986] and *Stander and Michel* [1989]. However, mechanistic details of the processes involved remain debatable.

While the strong directional alignment of c-axes had been widely documented for Arctic sea ice, very little documentation of this phenomenon existed for Antarctic sea ice prior to studies reported here and earlier [*Gow et al.,* 1981]. For this reason our characterization of land-fast sea ice in McMurdo Sound included extensive examination of c-axis alignment patterns conducted on azimuthally oriented ice samples in conjunction with instantaneous current measurements at the ice/water interface. Results of this work show that crystal texture and c-axis fabrics of columnar-congelation ice in McMurdo Sound are very similar, if not identical, to what is observed in the Arctic. Subsequent studies by *Jeffries et al.* [1993] and Jones [personal communication, 1996] of c-axis orientation patterns in columnar-congelation sea ice in McMurdo Sound substantially confirm the results reported here.

The overall pattern of c-axis alignments observed in October–November 1980 is indicated in Figure 2. Of the 26 first-year ice sites examined, 21 exhibited c-axis alignments ranging from moderate to strong. Here we follow *Weeks and Gow* [1978] in defining c-axis fabric strength in terms of the s_0 values of the horizontally-disposed c-axes. Such axes plot on or close to the perimeter of the Schmidt equal area net. As indicated in Figures 7 to 11, any significant alignment of the c-axes will be expressed as a clustering of the axes about a mean value. Fabrics

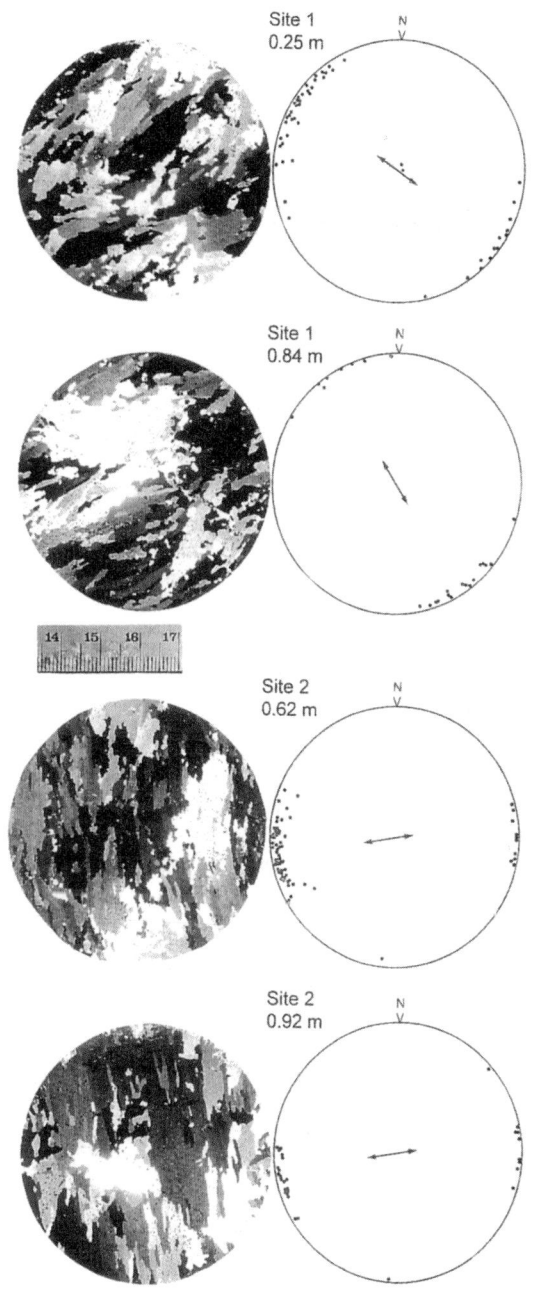

Figure 8. Ice-texture photographs and associated c-axis fabric plots of horizontal thin sections of columnar-congelation ice from sites 1 and 2 in McMurdo Sound. Note the slight (20°) swing in the c-axis alignment direction at Site 1. The smallest scale subdivisions measure 1 mm.

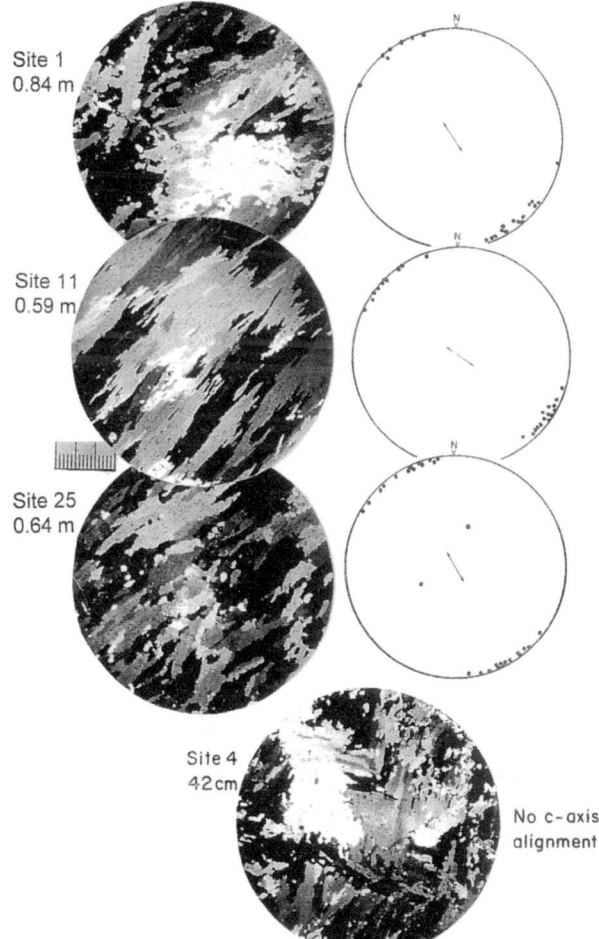

Figure 7. Ice-texture photographs, taken between crossed polarizers, and associated c-axis fabric plots of horizontal thin sections of columnar-congelation ice from four coring sites in McMurdo Sound. The scale beneath the site 25 photograph applies to all four sites; scale subdivisions measure 1 mm.

designated as strongly aligned are limited to those with s_0 values up to, but not exceeding 15°. Fabrics exhibiting s_0 values ranging between 15° and 30° are designated as moderately aligned. All fabrics with s_0 values exceeding 30° are assigned as weak or non-aligned. Only five sites in

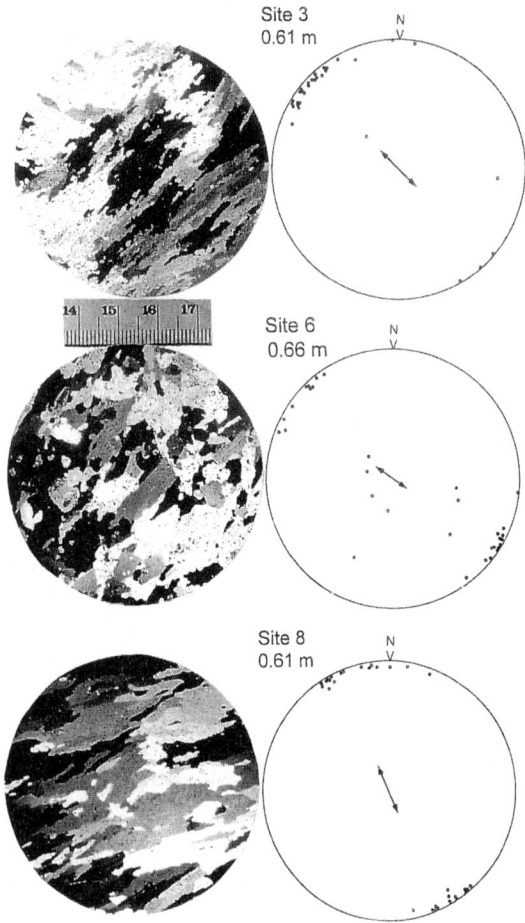

Figure 9. Ice-texture photographs and associated c-axis fabric plots of horizontal thin sections of columnar-congelation ice from sites 3, 6 and 8 in McMurdo Sound. The smallest scale subdivisions measure 1 mm.

extinguish optically, without recourse to precise measurements of the individual c-axes on the universal stage.

Examples of the crystalline textures and associated c-axis fabrics of horizontal thin sections of columnar ice from three sites, together with a section from a fourth site that exhibited no significant alignment of the c-axes are shown in Figure 7. Sutured crystal outlines characterize the texture of ice at sites 1 and 25, in contrast to the more sharply defined boundaries of crystals at site 11. However, the textures at all three sites exhibit a pronounced dimensional elongation of crystals perpendicular to the direction of predominantly strong c-axis alignments. A random arrangement of crystals is clearly evident in the section of ice from site 4.

Comparative textures and fabrics of horizontal thin sections of columnar ice from sites 1 and 2, respectively, are presented in Figure 8. Moderate to strong c-axis alignments are evident at both locations, with dimensional orienta-

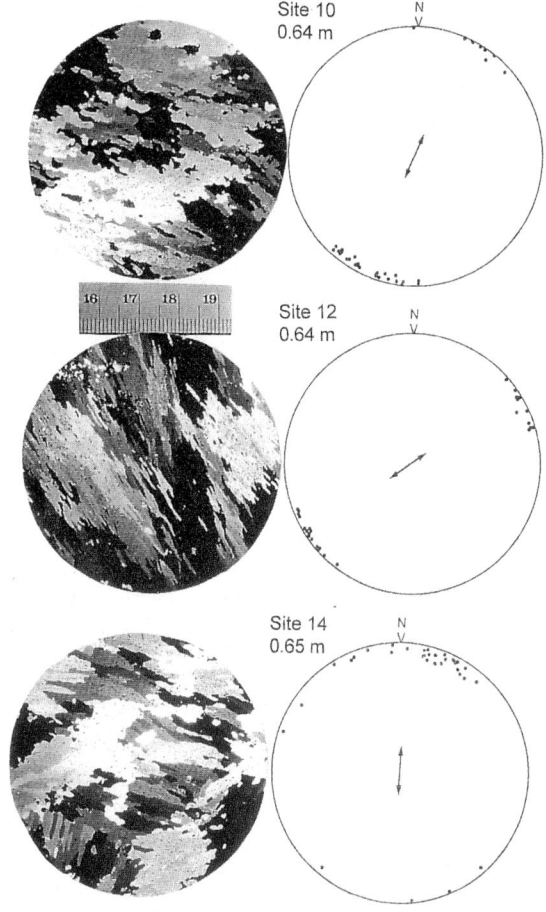

Figure 10. Ice-texture photographs and associated c-axis fabric plots of horizontal thin sections of columnar-congelation ice from sites 10, 12 and 14 in McMurdo Sound. The smallest scale subdivisions measure 1 mm.

McMurdo Sound displayed c-axis alignments that were weaker than moderate, based on the above spread angle (s_o) criterion (see Figures 2 and 3). Fabric strength and the extent of c-axis alignment relative to the ice thickness are indicated in Figure 6. The percentage of columnar-congelation ice exhibiting aligned c-axes at each site is given in Table 1. In general, c-axis alignments developed earliest at those sites displaying the strongest fabrics, usually within the top 0.60 m. The extent of this alignment was first noted in the series of horizontal thin sections cut from the blocks of ice obtained initially by chainsawing. Sections of deeper ice from the azimuthally oriented cores were also examined to determine the full extent and strength of the c-axis alignments. A semiquantitative approximation of the fabric alignment was also obtained from visual inspection of optical extinction patterns of the crystal aggregates in horizontal thin sections. This technique allows for the rapid determination of the range of angles over which crystals

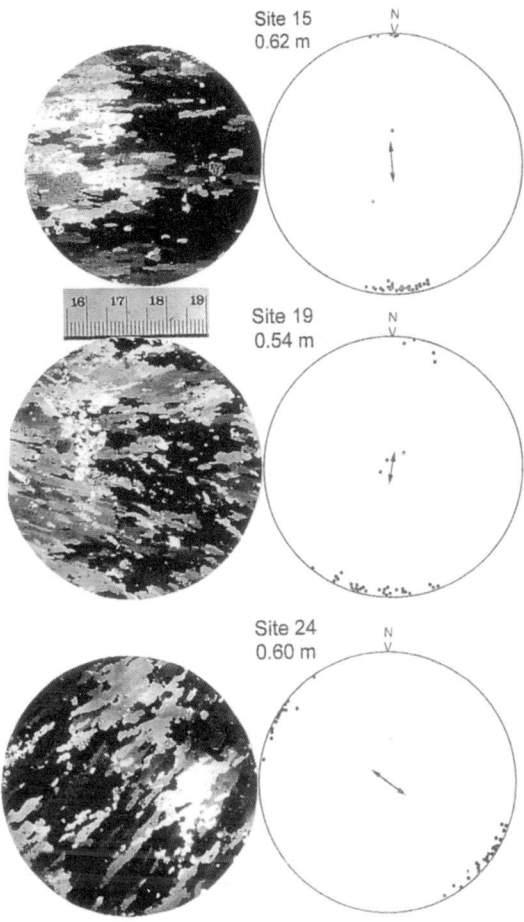

Figure 11. Ice-texture photographs and associated c-axis fabric plots of horizontal thin sections of columnar-congelation ice from sites 15, 19 and 24 in McMurdo Sound. The smallest scale divisions measure 1 mm.

tion of crystals also in evidence at site 2. A significant change (20°) in the alignment direction of the c-axes at site 1 is interpreted to indicate a correspondingly significant change in the current direction that occurred sometime during the period of growth from 0.25 to 0.84 m depth. This change in thickness likely occurred over a time interval of 5–6 weeks, based on a "best estimate calculation" of the freezing-degree index data for the 1980 winter. C-axis alignments at site 2 differ by less than 5°, well within the error limits of the measurements.

Texture/fabric diagrams for columnar ice from nine other sites in McMurdo Sound are shown in Figures 9, 10 and 11. The ice at most of these sites exhibited moderate to strong c-axis alignments and associated dimensional elongation of the crystals perpendicular to the predominant direction of c-axis alignment. Both the texture and the general nature of the c-axis alignments of the columnar component of McMurdo land-fast sea ice appear identical to its Arctic counterparts as described by *Weeks and Gow* [1978, 1980]. However, unlike the situation in mature Arctic sea ice, where c-axis alignments, once formed, invariably persist to the bottom of the ice sheet, c-axis alignments in McMurdo Sound sea ice usually degrade into weaker fabrics before the columnar-congelation ice becomes replaced by platelet ice. Congelation ice does not reappear except as an interstitial component of the platelet ice. This interstitial component is referred to as *interstitial congelation* ice by Jeffries et al. [1993] to distinguish it from platelet-free *congelation ice* of the upper layers of sea ice in McMurdo Sound. While it serves to consolidate the otherwise porous and permeable structure of the platelets, interstitial congelation ice does not appear to regenerate an aligned c-axis fabric during this phase of predominantly platelet ice growth. This lack of c-axis alignment can probably be attributed to the action of downward protruding platelets, hindering or even preventing any effective water-current control on c-axis orientation at the ice/water interface. *Jeffries et al.* [1993] also observed this phenomenon.

The weakening of c-axis alignments, occurring at some distance above the platelet ice at some sites, may be due to comingling of the earliest formed platelet crystals with columnar ice at a latter stage of its growth, or it might signal the onset of disturbed or changing current flow at the ice/water interface that preceded full-scale growth of the platelet ice. No attempt was made to separately catalogue co-mingled ice from the aligned columnar ice above or the platelet ice below.

The pattern of c-axes alignments depicted in Figure 2 generally agrees with the overall picture of late-winter surface water circulation patterns inferred from dynamic height topography supported by current meter measurements made at two sites in McMurdo Sound during October–November 1982 [*Lewis and Perkin*, 1985].

Attention has already been drawn to a measurable change (20°) in the c-axis alignment direction during growth of the columnar-congelation ice component at site 1. This change is interpreted as a response of the ice to a 20° change in the direction of the current at the freezing interface. Evidence of similar c-axis alignment changes in columnar-congelation ice at other sites was also obtained from thin section observations of optical extinction patterns of the crystal aggregate. These departures are also attributed to small-scale changes around an average long-term current direction.

At a number of sites, holes drilled for cores were used to insert a Marsh-McBirney current meter to measure both current velocities and direction directly beneath the ice sheet. The instrument consists of a transducer probe with cable, and a signal processor. The water velocity sensor design is based on the Faraday principle of electromagnetic conduction; it senses water flow in a plane normal to the longitudinal axis of the sensor. Readout resolution is 1 cm/s. Results of these essentially instantaneous current measurements, presented in Figure 12, show reasonable agreement between current direction and c-axes alignments at several sites that are consistent also with current directions inferred from coastal outlines, for example, sites 19, 15 and 28. At other locations however, as indicated in Figure 12, significant departures are observed between current directions measured directly under the ice in October–November 1980 and those inferred from c-axis alignments in ice formed several weeks or months earlier. Because of the land-fast nature of the sea ice in McMurdo Sound, any observed or inferred changes in current direction cannot be attributed to movement or repositioning of the ice during growth. However, such changes in current direction might be related to the columnar-congelation ice/platelet ice transition, implying some connection between ice type and changes in oceanic circulation patterns directly beneath the sea ice in McMurdo Sound. The seemingly fragile nature of platelet crystal accretion to the underside of the sea ice in McMurdo Sound would indicate that much of the platelet ice growth occurred during periods of relatively low current velocity.

According to *Lewis and Perkin* [1985], currents in McMurdo Sound exhibit tidal behavior and can be highly variable. For this reason caution needs to be exercised in evaluating results of instantaneous measurements of current flow directly beneath the sea ice cover. Nevertheless, our results show minimal current motion, not exceeding 1 cm/s at a number of sites, increasing to a maximum of 18 cm/s at site 2A and ranging from 12 to 17 cm/s at site 7. The 4 cm/s fluctuation at site 7 was measured over a period of about five minutes. Observations at site 28 are interesting in that currents measured just 1 meter apart, vertically, differed by 33° ± 5° with a corresponding minimal change in velocity of 1 cm/s.

Platelet Ice Formation

Platelet ice forms by consolidation of a network of platelet crystals by freezing of seawater in the voids between the platelet crystals. In this way platelet ice becomes a fully incorporated component of the sea ice sheet in McMurdo Sound. There appear to be two principal mechanisms for the crystallization of platelets in seawater: 1) by nucleation and growth of platelets directly to the bottom of an existing ice sheet and 2) by unrestrained nucleation of crystals in the water column, followed by uprise of the buoyant crystals to the underside of the ice sheet. Both mechanisms are discussed in detail by *Eicken and Lange* [1989] in reference to platelet ice formation in the southeastern Weddell Sea. The disposition of the crystals in McMurdo Sound platelet ice is consistent generally with mechanism 1; we obtained no compelling evidence for the existence of mechanism 2 such as horizontally layered platelet structure that would be expected of platelet crystals, formed in the water column, and subsequently transported by buoyancy forces to the underside of the ice sheet

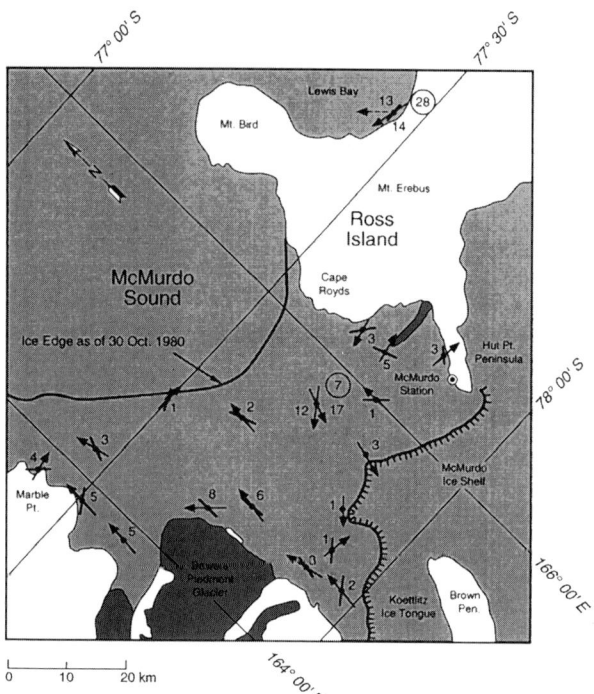

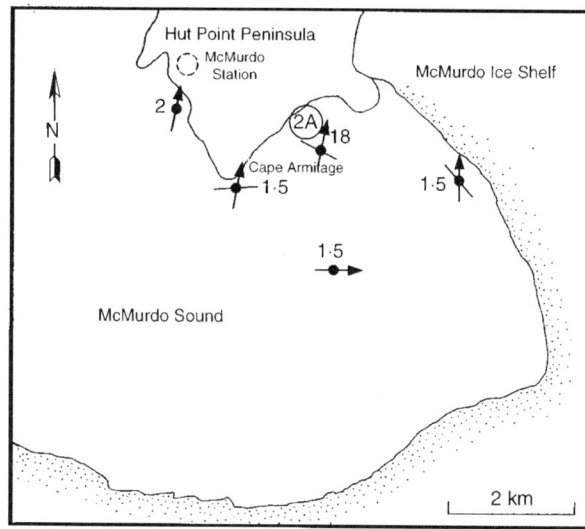

Figure 12. Comparison between c-axis alignment direction measurements of columnar-congelation ice formed earlier in the winter and current meter measurements (arrows) made directly beneath the ice cover at a number of sites in McMurdo Sound in October-November 1980. Numbers beside arrows are current velocities in cm/s. Site numbers were deleted to make for clearer presentation of current meter data.

[*Gow et al.*, 1987]. However, we frequently observed copious amounts of platelets rising to the tops of freshly drilled holes. These crystals are likely derived from a layer of platelets, as yet unconsolidated, adhering to the bottom of the ice sheet. These observations agree with those of *Crocker and Wadhams* [1989] and *Jeffries et al.* [1993] relevant to their studies of platelet ice growth in McMurdo Sound. *Eicken and Lange* [1989] also attribute part of platelet ice formation in southeastern Weddell Sea to consolidation of platelet crystals that grow attached to the bottom of the ice sheet.

Gow et al. [1987] also reported the occurrence of large wafer-like crystals in the bottoms of several multiyear floes in the western Weddell Sea. These crystals were so loosely bonded that it was often difficult to obtain competent cores when drilling this kind of ice. While bearing some resemblance to platelet ice in McMurdo Sound, it would appear that such loosely consolidated assemblages of crystals observed by *Gow et al.* [1987] in Weddell Sea pack ice may have more in common, structurally, with so-called underwater ice described by Serikov [1963], which he observed constituting the bulk of one-year-old sea ice forming off the east Antarctic coast near Mirny Station. According to Serikov, the platelets comprising the "underwater ice" nucleate in the water column beneath an existing ice sheet, prior to floating up to the bottom of the ice and freezing to it. The case for deep water formation of platelet crystals has been made by *Dieckmann et al.* [1986], who reported on the occurrence of an ice platelet mass (mistakenly identified by echosounder for a shoal of krill), trawled from 250 m depth near the Filchner Ice Shelf. Formation of this mass of platelets was attributed to crystallization in supercooled water streaming out from beneath the Filchner Ice Shelf. This at-depth-source of platelets is considered by *Eicken and Lange* [1989] to be an important contributor to platelet ice formation in the southeastern Weddell Sea.

In McMurdo Sound, in 1980, platelet ice made its first appearance only after substantial growth of normal columnar-congelation ice, of the order of 1.70–2.00 m thick. This late-stage appearance of platelet ice suggests that its formation is triggered by the onset of events or conditions in the water column that are inimical to the undisturbed growth of normal congelation ice. The initial appearance of platelet ice may vary from year to year as shown, for example, by *Crocker and Wadhams* [1989] on the basis of their own measurements performed in the winter of 1986 and on the observations made in 1911 and 1912 by *Wright and Priestley* [1922]. According to Crocker and Wadhams, crystals making up the platelets are formed when supercooled water comes in contact with the underside of the sea ice in McMurdo Sound. The source of this water is generally believed to be the adiabatically supercooled, low-density water that flows northward into the Sound from beneath the Ross Ice Shelf [*Lewis and Perkin*, 1985; *Crocker and Wadhams*, 1989; *Jeffries et al.*, 1993].

Crocker and Wadhams [1989] report that the first appearance of platelet ice in McMurdo Sound during 1986 occurred in July. They also cite the observations of *Wright and Priestley* [1922] of platelet crystal growth on manila ropes during the periods 5 August–12 October 1911, and 28 August–28 September 1912. On the assumption that platelet crystal growth to the bottom of the sea ice and on manila ropes are manifestations of the same process, then it would appear that 1) several months of normal congelation ice growth precedes the onset of platelet ice formation, 2) the date of the first appearance of platelet ice can vary significantly from year to year, and 3) the duration of platelet ice growth is also variable. Additionally, our observations and those of *Jeffries et al.* [1993] would indicate that no further growth of columnar-congelation ice occurs once platelet ice growth is established, except as ice filling the voids between the platelets. The amount of platelet ice formed can vary considerably from site to site, as indicated in Figure 6 and Table 1. Year to year variations in platelet ice thickness at a particular site can also be appreciable. For example, *Sullivan et al.* [1983] observed significant differences at their major study site near Cape Armitage: a virtual absence of identifiable platelet ice in 1980 and 1981, but a platelet layer thickness of up to 1.0 m in 1982. The importance of supercooled water to the whole process of platelet ice formation is further indicated by the virtual absence of platelet ice at site 28 located outside McMurdo Sound close to the shore in Lewis Bay, where it is essentially isolated from ice shelf water.

Platelet ice grows through the loss of heat through the water column, whereas the growth of columnar-congelation ice is controlled by heat loss by conduction through the overlying ice sheet. As noted by *Jeffries et al.* [1993], these two growth mechanisms collectively result in greater thicknesses of ice in McMurdo Sound than would otherwise occur with just congelation ice growth. An Arctic counterpart of sub-ice platelet ice growth beneath landfast first-year ice does not seem to exist or, if it does, it is yet to be demonstrated. However, platelet ice exhibiting structural characteristics very similar to that of Antarctic platelet ice has been observed in Arctic second-year and multiyear floes [*Eicken*, 1994; *Jeffries et al.*, 1995]. One suggested mechanism, first proposed by *Eicken* [1994], involves growth of platelets in under-ice melt ponds formed from the drainage of surface melt water to the undersides of floes. A second mechanism, suggested by *Jeffries et al.* [1995], invokes the operation of ice pumps in which ice that is melted off the deeper parts of floes and ridges is deposited as platelet ice at a higher level.

Examples of the variable nature of crystal textures observed in vertical thin section of platelet ice at six widely separated sites in McMurdo Sound are shown in Figures 13 and 14. The textures, dominated by blade-like crystal forms, are very similar to those reported by *Jeffries et al.* [1993]. Horizontal thin section photographs of platelet ice from five sites in McMurdo Sound are presented in Figure 14. The unique ice plate/brine layer substructure of the interstitial congelation ice crystals is evident in several of the sections.

the measured properties. Ice thicknesses ranged from 1.60 to 2.49 m. The mean thickness of 2.02 m is in generally good agreement with results obtained by other observers over the past 15 years. However, comparisons with measurements made during the 1960s indicate that a significant decrease of the annual ice thickness may have occurred in the intervening years. Snow thicknesses varied according to the proximity of the sea ice to the McMurdo Ice Shelf, ranging from 0.50 to 0.60 m near the shelf edge to only trace amounts in the more remote parts of the embayment. A strong correlation was also observed between the light-inhibiting nature of the snow cover and the extent of growth of visible algae in the basal ice, with little or no visible algae observed at locations where the snow cover exceeded 0.20 m.

Three major ice types were observed contributing to the structure of the sea ice in McMurdo Sound. The dominant ice type was columnar-congelation ice overlain by a layer of transition ice and underlain by platelet ice. The transition ice included granular ice at the top, consisting

Figure 13. Vertical thin section photographs taken between crossed polarizers of the variable crystal textures observed in basal platelet ice at sites 7, 9, 14, 17, 24 and 26 in McMurdo Sound.

CONCLUSIONS

Late winter observations of the physical and structural characteristics of land-fast sea ice in McMurdo Sound, Antarctica, have revealed significant variability in many of

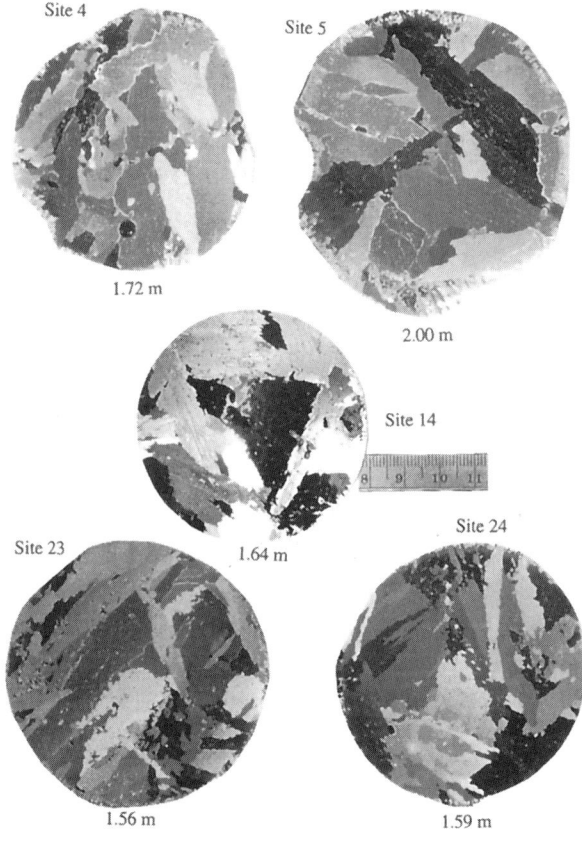

Figure 14. Horizontal thin section photographs taken between crossed polarizers of platelet-ice textures at sites 4, 5, 14, 23 and 24 in McMurdo Sound. The smallest subdivisions on the scale beside the site 14 thin section measure 1 mm. The same scale applies to all five sites.

mainly of frazil, overlying that section of initial congelation ice growth leading to crystal textures exhibiting substantially horizontal c-axes. Transition ice represented 15% of the total ice thickness. Continued growth of congelation ice generally resulted in the alignment of the c-axes within the horizontal plane. This constituted the columnar-congelation zone consisting of vertically elongated crystals 100 mm or more in length. Fully developed c-axis alignments did not generally occur until some distance into the columnar-congelation ice zone and usually weakened or terminated some distance above the platelet ice. Columnar-congelation ice comprised, on average, 72% of the total ice thickness, increasing to about 83% if the columnar-congelation component of the transition layer is included. Platelet ice comprised the basal ice at all first-year sampling sites; it represented, on average, 13% of the ice cover in McMurdo Sound.

Cores were carefully oriented in the azimuth to facilitate determination of c-axis alignments relative to measured and inferred current directions. Current action at the ice/water interface is now known to exercise a major effect in which the c-axes of the crystals become aligned parallel to the long-term current direction. In general, it was found that the c-axis alignments corresponded reasonably well with what is known of water circulation patterns in McMurdo Sound, and with current directions inferred from coastal outlines. However, currents measured beneath the ice during active growth of the platelet ice phase often deviated significantly from those inferred from c-axis alignments of columnar-congelation ice formed several weeks or months earlier. These observations would suggest that significant changes in the pattern of near-surface currents have occurred coincidentally with the onset of growth of platelet ice onto the bottom of the existing sea ice sheet. Platelet ice originates by the nucleation and growth of platelet crystals directly to the bottom of the ice. Actual consolidation to platelet ice occurs by freezing of water in the interstices of the platelet crystal network. In the particular case of McMurdo Sound, platelet ice formation is generally believed to be driven by adiabatic expansion and supercooling of low-density water emerging from beneath the Ross Ice Shelf. Considering the fragile nature of platelet crystals accreting to the underside of the ice, it seems likely that much of this in situ growth must have occurred during periods of reduced current velocity prior to consolidation of the platelet crystal structure by freezing of the interstitial water.

Bulk salinities averaged 6 $^o/_{oo}$, which is appreciably more saline than for ice of corresponding thickness and age in the Arctic. C-shaped profiles, linked to salinity enhancement at the tops and bottoms of cores, were observed at a number of sampling sites. Most profiles, however, exhibited iso-saline characteristics for the bulk of the ice sheet thickness.

Acknowledgments. We are indebted to the Office of Polar Programs, National Science Foundation, for financial support of this project. The support provided by Antarctic Services, Inc., and by VX-6, U.S. Navy, is gratefully acknowledged. We wish to thank Dr. Stephen Jones for access to his unpublished data on McMurdo Sound sea ice. The authors also acknowledge the support of the drafting and photo service sections of CRREL and Vicki Keating in the preparation of this paper.

REFERENCES

Ackley, S.F., Mass-balance aspects of Weddell Sea pack ice, *J. Glaciol.*, 24(90), 391–405, 1979.

Ackley, S.F., Sea-ice atmosphere interactions in the Weddell Sea using drifting buoys, in: *Sea Level, Ice and Climatic Change* (I. Allison, Ed.), International Association of Hydrological Sciences Publication No. 131, 177–191, 1981.

Ackley, S.F., A.J. Gow, K.R. Buck, and K.M. Golden, Sea ice studies in the Weddell Sea aboard USCGC *Polar Sea*, *Antarct. J. U.S.*, 15(5) 84–86, 1980.

Allison, I., and S. Qian, Characteristics of sea ice in the Casey region, ANARE *Research Notes* No. 28, 47–56, 1985.

Bennington, K.O., Some crystal growth features of sea ice, *J. Glaciol.*, 4(36), 669–688, 1963.

Campbell, K.J., and A.J. Orange, The electrical anisotropy of sea ice in the horizontal plane, *J. Geophys. Res.*, 79(33), 5059–5063, 1974.

Cherepanov, N.V., Structure of sea ice of great thickness, *Trudy Arktiki i Antarktiki Nauchno-Issled Institute*, 367, 13–18, 1964.

Cherepanov, N.V., Spatial arrangement of sea ice crystal structure, *Problemy Arktiki I Antarktiki*, 38, 176–181, 1971.

Cox, G.F.N., and W.F. Weeks, Salinity variations in sea ice, *J. Glaciol.*, 13(67), 109–120, 1974.

Crocker, G.B., and P. Wadhams, Modelling Antarctic fast-ice growth, *J. Glaciol.*, 35(119), 3–8, 1989.

Dieckmann, G., G. Rohardt, H. Hellmer, and J. Kipfstuhl, The occurrence of ice platelets at 250-m depth near the Filchner Ice Shelf and its significance for sea ice biology, *Deep-Sea Res.*, Part A, 33(2), 141–148, 1986.

Eicken, H., Structure of under-ice melt ponds in the central Arctic and their effects on the sea-ice cover, *Limnol. Oceanogr.*, 39(3), 682–694, 1994.

Eicken, H. and M.A. Lange, Development and properties of sea ice in the coastal regime of the southeastern Weddell Sea, *J. Geophys. Res.*, 94(C6), 8193–8206, 1989.

Gow, A.J., W.F. Weeks, J.W. Govoni, and S.F. Ackley, Physical and structural characteristics of sea ice in McMurdo Sound, *Antarct. J. U.S.* 16(5), 94–95, 1981.

Gow, A.J., S.F. Ackley, W.F. Weeks and J.W. Govoni, Physical and structural characteristics of Antarctic sea ice, *Ann. Glaciol.*, 3, 113–117, 1982.

Gow, A.J., S.F. Ackley, K.R. Buck and K.M. Golden, Physical and structural characteristics of Weddell Sea pack ice, *CRREL Report 87-14*, 1987.

Gow, A.J., D.A. Meese, D.K. Perovich, and W.B. Tucker, The anatomy of a freezing lead, *J. Geophys. Res.*, 95(C10), 18221–18232, 1990.

Gow, A.J., and W.B. Tucker, Physical and dynamic properties of sea ice in the polar oceans, *CRREL Monograph 91-1*, 1991.

Gow, A.J., and J.W. Govoni, An 80 year record of retreat of the Koettlitz Ice Tongue, McMurdo Sound, Antarctica, *Ann. Glaciol.*, 20, 237-241, 1994.

Heine, A.J., Ice breakout around the southern end of Ross Island, Antarctica, *N.Z.J. Geol. and Geophys.*, 6(3), 395–402, 1963.

Hendrickson, G., and R. Rowland, Strength studies on Antarctic sea ice, *CRREL Technical Report 157*, 1965.

Jacka, T.H., I. Allison, R. Thwaites, and J.C. Wilson, Characteristics of the seasonal sea ice of East Antarctica and comparisons with satellite observations, *Ann. Glaciol.*, 9, 85–91, 1988.

Jeffries, M.O., and W.F. Weeks, Fast ice properties and structure in McMurdo Sound, *Antarct. J. U.S.*, 26(5), 94–95, 1991.

Jeffries, M.O., and W.F. Weeks, Structural characteristics and development of sea ice in the western Ross Sea, *Antarctic Science*, 5, 63–75, 1992.

Jeffries, M.O., W.F. Weeks, R. Shaw, and K. Morris, Structural characteristics of congelation and platelet ice and their role in the development of Antarctic land-fast sea ice, *J. Glaciol.* 39(132), 223–238, 1993.

Jeffries, M.O., K. Schwartz, K. Morris, A.D. Veazey, H.R. Krouse, and S. Cushing, Evidence for platelet ice accretion in Arctic sea ice development, *J. Geophys. Res.*, 100 (C6), 10,905–10,914, 1995.

Kohnen, H. On the dc-resistivity of sea ice, *Z. für Gletscherkunde und Glazialgeologie*, 11(2) 143–154, 1976.

Kovacs, A., and R.M. Morey, Radar anisotropy of sea ice due to preferred azimuthal orientation of the horizontal c-axes of ice crystals, *J. Geophys. Res.* 83(12), 6037–6046, 1978.

Kovacs, A., A.J. Gow, J.H. Cragin, and R.M. Morey, The brine zone of the McMurdo Ice Shelf, Antarctica, *CRREL Report 82-39*, 1982.

Lange, M.A., S.F. Ackley, P. Wadhams, G.S. Dieckmann, and H. Eicken, Development of sea ice in the Weddell Sea, *Ann. Glaciol.*, 12, 92–96, 1989.

Langhorne, P.J., Laboratory experiments on crystal orientation in NaCl ice, *Ann. Glaciol.*, 4, 163–169, 1983.

Langhorne, P.J., and W.H. Robinson, Alignment of crystals in sea ice due to fluid motion, *Cold Reg. Sci. Technol.*, 12(2), 197–214, 1986.

Langway, C.C., Ice fabrics and the universal stage, *SIPRE Technical Report 62*, 1958.

Leventer, A., R.B. Dunbar, M.R. Allen, and R.Y. Wayper, Ice thickness in McMurdo Sound, *Antarct. J. U.S.*, 22(5), 94–96, 1987.

Leventer, A., and R.B. Dunbar, Recent diatom record of McMurdo Sound: Implications for history of sea ice extent, *Paleoceanography*, 3(3), 259–274, 1988.

Lewis, E.L., and R.G. Perkin, The winter oceanography of McMurdo Sound, Antarctica, *Antarct. Res. Ser.*, 43, 145–165, 1985.

Morris, K., and M.O. Jeffries, Ice thickness variability of the McMurdo Sound landfast ice runway, *Antarct. J. U.S.*, 27(5), 83–85, 1992.

Nakawo, M., and N.K. Sinha, Growth rate and salinity profile of first-year sea ice in the high Arctic, *J. Glaciol.*, 27(96), 315–330, 1981.

Paige, R.A., Crystallographic studies of sea ice in McMurdo Sound, Antarctica, *U.S. Naval Civil Engineering Laboratory, Technical Report R494*, 1966.

Paige, R.A., Sea ice on McMurdo Sound, Antarctica: Deep Freeze 67 thickness and temperature studies, *Naval Civil Engineering Laboratory, Technical Note N-927*, 1967.

Paige, R.A., Breakout of the McMurdo Ice Shelf, *Naval Civil Engineering Laboratory, Internal Report 7*, 1971.

Palmisano, A.C., S.T. Kottmeier, R.L. Moe, and C.W. Sullivan, Sea ice microbial communities. IV. The effect of light perturbation on microalgae at the ice-seawater interface in McMurdo Sound, Antarctica, *Marine Ecol. Prog. Ser.*, 21, 37–45, 1985.

Peyton, H.R., Sea ice strength, *Geophysical Institute, University of Alaska Report, UAG-182*, 1966.

Serikov, M.I., Structure of Antarctic sea ice, *Infor. Bull. of Soviet Antarctic Exped.*, 4(5), 265–266, 1963.

Smith, D.D., Ice lithologies and structure of ice island Arlis II, *J. Glaciol.*, 5(37), 17–38, 1964.

Stander, E., and B. Michel, The effect of fluid flow on the development of preferred orientations in sea ice: Laboratory experiments, *Cold Reg. Sci. Technol.*, 17(2), 153–161, 1989.

Sullivan, C.W., A.C. Palmisano, S. Kottmeier, S. McGrath-Grossi, R. Moe, and G.T. Taylor, The influence of light on development and growth of sea-ice microbial communities in McMurdo Sound, *Antarct. J. U.S.*, 18(3), 77–179, 1983.

Sullivan, C.W., A.C. Palmisano, S. Kottmeier, S. McGrath-Grossi, and R. Moe, The influence of light on growth and development of the sea-ice microbial community in McMurdo Sound. in: *Antarctic nutrient cycles and food webs (Proceedings of the Fourth SCAR Symposium on Antarctic biology)*, edited by W.R. Siegfried, P.R. Condy and R.M. Laws, pp. 78–83, 1985.

Untersteiner, N., Natural desalination and equilibrium salinity profile of perennial sea ice, *J. Geophys. Res.*, 73(4), 1251–1257, 1968.

Veazey, A.D., Development and variability of the structure and physical properties of land fast sea ice in McMurdo Sound, Antarctica, M.S. thesis, 168 pp., Univ. of Alaska, Fairbanks, May 1994.

Weeks, W.F., and S.F. Ackley, The growth, structure and properties of sea ice, *CRREL Monograph 82-1*, 1982.

Weeks, W.F., and A.J. Gow, Preferred crystal orientations in the fast ice along the margins of the Arctic Ocean, *J. Geophys. Res.*, 83(C10), 5105–5121, 1978.

Weeks, W.F., and A.J. Gow, Crystal alignments in the fast ice of Arctic Alaska, *J. Geophys. Res.*, 85(C2), 1137–1146, 1980.

Wright, C.S., and R.E. Priestley, *British (Terra Nova) Antarctic Expedition, 1910–1913. Glaciology*, London, Harrison and Sons, 1922.

(Received November 21, 1996; accepted March 24, 1997)

LINKING LANDFAST SEA ICE VARIABILITY TO MARINE ICE ACCRETION AT HELLS GATE ICE SHELF, ROSS SEA.

J.-L. Tison[1], R.D. Lorrain[1] and A. Bouzette[1], M. Dini[2], A. Bondesan[3], and M. Stiévenard[4]

Eleven first-year sea ice cores collected in the vicinity of Hells Gate Ice Shelf (Terra Nova Bay, Ross Sea) are analyzed for their textures, ice fabrics, salinities, chemical composition and oxygen-18 values. These are compared to a new data set of four 10-45 m long marine ice cores drilled close to the ice shelf front. During most of the winter, granular frazil ice accretion prevails at the bottom of the landfast first-year sea ice cover in front of Hells Gate Ice Shelf, eventually in alternation with a draped facies. At the end of the winter, platelet ice is more commonly found. These facies are thought to result from frazil ice production in the Ice Shelf Water outflow associated with a Deep Thermohaline Circulation (DTC) in which High Salinity Shelf Water (HSSW) formed in the Terra Nova Bay Polynya descends beneath the Nansen Ice Shelf. Bottom accretion at the base of the landfast first-year sea ice still occurs during the summer as a banded rectangular facies. It can form up to 54% of the ice core thickness. The seasonality of this banded facies accretion at the bottom of the landfast sea ice, the dominance of the banded marine ice facies closer to the ice shelf front and the chemical trends in the granular marine ice facies suggest that a link must exist between the banded facies' genesis and the tidal pumping of warm surface waters below the ice shelf during the summer.

INTRODUCTION

Ice shelves and ice streams are an essential key to the stability of ice sheets under climatic changes. Ice shelves play a buttressing role on the ice streams' outflow, through their restricted flow over ice rises and rumples as well as due to lateral boundaries. A major issue in this regard is the process of ice shelf stabilization through increased thickness resulting from marine ice accretion at the bottom (ice crystals formed in the water column below the ice shelf). Considerable amounts of marine ice are sometimes observed where they are not predicted to occur in such quantities by 2-D models applied along longitudinal profiles of the ice shelf/sub-ice shelf cavity system [*Hellmer and Jacobs*, 1992].

The mass balance of ice shelves at the ice-ocean interface is, however, a variable still difficult to appraise, partly because of poor accessibility. Firstly, only a few deep ice cores drilled through the entire ice shelf thickness (or nearly so) are available at present in the Antarctic [*Eicken et al.*, 1994; *Morgan*, 1972; *Oerter et al.*, 1994; *Oerter et al.*, 1992; *Ragle et al.*, 1960; *Zotikov et al.*, 1980]. Secondly, specific climatic conditions, namely strong katabatic wind regimes inducing sustained surface ablation, sometimes allow marine ice to outcrop at the surface of smaller antarctic

---

[1]Département des Sciences de la Terre et de l'Environnement, Université de Bruxelles, Brussels, Belgium
[2]Laboratorio di Geochimica Isotopica, Università di Trieste, Trieste, Italy
[3]Dipartimento di Geografia "G. Morandini", Università di Padova, Padova, Italy
[4]Laboratoire de Modélisation du Climat et de l'Environnement, Gif sur Yvette, France

Copyright 1998 by the American Geophysical Union

ice shelves and ice tongues providing sampling opportunities at minimal cost [*Gow and Epstein*, 1972; *Souchez et al.*, 1991; *Tison et al.*, 1993]. A third approach consists of the study of landfast sea ice forming at the front of ice shelves or in the large rifts that often dissect the ice shelf (ice tongue) area close to the front [*Jeffries et al.*, 1993; *Kipfstuhl*, 1991; *Tison et al.*, 1991]. These landfast sea ice cores show specific facies rarely described in drifting pack ice.

The purpose of this paper is to: (1) present a new data set on 10-45 m long marine ice cores from Hells Gate Ice Shelf (HGIS), that will be shown to change, at least to some extent, the perspectives obtained from previous work; (2) present two data sets on first-year landfast sea ice at HGIS and discuss the processes leading to the build-up of the landfast ice cover; (3) relate the spatial and temporal variability of the landfast first-year sea ice facies to the one observed in the marine ice cores on a larger time scale (typically several hundreds of years); and (4) suggest paths for future investigations of ice-ocean interactions in the area.

CURRENT SEA ICE/MARINE ICE FACIES NOMENCLATURE

At this stage, we feel it is useful to support our data with a summary of the present-day knowledge of the sea ice textural facies and associated genetic processes. The purpose of this section is primarily to unambiguously illustrate the terminology used in the following sections. It has often been a matter of debate in which circumstances should textural or genetic terminology be used, and many authors use both undiscriminately. Also, the same terminology is used for different facies by different authors and vice-versa. For example, as discussed below, there is a strong discrepancy between the textural term of "platelet ice" introduced by *Lange* [1988] and the actual texture of the ice described, where all crystals appear acicular (needle-like) both in horizontal and vertical thin sections. Finally, some textural genetic equivalencies are still controversial or poorly known. Therefore, the contents of Plate 1 and Table 1 should by no means be considered as an exhaustive classification of sea ice facies, a subject far beyond the scope of the present work. Marine ice studies are fairly recent and, therefore, no attempt has been made until now to develop a proper nomenclature. Initially, textural facies in marine ice cores should not significantly differ from those observed in sea ice given the similarity of environments. Furthermore, as shown later in the paper, ice facies described in marine ice are also showing up in landfast sea ice cores. For these reasons, Plate 1 groups the main sea- and marine-ice textural facies under the same nomenclature. It should, however, be kept in mind that the marine ice accreted at the bottom of large ice shelves undergoes large post-depositional cumulative strains with time that can lead to considerable alterations of the original texture. Similarly, the impact of shearing on the ice shelf margins and/or converging flow in smaller ice shelves (as is the case for HGIS) can also partially alter the textures in the marine ice, producing typical facies that should not be expected in sea ice (even if it is deformed at a later stage).

In Plate 1, we have selected small illustrative samples that we believe best reflect the descriptive terms. Text in blue relates to the main "classical" marine or sea ice types from the literature. Text in red refers to more recent discoveries in landfast sea ice or marine ice, mainly in the Ross Sea area. Table 1 summarizes, in a few simple words, possible textural-genetic connections. Both Plate 1 and Table 1 were compiled from multiple references in the literature [*Dieckmann et al.*, 1986; *Eicken and Lange*, 1989; *Eicken et al.*, 1994; *Gow et al.*, 1987; *Jeffries et al.*, 1993; *Kipfstuhl*, 1991; *Lange*, 1988; *Lange*, 1990; *Lorrain et al.*, 1997; *Oerter et al.*, 1992; *Tison et al.*, 1993; *Weeks and Ackley*, 1986]. In this paper we will use the "facies" terminology "sensu-lato", that is, to describe various textural arrangements, without any implication of their genesis.

Columnar ice, made of large, vertically elongated crystals with intracrystalline substructures is unanimously recognized as the unequivocal signature of *congelation* ice resulting from direct freezing of sea water, at the bottom of an ice floe, for example. It is also generally agreed that the term *granular* applies to any sea (marine ice) texture with equigranular crystals. When it is polygonal (sharp linear crystal contours) with no interlocking grains and often quite porous, it is the signature of *snow ice*, a facies inherited from sea water infiltration of the surface snow layer of the sea ice cover. Another granular facies closely related to the snow cover, but less frequently encountered, and described but not illustrated here, is *superimposed ice* [see, for example, *Kawamura et al.*, 1997]. When granular ice is polygonal, but clearly shows interlocking morphology of grain boundaries and eventually subgrains and polygonization features, it points to a considerable retexturing of the initial crystals through recrystallization under stress.

Frazil ice is a general genetic term for individual ice crystals formed in the water column. Although, following the WMO nomenclature [*WMO*, 1985], individ-

Table 1. Textural-Genetic Equivalences for Sea Ice and Marine Ice

Textural Facies	Sub-facies	Genetic term	Process
Columnar		*Congelation sea ice*	Direct freezing of sea water
Granular	Polygonal (not interlocking, voids)	*Snow ice*	Infiltration of sea water in the surface snow-cover
	Polygonal (interlocking, medium to coarse)	*Frazil ice*	Modified by dynamical recrystallization (cumulative strain during ice shelf spreading)
	Orbicular	*Frazil ice*	Individual ice crystals forming in the water column: • wind- and wave-induced (surface layer-max:50 cm) • double diffusion between two water masses • thermohaline convection at the base of a growing congelation layer • adiabatic supercooling in rising water masses
Platelet		*Frazil ice*	Adiabatic supercooling in rising water masses, and passive transport to the front of ice shelves
Draped		*Mixed Platelet/ Congelation*	
String-lined		*Frazil ice*	Modified by dynamical recrystallization under compressive stress
Banded	Rectangular	*Frazil ice*	Adiabatic supercooling in rising water masses
	Wave-like		

ual crystals are described as "fine spicules or plates of ice, suspended in water", it can result in more diverse textural facies both in sea ice and marine ice. It commonly occurs in sea ice as a *granular orbicular* texture, with rounded crystal boundaries both in vertical and horizontal thin sections. *Weeks and Ackley* [1986] gave a clear synthesis of the four major processes that contribute to frazil ice production in the polar oceans (see Table 1). The *wind- and wave-induced* process concerns sea ice exclusively. *Thermohaline convection at the base of a growing congelation ice layer* applies wherever congelation ice forms, that is both below sea ice and, at least theoretically, at the bottom of ice shelves.

Double-diffusion between two water masses occurs wherever a salinity gradient exists between water masses both at their pressure melting point. Finally, *adiabatic supercooling in rising water masses* requires the large thicknesses of ice shelves (marine ice), although it can eventually feed the landfast sea ice cover with individual frazil ice crystals.

Of particular interest is the *platelet facies*. Since it has generally only been observed close to ice shelf fronts it has been attributed to unrestrained frazil ice nucleation and growth resulting from adiabatic supercooling in ascending Ice Shelf Water as part of the Deep Thermohaline Convection (DTC) [e.g., *Eicken* and Lange, 1989; *Kipfstuhl*, 1991]. Nucleation and growth of plate-like frazil ice crystals directly to the bottom of an existing ice sheet from an adiabatically supercooled water mass has also been invoked [*Eicken and Lange*, 1989; *Gow et al.*, this volume]. Some authors [*Eicken et al.*, 1994] suggest that the platelet facies is the original texture of all marine ice formed in the DTC under large ice shelves, and that it is eventually later reduced in size and shape under deviatioric buoyancy stress, as thick layers of loose platelets accrete at the bottom. Platelet facies are indeed only observed in very small amounts and in particular occurrences in marine ice cores [*Tison et al.*, 1993]. Some debris-free granular marine ice facies observed at HGIS are, however, so fine grained that, in that case, a process of fragmentation under the deviatoric buoyancy stress is hard to envisage. Furthermore, these fine grained facies also occur as irregular pockets surrounded by populations with larger crystal sizes, and these structures are more easily explained in terms of heterogeneity of the loose frazil as it consolidates, rather than in terms of homogeneous large scale post-depositional deformation structures. *Bombosch and Jenkins* [1995] suggest, by modeling frazil ice production in Ice Shelf Water, that the platelet ice facies could be associated with the evolution of ice crystals passively entrained in Ice Shelf Water as it levels off and exits at some depth in front of the ice shelf.

Recent work has demonstrated that what is commonly named as the platelet facies shows in fact a considerable variability. *Lange* [1988], who initially introduced the terminology, describes this facies as predominantly made of "elongated grains" (spicules) both in vertical [*Lange*, 1988: Figure 5*a*] and horizontal [*Lange*, 1988: Figure 5*b*] thin sections. The author mentions that platelet ice is, however, often interspersed with regions of smaller equigranular grains. Further observations [*Jeffries and Weeks*, 1993; *Jeffries et al.*, 1993] pointed to the fact that the platelet facies generally occurs as a *mixed platelet/congelation facies* with a wide range of proportions between the two components. Figure 1*b* illustrates an example of the platelet ice facies comprised almost entirely of platelet crystals. Figure 1*a* shows an example where individual platelet crystals are less abundant. In that case, they are eventually less angular and show wavy uneven edges. This we call the *draped facies* following *Jeffries et al.* [1993] who pointed out the draped appearance of the platelet crystals. Other mixed facies like the mixed columnar/granular facies were also extensively described in the Weddell Sea [e.g. *Eicken and Lange*, 1989].

When granular orbicular marine ice is submitted to strong compressive stresses (for example between two coalescent continental floating ice flows) it develops fold structures underlined by rectangular crystals arranged in individual curved alignments. This particular facies was recently described by *Tison et al.* [1997] and called *string-lined facies*. It is not known presently if it could develop in strongly deformed pressure ridges of landfast sea ice.

Finally, as discussed further in this paper, a new marine ice facies was discovered in the frontal area of Hells Gate Ice Shelf. Typically it is composed of stacks of rectangular crystals of various sizes in vertical thin section, the limits of which are commonly underlined by trains of brine inclusions [*Lorrain et al.*, 1997; *Tison et al.*, 1993]. This facies was named "banded facies" and shows strong similarities with the bottom layer of some of the southernmost sea ice cores (52 and 54) described by *Gow et al.* [1987] in the Weddell Sea. These authors describe it as "wafer-like" crystals [*Gow et al.*, 1987: Figure 14 (left) and Figure 25] and note that "the crystals are so loosely bonded at the very bottom that it was often difficult to obtain competent core when drilling floes containing this kind of ice". Ice fabrics in these layers exhibited vertical c-axes and the authors suggested that this type of ice might represent underwater ice as described by Russian observers at Antarctic coastal locations [e.g.,*Serikov*, 1963]. We will see later that this facies is also found in the landfast sea ice accreting in front of HGIS. It should be noted (see Plate 1 and Figure 1) that the banded facies strongly differs from the other frazil ice facies associated with adiabatic supercooling in DTC, i.e., platelet and draped facies. It shows no interstitial congelation ice, and the crystals appear rectangular in vertical thin sections, and platy (with shaded extinctions) in horizontal thin sections (see also Plate 3 and Figures 11 and 13). Furthermore, c-axes are strongly concentrated in a single maximum roughly perpendicular to the plates instead of the random arrangement found in the platelet facies. As a matter of fact, this facies would probably better deserve the "platelet" denomination than the platelet facies which would, in turn, be better described as an "acicular facies" or "mixed acicular-congelation facies".

CURRENT STATE OF KNOWLEDGE OF SEA ICE VARIABILITY IN THE ROSS SEA

Recent work by *Jeffries and Weeks* [1991*a*; 1991*b*; 1993], *Jeffries et al.* [1993] and *Jeffries and Adolphs* [1997], has concentrated on sea ice properties in

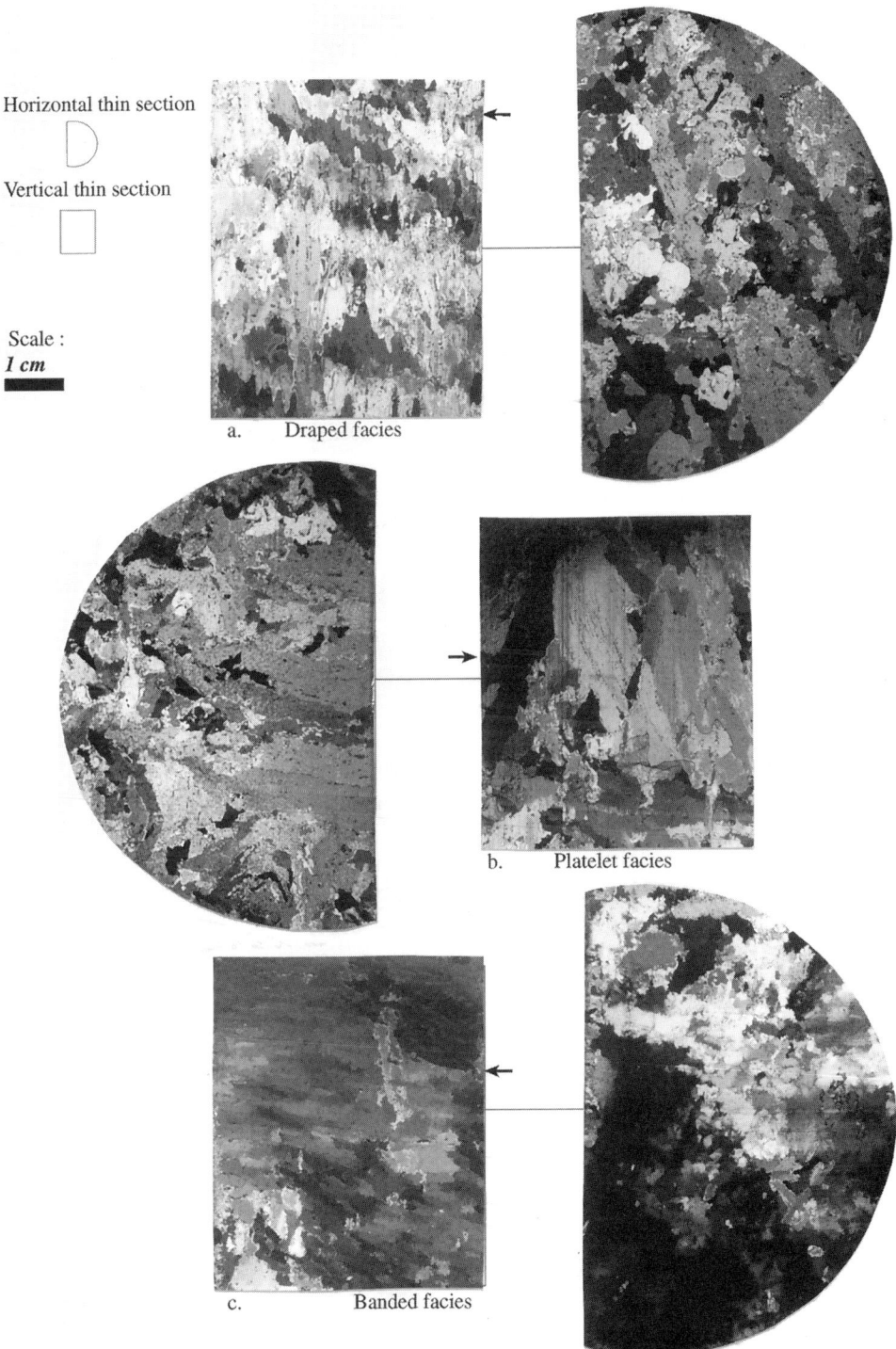

Fig. 1 : Comparison of the draped (a), platelet (b) and banded (c) facies in horizontal and vertical thin sections. The arrow on the vertical thin sections shows the location of the horizontal thin sections. All samples are from the landfast sea ice data sets at HGIS (a = core SA, depth : 212 cm; b = core SA, depth : 221 cm; c = core SE, depth : 175 cm).

McMurdo Sound and the western Ross Sea. *Jeffries and Weeks* [1993] sampled the sea ice cover in a transect from the Balleny Islands (62.50° S, 158.8° E) to McMurdo Station, via Cape Adare, Terra Nova Bay and Franklin Island (their Figure 1). Eighteen ice cores between 0.53 and 1.56 m long were sampled and analysed for their textural facies. Most of these cores were obtained quite far from the coast. Their texture is dominated by columnar (congelation) ice (22 to 100%) and frazil ice (5 to 74.4%). The latter is thought to have originated from a "pancake cycle" similar to the one described by *Wadhams et al.* [1987] and *Lange et al.* [1989] in the Weddell Sea. Attention should be drawn, however, to cores RS14 to RS16, that were retrieved from the South side of Drygalski Ice Tongue, i.e. close to a coastal floating ice body. Indeed, one of them (RS15) is the only one displaying platelet ice, and this in considerable amounts (64%). The authors drew attention on the fact that ISW was detected near the ice tongue and that the high proportion of platelet ice observed in the core might reflect adiabatic supercooling in this low density water outflow. On the other hand, core RS16 shows the highest percentage of frazil ice (88.5%), of which it is exclusively formed, if we discard "deformed ice" facies. Most recently, *Jeffries and Adolphs* [1997] have confirmed the strong contrast between pack ice and landfast sea ice characteristics in the Ross Sea, on an even wider scale. The pack ice floes they described were primarily built through dynamic processes and showed generally lesser thickness than the landfast ice for which thermodynamic processes are dominant.

Of even greater interest for studies of ice-ocean interactions under ice shelves is the investigation of landfast sea ice in McMurdo Sound (see detailed synthesis in *Gow et al.* [this volume]). *Jeffries et al.* [1993] describe sixteen sea ice cores between 1.26 and 2.34 m long sampled in the area. A typical structure of all these cores (illustrated in their Figure 6) consists of an upper congelation ice layer and lower layers of mixed congelation/platelet or platelet ice. Overall, the fast ice comprised 62.1% congelation ice, 28.3% platelet ice and 9.3% congelation/platelet ice. However the proportion of platelet ice might be underestimated due to bottom melting earlier in the summer. Apart from two exceptions, the basal layer of each core was platelet ice. However, the authors stressed clearly that the distinction between platelet and mixed platelet/congelation ice was quite arbitrary, since the platelet facies always showed interstitial congelation ice. The limit was arbitrarily taken at 50% between the "mixed" and the "pure" platelet facies. *Jeffries et al.* [1993] and *Jeffries and Weeks* [1993] also showed that fabric diagrams from platelet and platelet/congelation facies are bi-modal: c-axes concentrated in the equatorial circle on a Schmidt net correspond to congelation ice crystals whilst most of the platelet crystals from the same sample show a random distribution. It is also worth noting that a wide variety of platelet ice textures was observed in McMurdo Sound. The same authors mention that "...many of the platelets were less angular, more equidimensional and had wavy, uneven edges... densely packed accumulations of these wavy-edged platelets gave the ice a distinctive appearance resembling a draped fabric". Comparison of the percentage of each core containing platelet ice with contour lines of surface supercooling measured in 1982 by *Lewis and Perkins* [1986] supports the idea that the platelet ice originated by adiabatic supercooling of the Ice Shelf Water formed below the Ross Ice Shelf.

Finally, *Jeffries et al.* [1993] used the mean plate width in the congelation crystal sub-structure to compare growth rates of congelation ice at the top of the cores and of interstitial congelation ice in the mixed platelet/congelation facies. Surprisingly, they noticed no significant trend of decreasing growth rate with depth, as would be expected in a process where freezing is mainly controlled by heat conduction through the existing sea ice cover. This particular phenomenon was seen as the expression of a lower heat flux at the platelet-rich growth interface (where only interstitial water has to be frozen and is guaranteed to be at, or even below, the freezing point) than at the base of the unconsolidated platelet crystal layer.

The most recent description of McMurdo Sound landfast sea ice (based on a data set sampled in October-November 1980) by *Gow et al.* [this volume] generally supports the material published earlier. It shows in greater detail the relationship between the circulation patterns of near-surface currents and the strength of the ice fabric in the columnar-congelation ice. It also suggests from field and textural observations that the dominant mechanism of platelet ice accretion is nucleation and growth of plate-like frazil ice crystals directly to the bottom of the existing ice sheet.

HELLS GATE ICE SHELF : MORPHOLOGY, ICE DYNAMICS AND MASS BALANCE

Hells Gate Ice Shelf (Figure 2) is a small antarctic ice shelf located in Terra Nova Bay (Victoria Land, lat. 74°50'S, long. 163°50'E). It is derived from the larger

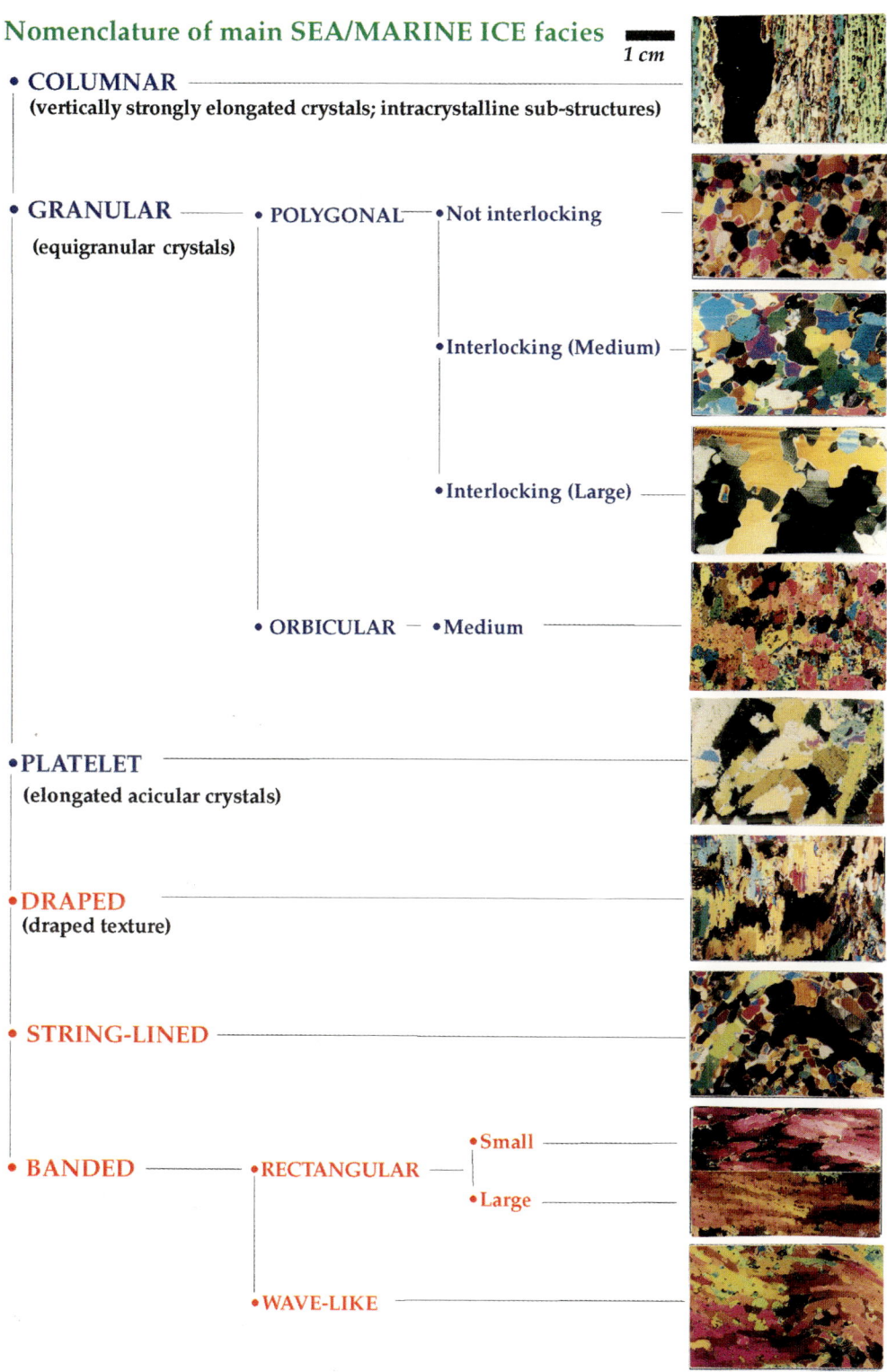

Plate 1 : Nomenclature for the main sea ice and marine ice facies discussed in this paper. The thin sections are in the vertical plane and the scale is valid for all pictures (see text for details).

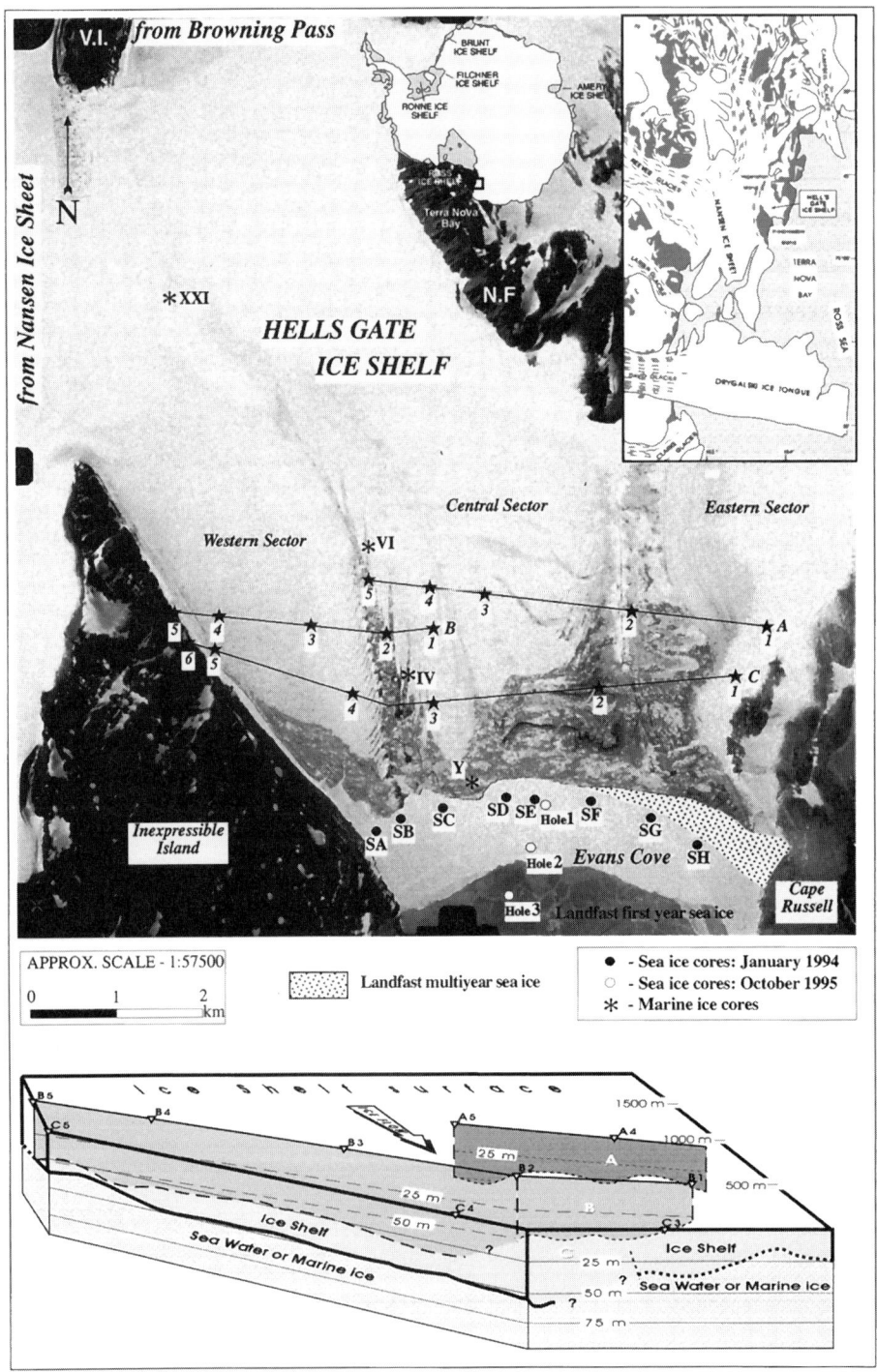

Fig. 2 : Location and main features of Hells Gate Ice Shelf showing the position of marine and sea ice cores and of the echosounding profiles from 1991 (top). The bottom drawing is a 3-D sketch of ice shelf thicknesses deduced from the 1991 echosoundings. Aerial photograph in the background is : Terra Nova Bay - 11.06.1985 - FL 152.27 - TMA-2851-V. V.I. = Vegetation Island; N.F. = Northern Foothills.

Nansen Ice Shelf (NIS) and reaches the sea near Cape Russell. HGIS extends from North to South for 16.6 km with a maximum width of 9.8 km. It is composed of three main sectors (western, central, eastern) separated by two medial moraines stretching roughly North-South.

Early radio-echo sounding (RES) profiles [*Souchez et al.*, 1991] obtained about 1.5 km inland from the ice front (Figure 2, bottom), showed a maximum ice shelf thickness of 70 m in the western sector reducing to only 20 m in some areas of the central sector. Furthermore, the interface picked-up by the RES was typically undulating in the central sector (A5-A3, B2-B1, C4-C3 in Figure 2 bottom), in phase with a clear-cut surface boundary between clear ice upstream and dark ice downstream. These data must, however, be treated carefully since, in several instances (especially in the central sector), the return signals were quite weak. It is, therefore, not known, in these cases, if the depth obtained corresponds to the ice-ocean interface, or to the continental meteoric ice-marine ice interface. Several arguments favor the latter hypothesis. First, seismic investigations [*F. Merlanti*, personal communication, May 1997] performed in the same area systematically provide higher ice thicknesses (22% higher in the western sector, 33% to 55% higher in the central sector). Secondly, isotopical ($\delta^{18}O$, δD) and crystallographic investigations of the surface clear-cut boundary indicate that it delimits continental meteoric ice (light gray, upstream) from marine ice (dark gray, downstream) [*Lorrain et al.*, 1997; *Tison et al.*, 1993]. Finally, core Y (Figure 2, top), obtained downstream of line C within a few tens of meters of the ice front, reached a depth of 44 m, clearly in contradiction of the RES data from C3-C2. For the same reasons, more recent RES data from further upstream are somewhat discordant [*Tison et al.*, 1997 and *A. Lozej*, personal communication]. However, measurements agree on a thickness of about 150 m some 6 kilometers inland from the front. A maximum value of 250 m has been measured nearly 10 kilometers from the front, in the central part of the flow between Vegetation Island and the Northern Foothills (Figure 2).

The specific lay-out of the meteoric ice/marine ice boundary near the front, the shape of the marine ice outcrops visible upstream in the eastern part of the central sector, the large scale foliation patterns and the RES return from the central sector all suggest that the ice shelf results from the coalescence of several individual ice flows of different sizes, marine ice accreting in the upstream area (near the grounding area or around a pinning point) acting as a "welding unit" between these various ice flows. This is a situation similar to the one detected, at a much larger scale, beneath the Ronne Ice Shelf by *Corr et al.* [1995]. Comparing RES data to precise GPS determination of ice thickness at Ronne Ice Shelf revealed large discrepancies due to considerable lateral accumulation of marine ice between the Rutford and Evans Ice Streams. These accumulations could only have formed as the two ice streams were ungrounding and joining together, upstream of the measured profiles.

Large scale morphological patterns (e.g. twin alignments of morainic dirt cones), marine ice outcropping patterns and ice foliation patterns also suggest a complex dynamical behavior of converging flow in the Hells Gate Ice Shelf. Side effects considerably affect the flow patterns in some areas [*Tison et al.*, 1997]. An oblique component of the flow in the western sector brings the ice in compression against the central sector, and results in shearing of part of the marine ice initially accreted between the two flows (Figure 3a). This results in the typical outcropping and foliation patterns of Figure 3b. At a smaller scale, this compressive regime affects the crystal structures both in the granular and banded facies (Figure 3c). The granular facies develops a preferential curved alignment of rectangular crystals ("string-lined" facies) and the initially rectangular crystals of the banded facies are re-arranged in folded structures. Similar features were successfully modeled and experimentally produced by *Wilson et al.* [1986] and *Wilson and Zhang* [1994]. More generally, the granular facies is frequently polygonal with interlocking crystal boundaries, a sign of crystal evolution under considerable cumulative strains (see section 2).

Horizontal surface velocities were estimated using two independent techniques by *Baroni* [1990], *Baroni et al.* [1991a], *Baroni et al.* [1991b], and *Frezzotti* [1993]. The western medial moraine is associated with marine ice forming at the southern tip of Vegetation Island (V.I. in Figure 2), where the flows from NIS and from Browning Pass meet. This moraine is composed of twin dirt cones that contain shells, worm tubes (serpulids) and sponges spicules, some in "living" position. Hypothesizing that these biogenic materials were incorporated into the marine ice at the ice-bedrock interface around Vegetation Island, dating of the shells [*Baroni*, 1990] provided estimates of surface velocities varying between 10.4 and 17.4 m a^{-1}. Similarly, on the eastern moraine, velocities of 3 to 3.7 m a^{-1} were obtained. These values are in good agreement with those calculated from comparison of aerial photographs taken in 1956 and 1985 giving 2.9 to 3.8 m a^{-1} and 8.8 to 11.9 m a^{-1} for the eastern and western sectors respectively.

Marine ice accretion rates are difficult to assess

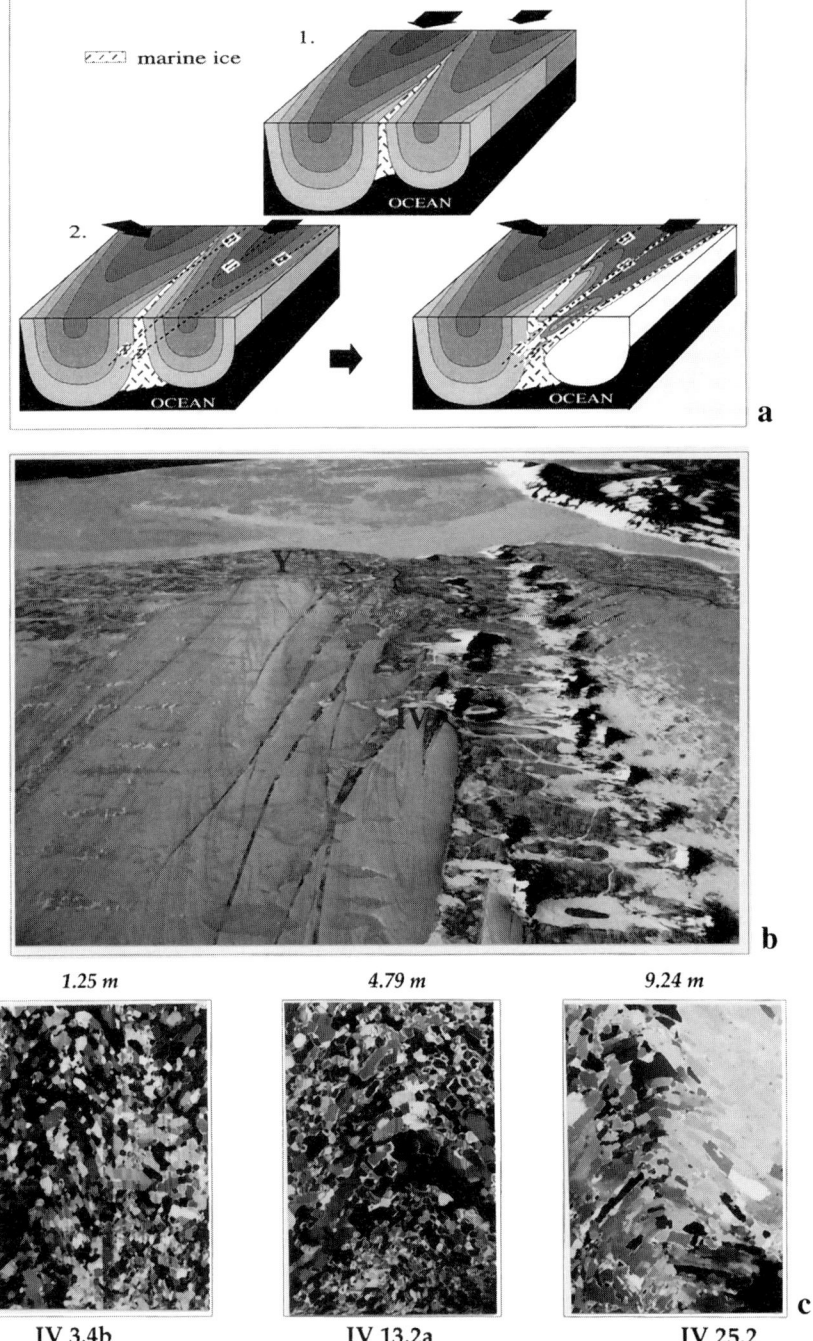

Fig. 3 : Deformational structures at Hells Gate Ice Shelf (see text for details): a) Diagram illustrating the process of shearing of marine ice at the junction of two individual ice flows with different flow directions (2); b) Oblique aerial photograph of the western-central sector of HGIS, close to the ice shelf front, showing the location of ice cores IV and Y. Note the similarity of patterns with diagram a; c) Three examples of folds and crystals bending in the orbicular and banded marine ice facies of core IV. Thin sections are shown at scale 1:1 in the vertical plane. Mean depths are shown on top and sample codes below.

from mass balance calculations, since these require surface strain rates measurements that are not available at present for HGIS. A crude estimate can, however, be made on the basis of shell dating and thickness estimates near the front. *Baroni* [1990] obtained a range of 0.122 to 0.250 m a^{-1}, depending on the method used to correct the ^{14}C ages BP. These values are, however, minimum estimates, since the shells might have been brought to the surface further upstream in the ablation area, where thicknesses are higher. Another calculation can be made considering the surface velocity range obtained from aerial photographs for the western moraine, and the (minimum) seismic ice thickness of 126 m [*F. Merlanti*, personal communication, May 1997] measured at the location of marine ice core XXI. As clearly seen on Figure 2, this ice core is located on the most upstream outcrops of the marine ice body that formed at the southern tip of Vegetation Island. A somewhat higher range of 0.43 to 0.58 m a^{-1} is obtained. These marine ice accretion rates are also higher than the surface ablation rates of 0.17 to 0.28 m a^{-1} calculated by *Baroni* [1990] from dirt cone build-up rates or 0.20 to 0.30 m a^{-1} estimated from bamboo poles ablation measurements [*A. Bondesan*, personal communication, June 1996]. The marine ice accretion rates estimated here are of the same order of magnitude as those modelled by *Jenkins and Bombosch* [1995] at Filchner-Ronne Ice Shelf for different frazil crystal sizes (about 0.5 m a^{-1} mean value), although these were occurring much further away from the grounding line in their case of a much larger ice shelf with expected much lower thickness gradients [*Paterson*, 1994: Figure 12.3]. Furthermore, the estimates given here at the location of ice core XXI should still be considered as minimum values, since accretion is thought to occur in the inverted depression associated with the merging of the two ice flows at the southern tip of Vegetation Island, as discussed above.

Finally, it is worth comparing marine ice accretion rates to an estimate of direct accretion rates from heat conduction through the ice shelf at the location of ice core XXI. Using a thickness of 126 m and a mean air temperature of -18°C [*Ronveaux*, 1992: Table 4] gives a value of 0.029 m a^{-1} for direct freezing of sea water at the interface. Therefore, marine ice accretion rates are an order of magnitude higher than for direct congelation of sea water at the bottom of 126 m of ice. It is only for marine ice accretion under less than 20 m of solid ice that the consolidation rate of the loose frazil will "keep pace" with the accretion rate, considering a porosity of 40%. A considerable delay will therefore exist between the accretion time and the consolidation time, especially for the lower units of the loose frazil ice accretion. This is of crucial importance, as will be discussed later.

MARINE ICE RECORDS AT HELLS GATE ICE SHELF

Previous Work

Taking advantage of specific climatic conditions that allow marine ice to outcrop at the surface of Hells Gate Ice Shelf, *Souchez et al.* [1991] and *Tison et al.* [1993] used shallow depth (2 m) surface ice cores on two longitudinal profiles (along flow lines) to study marine ice properties within a few kilometers of the ice shelf front. The latter authors have shown that two main types of marine ice exist at Hells Gate Ice Shelf, with contrasting textural and chemical properties revealing specific environments and depositional processes.

The first type is a granular facies, similar to the one described in other deeper marine ice cores with thicknesses ranging from 45 to 170 m (Amery Ice Shelf - [*N. Young*, personal communication, March 1995] and Filchner-Ronne Ice Shelf [*Eicken et al.*, 1994; *Oerter et al.*, 1994; *Oerter et al.*, 1992]). At Filchner-Ronne, crystals were equigranular in dimensions but showed a regular increase in size down the profile that was explained in part by the thermal history of this layer [*Eicken et al.*, 1994; *Oerter et al.*, 1994; *Oerter et al.*, 1992]. Conductivities (salinities) were very low (down to <0.03‰ salinity, i.e., up to 2 orders of magnitude lower than those measured in sea ice), decreasing downwards along the profile and eventually stabilizing in the lower part of the core. Stable isotopes were reckoned as fairly constant and reflecting equilibrium fractionation during freezing from sea water (slightly positive δ^{18}O values) both at Amery and Filchner-Ronne Ice Shelves. Sediment layers (aggregates of silt and clay-sized minerals and biogenic material) were found within the marine ice at Filchner Ronne Ice Shelf, close to the transition from meteoric ice.

In addition to the granular facies (not only showing very low salinities, but also strong chemical fractionation and random ice fabrics at HGIS), another facies, called "banded facies" (Plate 1 and Figure 1c), was observed in considerable amounts, with higher mean salinities, less chemical fractionation and strong single maximum fabrics. Statistical counting on a longitudinal profile suggested that there might be an increasing proportion of the banded facies as one gets closer to the ice shelf front. This was later confirmed by a systematic

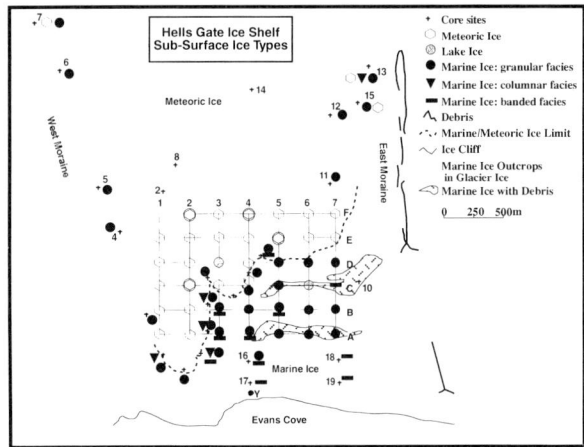

Fig. 4 : Sub-surface (2m depth) ice types at Hells Gate Ice Shelf (after [*Lorrain et al.*, 1997]).

grid survey of 60 shallow cores (Figure 4 and *Lorrain et al.* [1997]). From these data, together with observed shifts in isotopic compositions ($\delta^{18}O$-δD) of marine ice samples larger than the maximum possible equilibrium fractionation from sea water values, we conclude that active transfer of marine ice was occurring in a downstream direction at the bottom of Hells Gate Ice Shelf [*Souchez et al.*, 1991].

Of particular concern are the very low salinities observed in the granular marine ice facies. Authors agree on the fact that such levels can be produced neither by a simple process of equilibrium freezing (at very low rate) of loose frazil ice crystals in a matrix of sea water, nor by desalination processes at a later stage [*Eicken et al.*, 1994; *Tison et al.*, 1993]. *Kipfstuhl et al.* [1992] and *Eicken et al.* [1994] favor a process by which mechanical compaction of centimeter-sized platelet ice crystals results in reduction of crystal size, reduction of porosity and consolidation under the deviatoric buoyancy stress, with no significant in-situ ice growth. *Tison et al.* [1993] propose another explanation that links the low salinities of the consolidated granular facies to dilution processes of the "host" water in which accretion of the loose frazil ice crystals occurs. Dilution results from melting at the base of the ice shelf, and irregularities at the ice shelf/ocean interface favor "trapping" and relative isolation of both the loose crystals and the water of reduced density. This process allows for chemical fractionation under differential diffusion processes between the "host" water of the loose frazil and the ocean water below the interface. The whole mechanism, however, requires having both melting and frazil ice accretion occurring in areas nearby. One way to deal with this ap-

parent contradiction is to consider areas near the grounding line where melting by HSSW inflows in the DTC can occur and where large lateral irregularities must exist because of merging of individual ice flows (see discussion of RES in previous section). These would hold particularly for larger and thicker ice shelves like the Filchner-Ronne, but obviously not for HGIS where maximum ice thicknesses of 250 m are observed upstream of the southern tip of Vegetation Island. The heat source for melting was therefore left unresolved in previous marine ice studies at HGIS.

Location and Sampling of a New Data Set

During the IXth Italian Antarctic Expedition (1993-1994) about 60 shallow (2 m) to medium depth (10-45 m) ice cores where sampled at Hells Gate Ice Shelf, downstream of Vegetation Island, a local pinning point (Figures 2 and 4). We will concentrate here on a set of four cores (XXI, VI, IV, Y), drilled on the eastern flank of the western dirt cones moraine and aligned parallel to the flow lines, as deduced from the horizontal surface velocity field [*Frezzotti*, 1993].

The cores were retrieved using a modified version of the CRREL Rand Auger [*Rand and Mellor*, 1985], transferred into polyethylene bags, stored at -25°C at Terra Nova Station and shipped to Brussels at the same temperature for further processing. Vertical thin sections were cut along the total length of each core to allow observation of the different ice textures and to guide the isotopic sampling. The thin-sections were prepared using standard microtoming techniques and photographed between crossed polarizing filters. It should be noted, in this regard, that the subvertical striae visible in some of the thin sections illustrated in this paper are artefacts of the microtoming procedure. After reducing the diameter of each core by 5 mm to eliminate surface contamination, two samples, each with dimensions of 1x2x7.5 cm, were collected at depths corresponding to the top of each thin section, i.e., with a mean sample spacing of about 0.1 m. Non-contaminating polyethylene gloves were always used for handling, and cutting was performed with a clean band-saw previously tested as non-contaminating for the major ions of interest. The twin samples were then allowed to melt in polyethylene tubes, before their conductivity was measured using a Taccussel CD 810 conductimeter in a 25±0.01°C thermostatically-controlled bath. Salinities were deduced using calibration curves based on high precision ($\pm$ 10^{-3} ‰) IAPSO salinity standards provided by Ocean Scientific International. The overall precision of the method is ±

0.05‰. The other sample collected at the same depth was analyzed for Na, K, Ca and Mg with a Varian SpectrAA 300 atomic absorption spectrophotometer.

Samples were also collected in selected locations for co-isotopic analyses of deuterium (δD) and oxygen-18 ($\delta^{18}O$). Samples with a volume of 5 ml were collected with the band-saw, and those with a salinity above 0.5‰ were distilled under vacuum prior to the isotopic analyses by a procedure proven to conserve the isotopic signal [*Tison et al.*, 1993]. The measurements were performed at the Nuclear Research Center of Saclay (France). HDO and $H_2^{18}O$ concentrations are given in δ-units versus V.S.M.O.W. (Vienna Standard Mean Ocean Water) expressed in per mil. The accuracy of the measurement is ±0.5‰ in δD and ±0.1‰ in $\delta^{18}O$.

Results and Discussion

Plate 2 shows selected pictures of the textural characteristics of the studied ice cores. Cores XXI, VI and IV are each 10 m long and core Y is 45 m long. The pictures are full scale vertical thin sections photographed between crossed polarizers. Depths are given at the appropriate level on the right side of the pictures. The four cores were drilled on a longitudinal profile in the marine ice outcropping between the western and the central sectors (Figure 2 top). Given that the marine ice thickness is already 126 m at the location of core XXI, that the mean ablation rate is estimated to be between 0.20 and 0.30 m a^{-1} and that the surface velocities are between 9 and 12 m a^{-1} the cores are thought to adequately approximate what would be recorded with depth in a single ice core, close to the ice shelf front, supposing it had not been subjected to surface sublimation. Indeed, using the most conservative estimates for surface velocity (9 m a^{-1}) and for ablation rate (0.20 m a^{-1}) one can calculate that the amount of marine ice ablation between two successive core locations is 3 to 8 times higher than the core length at the most upstream site of each pair. Resulting profiles are therefore plotted on top of each other in Figures 5 and 6. It should be stressed, however, that this does not necessarily mean that the ice in core IV is younger than in core XXI, since this obviously depends on factors like the balance between horizontal velocities, surface ablation rates, marine ice accretion rates (of which we only have minimum estimates) and longitudinal extension of the marine ice depositional areas. Also, the dynamical characteristics of HGIS flow described above are likely to affect the original stratigraphy of the marine ice deposits to a certain extent, and this will have to be taken into account in the discussion below.

Figure 5 gives the smoothed salinity profile. Figure 6 shows the K/Mg ratio at the same resolution and with the same smoothing interval as for salinity. This K/Mg ratio was chosen since it is unlikely to be affected by salt precipitation at temperatures encountered in the HGIS environment, or by subsequent long-term storage in freezers. Indeed, *Richardson* [1976], in his study of the phase relations in sea ice as a function of temperature, indicates that $CaCO_3.6H_2O$ is the first salt to precipitate from sea water at a temperature close to the freezing point and that $Na_2SO_4.10H_2O$ and $CaSO_4.2H_2O$ soon follow, at temperatures of -8°C and -10°C respectively. Following the same author, concentrations in K and Mg ions in sea ice at sub-freezing temperatures under quasi-equilibrium conditions do not change until the temperature falls below -34°C.

Figure 7 shows the same data sets in K/Mg-salinity correlation graphs for each of the cores, compared to the range of sea water values of the K/Mg ratio (0.087-0.096) compiled from various authors [*Addison*, 1977; *Goldberg*, 1957; *Wilson*, 1975]. This reference level is also shown on Figure 6.

As mentioned before, selected samples were collected for co-isotopic measurements in δD-$\delta^{18}O$. A commonly used representation for these data is a δD-$\delta^{18}O$ diagram (Figure 8), where samples of natural waters from large unconfined reservoirs would lie on a Meteoric Water Line of equation $\delta D = 8\delta^{18}O + 10$ [*Dansgaard et al.*, 1973]. For the purpose of the discussion below, samples have been grouped in the two main marine ice textural facies.

Textural results from the studied cores (Plate 2) confirm the observations from previous work at HGIS. The two upstream cores (XXI and VI) show a granular facies exclusively, whilst core Y consists of 97% banded facies. The remaining 3% consists of granular ice, confined to discrete layers or pockets in the first 10 m of the core. Core IV, located about half-way between VI and Y (Figure 2), is of particular interest. Down to 5.35 m it shows a typical fine grained granular texture. It is also quite rich in fine insoluble particles (up to 0.8% debris in weight, excluding surface samples), as opposed to the other cores where virtually no debris was found. This is what would be expected for marine ice accreted close to a rock interface where suspension of mineral particles in the water can occur under tidal agitation. Below 6 m, core IV is exclusively banded, with the exception of the bottom 50 mm which consist of granular ice. Between 5.35 and 6 m, mixed

Plate 2 : Selected typical textural characteristics of the four marine ice cores, shown in vertical thin sections between crossed polarizers. The scale shown at the top is valid for all photographs. Depths in meters are indicated to the right. Selection has been made to give a complete overview of textural variability inside the cores, rather than a true representation of the frequency of occurrence of the different facies.

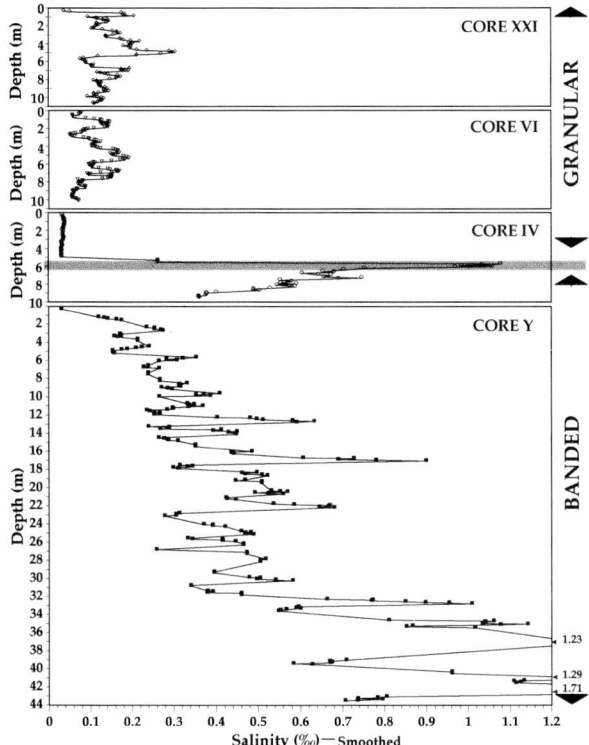

Fig. 5 : Smoothed salinity profiles in the four marine ice cores. The smoothing interval is 0.40 m. The shaded area shows the transition zone of mixed congelation-platelet ice facies in core IV (see text for details).

by the strong "noise" of unsmoothed salinity profiles. It would, therefore, be hazardous to discriminate between genuine and post-tectonic events in the fluctuation of salinity and K/Mg in each individual core of Figures 5 and 6, and we will only focus on general trends and mean values for each core and from core to core. Furthermore, besides dynamical disturbances, we have to keep in mind two other possible sources of alteration of the signals: surface modifications during the summer melt season, and interactions with the solid impurity content. The first process could be responsible for desalination of the top meter of the cores, as seems to be the case for all cores in Figure 5. It could also increase the K/Mg ratio through surface input of impurities of continental origin, since, as demonstrated by *Casella et al.* [in press] for snow accumulation in the Terra Nova Bay area (Mc Carthy Ridge and Styx Glacier), potassium is mainly of continental origin, whilst magnesium is mainly oceanic.

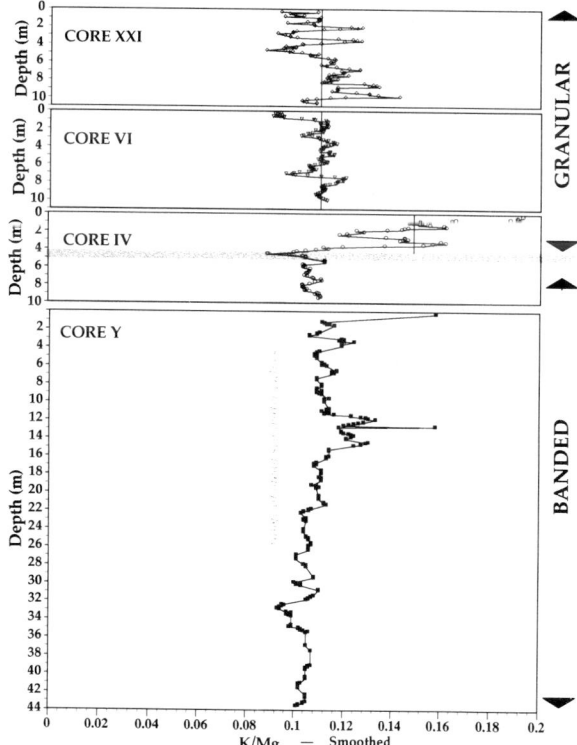

Fig. 6 : Smoothed K/Mg profiles (ratios calculated from concentrations expressed in meq/l units) in the four marine ice cores. The smoothing interval is 0.40 m. The dark shaded area shows the transition zone of mixed congelation-platelet ice facies in core IV (see text for details). The K/Mg range in sea water is also indicated in light gray (see text for details).

congelation-platelet ice texture is observed, a facies only described once in the marine ice literature [*Souchez et al.*, 1991], at the limit of meteoric and granular marine ice at HGIS. This spatial layout supports the assumption made above that the vertical juxtaposition of the four cores illustrates the large scale (kilometre) distribution of the textural facies in a vertical plane oriented parallel to the mean flow line. The assumption, however, probably does not hold at the smaller metre to centimetre scales, owing to the specific geometry of Hells Gate Ice Shelf. The sharp textural contrast occurring at 5.35 m and the re-occurrence of the granular facies at the bottom of core IV could, therefore, be the signature of shearing of a lower stratigraphic unit higher up in the core. The tilting of the crystals might be a further argument in this respect.

The occurrence of medium to small-scale disturbances in the individual cores' stratigraphy are quite likely to affect equally the distribution of their chemical characteristics, as suggested by the sharp salinity gradient in core IV between 5 and 6 m depth (Figure 5) and

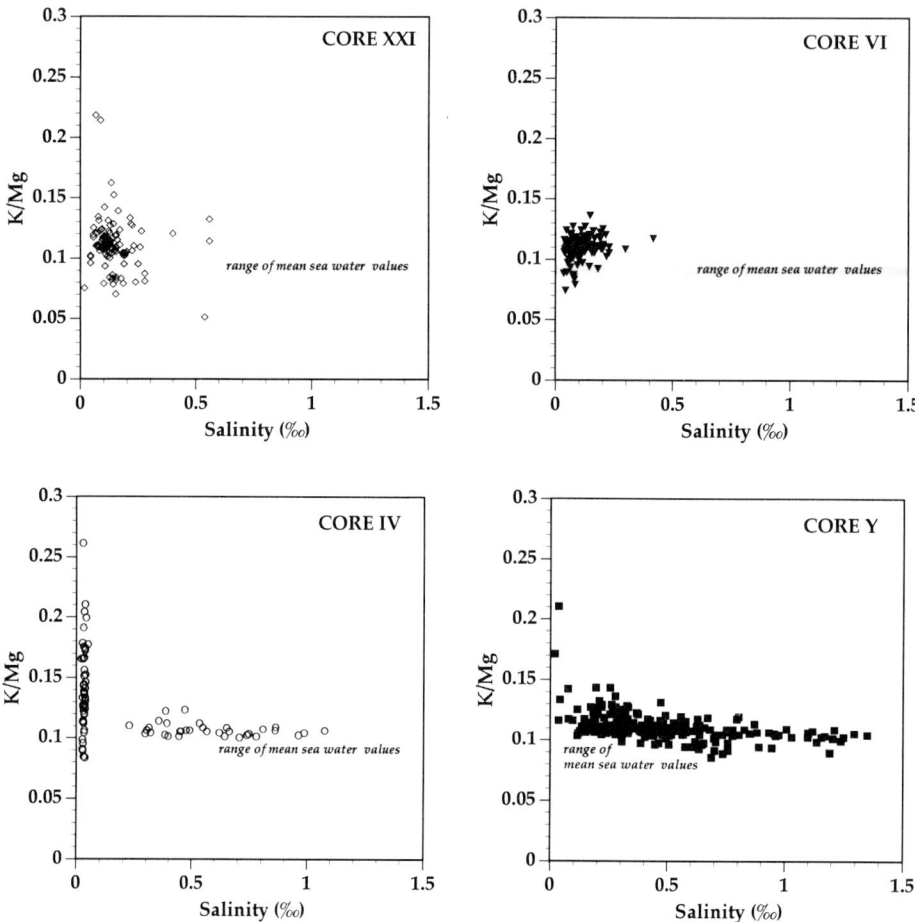

Fig. 7 : K/Mg-Salinity relationship in the four marine ice cores. Same conventions as in Figures 5 and 6.

Interactions with solid inclusions could also modify the chemical content of the ice samples through simple dissolution or, more likely, desorption effects from the surface of silt to clay-sized particles. This latter process, particularly, was shown to increase the alkali concentration with respect to that of the alkaline-earths in the solution, even after relatively short (minutes) contact times and for small amounts of particles [*Souchez et al.*, 1978]. This would, then, provide an artificial means of increasing the K/Mg in marine ice samples loaded with fine-grained solid impurities. Careful examination of the bottom of the sample tubes used for the chemical analyses performed on core IV revealed a thin deposit of insoluble residues. These were filtered, weighed with a precision balance (±0.1 mg) and expressed as debris/ice concentrations in weight percent. The results are plotted as a function of the K/Mg ratio in Figure 9. Except for three surface samples, all concentrations are low (below 0.8% in weight), and there is no significant correlation ($r^2 = 0.07$) between the two data sets. Contamination by the solid impurity content is, therefore, in this case, not likely to alter the conclusions of our study.

The salinity trend (Figure 5) in the granular facies corresponds to a decrease in a downstream direction, from maximum values of 0.3‰ in core XXI to values less than 0.03‰ in core IV. Both the trend and the minimum values are consistent with the observations in other deeper marine ice cores at Filchner Ronne Ice Shelf and Amery Ice Shelf [*Eicken et al.*, 1994; *Morgan*, 1972; *Oerter et al.*, 1992]. There is a sharp increase of the salinity when entering the mixed platelet-congelation horizon in core IV, with values of up to 1.1‰. In the banded facies of core Y, salinities are noticeably higher and universally increase with depth, especially below 32 m, where water was first

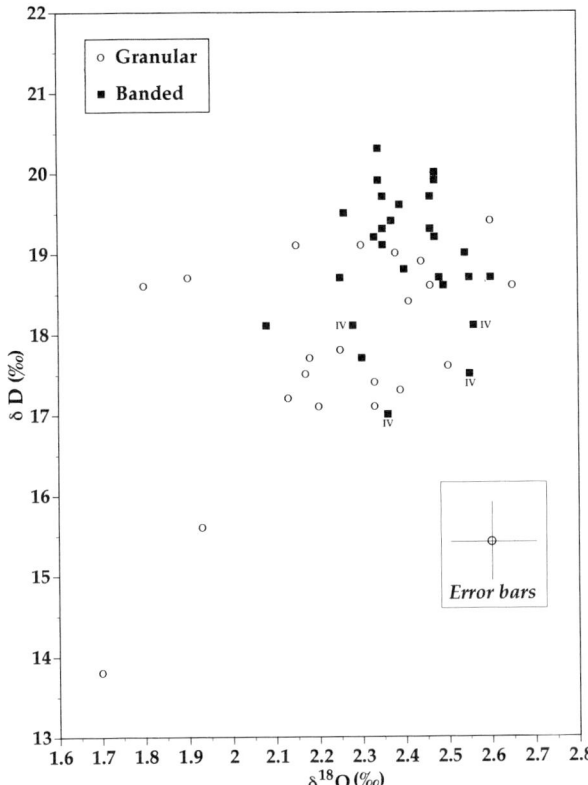

Fig. 8 : δD - δ^{18}O (‰) relationship for the two main marine ice facies.

observed to invade the drill hole. The water level rose slowly in the hole in the next 24 hours, and stabilized at 11 m depth, which is equivalent to the ice-shelf freeboard in front of the drill site. Conductivity measurements on two water samples taken in the hole at the beginning and at the end of the process were equivalent to respectively 3 and 0.9 times the conductivity of normal sea water. This indicates that the bottom part of the banded facies is still thoroughly permeable to sea water and that the "in-situ" interstitial water in these layers is considerably enriched by the brine exclusion process resulting from the consolidation of the upper levels. This supports the idea that, in this case, freezing is actively taking place in the consolidation process.

In previous work (Tison et al., 1993), the K/Mg ratio of marine ice samples was shown to be a good indicator of active chemical fractionation below the ice shelf. In particular, K/Mg - salinity relationships clearly showed a transition from saltier samples, with K/Mg ratios close to the sea water value, to low salinity samples with much higher K/Mg ratios. Simple experiments, where fresh-water ice blocks were melted in a sea water reservoir above its freezing point [*Tison et al.*, 1993: their Table 1 and Figure 5], were shown to result in a 30% reduction of the salinity and a 125% increase of the K/Mg ratio in the interface water layer.

This process, reflecting a selective diffusion mechanism from undiluted to diluted sea water had previously been demonstrated by *Ben-Yaakov* [1972] and shown to lead to an increase of K/Mg of up to 200% of its original value in sea water. Repeating the experiment described above with a gauge incising the bottom of the melting ice block, showed increased differences, thereby indicating accumulation of waters of least density in inverted depressions below the ice [*Tison et al.*, 1993]. It was then proposed that melting below an ice shelf would enhance dilution of the interstitial host water within the loose frazil accumulating underneath. It would also enhance chemical fractionation between sea water and this interstitial water, resulting in the original chemical signature of the granular facies.

Data presented in Figure 7 confirm this behavior. Core IV is especially illustrative, showing two distinct groups of samples: the low salinity-high K/Mg in the topmost granular facies, and the high salinity-low K/Mg in the lowermost banded facies and the transition zone. The granular facies of cores XXI and VI show a similar pattern of a dense cloud of samples with rather low salinities and moderate fractionation. The banded facies of core Y stretches between a similar cloud (although slightly more saline) and numerous saltier samples with sea water K/Mg values. The few low-salinity / high K/Mg samples in core Y can be shown to correspond to the discrete layers of granular crystals in the upper part of the core, and are thus not representative of the banded facies.

The new data set, however, differs in two major respects from the results described in *Tison et al.* [1993], suggesting that some of their conclusions have to be reconsidered. The maximum K/Mg value of 0.26 observed here is of the same magnitude as the maximum shift observed by *Ben-Yaakov* [1972] after three days of diffusion experiments. *Tison et al.* [1993], however, quote K/Mg values as high as 1.26 in coarse granular frazil from shallow (2 m) cores, and, therefore, call for repeated melting events at the ice-ocean interface to account for this cumulative fractionation. The fact that such very high values were only observed close to the surface, and did not occur in any of the deeper cores described in this paper, suggests possible surface contamination processes, even though the first 0.5 m of the cores were systematically discarded in *Tison et al.* [1993]. It is worth noting that, in these surface cores, the highest values were observed in the coarser textural

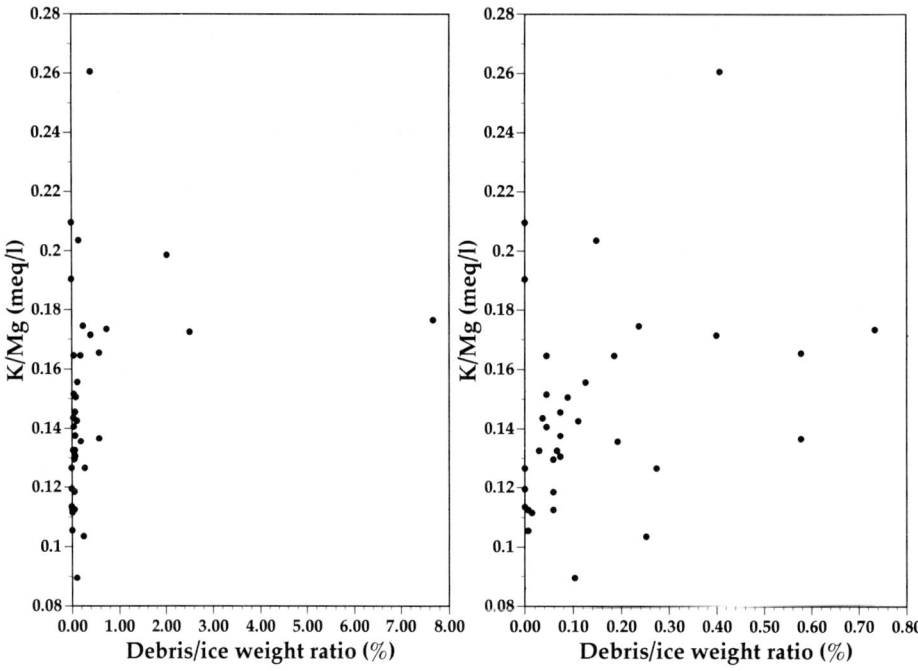

Fig. 9 : K/Mg ratio as a function of the debris/ice weight ratio for samples of the granular facies in core IV. K/Mg ratios are calculated from concentrations expressed in meq/l units.

horizons where recrystallization processes during the summer season were probably the strongest. Along the same line, the observations by these authors of decreasing K/Mg values with depth, suggesting a stronger chemical sorting in the older frazil (further away from the ice-ocean interface on the same vertical), is probably only mirroring a decreasing effect of alteration in the surface layers.

On the contrary, the behavior of the K/Mg ratio in Figure 6 seems to indicate a stronger chemical fractionation with depth in the granular facies. This is clear if one compares the mean values of cores XXI and VI to the mean value of core IV. On the other hand, although cores XXI and VI show the same mean values (due to homogenizing dynamical effects) there is a trend towards larger K/Mg ratios with depth.

Souchez et al. [1991] have shown, on co-isotopic grounds, that marine ice transfer occurs at the bottom of HGIS. Now that the spatial distribution of the two main marine ice facies is clearly demonstrated, and that core IV suggests an episode of direct freezing under heat conduction through the ice shelf (mixed platelet-congelation transition) occurring between the successive accumulation of these two facies, it is tempting to see the banded facies as forming in a water partially diluted by the melting of the granular facies. Could this process be traced by the co-isotopic signature of the two facies? This is a difficult task to achieve, if one keeps in mind all the cumulative steps involved in the build up of a consolidated marine ice layer. The final isotopic signature will depend on a number of factors: the signal of the "parent water" for individual frazil ice crystals, the apparent fractionation coefficient, the signal of the "host water" where individual frazil ice crystals accumulate, the porosity of the loose frazil (proportion of water to ice) and the efficiency of fractionation during the consolidation process by freezing of the "host water". The slight increase of the isotopic signal in the water resulting from mixing of the granular ice meltwater with sea water, that will eventually form the parent water for individual crystals of the banded facies, could easily become lost in this multiple step process. Nevertheless, Figure 8 plots the δD-$\delta^{18}O$ relationship for samples from the two facies in cores XXI, IV and Y. Although, as foreseen, all the samples lie in the area of a few error bars, the pattern suggests that samples from the banded facies are slightly enriched with regards to those from the granular facies, especially if one does not consider the four samples from the banded facies occurring just below the transition zone in core IV.

New insights from the data set on deeper marine ice cores at HGIS can be summarized as follows. The existence of two main facies of marine ice has been confirmed. Besides the widespread granular facies described in all other marine ice cores, a banded facies is dominant closer to the ice shelf front. In places where both facies are present on top of each other, a transition zone exists showing a mixed platelet-congelation facies indicating an episode of direct freezing of sea water at the interface driven by heat conduction through the ice shelf. The granular facies is characterized by low salinities and moderate to high chemical fractionation. The trend is to a decrease in salinity and increase in fractionation downstream on the profile (i.e., as one gets closer to the ice shelf front), suggesting that the heat source requested for melting and dilution processes, hypothesized by *Tison et al.* [1993], is to be found in the frontal area of HGIS. A potential candidate for this is the tidal pumping in summer of warmer water of the coastal currents, described by *Jacobs et al.* [1992] as "circulation mode-3". This circulation-mode affects shallow ice shelf bases and walls within 100 km of the ice front and leads to high melting rates, involving about 35% of the total annual net loss of Antarctic ice shelves [*Jacobs et al.*, 1992].

LANDFAST FIRST-YEAR SEA ICE RECORDS AT HELLS GATE ICE SHELF

In this section we focus on the information that can be gained from the study of the short-term depositional pattern in landfast first-year sea ice in front of HGIS and discuss how these can shed some light on the genesis of the banded marine ice facies in the context of ice-ocean interaction processes.

The Data Set

In the course of January 1994, during the IXth Italian Antarctic Expedition, eight landfast first-year sea ice cores were collected. The cores were equally distributed on a west-east transect of Evans Cove (Figure 2) roughly following the limit of the remaining sea ice at the time, within a distance of 50-150 m from the ice shelf front (to the West) or from the multiyear landfast sea ice (to the East). For the western cores (SA to SC), no significant amount of loose ice crystals was found in the drill holes and the bottom of the cores was quite solid. In contrast, in cores SD to SH, a thick layer of about 3-4 m (rough estimate from the sea ice cover freeboard measured in the hole, supposing a porosity of about 50% i.e., a density of 0.45 for the loose frazil) of loose frazil ice crystals was found at the bottom of the solid sea ice cover. These frazil ice crystals were rather small and plate-like, up to 10 mm in diameter and about 1 mm thick and typically of rectangular shape.

To complement the textural results from 1994, three additional first-year sea ice cores were sampled at the end of the winter period (October 1995) during the XIth Italian Antarctic Expedition. These were located roughly on a North-South transect respectively 200m, 700m and 1500m from the ice shelf front in its central part (Figure 2: Holes 1, 2, 3). Again, large amounts of loose frazil crystals were accumulating at the bottom of the ice cover. These were much larger than those observed in the summer, commonly reaching 80 to 100 mm in diameter and a thickness of about 2 mm. Individual frazil ice crystals were often welded together in "packs" of up to 10 units.

Sampling Procedure and Analytical Treatment

The sea ice cores were sampled and processed as described before for marine ice as far as crystallographic investigations and salinity measurements are concerned. Using textural profiles in conjunction with salinity profiles from all ice cores, a total of 70 samples were selected for $\delta^{18}O$ measurements. The isotopic measurements were performed at the University of Trieste using a Finnigan Delta-S mass spectrometer. Precision of the measurement is ± 0.08‰. In selected cores, c-axis measurements were performed using a four-axis universal stage and standard procedures developed by *Langway* [1958]. The data were plotted on the lower hemisphere of a Schmidt (equal-area) net.

Results

Textures and fabrics in the October 1995 profiles (end of winter). Figure 10 summarizes the textural data of the three cores sampled at the end of the winter on the North-South transect. In this and the following figures and descriptions, a depth scale in centimeters will be used, in accordance with the scale of the textural and chemical variability. All the cores show the same major characteristics. First of all, no congelation ice is found in the cores, unlike the top of the sea ice cores from McMurdo Sound [*Jeffries et al.*, 1993; *Gow et al.*, this volume]. Most of the winter accretion consists of granular frazil, with some limited occurrences of the draped (mixed congelation/platelet) facies. Platelet accretion is also present, but limited to the end of the

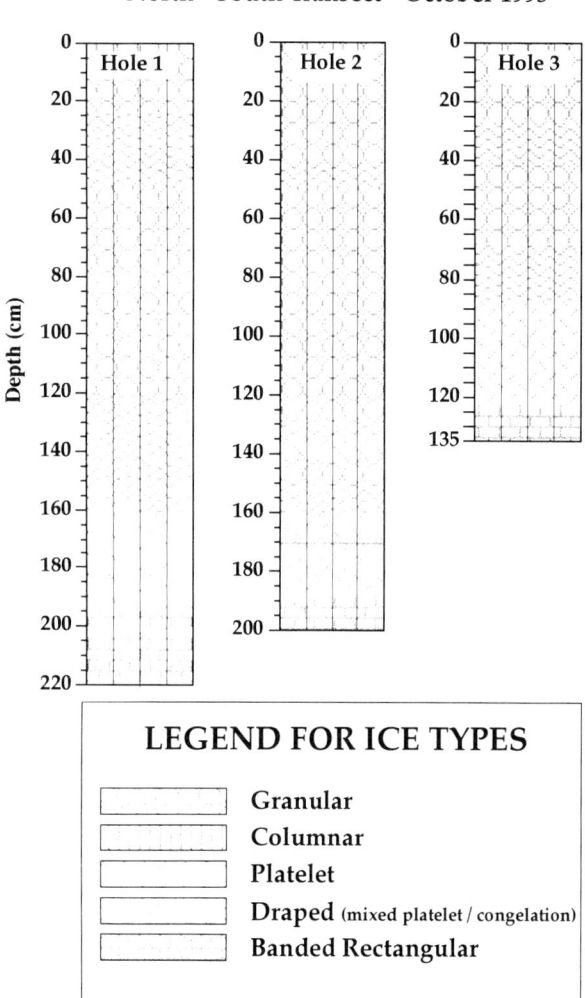

Fig. 10 : Textural profiles from the October 1995 (north-south transect) data set. See Plate 1 for thin sections illustrations of the ice types.

winter season in all cores. However, platelet ice does not form the very bottom of the core, where small amounts (max. 20 cm) of the banded rectangular facies occur.

Plate 3 and Figure 11 illustrate in more detail the textural and crystallographic characteristics of the cores at "Hole 1" and "Hole 2". Of major interest is the contrast between the granular facies forming the surface layer and the one occurring further below at 50-125 cm depth. The first is orbicular, richer in air/brine inclusions, shows heterogeneous colors between crossed polarizers and a fabric close to random. The latter has fewer bubble inclusions and these are of larger size.

Crystals are also slightly elongated in the vertical plane (see also Figure 13). Both in vertical and horizontal thin sections, the color is quite uniform, indicating a strongly oriented pattern, as confirmed by the c-axes clustering. Even more significant is the type of clustering, which is close to a strong maximum in the vertical plane, an unusual feature in sea ice.

Platelet ice and draped mixed congelation/platelet facies show essentially random fabrics in accordance with those measured by *Jeffries et al.* [1993] and *Jeffries and Weeks* [1993], although we did not perform separate plotting of platelet and congelation crystals to show the bimodal nature of this arrangement. Finally, the bottom, banded rectangular facies shows strongly aligned vertical c-axes.

Textures in the January 1994 profiles (mid-summer). Textural profiles from mid-summer 1994 (Figure 12) exhibit, in their top sections, similar features to those observed two years later (October 1995), with the exception that more draped facies events occurred during the course of the winter. There is, however, for the bottom part of the cores, a strong contrast both in time (as compared to the October 1995 cores) and space (comparing western SA-SB-SC and eastern SD-SE-SF-SG-SH cores). Platelet accretion marks the end of the winter season in the western cores. As in the McMurdo Sound samples, the relative thickness of this bottom platelet layer might be underestimated, given possible ablation in the course of the summer. In contrast to the other sites, virtually no platelet facies are observed in core SD to SH, either because it never formed, or because it melted away before accretion of the banded rectangular facies.

Probably the most striking feature of this West-East mid-summer transect is the well developed (up to 1.40 m consolidated) banded rectangular facies at the base of the eastern (SD to SH) cores. This layer is neither present (or barely so) at the same location at the end of the winter season (compare "Hole 1" in Figure 10 to "SE" in Figure 12), nor at the same moment in the western sector of Evans Cove.

As an example, Figure 13 illustrates the typical textural contrast between the granular frazil ice occurring between 75 and 85 cm depth and the banded rectangular facies accreted at the bottom in core SE. The former shows appreciable elongation in the vertical plane resulting in U-shaped crystals and uniform tints in horizontal thin section, indicating near-vertical c-axes clustering. The latter displays obvious strongly elongated rectangular shapes with intracrystalline brine inclusions in vertical thin sections, and appears as large flattened

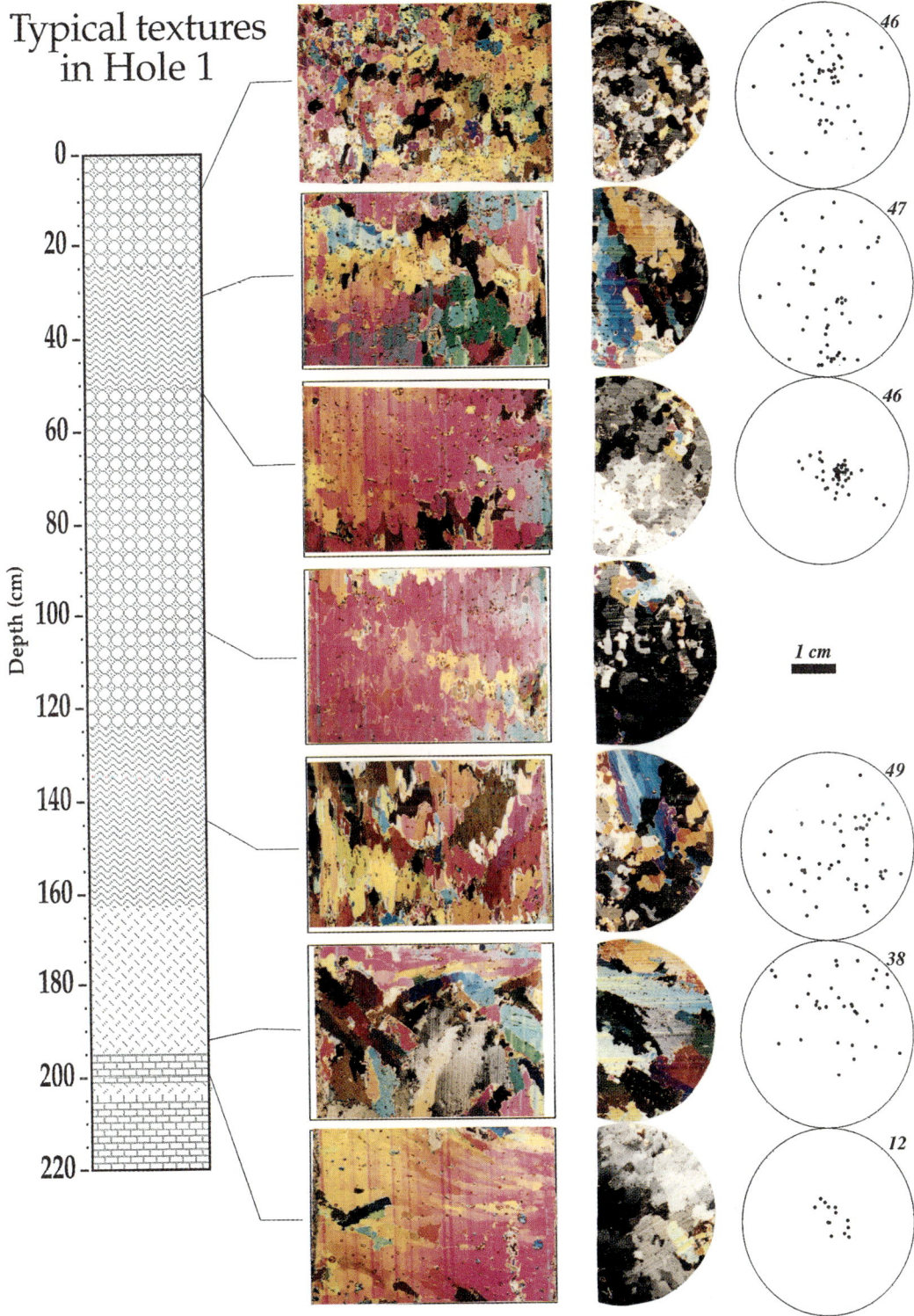

Plate 3 : Textures and fabrics in Hole 1 (October 1995). Rectangular photographs are vertical thin sections and circular photographs show horizontal thin sections at equivalent depths. The scale is valid for all pictures. C-axes are shown in the horizontal plane and plotted in the lower hemisphere of a Schmidt net and the number of c-axes measured is specified. Legend of ice types as in Figure 10.

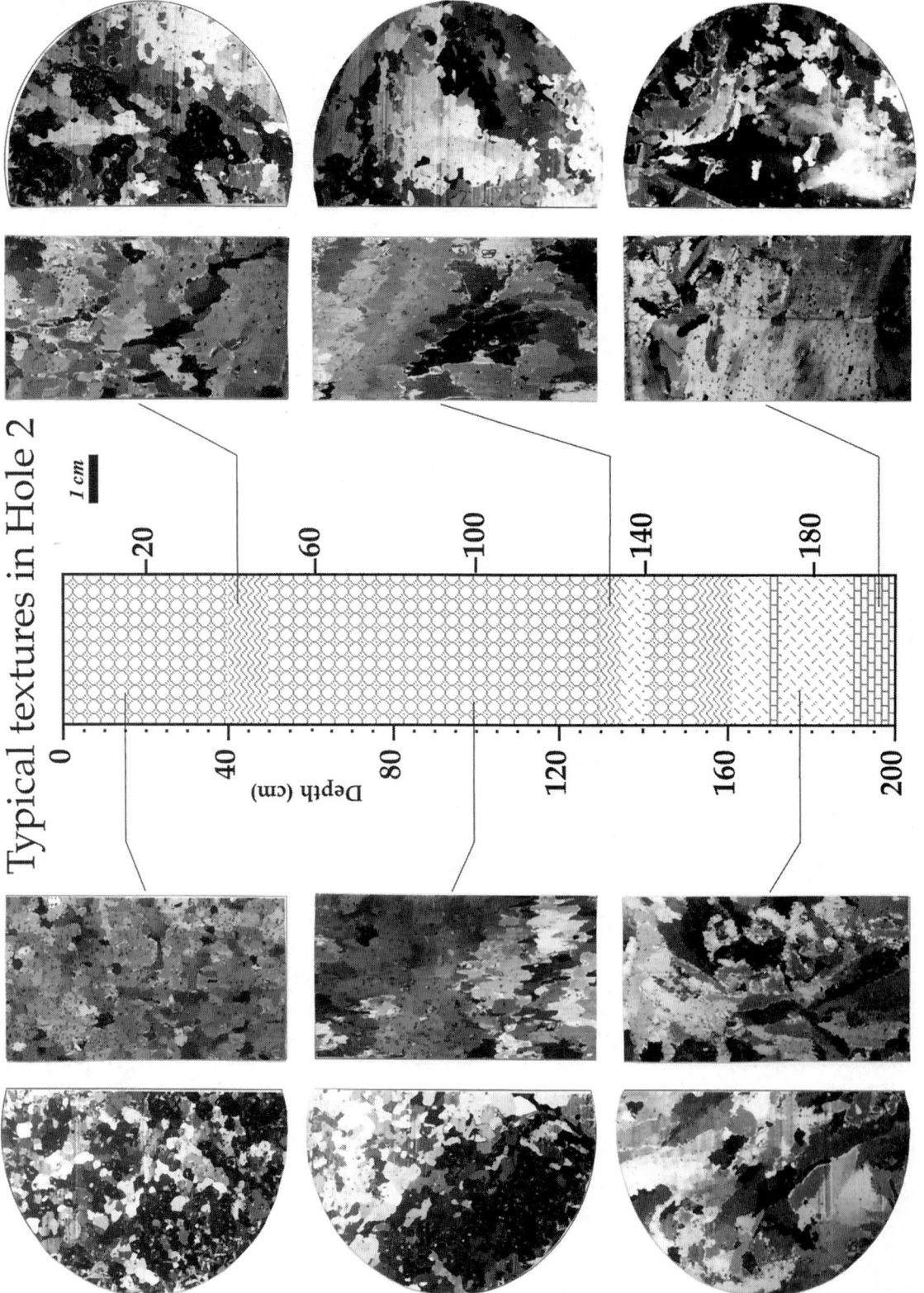

Fig. 11 : Textures in Hole 2 (October 1995). Conventions as in Plate 3.

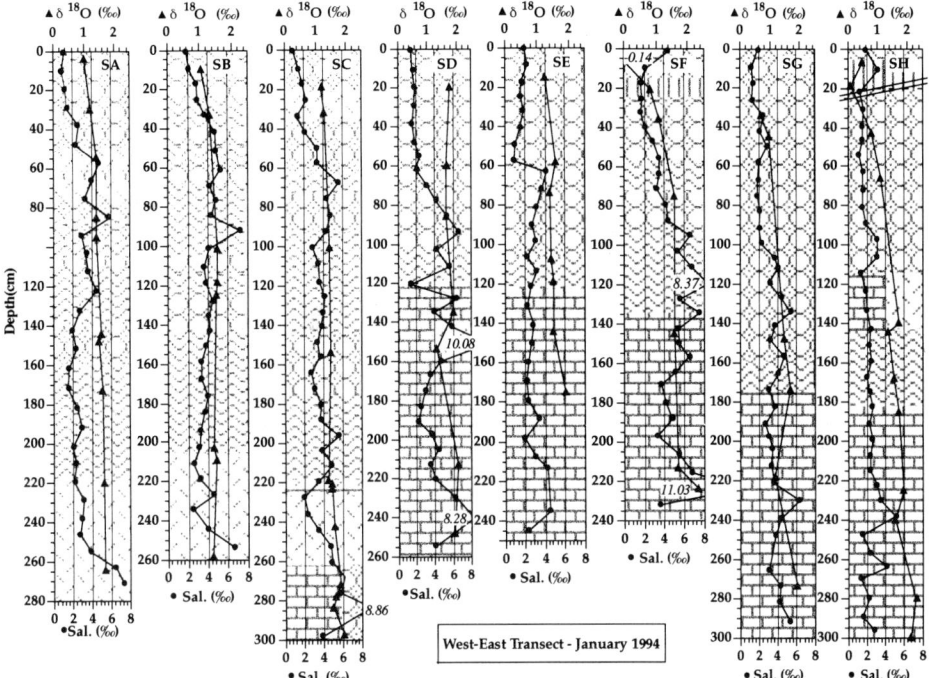

Fig. 12 : Textural, salinity and $\delta^{18}O$ profiles from the January 1994 data set. See Plate 1 for thin sections illustrations of the ice types. The sloping lines in the top part of core SH refer to deformational structures. Legend of ice types as in Figure 10.

crystals displaying near vertical c-axes orientations in horizontal thin sections.

Salinity and $\delta^{18}O$ profiles in the 1994 data set. Figure 12 displays all the salinity and $\delta^{18}O$ results in the landfast sea ice cores from the January 1994 (midsummer) data set. Figures 14a and 14b are enlarged graphs for selected stations, to ease the comparison between western (SA-SB - Figure 14a) and eastern (SE-SF - Figure 14b) cores. In the top 40-90 centimeters salinity rises from 1-2‰ to 4-8‰, whilst $\delta^{18}O$ steadily increases from a minimum of -0.14‰ to about +1.5‰. Further down (40-90cm to 120-140cm), salinity fluctuates in the 4-8‰ range, whilst $\delta^{18}O$ stabilizes. In this unit, there is no obvious correlation between textural (granular vs. mixed platelet/congelation) and compositional properties. In most of the lower half of the cores, salinity steadily decreases whilst $\delta^{18}O$ remains quite constant in cores that do not show a banded facies accreted at the bottom (SA and SB). In all other cores with a banded facies at the bottom, the $\delta^{18}O$ signal rises slightly. Finally, a general salinity increase is observed in the bottom 40-60 centimeters of most cores, although it is not so clear-cut in those cores consisting of banded rectangular ice (SC, SE, SG and SH).

Discussion

Previous work on marine ice has already underlined the potential difficulties in interpreting salinity-stable isotope data in consolidated frazil ice (see for example *Tison et al.* [1993], *Eicken et al.* [1994] and discussion in previous sections). In congelation sea ice, resulting from direct freezing of the ocean water reservoir at the base of an existing sea ice cover, salinity and $\delta^{18}O$ essentially depend on two factors : the signal of the "parent" ocean water and the freezing rate. If one assumes constant parent water properties and limited desalination processes, the profiles will essentially reflect fluctuations of the freezing rate with depth. A decreasing freezing rate will enhance expulsion of brines (lowering bulk ice salinity) and increase isotopic fractionation (increasing ice δ-values) as one approaches equilibrium. Numerous studies of salinity profiles in first-year congelation sea ice (see, for example, *Nakawo and Sinha* [1981] and synthesis in *Weeks and Ackley* [1986]) display a typical c-shaped profile that has been satisfactorily modeled, combining initial salt entrapment with subsequent drainage processes [*Cox and Weeks*, 1988]. Apart from the relatively enriched top (frazil and brine-

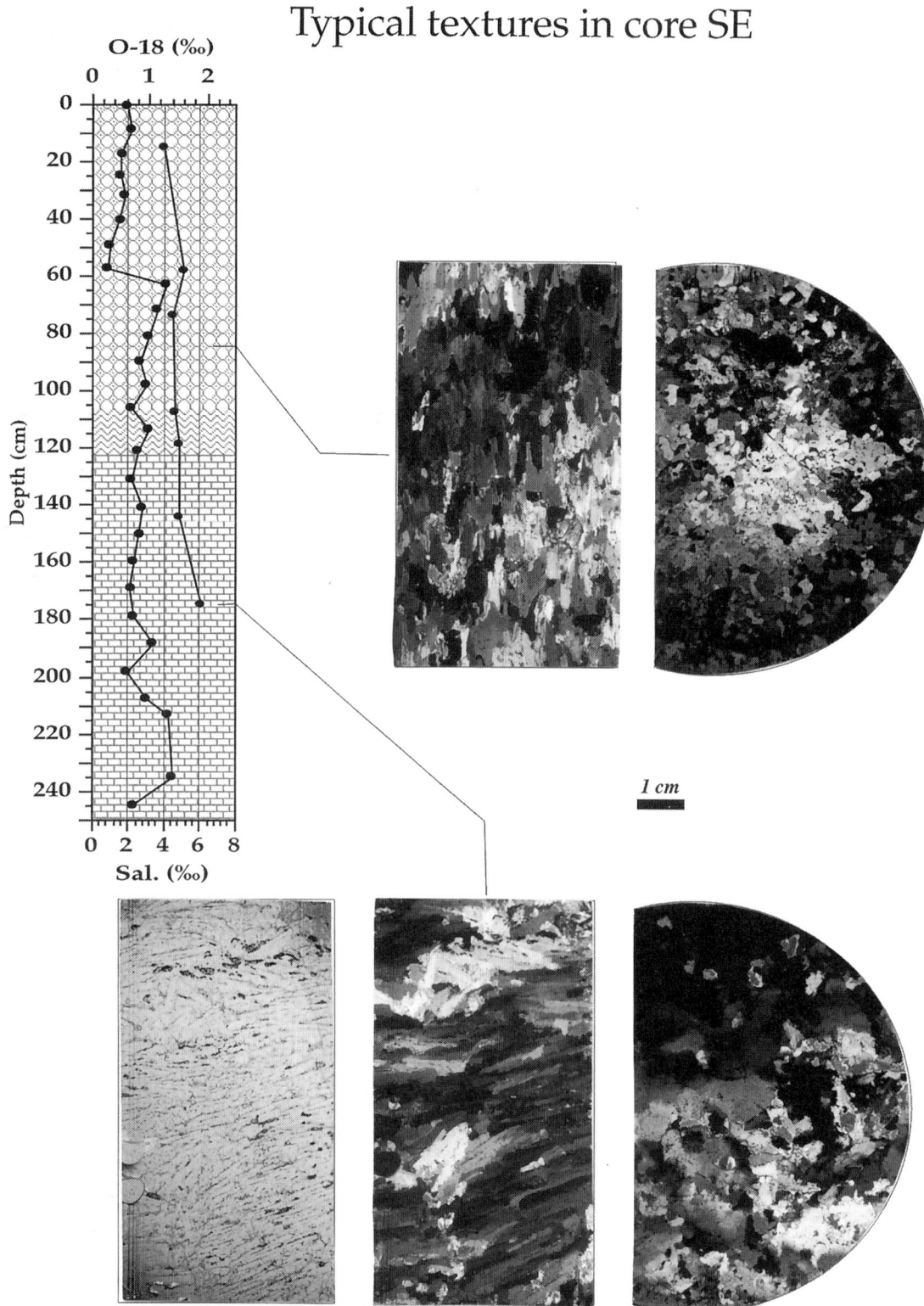

Fig. 13 : Ice core structure/stratigraphy and ice textures in core SE (January 1994). Same conventions as for Plate 3. The bottom left photograph is taken in transmitted light.

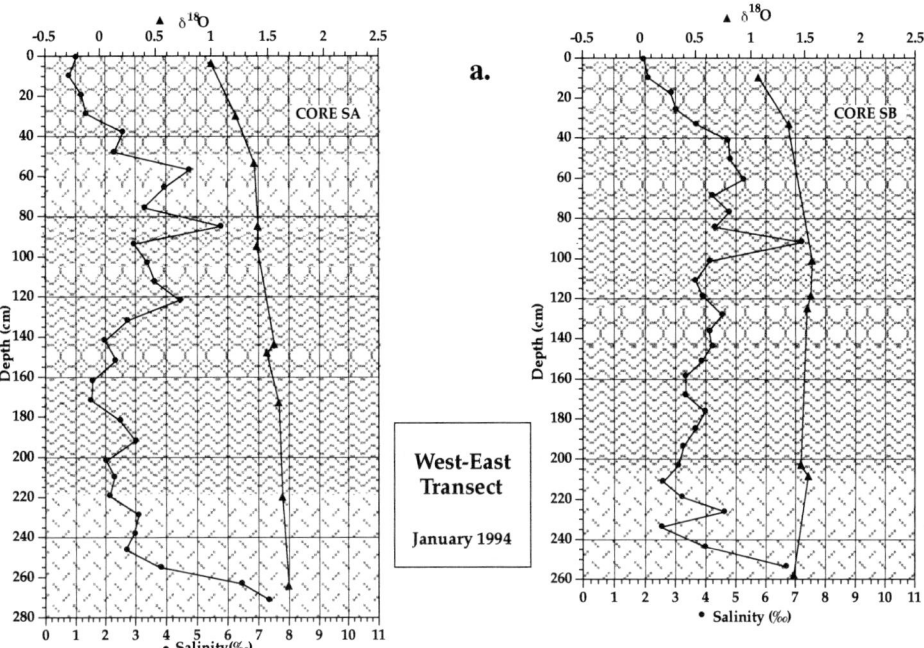

Fig. 14a : Enlarged structure/stratigraphy/texture diagrams and salinity and $\delta^{18}O$ profiles for cores SA-SB. Legend of ice types as in Figure 10.

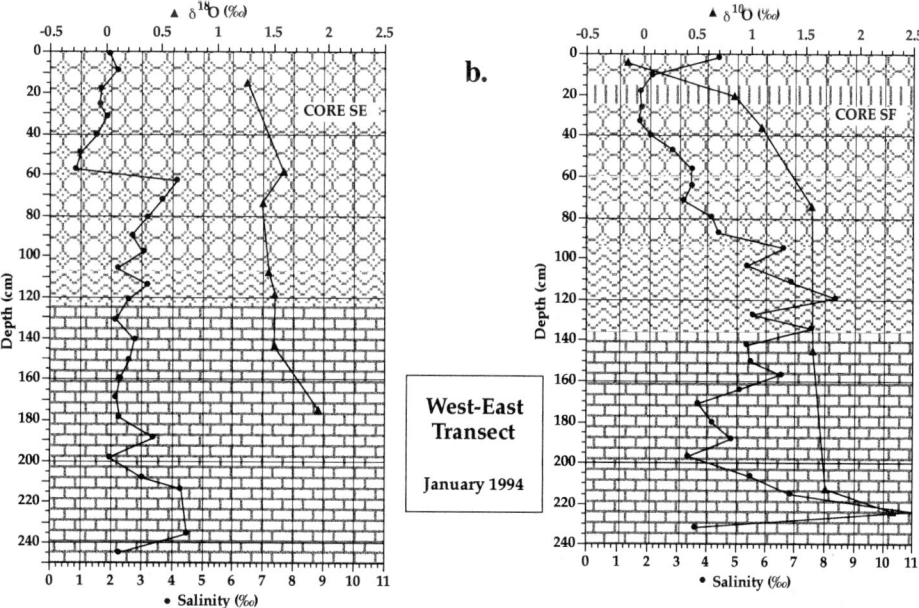

Fig. 14b : Enlarged structure/stratigraphy/texture diagrams and salinity and $\delta^{18}O$ profiles for cores SE-SF. Legend of ice types as in Figure 10.

rich snow ice, rapid congelation growth rates in the top 0.10 m, expulsion of brine onto the surface following freezing) and bottom (capillary retention of brine in the highly permeable skeletal layer [*Gow et al.*, 1990]), the profile will reflect the dependance of the freezing rate on the balance between the trend of the ambient air temperature and the insulating effect of the ice. This is clearly seen, for example, in a detailed case study of a freezing lead in the Arctic by *Gow et al.* [1990: Figures. 2, 4a and 4c]. When desalination processes are important, as in arctic multiyear sea ice, for example, this initial freezing effect can be completely altered and the c-shaped profile turned into a monotonically increasing salinity with depth (e.g., Figure 65 in *Weeks and Ackley* [1986]).

These relatively simple patterns are not valid for frazil/platelet/mixed sea ice, unless one develops a set of hypotheses on the variables in play (see discussion in *Eicken* [this volume]). Indeed, one has to remember that the formation process is threefold: formation of individual ice crystals in a "parent" water, accretion of the loose ice crystals in a "host" water and, finally, consolidation by freezing of the interstitial "host water". In such a process, the resulting isotopic signal will depend on many factors such as the signal of the parent water (with poorly know fractionation efficiency), the signal of the host water, the freezing rate of the host water, and the grain size and porosity of the loose frazil. The salinity will depend on the same factors, except perhaps the signal of the parent water, if one considers that frazil ice crystals are chemically pure.

It is difficult, with the two data sets at hand, to discriminate inter-annual from seasonal variability in the cores' facies. However, the contrast between eastern and western cores in January 1994, the strong textural similarity between cores SA, SB and SC and those from October 1995, and the occurrence of large amounts of loose frazil below the eastern side of the sea ice cover in mid-summer, suggest the dominance of seasonal effects on the main facies contrasts. Therefore, a comparison of various salinity, $\delta^{18}O$ and textural profiles from different periods of the year and at two year intervals enables us to propose a typical sequence of first-year sea ice accretion in Evans Cove at the front of HGIS, keeping in mind the potential complications discussed above.

Consolidation of wind- and wave-induced frazil ice produced in surface sea water, showing dilution effects by continental meltwaters (lower salinity, lower $\delta^{18}O$), forms the initial sea ice cover. The crystallographic and textural contrasts between the granular facies forming the surface layer and the one occurring further below suggest different origins. The richness in air/brine inclusions and the random c-axes fabric in the surface layer are consistent with a wind- and wave-induced frazil ice forming at the beginning of the winter, in highly agitated waters subject to strong katabatic winds. The prevailing north west origin of these winds (see records at Automatic Weather Station Manuela: Figure 37 in *Ronveaux* [1992]) however prevents fully efficient export of the new frazil ice crystals formed in Evans Cove towards the Terra Nova Bay Polynya. Consolidation of the initial cover can thus occur. The salinity and $\delta^{18}O$ signatures in this 40-90cm thick surface layer reflect a dominant influence of the host water signal, since both variables increase with depth. This is compatible with the surface water conditions close to the ice shelf at the end of the summer. This water is likely to be slightly diluted and depleted in heavy isotopes from contribution of melted meteoric ice that is seen to accumulate in numerous ponds at the surface of HGIS. This meltwater is subsequently drained by supraglacial streams joining in a waterfall at the ice shelf front. A dominant influence of the freezing rate would, in contrast, decrease salinity and increase $\delta^{18}O$ with depth. The salinity increase with depth could eventually reflect a post-genetic desalination process. However, in that case, the isotopic signal should not be significantly altered, since there is no fractionation on ice melting [*Friedman et al.*, 1964; *Moser and Stichler*, 1980]. The salinity anomalies in the first 20cm of cores SF and SH are probably associated with deformational processes (rafting), as also suggested by the occurrence of oblique textural discontinuities that were an obvious exception in the very top part of these cores. Finally, it should be stressed that the thickness of this surficial layer might be underestimated, given the strong surface sublimation that probably dominates because of the steady katabatic winds that prevail in the area (see previous discussion).

In more than half of the cores, the surface layer is underlain by a platelet or a draped facies. Then, the lower granular facies develops, eventually in alternation with the draped facies. This granular facies shows a slight elongation of the crystals in the vertical plane, denoting partial influence of a downward congelation process. However, instead of the usual concentration of c-axes in the horizontal plane typical of congelation sea ice, it displays a strong clustering along the vertical. This suggests that most of the winter accretion in front of HGIS consists of small individual discs of frazil ice crystals (with c-axes perpendicular to the disc), gently settling upward at the base of the existing sea ice cover in a calm environment, allowing orderly packing. Sweeping of wind- and wave- induced frazil ice crystals formed in the polynya down to a maximum 1.70 m underneath the already existing fast ice cover in Evans Cove is improbable because of the steady south-eastward blowing katabatic winds. These are constantly "skimming" away the thin surficial buoyant layer of newly formed frazil. Moreover, the thickness of this facies tends to increase towards the ice shelf front in the 1995 cores (Figure 10). Dynamical thickening through rafting is equally precluded given the maximum depth observed for this facies and the absence of typical signatures of the process (see, for example, Eicken [this volume] and discussion above). Also (see processes in Table 1), (1) wind- and wave- induced frazil would build a facies similar to the surface layer described above, (2) no congelation ice layer is present to sustain small scale thermohaline convection and (3) there is no obvious source for diluted waters in the environment to induce double-diffusion processes. The most plausible source for these frazil ice crystals is, therefore, active supercooling in the Deep Ice Shelf Water adiabatically rising below HGIS during the winter. As shown in the next section, the structure of the water column in front of HGIS confirms the presence of large amounts of Deep Ice Shelf Water (DISW). Further consolidation of the host water leads to elongation of the crystals, which, however, retain their original crystallographic signature. In the top part of this second unit, salinity fluctuates around a mean value and $\delta^{18}O$ stabilizes. This probably reflects the combination of Ice Shelf Water characteristics for the "parent" water of the loose frazil and the onset of constant more saline winter "host" water characteristics. Further down, in the lower half of the winter accretion of cores SA and SB (120-220cm), salinity steadily decreases and $\delta^{18}O$ remains quite constant. This could either reflect a progressive reduction of the freezing rate or a progressive change in the host water characteristics towards dilution. In the first case, the rise of the isotopic signal in the freezing interstitial water could be "masked" by the background (up to 60%) signal of the loose individual crystals. This however would be in contradiction to the "a_0" (plate width in individual congelation ice crystals) measurements of Jeffries et al. [1993] in interstitial congelation ice from the lower layers of the McMurdo Sound sea ice cover, which indicate no decrease of the freezing rate with depth. The other alternative, although rather improbable in the middle of the winter, would be the scenario of a rising $\delta^{18}O$ in a diluted host water as suggested for the bottom part of cores SD to SH discussed below.

The end of the winter season is dominated by platelet accretion (Hole 1 to 3 and SA to SC). This facies is nearly absent in the eastern cores, where the winter accretion is thinner. It is not clear, however, if the lack of platelet ice in the eastern cores results from a lower accretion of platelet ice compared to the western zone, or increased bottom melting in the beginning of the summer season.

Bottom accretion still occurs during the summer as a banded rectangular facies. This facies does not exist in front of the western sector (west of the western medial moraine) and forms between 44% and 54% of all consolidated ice cores in front of the central and eastern sector. There also, a thick layer (several meters) of unconsolidated loose rectangular frazil ice crystals exists at the base of the sea ice cover from the end of the winter through most of the summer. This banded rectangular facies is thus actively building preferentially in the eastern sector in the first half of the summer. The $\delta^{18}O$ signal increases slightly as the banded facies starts to build up. This last observation is consistent with the suggestion made in the previous section that warm surface water is forced under HGIS by tidal action and partially melts marine ice accreted at the bottom. The meltwater produced is likely to increase the δ-value of the resultant sub-ice shelf top water layer, since marine ice has a slightly positive signature. This water mass will be the host water (and possibly the parent water, if further growth or new crystal growth is allowed by adiabatic supercooling) for the loose crystals that will form the banded rectangular facies in the landfast sea ice, resulting in the higher $\delta^{18}O$ values observed.

Another way to visualize this process is to plot the granular frazil from winter accretion in the sea ice and the banded rectangular frazil from summer accretion, on a salinity-$\delta^{18}O$ diagram (Figure 15). The banded rectangular frazil is clearly shifted towards higher salinities and higher $\delta^{18}O$ values. This trend is similar to the one observed for co-isotopic values in the marine ice cores (Figure 8). However, $\delta^{18}O$ ranges in the sea ice are, in both facies, shifted towards less positive values. This can be clearly related to differences in freezing rates of the host water, which must be significantly higher below 1.50 m of first-year sea ice than below several tens of meters of shelf ice. Higher salinities in the banded rectangular facies probably result partly from grain sizes, shapes, and hence, porosity.

Finally, the sudden increase of salinity at the very base of nearly all profiles is typical of the bottom of sea ice that has not yet undergone the rapid desalination which often occurs in the first few weeks after initial

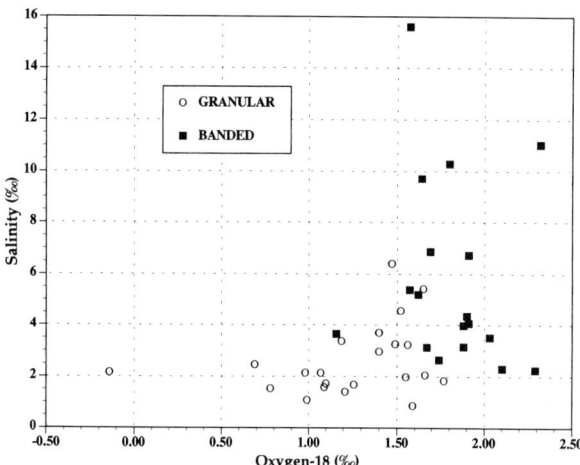

Fig. 15 : $\delta^{18}O$ versus salinity for samples of the granular and banded rectangular facies.

freezing [*Nakawo and Sinha*, 1981]. In this case it could also result from higher porosity of the bottom layers. This situation is, however, less clear in those cores where the banded rectangular facies exists at the bottom, which probably reflects brine loss from the loosely consolidated bottom layer of these ice cores from the eastern part of Evans Cove during sampling.

LINKING MARINE ICE AND SEA ICE RECORDS IN AN ICE/OCEAN INTERACTION PERSPECTIVE

The granular marine ice facies observed at HGIS must result from adiabatic supercooling in Ice Shelf Water produced in a Deep Thermohaline Circulation (DTC), as it occurs beneath larger ice shelves [*Jacobs et al.*, 1992]. Adiabatic decompression in recently opened bottom crevasses at the grounding line could only account for the production of negligible amounts of granular ice. A rough calculation can be made hypothesizing that all the supercooling occurring within a sea water mass at local pressure melting point as it is adiabatically rising in a bottom crevasse is compensated by ice crystal growth. For example, a 400 m water column, with a "mean" adiabatic rise of 200 m, would result in the production of, at most, 1 m of loose frazil with a 50% porosity. The only oceanic profile presently available in Evans Cove [*Fabiano et al.*, 1991], shows Ice Shelf Water from 100 m deep to about 100 m above the sea floor, which consists of a narrow trough 700 m deep in the central part of Evans Cove, with depth increasing inland. The maximum thickness of HGIS as far as half way to the northern tip of Vegetation Island is only 250 m, which implies a larger-scale circulation pattern in which HSSW produced elsewhere, e.g. in the Terra Nova Bay Polynya, can access the Nansen Ice Sheet grounding lines, perhaps beneath Reeves and Priestley Glaciers (Figure 2). Part of the Ice Shelf Water produced in those locations could recirculate along the eastern and western flanks of Vegetation Island, carrying with it loose frazil ice crystals formed by adiabatic supercooling. These crystals would accumulate in the depression formed at the southern tip of Vegetation Island, between the two merging flows as suggested previously.

Remnants of frazil ice crystals formed in this manner (hereafter DTC-granular facies) escape at the front of the ice shelf, where they accrete under the sea ice during the winter to form the granular facies with vertical c-axes fabrics. In all first-year sea ice cores that are thought to have retained the whole of the winter accretion (Hole 1, 2 and 3; SA, SB, SC), platelet ice bounds the lower limit of this winter accretion, while the upper limit (boundary with the initial wind- and wave- frazil ice accretion where it exists) consists of either the platelet (SA, SC) or the draped mixed congelation/platelet facies (Hole 1, 2 and 3). We postulate that these observations can be interpreted as the expression of variable intensities of the DTC. *Bombosch and Jenkins* [1995], show that the large size of crystals in the platelet facies reflects slower frazil ice crystal growth rates in the Ice Shelf Water as it levels off and eventually escapes at depth in front of the ice shelf. Transition from a small grained DTC-granular facies to a platelet or a draped facies in the landfast first-year sea ice could thus reflect variations in the degree of supercooling of the ISW where these crystals were formed. In early winter, enhanced freezing at the polynya surface is likely to increase HSSW production and speed up the DTC. At that time, the density contrast between ISW and HSSW should be moderate and supercooling weak. This would favor formation of the platelet facies or, probably more often, the draped mixed platelet-congelation facies, since congelation ice growth rates would still be competitive with the platelet crystal accretion rates beneath a thin sea ice cover.

In the mid-winter, the DTC may accelerate, resulting in higher density contrasts between ISW and HSSW and stronger supercooling producing the fine grained matrix of the DTC-granular sea ice facies. At the end of the winter, a slower DTC would again favor larger crystal growth. The heat sink through the sea ice cover is also reduced, along with the consolidation rate. The latter part of this proposed cycle would result in the platelet ice facies observed at the bottom of the winter accretion of the cores where summer melting at the base of the ice was less important. In addition, large scale fluctuations in the HSSW production related to katabatic wind activity during the winter could explain the alternations of the draped and the DTC-granular facies.

Several hypothesis can be formulated for the genesis of the banded marine and sea ice facies:

Post-depositional deformation processes leading to recrystallization of the original textures and fabrics. This is unlikely, as the banded facies are spatially confined to the frontal zone, and not developed in cores XXI and VI, where surface morphology and foliation show that the deformation is quite active. Also, discs or plates are seen to accrete in the water column below the sea ice cover and thin sections of these half frozen bottom deposits are identical to the banded facies observed at the base of the solid level sea ice cover above, where no evidence of deformation was found.

Expression of a spatially changing environment for the frazil ice crystals formed in the DTC, as the geometry of the base of the ice shelf switches from highly irregular (with transverse inverted depressions between individual flows close to the grounding line or around pinning points) towards a smoother interface closer to the front. In the first case, rapid lateral ascension of frazil crystals along steep slopes would favor the production of small crystals accreting in a random fashion at the base of the ice shelf. In the latter case, slower ascension further away from the initial production site would favor slower growth of larger crystals, accreting in an orderly fashion at the bottom of the ice shelf. The lower growth rates would result in higher isotopic values because of the higher fractionation efficiency. However, granular marine ice is observed up to several hundred meters downstream of the contact line between meteoric ice and marine ice (Figure 4). Moreover, if this hypothesis is correct, only the banded facies should be found in the first-year landfast sea ice in front of the ice shelf.

Expression of a temporally changing environment. The seasonal character of the banded facies accretion (summer only) suggests that it is linked to seasonal processes in the water column in front of or below the ice shelf. We have seen that the lower salinity and enhanced chemical fractionation in the granular marine ice consolidating closer to the ice shelf front suggest tidal forcing of warm surface waters below HGIS during the summer. These waters would favor partial melting of the loose granular frazil and produce a water mass at local pressure melting point showing lower salinities and higher isotopic signatures. This water mass, perhaps mixed with water carrying DTC-granular frazil, would eventually exit at the front. The marine and sea ice banded facies, then represent either transformed granular frazil ice crystals initially formed in the DTC, or new crystals formed by adiabatic supercooling, or both.

CONTRASTS WITH SEA ICE ACCRETION IN MCMURDO SOUND

The first-year sea ice cover at HGIS differs from the one observed in McMurdo Sound [*Jeffries et al.*, 1993; *Gow et al.*, this volume] in several aspects. Firstly, no thick (about 1 m) layer of congelation ice exists at the top of the cores in Evans Cove. Congelation ice occurs in limited amounts (10-15 cm) in only two cores (SF and SH), with both cores showing evidence of deformation in their top sections. Strong sublimation under katabatic winds could be responsible for ablation of a surficial congelation ice layer, but this is improbable for most of the cores, given the general occurrence of wind- and wave-induced frazil ice in the surficial layer. This contrasting situation simply reflects the calm conditions of landfast sea ice growth in McMurdo Sound in 1990, as suggested by *Jeffries et al.* [1993], as opposed to initial turbulent growth conditions occurring in Evans Cove.

Secondly, winter accretion is dominated more by the DTC-granular facies than by the congelation and platelet ice facies. This could be the expression of different geometries for HGIS and the Ross Ice Shelf (near McMurdo Sound). The length of HGIS is indeed quite small and, therefore, the granular frazil possibly "escapes" more easily from under the ice shelf, whilst it accretes much more upstream at the base of the Ross Ice Shelf. Platelet accretion is limited during the winter and more abundant at the end of the season at HGIS, but this is not necessarily in contradiction to the situation in McMurdo Sound. As a matter of fact, it is the case that the platelet observed at the base of all McMurdo ice cores accretes in the second half of the winter, and that congelation ice growth dominates in the beginning of the winter. Ice growth calculations based on freezing degree days indicate that platelet ice crystals begin to appear at the base of the existing columnar ice layer in McMurdo Sound at some time during the period July-September [*Gow et al.*, this volume; *Veazey*, 1994].

Finally, as underlined before, a thick layer of banded rectangular frazil accretes at the bottom of the sea ice at HGIS during the summer, and it is not observed in McMurdo Sound. Since warm surface waters have been detected in the McMurdo Sound area (see, for example, *Jacobs et al.* [1989: Stations 1, 70, 71]), contrasts in ocean circulation patterns, size and geometry of the ice shelf, or unavailability of loose DTC-granular frazil close to the ice shelf front are amongst possible reasons for this discrepancy.

CONCLUDING REMARKS

Our data set on landfast first-year sea ice from HGIS has confirmed that the variability of sea ice facies increases as one gets close to ice shelves. Besides the main columnar and granular facies associated respectively with congelation ice, wind- and wave-induced frazil ice or snow ice usually found in drifting sea ice (eventually as mixed facies), other facies develop that can be linked to ice-ocean interactions below ice shelves. The platelet ice facies, already shown before to be linked to the proximity of ice shelves, can turn into a draped mixed congelation-platelet facies depending on the balance between the accretion rate of the individual acicular platelet ice crystals, formed in the supercooled Ice Shelf Water under circulation mode-1, and the congelation freezing rate at the bottom of the existing sea ice cover. In addition, the Deep Thermohaline Convection can sustain winter accretion of a granular facies that differs from the surficial wind- and wave-induced facies by its much lower bubble content and its slightly vertically elongated crystals showing a strong vertical maximum of c-axes. Finally, bottom accretion still occurs during the summer as a banded facies which is thought to reflect interactions of the ice shelf base with a circulation mode-3 under HGIS.

Annual rhythmicity in landfast first-year sea ice records at HGIS, therefore allow us to investigate the variability within and between the two main ocean circulation modes below the ice shelf. Surprisingly, this rhythmicity is not recorded in the 45m-deep marine ice core Y close to the ice shelf front, where the banded facies forms more than 95% of the total core. A logical reason for this is the strong contrast in thermodynamical regimes between the bottom of the ice shelf and the base of the first-year sea ice cover. To be recorded, the seasonal variability of the frazil ice accretion needs to be "frozen-in". Obviously, the freezing rate at the bottom of an ice shelf is one to two order of magnitude less than the one at the base of a sea ice cover. Therefore, the winter accretion below the ice shelf and close to the front (where circulation mode-3 should be the most active) is easily remobilized and melted away during the next summer. Marine ice cores, therefore, mainly record the main spatial differentiation of the two ocean circulation modes on time scales of hundreds of years. This, in turn, highlights the specific input of landfast first-year sea ice studies as a useful complement of coupled oceanographical-shelf ice research in the vicinity of floating ice bodies.

Hells Gate Ice Shelf is a small and narrow ice shelf subject to a strong katabatic wind regime. We have shown that these characteristics makes it easier to study marine ice properties at minimal cost but, at the same time, in a peculiar dynamical setting that enhances the complexity of data interpretation. This also raises the question as to the degree to which the findings discussed here can be applied to large floating ice bodies? First of all, HGIS shows geometries of granular marine ice accretion, between the individual ice flows forming the ice shelf, that are similar to those on the Ronne Ice Shelf [*Corr et al.*, 1995]. This stresses the importance of considering 3-D modelling of the DTC, since local transverse gradients of the ice shelf base might exist close to grounding lines and around pinning points, that are much steeper than the longitudinal gradient considered in 2-D modelling exercises [*Bombosch and Jenkins*, 1995; *Hellmer and Jacobs*, 1992; *Jenkins and Bombosch*, 1995]. Recent developments along these lines [*Grosfeld et al.*, 1997; *Williams et al.*, in press] stress the complexity of marine ice accumulation patterns below ice shelves. On the other hand, HGIS should really be considered as part of the larger Nansen Ice Sheet. Although this whole floating ice mass is still smaller than the Filchner-Ronne or the Amery ice shelves, it clearly shows evidence of a similar 3-D oceanic circulation in which Deep Thermohaline Convection, Ice Shelf Water production and marine ice accretion take place. Recent radio echo sounding profiles, detailed surface velocity and GPS thickness measurements at Nansen Ice Sheet demonstrate that large marine ice production rates are required in order to sustain a steady state profile [*M. Frezzotti*, personal communication, June 1996]. Extensive marine ice outcrops are clearly seen at the NIS surface and are currently under investigation by the authors. Finally, accretion of the banded marine ice facies in the frontal zone of HGIS (and at the bottom of the landfast sea ice) might also be considered as a peculiar process, essentially restricted to the study case. Indeed, no deep marine ice cores from the major antarctic ice shelves has ever shown this type of facies, and the frontal zone of ice shelves has always been considered as subject to a strong basal melting regime (e.g. [*Jacobs et al.*, 1992]). It should be noted, however, that none of the deep marine ice cores has been drilled within a few tens of kilometers of the margins of ice shelves and that recent 3-D modelling of ice-ocean interactions in sub-ice shelf cavities reproduce a fringe of net basal accretion near these ice shelf margins in a number of realistic scenarios [*Grosfeld et al.*, 1997; *Williams et al.*, in press]. The similarity between the banded facies and the "wafer-like" facies described by *Gow et al.* [1987] in the Weddell Sea also suggests that our observations in the Ross Sea might be valid for other ice shelf areas around Antarctica.

Oceanographic measurements at the ice shelf front are obviously a crucial "missing link" in this study. Future work should focus on coupled oceanographic and landfast sea ice studies both in summer and winter. The development of experimental procedures to better understand the controlling factors on the shape and size of frazil ice crystals, and to support interpretation of the observed landfast sea ice facies variability in the field are also essential.

Acknowledgments. This paper is a contribution to the Belgian Antarctic Program (SSTC-Science Policy Office). The authors are greatly indebted to the "Programma Nazionale di Ricerche in Antartide" (PNRA) and to the logistic team of Terra Nova Bay for unconditional and efficient support in the field. Thanks are also due to Dr. J. Jouzel and Prof. A. Longinelli for allowing use of the facilities in their respective Laboratories. M. Dini benefited from a research scholarship from the University of Trieste. J.-L. Tison is a Research Associate of the Belgian National Fund for Scientific Research (F.N.R.S.). The authors especially wish to thank H. Eicken, S. Jacobs and three anonymous referees for constructive and detailed comments on earlier versions of the manuscript. M. Naessens is acknowledged for analytical work on some of the sea ice cores.

REFERENCES

Addison, J.R., Impurity concentrations in sea ice., *J. Glaciol.*, *18*(78), 117-127, 1977.

Baroni, C., The Hells Gate and Backstairs Passage Ice Shelves, Victoria Land, Antartica, *Mem. Soc. Geol. Ital.*, *43*, 123-144, 1990.

Baroni, C., M. Frezzotti, C. Giraudi, and G. Orombelli, Ice flow and surficial variation inferred from satellite image and aerial photograph analysis of Larsen Ice Tongue, Hells Gate and Nansen Ice Shelves (Victoria Land, Antarctica), *Mem. Soc. Geol. Ital.*, *46*, 69-80, 1991a.

Baroni, C., B. Stenni, and P. Iacumin, Oxygen isotopic composition of ice samples from the Hells Gate and Backstairs Passage ice shelves (Victoria Land, Antarctica) : Evidence of bottom freezing, *Mem. Soc. Geol. Ital.*, *46*, 45-48, 1991b.

Ben-Yaakov, S., Diffusion of sea water ions - I. Diffusion of sea water into a dilute solution, *Geochim. Cosmochim. Acta*, *36*, 1395-1406, 1972.

Bombosch, A., and A. Jenkins, Modeling the formation and deposition of frazil ice beneath Filchner-Ronne Ice Shelf., *J. Geophys. Res.*, *100*(C4), 6983-6992, 1995.

Casella, F., R. Udisti, and G. Piccardi, The oceanic source contribution to the snow composition, as function of elevation, at two Antarctic coastal stations, *Terra Antartica*, in press.

Corr, H., M. Popple, and A. Robinson, Airborne radio echo investigations of a marine ice body. *Filchner-Ronne Ice Shelf Programme (FRISP) Report 8*, AWI - Bremerhaven, 14-17, 1995.

Cox, G.P.N., and W.F. Weeks, Numerical simulations of the profile properties of undeformed first-year sea ice during the growth season, *J. Geophys. Res.*, *93*(10), 12449-12460, 1988.

Dansgaard, W., S.J. Johnsen, H.B. Clausen, and N. Gundestrup, Stable isotope glaciology, *Medd. Grønl.*, *197*(2), 1-53, 1973.

Dieckmann, G.S., G. Rohardt, H. Hellmer, and J. Kipfstuhl, The occurrence of ice platelets at 250m depth near the Filchner Ice Shelf and its significance for sea ice biology., *Deep-Sea Res.*, *33*, 141-148, 1986.

Eicken, H., Factors determining microstructure, salinity and stable-isotope composition of Antarctic sea ice : Deriving modes and rates of ice growth in the Weddell Sea, *Antarct. Res. Ser.*, this volume.

Eicken, H., and M.A. Lange, Development and properties of sea ice in the coastal regime of the Southeastern Weddell Sea, *J. Geophys. Res.*, *94*(C6), 8193-8206, 1989.

Eicken, H., H. Oerter, H. Miller, W. Graf, and J. Kipfstuhl, Textural characteristics and impurity content of meteoric and marine ice in the Ronne Ice Shelf, Antarctica, *J. Glaciol.*, *40*(135), 386-398, 1994.

Fabiano, M., P. Povero, G. Catalano, and F. Benedetti, Hydrological data collected during the biological, chemical and geological sampling in Terra Nova Bay, *Nat. Sc. Com., Ocean. Camp. 1989-90, Data Rep.*, 35-71, 1991.

Frezzotti, M., Glaciological study in Terra Nova Bay, Antarctica, inferred from remote sensing analysis, *Ann. Glaciol.*, *17*, 63-71, 1993.

Friedman, I., A.C. Redfield, B. Schoen, and J. Harris, The variation of the deuterium content of natural water in the hydrologic cycle, *Rev. Geophys.*, *2*(1), 177-189, 1964.

Goldberg, E.D., Biogeochemistry of trace elements., *Geol. Soc. Amer. Mem.*, *67*, 345-358, 1957.

Gow, A.J., S.F. Ackley, K.R. Buck, and K.M. Golden, Physical and structural characteristics of Weddell Sea pack ice, *CRREL Report*, *87-14*, pp. 70, 1987.

Gow, A.J., S.F. Ackley, J.W. Govoni, and W.F. Weeks, Physical and structural properties of land-fast sea ice in

McMurdo Sound, Antarctica, *Antarct. Res. Ser.*, this volume.

Gow, A.J., and S. Epstein, On the use of stable isotopes to trace the origins of ice in a floating ice tongue, *J. Geophys. Res.*, 77, 6552-6557, 1972.

Gow, A.J., D.A. Meese, D.K. Perovich, and W.B. Tucker III, The anatomy of a freezing lead, *J. Geophys. Res.*, 95(C10), 18221-18232, 1990.

Grosfeld, K., R. Gerdes, and J. Determann, Thermohaline circulation and interaction between ice shelf cavities and the adjacent open ocean, *J. Geophys. Res.*, 102(C7), 15595-15610, 1997.

Hellmer, H.H., and S.S. Jacobs, Ocean interactions with the base of the Amery Ice Shelf, Antarctica, *J. Geophys. Res.*, 97(20), 305-320, 1992.

Jacobs, S.S., W.E. Haines, J.J.L. Ardai, and P.A. Mele, Ross Sea oceanographic data, 1983-1987, *Lamont-Doherty Geological Observatory of Columbia University, New York*, 1989.

Jacobs, S.S., H.H. Helmer, C.S.M. Doake, A. Jenkins, and R.M. Frolich, Melting of ice shelves and the mass balance of Antarctica., *J. Glaciol.*, 38(130), 375-387, 1992.

Jeffries, M.O., and W.F. Weeks, Fast-ice properties and structure in McMurdo Sound, *Antarct. J. U.S.*, 26(5), 94-95, 1991a.

Jeffries, M.O., and W.F. Weeks, Summer pack-ice properties and structure in the western Ross Sea, *Antarct. J. U.S.*, 26(5), 95-97, 1991b.

Jeffries, M.O., and W.F. Weeks, Structural characteristics and development of sea ice in the western Ross Sea, *Antarct. Sci.*, 5(1), 63-75, 1993.

Jeffries, M.O., W.F. Weeks, R. Shaw, and K. Morris, Structural characteristics of congelation and platelet ice and their role in the development of Antarctic land-fast sea ice, *J. Glaciol.*, 39(132), 223-238, 1993.

Jeffries, M.O., and U. Adolphs, Early winter ice and snow thickness distribution, ice structure and development of the western Ross Sea pack ice between the ice edge and the Ross Ice Shelf, *Antarct. Sci.*, 9(2), 188-200, 1997.

Jenkins, A., and A. Bombosch, Modeling the effects of frazil ice crystals on the dynamics and thermodynamics of Ice Shelf Water plumes., *J. Geophys. Res.*, 100(C4), 6967-6981, 1995.

Kawamura, T., K.I. Oshima, T. Takizawa, and S. Ushio, Physical, structural, and isotopic characteristics and growth processes of fast sea ice in Lützow-Holm Bay, Antarctica, *J. Geophys. Res.*, 102(C2), 3345-3355, 1997.

Kipfstuhl, J., Zur entstehung von unterwassereis und das wachstum und die energiebilanz des meereises in der Akta Bucht, Antarktis. On the formation of underwater ice and the growth and energy budget of the sea ice in Atka Bay, Antarctica, *Ber. Polarforsch.*, 85, 1991.

Kipfstuhl, J., G. Dieckmann, H. Oerter, H. Hellmer, and W. Graf, The origin of green icebergs in Antarctica., *J. Geophys. Res.*, 97(C12), 20319-20324, 1992.

Lange, M.A., Basic properties of Antarctic sea ice as revealed by textural analysis of ice cores, *Ann. Glaciol.*, 10, 95-101, 1988.

Lange, M.A., Development and physical properties of sea ice in the Weddell Sea, Antarctica, in *Sea Ice Properties and Processes, CRREL Monograph 90-1*, 22-40, 1990.

Lange, M.A., S.F. Ackley, P. Wadhams, G.S. Dieckmann, and H. Eicken, Development of sea ice in the Weddell Sea, Antarctica, *Ann. Glaciol.*, 12, 92-96, 1989.

Langway, C.C.J., Ice fabrics and the Universal stage, *CRREL Tech. Rep. 62*, pp. 16, 1958.

Lewis, E.L., and R.G. Perkin, Ice pumps and their rates, *J. Geophys. Res.*, 91(10), 11.756-11.762, 1986.

Lorrain, R., J.-L. Tison, A. Bondesan, D. Ronveaux, and M. Meneghel, Preliminary results from 60 shallow cores and from one 45-m deep marine ice core at Hells Gate Ice Shelf, Victoria Land, Antarctica, *Terra Antartica Reports*, 1, 19-24, 1997.

Morgan, V.I., Oxygen isotope evidence for bottom freezing on the Amery Ice Shelf, *Nature*, 238, 392-394, 1972.

Moser, H., and W. Stichler, Environmental isotopes in ice and snow, in Handbook of Environmental Isotope Geochemistry, in *The terrestrial environment*, edited by P. Fritz, and J. Fontes, pp. 141-178, Elsevier, New York, 1980.

Nakawo, M., and N. Sinha, Growth rate and salinity profiles of first-year sea ice in the high Arctic, *J. Glaciol.*, 27, 315-330, 1981.

Oerter, H., H. Eicken, J. Kipfsthul, H. Miller, and W. Graf, Comparison between ice core B13 and B15, *Filchner-Ronne Ice Shelf Programme (FRISP) Report 7*, AWI - Bremerhaven, 29-36, 1994.

Oerter, H., J. Kiptstuhl, J. Determann, H. Miller, D. Wagenbach, A. Minikin, and W. Graf, Evidence for basal marine ice in the Filchner-Ronne Ice Shelf, *Nature*, 358, 399-401, 1992.

Paterson, W.S.B., *The Physics of Glaciers*, 480 pp., Pergamon, Oxford, 1994.

Ragle, R.H., B.L. Hansen, A. Gow, and R.W. Patenaude, Deep core drilling in the Ross Ice Shelf, Little America V, Antarctica, *CRREL Technical Report, 70*, pp. 10, 1960.

Rand, J., and M. Mellor, Ice-coring augers for shallow depth sampling, *CRREL Rep. 85-21*, pp.22, 1985.

Richardson, C., Phase relationships in sea ice as a function of temperature, *J. Glaciol.*, 17(77), 507-519, 1976.

Ronveaux, D., The dynamics of a small antarctic ice shelf as indicated by an ice composition study, thèse thesis, Université Libre de Bruxelles, 1992.

Serikov, M.I., Structure of Antarctic sea ice, *Information Bulletin of the Soviet Antarctic Expedition*, 4(5), 265-266, 1963.

Souchez, R., M. Lemmens, R. Lorrain, and J.-L. Tison, Pressure-melting within a glacier indicated by the chemistry of regelation ice, *Nature*, 273, 454-456, 1978.

Souchez, R., M. Meneghel, J.-L. Tison, R. Lorrain, D. Ronveaux, C. Baroni, A. Lozej, I. Tabacco, and J. Jouzel, Ice composition evidence of marine ice transfer along the bottom of a small Antarctic ice shelf, *Geophys. Res. Lett.*, *18*(5), 849-852, 1991.

Tison, J.-L., A. Bondesan, G. Delisle, A. Lozej, F. Merlanti, and L. Janssens, A dynamical approach to explain ice structures and complex morainic genesis on a partially grounded ice shelf (Hells Gate Ice Shelf - Victoria Land, Antarctica), *Terra Antartica Reports*, 1, 33-38, 1997.

Tison, J.-L., E.M. Morris, R. Souchez, and J. Jouzel, Stratigraphy, stable isotopes and salinity in multi-year sea ice from the rift area, south George VI Ice Shelf, Antarctic Peninsula, *J. Glaciol.*, *37*(127), 357-367, 1991.

Tison, J.-L., D. Ronveaux, and R. Lorrain, Low salinity frazil ice generation at the base of a small antarctic ice shelf, *Antarct. Sci.*, *5*(3), 309-322, 1993.

Veazey, A.D., Development and variability of the structure and physical properties of landfast sea ice in McMurdo Sound, M.S. Thesis, University of Alaska Fairbanks, 1994.

Wadhams, P., M.A. Lange, and S.F. Ackley, The ice thickness distribution across the Atlantic sector of the Antarctic Ocean in midwinter, *J. Geophys. Res.*, *92*, 14535-14552, 1987.

Weeks, W.F., and S.F. Ackley, The growth, structure and properties of sea ice, in *The geophysics of sea ice*, edited by N. Untersteiner, pp. 9-164, Martinus Nyhoff Publ., Dordrecht (Nato ASI Series B, Physics), 1986.

Williams, M.J.M., R.C. Warner, and W.F. Budd, The effects of ocean warming on melting and ocean circulation under the Amery Ice Shelf, East Antarctica, *Ann. Glaciol.*, in press.

Wilson, C.J.L., J.P. Burg, and J.C. Mitchell, The origin of kinks in polycrystalline ice, *Tectonophysics*, *127*(1-2), 27-48, 1986.

Wilson, C.J.L., and Y. Zhang, Comparison between experiment and computer modelling of plane strain simple shear ice deformation., *J. Glaciol.*, *40*(134), 46-55, 1994.

Wilson, T.R.S., Salinity and the major elements of sea water, edited by Riley and Skirrow, pp. 365-413, Academic Press, London, 1975.

World Meteorological Organization, WMO *Sea Ice Nomenclature Terminology, Codes and Illustrated Glossary*, WMO/DMM/BMO 259-TP-145, Secretariat of the WMO, Genova, 1985.

Zotikov, I.A., V.S. Zagorodnov, and J.V. Raikovsky, Core drilling through the Ross Ice Shelf (Antarctica) confirmed basal freezing, *Science*, *207*(4438), 1463-1465, 1980.

A. Bouzette, R. D. Lorrain and J.-L. Tison, Département des Sciences de la Terre et de l'Environnement, Laboratoire de Glaciologie, Faculté des Sciences (CP 160/03), Université Libre de Bruxelles, 50, av. F.D. Roosevelt, 1050 Bruxelles, Belgique.

M. Dini, Laboratorio di Geochimica Isotopica, Dipartimento di Scienze della Terra, Università di Trieste, via E. Weiss, 6, 34127 Trieste, Italy.

A. Bondesan, Dipartimento di Geografia G. Morandini, Università di Padova, via del Santo, 26, I-35123, Padova, Italy.

M. Stiévenard, Laboratoire de Modélisation du Climat et de l'Environnement, CEA/DSM, Bâtiment 709, Orme des Merisiers, CE Saclay, F-91191, Gif-sur-Yvette, CEDEX France.

(Received June 23, 1997;
accepted August 22, 1997)